Y0-DLC-430

ENVIRONMENTAL SCIENCE

SUSTAINING YOUR WORLD

G. TYLER MILLER • SCOTT E. SPOOLMAN

ON ASSIGNMENT National Geographic photographer Carsten Peter accompanied a group of explorers to document the untouched subterranean beauty of Vietnam's Hang Son Doong. The enormous cave is part of a national park that was named a World Heritage site in 2003. Although longer and deeper caves have been discovered, none are as large as Hang Son Doong, which could easily fit several 40-story skyscrapers.

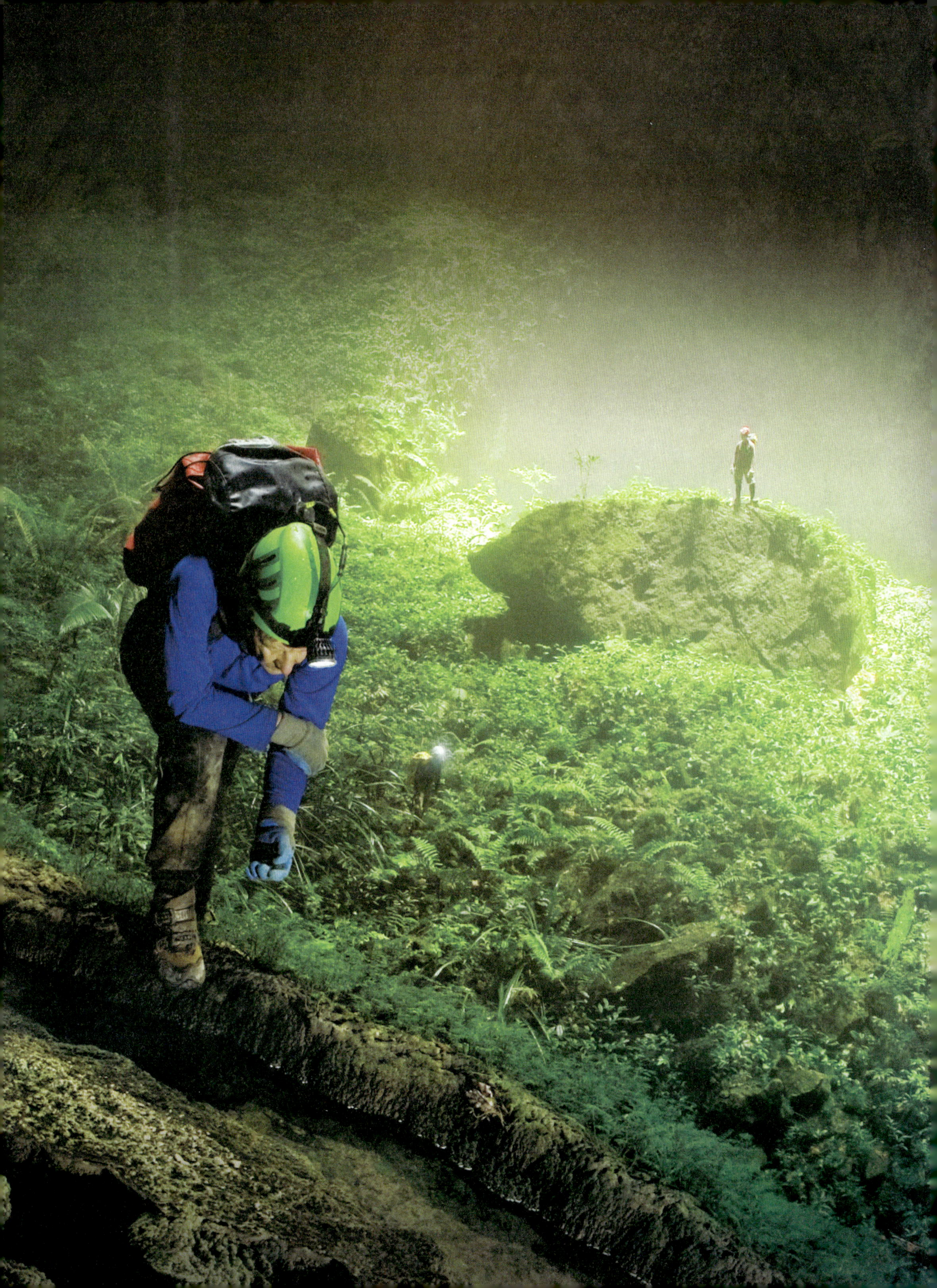

NATIONAL GEOGRAPHIC | CENGAGE

Acknowledgments

Grateful acknowledgment is given to the authors, artists, photographers, museums, publishers, and agents for permission to reprint copyrighted material. Every effort has been made to secure the appropriate permission. If any omissions have been made or if corrections are required, please contact the Publisher.

Photographic Credits

Cover © Carsten Peter/National Geographic Creative. Acknowledgments and credits continue on page 694.

Printed in the United States of America

Print Number: 03
Print Year: 2024

Copyright © 2023 Cengage Learning, Inc.

ALL RIGHTS RESERVED. No part of this work covered by the copyright herein may be reproduced or distributed in any form or by any means, except as permitted by U.S. copyright law, without the prior written permission of the copyright owner.

"National Geographic," "National Geographic Society," and the Yellow Border Design are registered trademarks of the National Geographic Society® Marcas Registradas.

For product information and technology assistance, contact us at Customer & Sales Support,
888-915-3276

For permission to use material from this text or product, submit all requests online at
www.cengage.com/permissions

Further permissions questions can be emailed to
permissionrequest@cengage.com

National Geographic Learning | Cengage
200 Pier 4 Blvd., Suite 400
Boston, MA 02210

National Geographic Learning, a Cengage company, is a provider of quality core and supplemental educational materials for the PreK–12, adult education, and ELT markets. Cengage is a leading provider of customized learning solutions with employees residing in nearly 40 different countries and sales in more than 125 countries around the world. Find your local representative at **NGL.Cengage.com/RepFinder**.

Visit National Geographic Learning online at **NGL.Cengage.com**.

ISBN: 978-0-357-54184-5

BRIEF CONTENTS

Detailed Contents vi

Preface xx

▮ Into the Okavango with National Geographic 02

ECOLOGY AND ECOSYSTEMS   12

CHAPTER 1 The Environment and Sustainability 14

CHAPTER 2 Science, Matter, Energy, and Systems 44

CHAPTER 3 Ecosystem Dynamics 66

ENGINEERING PROJECT 1 Design a Method for Treating Contaminated Soil 100

BIODIVERSITY   102

CHAPTER 4 Biodiversity and Evolution 104

CHAPTER 5 Species Interactions, Ecological Succession, and Population Control 132

CHAPTER 6 Ecosystems and Climate 156

ENGINEERING PROJECT 2 Design a System to Assess a Local Species 194

▮ Partners in Sustainability: Nature Museums and Preserves 196

SUSTAINING BIODIVERSITY   206

CHAPTER 7 Saving Species and Ecosystem Services 208

CHAPTER 8 Sustaining Biodiversity: An Ecosystem Approach 242

ENGINEERING PROJECT 3 Design a Solar Cooker 276

ENVIRONMENTAL QUALITY   278

CHAPTER 9 Food, Soil, and Pest Management 280

CHAPTER 10 Water Resources and Water Pollution 326

CHAPTER 11 Geology and Nonrenewable Mineral Resources 360

CHAPTER 12 Nonrenewable Energy Resources 390

CHAPTER 13 Renewable Energy Resources 420

ENGINEERING PROJECT 4 Design a Wind-Powered Generator 456

▮ Citizen Science: Of the People, By the People, For the People 458

ENVIRONMENTAL CONCERNS   468

CHAPTER 14 Human Population and Urbanization 470

CHAPTER 15 Environmental Hazards and Human Health 498

CHAPTER 16 Air Pollution, Climate Change, and Ozone Depletion 530

CHAPTER 17 Solid and Hazardous Waste 576

CHAPTER 18 Environmental Economics, Politics, and Worldviews 612

ENGINEERING PROJECT 5 Design a Carbon-Capturing Device 642

v

CONTENTS

Into the Okavango with
National Geographic.................. 02

■ **NATIONAL GEOGRAPHIC** Featured Explorers.......... xvi
■ **NATIONAL GEOGRAPHIC** Learning Framework xix
PREFACE xx
■ **SPECIAL FEATURE**
Into the Okavango with National Geographic........... 02

Unit 1 Ecology and Ecosystems 12

CHAPTER 1
THE ENVIRONMENT AND SUSTAINABILITY 14
■ **EXPLORERS AT WORK** Reconnecting People with Nature, with Juan Martinez................................. 16
CASE STUDY The Greening of American Schools 18
1.1 What Are Some Key Factors of Sustainability? 19
1.2 How Are Our Ecological Footprints Affecting Earth?.. 23
1.3 What Causes Environmental Problems and Why Do They Persist? 28
1.4 What Is an Environmentally Sustainable Society?.. 36
■ **EXPLORERS AT WORK** Dennis Dimick................ 36

TYING IT ALL TOGETHER The Greening of the American Mindset.. 40
Chapter 1 Summary..41
Chapter 1 Assessment 42

CHAPTER 2
SCIENCE, MATTER, ENERGY, AND SYSTEMS 44
■ **EXPLORERS AT WORK** New Paths to Engaging Science, with Nalini Nadkarni.. 46
CASE STUDY Experimenting with a Forest.................... 48
2.1 What Do Scientists Do? 49
   Engineering Focus: What Is Engineering? 50
   ■ **EXPLORERS AT WORK** John Francis................... 51
2.2 What Is Matter? 55
2.3 What Is Energy? 58
2.4 What Are Systems? 61
TYING IT ALL TOGETHER The Hubbard Brook Forest Experiment..62
Chapter 2 Summary... 63
Chapter 2 Assessment 64

vi   TABLE OF CONTENTS

CHAPTER 3
ECOSYSTEM DYNAMICS 66

■ **EXPLORERS AT WORK** Eco-Paradise at Serra Bonita, with Vitor Becker 68

CASE STUDY Disappearing Tropical Rain Forests 70

3.1 What Are Earth's Major Spheres, and How Do They Support Life? 71
 Science Focus: Nutrient Cycling 72

3.2 What Are the Major Ecosystem Components? 74
 Engineering Focus: Nature's Cleanup Crew 79

3.3 What Happens to Energy in an Ecosystem? 81

3.4 What Happens to Matter in an Ecosystem? 84
 Science Focus: Water 87

3.5 How Do Scientists Study Ecosystems? 92
 ■ **EXPLORERS AT WORK** Thomas E. Lovejoy 94
 Science Focus: Testing Planetary Boundaries 95

TYING IT ALL TOGETHER The Energy Cost of Cutting Rain Forests 96

Chapter 3 Summary 97

Chapter 3 Assessment 98

ENGINEERING PROJECT 1
Design a Method for Treating Contaminated Soil 100

Unit 2 Biodiversity 102

CHAPTER 4
BIODIVERSITY AND EVOLUTION 104

■ **EXPLORERS AT WORK** Fighting a Deadly Frog Disease, with Anna Savage 106

CASE STUDY Amphibians at Risk 108

4.1 What Is Biodiversity and Why Is It Important? 109
 Science Focus: Insects and Ecosystem Services 114

4.2 What Roles Do Species Play in Ecosystems? 115

4.3 How Does Life on Earth Change Over Time? 120
 ■ **EXPLORERS AT WORK** Edward O. Wilson 121

4.4 What Factors Affect Biodiversity? 124
 Science Focus: Geological Processes and Biodiversity 126

TYING IT ALL TOGETHER Piecing Together the Puzzle of Chytrid Disease 128

Chapter 4 Summary 129

Chapter 4 Assessment 130

TABLE OF CONTENTS   vii

CONTENTS

Partners in Sustainability
Nature Museums and Preserves 196

CHAPTER 5
SPECIES INTERACTIONS, ECOLOGICAL SUCCESSION, AND POPULATION CONTROL 132

■ **EXPLORERS AT WORK** Sea Otters Step Up, with Jim Estes 134

CASE STUDY The Southern Sea Otter: A Species in Recovery 136

5.1 How Do Species Interact? 137

5.2 How Do Ecosystems Respond to Changing Conditions? 142

5.3 What Limits the Growth of Populations? 145
■ **EXPLORERS AT WORK** Zeb Hogan 146

TYING IT ALL TOGETHER Southern Sea Otters Face an Uncertain Future 152

Chapter 5 Summary 153

Chapter 5 Assessment 154

CHAPTER 6
ECOSYSTEMS AND CLIMATE 156

■ **EXPLORERS AT WORK** Saving the Last Wild Places in the Ocean, with Enric Sala 158

CASE STUDY The Importance of Coral Reefs 160

6.1 What Factors Influence Climate? 161
Science Focus: Greenhouse Gases and Climate 165

6.2 What Are the Major Types of Terrestrial Ecosystems? 167
Science Focus: Desert Adaptations 169
■ **EXPLORERS AT WORK** Alizé Carrère 175

6.3 What Are the Major Types of Marine Ecosystems? 180

6.4 What Are the Major Types of Freshwater Systems? 186

TYING IT ALL TOGETHER Coral Reef Recovery 190

Chapter 6 Summary 191

Chapter 6 Assessment 192

ENGINEERING PROJECT 2
Design a System to Assess a Local Species 194

■ SPECIAL FEATURE
Partners in Sustainability: Nature Museums and Preserves 196

viii TABLE OF CONTENTS

Unit 3 Sustaining Biodiversity 206

CHAPTER 7
SAVING SPECIES AND ECOSYSTEM SERVICES 208

■ **EXPLORERS AT WORK** A Focus on Saving Honeybees, with Anand Varma 210

CASE STUDY A Honeybee Mystery 212

7.1 What Causes Extinction and What Are Its Impacts? 213
  ■ **EXPLORERS AT WORK** Joel Sartore 217

7.2 What Role Do Humans Play in the Loss of Species and Ecosystem Services? 221
  Science Focus: Honeybee Losses 223

7.3 How Can We Sustain Wild Species and Ecosystem Services? 231
  Science Focus: Wildlife Forensics 235

TYING IT ALL TOGETHER Colony Collapse Disorder 238

Chapter 7 Summary 239

Chapter 7 Assessment 240

CHAPTER 8
SUSTAINING BIODIVERSITY: AN ECOSYSTEM APPROACH 242

■ **EXPLORERS AT WORK** Protecting Biodiversity One Bird at a Time, with Çağan Şekercioğlu 244

CASE STUDY Costa Rica—A Global Conservation Leader 246

8.1 How Can Forests Be Better Managed? 247
  Science Focus: What Are Ecosystem Services Worth? 248

8.2 How Can Grasslands Be Better Managed? 255

8.3 How Can Protected Lands Be Better Managed? 256

8.4 How Does the Ecosystem Approach Help Protect Terrestrial Biodiversity? 262
  ■ **EXPLORERS AT WORK** Kamaljit S. Bawa 264

8.5 How Does the Ecosystem Approach Help Protect Aquatic Biodiversity? 266

TYING IT ALL TOGETHER Rates of Tropical Forest Loss 272

Chapter 8 Summary 273

Chapter 8 Assessment 274

ENGINEERING PROJECT 3
Design a Solar Cooker 276

TABLE OF CONTENTS   ix

CONTENTS

Unit 4 Environmental Quality 278

CHAPTER 9
FOOD, SOIL, AND PEST MANAGEMENT 280
- **EXPLORERS AT WORK** Sustainable Agriculture: Is It Possible? with Jennifer Burney 282
- **CASE STUDY** Growing Power—An Urban Food Oasis 284
- 9.1 What Is Food Security? 285
- 9.2 How Is Food Produced? 290
- 9.3 How Are Environmental Issues and Food Production Connected? 302
  - Science Focus: Soil Science 303
  - **EXPLORERS AT WORK** Jerry Glover 306
- 9.4 How Can Society Manage Agricultural Pests More Sustainably? 311
- 9.5 What Are Sustainable Solutions for Food Production? 314
- **TYING IT ALL TOGETHER** Sustainability Starts With You 322
- Chapter 9 Summary 323
- Chapter 9 Assessment 324

CHAPTER 10
WATER RESOURCES AND WATER POLLUTION 326
- **EXPLORERS AT WORK** Rescuing the Colorado River Delta, with Osvel Hinojosa Huerta 328
- **CASE STUDY** The Colorado River Story 330
- 10.1 Why Is Fresh Water in Short Supply? 331
- 10.2 How Can People Increase Freshwater Supplies? 334
  - Science Focus: Aquifer Depletion in the United States 335
- 10.3 How Can People Use Fresh Water More Sustainably? 338
  - **EXPLORERS AT WORK** Sandra Postel 341
- 10.4 How Can People Reduce Water Pollution? 342
  - Engineering Focus: Working with Nature to Treat Sewage 354

- **TYING IT ALL TOGETHER** Saving the Colorado River 356
- Chapter 10 Summary 357
- Chapter 10 Assessment 358

CHAPTER 11
GEOLOGY AND NONRENEWABLE MINERAL RESOURCES 360
- **EXPLORERS AT WORK** Better Batteries for Electric Cars, with Yu-Guo Guo 362
- **CASE STUDY** The Importance of Rare Earth Metals 364
- 11.1 How Do Geological Processes Relate to Society and the Environment? 365
- 11.2 What Are Earth's Mineral Resources and How Long Might Reserves Last? 370
- 11.3 What Are the Effects of Using Mineral Resources? 376
  - **EXPLORERS AT WORK** Bob Ballard 377
- 11.4 How Can Society Use Mineral Resources More Sustainably? 383
  - Science Focus: The Nanotechnology Revolution 384
- **TYING IT ALL TOGETHER** Rare Earth Metals and Sustainability 386
- Chapter 11 Summary 387
- Chapter 11 Assessment 388

CHAPTER 12
NONRENEWABLE ENERGY RESOURCES 390
- **EXPLORERS AT WORK** Hunting for Methane Leaks, with Katey Walter Anthony 392
- **CASE STUDY** Fracking for Oil and Gas 394
- 12.1 What Is Net Energy and Why Is It Important? 395
- 12.2 What Are the Advantages and Disadvantages of Using Fossil Fuels? 397
- 12.3 What Are the Advantages and Disadvantages of Using Nuclear Power? 408
  - Science Focus: Nuclear Fission Reactors 409
  - **EXPLORERS AT WORK** Leslie Dewan 412

x TABLE OF CONTENTS

Citizen Science:
**Of the People, By the People,
For the People**.................................... 458

TYING IT ALL TOGETHER Trends in Energy Use 416

Chapter 12 Summary... 417

Chapter 12 Assessment 418

CHAPTER 13
RENEWABLE ENERGY RESOURCES 420

▪ **EXPLORERS AT WORK** The Boiling River, with
Andrés Ruzo... 422

CASE STUDY The Potential for Wind Power in the
United States 424

13.1 Why Is Energy Efficiency an Important
Energy Resource?... 425

13.2 What Are Sources of Renewable Energy? 433
▪ **EXPLORERS AT WORK** Xiaolin Zheng 437

13.3 How Can Society Transition to a More
Sustainable Energy Future?................................... 449
Engineering Focus: Green City............................. 450

TYING IT ALL TOGETHER Wind Power and
Sustainability 452

Chapter 13 Summary... 453

Chapter 13 Assessment 454

ENGINEERING PROJECT 4
Design a Wind-Powered Generator 456

▪ **SPECIAL FEATURE**
Citizen Science: Of the People, By the People,
For the People... 458

Unit 5 Environmental Concerns 468

CHAPTER 14
HUMAN POPULATION AND URBANIZATION 470

▪ **EXPLORERS AT WORK** Healthy Planet, Healthy
People, with Christopher Golden 472

CASE STUDY Population 7.9 Billion.............................474

14.1 How Many People Can Earth Support? 475
Science Focus: The Population Debate............... 476

14.2 What Factors Influence the Size of the
Human Population?... 477

14.3 What Are the Effects of Urbanization on
the Environment? 482

14.4 How Can Cities Become More Sustainable?...... 489
▪ **EXPLORERS AT WORK** Caleb Harper 491
Engineering Focus: Curitiba's
Transportation System 492

TYING IT ALL TOGETHER
Analyzing Population Growth....................................... 494

Chapter 14 Summary... 495

Chapter 14 Assessment 496

TABLE OF CONTENTS   xi

CONTENTS

CHAPTER 15
ENVIRONMENTAL HAZARDS AND HUMAN HEALTH 498

- **EXPLORERS AT WORK** Health Testing Made Simple, with Hayat Sindi 500

CASE STUDY Mercury's Toxic Effects 502

15.1 What Are the Major Types of Health Hazards? 503

15.2 How Do Biological Hazards Threaten Human Health? 504
   Science Focus: Antibiotic Resistance 505

15.3 How Do Chemical Hazards Threaten Human Health? 511
   Science Focus: The BPA Controversy 514

15.4 How Can People Evaluate Risks from Chemical Hazards? 515

15.5 How Do People Perceive and Avoid Risks? 522
   EXPLORERS AT WORK Dan Buettner 524

TYING IT ALL TOGETHER What Makes a Healthy Community? 526

Chapter 15 Summary 527

Chapter 15 Assessment 528

CHAPTER 16
AIR POLLUTION, CLIMATE CHANGE, AND OZONE DEPLETION 530

- **EXPLORERS AT WORK** Conversing with Glaciers, with Erin Pettit 532

CASE STUDY Melting Ice in Greenland 534

16.1 What Are the Major Air Pollution Problems? 535
   Science Focus: Indoor Air Pollution 544

16.2 What Are the Effects of Climate Change? 546
   Science Focus: Using Models to Predict Change 552
   EXPLORERS AT WORK Sylvia Earle 561

16.3 How Can People Slow Climate Change? 562
   Engineering Focus: Carbon Capture 565

16.4 How Can People Reverse Ozone Depletion? 569

TYING IT ALL TOGETHER Melting Ice and Sustainability 572

Chapter 16 Summary 573

Chapter 16 Assessment 574

CHAPTER 17

SOLID AND HAZARDOUS WASTE 576

- **EXPLORERS AT WORK** Waging War Against Food Waste, with Tristram Stuart 578

CASE STUDY E-Waste—An Exploding Problem 580

17.1 What Are Problems Related to Solid and Hazardous Waste? 581

17.2 How Should Society Deal with Solid Waste? 587
   Science Focus: Bioplastics 592
   - **EXPLORERS AT WORK** T.H. Culhane 595

17.3 How Should Society Deal with Hazardous Waste? 596

17.4 How Can Society Transition to a Low-Waste Economy? 604

TYING IT ALL TOGETHER MSW Ecological Footprint 608

Chapter 17 Summary 609

Chapter 17 Assessment 610

CHAPTER 18

ENVIRONMENTAL ECONOMICS, POLITICS, AND WORLDVIEWS 612

- **EXPLORERS AT WORK** Adventurers and Scientists for Conservation, with Gregg Treinish 614

CASE STUDY The United States, China, and Sustainability 616

18.1 How Are Economic Systems Related to the Biosphere? 617
   Science Focus: Biosphere 2: A Lesson In Humility 619

18.2 How Can People Use Economic Tools to Address Environmental Problems? 620

18.3 How Can Society Enact More Just Environmental Policies? 625
   - **EXPLORERS AT WORK** Jane Goodall 633

18.4 How Can Society Live More Sustainably? 635

TYING IT ALL TOGETHER A Sustainable World 638

Chapter 18 Summary 639

Chapter 18 Assessment 640

ENGINEERING PROJECT 5
Design a Carbon-Capturing Device 642

APPENDIX 1 Science Safety Guidelines 644
APPENDIX 2 Measurement Units 648
APPENDIX 3 Periodic Table of Elements 649
APPENDIX 4 Countries of the World 650
GLOSSARY 652
GLOSARIO 666
INDEX 682
ACKNOWLEDGMENTS 694

About the Authors

G. TYLER MILLER

G. Tyler Miller has written 62 textbooks for introductory courses in environmental science, basic ecology, energy, and environmental chemistry. Since 1975, Miller's books have been the most widely used textbooks for environmental science in the United States and throughout the world. They have been used by almost 3 million students and have been translated into eight languages.

Miller has a professional background in chemistry, physics, and ecology. He has a PhD from the University of Virginia and has received two honorary doctoral degrees for his contributions to environmental education. He taught college for 20 years, and developed one of the nation's first environmental studies programs, before deciding to write environmental science textbooks full-time in 1975.

He describes his hopes for the future as follows.

"If I had to pick a time to be alive, it would be the next 75 years. Why? First, there is overwhelming scientific evidence that we are in the process of seriously degrading our own life-support system. In other words, we are living unsustainably. Second, within your lifetime we have the opportunity to learn how to live more sustainably by working with the rest of nature, as described in this book.

"I am fortunate to have three smart, talented, and wonderful sons—Greg, David, and Bill. I am especially privileged to have Kathleen as my wife, best friend, and research associate. It is inspiring to have a brilliant, beautiful (inside and out), and strong woman who cares deeply about nature as a lifemate. She is my hero. I dedicate this book to her and to the earth."

SCOTT E. SPOOLMAN

Scott Spoolman has more than 30 years of experience in educational publishing. He has worked with Tyler Miller first as a contributing editor and then as coauthor of *Living in the Environment, Environmental Science*, and *Sustaining the Earth*. With Norman Myers, he coauthored *Environmental Issues and Solutions: A Modular Approach.*

Spoolman holds a master's degree in science journalism from the University of Minnesota. He has authored numerous articles in the fields of science, environmental engineering, politics, and business. He has also worked as a consulting editor in the development of over 70 college and high school textbooks in fields of the natural and social sciences.

In his free time, he enjoys exploring the forests and waters of his native Wisconsin along with his family—his wife, environmental educator Gail Martinelli, and his children, Will and Katie.

Spoolman has the following to say about his collaboration with Tyler Miller.

"I am honored to be working with Tyler Miller as a coauthor to continue the Miller tradition of thorough, clear, and engaging writing about the vast and complex field of environmental science. I share Tyler Miller's passion for ensuring that these textbooks and their multimedia supplements will be valuable tools for students and instructors. To that end, we strive to introduce this interdisciplinary field in ways that will be informative and sobering, but also tantalizing and motivational.

"If the flip side of any problem is indeed an opportunity, then this truly is one of the most exciting times in history for students to start an environmental career. Environmental problems are numerous, serious, and daunting, but their possible solutions generate exciting new career opportunities. We place high priorities on inspiring students with these possibilities, challenging them to maintain a scientific focus, pointing them toward rewarding and fulfilling careers, and in doing so, working to help sustain life on the earth."

From the Authors

My Environmental Journey — G. Tyler Miller

My environmental journey began in 1966 when I heard a lecture on population and pollution problems by Dean Cowie, a biophysicist with the U.S. Geological Survey. It changed my life. I told him that if even half of what he said was valid, I would feel ethically obligated to spend the rest of my career teaching and writing to help students learn about the basics of environmental science. After spending six months studying the environmental literature, I concluded that he had greatly underestimated the seriousness of these problems.

I developed an undergraduate environmental studies program and in 1971 published my first introductory environmental science book, an interdisciplinary study of the connections between energy laws (thermodynamics), chemistry, and ecology. In 1975, I published the first edition of Living in the Environment. Since then, I have completed multiple editions of this textbook, and of three others derived from it, along with other books.

Beginning in 1985, I spent ten years in the deep woods living in an adapted school bus that I used as an environmental science laboratory and writing environmental science textbooks. I evaluated the use of passive solar energy design to heat the structure; buried earth tubes to bring in air cooled by the earth (geothermal cooling) at a cost of about $1 per summer; set up active and passive systems to provide hot water; installed an energy-efficient instant hot water heater powered by LPG; installed energy-efficient windows and appliances and a composting (waterless) toilet; employed biological pest control; composted food wastes; used natural planting (no grass or lawnmowers); gardened organically; and experimented with a host of other potential solutions to major environmental problems that we face.

I also used this time to learn and think about how nature works by studying the plants and animals around me. My experience from living in nature is reflected in much of the material in this book. It also helped me to develop the six simple factors of sustainability that serve as the integrating theme for this textbook and to apply these factors to living my life more sustainably.

I came out of the woods in 1995 to learn about how to live more sustainably in an urban setting where most people live. Since then, I have lived in two urban villages, one in a small town and one within a large metropolitan area.

Since 1970, my goal has been to use a car as little as possible. Since I work at home, I have a "low-pollute commute" from my bedroom to a chair and a laptop computer. I usually take one airplane trip a year to visit my sister and my publisher.

As you will learn in this book, life involves a series of environmental trade-offs. I hope you will join me in striving to live more sustainably and sharing what you learn with others. It is not always easy, but it sure is fun.

Cengage Learning's Commitment to Sustainable Practices

We the authors of this textbook and Cengage Learning, the publisher, are committed to making the publishing process as sustainable as possible. This involves four basic strategies:

- *Using sustainably produced paper whenever possible.* The book publishing industry is committed to increasing the use of recycled fibers, and Cengage Learning is always looking for ways to increase this content. Cengage Learning works with paper suppliers to maximize the use of paper that contains only wood fibers that are certified as sustainably produced, from the growing and cutting of trees all the way through paper production.

- *Reducing resources used per book.* The publisher has an ongoing program to reduce the amount of wood pulp, virgin fibers, and other materials that go into each sheet of paper used. New, specially designed printing presses also reduce the amount of scrap paper produced per book.

- *Recycling.* Printers recycle the scrap paper that is produced as part of the printing process. Cengage Learning also recycles waste cardboard from shipping cartons, along with other materials used in the publishing process.

- *Process improvements.* In years past, publishing has involved using a great deal of paper and ink for the writing and editing of manuscripts, copyediting, reviewing page proofs, and creating illustrations. Almost all of these materials are now saved through use of electronic files. Very little paper and ink were used in the preparation of this textbook.

Program Reviewers

Debra L. Bloomquist, Ph.D.
University of Toledo
Toledo, Ohio

Isaac E. Chandler
Teacher, Science Department
Suwannee High School
Live Oak, Florida

Amanda Gilbert, Ph.D.
University of Toledo
Toledo, Ohio

Donna L. Governor, Ph.D.
Teacher, Science
North Forsyth High School
Cumming, Georgia

Ronald Hochstrasser
Environmental Science Instructor
Sycamore Community High School
Cincinnati, Ohio

Michelle Joyce
Teacher, Science & Mathematics
Palmetto Ridge High School
Naples, Florida

Robert J. Seiple
Teacher, Science Department
Sycamore High School
Montgomery, Ohio

NATIONAL GEOGRAPHIC

Program Consultant

Featured Explorers

Christopher Thornton
Vice-Chair, Committee for Research and Exploration

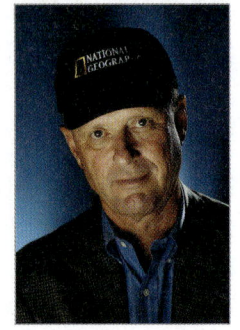

Robert Ballard
Ocean Explorer
National Geographic Explorer-in-Residence

Kamaljit S. Bawa
Conservation Biologist
National Geographic Grantee

Vitor Osmar Becker
Conservationist
National Geographic Grantee

Steve Boyes
Conservation Biologist
National Geographic Emerging Explorer

Dan Buettner
Author
National Geographic Fellow

Jennifer Burney
Environmental Scientist
National Geographic Emerging Explorer

Alizé Carrère
Cultural Ecologist
National Geographic Young Explorer

T.H. Culhane
Urban Planner
National Geographic Blackstone Innovation Challenge Grantee

Leslie Dewan
Nuclear Engineer
National Geographic Emerging Explorer

xvi

Featured Explorers continued

Dennis Dimick
Conservationist
National Geographic
Environmental Editor

Sylvia Earle
Oceanographer
National Geographic
Explorer-in-Residence

Jim Estes
Ecologist and
Evolutionary Biologist
National Geographic
Grantee

John Francis
Marine Biologist
National Geographic
Grantee and
Vice President for
Research, Conservation,
and Exploration

Christopher Golden
Ecologist and
Epidemiologist
National Geographic
Emerging Explorer

Jerry Glover
Agroecologist
National Geographic
Emerging Explorer

Jane Goodall
Primatologist
National Geographic
Explorer-in-Residence

Yu-Guo Guo
Chemist
National Geographic
Emerging Explorer

Caleb Harper
Urban Agriculturalist
National Geographic
Emerging Explorer

Zeb Hogan
Ecologist and
Photographer
National Geographic
Fellow

Osvel Hinojosa Huerta
Conservationist
National Geographic
Emerging Explorer

Thomas E. Lovejoy
Tropical and
Conservation Biologist
National Geographic
Fellow

Juan Martinez
Environmentalist
National Geographic
Emerging Explorer

Nalini Nadkarni
Biologist and Forest
Ecologist
National Geographic
Grantee

Erin Pettit
Glaciologist and
Geophysicist
National Geographic
Emerging Explorer

xvii

Featured Explorers continued

Sandra Postel
Freshwater Conservationist
National Geographic Fellow

Andrés Ruzo
Geologist
National Geographic Young Explorer

Enric Sala
Marine Ecologist
National Geographic Explorer-in-Residence

Joel Sartore
Photographer
National Geographic Fellow

Anna Savage
Evolutionary Geneticist
National Geographic Young Explorer

Çağan Hakkı Şekercioğlu
Ornithologist and Conservation Ecologist
National Geographic Emerging Explorer

Shah Selbe
Engineer and Conservation Technologist
National Geographic Emerging Explorer

Hayat Sindi
Science Entrepreneur
National Geographic Emerging Explorer

Tristram Stuart
Author and Campaigner
National Geographic Emerging Explorer

Gregg Treinish
Adventurer and Conservationist
National Geographic Emerging Explorer

Anand Varma
Natural History Photographer
National Geographic Young Explorer

Katey Walter Anthony
Aquatic Ecologist and Biogeochemist
National Geographic Emerging Explorer and Blackstone Innovation Challenge Grantee

Edward O. Wilson
Biologist and Author
National Geographic Hubbard Award recipient

Xiaolin Zheng
Nanoscientist
National Geographic Emerging Explorer

xviii

The National Geographic Learning Framework

Environmental Science: Sustaining Your World is one of the first programs to feature the National Geographic Learning Framework. The Learning Framework is an educational foundation based on research and perspectives from diverse fields of knowledge. It recognizes the distinct core principles and focus areas established at National Geographic along with the values held by families, communities, educators, and cultures.

The Learning Framework provides a common language that defines learning along three dimensions: the Attitudes, Skills, and Knowledge (A.S.K.) of "explorers" of all types, from the National Geographic Explorers featured in this program, to curious students exploring the world around them. The Learning Framework provides a way to ensure that students, educators, and the National Geographic Society are working toward common learning goals, and informs how we measure the impact of National Geographic products and resources.

The National Geographic Learning Framework is part of the Take Action or Citizen Science challenge included at the end of every chapter. It provides students with the opportunity to engage with the science content through the dimensions of the Learning Framework.

The Learning Framework's A.S.K. Dimensions

The chart below outlines the Attitudes, Skills, and Knowledge (A.S.K.) dimensions of the National Geographic Learning Framework as it applies to this Environmental Science program.

ATTITUDES	SKILLS	KNOWLEDGE
CURIOSITY. An explorer remains curious about how the world works throughout his or her life. An explorer is adventurous, seeking out new and challenging experiences. **RESPONSIBILITY.** An explorer has concern and empathy for the welfare of other people, cultural resources, and the natural world. An explorer is respectful, considers multiple perspectives, and honors others regardless of differences. An explorer is a global citizen. **EMPOWERMENT.** An explorer acts on curiosity, respect, responsibility, and adventurousness and persists in the face of challenges.	**OBSERVATION.** An explorer notices and documents the world around her or him and is able to make sense of those observations. **COMMUNICATION.** An explorer is a storyteller, communicating experiences and ideas effectively through language and media. An explorer has literacy skills, interpreting and creating new understanding from spoken language, writing, and a wide variety of visual and audio media. **COLLABORATION.** An explorer works effectively with others to achieve goals. **PROBLEM SOLVING.** An explorer is able to generate, evaluate, and implement solutions to problems. An explorer is a capable decision-maker—able to identify alternatives and weigh trade-offs to make a well-reasoned decision.	*In addition to the skills and attitudes of an explorer, people need to understand how our ever-changing and interconnected world works to function effectively and act responsibly as a global citizen. Critical knowledge required of explorers can be expressed through the four National Geographic key focus areas:* **OUR HUMAN STORY.** Exploring where we came from, how we live today, and where we may find ourselves tomorrow. **OUR LIVING PLANET.** Understanding the amazing, intricate, and interconnected systems of the changing planet we live on, its geography, and how the planet impacts living organisms and cultures. **CRITICAL SPECIES.** Revealing, celebrating, and helping to protect the amazing and diverse species with which we share our world. Appreciating the impact human history has had on other species. **NEW FRONTIERS.** Throughout history, searching every day for the "new" and the "next," using the latest technology and science to go places no one has ever been and find answers no one has ever found.

PREFACE

"Students who can begin early in their lives to think of things as connected, even if they revise their views every year, have begun the life of learning." —Mark Van Doren

Why Is It Important to Study Environmental Science?

Welcome to **environmental science**—an interdisciplinary study of how Earth works, how we interact with Earth, and how we can deal with the environmental problems we face. Because environmental issues affect every part of your life, the concepts, information, and issues discussed in this book and the course you are taking will be useful to you now and throughout your life.

Understandably, we are biased, but *we strongly believe that environmental science is the single most important course that you could take*. What could be more important than learning about Earth's life-support system, how our choices and activities affect it, and how we can reduce our growing environmental impact? Overwhelming evidence indicates that we will have to learn to live more sustainably by reducing our degradation of the planet's life-support system so that Earth can provide for future generations as well as it has for us. We hope this book will inspire you to become involved in this change in the way we view and treat Earth, which sustains us, our economies, and all other living things.

Improve Your Study and Learning Skills

You can get the most from this book and your course by improving your study and learning skills. Here are some suggestions for doing so.

- **Get organized.**
- **Make daily to-do lists.** Put items in order of importance, focus on the most important tasks, and assign a time to work on these items. Shift your schedule as needed to accomplish the most important items.
- **Set up a study routine in a distraction-free environment.** Develop a daily study schedule and stick to it. Study in a quiet, well-lit space. Take breaks every hour or so. During each break, take several deep breaths and move around; this will help you to stay more alert and focused.
- **Avoid procrastination.** Do not fall behind on your reading and other assignments. Set aside a particular time for studying each day and make it a part of your daily routine.
- **Make molehills out of mountains.** It can be difficult to read an entire chapter or book, write an entire paper, or cram for a test within a short period of time. Instead, break these large tasks (mountains) down into a series of small tasks (molehills). Each day, read a few pages of the assigned book or chapter, write a few paragraphs of the paper, and review what you have studied and learned.
- **Ask and answer questions as you read.** For example, "What is the main point of a particular section or paragraph?" Relate your own questions to the key question and core ideas addressed in each chapter lesson.
- **Focus on key terms.** Use the glossary in the back of this book to look up the meaning of terms or words you do not understand. This book shows all key terms in bold type the first time they are defined. The Chapter Assessment questions at the end of each chapter review some of the chapter's key terms. Flash cards for testing your mastery of key terms are available within MindTap, or you can make your own.
- **Interact with what you read.** Keep a notebook to paraphrase key sentences. Note page numbers of important paragraphs, diagrams, or sections you want to return to. Use the book's index to locate specific information. The MindTap edition supports extensive note-taking features such as highlighting, bookmarking, and adding your own notes to the text.
- **Review to reinforce learning.** Before class, review the material you learned in the previous class session and read the assigned material.
- **Become a good note taker.** Learn to write down the main points and key information from any lecture using your own shorthand system. Review, fill in, and organize your notes as soon as possible after each class.
- **Check what you have learned.** Within every lesson are multiple Checkpoint questions for you to check your understanding. The Lesson Assessments at the end of each lesson will help you gauge your understanding of the core ideas. We suggest that you try to answer the Checkpoint and Lesson Assessment questions as you advance through the chapters.

- ***Write out answers to questions to focus and reinforce learning.*** Write down your answers to the critical thinking questions in the Explorers at Work and Science and Engineering Focuses. These questions, along with the As You Read in the Case Study, are designed to inspire you to think critically about key ideas and connect them to other ideas and to your own life. Also, write down your answers to all Chapter Assessment questions. You can find additional quizzes in MindTap. Save your answers for review and test preparation.
- ***Use the buddy system.*** Study with a friend or become a member of a study group to compare notes, review material, and prepare for tests. Explaining something to someone else is a great way to focus your thoughts and reinforce your learning. Attend any review sessions offered by your teacher.
- ***Learn your teacher's test style.*** Does your teacher emphasize multiple-choice, fill-in-the-blank, true-or-false, factual, or essay questions? How much of the test will come from the textbook and how much from lecture material? Adapt your learning and studying methods to this style.
- ***Become a good test taker.*** Avoid cramming. Eat well and get plenty of sleep before a test. Arrive on time or early. Calm yourself and increase your oxygen intake by taking several deep breaths. (Do this also about every 10–15 minutes while taking the test.) Look over the test and answer the questions you know well first. Then work on the harder ones. Use the process of elimination to narrow down the choices for multiple-choice questions. For essay questions, organize your thoughts before you start writing. If you have no idea what a question means, make an educated guess. You might earn some partial credit and avoid getting a zero. Another strategy for getting some credit is to show your knowledge and reasoning by writing something like this: "If this question means so-and-so, then my answer is _____?

Improve Your Critical Thinking Skills

Critical thinking involves developing skills to analyze information and ideas, judge their validity, and make decisions. Critical thinking helps you distinguish between facts and opinions, evaluate evidence and arguments, and take and defend informed positions on issues. It also helps you integrate information, see relationships, and apply your knowledge to new situations. Here are some basic skills for learning how to think more critically.

- ***Question everything and everybody.*** Be skeptical, as any good scientist is. Do not believe everything you hear and read, including the content of this book, without evaluating the information. Seek other sources and opinions.
- ***Identify and evaluate your personal biases and beliefs.*** Each of us has biases and beliefs taught to us by our parents, teachers, friends, role models, and our own experience. What are your basic beliefs, values, and biases? Where did they come from? What assumptions are they based on? How sure are you that your beliefs, values, and assumptions are right—and why are they right? According to the American psychologist and philosopher William James, "A great many people think they are thinking when they are merely rearranging their prejudices."
- ***Be open-minded and flexible.*** Be open to considering different points of view. Suspend judgment until you gather more evidence, and be willing to change your mind. Recognize that there may be a number of useful and acceptable solutions to a problem and that very few issues are either black or white. Try to take the viewpoints of those you disagree with to better understand their thinking. There are trade-offs involved in dealing with any environmental issue, as you will learn in this book.
- ***Be humble about what you know.*** Some people are so confident in what they know that they stop thinking and questioning. To paraphrase American writer Mark Twain, "It's what we know is true, but just ain't so, that hurts us."
- ***Find out how the information related to an issue was obtained.*** Are the statements you heard or read based on firsthand knowledge and research or on hearsay? Are unnamed sources used? Is the information based on reproducible and widely accepted scientific studies or on preliminary scientific results that may be valid but need further testing? Is the information based on a few isolated stories or experiences or on carefully controlled studies that have been reviewed by experts in the field involved? Is it based on unconfirmed and doubtful scientific information or beliefs?

PREFACE

- ***Question the evidence and conclusions presented.*** What are the conclusions or claims, based on the information you're considering? What evidence is presented to support them? Does the evidence actually support them? Is there a need to gather more evidence to test the conclusions? Are there other, more reasonable conclusions?
- ***Try to uncover differences in basic beliefs and assumptions.*** On the surface, most arguments or disagreements involve differences of opinion about the validity or meaning of certain facts or conclusions. Scratch a little deeper and you will find that many disagreements are based on different (and often hidden) assumptions concerning how we look at and interpret the world around us. Uncovering these basic differences can allow the parties involved to understand one another's viewpoints and to "agree to disagree" about their basic assumptions, beliefs, or principles.
- ***Try to identify and assess any motives on the part of those presenting evidence and drawing conclusions***. What is their expertise in this area? Do they have any unstated assumptions, beliefs, biases, or values? Do they have a personal agenda? Can they benefit financially or politically from acceptance of their evidence and conclusions? Would investigators with different basic assumptions or beliefs take the same data and come to different conclusions?
- ***Expect and tolerate uncertainty.*** Recognize that scientists cannot establish absolute proof or certainty about anything. However, the results of science can have a high degree of certainty and reliability.
- ***Check the arguments you hear and read for logical errors and debating tricks.*** Here are six of many examples of such debating tricks. First, attack the presenter of an argument rather than the argument itself. Second, appeal to emotion rather than facts and logic. Third, claim that if one piece of evidence or one conclusion is false, then all other related pieces of evidence and conclusions are false. Fourth, say that a conclusion is false because it has not been scientifically proven. (Scientists never prove anything absolutely, but they can often establish high degrees of certainty.) Fifth, inject irrelevant or misleading information to divert attention from important points. Sixth, present only either/or alternatives when there may be a number of options.
- ***Do not believe everything you read on the Internet.*** The Internet is a wonderful and easily accessible source of information. It includes alternative explanations and opinions on almost any subject or issue—much of it not available in the mainstream media and scholarly articles. Blogs of all sorts have become a major source of information, more important than standard news media for some people. However, because the Internet is so open, anyone can post anything they want to some blogs and other websites with no editorial control or review by experts. As a result, evaluating information on the Internet is one of the best ways to put into practice the principles of critical thinking discussed here. Use and enjoy the Internet, but think critically and proceed with caution. For example, be wary of "clickbait" articles. These articles use compelling headlines and imagery as lures to attract readers to their advertisements, but their claims are not necessarily supported by sufficient evidence.
- ***Develop principles or rules for evaluating evidence.*** Develop a written list of principles to serve as guidelines for evaluating evidence and claims. Continually evaluate and modify this list on the basis of your experience.
- ***Become a seeker of wisdom, not a vessel of information.*** Many people believe that the main goal of their education is to learn as much as they can by gathering more and more information. We believe that the primary goal is to learn how to sift through mountains of facts and ideas to find the few nuggets of wisdom that are the most useful for understanding the world and for making decisions. This book is full of facts and numbers, but they are useful only to the extent that they lead to an understanding of key ideas, connections, concepts, and scientific laws and theories. The major goals of the study of environmental science are to find out how nature works and sustains itself (environmental wisdom) and to use principles of environmental wisdom to help make human societies and economies more sustainable, more just, and more beneficial and enjoyable for all. As writer Sandra Carey observed, "Never mistake knowledge for wisdom. One helps you make a living; the other helps you make a life."

To help you practice critical thinking, we have supplied questions throughout this book. A good way to improve your critical thinking skills is to compare your answers with those of your classmates and to discuss how you arrived at your answers.

Use the Learning Tools We Offer in This Book

In every chapter, we have included a number of tools that are intended to help you improve your learning skills and apply them. First, consider the **Key Questions** list at the beginning of each chapter. You can use these to preview a chapter and to review the material after you've read it.

Each **Explorers at Work** profiles the positive, inspiring, and creative problem-solving thinkers and scientists of National Geographic. Little by little, day by day, they are changing the world. Read how they are doing it, and then think about what you could do now and in the future.

A **Case Study** anchors a chapter's big ideas to a real-world example. At the end of each Case Study is an *As You Read* note that challenges you to connect the chapter concepts to your own knowledge or life experiences. Learning objectives are found in the **Core Ideas and Skills** box at the start of every lesson. The **Key Terms** box gives you an at-a-glance list of the important words you will encounter that you can look up in the **English** or **Spanish Glossary**. Our six factors of sustainability summarize the overarching theme of this book. The six factors are introduced in Figures 1-2 and 1-3 and we refer back to them throughout the book.

Consider This boxes call out some of the often surprising connections between environmental problems or processes and the products we use or the activities we partake in. They are intended to get you to think carefully about the choices we take for granted, and about how our activities might affect the environment.

All chapters include big, bold **photographs**. "Seeing is believing," and the right photos can inspire curiosity and help promote greater wisdom through knowledge. Many photographs in these pages were taken by today's best photographers. We encourage you study their images and think about what they might mean. Every photograph was chosen for a reason.

Following the last lesson is a **Tying It All Together**, which guides you through the analysis of data or information related to the real-world challenges discussed in the Case Study and throughout the chapter. Each Tying It All Together is an opportunity to improve your skills in the areas of **Science, Technology, Engineering, and Math (STEM)**. We recommend that you work in teams to solve these STEM challenges. After the Tying It All Together, a **Chapter Summary** reviews what we consider to be the big ideas and important details from each lesson.

Finally, we included a **Chapter Assessment** with questions that are organized into *Key Terms*, *Key Concepts*, and *Critical Thinking*. These questions give you practice applying chapter material to the real world and your own life. The Chapter Assessments end with a **Chapter Activities** section made up of two parts. *Part (A)* is a **Develop Models**, **Experiment**, or **Investigate** activity. These offer you another chance to hone your STEM skills and deepen your understanding of the chapter content. *Part (B)* is a **Citizen Science** or **Take Action** activity. The Citizen Science activities encourage you to get involved in an ongoing project or National Geographic initiative. Alternatively, the Take Action activities are more about using scientific knowledge to inform environmental action.

Each unit ends with a guided **STEM Engineering Project** in which you get to take the wheel as the problem-solver working on a real environmental issue. An Engineering Project may take several weeks to complete and involve careful planning and collaboration. In your role as an engineer, you will get to put to use your math and technology skills, apply your scientific knowledge, and be creative.

Know Your Own Learning Preference

People prefer to learn in different ways. It can be helpful to know your own learning preference. For example, you may find reading and viewing illustrations and diagrams to be the most enjoyable way to learn. Or, you may enjoy learning more by listening and discussing, in which case you might also benefit from reading aloud while studying and using an audio recorder in lectures for study and review. If you like to approach learning logically, you might enjoy using concepts and logic to uncover and understand a subject rather than relying mostly on memory.

This book and the related MindTap edition contain plenty of tools for all learners. The animations and videos in MindTap provide visual and auditory support for many of the concepts presented. In addition, online features such as an easy-to-use note-taking feature and flash cards help you remember

PREFACE

key terms and concepts. This is a highly visual book with hundreds of photographs and diagrams carefully selected to illustrate important concepts and processes. Podcasts featuring interviews of National Geographic Explorers add context to environmental issues. And for those of you who enjoy learning through logic, the Key Questions and Core Ideas and Skills provide a logical progression through each lesson and help you connect concepts and major principles. We urge you to become aware of your own learning preferences to make the most of these tools.

This Book Presents a Positive, Realistic Environmental Vision of the Future

Our goal is to present a positive vision of our environmental future based on realistic optimism. To do so, we strive not only to present the facts about environmental issues, but also to give a balanced presentation of different viewpoints. We consider the advantages and disadvantages of various technologies and proposed solutions to environmental problems. We argue that environmental solutions usually require trade-offs among opposing parties, and that the best solutions are win-win solutions. Such solutions are achieved when people with different viewpoints work together to come up with a solution that both sides can live with. And we present the good news as well as the bad news about efforts to deal with environmental problems.

One cannot study a subject as important and complex as environmental science without forming conclusions, opinions, and beliefs. However, we argue that any such results should be based on use of critical thinking to evaluate conflicting positions and to understand the trade-offs involved in most environmental solutions. To that end, we emphasize critical thinking throughout this textbook, and we encourage you to develop a practice of thinking critically about everything you read and hear, both in school and throughout your life.

Help Us Improve This Book

Researching and writing a book that covers and connects the numerous major concepts from the wide variety of environmental science disciplines is a challenging and exciting task. Almost every day, we learn about some new connection in nature. However, in a book this complex, there are bound to be some errors—some typographical mistakes that slip through and some statements that you might question, based on your knowledge and research. We invite you to contact us to correct any errors you find, point out any bias you see, and suggest ways to improve this book. Please e-mail your suggestions to Tyler Miller at mtg89@hotmail.com or Scott Spoolman at spoolman@tds.net.

Now start your journey into this fascinating and important study of how Earth's life-support system works and how we can leave our planet in a condition at least as good as what we now enjoy.

Online Learning Solutions and Resources

You may have a variety of digital materials available to you to help you take your learning experience beyond this physical book.

- **MindTap Environmental Science.** MindTap is a digital learning platform that works alongside your other resources to deliver course content across the range of electronic devices in your life. MindTap is built on an app model allowing better digital collaboration and delivery of content across many different products and resources, some of which you may already be using.

ENVIRONMENTAL SCIENCE

SUSTAINING YOUR WORLD

NATIONAL GEOGRAPHIC | CENGAGE

G. TYLER MILLER • SCOTT E. SPOOLMAN

NATIONAL GEOGRAPHIC

Into the
Okavango
with National Geographic

The Beating Heart of Our Planet: The Okavango Delta

Landlocked on the continent of Africa in the northwestern part of Botswana, the Okavango Delta is one of the richest and most beautiful wildlife wildernesses on the planet. Unlike most deltas, it is entirely inland. Each winter, instead of flowing into an ocean or sea, the Okavango River's life-giving waters spill out across a flat, thirsty landscape of grassy savannas. Its waters seem to stir up life from the very dust and inspire long, epic migrations of some of Earth's largest land mammals. The complex patterns that have evolved around the ebb and flow of this miraculous region are part of the pulse of a whole continent. But the *catchment areas* (land areas from which rainfall flows into a river or body of water) in Angola and Namibia have remained unprotected and relatively unexplored ... until now.

Visible from space, the Okavango Delta is the size of the state of Texas. It's one of Africa's last truly wild landscapes, like the Sahara, the Serengeti, and the Congo. Some 100,000 elephants roam free across the land. Lions, leopards, hyenas, rhinos, cheetahs, crocodiles, and wild dogs also thrive here, as well as nearly 500 bird and over 1,000 plant species.

Since 2011, the Into the Okavango expedition team, made up of indigenous Bayei river bushmen, National Geographic Explorers, and numerous engineers, artists, and scientists, has been making visits to the remote delta in one of the first "live-data" expeditions. They constantly record and upload data from the field to their website via satellite. The data is also available through a public API, or application programming interface, that allows anyone to analyze and examine the information. "We're connecting society with the wilderness," explains conservation biologist and expedition leader Steve Boyes.

When the team is in the Okavango, state-of-the-art sensors record personal data about expedition members every 10 seconds, including heart rate and GPS positional data. Team members also post water quality data and document animal and bird sightings with photos from the field. These data will serve as a baseline to which data from future studies can be compared. It's all part of an effort to measure the natural conditions of the environment and how humans continue to affect the Okavango ecosystem.

An African elephant ambles across the flooded savanna, which is a seasonal refuge for thousands of elephants and other animals.

Into the Okavango:
An Explorer's Interview

Shah Selbe is an engineer, a conservation technologist, and a National Geographic Emerging Explorer. He is also an important member of the Into the Okavango expedition team. National Geographic Learning interviewed Dr. Selbe about biodiversity, his experiences in the Okavango, and what we can do to help protect the environment.

National Geographic Learning: Shah, why do you believe Earth's biodiversity is at risk?

Shah Selbe: For most of the time that humans have inhabited this planet, their impact on Earth's ecosystems has been relatively minimal. Although there has been proof of human activity driving certain species to extinction, it wasn't until recent industrialization that larger-scale change resulted. Since then, we have started to see large shifts in previously stable ecosystems coming from habitat loss, overhunting, excessive fishing, or pollution. The extinction rate of species has been observed to be as much as 1,000 times what we would consider normal, leading scientists to consider this the sixth mass extinction event in Earth's history. Many people believe we have entered a new era during which geologically significant conditions and processes are being profoundly altered by human activities. The most commonly known outcome of this era is climate change.

NGL: Could you give a specific example of how biodiversity is impacted by environmental problems?

Selbe: There are a number of ways that environmental problems can impact biodiversity. Climate change is driven by carbon dioxide emissions that come from our industrial processes, coal power plants, and automobiles. As the amount of carbon dioxide in the air increases, the oceans compensate by absorbing more of it. Current estimates put 30–40% of the carbon dioxide that humans release dissolving into our oceans, rivers, and lakes. This impacts biodiversity because the carbon dioxide reacts with the water to make the oceans more acidic. This is particularly problematic for organisms like coral, plankton, oysters, and mussels. These species help build habitat for many other species,

Shah Selbe poses with a giraffe's skull that he found. The hornlike growths on top of the skull are called *ossicones*.

play key roles in the food chain, and help clean pollutants from the water.

NGL: What are one or two meaningful solutions that you believe could significantly impact Earth's environmental problems?

Selbe: One of the greatest things we can do to help environmental impact is create protected areas or reserves. By creating areas where destructive activities (like commercial fishing, mining, or development) are not allowed, it helps to alleviate stress from the ecosystem and allow wildlife populations to rebound. We have seen how truly successful this can be, but only if it is actually protected. Some of the land we have most successfully protected in the United States is our national parks, which are some of the most beautiful parts of Earth and have become a point of pride.

Additionally, the emerging field of conservation technology has the potential to help us solve some of Earth's environmental problems. We now have the technologies that can help us learn more about the planet and measure changes accurately enough to do something about it, as we are doing in the Okavango Delta. Thanks to the innovation we have seen in computers and communication systems (like cellular

Giraffes search for food in the Okavango. With a population in the wild of more than 100,000 individuals, giraffes are not considered threatened. Like most megafauna, however, their population is declining.

INTO THE OKAVANGO WITH NATIONAL GEOGRAPHIC

networks and the Internet), we can now have a profound impact in protecting these places, sharing the beauty of them with people everywhere, and making the most out of the limited resources that scientists and conservationists have.

NGL: What is one of the most significant environmental challenges to biodiversity in the Okavango Delta?

Selbe: One of the biggest challenges to the Okavango Delta has to do with the source of the floodwaters—the inlet waters. The Delta starts in Angola, flows through Namibia, and then fans out into a massive delta in the middle of the Kalahari Desert. The Okavango is home to critically important species of elephants, hippos, lions, and many of those other charismatic megafauna (very large animals) that draw so many of us to sub-Saharan Africa. If anything were to happen to that inlet water due to mining activity, pollutants, or building of dams, then it could fundamentally change the floods that supply water to those animals.

Protection of those inlet waters would be the most important way to ensure healthy biodiversity in the Okavango. One of the best ways to do that would be to declare the areas upstream as protected reserves as well. This could help to limit development and stop pollutants from entering the water. It would also help stop poaching and provide wildlife corridors for the animals of the Delta to use. (Wildlife corridors are natural regions connecting animal populations that are separated by human activities such as urban or agricultural land development.) If we can extend this sort of protection through Namibia and Angola, it could have incredible implications on the animals of the region.

As the third-largest land mammal on Earth, the hippopotamus is not to be taken lightly. Bulls are highly territorial in water. Pods, or groups of hippopotamuses can be among the most dangerous animals in the Okavango.

NGL: How do the efforts of you and the other members of the Okavango Delta team help protect biodiversity?

Selbe: Our team works to protect biodiversity in a number of ways. First, we conduct a yearly wildlife census on mekoro (dugout canoes, or canoes made from a hollowed tree trunk) because that mode of transportation allows us to get an unparalleled idea of species numbers and behaviors. During that census, we take habitat photographs and collect environmental data (water quality, air quality) that we can correlate with our wildlife sightings and GPS coordinates. As the conservation technologist, I also create environmental monitoring stations, which I call "the Internet of Earth things," that send back data in real-time and allow us to monitor the water even when we aren't in the field.

NGL: What's the most exciting adventure you've had in the Okavango Delta region?

Selbe: Deep in the wild during our expeditions, we encounter exciting adventures on a daily basis. We have been charged by hippos and elephants, had hyenas surround our tents, rowed over massive crocodiles, and watched lions walk through our camp. We are experienced in dealing with these sorts of scenarios and we are careful, but they are each adventures unto themselves. The most scary thing is dealing with the hippos. They are fiercely territorial creatures and will charge you full speed if you get too close. When a semiaquatic 3,000-pound animal charges at you at full speed while you are in a canoe, you remember that vividly. Thankfully we have always ended up safe in the end.

NGL: Your Okavango Delta "team" is diverse and multi-talented. Could you explain how people with a wide range of specialties and backgrounds play a role in your efforts?

Selbe: The entire birth of our Into the Okavango project came from an interdisciplinary approach. Steve Boyes, the expedition leader, is a charismatic conservation biologist and one of the most inspiring

Local members of the expedition taught Boyes, Selbe, and others the basics of how to pole the mekoro—dugout canoes that have been the Bayei's preferred mode of transportation for centuries. Today most mekoro are made of fiberglass to conserve trees.

advocates of protecting the delta. Jer Thorp, the data artist, is a brilliant individual who has fundamentally changed how we visualize and share the data we collect. As the conservation technologist, I bring in an understanding of sensors and hardware that can profoundly change how we do science in the region. Every member of our team brings capabilities like this into the effort. It is very inspiring to be surrounded by such incredible people!

NGL: How does technology play a role in the work you do in the Okavango?

Selbe: A large part of my job is figuring out how to use new technologies to protect these places. It can vary, from sensors to drones to smartphone apps, all depending on the problem that needs to be solved. I was one of the first to start working in this field, so my work has developed alongside other conservationists. There is plenty of potential for it to have profound impacts on how we protect the planet, provided that we do it sustainably and focus on helping the communities that live near the implementation areas.

NGL: How does forming relationships with people help you accomplish your goals in the Okavango?

Selbe: All of the work we do is done side-by-side with the people of the delta. Our guides on the mokoros are members of the Bayei tribe, people who call the delta home. We work closely with local scientists, politicians, conservationists, and tourist industries to support our project and build an optimistic future for the Okavango. Without those relationships, we wouldn't be able to do any of the work we do.

NGL: When you are in the Okavango, what do you miss the most about home?

Selbe: When we are camping so far away from civilization, we start to miss pretty basic stuff that we normally take for granted. I think, for me, that tends to be hamburgers and tacos! You start to crave those pretty badly when being away from them for so long. I also miss my family since it is pretty difficult to find time to talk with them when you are beyond any cellular signal.

NGL: What is the most enjoyable part of doing fieldwork?

Selbe: I love everything about being in the field. There is a certain amount of resourcefulness that you have to adopt in order to make things work out there. It is daily problem-solving that also involves innovating on the spot to meet your science objectives or stay safe. There is also a camaraderie that comes from being out with the team that you can't quite experience unless you go through something like that. It is incredible.

NGL: What can a high school student do to help protect biodiversity and the environment?

Selbe: Everyone can help protect the environment. The easiest way is to start to learn more about the products you use and foods that you eat, and to try to make better choices that have less of an impact on the environment. There is quite a lot of power in market pressure, so if enough of us stop using something that is environmentally damaging, companies may stop making it.

There are also many volunteer opportunities available for both biodiversity and environmental protection, some of which can be done from home (like tracking data). And it always helps to learn more about the world through science so that you have a better understanding of sustainability.

> "We have unprecedented opportunities to improve the world. But only if we act in time."
> —SHAH SELBE

A group of Hambukushu women fish with baskets. Like the Bayei, the Hambukushu people are an indigenous tribe of the Okavango.

INTO THE OKAVANGO WITH NATIONAL GEOGRAPHIC

Environmental Science
IN ACTION

When an international team of scientists, filmmakers, engineers, and adventurers arrives in a remote region of a country like Angola to monitor and protect the environment, they attract some attention. Social media lit up on May 25, 2015, when an astronaut in the International Space Station captured a stunning photograph of the Okavango Delta system and tweeted a simple message to the team: "Wishing you luck." The world needs more people on a mission of conservation and sustainability—people like those on the Into the Okavango expedition team.

As they gather important data about the wildlife, water, and environment, the team also forges personal relationships with the people who depend on the water of this region. They hear their concerns, stories, and opinions. "We've met with both of the governors of the provinces we're passing through," says expedition leader, Steve Boyes. "We have support from the president of Angola and the ambassadors of the U.S., the U.K., South Africa, and Namibia." It's important to note that the conservation work Boyes and his colleagues are doing is supported by governmental officials and local citizens, in part because of the potential for bringing tourism to the region, primarily along the Okavango River's catchments.

But while the Okavango Delta itself is a protected region, its headwaters, which originate in the highlands of Angola, are not. That part of the catchment, the Cuito River, flows through an area riddled with dangerous land mines, buried in the ground during the country's decades-long civil war. The fighting in this region ended in 2002, but removing the mines has been a slow and challenging process. Yet, as more of the mines are removed, local people are able to return to their native lands. Although this has been a good thing for people, there have been unintended consequences.

Backlit by the sun, a Bayei bushman in a mokoro peers into the calm floodwaters.

When Angolans move back into their homelands, they clear trees from the region to farm the land to support their families. But the trees they are removing in the headwaters region are the reason the Okavango Delta exists. Without these woodland trees, the waterways will stop flowing and the delta will dry up. In essence, the land mines were protecting the landscape from this deforestation, and now the land mines are being removed. Says Boyes, "As they de-mine and open up a region, you can see where people are clearing the forest to grow cassavas. That's causing a lot of damage. It's adding a lot of urgency to our work."

Now it's up to conservationists like Boyes and the Into the Okavango expedition to gather valuable data and provide solutions that the Angolan government can implement in order to establish sustainable environmental protection. "We need to look at better land management. We need to look at critical ecosystems we can protect. We're documenting the habitat, the unique species, and issuing opinions to the government on potential tourism possibilities, game reserves, national parks, protected areas." In other words, they are working hard, but there is much more work to be done to save the Okavango Delta.

In the Okavango, Shah Selbe's tools range from an advanced water-quality sensor probe with solar panels to a basic pen and notebook.

INTO THE OKAVANGO WITH NATIONAL GEOGRAPHIC 11

UNIT 1
ECOLOGY AND ECOSYSTEMS

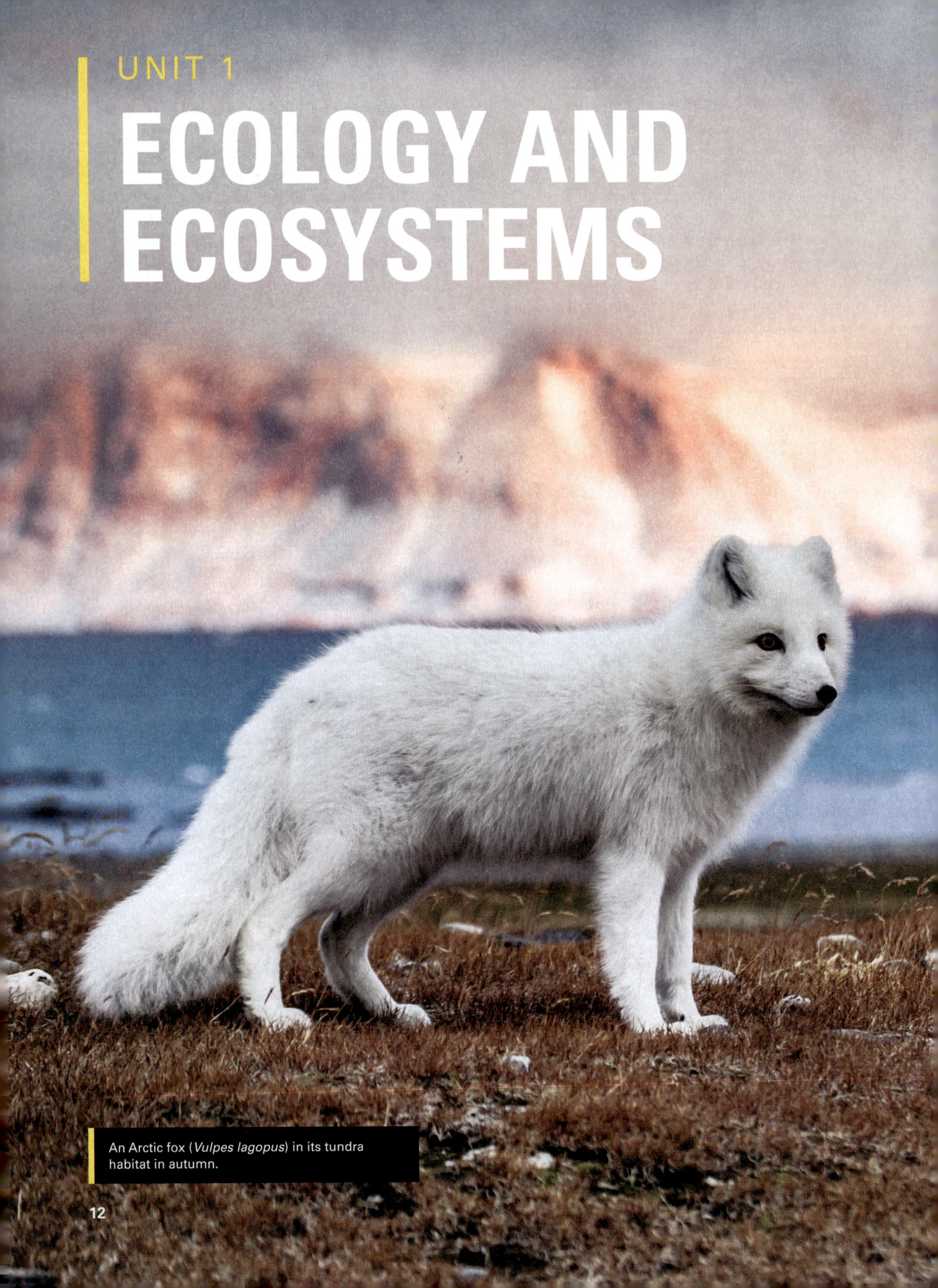

An Arctic fox (*Vulpes lagopus*) in its tundra habitat in autumn.

The Arctic fox (*Vulpes lagopus*) is an example of an animal uniquely adapted to its ecosystem. These adept hunters' relatively small size and thick fur protect them from the frigid winters of the Arctic tundra. They can withstand external temperature differences of 90–100°C (160–180°F) from their internal body temperatures. Their coats change color seasonally from white to brown, camouflaging them from predators and making it easier for them to hunt prey. Though the Arctic fox is not endangered globally, some local communities are threatened by overhunting, competition with other fox species, and loss of habitat due to climate change.

Because of this, legislatures in affected regions are implementing hunting ordinances as well as other regulations aimed at limiting the decline of Arctic fox populations.

In this unit, you will learn about ecology and the complex interactions between different parts of the Earth system. You will examine why global citizens find it important to assure the sustainability of the planet's ecosystems and how maintaining our planet's future hospitability depends on the preservation of habitats and resources.

CHAPTER 1
THE ENVIRONMENT AND SUSTAINABILITY

CHAPTER 2
SCIENCE, MATTER, ENERGY, AND SYSTEMS

CHAPTER 3
ECOSYSTEM DYNAMICS

ENGINEERING PROJECT 1: Design a Method for Treating Contaminated Soil

THOUSANDS OF PEOPLE rely on fish from reefs in the Coral Triangle—a geographic region home to three-quarters of the world's coral species. Conservation groups have worked hard to protect the high biodiversity in these waters. Their efforts and the local fishing practices mean that the natural resources here will be able to support the local fishing communities for generations to come.

CHAPTER 1
THE ENVIRONMENT AND SUSTAINABILITY

KEY QUESTIONS

1.1 What are some key factors of sustainability?

1.2 How are our ecological footprints affecting Earth?

1.3 What causes environmental problems and why do they persist?

1.4 What is an environmentally sustainable society?

NATIONAL GEOGRAPHIC | EXPLORERS AT WORK

Reconnecting People with Nature

with National Geographic Explorer Juan Martinez

Ten years ago no one would have predicted that Juan Martinez would become a leader in making nature accessible to urban and at-risk teens, least of all Juan himself. Juan grew up in South Los Angeles, in a poor neighborhood known for gang violence. In fact, he was headed for a different life, possibly among the city's gangs. In his neighborhood, you did what you had to do to survive and support your family.

A failing grade in a high school science course made all the difference. Martinez's teacher recognized his potential and gave him an ultimatum—stay in detention all year or join the Eco-Club. Anything was better than detention, so Juan soon found himself planting jalapeño seeds in the club's garden. He eventually earned a chance to join a two-week trip to Wyoming's Teton Science Schools. It was his first real exposure to the dramatic topography of the outdoors. Martinez admits that "ten years later, I still can't find words to describe the first moment I saw those mountains rising up from the valley."

Martinez's experiences in nature changed his life in ways that still amaze him. "I went from hating everyone and everything to being part of organizations that are all about building supportive, thriving communities. Nature was my bridge between those two totally different worldviews." He became determined to share that experience of transformation with others.

Today, Juan is a leader in providing outdoor learning experiences of all kinds to young people across the country. His activities range from working side-by-side with local gardeners and taking urban young people into the wilderness to attending meetings on nature issues at the White House. He has advised the U.S. Department of the Interior on plans to create a youth conservation corps. He also heads up the Natural Leaders Network of the Children & Nature Network. This organization links environmental organizations with corporations, government, and education, all with the goal to connect young people with the natural environment.

It's easy to focus on environmental disasters or the sometimes depressing news on climate change, endangered species, or habitat loss. Instead, Martinez and the Natural Leaders Network make a point of emphasizing the benefits of engaging with the natural world and building a sustainable life. Martinez understands firsthand the deeply positive effect that connecting with the outdoors can have on an individual—but he also recognizes how meaningful it can be to help others find those same connections. He says, "When I think back to the people who were here to step in for me at a critical moment, I want to make that same kind of difference." Reaching young people in classrooms and on the streets, Juan Martinez is a powerful voice for a greener generation.

Thinking Critically

Infer Juan Martinez works on a daily basis to get young people outdoors and hands-on with nature. In what ways might his work help support the drive for a more sustainable planet, both today and in the future?

Juan Martinez works to connect people—especially youth—with the outdoors. Experiences in nature changed his life for the better and he hopes to do the same for others.

17

CASE STUDY
The Greening of American Schools

Sustainability is the capacity of Earth's natural systems and human cultural systems to survive and flourish into the long-term future. To ensure sustainability, humans must conserve resources and maintain environmental conditions necessary for health and survival, both now and in the future. Sustainability is the big idea of this book.

Many schools today are opting for more sustainable facilities and include environmental and sustainability topics in their curricula. In constructing a new K–12 building in 2012, the Milton-Union School District in rural West Milton, Ohio, put several sustainability measures into practice. These "green" features included:

- Large exterior windows in each classroom to reduce the need for artificial lighting.
- Sprayed polyurethane foam insulation that provides a barrier without gaps to help regulate the indoor climate.
- High-efficiency chillers to create ice at night when electricity rates are lower. The ice then cools the building during the day.
- A wind turbine, a solar cell array, and a solar thermal system that heats city water to reduce reliance on natural gas.
- A rainwater harvesting system that stores 283,900 liters (75,000 gallons) of water underground (about the volume of 960 average-sized bathtubs). The rainwater is used to flush all toilets and urinals.
- A long ditch that filters stormwater runoff as it soaks into the ground.

During construction of Milton-Union School District, about 76 percent of "waste" was sent to recycling or repurposing centers. Builders chose materials with high levels of recycled content. And now that school is underway, students are in charge of recycling and waste reduction programs.

The sustainability effort goes beyond the school's infrastructure too. Bus routes are designed for efficiency. Ten percent of the students walk or bike to school. Students receive preferred parking if they drive fuel-efficient cars. The cafeteria uses local food whenever possible. The school grounds include woods and open prairie. Landscaping incorporates water-efficient native plant species.

Curricula at all levels at Milton-Union include environmental and sustainability themes. Preschoolers and kindergarteners visit the rain gardens and woods for art activities that focus on the relationship between plants, insects, and animals on the school campus. Elementary school students investigate living and nonliving things, including the complex habitats of various plants and animals and their special adaptations. High school students focus on biodiversity with emphasis on the impact that humans have on the environment.

This chapter introduces you to environmental science and sustainability. You will learn about today's environmental problems, how they developed, and how humankind can reach the ultimate goal of an environmentally sustainable society.

As You Read Think about other ways people and businesses are increasing energy efficiency and reducing their environmental impact.

FIGURE 1-1 ▼
Sustainable aspects of Milton-Union School District include a wind turbine and "green" landscaping.

1.1 What Are Some Key Factors of Sustainability?

CORE IDEAS AND SKILLS

- Discuss the study of environmental science and its goals.
- Describe the concept of sustainability and its significance to environmental science.
- Understand the link between ecosystem services and natural resources.

KEY TERMS

environment
environmental science
ecology
ecosystem
environmentalism
sustainability
natural capital
natural resource
inexhaustible resource
renewable resource
nonrenewable resource
ecosystem service

Environmental Science Is a Study of Connections in Nature

The **environment** is everything around you. It includes all the living things (such as plants, animals, and people) and the nonliving things (such as air, water, and sunlight) with which you interact. With each breath you take, for example, you are interacting with the environment. Despite humankind's many scientific and technological advances, all people depend on Earth's resources for clean air and water, food, shelter, energy, and fertile soil.

Environmental science is an interdisciplinary study of how humans interact with the environment. It includes information and ideas from engineering as well as natural sciences such as biology, chemistry, geology, meteorology, and climatology. It also includes ideas from social sciences such as geography, economics, and political science. Environmental science even includes the study of ethics, or moral judgments of right and wrong that influence human behavior.

The study of environmental science involves three fundamental goals.

1. Learn how life on Earth has survived and thrived.
2. Understand how humans interact with the environment.
3. Find ways to deal with environmental problems and live more sustainably.

An important part of environmental science is ecology (Chapter 3). **Ecology** is the branch of biology that focuses on how living things interact with the living and nonliving parts of their environment. Each organism, or living thing, in the environment belongs to a species. A species is a group of organisms having a unique set of characteristics that set it apart from other groups (Chapter 4).

A major focus of ecology is the study of ecosystems. An **ecosystem** is one or more communities of different species interacting with one another and with the chemical and physical factors of their nonliving environment. For example, a forest ecosystem consists of plants, animals, and organisms that decompose organic materials, all interacting with one another and the chemicals in the forest's air, water, and soil.

Environmental science and ecology should not be confused with environmentalism. **Environmentalism** is not a scientific field of study, but rather a social movement dedicated to protecting Earth and its resources. Environmentalism is practiced more in the realms of politics and ethics than science. However, the two are related. The findings of environmental scientists can provide evidence to back or refute environmentalists' claims and activities.

checkpoint What is environmental science?

Scientific Factors of Sustainability

Earth is a remarkable example of a sustainable system. Life has existed on Earth for about 3.8 billion years. During this time, the planet has experienced several catastrophic environmental changes: gigantic meteorite impacts, ice ages lasting hundreds of millions of years, warming periods during which sea levels rose by hundreds of feet, and five mass extinctions—each wiping out more than half of the world's existing species. Despite these dramatic environmental changes, an astonishing variety of life has survived.

Modern humans have been around for a mere 200,000 years—less than the blink of an eye relative to the 3.8 billion years life has existed on Earth. Scientific evidence suggests that during that short time, humans have seriously degraded the natural systems that support life. Thus, the newest challenge for the human species is to learn how to live more sustainably.

FIGURE 1-2 ▼
Three scientific factors of sustainability help explain how the planet has sustained such a huge variety of life for billions of years, despite drastic environmental changes.

SCIENTIFIC FACTORS OF SUSTAINABILITY	
Solar energy (radiant energy from the sun)	The sun's energy warms the planet and provides the energy that plants use to produce their own food, which in turn provides energy for life's functions. The sun's energy indirectly powers the wind and flowing water people use to produce electricity.
Biodiversity	Biodiversity is variety in the form of species, genes, ecosystems, and environmental interactions. Biodiversity provides ways for organisms to adapt to changing environmental conditions, including gradual or catastrophic changes.
Nutrient cycling	The nutrients, or chemicals, required to sustain life are circulated from the environment through various organisms and returned to the environment in a process called nutrient cycling or chemical cycling.

FIGURE 1-3 ▼
Three social factors of sustainability can help humankind make a transition to a more environmentally and economically sustainable future.

SOCIAL FACTORS OF SUSTAINABILITY	
Economics	Economics is the social science concerned with the production and consumption of goods and services in the creation and transfer of wealth. Some economists urge manufacturers to include the environmental and health costs of producing goods and services in their market prices. This practice is called full-cost pricing. This would give consumers information about the environmental impacts of products.
Political science	Political science is the study of systems of government and the analysis of political behaviors. Political scientists often look for win-win solutions to environmental problems that will benefit the largest number of people as well as the environment.
Ethics	Ethics, a branch of philosophy, is the systematic study of right and wrong actions. Ethicists argue that society should leave the planet in a condition that is as good as or better than it is in now. This is our responsibility to future generations.

Sustainability is the capacity of Earth's natural systems that support life—including human social systems—to maintain stability or to adapt to changing environmental conditions indefinitely. Three major factors have played key roles in the long-term sustainability of life, as summarized in Figure 1-2. Our understanding of these factors is based on research in the natural sciences. Therefore, this text refers to them as "scientific factors" of sustainability.

checkpoint Which sciences support the three scientific factors of sustainability?

Social Factors of Sustainability

Historical and ongoing human experiences related to environmental problems, solutions, and trade-offs help reveal the social factors behind sustainability. These social factors are based on studies from fields in the social sciences, especially economics and political science, and from ethics. These factors are summarized in Figure 1-3.

Understanding the social factors of sustainability helps decision makers determine what courses of action they should take, given known scientific facts. The natural sciences alone cannot provide answers to all the environmental challenges society faces. You will explore the factors of sustainability further in this chapter and apply them throughout this book.

checkpoint What is full-cost pricing?

Sustainability and Natural Capital

A necessary part of sustainability is the availability of natural capital. **Natural capital** consists of the natural resources and ecosystem services that keep humans and other species alive and that support human economies (Figure 1-6).

Natural Resources Materials and energy sources in nature that are essential or useful to humans are called **natural resources**. Resources such as surface water, trees, and edible wild plants are directly available for use. Other resources such as petroleum, minerals, wind, and underground water become useful to humans with more effort and technological knowledge. Natural resources fall into three categories: inexhaustible resources, renewable resources, and nonrenewable resources. Figure 1-4 defines these terms and gives examples of each category.

Ecosystem Services **Ecosystem services** are the natural services—provided by healthy ecosystems—that support life and human economies at no monetary cost (Figure 1-6). For example, forests help purify air and water, regulate climate, reduce soil erosion, and provide millions of species with a place to live. Examples of key ecosystem services include:

- nutrient cycling
- purification of air and water
- renewal of topsoil
- pollination
- pest control

A vital ecosystem service that affects all other ecosystem services is nutrient cycling, which is one of the key scientific factors of sustainability described in Figure 1-2. Without nutrient cycling in topsoil there would be no plants, no pollinators (another ecosystem service), and no humans or other land animals. The absence of nutrient cycling in topsoil would also disrupt the ecosystem service of purifying air and water.

Without solar energy, Earth would be without ecosystem services and most natural resources. The nutrient cycling that supports so many ecosystem services is also powered, directly or indirectly, by the sun. Humankind and its economies depend on the Earth-sun system for Earth's natural capital—the natural resources and ecosystem services that sustain life.

FIGURE 1-4

Natural Resources	
Inexhaustible resources	Solar energy is an **inexhaustible resource** because its continuous supply is expected to last for at least 6 billion years, until the sun dies. Energy from the sun also powers Earth's inexhaustible winds and flowing waters, which people can use to produce electricity.
Renewable resources	A **renewable resource** is any resource that can be replenished by natural processes within hours to centuries. As long as people do not use the resource faster than natural processes can replace it, the resource will be renewable. Examples of renewable resources include forests, grasslands, wildlife, fertile topsoil, clean air, and fresh water. The highest rate at which people can use a renewable resource indefinitely without reducing its available supply is called its sustainable yield.
Nonrenewable resources	**Nonrenewable resources** exist in a fixed amount, or stock. They take millions to billions of years to form. On the human time scale, these resources will be used more quickly than they can be replaced. Examples of nonrenewable resources are fossil fuel energy resources (such as oil, coal, and natural gas), metallic mineral resources (such as copper and aluminum), and nonmetallic mineral resources (such as salt and sand).

FIGURE 1-5
How do you think a forest ecosystem differs from a river or mountain ecosystem? What natural resources can you identify in this photograph?

FIGURE 1-6
Natural Capital
Natural capital consists of natural resources (blue) and ecosystem services (orange) that support and sustain Earth's life as well as human economies.

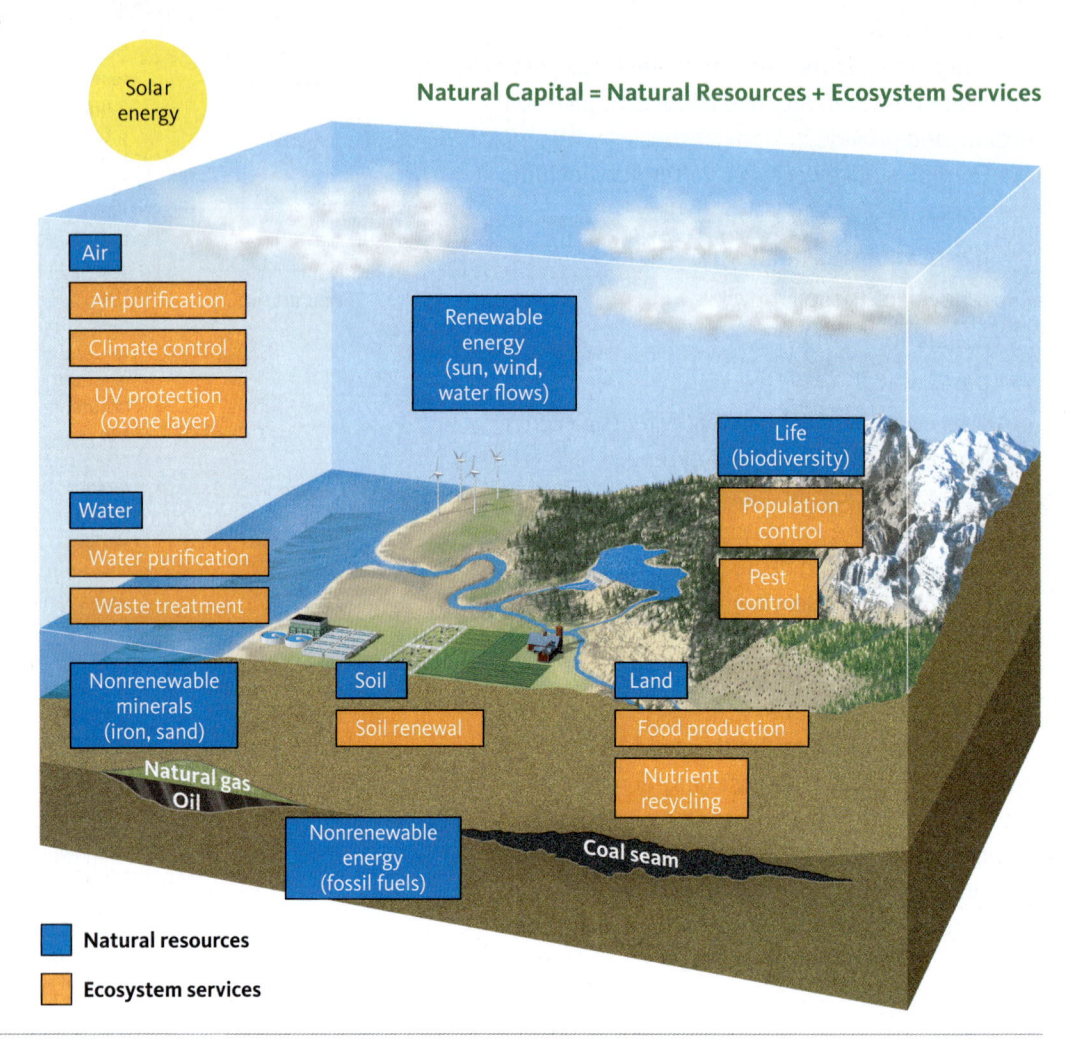

Natural Capital = Natural Resources + Ecosystem Services

Activities That Degrade Natural Capital
Human activities can severely degrade natural capital, which affects sustainability. Humans do this by using renewable resources faster than they can be restored naturally. Some human activities overload Earth's renewable air and water systems with pollution and wastes. For example, people in many areas of the world are replacing forests with crop plantations that require large inputs of energy, water, fertilizer, and pesticides. Many people dump chemicals and wastes into rivers, lakes, and oceans faster than they can be cleansed through natural processes. Many of the plastics and other synthetic materials people use poison wildlife and disrupt nutrient cycles because they cannot be broken down and used as nutrients by other organisms.

Solutions to Harmful Activities
Once harmful activities are identified, people can work together to create solutions that protect Earth's natural capital and guarantee that it is used sustainably. Environmental scientists research problems such as the degradation of forests and other sources of natural capital. Social scientists look for economic and political solutions.

One solution to reduce the loss of forests is to stop burning or cutting down mature forests. This cannot be done unless governments educate their citizens about the ecosystem services forests provide, reduce poverty, provide or develop other ways to meet people's resource needs, and pass laws to protect the forests. Conflicts can arise when such protections have a negative economic effect on groups of people or certain industries. Dealing with such conflicts often involves both sides making compromises or trade-offs. For example, a timber company may be allowed to cut a portion of the forest as long as it plants new trees to replace the ones that are harvested. In exchange for

22 CHAPTER 1 THE ENVIRONMENT AND SUSTAINABILITY

FIGURE 1-7
Logs stand five stories high in Alberta, Canada. How is clearing of forests an example of natural capital degradation?

the cost of replanting, the government may subsidize (pay part of the cost of) planting new trees.

Each individual—including you—plays an important role in the drive to become a more sustainable society. History has shown that most of the significant changes in human systems have begun through the collective actions of individuals.

checkpoint What are some human activities that degrade natural capital?

1.1 Assessment

1. **Classify** Is solar energy an ecosystem service? Why or why not?
2. **Contrast** What is the difference between ecosystem services and natural resources?
3. **Apply** Give an example of an ecosystem service that is not mentioned in the text and defend your answer.

CROSSCUTTING CONCEPTS

4. **Stability and Change** Life has existed on Earth for about 3.8 billion years despite gradual and sudden environmental changes. How do the scientific factors of sustainability help explain how life has continued to survive and evolve?

1.2 How Are Our Ecological Footprints Affecting Earth?

CORE IDEAS AND SKILLS
- Recognize some major environmental problems that lead to natural capital degradation.
- Describe the purpose of the ecological footprint and IPAT models.

KEY TERMS

environmental degradation
pollution
point source
nonpoint source
ecological footprint

Good News: Many People Have a Better Quality of Life

As the world's dominant animal, humans have an immense power to degrade or sustain Earth's life support system. Collectively, people have the power to decide such matters as whether a forest is preserved or cut down, or whether a river is dammed or allowed to flow. People's actions will have an impact on the temperature of the atmosphere, the temperature and the acidity of the ocean, and which species survive or become extinct. At the same time, creative thinking, economic growth, scientific research, grassroots political pressure by citizens, and regulatory laws have improved the quality of life for many of Earth's people, especially in the United States and most other more-developed countries.

LESSON 1.2

Humans have developed an amazing array of useful materials and products. We have learned how to use wood, fossil fuels, the sun, wind, the nuclei of certain atoms, and Earth's heat (geothermal energy) to supply us with enormous amounts of energy. We live and work in artificial environments in the form of buildings and cities. We have invented computers to extend our brains, robots to perform repetitive tasks with great precision, and electronic networks to enable instantaneous global communication.

Globally, life spans are increasing, infant mortality is decreasing, education is on the rise, some diseases are being conquered, and the population growth rate has slowed. The food supply is generally more abundant and safer, air and water are getting cleaner in many parts of the world, and exposure to toxic chemicals is more avoidable. People have protected some endangered species and ecosystems, and forests are growing back in some areas.

Scientific research and technological advances financed by affluence helped achieve these improvements in life and environmental quality. Education also spurred many citizens to insist that businesses and governments work toward improving environmental quality.

People have learned a lot about how Earth works and how to identify environmental problems, their causes, and their solutions. With the internet and mobile phones, humans are a connected species with growing access to information that could help society shift to a more sustainable path.

checkpoint Give at least three reasons why many people around the world have an improved quality of life.

Humans Are Living Unsustainably

According to a large body of scientific evidence, people are living unsustainably. People continually waste, deplete, and degrade much of Earth's natural capital—a process known as **environmental degradation** or natural capital degradation (Figure 1-8). Biologist Garrett Hardin (1915–2003) called such degradation the "tragedy of the commons." Some renewable resources are not owned by anyone and can be used by almost anyone. Examples include the atmosphere, the ocean, and its fish. Degradation occurs because each user reasons, "The little bit that I use or pollute is not enough to matter." When the level of use is small, this logic works. Eventually, however, the total effect of large numbers of people trying to exploit a widely available or shared resource can degrade the resource, eventually exhausting or ruining it. Then no one benefits and everyone loses. That is the tragedy.

FIGURE 1-8
Natural Capital Degradation Climate change, environmental pollutants in water and air, and soil erosion are examples of natural capital degradation.

Degradation of Normally Renewable Natural Resources

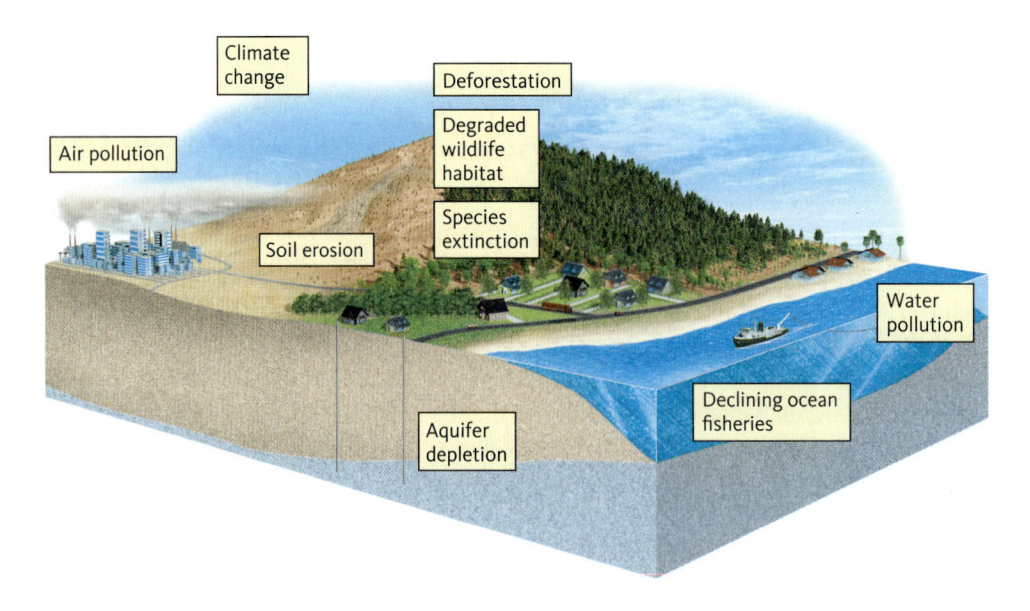

- Climate change
- Air pollution
- Soil erosion
- Deforestation
- Degraded wildlife habitat
- Species extinction
- Aquifer depletion
- Water pollution
- Declining ocean fisheries

FIGURE 1-9
Pollution, in the form of plastic waste, washes up on a beach in Venezuela.

In 2005, the United Nations (UN) released a report titled *Millennium Ecosystem Assessment*. It is the product of a four-year study conducted by 1,360 experts from 95 countries. According to this study, human activities have overused about 60% of Earth's ecosystem services. (See the orange boxes in Figure 1-6.) Most of this has occurred since 1950, which suggests the pace of degradation has increased. The UN's summary warned that "human activity is putting such a strain on the natural functions of Earth that the ability of the planet's ecosystems to sustain future generations can no longer be taken for granted." The problems are extremely complex, but scientific, economic, and political solutions could be implemented within the near future.

Countries, like individuals, differ in their resource use and environmental impact. The world's states can be classified economically as either "more developed" or "less developed," based primarily on their average income per person. Distinguishing between countries in terms of their economic activity is useful for studying the resource use and environmental impacts of countries and regions.

More-Developed Countries Industrialized countries with high average incomes account for only 16% of the world's population. They include the United States, Canada, Japan, Australia, and most European countries.

Less-Developed Countries All other nations, in which 84% of the world's people live, are considered less developed economically. Most of these countries are in Africa, Asia, and Latin America. Some less-developed countries are further classified as middle-income, moderately developed countries such as China, Brazil, Thailand, and Mexico. Others are low-income, least-developed countries, including Congo and Haiti.

checkpoint What is the tragedy of the commons?

Pollution Comes from a Number of Sources

A major environmental problem is pollution. **Pollution** is contamination of the environment by any chemical or agent, such as noise or thermal energy, at levels considered harmful to the health, survival, or activities of organisms. Some pollution occurs naturally. Volcanic eruptions are a natural source of pollutants that contaminate the air and water. Other pollution is the result of human activities such as energy production, industry, agriculture, and transportation.

Pollutants originate from point sources and nonpoint sources. According to the Environmental Protection Agency (EPA), a **point source** of pollution is "any single identifiable source of pollution from which pollutants are discharged." Examples are air pollution from the exhaust pipes of motor vehicles, chimneys, and smokestacks and water pollution from the drainage pipes and ditches of industrial and sewage treatment plants. **Nonpoint source** pollution consists of pollutants that come from many diffuse sources that are hard to pinpoint. Examples are runoff of water pollutants from croplands, residential areas, clear-cut forests, and construction sites, as well as deposition of pollutants on land from the atmosphere.

People have tried to deal with pollution in two very different ways. One approach is pollution cleanup, which involves cleaning up or diluting pollutants that have already been, or are being, released. For example, devices in smokestacks and automobile exhaust systems can remove various types of pollutants. Polluted water can be treated to remove pollutants before they are discharged.

The other approach is pollution prevention, which focuses on reducing or eliminating the production of pollutants. For example, governments can enact pollution control laws that ban or minimize the emissions of certain pollutants. Scientists and engineers can come up with new designs for automobile engines, power plants, or industrial processes that do not produce certain pollutants. According to many environmental scientists, pollution prevention is an important step toward a more sustainable future because, in the long run, it is more effective and less expensive than cleanup.

checkpoint What is the difference between point source and nonpoint source pollution?

Modeling Environmental Impacts

Overusing or misusing renewable resources can result in natural capital degradation, pollution, and waste (Figure 1-9). The extent of this harmful environmental impact is called an **ecological footprint**. This is the amount of land and water needed to supply an individual or a population with renewable resources. It is also a measure of the amount of land and water required to absorb and recycle the wastes and pollution produced by such resource use. The size of the footprint is proportional to the amount of renewable resources consumed.

FIGURE 1-10

Human Ecological Footprint The human ecological footprint has an impact on about 83% of Earth's total land surface.

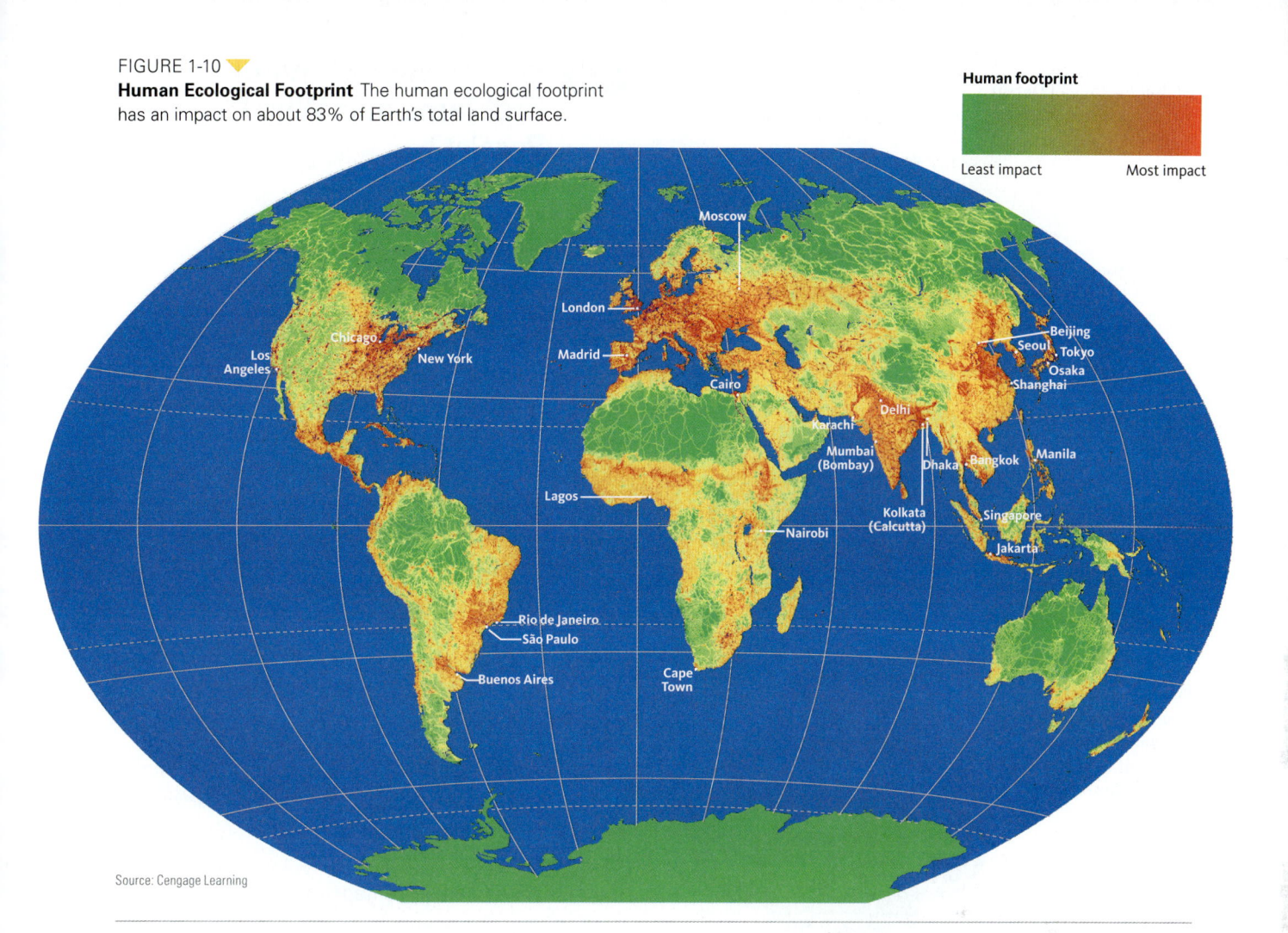

Human footprint

Least impact — Most impact

Source: Cengage Learning

You can find ecological footprint models on the internet and use them to determine your own ecological footprint. The per capita ecological footprint is the average ecological footprint of an individual in a given country or area. Figure 1-10 shows the human ecological impact in different parts of the world.

If the total ecological footprint for a city, country, or the world is larger than its biological capacity to replenish resources and absorb the resulting wastes and pollution, the area is said to have an ecological deficit. In other words, an ecological deficit results when people are living unsustainably by depleting natural capital. Ecological footprint data and models have been in use since the mid-1990s. Though imperfect, they provide useful rough estimates of individual, national, and global environmental impacts. Globally, humans are running up a huge ecological deficit that is expected to increase (Figure 1-11).

Another environmental impact model was developed in the 1970s by scientists Paul Ehrlich and John Holdren. The IPAT model shows that the environmental impact (I) of human activities is the product of three factors: population size (P), affluence (A), and the beneficial and harmful environmental effects of technologies (T). The following equation summarizes the IPAT model:

Impact (I) = Population (P) × Affluence (A) × Technology (T)

While the ecological footprint model emphasizes the use of renewable resources, the IPAT model includes the environmental impact of using both renewable and nonrenewable resources. The T factor can be either harmful or beneficial. Some forms of technology such as polluting factories, gas-guzzling motor vehicles, and coal-burning power plants increase our harmful environmental impact by raising the T factor. The goal is to decrease the harmful T factor with use of pollution control and prevention technologies, fuel-efficient cars, and wind turbines and solar cells that generate electricity with a low environmental impact. In doing so, the ecological footprint will be reduced.

LESSON 1.2

FIGURE 1-11

Global Ecological Deficit Each scenario shows the estimated number of Earths needed to support humankind's ecological footprint.

Needed today

Needed in 2030

Needed today if everyone on Earth had the same ecological footprint as the average American

Source: World Wide Fund for Nature's *Living Planet Report 2012*

checkpoint Give some examples of how ecological footprints are increasing.

1.2 Assessment

1. **Explain** Why is nonpoint source pollution more difficult to control than point source pollution?
2. **Research** Find a reliable ecological footprint calculator and use it to determine the size of your own ecological footprint. How does your footprint compare to the average? How do you think the calculator's accuracy could be improved?

CROSSCUTTING CONCEPTS

3. **Scale, Proportion, and Quantity** In 2012, the total ecological footprint of China was only just below that of the European Union. But the ecological footprint of the average European was nearly three times greater than that of the average Chinese person's. Explain how this is possible.

1.3 What Causes Environmental Problems and Why Do They Persist?

CORE IDEAS AND SKILLS

- Discuss the major causes of environmental problems.
- Describe the different forms of environmental degradation that can result from affluence and from poverty.
- Explain how one's environmental worldview affects one's attitude toward living sustainably.
- Compare the three major categories of environmental worldviews.

KEY TERMS

| exponential growth | environmental ethics | life-centered worldview |
| environmental worldview | human-centered worldview | Earth-centered worldview |

The Human Population Is Growing at a Rapid Rate

According to a significant number of environmental and social scientists, population growth is one of the major causes of environmental problems (Figure 1-12). **Exponential growth** occurs when a quantity increases at a fixed percentage per unit of time, such as 2% a year. Exponential growth starts slowly, but after a few doublings, grows to enormous numbers because each doubling is twice the total of all earlier growth. For example, a sheet of paper is very thin, but when folded in half, the thickness doubles. Now imagine if you doubled the thickness of the folded sheet of paper—then it would be four times thicker. If you repeated this process just 50 times the paper would be thick enough to almost reach the sun, which is 150 million kilometers (93 million miles) from Earth.

The human population has been growing exponentially, though the rate has slowed in recent years. In 2015, the rate of growth was 1.16%. By 2021, the rate of growth was 1.03%. Today there are more than 7.9 billion people on the planet. By 2050, the population could reach 9.7 billion. Because of recent declines in the growth rate, it is difficult to forecast exactly how the global population will change in the future. There are many predictions as to why the rate is decreasing, but one fact is certain: the population in 2050 will be significantly greater than it is today.

28 CHAPTER 1 THE ENVIRONMENT AND SUSTAINABILITY

FIGURE 1-12
Growing global populations affect already large cities. In Hong Kong, city planners build taller structures to support the increasing population within the city's limited space.

No one knows how many people Earth can support indefinitely. No one knows the level of average resource consumption per person that will seriously degrade the planet's natural capital. However, humanity's large and expanding ecological footprint and the resulting widespread natural capital degradation are disturbing warning signs. Some scientists argue that such severe environmental degradation could be reduced by slowing population growth and the leveling off the population in coming decades. Some ways to do this include reducing poverty through economic development, promoting family planning, and elevating the status of women, as discussed in Chapter 14.

checkpoint What is a population's doubling time?

Affluence and Unsustainable Resource Use

The lifestyles of the world's expanding population of consumers are built on growing affluence as more people earn higher incomes. Affluent societies often include high levels of resource consumption.

> **CONSIDER THIS**
>
> **You can calculate doubling time with the "rule of 70."**
>
> The approximate doubling time of the human population can be calculated by using the rule of 70. (You can apply this rule to any quantity that is increasing exponentially.)
>
> *doubling time (years) = 70 ÷ annual growth rate (%)*
>
> The world's population is growing at about 1% per year. At this rate, about how long will it take for the human population to double?

This improvement in quality of life can increase environmental degradation, wastes, and pollution unless individuals live more sustainably.

The effects can be dramatic. The World Wildlife Fund (WWF) estimates that the United States, with only 4.2% of the global population, is responsible for about 15% of the global ecological footprint. Only

FIGURE 1-13 ▼
Exponential Growth The J-shaped curve represents past exponential world population growth, with projections to 2100. As the rate of growth has slowed, the J-shaped curve of growth changes to an S-shaped curve. (This figure is not to scale.)

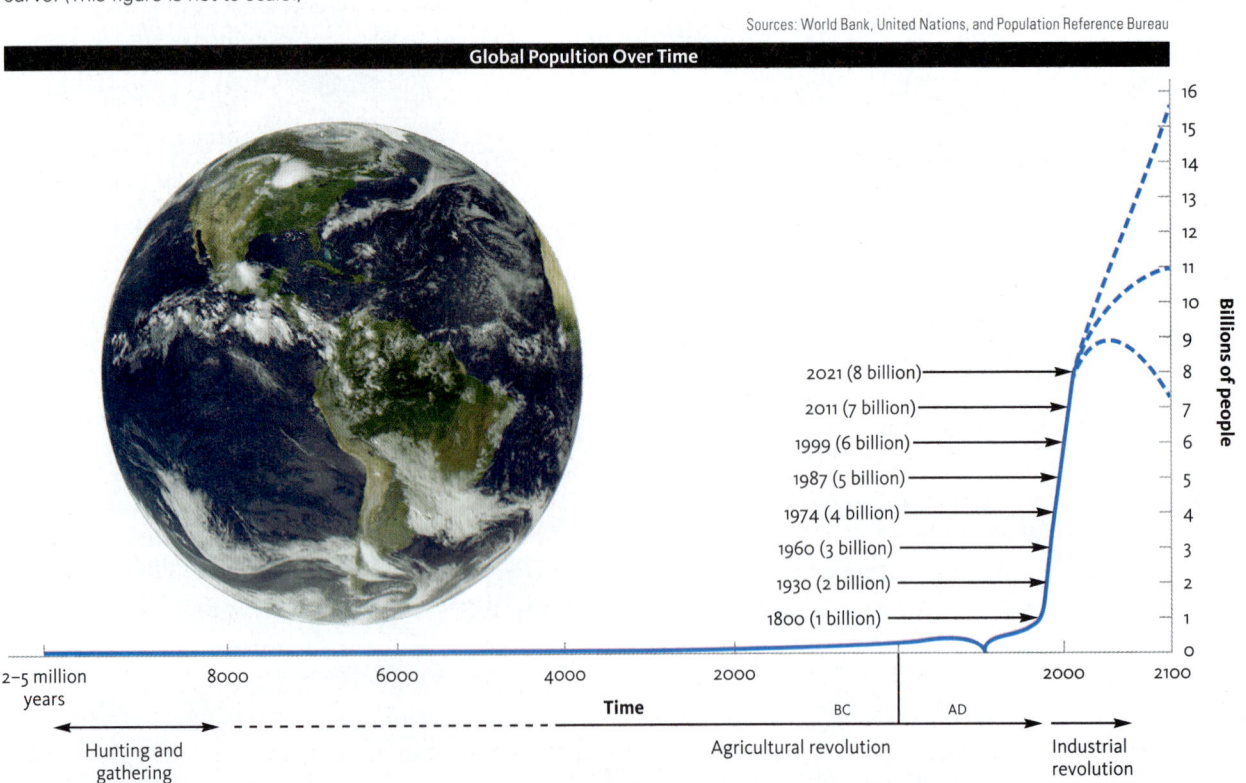

Sources: World Bank, United Nations, and Population Reference Bureau

Global Population Over Time

- 2021 (8 billion)
- 2011 (7 billion)
- 1999 (6 billion)
- 1987 (5 billion)
- 1974 (4 billion)
- 1960 (3 billion)
- 1930 (2 billion)
- 1800 (1 billion)

Time: 2–5 million years | 8000 | 6000 | 4000 | 2000 BC | AD | 2000 | 2100

Hunting and gathering → Agricultural revolution → Industrial revolution five countries make up half of the world's ecological footprint. The WWF has projected humans would need five Earths if everyone lived as the average American does (Figure 1-11).

Higher resource consumption contributes to air pollution and water pollution from factories and motor vehicles. It also contributes to land degradation from mining raw materials used to make the products consumers buy. Another downside to affluence is that it enables consumers to obtain resources from almost anywhere in the world without seeing the harmful effects of their high-consumption lifestyles.

On the other hand, affluence can allow for widespread and better education, which can lead people to become more concerned about environmental quality. Affluence also makes more money available for developing technologies to reduce pollution, environmental degradation, and waste.

As a result, in the United States and other affluent countries, the air is clearer, drinking water is purer, and most rivers and lakes are cleaner than they were in the 1970s. The food supply is more abundant and safer, and the incidence of life-threatening infectious diseases has been greatly reduced. Life spans are longer, and some endangered species are being saved from extinction caused by human activities.

checkpoint How can affluence affect one's environmental impact?

Poverty Has Harmful Environmental and Health Effects

Poverty is a condition in which people lack enough money to fulfill their basic needs for food, water, shelter, health care, and education. According to the World Bank, about one in every four people, or 1.8 billion people, live on less than $3.10 a day. Of this 1.8 billion, nearly 900 million people—more than three times the U.S. population—live in extreme poverty, or the equivalent of less than $1.90 a day. This is less than many people spend for a bottle of water or a cup of coffee. Fortunately, the percentage of the world's population living in extreme poverty decreased from 52% in 1981 to 9.2% in 2021.

Poverty causes a number of harmful environmental and health effects. The daily lives of the world's poorest people center on getting enough food, water, and fuel for cooking and heating to survive. These individuals are too desperate for short-term survival to worry about long-term environmental quality or sustainability. Thus, collectively, they may be forced to degrade forests, topsoil, and grasslands, and deplete fisheries and wildlife populations to stay alive. In some places, governments and independent organizations have helped poverty stricken populations achieve healthy living conditions that minimize harm to the environment.

Environmental degradation can have severe health effects on poor people. One problem is life-threatening malnutrition. Malnutrition, a lack of protein and other nutrients needed for good health, results from environmental degradation that interferes with food production.

Another issue is illness caused by limited access to adequate sanitation and clean drinking water. More than one-third of the world's people have no bathroom facilities and are forced to use backyards, alleys, ditches, and streams. As a result, one of every nine of the world's people gets water for drinking, washing, and cooking from sources polluted by human or animal feces.

According to the World Health Organization, air pollution also threatens millions of people in less-developed countries. Outdoor air pollution causes more than 1 million deaths per year in China. Globally, about 4 million people die every year from indoor air pollution, mostly from the smoke from open fires or stoves used for heating and cooking.

checkpoint How can poverty affect one's environmental impact?

Prices of Goods and Services Exclude Environmental and Health Costs

Another basic cause of environmental problems has to do with how the marketplace prices goods and services. Companies using resources to provide goods for consumers generally are not required to pay for most of the harmful environmental and health costs of supplying those goods.

For example, timber companies pay the cost of clear-cutting forests. They do not pay for the resulting environmental degradation and loss of wildlife habitat. One impact of clearing a forest is soil erosion. This reduces the humidity of nearby forests that have not been cleared, which in turn can lead to forest fires. The primary goal of a company is to maximize profits for its owners or stockholders. Companies are not inclined to add environmental costs to their prices voluntarily. Because the prices

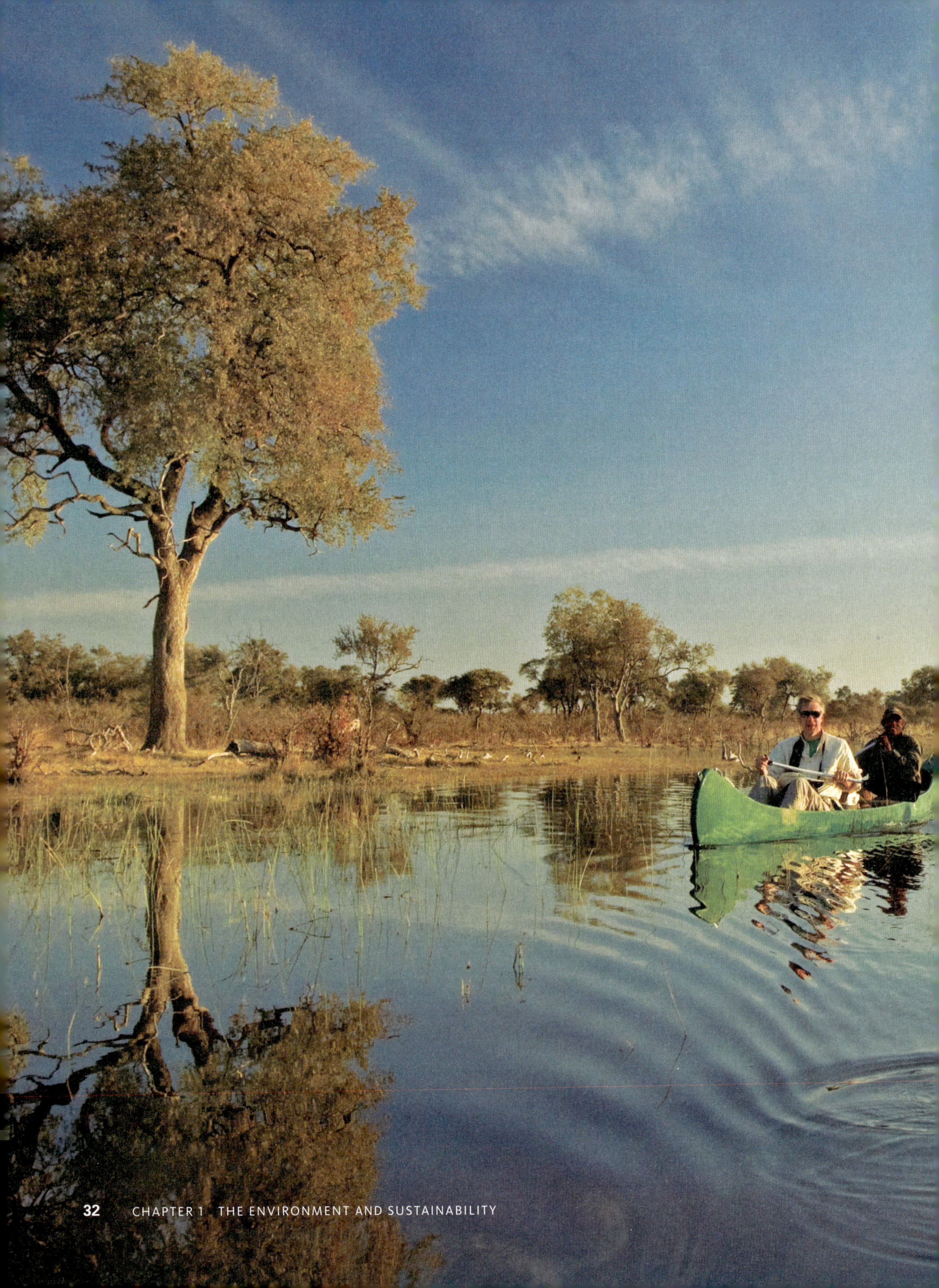

FIGURE 1-14
ON ASSIGNMENT National Geographic Explorer Beverly Joubert photographed her husband, Dereck Joubert (also an NG Explorer), on the Selinda Spillway in Botswana. Beverly and Dereck, who are award-winning filmmakers, lead a group of ecotourists in search of elephants. Ecotourism provides tourist dollars for locals and opportunities for the tourists to connect with nature.

of goods and services do not include most of their harmful environmental and health costs, consumers and decision makers have no effective way to evaluate these harmful effects.

Another problem may arise when governments give companies subsidies. For example, states may lend money without interest to lure new businesses into the area. Cities may exempt businesses from paying property taxes for multiple years so companies will build in the city. This helps to create jobs and stimulate economies, but environmentally harmful subsidies encourage depletion and degradation of natural capital. An example of an environmentally beneficial subsidy is exempting a business from sales tax when it buys new pollution control equipment.

People could live more sustainably and increase their beneficial environmental impact if the harmful environmental and health costs of the goods and services were included in the market prices of the goods they buy. Such full-cost pricing is a powerful tool and is part of the economic social factor of sustainability shown in Figure 1-3. Economic sustainability could be achieved in multiple ways. One way is to shift from harmful government subsidies to environmentally beneficial subsidies. Another solution is to heavily tax companies that cause pollution and waste.

checkpoint How can some government subsidies cause environmental harm and endanger human health?

People Are Increasingly Isolated from Nature

Today, more than half of the world's people live in urban areas. This shift from rural to urban living is continuing at a rapid pace. Urban environments and the increased use of cell phones, computers, and other electronic devices are isolating people, especially children, from the natural world.

Some argue this has led to a phenomenon known as nature deficit disorder. People with this disorder may suffer from stress, anxiety, depression, and other problems. Children and adults can gain many benefits from outdoor activities. Experiencing nature can lead to better health, reduced stress, improved mental abilities, and increased imagination and creativity. It also can provide a sense of wonder and connection to life on Earth. As described in the feature at the beginning of this chapter, the benefits of interacting with nature inspire National Geographic Explorer Juan Martinez to find ways to build connections between people and their environment.

checkpoint What are the possible effects of nature deficit disorder?

People Have Different Views About Environmental Problems and Their Solutions

People's opinions differ on the nature and seriousness of the world's environmental problems as well as how to solve them. These differences arise mostly because of varying environmental worldviews. Your **environmental worldview** is your set of assumptions and values concerning the natural world and what you think your role in managing it should be. **Environmental ethics**, the study of varying beliefs about what is right and wrong with how people treat the environment, provides useful tools for examining worldviews. Here are some important ethical questions relating to the environment:

FIGURE 1-15
This woman in New Delhi, India, is taking apart circuit boards from electronics and sorting the pieces so they can be sold and recycled. With the usage of electronics comes the need to dispose of products at the end of their working life.

- Why should people care about the environment?
- Are humans the most important species on the planet, or are they just another one of Earth's millions of forms of life?
- Do people have an obligation to see that their activities do not cause the extinction of other species? If so, should people try to protect all species or only some? How does society decide which ones to protect?
- Does the current generation have an ethical obligation to pass the natural world on to future generations in a condition that is as good as or better than what they inherited?
- Should every person be entitled to equal protection from environmental hazards regardless of race, gender, age, national origin, income, social class, or any other factor? (This is the central ethical and political issue for what is known as the environmental justice movement.)
- Should individuals and society as a whole seek to live more sustainably and, if so, how?

People with different environmental worldviews can take the same data, be logically consistent with them, and arrive at quite different answers to questions about environmental sustainability. This happens because individuals start with different assumptions and moral, ethical, or religious beliefs. What follows is a brief introduction to the three major categories of environmental worldviews: human-centered, life-centered, and Earth-centered. Environmental worldviews are discussed in greater detail in Chapter 18.

Human-Centered Worldview A **human-centered worldview** sees the natural world as a support system for human life. It is divided into the planetary management worldview and the stewardship worldview. Both worldviews hold that humans are separate from and in charge of nature and that society should manage Earth for the benefit of humans. The stewardship worldview adds that people have a responsibility to be caring and responsible managers, or stewards, of the planet.

Life-Centered Worldview According to the **life-centered worldview**, all species have value in fulfilling their particular role within the biosphere, regardless of their potential or actual use to society. Most individuals with a life-centered worldview believe society has an ethical responsibility to avoid hastening the extinction of species through human activities.

Earth-Centered Worldview The **Earth-centered worldview** maintains that people are part of, and dependent on, nature. It maintains that Earth's natural capital exists for all species, not just for humans. According to this view, economic success and the long-term survival of cultures and species depend on learning how life on Earth has sustained itself for billions of years. Lessons from nature should influence how people think and act.

checkpoint Is your environmental worldview human-centered, life-centered, or Earth-centered? What makes you describe it that way?

1.3 Assessment

1. **Apply** Give an example of a technology that reduces some of the environmental degradation that results from affluence. Give another example of a technology or practice that reduces some of the environmental degradation associated with poverty.

2. **Evaluate** Which category of environmental worldview do you think would be most likely to lead to a sustainable future if it were widely accepted, and why?

CROSSCUTTING CONCEPTS

3. **Cause and Effect** Some people view the rapid population growth in less-developed nations as the primary cause of environmental problems. Others see the high rate of resource use per person in more-developed countries as the problem. Which factor do you think is more significant? Why?

SCIENCE AND ENGINEERING PRACTICES

4. **Using Math** Use the rule of 70 to estimate the population doubling times of Niger and the United States based on the data below. How many times will Niger's population have to double to surpass the United States in population? (Extra credit: In what year will that happen?)
   - **Niger:** Pop. 24.2 million; 3.8% growth rate (2020)
   - **U.S.:** Pop. 330 million; 0.4% growth rate (2020)

LESSON 1.3

1.4 What Is an Environmentally Sustainable Society?

CORE IDEAS AND SKILLS

- Describe the rise of environmental conservation and protection in the United States.
- Explain the basic difference between preservation and conservation.
- Discuss how people can achieve an environmentally sustainable society.

KEY TERMS

environmentally sustainable society

natural income

Environmental Conservation and Protection in the United States

When European colonists first arrived in what is now the United States, they viewed North America as a land with inexhaustible resources and a wilderness to be conquered and managed for human use. Settlers cleared forests to build settlements. They plowed grasslands to plant crops, and mined for gold, lead, and other minerals.

Early Conservation Efforts In 1864, George Perkins Marsh, a scientist and member of Congress from Vermont, questioned the idea that America's resources were inexhaustible. He used scientific studies to show how the rise and fall of past civilizations were linked to the misuse of their soils, water supplies, and other resources. Marsh was one of the founders of the U.S. conservation movement.

Early in the 20th century, this movement split into two factions over how to use public lands. The preservationist view, led by naturalist John Muir, wanted wilderness areas on some public lands to be left untouched so they would be preserved indefinitely. He was largely responsible for establishing Yosemite National Park in 1890. In 1892, he founded the Sierra Club, which is to this day a political force working on behalf of the environment. Muir called for setting aside some public lands as protected wilderness, an idea that was not enacted into law until 1964.

The conservationist view was led by Theodore Roosevelt and Gifford Pinchot. Roosevelt was president of the United States, and Pinchot was the first chief of the U.S. Forest Service. They believed all public lands should be managed wisely

---

NATIONAL GEOGRAPHIC | EXPLORERS AT WORK
Dennis Dimick Conservationist and Former National Geographic Executive Environment Editor

Dennis Dimick is passionate about the power of storytelling to inspire people to care about the planet. In his role as environmental editor at National Geographic Society's flagship publication, Dimick helped plan stories for magazine issues and worked with senior writers and photographers in crafting their coverage and reportage of environmental stories. He helped to produce several major features, including a series on global population and a later series on global food security. He also oversaw a three-story series on climate change.

Deep journalism on difficult environmental problems such as climate change often requires gathering and analyzing data in collaboration with scientists and research teams. The process can generate a lot of complex information. Dimick worked to synthesize and distill this information into compelling and scientifically sound narratives for the general reader of *National Geographic* magazine (NGM).

In November 2015, NGM published an issue focused entirely on climate change. In an editorial piece on the very last page, Dimick has the final word: "Every day we learn more about Earth's atmosphere, water, land—and our impact upon it. The health of a tiny blue planet is at risk. What will we do with what we know?"

At the end of 2015, Dimick retired from the National Geographic Society and joined the board of directors for the non-profit educational organization The Society of Environmental Journalists.

FIGURE 1-16
John Muir (1838–1914) led the preservationist school of environmental conservation.

and scientifically, primarily to provide resources for people. Roosevelt's term of office (1901–1909) has been called the country's Golden Age of Conservation. He established more than 50 national wildlife reserves and more than tripled the size of the national forest reserves.

Conservation efforts in the United States continued to straddle these two schools of thought. Aldo Leopold, a wildlife manager, professor, writer, and conservationist, personified this dual approach. Trained in the conservation school, he shifted toward the preservation school. He became a pioneer in forestry, soil conservation, wildlife ecology, and wilderness preservation. In 1935, he helped found the U.S. Wilderness Society. Through his writings, especially his 1949 book, *A Sand County Almanac*, he laid the groundwork for the field of environmental ethics. He argued that the role of the human species should be to protect nature, not conquer it.

Later in the 20th century, it was necessary to broaden the concept of resource conservation. The concept included preservation of the quality of the planet's air, water, soil, and wildlife. A prominent pioneer in that effort was Rachel Carson, whose book *Silent Spring* was published in 1962. Carson's book documented the pollution of air, water, and wildlife from the widespread use of harmful pesticides such as DDT. This influential book heightened public awareness of pollution problems and led to the regulation of several dangerous pesticides.

Between 1940 and 1970, the United States underwent rapid economic growth and industrialization. The byproduct of industrialization was increased air and water pollution and growing mounds of solid and hazardous wastes. Air pollution was so bad in many cities that drivers had to use their car headlights during the daytime. Thousands died each year from the harmful effects of air pollution. The Cuyahoga River, running through Cleveland, Ohio, was so polluted with oil and other flammable pollutants that it caught fire several times. A devastating oil spill off the California coast took place in 1969. Well-known wildlife species such as the American bald eagle, grizzly bear, whooping crane, and peregrine falcon became endangered.

The Decade of the Environment: 1970s Growing publicity over these problems led the American public to demand government action. When the first Earth Day was held on April 22, 1970, 20 million people in more than 2,000 communities and on college campuses attended rallies to demand improvements in environmental quality. Earth Day led to the creation of the Environmental Protection Agency (EPA) in 1970 and to passage of most of the environmental laws now in place in the United States. The 1970s are known as the decade of the environment. The United States led the world in expanding environmental awareness,

FIGURE 1-17
President Theodore Roosevelt (1858–1919) was a leader in the conservationist school of environmental conservation.

LESSON 1.4   37

developing wildlife conservation, and strengthening environmental protection.

Since 1970, many grassroots environmental organizations have sprung up to help deal with environmental threats. Interest in environmental issues grew on many college campuses, resulting in the expansion of environmental studies courses and programs. Individuals became more aware of critical, complex, and largely invisible environmental issues. These included losses in biodiversity, ocean pollution and acidification, atmospheric warming, and the threat of climate change.

The 1980s and After In the 1980s there was a backlash against environmental laws and regulations led by some corporate leaders and members of Congress. They argued that environmental laws were hindering economic growth. They pushed to weaken or eliminate many environmental laws passed during the 1970s and eliminate the EPA. The 1990s saw increasingly sophisticated "disinformation" campaigns, funded by powerful business interests, that were meant to confuse or mislead the public on important environmental issues. Many of these campaigns continue today. Since the 1990s, environmental leaders and their supporters have spent much of their time and resources fighting

FIGURE 1-18
Aldo Leopold (1887–1948) became a leading conservationist. With more than 350 books and articles published, he has been called "probably the most quoted voice in the history of conservation."

FIGURE 1-19
Rachel Carson (1907–1964) alerted society to the harmful effects of the widespread use of pesticides.

efforts to discredit the environmental movement. Some analysts call for the United States to regain and strengthen its global role in improving environmental quality and shifting to a more environmentally sustainable society and economy.

checkpoint What led more Americans to demand action on environmental issues in the 1970s?

Achieving an Environmentally Sustainable Society

An **environmentally sustainable society** protects natural capital and lives off its income. Most environmental scientists say the ultimate goal of humankind should be to achieve an environmentally sustainable society. This society is one that meets the current and future basic resource needs of its people. Meeting people's needs should be done in a just and equitable manner without compromising the ability of future generations to meet their basic resource needs. This is in keeping with the social factors of sustainability.

Imagine that you win $1 million in a lottery. Suppose you invest this money (your capital) and earn 10% interest per year. If you live on just the interest income made by your capital, you will have a sustainable annual income of $100,000. You can spend $100,000 each year indefinitely and not deplete your capital. However, if you consistently spend more than your income, you will deplete your capital. Even if you spend just $10,000 more per year while still allowing the interest to accumulate, your money will be gone within 18 years.

The lesson here is an old one: Protect your capital and live on the income it provides. Deplete or waste your capital, and you will move from a sustainable to an unsustainable lifestyle.

The same lesson applies to using Earth's natural capital. This natural capital is a global trust fund of free natural resources and ecosystem services that are available to people now and in the future, and to all of Earth's other species. Living sustainably means living on **natural income**, which is the portion of renewable resources that can be used sustainably. By preserving and replenishing the natural capital that supplies this natural income, people can find ways to reduce their ecological footprint while expanding their beneficial environmental impact.

Although it may take hundreds or thousands of years, ecologists note that ecosystems can recover from much of the environmental degradation caused by human activities. But making a shift toward a more sustainable future involves some tough challenges. Environmental problems are so complex and widespread that it may seem hopeless, but that is not true. There is plenty of reason to hope—and to act. For instance, consider these two pieces of good news from the social sciences. First, research suggests it takes only 5–10% of the population of a community, a country, or the world to bring about major social and environmental change. Second, this research also shows that change can occur much faster than most people believe.

Anthropologist Margaret Mead famously summarized the potential for social change: "Never doubt that a small group of thoughtful, committed citizens can change the world. Indeed, it is the only thing that ever has." Engaged citizens in communities and schools around the world are proving Mead right. Join them.

checkpoint What is an environmentally sustainable society?

FIGURE 1-20
Julia "Butterfly" Hill lived in a redwood tree in California for two years to prevent a logging company from cutting down the tree. Her efforts and those of her supporters ultimately protected the tree from being logged.

1.4 Assessment

1. **Explain** What is the connection between natural capital, natural income, and an environmentally sustainable society?

2. **Summarize** Name the early individuals who had a major impact on the American public's awareness of the environment and sustainability, and explain their accomplishments.

3. **Classify** Into which category of environmental worldview would you put the early preservationists? Into which category would you put the early conservationists?

4. **Evaluate** In the 1970s, public opinion favored cleaning up the environment. Since the 1980s, some members of Congress and industry leaders have fought against environmental laws that regulate the amount of pollution that can be added to the environment. Are they working in the best interests of the public or of corporations?

SCIENCE AND ENGINEERING PRACTICES

5. **Engaging in Argument** Work in teams to collect data on the environmental impact of a product people rely on or use regularly. Construct an argument about its environmental impact. Find out what materials it is made from and how they are obtained, where the product is made versus where it is used, how it benefits society, and other factors that determine its environmental impact.

LESSON 1.4   39

TYING IT ALL TOGETHER STEM
The Greening of the American Mindset

As you read in this chapter and will continue to explore throughout this book, society faces many serious environmental problems. This book is about understanding these problems and finding solutions to them. A key to most solutions is to design economic and social systems, as well as individual lifestyles, that do not run counter to or interfere with the factors of sustainability. Humans can use scientifically informed social strategies to try to slow the losses of biodiversity, sharply reduce waste and pollution, switch to more sustainable sources of energy, and promote more sustainable forms of agriculture and other uses of land and water. People can also use their knowledge to sharply reduce poverty and slow population growth if necessary.

The Case Study at the beginning of the chapter highlighted ways that one school district is using scientific knowledge and making changes to be more sustainable. Individuals—like yourself and your classmates—can encourage changes like those seen at Milton-Union School District. Individuals can also make changes in their own lifestyles.

You and your fellow students are members of the 21st century's transition generation. You will play a major role in deciding if society creates a sustainable future or stays on a path of environmental degradation. To achieve sustainability, you will have to confront the urgent challenges posed by the environmental problems discussed in this book.

You may also need to make tough decisions personally and address your own lifestyle. It is an exciting time to be alive as people strive to develop a more sustainable relationship with this planet that is our only home.

In Lesson 1.2 you calculated your own ecological footprint. With that exercise in mind, answer the following questions.

1. What are some ways you could reduce your ecological footprint? How easy or difficult would it be for you to implement the changes necessary to reduce your footprint?
2. Look at this data from a sample utility bill. Calculate the amount of natural gas and electricity that this household would use over the course of a month and the course of a year.

> **Average Daily Gas Usage:** 1.97 therms
> **Average Daily Electricity Usage:** 28.19 kilowatt hours

3. Visit the EPA's Greenhouse Gas Equivalency Calculator website. Enter your answers from item 2 into the calculator.
   a. Compare the emissions from the sample bill to the emission equivalents of tons of waste sent to the landfill.
   b. Compare the emissions from the sample bill to the emission equivalents of garbage trucks of waste recycled instead of landfilled.
4. Compare your answers from item 3. How do those answers support the statement that recycling is a more sustainable solution to waste than sending it to the landfill, even though recycling still generates emissions?
5. Find out how much natural gas and electricity your household uses in any given month. How does your carbon emission footprint compare to the sample bill? How does it compare to your classmates' footprints? What do you think are factors that could influence utility emissions for a household?
6. Carbon offsets are actions that make up for the harmful effects of emitting carbon dioxide and other harmful gases. For example, if you burn gasoline driving in your car, you emit carbon dioxide. To offset these emissions, you can plant trees that capture carbon dioxide. Depending on the number, size, type, and age of tree you plant, your emissions can be fully or partially offset. Using the EPA website again, how many acres of forest would it take to offset (sequester) the natural gas and electrical emissions the sample household emits in one year?
7. How could you reduce your carbon footprint based on your household's natural gas and electricity use? How would this affect your overall ecological footprint?

CHAPTER 1 SUMMARY

1.1 What are some key factors of sustainability?

- Environmental science includes three goals. The first is to learn how life on Earth has survived and thrived. The second is to understand how humans interact with the environment. The third is to find ways to deal with environmental problems and live more sustainably.
- There are six factors of sustainability. Three are based on the natural sciences and three are based on social sciences. Sustainability is comprised of a combination of factors: solar energy, biodiversity, nutrient cycling, economics, political science, and ethics.
- Natural capital is comprised of natural resources (inexhaustible, renewable, and nonrenewable) and ecosystem services. Whether or not sustainability is achieved depends on how humans use natural capital.

1.2 How are our ecological footprints affecting Earth?

- Humans are living unsustainably by causing natural capital degradation. The tragedy of the commons helps explain why people may degrade natural resources.
- Every person or group of people produces an ecological footprint, which may or may not be harmful. An ecological footprint is equivalent to the land and water needed to produce the natural resources consumed by a person or population and to absorb their wastes.

1.3 What causes environmental problems and why do they persist?

- Major causes of environmental problems are population growth, wasteful and unsustainable resource use, poverty, avoidance of full-cost pricing, increasing isolation from nature, and conflicting environmental worldviews.
- People have different environmental worldviews, which reflect how they think the world works and what their role in the world should be.

1.4 What is an environmentally sustainable society?

- Many individuals have had a major impact on environmental awareness and sustainability. The first Earth Day is credited with increasing awareness about environmental issues in the United States in the 1970s.
- Living sustainably means living off Earth's natural income without depleting or degrading the natural capital that supplies it.

MindTap If you have been provided with access to a MindTap course, additional resources are available at login.cengage.com.

CHAPTER 1 ASSESSMENT

Review Key Terms

Select the term that best fits the definition. Not all key terms will be used.

> Earth-centered worldview
> ecological footprint
> ecology
> ecosystem
> ecosystem service
> environment
> environmental degradation
> environmental ethics
> environmental science
> environmental worldview
> environmentalism
> environmentally sustainable society
> exponential growth
> human-centered worldview
> inexhaustible resource
> life-centered worldview
> natural capital
> natural income
> natural resource
> nonpoint source
> nonrenewable resource
> point source pollution
> renewable resource
> sustainability

1. Materials and energy sources in nature that are essential or useful to humans
2. The natural services that support life and human economies at no monetary cost, such as pollination and the purification of air and water
3. A social movement dedicated to sustaining Earth's systems that support life
4. Something that cannot be used up or depleted
5. Asserts that humans are part of, and dependent on, nature and that Earth's life-support system exists for all species, not just for humans
6. Harmful contamination of the environment by a chemical or agent entering it over a diffuse area
7. The study of varying beliefs about what is right and wrong with how people treat the environment; provides useful tools for examining worldviews
8. The natural resources and ecosystem services that keep humans and other species alive and that support human economies

Review Key Concepts

9. Distinguish among environmentalism, ecology, and environmental science.
10. What are the six factors of sustainability?
11. Describe the sun's role in the sustainability of Earth's environment.
12. What role does biodiversity play in the sustainability of Earth's environment?
13. What is nutrient cycling, and why is it considered both a scientific factor of sustainability and an ecosystem service?
14. Compare and contrast inexhaustible, renewable, and nonrenewable resources. Give two examples of each type of resource.
15. What are some things that people can do to reduce degradation of natural capital?
16. According to Juan Martinez, what are some life skills that can be practiced by connecting with nature?
17. What are some factors that would need to be considered in full-cost pricing of gasoline that is used to power vehicles?
18. Some ethicists argue it is our responsibility to leave Earth in as good condition as it is in now, or in better condition. Do you agree? Explain why or why not.

Think Critically

19. What do you think are the two most unsustainable components of your lifestyle? What are two sustainable components? List two ways in which you could make your lifestyle more sustainable.
20. Explain why you agree or disagree with the following statements:
    a. Stabilizing population is not desirable because, without more consumers of products, economic growth would stop.
    b. We will never run out of resources. Technology will lead to substitutes and waste reduction.
    c. We can shrink our ecological footprints and improve the environment at the same time.
21. How does the statistic that the United States makes up 15% of the world's ecological footprint make you feel? How could you and others fix this problem?
22. Explain why you agree or disagree with each of the following statements:
    a. Humans are in charge of Earth.
    b. The value of other forms of life depends only on whether they are useful to humans.
    c. All forms of life have an inherent right to exist.

23. What are the basic beliefs within your environmental worldview? Are these beliefs consistent with the answers you gave to number 22? Are your actions that affect the environment consistent with your environmental worldview? Explain.

    At the end of this course, return to your answer to see whether your environmental worldview has changed.

Chapter Activities

A. Develop Models: Ecological Footprints STEM

If an ecological footprint is larger than the biological capacity of a region to replenish its renewable resources and absorb the resulting waste and pollution, there is an ecological deficit. If the reverse is true, the region has an ecological credit or reserve. In this activity you will examine the average, per-person biological capacity and ecological footprint of different countries and the world.

Materials
sheets of paper (8.5 x 11)
ruler
tape
scissors

1. Divide into 12 groups, each representing a different country listed in Figure 1-21.

2. Use the suggested materials to show the amount of land area (in hectares) required to support one person in your group's country.

   1 sheet of paper = 1 hectare = 10,000 m^2

3. When your land-area representation is complete, label it with the name of your country. Also include the following information:
   a. the number of hectares per person that it represents
   b. the number of football fields (including end zones) that could fit inside
   c. the country's ecological deficit or credit (Hint: For World, the ecological deficit is −0.9 hectare.)

4. Some countries with an ecological credit have a per capita footprint that is larger than the per capita footprint of countries with an ecological deficit. What factors do you think are used in calculating a country's ecological impact?

FIGURE 1-21

Source: World Wide Fund for Nature's *Living Planet Report 2012*

Place	World	United States	Canada	Mexico	Brazil	South Africa	Saudi Arabia	Israel	Germany	Russia	India	China	Australia
Per capita biological capacity (hectares/person)	1.8	3.9	14.9	1.4	9.6	1.2	0.7	0.3	2.0	6.6	0.5	0.9	14.6
Per capita ecological footprint (hectares/person)	2.7	7.2	6.4	3.3	2.9	2.6	4.0	4.0	4.6	4.4	0.9	2.1	6.7

B. Take Action

National Geographic Learning Framework
Attitudes | Empowerment
Skills | Communication
Knowledge | Our Living Planet

What features or practices could be implemented at your school to promote environmental sustainability? Identify one or two ways in which your school could be more sustainable. Write a letter or email to your principal or superintendent. In your letter, use as much real data as you can find. Include how the benefits of implementing the new feature or practice will outweigh the costs.

CHAPTER 2
SCIENCE, MATTER, ENERGY, AND SYSTEMS

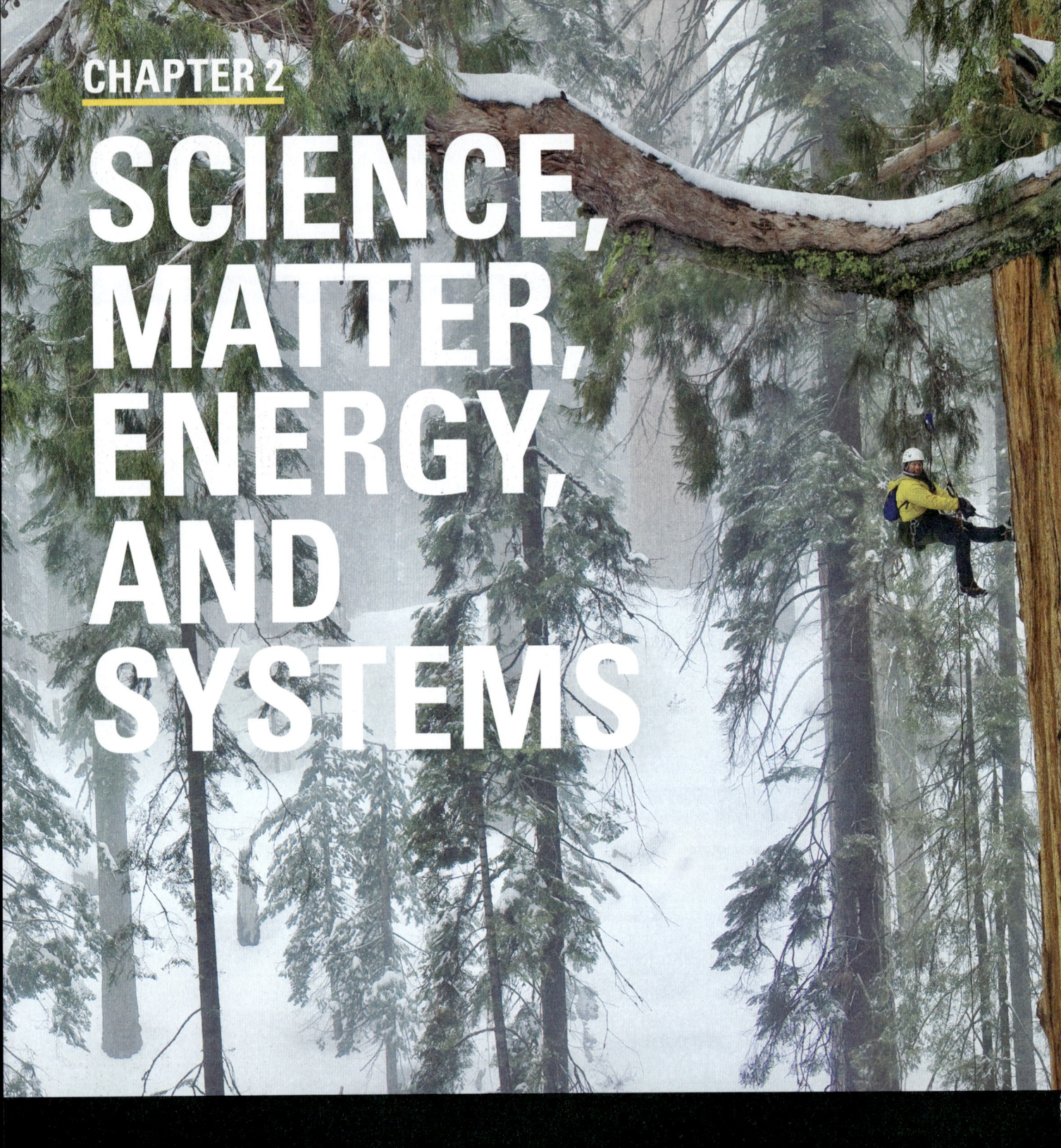

THE PRESIDENT is the nickname given to this giant sequoia, towering about 75.25 meters (247 feet) high. This forest giant consists of the same basic types of particles that make up all life—namely, long chains of carbon. Where did the sequoia get all that carbon? The enormous tree absorbed most of its carbon directly from the atmosphere. Understanding exchanges in matter is key to understanding ecosystems.

KEY QUESTIONS

2.1 What do scientists do?

2.2 What is matter?

2.3 What is energy?

2.4 What are systems?

Researchers measure a 3,200-year-old giant sequoia in California's Sequoia National Park

NATIONAL GEOGRAPHIC | EXPLORERS AT WORK

New Paths to Engaging Science

with National Geographic Explorer Nalini Nadkarni

Nalini Nadkarni has all the academic credentials anyone could ask for. She's a Ph.D. biologist and forest ecologist at the University of Utah, where she's also Director of the Center for Science and Mathematics Education. She's accrued an impressive record of fieldwork, using mountain-climbing techniques, construction cranes, walkways, and hot air balloons to explore the canopies of rain forests on four continents.

But Dr. Nadkarni's real passion is finding ways to bring that important work to people outside the scientific community. She suggests that "if you want to go beyond that small percentage of people who are already environmentally and scientifically aware, you have to make your work somehow link with a passion, interest, or profession of someone who isn't interested in science or nature."

Nadkarni has found some very creative ways to reach new audiences. She invited George "Duke" Brady, a California rap singer, to join her on a tree-climbing expedition. Nadkarni believes Brady's resulting rap about his experience appealed to students in a way that more traditional approaches may not have. That's just one example. Here's another: Treetop Barbie. Combining science and fashion, this popular doll wears the clothing of a forest ecologist and comes with a field guide to canopy plants and animals.

Nalini Nadkarni's most ambitious effort has involved one of the last places you might expect—the prisons of the Washington State Department of Corrections. The Sustainability in Prisons Project helps incarcerated men and women contribute to conservation and ecological research. The benefits go both ways, Nadkarni insists. The prisoners have a purpose and learn useful skills on the job as they do scientific work, and the scientific community benefits from their projects and the data they gather.

Some of those projects involve time-consuming and meticulous procedures. For example, inmates raise frogs, rear butterflies, catalog moss species, and practice beekeeping. Inmates volunteer for such projects, Nadkarni explains, because they can contribute to society despite being confined. In turn, the inmates hear guest lectures and acquire job skills.

The prisons participating in the Sustainability project often begin to follow more sustainable practices, such as recycling, composting, and organic gardening. Nadkarni covered the walls of an exercise room in one prison with a picture of trees. To everyone's surprise, violence in that unit decreased. Nadkarni's conclusion? "There isn't a person on Earth who couldn't use a connection with nature."

Nadkarni has found some unique ways to bring science to communities that need it. She shows us that all the content knowledge in the world isn't worth much without critical thinking and creativity.

Thinking Critically

Evaluate What factors do you think would be involved in implementing programs similar to the Sustainability in Prisons Project? List some possible pros and cons.

Nadkarni prepares to climb a tree in Costa Rica's Monteverde Cloud Forest Reserve.

CASE STUDY
Experimenting with a Forest

Suppose a logging company plans to cut down all of the trees on a hillside near your home. You are very concerned and want to know about the possible harmful environmental effects.

One way to learn about such effects is to conduct a controlled experiment. To conduct an experiment, scientists begin by identifying key variables, such as water loss and soil nutrient content. Then, they set up two groups. One is the experimental group, in which a chosen variable is changed in a known way. The other is the control group, in which the chosen variable is not changed. Then they compare the results from the two groups.

In 1963, botanist F. Herbert Bormann, forest ecologist Gene Likens, and their colleagues began carrying out such a controlled experiment. Their goal was to compare the loss of water and soil nutrients from an area of uncut forest (the control site) with one that had been stripped of its trees (the experimental site).

The researchers built V-shaped concrete dams across the creeks at the bottoms of several forested valleys in the Hubbard Brook Experimental Forest in New Hampshire. The dams were designed so that all surface water leaving each valley had to flow across an area where scientists could measure its volume and dissolved nutrient content.

First, the researchers measured the amounts of water and dissolved soil nutrients flowing from an undisturbed forested area in one of the valleys (the control site) (Figure 2-1, left). These measurements showed that an undisturbed mature forest is very efficient at storing water and retaining nutrients in its soils.

Next, they set up an experimental site (Figure 2-1, right). One winter, they cut down all the trees and shrubs in that valley, left them where they fell, and sprayed the area with herbicides to prevent the regrowth of vegetation.

The researchers then compared the outflow of water and nutrients in this experimental site with those in the control site.

The scientists found that, with no plants to help absorb and retain water, the amount of water flowing out of the deforested valley increased by 30–40%. As this excess water ran rapidly over the ground, it eroded soil and carried dissolved nutrients, such as nitrates, out of the topsoil in the deforested site (Figure 2-1). Overall, the loss of key soil nutrients from the experimental forest was six to eight times that in the nearby uncut control forest.

In this chapter, you will learn how scientists study nature and about the matter and energy that make up the world. You will also learn about scientific laws that govern the changes to matter and energy. And you will learn the important difference between a scientific hypothesis and a scientific theory.

As You Read Think about how ecologists might use basic chemistry and the laws of matter and energy to study ecosystems.

FIGURE 2-1

Nitrate Results
This graph shows the concentrations of nitrate ions in water from the forested watershed (control site) and the deforested watershed (experimental site).

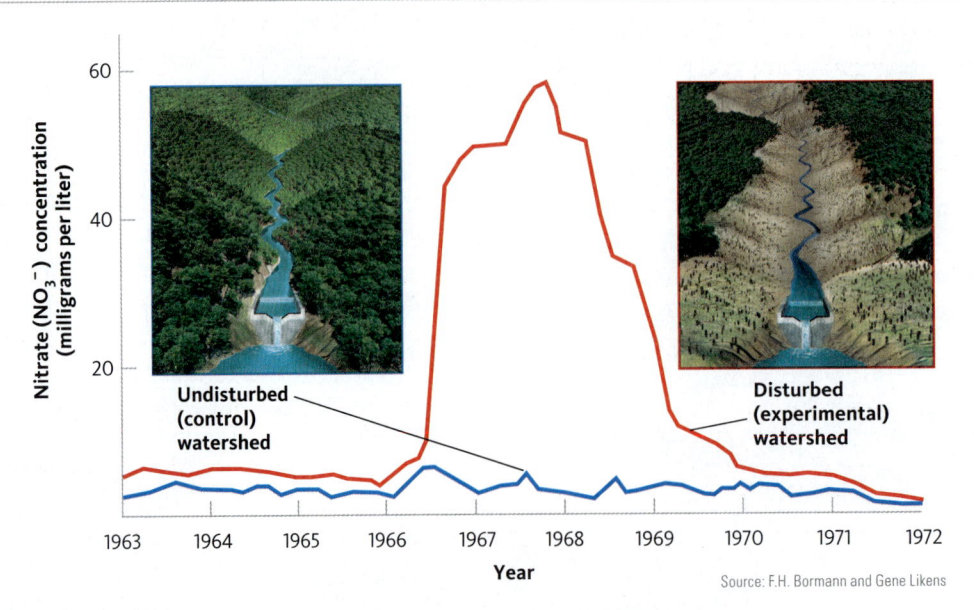

Nitrate (NO_3^-) concentration (milligrams per liter)

Undisturbed (control) watershed

Disturbed (experimental) watershed

Year

Source: F.H. Bormann and Gene Likens

48 CHAPTER 2 SCIENCE, MATTER, ENERGY, AND SYSTEMS

2.1 What Do Scientists Do?

CORE IDEAS AND SKILLS

- Describe the scientific methods and the importance of observation, experimentation, and models.
- Recognize the importance of evidence, hypotheses, theories, and laws in science.
- Understand the benefits and limitations of science.

KEY TERMS

science
scientific method
scientific hypothesis
data
model
scientific theory
peer review
scientific law

Scientists Use a Variety of Methods

Science is a broad field of study focused on discovering how nature works and using that knowledge to describe what is likely to happen in nature. It is based on the assumption that events in the physical world follow orderly cause-and-effect patterns that can be understood through careful observation, measurements, and experimentation.

Scientists use several **scientific methods**, or practices, to advance knowledge and understanding of how the natural world works. Figure 2-2 summarizes these practices. While the immediate goal of science is to build knowledge of the natural world, that knowledge can be applied in a number of ways. For example, engineers use scientific knowledge to design solutions that improve society and protect the planet and its resources.

In the Case Study, Bormann and Likens used scientific practices to find out how clearing a forest affects its ability to store water and retain soil nutrients. They asked questions, did research, collected information, and proposed a hypothesis. A **scientific hypothesis** is a possible and testable answer to a scientific question or explanation of what scientists observe in nature. Hypotheses can be written as *"if, then"* statements. Bormann and Likens came up with the following hypothesis: *If* land is cleared of its vegetation and exposed to rain and melting snow, *then* the land retains less water and loses nutrients.

Next, Bormann and Likens designed experiments to collect **data**, or information, to test their hypothesis. In an experiment, researchers try to keep all variables the same between a control group and an experimental group except for the variable they are testing—in this case, deforestation. Controlled experiments are the only way to show that one variable causes another. Correlation studies, on the other hand, can show that two variables are related, but they can't show how they are related.

FIGURE 2-2
Scientific Practices This table summarizes the main practices that scientists use to advance knowledge. Scientists may use these practices in any order. Some scientists use all of the practices, while others focus on only one or a few.

THE PRACTICES OF SCIENCE
Asking questions
Developing and using models
Planning and carrying out investigations
Analyzing and interpreting data
Using math
Forming explanations
Forming arguments from evidence
Obtaining, evaluating, and communicating information

Bormann and Likens tested their hypothesis twice. The first set of data recorded the amount of nitrogen in the runoff. They repeated their experiment to determine the amount of phosphorus in the water. The experimenters wrote scientific articles describing their research. Other scientists evaluated their data and conclusions. These reviews and further research supported their hypothesis.

Models are often also used to conduct experiments or form explanations. A **model** is a physical or mathematical representation of a structure or system. Data from the research carried out by Bormann and Likens and others was fed into models, which also supported their hypothesis.

A well-tested and widely accepted scientific hypothesis or a group of related hypotheses is called a scientific theory. A **scientific theory** is one of the most important and certain results of science and is based on a large body of evidence. The research conducted by Bormann and Likens and other scientists led to the scientific theory that trees and other plants hold soil in place and retain water and nutrients needed to support the plants.

checkpoint What is a scientific hypothesis?

LESSON 2.1

ENGINEERING FOCUS 2.1

WHAT IS ENGINEERING?

Engineering is all about defining problems and designing and testing solutions to the problems. In fact, anyone who applies science and math to solve a practical problem is using the practices of engineering (Figure 2-3).

Whereas the result of science is an advancement in human knowledge, the result of engineering is a new or improved product. A product can be a structure such as a dam, a technology such as a new software application, or even a new procedure or method.

The outputs of science and engineering feed into one another. Engineers use knowledge gained by scientists—and scientists use tools developed by engineers. New or improved tools and methods can lead to new knowledge, and new knowledge leads to the development of new tools.

Engineering's many specialties fall into different broad categories, including mechanical, electrical, civil, chemical, and biological engineering. One exciting new field of engineering is conservation technology, which uses technology to solve ecological problems (Figure 2-4).

Thinking Critically
Apply Describe an example in which a new tool or technology led to new scientific knowledge.

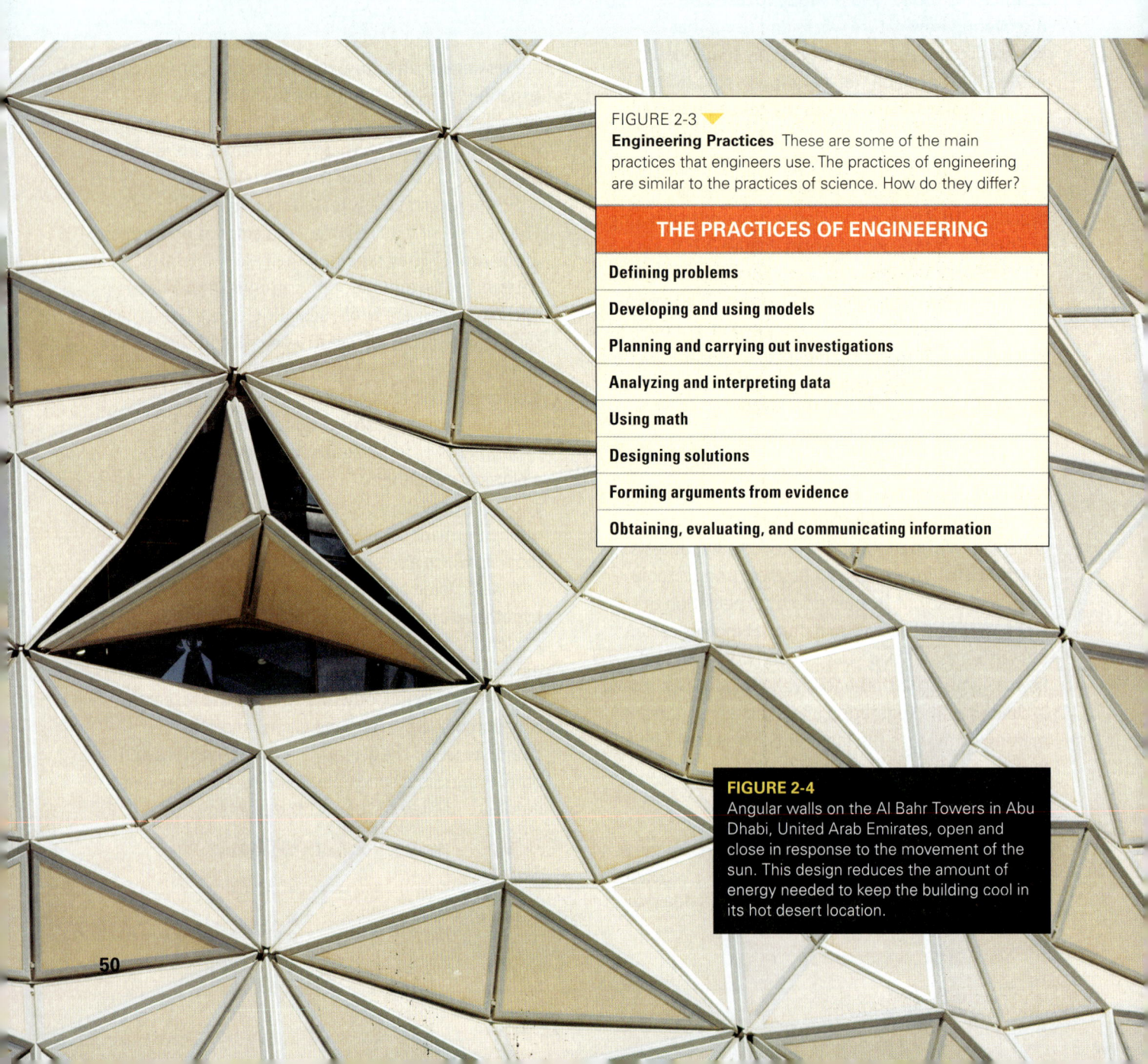

FIGURE 2-3
Engineering Practices These are some of the main practices that engineers use. The practices of engineering are similar to the practices of science. How do they differ?

THE PRACTICES OF ENGINEERING
Defining problems
Developing and using models
Planning and carrying out investigations
Analyzing and interpreting data
Using math
Designing solutions
Forming arguments from evidence
Obtaining, evaluating, and communicating information

FIGURE 2-4
Angular walls on the Al Bahr Towers in Abu Dhabi, United Arab Emirates, open and close in response to the movement of the sun. This design reduces the amount of energy needed to keep the building cool in its hot desert location.

NATIONAL GEOGRAPHIC | EXPLORERS AT WORK
John Francis Marine Biologist

Virtually all environmental scientists have a deep personal connection to the natural world. How can young people establish such a connection, especially as they spend more time indoors and in front of devices? National Geographic Explorer and marine biologist John Francis is leading Americans back into the wonders of the natural world.

When he was 19, John Francis began studying marine mammals on remote islands in the Americas. After earning his Ph.D. and conducting research for the Smithsonian Institution, he was awarded two National Geographic Society grants. The grants enabled him to study and film a rare fur seal that lives only on an isolated island off Chile. Dr. Francis's passion for ecology and filmmaking led to his role as a producer of wildlife films with National Geographic Television. During his six years as a producer, he made films covering everything from chimps and tigers to whales and sharks.

Dr. Francis has also served as the Vice President for Research, Conservation, and Exploration at the National Geographic Society. In collaboration with the National Park Service (NPS), he developed a ten-year series of BioBlitz events held in national parks around the country, starting in 2007. During a BioBlitz, teams of volunteer scientists, families, students, teachers, and other community members work to identify as many species of organisms as possible in a particular area in a 24-hour period. The BioBlitz provides a "snapshot" of species diversity and helps bring people closer to nature.

The ninth NGS-NPS BioBlitz was held in the Hawai´i Volcanoes National Park and involved more than a thousand participants. The event included a celebration of local culture to acknowledge the connections among land, history, and culture. The resulting inventory from the Hawai´i blitz included observations of 22 species never before recorded in Volcanoes National Park. These events have served as models for BioBlitzes around the world.

FIGURE 2-5
An i'iwi bird perches on a Hawaiian raspberry branch. The scientific practice of obtaining information was the focus of the ninth annual BioBlitz in Hawai'i Volcanoes National Park.

LESSON 2.1

FIGURE 2-6
ON ASSIGNMENT National Geographic Photographer Tim Laman snapped this shot of a moth scientist using a light trap in a New Guinea rain forest. Which of the science practices do you think the scientist is engaged in? Which of the engineering practices might he also use?

52 CHAPTER 2 SCIENCE, MATTER, ENERGY, AND SYSTEMS

LESSON 2.1

Scientific Inquiry Advances Human Knowledge

Scientific inquiry is based on values that help advance scientific knowledge. Some of the values that support good science are logic, critical thinking, objectivity, open-mindedness, and honest reporting. Science is a worldwide effort conducted by people from many nations and cultures. Many scientists work collaboratively on teams. They almost always build from an existing body of scientific research and knowledge. For example, Bormann and Likens likely consulted many prior studies and used existing nutrient-testing methods and dam designs.

Scientists also review each other's work in a process called peer review. **Peer review** involves scientists publishing details of the methods they used, the results of their experiments, and the reasoning for their interpretations. Other scientists in the same field (their peers) evaluate their work. Scientific knowledge advances in this self-correcting way, with scientists questioning and confirming the data and hypotheses of their peers. Scientific findings that are repeatable are considered *reliable*.

checkpoint What is the purpose of peer review?

Theories and Laws Are the Most Certain Results of Science

Scientific theories are based on hypotheses that have been tested widely, are supported by extensive evidence, and are accepted as useful explanations of phenomena by most scientists in a field of study. Scientific theories can change and evolve as new scientific knowledge becomes available.

Another important and reliable outcome of science is a **scientific law**—a well-tested and widely accepted description of observations that have been repeated many times in a variety of conditions. One example is the law of gravity. After making many thousands of observations and measurements of objects falling from different heights, scientists developed the following scientific law: All objects fall to Earth's surface at predictable speeds.

The difference between a scientific law and a scientific theory is that a law describes an observable phenomenon, while a theory is a well-tested and generally accepted scientific explanation as to why that phenomenon occurs.

checkpoint What is the difference between a scientific theory and a scientific law?

Science Has Limitations

Scientific inquiry has limitations in explaining and predicting natural phenomena. For example, scientific research cannot prove anything absolutely. Some degree of uncertainty in scientific measurements, observations, and models will always exist—even if that uncertainty is very small. Uncertainty also results from the fact that science requires the use of statistical tools. For example, there is no way to measure accurately how many metric tons of soil are eroded annually worldwide. Instead, scientists use statistical sampling and mathematical methods to estimate such numbers.

Scientists don't use the word *proof* in the same way as many nonscientists because it can falsely imply "absolute proof." For example, most scientists would not say, "Science has proven that cigarettes cause lung cancer." Instead they might say, "Overwhelming evidence from thousands of studies indicates that people who smoke regularly for many years have a greatly increased chance of developing lung cancer."

Another limitation of science is that of the scientists themselves. Like all humans, they are not totally free of bias about their own results and hypotheses. The self-imposed standards of evidence required through peer review help protect against personal bias and falsifying scientific results.

Despite these limitations, science is the most useful way of learning about how nature works and predicting how it might behave in the future.

checkpoint What are two limitations of science?

2.1 Assessment

1. **Identify** Describe one example of each of the scientific practices (Figure 2-2) in the Hubbard Brook experiment.
2. **Explain** What are the benefits and limitations of scientific models?
3. **Synthesize** Describe how engineering was used in the Hubbard Brook experiment.
4. **Contrast** How does a scientific hypothesis differ from a scientific theory?

CROSSCUTTING CONCEPTS

5. **Cause and Effect** Do the results of the Hubbard Brook experiment show a cause-and-effect relationship? Explai

2.2 What Is Matter?

CORE IDEAS AND SKILLS
- Define matter and identify its building blocks.
- Distinguish between physical and chemical changes.
- Describe the law of conservation of matter.

KEY TERMS

matter	ion	chemical reaction
element	pH	law of conservation of matter
compound	organic compound	
atom	physical change	
isotope	chemical change	
molecule		

Matter Consists of Elements and Compounds

Matter is anything that has mass and takes up space. Matter commonly exists in one of three physical states—solid, liquid, or gas—at a given temperature and pressure. Chemically, matter can be an element or a compound.

Elements An **element** is a type of matter with a unique set of properties that cannot be broken down into simpler substances by chemical means. Chemists refer to each element with a unique symbol, such as C for carbon and Ca for calcium. The known elements are arranged on the basis of their chemical behavior. The chart of these elements is called the periodic table of elements (as shown in Appendix 3). The periodic table currently contains 118 elements. Some of these elements have only been onservable in laboratory experiments.

Most matter consists of **compounds**, which are combinations of two or more different elements held together in fixed proportions. For example, water is a compound that consists of the elements hydrogen (H) and oxygen (O).

The **atom** is the basic building block of matter. An atom is the smallest unit of matter into which an element can be divided and still have its distinctive chemical properties. Atoms are incredibly small. More than 3 million hydrogen atoms could sit side by side on the period at the end of this sentence.

Each atom contains certain numbers of three subatomic particles (Figure 2-7).

- Neutrons—particles with no charge that are located in the nucleus (center) of the atom
- Protons—particles with a positive (+) charge that are located in the nucleus of the atom
- Electrons—smaller particles with a negative (−) charge in rapid motion outside the nucleus

Each element has a unique atomic number equal to the number of protons in the nucleus of its atom. Carbon (C) has 6 protons in its nucleus, so it has an atomic number of 6. Uranium (U) has 92 protons in its nucleus. It has an atomic number of 92. The mass number of an atom is the total number of both the protons and the neutrons in its nucleus. Most of an atom's mass is concentrated in its nucleus. Because electrons have so little mass compared with protons and neutrons, the mass number does not include electrons.

Atoms of a particular element can vary in the number of neutrons they contain, and therefore, in their mass numbers. The forms of an element that have the same atomic number but different mass numbers are called **isotopes**. For example, the three most common isotopes of carbon are carbon-12, carbon-13, and carbon-14.

Molecules A second building block of matter is the **molecule**, a combination of two or more atoms of the same or different elements. The atoms are held together by forces known as chemical bonds. For example, a water (H_2O) molecule consists of two atoms of hydrogen (H_2) bound to one atom of oxygen (O).

Ions A third building block of some types of matter is an **ion**—an atom or group of atoms with a positive or negative charge (+ or −). Chemists use a superscript after the symbol of an ion to indicate the number of positive or negative electrical charges it has. For example, sodium ions (Na^+) have a net positive one charge. Chloride ions (Cl^-) have a net negative one charge. In this chapter's Case Study,

FIGURE 2-7 ▼
Atom This simplified model depicts a carbon-12 atom. The atom's nucleus consists of six protons, each with a positive electrical charge, and six neutrons with no electrical charge. Six negatively charged electrons move rapidly outside its nucleus.

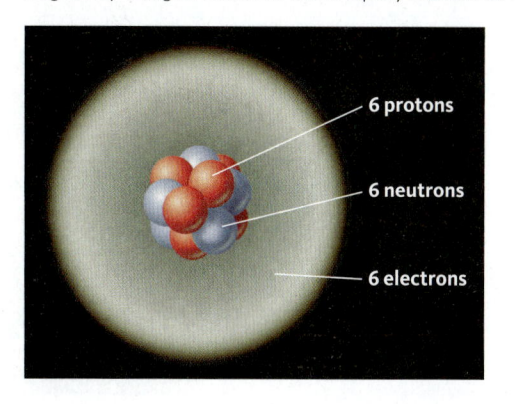

Bormann and Likens measured and compared the loss of nitrate ions (NO_3^-) from disturbed and undisturbed forests in their experiment (Figure 2-1). The nitrate ion is a nutrient that is essential for plant growth.

> **CONSIDER THIS**
>
> **You are composed of many of the same elements that are found in Earth's crust.**
>
> In addition to water, your body is made of calcium, phosphorus, potassium, sulfur, sodium, chlorine, magnesium, and iron. How do these elements get from rocks (Figure 2-9) into your body? They are broken down into soil, taken up by plants, and absorbed from your food. You will learn more about the cycling of matter in Chapter 3.

Ions are important for measuring a substance's acidity in a water solution. Acidity is based on the comparative amounts of hydrogen ions (H^+) and hydroxide ions (OH^-) in the solution. Scientists use the **pH** scale as a measure of acidity.

Pure water (not tap water or rainwater) has an equal number of H^+ and OH^- ions. Pure water is considered a neutral solution and has a pH of 7. A solution that has more H^+ ions than OH^- ions is an acidic solution, or an acid. Acids have a pH less than 7. A solution that has more OH^- ions than H^+ ions is a basic solution (Figure 2-8). It may also be called an alkaline solution or a base. A base has a pH greater than 7.

checkpoint How do elements and compounds differ?

FIGURE 2-8

The pH Scale The pH scale describes the acidity of solutions.

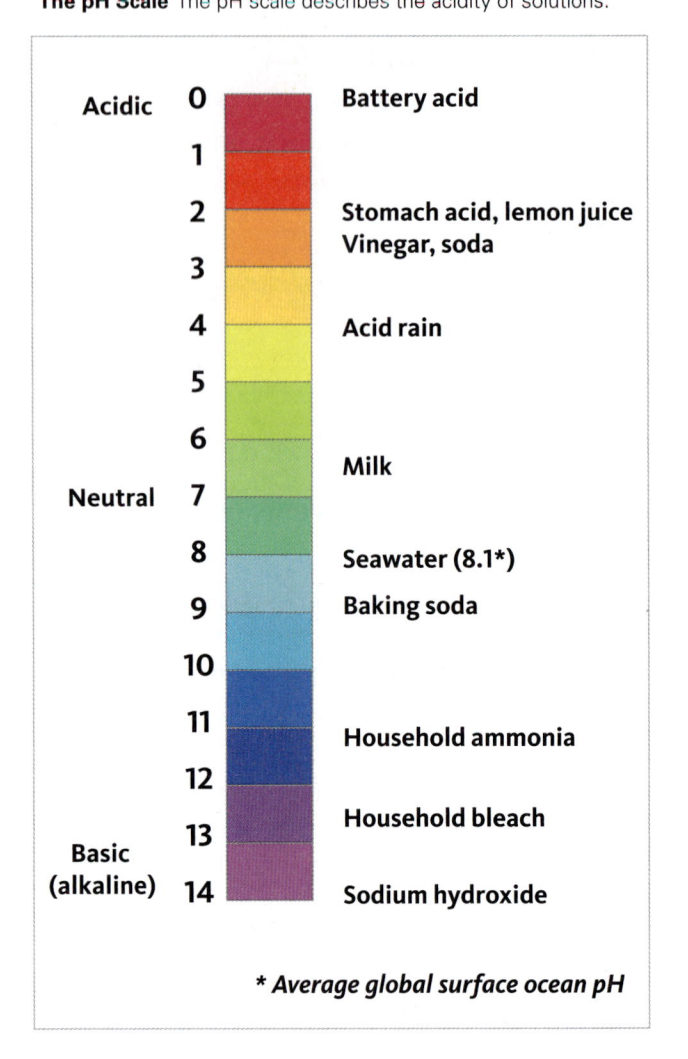

pH		Example
Acidic 0		Battery acid
1		
2		Stomach acid, lemon juice
3		Vinegar, soda
4		Acid rain
5		
6		Milk
Neutral 7		
8		Seawater (8.1*)
9		Baking soda
10		
11		Household ammonia
12		
13		Household bleach
Basic (alkaline) 14		Sodium hydroxide

Average global surface ocean pH

FIGURE 2-9

From Rocks to Organisms The mineral apatite is composed of the elements calcium, phosphorus, oxygen, hydrogen, fluorine, and chlorine. Apatite is a main source of nutrients for plants in soil.

Molecules of Life

In chemistry, an **organic compound** is a carbon-based compound. More specifically, an organic compound is a compound composed of at least two carbon atoms combined with atoms of one or more other elements. The exception is methane (CH_4), which has only one carbon atom. Plastics, table sugar, vitamins, aspirin, penicillin, and most of the chemicals in your body are organic compounds.

The millions of known organic (carbon-based) compounds include hydrocarbons and simple carbohydrates. Hydrocarbons are compounds of carbon and hydrogen atoms. Hydrocarbons are essential to living things. Organic compounds also include simple carbohydrates (simple sugars) made up of carbon, hydrogen, and oxygen atoms. An example is glucose ($C_6H_{12}O_6$), which most plants and animals break down in their cells to obtain energy.

Several types of larger and more complex organic compounds are called polymers. Polymers form when a number of simple organic molecules (monomers) are linked together by chemical bonds. Imagine monomers like rail cars linked in a freight train. The simple carbohydrate glucose is a monomer. Glucose can be found in complex carbohydrates such as starch and cellulose, as shown in Figure 2-10.

In addition to complex carbohydrates, three other groups of polymers essential to life include proteins, nucleic acids, and lipids. Proteins are polymers made up of amino acids and play many vital roles in the body. Nucleic acids are polymers formed by monomers called nucleotides. Genetic information is coded in the nucleic acids RNA and DNA. Lipids, which include fats and waxes, make up cell membranes and are also critical to nervous system function.

checkpoint Why is glucose classified as an organic compound?

Physical and Chemical Changes

Matter can undergo physical or chemical changes. When matter undergoes a **physical change**, there is no change in its chemical composition. A piece of aluminum foil cut into small pieces is still aluminum foil. When solid water (ice) melts and liquid water boils, the resulting liquid water or water vapor remain as H_2O molecules.

When a **chemical change**, or **chemical reaction**, takes place, there is a change in the chemical composition of the substances involved. Chemists use a chemical equation to show how chemicals are rearranged in a chemical reaction. For example, coal is made up almost entirely of the element carbon (C). When coal is burned completely in a power plant, the solid carbon in the coal combines with oxygen (O_2) from the atmosphere. This forms the gaseous compound carbon dioxide (CO_2). Figure 2-11 shows three different ways to represent this reaction.

Law of Conservation of Matter Elements and compounds can change from one physical or chemical form to another. However, atoms are never created or destroyed in the process. As the result of a physical change, a substance may have different physical properties, but it is still composed

FIGURE 2-11
Chemical Reaction Chemical reactions can be modeled with words, symbols, or illustrations.

Reactant(s) → Product(s)

Carbon + Oxygen → Carbon dioxide + Energy

C + O_2 → CO_2 + Energy

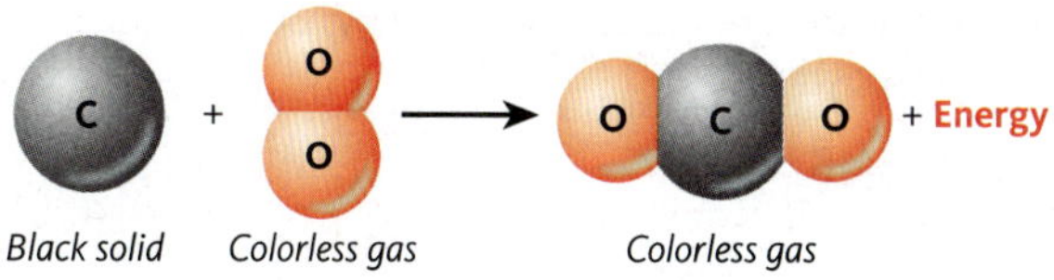

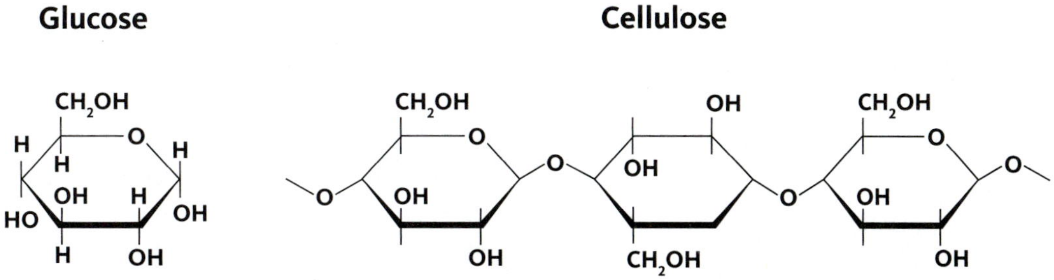

Black solid + Colorless gas → Colorless gas + Energy

Glucose

Cellulose

FIGURE 2-10
Carbon-based Molecules Glucose monomers (left) make up complex carbohydrate polymers, such as cellulose (right).

LESSON 2.2

of the same atoms, ions, or molecules (such as liquid water and ice). A chemical change results in different chemical combinations of atoms, ions, or molecules, forming different substances. The **law of conservation of matter** states that whenever matter undergoes a physical or chemical change, no atoms are created or destroyed.

checkpoint What is the difference between a physical change and a chemical change?

2.2 Assessment

1. **Identify** Name and describe three particles that make up atoms.
2. **Apply** Give an example of a physical change and an example of a chemical change to water.
3. **Generalize** How could you test the law of conservation of matter?
4. **Predict** Ground limestone can be added to soil to make it less acidic. Would you expect the pH of limestone in a water solution to be less than or greater than 7?

SCIENCE AND ENGINEERING PRACTICES

5. **Communicating Information** Explain how Figure 2-7 would be different if it illustrated carbon-14 instead of carbon-12.

2.3 What Is Energy?

CORE IDEAS AND SKILLS
- Recognize the different forms of energy.
- Understand the first and second laws of thermodynamics.

KEY TERMS

energy
kinetic energy
thermal energy
electromagnetic
 radiation
potential energy
first law of
 thermodynamics
second law of
 thermodynamics

Energy Comes in Many Forms

Suppose you find this book on the floor and you pick it up and put it on your desk. To do this you have to use a certain amount of work to move the book from one place to another. In scientific terms, work is done when any object is moved a certain distance (work = force × distance). **Energy** is the capacity to do work. There are two major types of energy: energy due to motion (called kinetic energy) and energy that is stored (called potential energy).

Kinetic Energy Matter that is moving has **kinetic energy**. Some examples of kinetic energy are flowing water, a speeding car, electricity (electrons flowing through a wire or other conducting material), and wind (a mass of moving air). Heat, or **thermal energy**, is another form of kinetic energy. Heat is the total kinetic energy of all the moving atoms, ions, or molecules in a sample of matter. The hotter an object is, the faster the motion of the atoms, ions, or molecules that make up the object. Temperature is a measure of the average heat or thermal energy of the atoms, ions, or molecules in a sample of matter. When two objects at different temperatures make contact with each other, heat, or thermal energy, flows from the warmer object to the cooler object. You probably have observed this when you have touched a hot dish. Heat from the dish flows into the cooler skin on your hand.

Electromagnetic radiation is another form of kinetic energy. Energy in electromagnetic radiation travels as a wave. This occurs as a result of changes in electrical and magnetic fields. Different wavelengths of electromagnetic radiation carry different amounts of energy (Figure 2-12). A wavelength is measured by the distance between successive peaks or troughs in the wave. Visible light, ultraviolet radiation, and radio waves are all examples of electromagnetic energy.

Electromagnetic radiation from the sun, or solar energy, is the major source of energy that fuels processes on Earth. This includes plant growth, winds, ocean currents, and climate systems. Solar energy is one of the key factors of sustainability discussed in Chapter 1. Without solar energy, Earth would be frozen and life as we know it would not exist.

Potential Energy Another major type of energy is potential energy. **Potential energy** is stored and potentially available for use. Potential energy can change into kinetic energy. For example, a book held in your hand has potential energy. The book has the potential to gain kinetic energy if you were to drop it. The chemical energy in the molecules of food you eat is also potential energy. Water in a reservoir behind a dam has potential energy (Figure 2-13). When the dam is opened and the water flows, potential energy transforms to kinetic energy. When a car engine burns gasoline, the potential energy stored in the chemical bonds of the gasoline molecules changes into kinetic energy that moves the car.

FIGURE 2-12

Electromagnetic Spectrum The electromagnetic spectrum consists of a range of electromagnetic waves, which differ in wavelength and energy content.

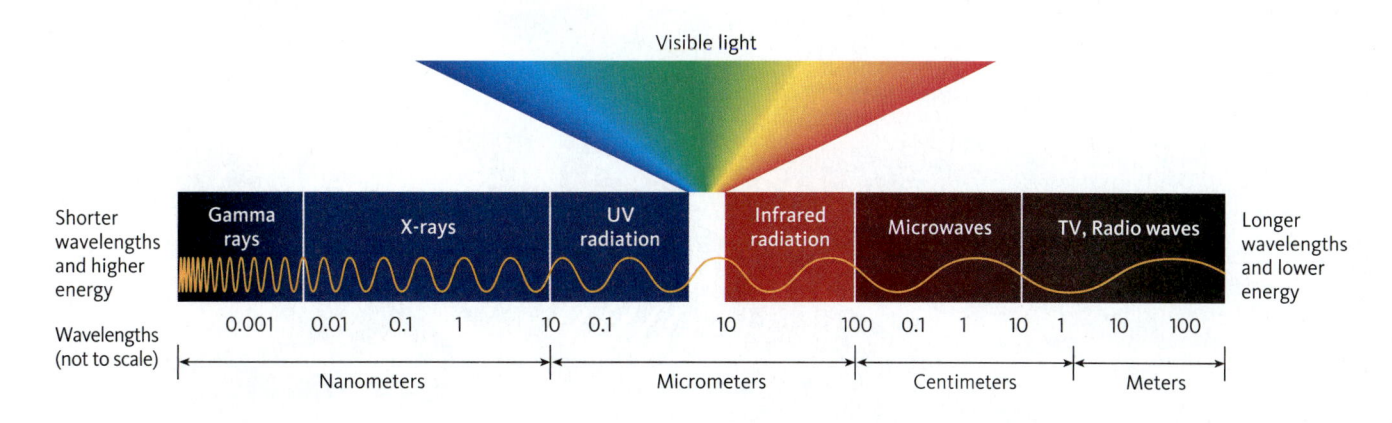

Energy Quality

Energy quality is a measure of the capacity of energy to do useful work. High-quality energy is concentrated energy that has a high capacity to do useful work. Examples are high-temperature thermal energy, concentrated sunlight, high-speed wind, and the energy released when people burn wood, gasoline, natural gas, or coal.

By contrast, low-quality energy is so spread out that it has little capacity to do useful work. For example, the energy in the exhaust molecules from the burnt gasoline in a car is far too spread out for humans to use to move or heat things.

checkpoint What are the two major types of energy?

Energy Changes Are Governed by Two Scientific Laws

First Law of Thermodynamics From millions of observations and measurements of energy changing from one form to another, scientists summarized their findings in the law of conservation of energy, also known as the **first law of thermodynamics**. According to this scientific law, whenever energy is converted from one form to another in a physical or chemical change, no energy is created or destroyed. The total amount of energy does not change.

Second Law of Thermodynamics Thousands of experiments have shown that whenever energy is transformed from one type to another in a physical or chemical change, the result is lower-quality energy. The energy is less usable after the change than before the change. This is a statement of the **second law of thermodynamics**. The "lost" usable energy usually takes the form of heat, which spreads into the environment and cannot be used to do much work. The random motion of air or water molecules in the environment further disperses this heat.

In other words, when energy changes from one form to another, it always goes from a more useful to a less useful form. This means that people can't recycle or reuse high-quality energy to perform useful work. Once the high-quality energy in a serving of food, a tank of gasoline, or a chunk of coal is released, it is degraded to low-quality energy and disperses as heat into the environment.

checkpoint What is the difference between the first and second laws of thermodynamics?

2.3 Assessment

1. **Identify** Give three examples of kinetic energy and three examples of potential energy.
2. **Summarize** What is the second law of thermodynamics?
3. **Classify** Why is thermal energy classified as kinetic energy?
4. **Explain** Why can't the high-quality energy released from a lump of burning coal be recycled?

SCIENCE AND ENGINEERING PRACTICES

5. **Constructing Explanations** Suppose someone claims they have invented a new automobile engine that can produce more energy than is found in the fuel used to run it. Apply scientific principles to evaluate the person's claim.

LESSON 2.3   59

FIGURE 2-13
When water in a reservoir flows through channels in a dam, its potential energy transforms into kinetic energy. The powerful kinetic energy transfers to a turbine inside the dam, producing electrical energy.

2.4 What Are Systems?

CORE IDEAS AND SKILLS
- Identify the key components of a system.
- Describe the ways in which systems respond to change.

KEY TERMS
system
feedback loop
ecological tipping point

Systems and System Models

A **system** is a set of components that function and interact in some regular way. A cell, the human body, a forest, a river, a dam, an economy, and Earth are all examples of systems.

Systems have the following key components: inputs, throughputs, and outputs of matter, energy, and information (Figure 2-14). A system can become unsustainable if the throughputs are greater than the ability of the environment to provide the required inputs or to absorb or dilute the system's outputs.

Most systems are affected by feedback. Feedback is matter, energy, or information that, when fed back into the system as input, increases or decreases a change to the system. Input that increases a change to a system is called positive feedback. Input that decreases a change to a system is called negative feedback. A **feedback loop** occurs when an output of matter, energy, or information is fed back into the system as an input and leads to changes in that system.

FIGURE 2-14
Inputs, Throughputs, and Outputs This greatly simplified model shows the flow of matter, energy, and information into and out of a system.

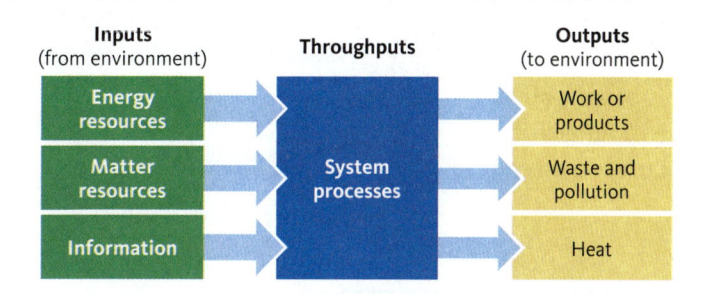

FIGURE 2-15
Negative Feedback Like a thermostat regulating the temperature in a home, a negative feedback loop has a stabilizing effect on a system.

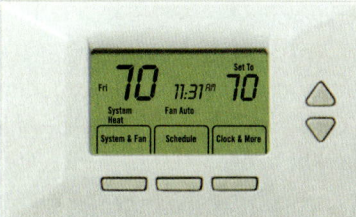

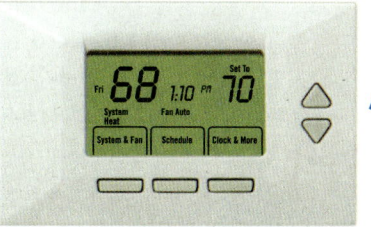

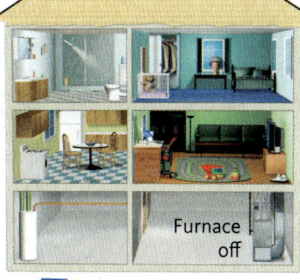

LESSON 2.4 61

Positive Feedback Loops A positive feedback loop causes a system to change further in the same direction. The "snowball effect" is an informal description of a positive feedback loop. As a hypothetical snowball rolls down a hill, it accumulates more snow, which increases its momentum. As the snowball gains momentum, it rolls faster, gaining more snow. The Hubbard Brook forest experiment in the Case Study is an example of a positive feedback loop. As vegetation was lost, nutrients washed away, causing more vegetation to be lost, and so on.

When a natural system becomes locked into a positive feedback loop, it can reach an **ecological tipping point**. Beyond this point, the system can change so drastically that it suffers from severe degradation or collapse.

Negative Feedback Loops A negative, or corrective, feedback loop causes a system to change in the opposite direction from which it is moving. A thermostat is a simple example of a negative feedback loop (Figure 2-15). When a furnace is running, the temperature in a house increases. The temperature information feeds back into the system. When the temperature reaches a preset temperature, the furnace turns off. Because the input causes the system to stop or decrease, the thermostat is an example of a negative feedback loop.

checkpoint What is a feedback loop?

2.4 Assessment

1. **Apply** Give three examples of systems in nature that are not discussed in this chapter.
2. **Explain** What can cause a system to become unsustainable?
3. **Analyze** Your classroom can be thought of as a system. Draw a model of the system. Include inputs, throughputs, and outputs.

CROSSCUTTING CONCEPTS

4. **Stability and Change** Describe and diagram a positive feedback loop that is not discussed in this chapter.

TYING IT ALL TOGETHER STEM
The Hubbard Brook Forest Experiment

In the controlled experiment discussed in this chapter's Case Study, the clearing of a mature forest degraded some of its natural capital. Specifically, the loss of trees and other vegetation altered the ability of the forest to retain and recycle water and other critical plant nutrients. Key nutrients that would normally have been recycled were lost through erosion.

FIGURE 2-16 ▼

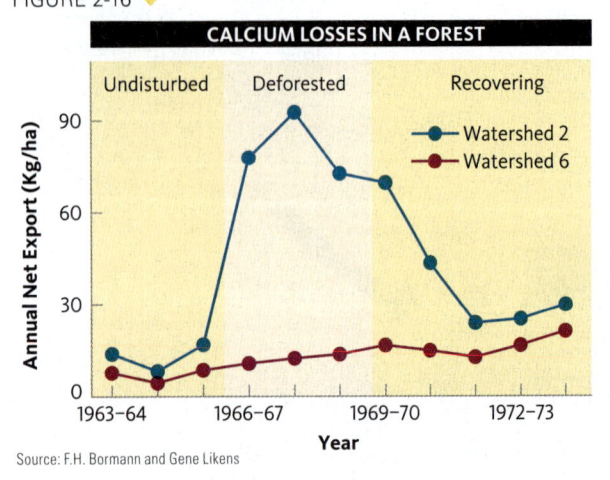

CALCIUM LOSSES IN A FOREST

(Graph showing Annual Net Export (Kg/ha) vs Year from 1963-64 to 1972-73, with phases labeled Undisturbed, Deforested, and Recovering. Two lines: Watershed 2 and Watershed 6.)

Source: F.H. Bormann and Gene Likens

Use the graph to help you answer the questions that follow.

1. In what year did the loss of calcium from the experimental site begin a sharp increase? What type of feedback loop most likely caused the sharp increase?
2. In what year did calcium losses peak? Why might calcium losses have reached a peak? What might have caused calcium losses to go back down again?
3. In what year were the calcium losses from the two sites closest together? During the span of the study, did they ever get that close again?
4. Experiments are difficult to conduct in the field because it is hard to control every variable. What are the benefits of conducting a controlled experiment?
5. Describe another experiment, either in a lab or the field, whose results could strengthen the findings of the Hubbard Brook experiment.

CHAPTER 2 SCIENCE, MATTER, ENERGY, AND SYSTEMS

CHAPTER 2 SUMMARY

2.1 What do scientists do?

- Scientists work to gain a better understanding of the natural world. Scientists engage in scientific practices such as asking questions, developing and using models, and planning and carrying out investigations. Scientific investigations often involve the use of controlled experiments to test hypotheses and gather data.
- Scientific theories have been tested widely, are supported by extensive evidence, and are accepted by most scientists in a field as being useful explanations of how some aspect of the natural world works. Scientific laws are descriptions of things that happen repeatedly and in the same way in nature.

2.2 What is matter?

- Matter is anything that has mass and takes up space. It can be an element or a compound. An element is a type of matter with a unique set of properties that cannot be broken down into simpler substances by chemical means. A compound is a combination of two or more different elements bonded together in fixed proportions.
- Isotopes are elements with the same number of protons but different numbers of neutrons and therefore different mass numbers. Ions are atoms or groups of atoms with a charge. Acidity is based on the relative concentrations of hydrogen ions and hydroxide ions in a solution.
- An organic compound is a carbon-based compound. Hydrocarbons are organic compounds of carbon and hydrogen atoms. Organic compounds also include simple carbohydrates such as glucose. Other important organic compounds include polymers. Four types of polymers are essential to life: complex carbohydrates, proteins, nucleic acids, and lipids.
- Matter can undergo physical or chemical changes. Physical changes do not change the chemical makeup of a substance. Chemical changes (chemical reactions) result in one or more new substances with different chemical properties than the original substance or substances. The law of conservation of matter states that whenever matter undergoes a physical or chemical change, no atoms are created or destroyed.

2.3 What is energy?

- Energy is the capacity to do work. Energy can be kinetic or potential. Kinetic energy is the energy associated with motion. Moving objects, electricity, thermal energy, and electromagnetic radiation are forms of kinetic energy. Potential energy is the stored energy associated with an object or atom's position. Chemical energy and the energy stored in an object's height above Earth are examples of potential energy.
- The first law of thermodynamics states that when energy is converted from one form to another in a physical or chemical change, no energy is created or destroyed. The second law of thermodynamics states that when energy is converted from one form to another in a physical or chemical change, the energy is converted to lower-quality or less usable energy.

2.4 What are systems?

- A system is a set of components that functions in some regular way. System components include inputs, throughputs, and outputs. Feedback loops can occur in systems when an output feeds back into the system as an input.

MindTap — If you have been provided with access to a MindTap course, additional resources are available at login.cengage.com.

CHAPTER 2 ASSESSMENT

Review Key Terms

Select the key term that best fits each definition. Some terms will not be used.

atom
chemical change
chemical reaction
compound
data
ecological tipping point
electromagnetic radiation
element
energy
feedback loop
first law of thermodynamics
ion
isotope
kinetic energy
law of conservation of matter
matter
model
molecule
organic compound
peer review
pH
physical change
potential energy
science
scientific hypothesis
scientific law
scientific method
scientific theory
second law of thermodynamics
system
thermal energy

1. A field of study focused on discovering how nature works
2. Involves practices including asking questions and constructing explanations
3. One of two or more forms of an element with the same atomic number but different mass numbers
4. Stored energy
5. Anything that has mass and takes up space
6. When a sample of matter undergoes this type of change, there is no change in its chemical composition
7. A wave of changing electrical and magnetic fields
8. A testable answer to a scientific question or explanation
9. The energy of the moving particles that make up matter
10. The capacity to do work
11. The smallest unit of matter into which an element can be divided and keep its distinctive chemical properties
12. A well-tested and widely accepted explanation of some aspect of the natural world
13. A representation or simulation of a system
14. States that energy cannot be created or destroyed in a physical or chemical change
15. A measure of acidity
16. Contains at least two carbon atoms combined with atoms of one or more other elements
17. Occurs when an output feeds into a system as input and causes changes in that system

Review Key Concepts

18. Describe three practices that scientists use to learn about nature.
19. What is matter?
20. What are elements and compounds?
21. What is pH?
22. Describe an example of a chemical change.
23. Why are organic compounds important to living organisms?
24. Define the law of conservation of matter.
25. What is the difference between kinetic energy and potential energy?
26. Explain why thermal energy is classified as kinetic energy.
27. What is the law of conservation of energy, also known as the first law of thermodynamics?
28. What is the second law of thermodynamics?
29. What are the main components of a system?
30. Describe an example of a negative feedback loop.

Think Critically

31. What ecological lesson can be gained from the experiment on the clearing of forests described in the Case Study that opened the chapter?
32. Suppose you observe that all the fish in a pond have disappeared. How might you use the scientific practices to determine the cause of this fish kill?
33. Reread the caption for the hydroelectric power plant (Figure 2-13). Infer the difference between an energy *transformation* and an energy *transfer*.

34. Use the second law of thermodynamics to explain why we can use oil only once as a fuel. Or, in other words, why can't we recycle its potential energy?

35. Because of the law of conservation of matter, there is no way to truly "get rid of" wastes. Why is the world not filled with waste matter?

Chapter Activities

A. Experiment: Runoff STEM

As described in the Case Study, ground cover affects the amount, quality, and composition of runoff. In this lab, you will create a model to help you conduct a controlled experiment similar to the larger study led by Bormann and Likens. During the experiment, you will use your model to answer the question, "How does ground cover affect the amount, quality, or composition of water runoff?"

Materials

3 plastic bottles	3 plastic cups
scissors	watering can
soil	string
twigs, bark, leaves, and/or plant debris	plants, seedlings, or pieces of sod

1. With the bottles lying on their side, cut each one lengthwise, leaving the bottom and opening of the bottle intact.

2. Place equal amounts of soil in each of the 3 bottles. Press down firmly and ensure that the soil is below the opening of the bottle.

3. Punch two holes in each of the cups and tie the string to create a handle. This will be used to hang over the bottle opening and collect the "runoff."

4. Look at the materials provided. As a group, discuss which two types of ground cover you would like to use in your experiment. Remember that you need a control, so the third bottle will only contain the soil.

5. Write your hypothesis. Identify your independent and dependent variables.

6. Place the ground covers chosen in the two remaining bottles and label them accordingly.

7. Hang the cups from the necks of the three bottles with the various ground covers in them.

8. Using the watering can, filled to a predetermined line, pour the water onto the soil at the end opposite to the opening and make observations about the water that flows into the cups.

Questions

1. What differences did you notice in the water collected in the cups? Why do you think these differences occurred?

2. Compare and contrast this model with the experiment done by Bormann and Likens. What conclusions could you draw from this model about the role of ground coverings in nature?

B. Citizen Science

National Geographic Learning Framework
Attitudes | Responsibility
Skills | Observation
Knowledge | Our Living Planet

Chemical conditions are critical factors affecting water quality. Acidity and the amount of dissolved ions, including nitrates and phosphates, must stay within certain ranges to support living things. The U.S. EPA and other organizations offer test kits that citizens can use to help monitor local bodies of water. For example, pH strips can be used to test for acidity and nitrate strips test nitrate ion concentration.

Join a citizen science group to monitor the quality of a local lake, estuary, stream, or wetland. Gather in small groups with your classmates to share your experiences.

CHAPTER 2 ASSESSMENT   65

CHAPTER 3
ECOSYSTEM DYNAMICS

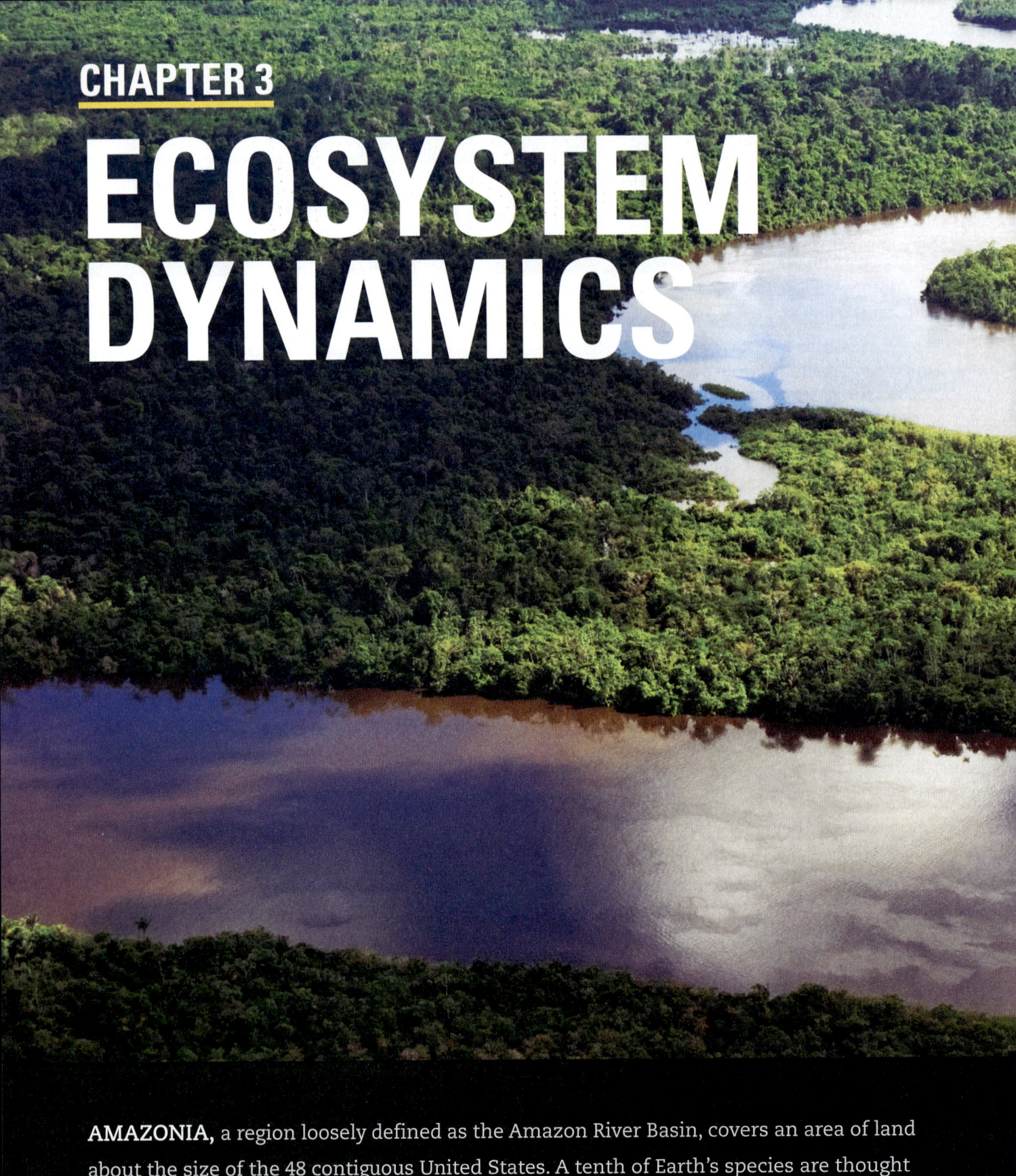

AMAZONIA, a region loosely defined as the Amazon River Basin, covers an area of land about the size of the 48 contiguous United States. A tenth of Earth's species are thought to live in Amazonia, which includes half of the planet's tropical rain forests. People have lived in this region for at least 13,000 years. In the past 50 years, however, human activity has destroyed close to 20 percent of Amazonia's rain forest.

KEY QUESTIONS

3.1 What are Earth's major spheres, and how do they support life?

3.2 What are the major ecosystem components?

3.3 What happens to energy in an ecosystem?

3.4 What happens to matter in an ecosystem?

3.5 How do scientists study ecosystems?

The Nanay River in Peru is one of the Amazon River's many tributaries.

NATIONAL GEOGRAPHIC | EXPLORERS AT WORK

Eco-Paradise at Serra Bonita

with National Geographic Explorer Vitor Becker

Vitor Becker takes a few moments each day to do something that hardly anyone else in the world can do: He feeds a buzzing flock of tiny hummingbirds from the palm of his hand. As you might imagine, this isn't something that happens in a suburban backyard, but at Serra Bonita, a nearly 2,225 hectare (5,000 acre) nature reserve named for the Serra Bonita Mountain in the Atlantic Forest of Brazil.

About the Atlantic Forest: There's good news and bad news. It's one of the most diverse biomes in the world. It is also one of the most destroyed in Brazil—only 8% of its original forests remain. Still, the region continues to have very high species diversity. In fact, protected areas like Serra Bonita are a refuge for thousands of species not found anywhere else in the world. This lush rain forest and the species it nurtures lead Becker to call Serra Bonita an "eco-paradise."

Dr. Becker studied forestry and trained as an entomologist, which is a scientist who studies insects. He, his wife, and their daughter—with help from National Geographic and many others—all work to maintain the reserve. Think about these numbers: More than 350 bird species, roughly 1,200 vascular plant species, and more than 70 frog species have been identified at Serra Bonita. Protecting them is no small task. The Brazilian rain forest is often a difficult place for conservationists to work because of illegal logging operations and other threats.

Despite that, the family plans to expand the reserve—and is committed to education too. The research center at Serra Bonita houses laboratories, collection rooms, and a library. The center supports many research projects that yield new information nearly every day.

Hummingbirds aren't Becker's only concern. There are also the thousands of moth species he has identified at Serra Bonita. Check out online videos on Serra Bonita, and you'll find howler monkeys sitting on his shoulders. And, though you won't see them in the videos, Becker knows puma prowl through the reserve, now free of threat from hunters.

The Serra Bonita website offers ideas for how to get involved in this great project, but there are many ways to take up the causes of ecosystem conservation and species preservation. You could start by asking questions about your own region. What plants and animals live there, and are they thriving—or just barely surviving? Consider starting a citizen science project to identify species in the area. Look for ways you can help and then set out to do it.

Thinking Critically
Draw Conclusions Even though only 8% of the original forests remain, the Atlantic Forest is still considered one of the most diverse regions on Earth. Can you conclude from these facts that the loss of forests has had little effect so far on the number of species found there? Why or why not?

Vitor Becker researches hummingbirds and insects at Serra Bonita, a nature reserve in the Atlantic Forest. Today only 8% of the Atlantic Forest remains—another reason why protecting places like Serra Bonita is so important.

CASE STUDY
Disappearing Tropical Rain Forests

Tropical rain forests support an incredible variety of life. They cover only about 7% of Earth's land surface but contain half of the plant and animal species found on land. These lush forests are warm and humid year-round because of their daily rainfall and nearness to the Equator. The biodiversity of tropical rain forests makes them an excellent natural laboratory for the study of ecosystems. An ecosystem is one or more communities of organisms that interact with one another and their nonliving environment.

To date, human activities have destroyed or disturbed more than half of Earth's tropical rain forests. People continue clearing the forests to grow more crops, graze more cattle, and build more settlements (Figure 3-1). Ecologists warn that without protection, most of the forests will be gone or severely damaged by the end of this century. The preservation efforts of individuals like National Geographic Explorer Vitor Becker help combat the degradation of these ecosystems.

Removing tropical rain forests reduces Earth's vital biodiversity, or the planet's variety of species and the habitats where they live. (See Chapter 4 for more about biodiversity.) Destroying the habitats of plant and animal species often results in their extinction. When the forest loses a key species, it can have a ripple effect that leads to the loss of other species.

Destroying tropical rain forests also accelerates atmospheric warming. Without tropical rain forests, the atmosphere warms, leading to climate change. Why? Eliminating large areas of trees means there are fewer plants to remove carbon dioxide (CO_2) during photosynthesis. Carbon dioxide is a gas that contributes to atmospheric warming. (See Chapter 16 for more about climate change.)

Large-scale loss of tropical rain forests can also change regional weather patterns in ways that prevent the forest from returning. When this irreversible tipping point is reached, rain forests become dryer, less diverse tropical grasslands. The presence of grasslands decreases rainfall in nearby forests, which further weakens them.

In this chapter, you will examine the living and nonliving components of ecosystems and how they function. You will learn how ecosystems support life, how ecologists study the interactions within and among different ecosystems, and the importance of maintaining ecosystem integrity. Healthy tropical rain forests like the one preserved at Serra Bonita are examples of sustainably functioning ecosystems.

As You Read Think about an ecosystem where you live. Consider what makes the ecosystem unique, as well as what it might have in common with a rain forest ecosystem. Which part of the ecosystem is most damaged or threatened? How might this affect the other parts of the ecosystem and its long-term sustainability?

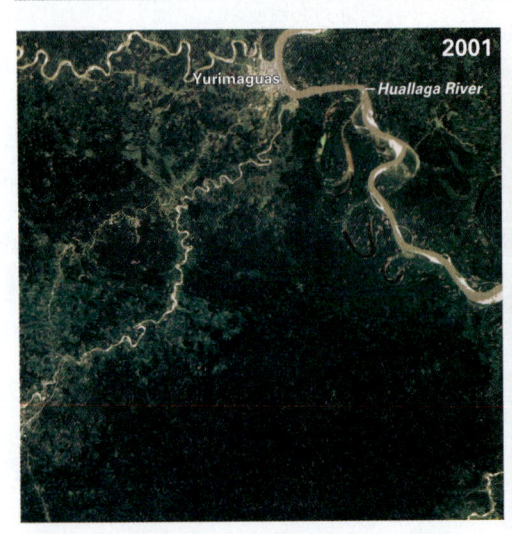

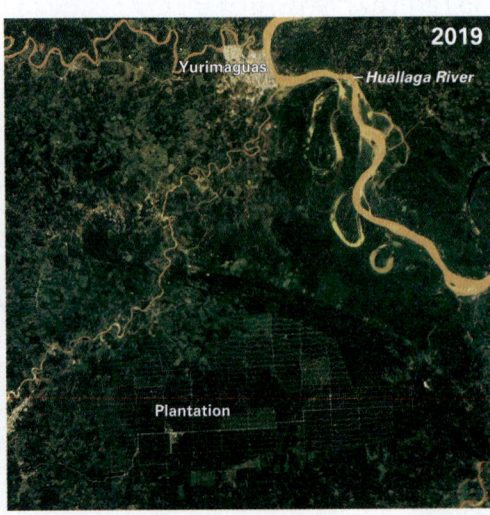

◀ FIGURE 3-1 Satellite images taken of the same area in 2001 (left) and 2019 (right) show the loss of tropical rain forest near the Peruvian city of Yurimaguas. A large swath of forest has been cleared for an oil palm plantation.

3.1 What Are Earth's Major Spheres, and How Do They Support Life?

CORE IDEAS AND SKILLS

- Describe the four major spheres that support life on Earth.
- Understand how nutrients cycle and energy flows through ecosystems.

KEY TERMS

geosphere	stratosphere	greenhouse
atmosphere	hydrosphere	effect
troposphere	biosphere	

Earth's Spheres Function As a Life-Support System

Earth's "life-support system" is based on the interaction among four planetary systems, or spheres. These spheres are called the atmosphere, hydrosphere, geosphere, and biosphere (Figure 3-2). The natural capital (resources and ecosystem services) on which life depends is the product of Earth's spheres and energy from the sun.

The Geosphere The **geosphere** consists of Earth's core, mantle, and thin outer crust—all the material above and below the surface that forms the planet's mass. Without its large mass, Earth would not have the gravitational force needed to keep the atmosphere from escaping into space. The geosphere's upper crust contains nutrients organisms need to live, grow, and reproduce (Science Focus 3.1). The crust also includes nonrenewable fossil fuels—coal, oil, and natural gas—and mineral resources.

The Atmosphere Held to Earth by gravity, the **atmosphere** is an envelope of gases surrounding the planet (Figure 3-4). If Earth were the size of a basketball, the atmosphere would be about the thickness of a sheet of paper. This thin blanket of gases shields the planet from meteors and blocks most of the sun's harmful ultraviolet (UV) radiation. It also helps regulate Earth's climates, allowing surface temperatures to be suitable for life to exist in the **troposphere**, the lowest layer of the atmosphere.

The troposphere is the layer in which weather occurs. It is also the only layer in which terrestrial organisms can survive. Thickest at the Equator, the troposphere extends up to 19 kilometers (12 miles) above sea level. At the Poles, the troposphere extends up to 6 kilometers (4 miles). Life in the atmosphere has evolved to tolerate the temperature ranges and composition of gases found only within the troposphere.

The **stratosphere** is the atmospheric layer above the troposphere. Although nothing lives in the stratosphere, this layer has a direct impact on life at the surface. The lower stratosphere contains a relatively high concentration of ozone (O_3), which is called the ozone layer. The ozone layer absorbs more than 95% of the sun's harmful UV radiation. As a result, it acts as a global sunscreen that allows life to exist on Earth's surface.

Three more atmospheric layers extend for hundreds of kilometers beyond the stratosphere: the mesosphere, the thermosphere, and the exosphere. Together, these five layers of the atmosphere protect Earth from the extremes of space.

FIGURE 3-2

Earth's Spheres Earth consists of a land sphere (geosphere), an air sphere (atmosphere), a water sphere (hydrosphere), and a life sphere (biosphere).

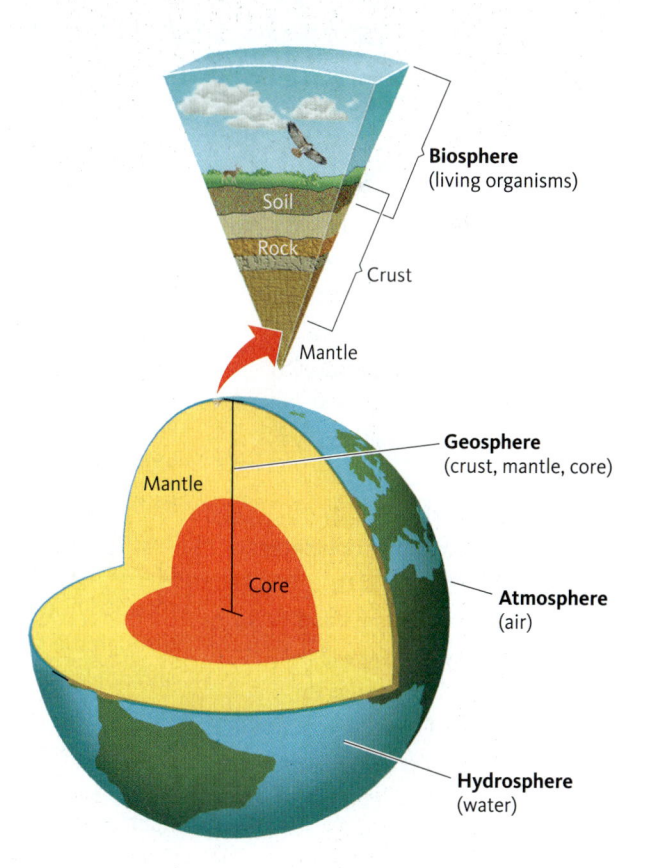

LESSON 3.1   71

FIGURE 3-3
Hundreds of macaws make up for the lack of sodium in their diet by eating small amounts of salty clay at this avian salt lick in Manú National Park, Peru. Sodium is an essential mineral nutrient.

SCIENCE FOCUS 3.1

NUTRIENT CYCLING

Life on Earth depends on two processes: the one-way flow of energy from the sun and the cycling of matter through the biosphere. This is in keeping with the solar energy and nutrient cycling factors of sustainability described in Chapter 1.

The life-sustaining energy of nearly every ecosystem originates with sunlight, converted to chemical energy by plants and other producers. As you will learn in Lesson 3.3, without the continual input of energy from the sun, nearly every ecosystem would quickly run out of the energy needed for life.

As energy flows through ecosystems, it fuels the building up and breaking down of chemical compounds (Lesson 3.4). The resulting atoms, ions, and molecules form the planet's living organisms and the nutrients they need to survive.

Earth does not get significant inputs of matter from space, so its supply of nutrients is fixed. Nutrients must be recycled to support each successive generation of organisms.

Carbon, oxygen, nitrogen, and phosphorus are among these recycled nutrients. They pass through the living (biosphere) and nonliving (atmosphere, hydrosphere, geosphere) parts of ecosystems. Water helps cycle these important nutrients and is itself an essential nutrient.

What makes something a nutrient? A *nutrient* is any matter that an organism needs to survive and function. Feeding is the means by which organisms obtain most of their nutrients.

Macronutrients are nutrients that organisms need in large amounts. They form the bulk of the foods you eat. Proteins, fats, and carbohydrates are examples of macronutrients. *Micronutrients*—vitamins and minerals—are nutrients that organisms need in very small amounts.

Vitamins are considered "organic" compounds because they contain carbon. Minerals, which do not contain carbon, are "inorganic" compounds. Some important mineral nutrients include calcium, zinc, potassium, and iron.

Thinking Critically
Infer How would nutrient cycling be affected if all of Earth's producers died off?

FIGURE 3-4
The Atmosphere On average, the troposphere layer is 78% nitrogen and 21% oxygen. The remaining 1% is mostly argon, water vapor, carbon dioxide, and other gases. The amount of water vapor may increase to more than 4% depending on altitude and air temperature.

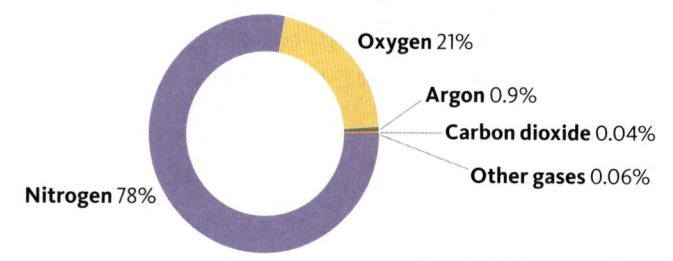

Oxygen 21%
Argon 0.9%
Carbon dioxide 0.04%
Other gases 0.06%
Nitrogen 78%

Source: National Aeronautics and Space Administration

FIGURE 3-5
The Hydrosphere At any given time, approximately 96.5% of Earth's water molecules exist in the ocean. What percentage of the hydrosphere consists of glaciers and ice caps? (Hint: The answer is not 68.7%.)

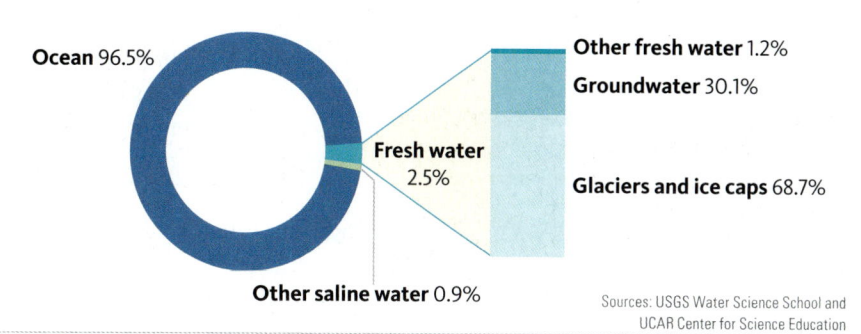

Ocean 96.5%
Other fresh water 1.2%
Groundwater 30.1%
Fresh water 2.5%
Glaciers and ice caps 68.7%
Other saline water 0.9%

Sources: USGS Water Science School and UCAR Center for Science Education

The Hydrosphere Glaciers, lakes, rivers, aquifers, water vapor, clouds, and the ocean are all part of the hydrosphere (Figure 3-5). The **hydrosphere** includes all of the gaseous, liquid, and solid water on or near Earth's surface. The distribution of water is dominated by the ocean, which covers about 71% of Earth's surface. The ocean contains about 96.5% of Earth's total supply of water. Less than 3% of Earth's water is available as fresh water, and most of that is frozen in polar ice caps and glaciers.

The Biosphere If Earth were an apple, the biosphere would be no thicker than the apple's skin. The **biosphere** consists of the parts of the atmosphere, hydrosphere, and geosphere where life exists. It is the living part of every ecosystem. One important goal of environmental science is to understand the key interactions that occur within this thin layer of air, water, soil, and organisms. Another is to understand how the biosphere is impacted by human activities.

checkpoint In which layer of the atmosphere do you live?

Earth's Spheres Interact

Through a process known as the **greenhouse effect**, solar energy warms the troposphere as it reflects from Earth's surface (geosphere) and interacts with carbon dioxide (CO_2), methane (CH_4), water vapor (from the hydrosphere and biosphere), and other greenhouse gases (atmosphere). These interactions are part of Earth's life-support system. Without the greenhouse effect, Earth would be too cold to support life as we know it.

Interactions among the spheres clean Earth's water and air. As plants absorb water and water transpires, or evaporates from their leaves, pollutants are absorbed. Animals such as clams and mussels filter impurities from bodies of water, and microorganisms in water and soil can break down many contaminants. As water evaporates from Earth's surface into the atmosphere, particles that make the water impure are left behind.

Forests play an important role in purifying air. Trees can absorb air-polluting gases near the surface. A single tree can produce enough oxygen for two people to breathe for a year. The same tree might absorb about 4.5 kilograms (10 pounds) of pollutants in a year.

checkpoint What are some of the gases that help produce the greenhouse effect?

3.1 Assessment

1. **Recall** What are Earth's four major spheres that support life?
2. **Explain** What is the greenhouse effect and why is it important to life on Earth?
3. **Generalize** What is Earth's "life-support system"?

CROSSCUTTING CONCEPTS

4. **Systems and System Models** Use one or more examples from everyday life to explain how Earth's four major spheres interact.

3.2 What Are the Major Ecosystem Components?

CORE IDEAS AND SKILLS

- Describe trophic levels and how they can be represented in a conceptual model.
- Explain the roles of producers, consumers, and decomposers in an ecosystem.
- Identify the different ways in which energy and matter are transformed in an ecosystem.
- Summarize the processes of photosynthesis and cellular respiration.

KEY TERMS

trophic level, producer, photosynthesis, consumer, primary consumer, herbivore, secondary consumer, tertiary consumer, carnivore, omnivore, decomposer, detritivore, aerobic respiration, anaerobic respiration

Ecosystems Have Several Important Components

Ecology is the branch of biology that focuses on how organisms interact with one another and their physical environment (Lesson 1.1). Scientists classify matter into levels of organization ranging from atoms to galaxies. Ecologists study interactions within and among several of these levels—from molecules to the biosphere. See Figure 3-6 for a definition of each level of organization.

The biosphere and its ecosystems are made up of living (biotic) and nonliving (abiotic) components. Examples of nonliving components are water, air, rocks, nutrients, thermal energy (heat), and sunlight. Living components include plants, animals, microbes, and all other organisms. Figure 3-7 is a simplified model of some of the living and nonliving components of a terrestrial ecosystem.

Ecologists assign each organism in an ecosystem to a feeding level called a **trophic level**. An organism's trophic level depends on (a) whether it makes food or finds food, and (b) if it finds food, what its feeding behavior is. Organisms are classified as producers or consumers by whether they make (produce) or find (consume) food.

Producers such as plants make the food they need from compounds in soil, carbon dioxide in air, and water—using the energy of sunlight. In the process known as **photosynthesis**, producers change radiant energy (sunlight) into chemical energy stored primarily in glucose. By harnessing the energy of light, producers can convert inorganic molecules of carbon dioxide and water into organic molecules such as glucose. Glucose ($C_6H_{12}O_6$) is an important building block of many energy-rich carbohydrates that are necessary for life. The chemical equation for photosynthesis is:

$$\text{carbon dioxide} + \text{water} + \text{light energy} \rightarrow \text{glucose} + \text{oxygen}$$
$$6\,CO_2 + 6\,H_2O + \text{light energy} \rightarrow C_6H_{12}O_6 + 6\,O_2$$

FIGURE 3-6

Ladder of Matter Ecology includes all of these levels of the organization of matter.

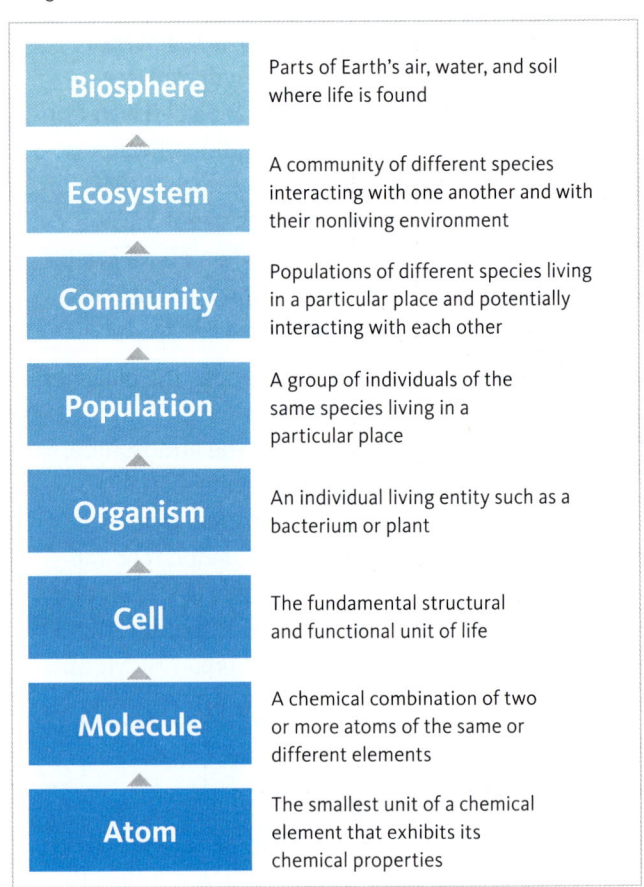

Biosphere	Parts of Earth's air, water, and soil where life is found
Ecosystem	A community of different species interacting with one another and with their nonliving environment
Community	Populations of different species living in a particular place and potentially interacting with each other
Population	A group of individuals of the same species living in a particular place
Organism	An individual living entity such as a bacterium or plant
Cell	The fundamental structural and functional unit of life
Molecule	A chemical combination of two or more atoms of the same or different elements
Atom	The smallest unit of a chemical element that exhibits its chemical properties

FIGURE 3-7

Matter on the Move Arrows in this simplified ecosystem trace the movement of matter through key living (biotic) and nonliving (abiotic) components.

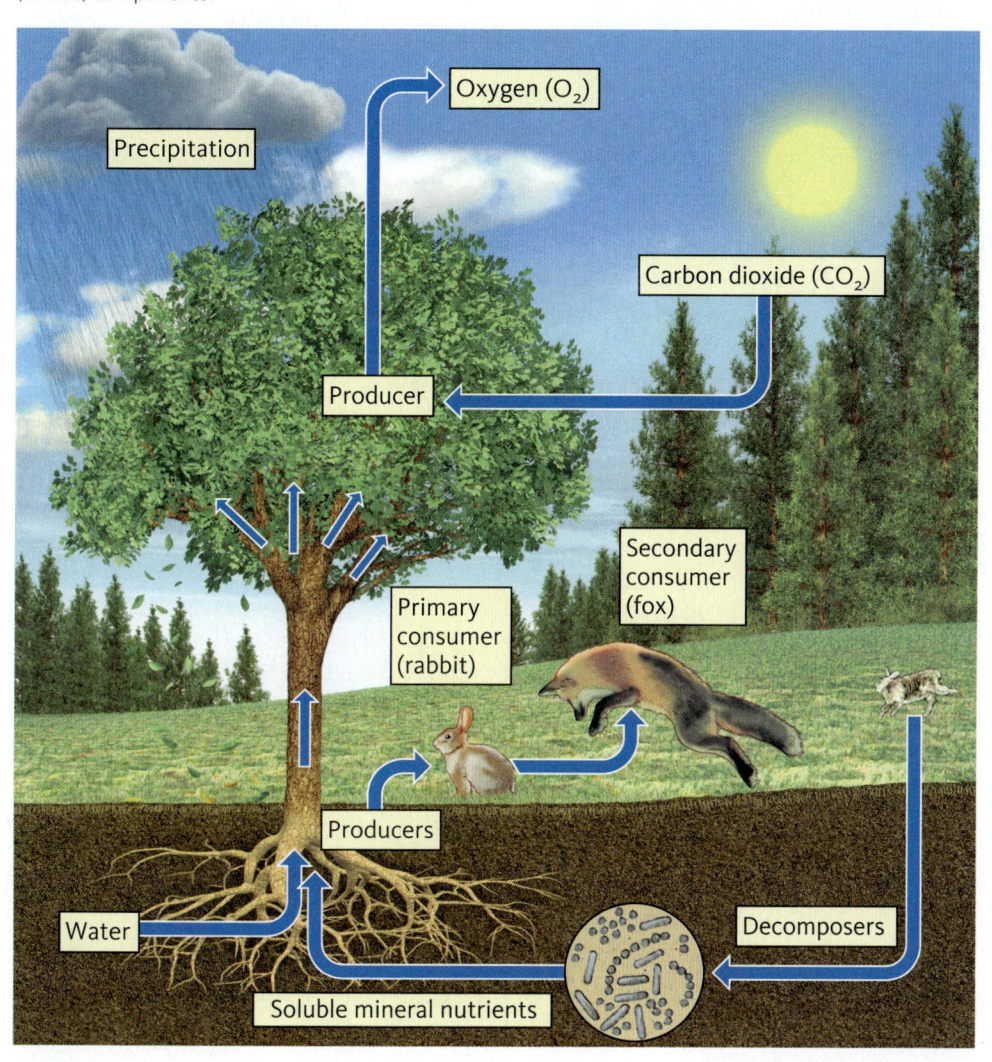

On land, most producers are green plants such as trees and grasses. In freshwater and ocean ecosystems, algae and aquatic plants growing near shorelines are the major producers. In open water, the dominant producers are phytoplankton—mostly microscopic organisms that float or drift in the water.

In contrast to producers, **consumers** are organisms that cannot produce their own food. They obtain food and energy by eating producers or other consumers or by feeding on their wastes or remains.

Primary consumers, or **herbivores**, are organisms that eat mostly green plants or algae. Examples of herbivores are caterpillars, giraffes, and zooplankton, which are tiny sea animals that feed on phytoplankton. **Secondary consumers** are animals that feed on primary consumers. **Tertiary** (or higher-order) **consumers** feed on both primary and secondary consumers.

Among the secondary and tertiary groups are carnivores and omnivores. **Carnivores** feed mostly on other animals. Some carnivores, including spiders, lions, and most small fishes, are secondary consumers. Others, such as tigers, hawks, and killer whales (orcas), are tertiary consumers.

Omnivores, such as pigs, rats, and humans, eat both plants and animals. Like carnivores, omnivores may be secondary or tertiary consumers.

All consumers in an ecosystem rely on the ecosystem's producers for their ability to make energy available to other organisms via photosynthesis. If producers were eliminated from an ecosystem, other organisms would run out of food.

LESSON 3.2

FIGURE 3-8
ON ASSIGNMENT National Geographic photographer Frans Lanting snapped this image of deforestation while flying above the edge of a lush rain forest in Brazil's Iguaçu National Park. The stark agricultural landscape reveals the park's boundary—and the loss of habitat when crops replace trees.

LESSON 3.2

Decomposers are consumers that get their nutrients by breaking down (decomposing) nonliving organic matter such as leaf litter, fallen trees, and dead animals. In the process of obtaining their own food, decomposers release nutrients from the wastes or remains of plants and animals. The process of decomposition returns nutrients to soil and water, making them available to the ecosystem. Most decomposers are bacteria and fungi. **Detritivores**, or detritus feeders, get their nourishment by consuming detritus, or freshly dead organisms, before they are fully decomposed. Detritus feeders include earthworms, some insects, hyenas, and vultures.

In natural ecosystems, decomposers and detritivores eliminate the build up of plant litter, animal wastes, and dead plants and animals. In doing so, they are the key to nutrient cycling. For example, decomposers and detritivores can transform a fallen tree into wood particles and, ultimately, simple inorganic molecules that producers absorb as nutrients (Figure 3-9). In this way, many nutrients that make life possible are continually recycled.

checkpoint What does an organism's trophic level indicate about that organism?

FIGURE 3-9

Detritivores and Decomposers Various detritivores and decomposers (mostly fungi and bacteria) "feed on" or digest parts of a log. They eventually convert its complex organic chemicals into simpler inorganic nutrients that can be used by producers.

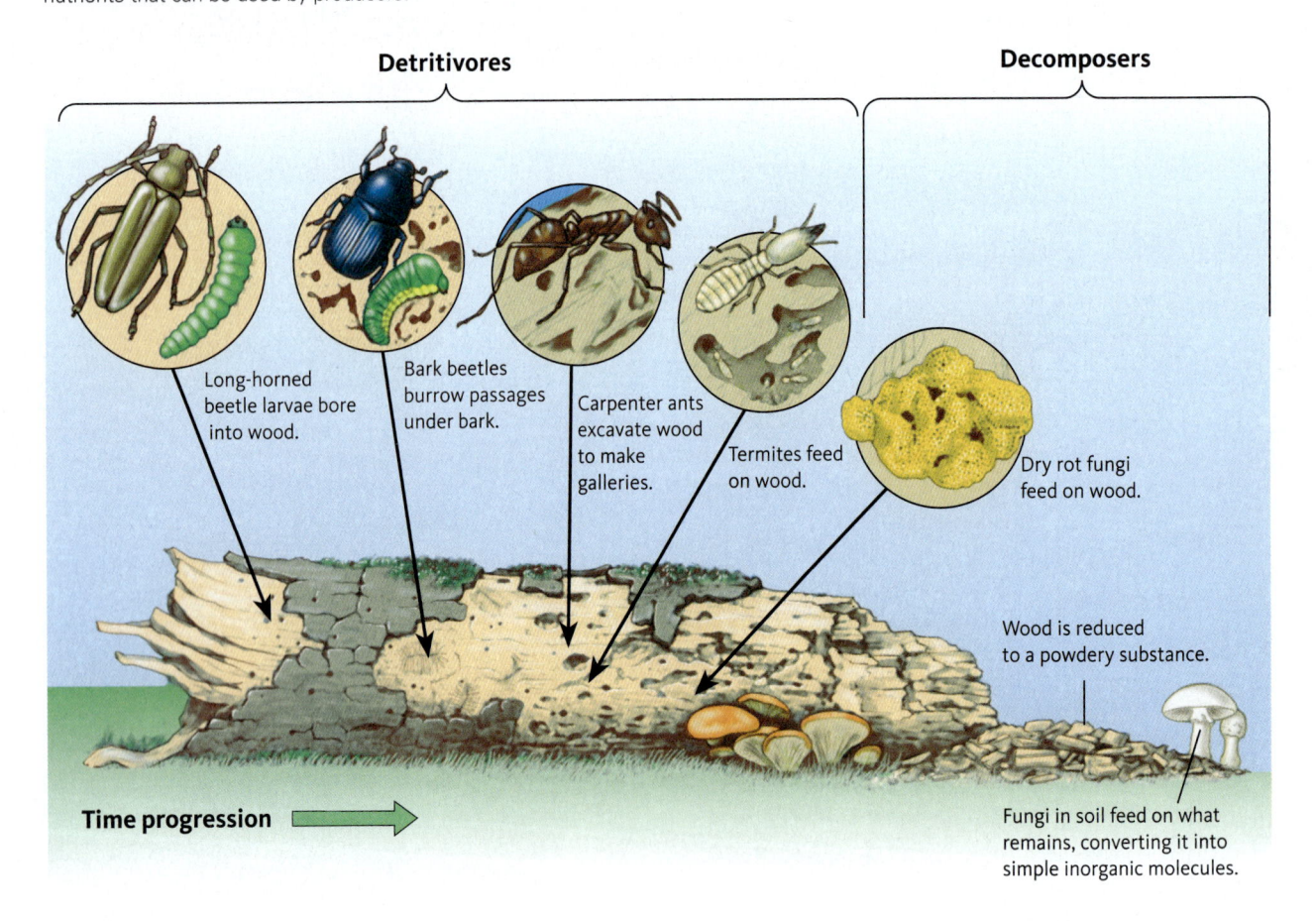

Detritivores

- Long-horned beetle larvae bore into wood.
- Bark beetles burrow passages under bark.
- Carpenter ants excavate wood to make galleries.
- Termites feed on wood.

Decomposers

- Dry rot fungi feed on wood.
- Wood is reduced to a powdery substance.
- Fungi in soil feed on what remains, converting it into simple inorganic molecules.

Time progression →

78 CHAPTER 3 ECOSYSTEM DYNAMICS

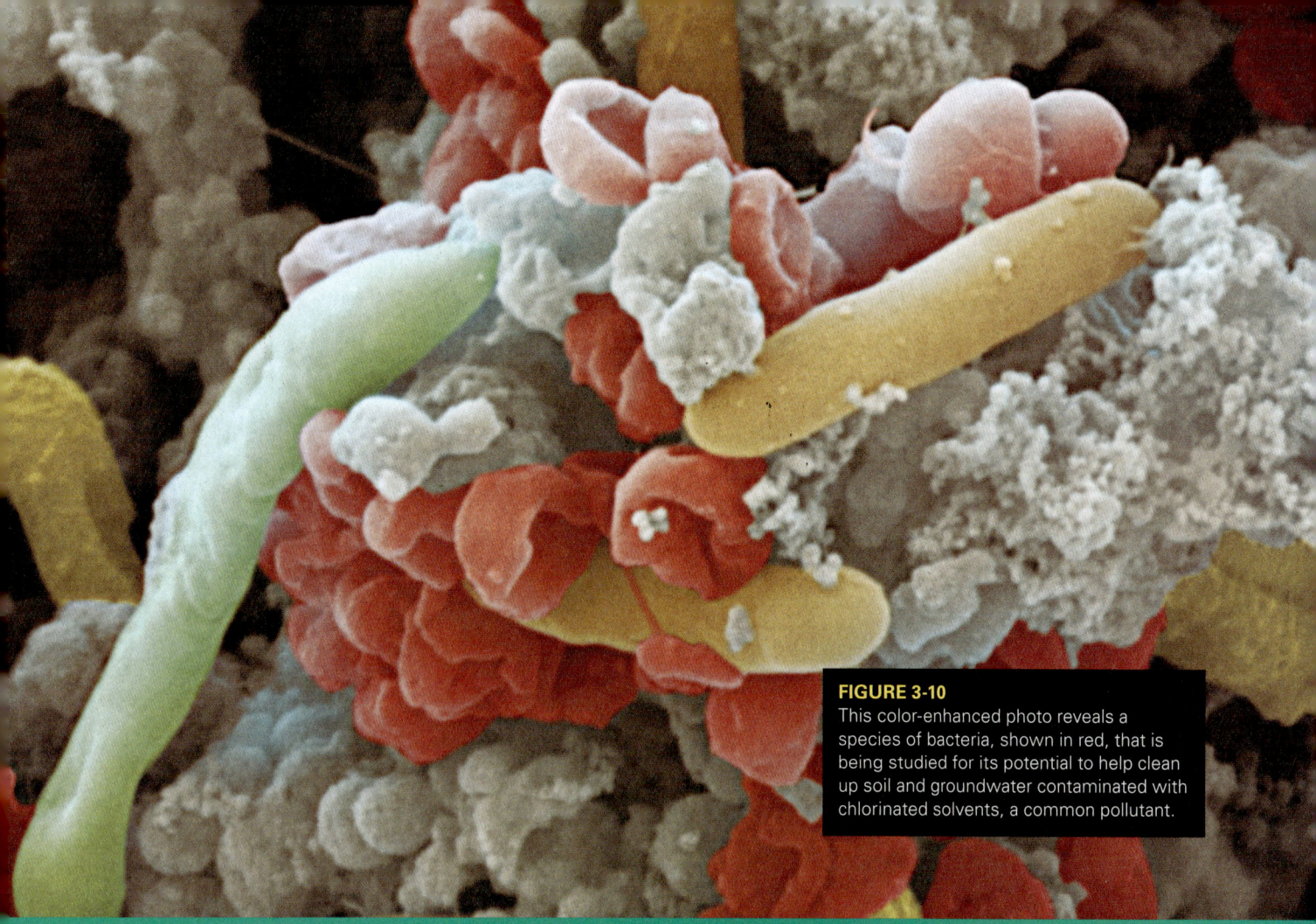

FIGURE 3-10
This color-enhanced photo reveals a species of bacteria, shown in red, that is being studied for its potential to help clean up soil and groundwater contaminated with chlorinated solvents, a common pollutant.

ENGINEERING FOCUS 3.2

NATURE'S CLEANUP CREW

The word *microbe*, or *microorganism*, is a catchall term for thousands of species of bacteria, protozoa, fungi, and floating phytoplankton. Microbes play key roles as decomposers throughout the entire biosphere.

Bacteria and fungi in the soil and oceans decompose organic wastes into inorganic nutrients such as nitrogen and phosphorus. The nutrients are then taken up by plants that are then eaten by consumers. Within your own intestinal ecosystem, trillions of bacteria are busily breaking down the food you eat.

Scientists and engineers have learned how to use microbes to break down pollutants in oil spills and toxic waste leaks. *Bioremediation* is the use of microbes or other decomposers to clean up polluted sites.

But using microbes for environmental cleanup can be challenging. Toxins can be hard to digest, and microbes must have genes that enable them to do this job. However, these same microbes may not be able to survive the environmental conditions at a cleanup site. Likewise, microbes suited to survive in such an environment may not be able to digest toxins.

Environmental engineers are solving this problem by creating custom-made bacteria with the genes needed for a specific job. Bioengineers have learned how to transfer the desired genes from one species of bacteria to another. The target species is usually one that naturally occurs in the sort of environmental conditions that are found at the cleanup site. The resulting combination of traits—an appetite for oil or other pollutants contaminating a given site, and the ability to flourish under the natural conditions found there—makes for much more resilient bacteria.

Thinking Critically
Evaluate What are some possible risks of introducing genetically engineered bacteria into the environment?

LESSON 3.2

FIGURE 3-11

Component Interactions The main components of an ecosystem are energy, matter, and organisms. Nutrient cycling and the flow of energy—first from the sun, then through organisms and into the environment as heat—link these components.

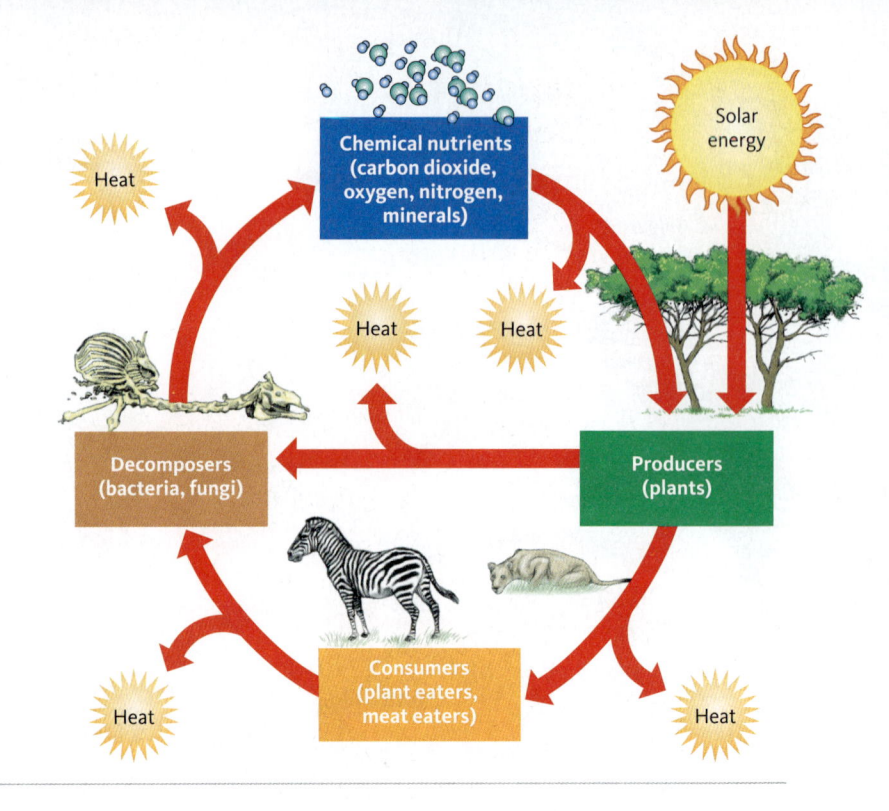

Cellular Respiration

Organisms use the chemical energy stored in glucose and other organic compounds to fuel their life processes. In most cells, this energy is released by **aerobic respiration**, which uses oxygen and glucose to produce energy. Carbon dioxide and water are the by-products of this reaction. The chemical equation for aerobic respiration is:

glucose + oxygen → carbon dioxide + water + energy

$$C_6H_{12}O_6 + 6\,O_2 \rightarrow 6\,CO_2 + 6\,H_2O + \text{energy}$$

Although the detailed steps differ, the net chemical change for aerobic respiration is the opposite of that for photosynthesis. As cells respire, the reaction also produces some thermal energy, which is eventually lost to the environment as heat.

Decomposers such as yeast and some bacteria get the energy they need by breaking down glucose and other organic compounds in the absence of oxygen. This form of cellular respiration is called **anaerobic respiration**, or fermentation. The by-products of anaerobic respiration are not carbon dioxide and water. Rather, they are compounds such as methane gas (CH_4, the main component of natural gas), ethyl alcohol (C_2H_6O), acetic acid ($C_2H_4O_2$, the key component of vinegar), and hydrogen sulfide (H_2S, a highly poisonous gas).

Anaerobic respiration also occurs temporarily in oxygen-starved muscle cells, a by-product of which is lactic acid ($C_3H_6O_3$). All organisms—including producers—get their energy from aerobic or anaerobic respiration. Only producers, however, carry out photosynthesis.

To summarize, ecosystems and the biosphere are sustained by the one-way energy flow from the sun through these systems and the nutrient cycling of key materials within them (Figure 3-11).

checkpoint What is the difference between aerobic and anaerobic respiration?

3.2 Assessment

1. **Contrast** How are photosynthesis and cellular respiration different?
2. **Infer** Could an ecosystem function without decomposers? Why or why not?

SCIENCE AND ENGINEERING PRACTICES

3. **Use Models** How would you revise Figure 3-7 to account for tertiary consumers, photosynthesis, aerobic respiration, and anaerobic respiration?

CROSSCUTTING CONCEPTS

4. **Energy and Matter** Explain why a natural ecosystem is both an open system and a closed system.

3.3 What Happens to Energy in an Ecosystem?

CORE IDEAS AND SKILLS

- Identify the role of food chains and food webs in an ecosystem.
- Describe the flow of energy through an ecosystem.
- Explain the difference between gross primary productivity and net primary productivity.

KEY TERMS

food chain
gross primary productivity (GPP)
food web
net primary productivity (NPP)

Food Chains and Food Webs

Food chains and food webs describe how energy flows through ecosystems. A sequence of organisms that serves as a source of nutrients or energy for the next level of organisms is called a **food chain**. Figure 3-13 illustrates a simplified food chain.

Organisms at each trophic level obtain high-quality chemical energy from their food. However, about 90% of the chemical energy is lost at each link in the food chain. Why is this transfer of energy so inefficient? As organisms live and grow and their cells respire, the chemical energy obtained through food is converted to other forms of energy. According to the laws of thermodynamics (Lesson 2.3), when energy is transformed (as in a food chain), there is an automatic loss of energy "quality," with most of it flowing into the environment as low-quality thermal energy (heat). At the same time, total energy is conserved, or does not change. Thus, there is less high-quality energy left to support large numbers of top predators such as tigers or hawks (Figure 3-12).

Food webs offer another way to describe the flow of energy through ecosystems. A food web is a complex network of interconnected food chains. Food webs are useful in studies at the ecosystem level. For example, scientists studying the effect of decreasing killer whale populations on marine ecosystem health may refer to food webs similar to the one illustrated in Figure 3-15. Food chains and food webs show how producers, consumers, and decomposers are connected to one another as energy flows through trophic levels in an ecosystem.

checkpoint What is the difference between a food chain and a food web?

Primary Productivity

Scientists measure the rates at which ecosystems produce chemical energy to compare ecosystems and understand how they interact. **Gross primary productivity (GPP)** is the rate at which an ecosystem's producers convert radiant energy into chemical energy. This energy is stored in compounds in their bodies. To stay alive, grow, and reproduce, producers must use some of their stored chemical energy for cellular respiration.

Net primary productivity (NPP) is the rate at which producers use photosynthesis to produce and

FIGURE 3-12
Only a small fraction of energy produced at the lowest trophic level is available to top predators such as this black-collared hawk.

LESSON 3.3

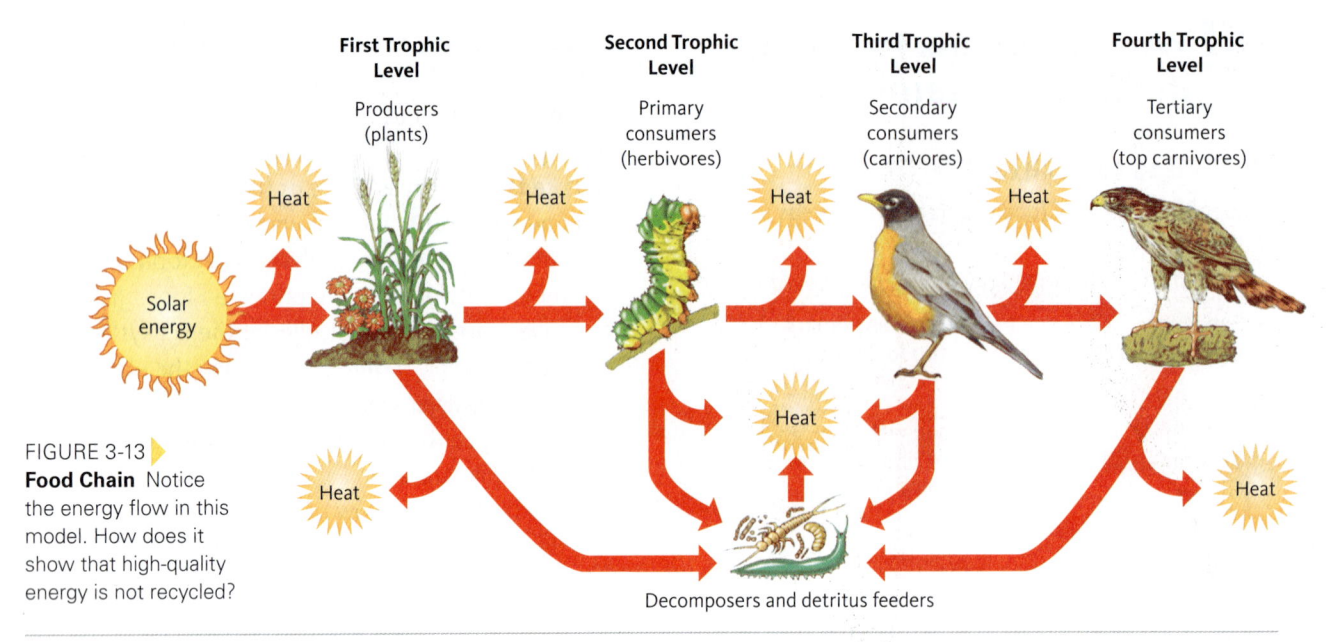

First Trophic Level
Producers (plants)

Second Trophic Level
Primary consumers (herbivores)

Third Trophic Level
Secondary consumers (carnivores)

Fourth Trophic Level
Tertiary consumers (top carnivores)

Decomposers and detritus feeders

FIGURE 3-13 ▶
Food Chain Notice the energy flow in this model. How does it show that high-quality energy is not recycled?

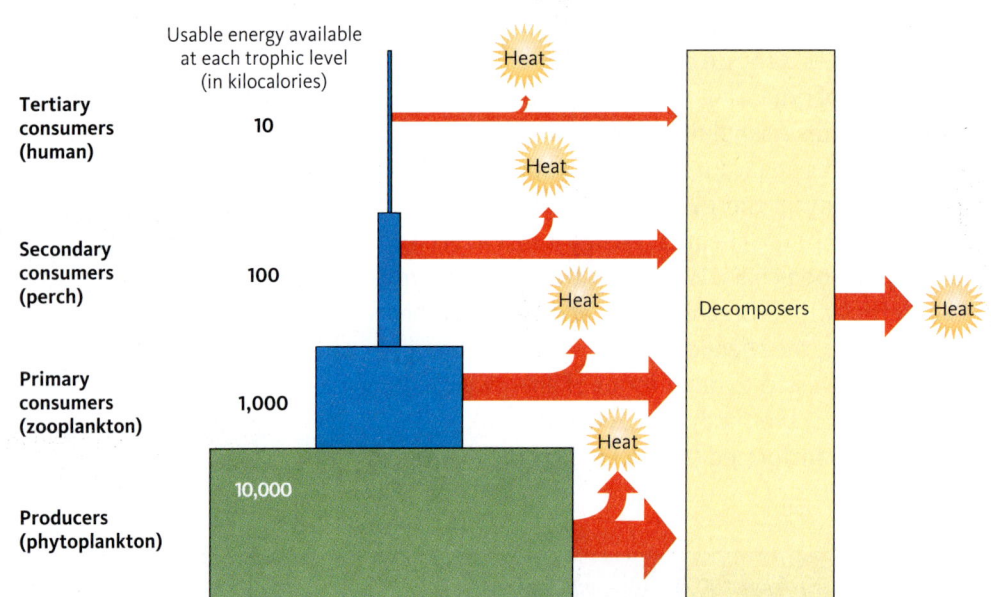

FIGURE 3-14 ▶
Energy Pyramid Energy pyramids model the upward flow of energy through trophic levels. The width of each bar represents usable energy available at each level. Bars become increasingly narrow due to decreases in usable energy available. The number of organisms supported by available energy decreases as well. Energy pyramids are not drawn to scale.

Usable energy available at each trophic level (in kilocalories)

Trophic level	Energy
Tertiary consumers (human)	10
Secondary consumers (perch)	100
Primary consumers (zooplankton)	1,000
Producers (phytoplankton)	10,000

store chemical energy, minus the rate at which they use some of this stored chemical energy through cellular respiration (Figure 3-14). In other words, NPP is the difference between gross primary productivity and cellular respiration. NPP is a measure of the rate at which producers make chemical energy potentially available to the consumers in an ecosystem.

Ecosystems vary in their NPP due to factors such as solar energy input, temperature, carbon dioxide and moisture levels, and nutrient availability. These factors influence the presence and function of photosynthesizers in an ecosystem. Tropical rain forests have a high NPP and collectively are large contributors to Earth's overall NPP. They have a great abundance and variety of plants to support a large biomass of consumers. By contrast, the open ocean has a low NPP. Yet it is more productive annually than any other ecosystem due to its enormous volume and huge numbers of phytoplankton and other producers.

Only the plant matter represented by NPP is available as nutrients for consumers. Thus, *the planet's NPP ultimately limits the number of consumers (including humans) that can survive*. When human activities damage the most highly productive ecosystems, Earth's total productivity is reduced. So is the total number of consumers it can support.

checkpoint What happens to energy as it flows through food chains and food webs?

CHAPTER 3 ECOSYSTEM DYNAMICS

FIGURE 3-15

Food Web A simplified food web shows some of the feeding relationships among marine organisms in the Southern Hemisphere. The shaded middle area is a food chain within the more complex food web. Many more species, including an array of decomposer and detritus feeder organisms, are not shown in this model.

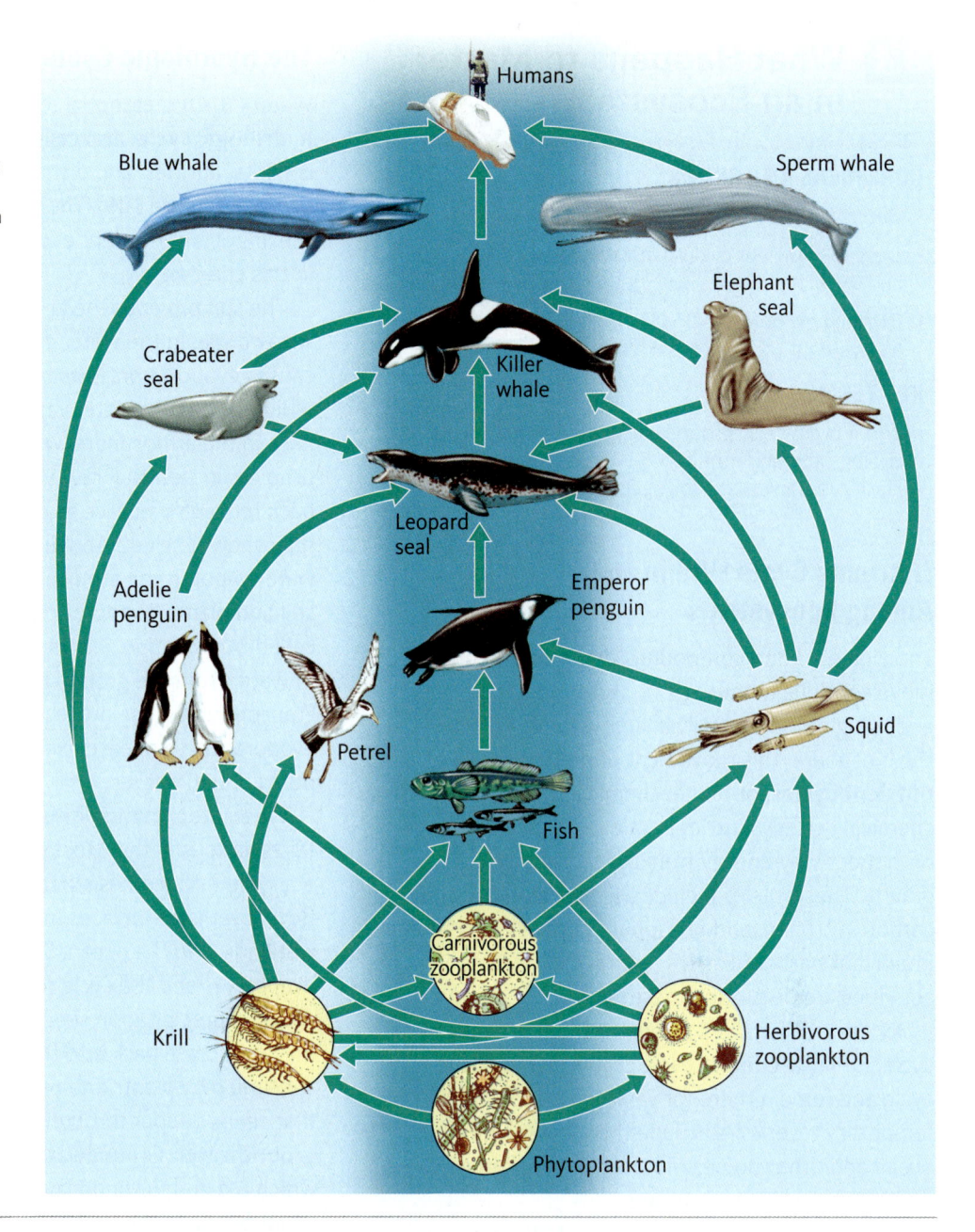

3.3 Assessment

1. **Identify Main Ideas** What is the ecological role of food chains and food webs?
2. **Summarize** What is the difference between gross primary productivity (GPP) and net primary productivity (NPP)?
3. **Synthesize** The plants in a red-tailed hawk's food chain produce 3 million kcal of chemical energy per day. How much energy is available to the hawk's trophic level, assuming 90% loss at each level? (Hint: Refer to Figure 3-13.)

SCIENCE AND ENGINEERING PRACTICES

4. **Developing and Using Models** Create a simplified model of a food chain in your region. Include the names of the organisms and their relationship to each other. Indicate the flow of energy, starting with the sun and including producers, consumers, and decomposers.

SCIENCE AND ENGINEERING PRACTICES

5. **Constructing Explanations** Explain how diagrams can be useful for studying smaller-scale mechanisms within the larger ecosystem. Discuss the limitations of such models.

LESSON 3.3 83

3.4 What Happens to Matter in an Ecosystem?

CORE IDEAS AND SKILLS

- Describe the hydrologic cycle.
- Describe nutrient cycles within and among ecosystems and the biosphere.
- Explain how human activities impact nutrient cycles in ecosystems.

KEY TERMS

nutrient cycle groundwater nitrogen cycle
hydrologic cycle aquifer phosphorus cycle
surface runoff carbon cycle

Nutrients Cycle Within and Among Ecosystems

The elements and compounds that make up nutrients move continually through air, water, soil, rock, and living organisms within ecosystems. Within the biosphere, this movement of matter occurs in **nutrient cycles**, or biogeochemical cycles (life-earth-chemical cycles). Nutrient cycles are driven directly or indirectly by energy from the sun and by Earth's gravity. These cycles include the hydrologic (water), carbon, nitrogen, and phosphorus cycles. They are important parts of Earth's natural capital. Yet, human activities are disrupting these cycles.

As a nutrient moves through a biogeochemical cycle, it may accumulate in a certain stage of the cycle and remain there for varying periods. Such temporary reservoirs include the atmosphere, the ocean and other bodies of water, underground deposits, and living organisms.

The diagrams on these pages relate to many key lessons of this book, and you may wish to revisit them regularly during your study of environmental science. For example, you might review the various cycles shown when you study forms of pollution or overuse of natural resources. They will help you understand the impacts human activities have on Earth's life-support system and possible solutions for maintaining sustainability on Earth.

checkpoint Why do you think it is important to understand biogeochemical cycles?

The Hydrologic Cycle

Water (H_2O) is essential to life on Earth. The **hydrologic cycle**, also called the water cycle, collects, purifies, and distributes Earth's fixed supply of water (Figure 3-16). The water cycle facilitates all of the important nutrient cycles that are discussed later in this chapter.

The sun provides the energy needed to power the water cycle. In the water cycle, incoming solar energy causes evaporation. *Evaporation* is the conversion of liquid water to water vapor. Most water vapor rises in the atmosphere, where it condenses into droplets in clouds. Gravity then draws the water back to Earth's surface as *precipitation*, such as rain, snow, or sleet. Above land about 90% of the water vapor in the atmosphere evaporated from soil and plants. Evaporation from plant surfaces is called *transpiration*. Plants draw enormous amounts of water from the ground through their roots. Transpiration is the process by which plants use evaporation to release excess water through tiny pores in their leaves.

When precipitation returns to Earth's surface, it takes various paths. Most precipitation falling on land ecosystems becomes **surface runoff**. Surface runoff flows over land surfaces into streams, rivers, lakes, wetlands, and the ocean. Some of that water then evaporates and the cycle repeats.

Some precipitation seeps into the soil. This water may evaporate back into the atmosphere or be consumed by plants and other organisms. Water that seeps deeper through soil is known as **groundwater**. Groundwater collects in **aquifers**, which are underground layers of sand, gravel, and water-bearing rock.

Water easily dissolves many compounds, which means it can be easily polluted. Throughout the hydrologic cycle, several natural processes purify water by drawing out pollutants. For example, when water evaporates, dissolved solids, including pollutants, are left behind. The hydrologic cycle can be viewed as a natural cycle of water quality renewal—an important and free ecosystem service. Without the hydrologic cycle's purification processes, humans and other species would rapidly run out of drinkable water.

FIGURE 3-16 ▼

Hydrologic Cycle This illustration shows a simplified model of the hydrologic, or water, cycle. Water circulates in various physical forms within the atmosphere, geosphere, hydrosphere, and biosphere. The red arrows and boxes identify major effects of human activities on this cycle.

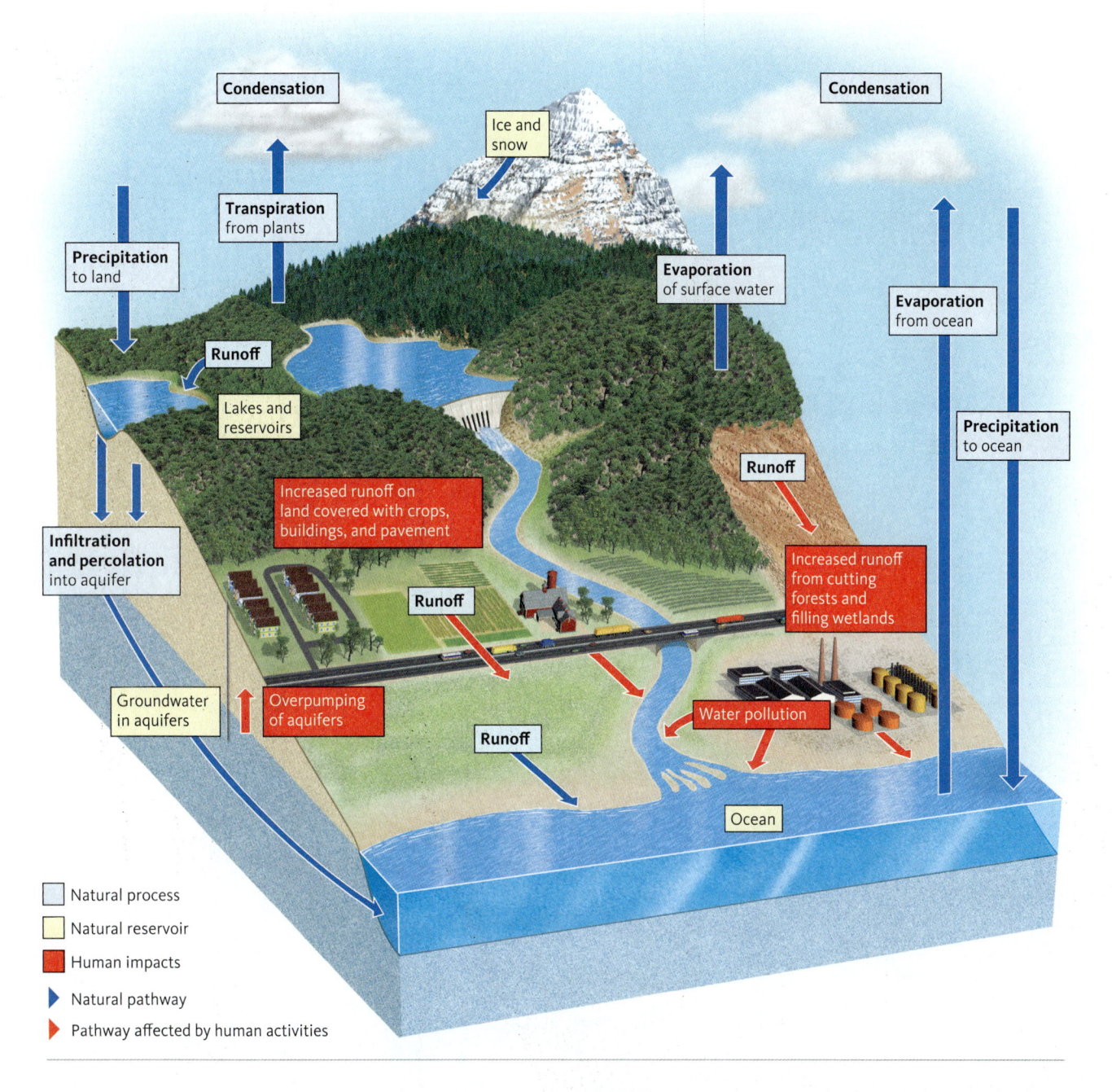

Condensation · Ice and snow · Condensation · Transpiration from plants · Evaporation of surface water · Evaporation from ocean · Precipitation to land · Runoff · Lakes and reservoirs · Precipitation to ocean · Runoff · Increased runoff on land covered with crops, buildings, and pavement · Infiltration and percolation into aquifer · Runoff · Increased runoff from cutting forests and filling wetlands · Groundwater in aquifers · Overpumping of aquifers · Runoff · Water pollution · Ocean

- ☐ Natural process
- ☐ Natural reservoir
- ☐ Human impacts
- ▶ Natural pathway
- ▶ Pathway affected by human activities

Only about 0.024% of Earth's vast water supply is available to humans and other species as liquid fresh water. This small fraction is further reduced when human activities pollute freshwater sources. Fresh water is found in accessible groundwater deposits and in surface water from lakes, rivers, and streams. Some groundwater deposits are too deep to extract affordably. The rest of the planet's water is too salty to drink or is permanently frozen in glaciers.

Human Impacts Humans alter the water cycle in three primary ways. (See the red arrows and boxes in Figure 3-16.)

First, people drain and fill wetlands for farming and urban development. Left undisturbed, wetlands provide the ecosystem service of flood control. Wetlands act like sponges to absorb and hold overflows of water from drenching rains or rapidly melting snow.

LESSON 3.4

FIGURE 3-17
ON ASSIGNMENT National Geographic photographer Peter McBride documents a canoeist's struggle through a shallow pool of garbage and muddy froth. This photo was taken at the end of the Colorado River, just inside Mexico.

SCIENCE FOCUS 3.3

WATER

Without water, Earth would be a lifeless planet. Water's unique properties make it one of nature's most extraordinary compounds. Here are a few of the reasons why water is so wondrous.

Water exists as a liquid over a wide range of temperatures. At first glance, this may not seem important. But what if liquid water had a narrower temperature range between freezing and boiling like so many other liquids? The ocean would have frozen solid or boiled away long ago.

Liquid water has a high heat capacity. In other words, water can store a large amount of thermal energy. It takes a lot more energy to raise the temperature of water than it does to raise the temperature of most other liquids. This property of water helps organisms regulate body temperature and plays a critical role in moderating Earth's climate.

Liquid water dissolves more substances than any other liquid. For this reason, water is often called the "universal solvent." In nutrient cycling, water is like the vehicle in which nutrients travel. Water carries dissolved nutrients into the tissues of living organisms and flushes waste products from those tissues. (More than half of your body mass is water.) It helps remove and dilute the water-soluble wastes of civilization. Unfortunately, this property makes water susceptible to pollution.

Water expands when it freezes. Ice floats on water because it has a lower density (mass per unit of volume) than its liquid form. Otherwise, lakes and streams in cold climates would freeze solid, killing virtually all of the aquatic life. This special property fractures rocks in a phenomenon called ice wedging. Thus, water plays a major role in shaping landscapes and forming soil.

Thinking Critically

Infer The expansion of water when it freezes plays a major role in shaping landscapes and forming soil. Which other property described above also plays a major role in altering landscapes?

Second, people withdraw fresh water from rivers, lakes, and aquifers, often at rates faster than natural processes can replace it. As a result, some aquifers are being depleted and several rivers no longer flow to the ocean.

Third, people clear vegetation from land for agriculture, mining, road building, and other activities. They cover much of the cleared land with buildings, concrete, and asphalt. This increases runoff and reduces infiltration that normally recharges groundwater supplies.

checkpoint How does energy from the sun drive the hydrologic cycle?

The Carbon Cycle

Carbon is the basic building block of the carbohydrates, fats, proteins, DNA, and all other organic compounds required for life. Carbon is found in every cell of your body. It is part of the carbohydrate molecules produced through photosynthesis and eaten or decomposed by consumers. In the **carbon cycle** (Figure 3-18), different compounds of carbon circulate through the biosphere, atmosphere, and parts of the geosphere and hydrosphere.

A key component of the carbon cycle is carbon dioxide (CO_2) gas. Carbon dioxide makes up only about 0.04% of the volume of the atmosphere and is also dissolved in water. The amount of carbon dioxide (along with water vapor) has a big effect on global temperatures because of the greenhouse effect (Lesson 3.1).

On land, photosynthesis by producers moves carbon from the atmosphere to the biosphere. In marine environments, producers remove carbon from water. Meanwhile, the cells of oxygen-consuming producers, consumers, and decomposers (both terrestrial and aquatic) carry out aerobic respiration. As you learned in Lesson 3.2, the by-product of aerobic respiration is water and CO_2. Together, the processes of photosynthesis and aerobic respiration circulate carbon through the biosphere.

FIGURE 3-18

Carbon Cycle This simplified model shows the circulation of various chemical forms of carbon in the global carbon cycle.

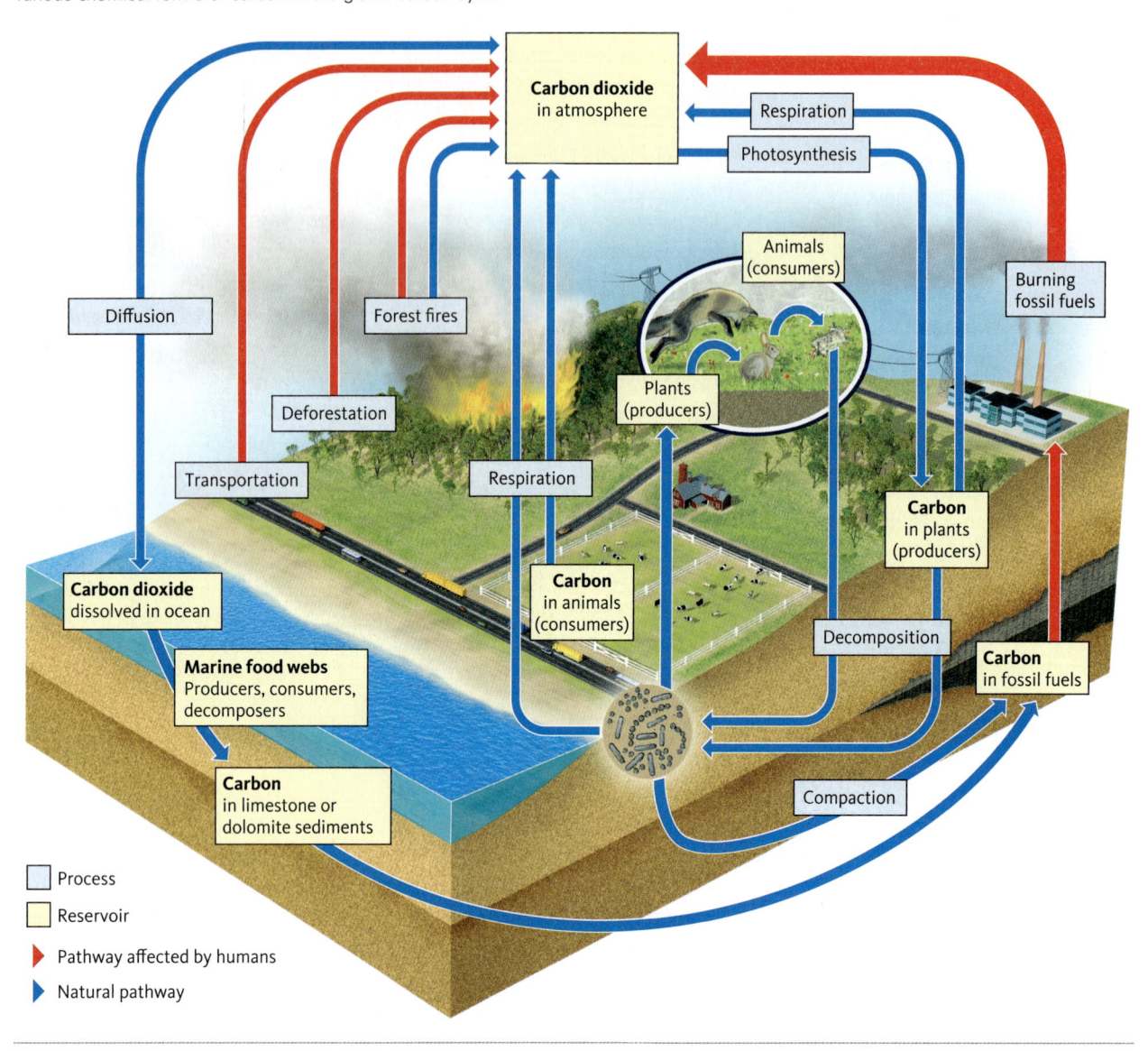

- Process
- Reservoir
- ▶ Pathway affected by humans
- ▶ Natural pathway

On land, decomposers release some of the carbon stored in the bodies of dead organisms back into the air as CO_2. Carbon dioxide can remain in the atmosphere for 100 years or longer. In water, decomposers release carbon that can be stored as insoluble minerals in bottom sediment for much longer periods. In fact, marine sediments are Earth's largest store of carbon.

Over millions of years, the carbon in deeply buried marine deposits of dead plant matter and algae were converted into carbon-containing fossil fuels. The high pressure from the weight of overlying sediments and heat released during the decomposition of dead matter formed coal, oil, and natural gas (fossil fuels).

Human Impacts Humans are altering the carbon cycle mostly by adding large amounts of carbon dioxide to the atmosphere. (See the red arrows and boxes in Figure 3-18.) In the past few hundred years, humans have extracted and burned huge quantities of fossil fuels that took millions of years to form. This has resulted in the release of tremendous quantities of CO_2 into the atmosphere. Humans also alter the cycle by clearing carbon-absorbing vegetation from forests, especially tropical forests, faster than it can grow back (Case Study). These alterations contribute to environmental problems that affect the atmosphere and ocean.

checkpoint Why is carbon essential to your survival?

FIGURE 3-19 ▼

Nitrogen Cycle Various chemical forms of nitrogen circulate in this simplified model of the nitrogen cycle. Red arrows indicate the major harmful human impacts. (Yellow box sizes do not represent relative reservoir sizes.)

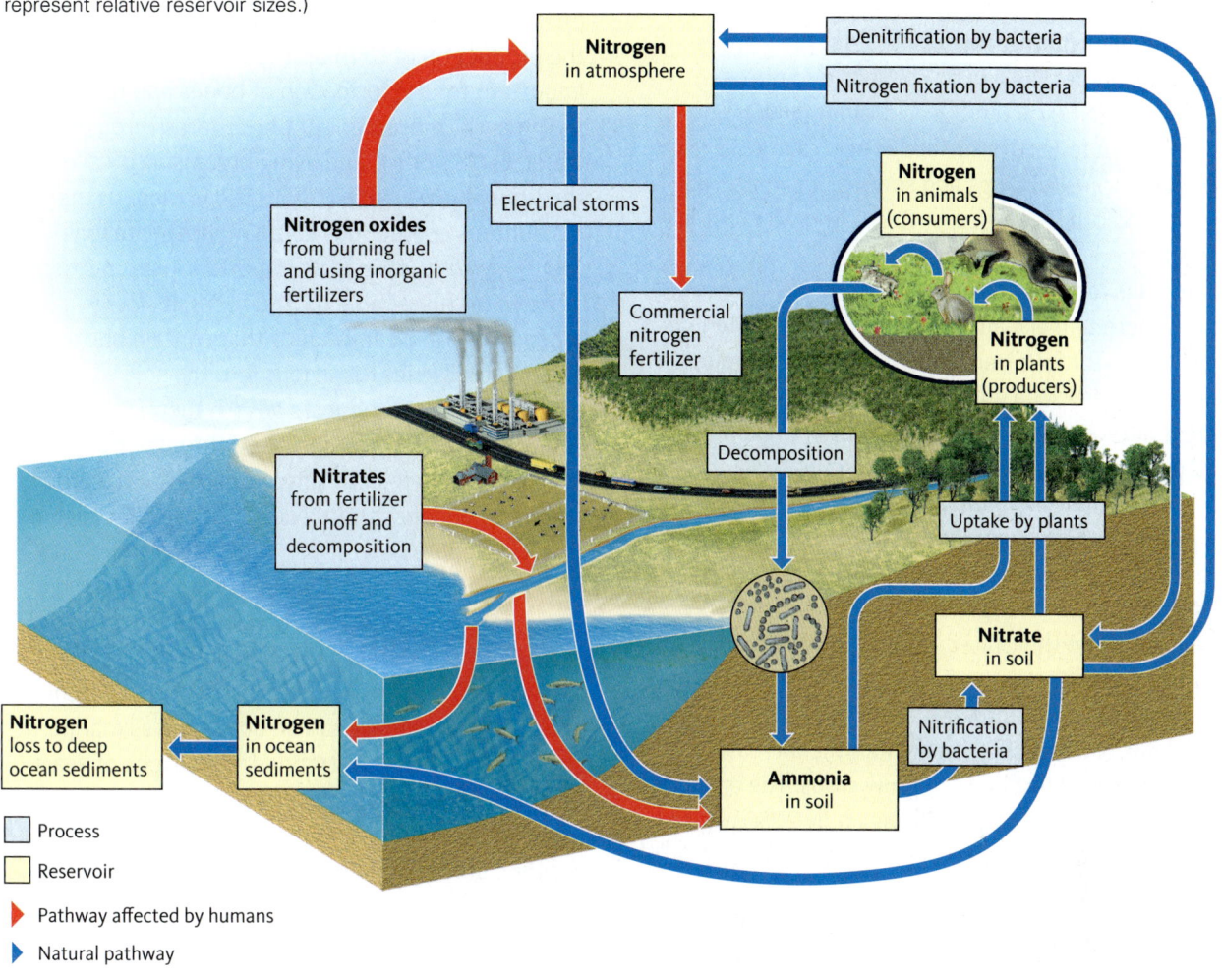

The Nitrogen Cycle

Nitrogen gas (N_2) makes up 78% of the volume of the atmosphere. Nitrogen is a crucial component of proteins, many vitamins, and DNA. Despite its abundance and importance to life, nitrogen cannot be absorbed and used directly as a nutrient by plants or other organisms. It becomes usable by producers only in the form of compounds such as ammonia (NH_3) and ammonium ions (NH_4^+).

These compounds are created within the **nitrogen cycle** (Figure 3-19). They result from reactions involving either lightning or specialized bacteria found in topsoil and aquatic ecosystems. Other bacteria convert most of the NH_3 and NH_4^+ in the topsoil to nitrate ions (NO_3^-), which the roots of plants take up. Plants use these forms of nitrogen to produce the proteins, nucleic acids, and vitamins necessary for their own survival and that of other organisms. Animals that eat plants absorb these nitrogen-containing compounds, as do detritivores and decomposers.

Organisms return nitrogen-rich organic compounds to the environment in their wastes and cast-off particles of matter such as leaves, skin, or hair. When organisms die, their bodies are decomposed or eaten by detritus feeders. In both instances, specialized bacteria break down the remains into simpler chemicals such as nitrate ions (NO_3^-), ammonia (NH_3) and ammonium ions (NH_4^+). Bacteria then convert such chemicals to N_2 gas, which returns to the atmosphere to begin the nitrogen cycle again.

LESSON 3.4

Human Impacts Human activities impact the nitrogen cycle in several ways. (See the red arrows and boxes in Figure 3-19.) Nitric oxide (NO) is added to the atmosphere as a product of combustion when humans burn gasoline and other fuels. In the atmosphere, NO can be converted to nitrogen dioxide gas (NO_2) and nitric acid vapor (HNO_3), which return to Earth's surface as acid rain. Acid rain damages stone buildings and statues. It can also kill forests and other plant ecosystems, and wipe out life in ponds and lakes.

Humans remove nitrogen (N_2) from the atmosphere to make ammonia (NH_3) and ammonium ions (NH_4^+) for fertilizers. In addition, humans alter the nitrogen cycle in aquatic ecosystems by adding excess nitrates (NO_3^-). These nitrates contaminate bodies of water through agricultural runoff of fertilizers, animal manure, and discharges from municipal sewage treatment systems. This can cause excessive growth of algae that reduce oxygen levels and cause stress or death of aquatic organisms.

According to the United Nations, during the last century, human activities have more than doubled the total amount of nitrogen entering Earth's biosphere. Most of this comes from the increased use of inorganic fertilizers to grow crops. The amount released is projected to double again by 2050, which would seriously alter the nitrogen cycle.

> **CONSIDER THIS**
>
> **Nutrient cycles connect past, present, and future forms of life.**
>
> Some of the carbon atoms in your skin may once have been part of an oak leaf, a dinosaur's skin, or a layer of limestone rock deep in the ocean. Your great-grandmother, George Washington, or a hunter-gatherer who lived 25,000 years ago may have breathed some of the nitrogen molecules you just inhaled. The hydrogen and oxygen atoms that formed the water you drank today may have flowed in the Nile River in Egypt thousands of years ago or floated in a cloud over the Pacific Ocean only two months ago.

checkpoint How can the release of excess nitrates into bodies of water affect aquatic organisms?

The Phosphorus Cycle

Phosphorus (P) is an element that is essential for living things. It is contained in ATP, a compound that provides energy for life processes. It is necessary for the production of DNA and cell membranes, and is important for the formation of bones and teeth.

The cyclic movement of phosphorus through water, Earth's crust, and living organisms is called the **phosphorus cycle** (Figure 3-20). Most of the phosphorus compounds contain phosphate ions (PO_4^{3-}), which serve as an important plant nutrient.

As water runs over exposed rocks, it slowly erodes inorganic compounds that contain phosphate ions. Water carries these ions into the soil, where plants and other producers absorb them. Phosphate compounds are then transferred through food webs from producers to consumers. Unlike water, carbon, and nitrogen, phosphorus does not cycle through the atmosphere.

The phosphorus cycle is slow compared to the water, carbon, and nitrogen cycles. As phosphate is eroded from exposed rocks, much of it is carried in rivers and streams to the ocean. When it reaches the ocean, phosphate can be deposited as marine sediments and remain trapped for millions of years. Over time, geological processes uplift and expose some of these seafloor deposits. The exposed rocks are then eroded, freeing up the phosphorus to re-enter the cycle.

Most soils contain little phosphate, which limits plant growth on land. For this reason, people often fertilize soil by adding phosphorus (as phosphate compounds mined from the ground). Under natural conditions, low levels of phosphorus also limit the growth of producer populations in many freshwater environments. Phosphate compounds are only slightly soluble in water, so they release fewer phosphate ions to aquatic producers that need them as nutrients.

Human Impacts Human activities, including the mining of large amounts of phosphate to make fertilizer, disrupt the phosphorus cycle. (See the red arrows in Figure 3-20.) By clearing tropical forests, humans expose the topsoil to greater erosion, which reduces phosphate levels in the soil.

Phosphate fertilizer is then added to make up for the lost phosphate. Eroded topsoil and fertilizer washed from fertilized crop fields, lawns, and golf courses carry large quantities of phosphate ions into streams, lakes, and oceans. Phosphates stimulate the growth of producers such as algae. Similar to nitrogen-rich runoff, phosphate-rich runoff from the land often causes huge increases in algae populations. As the algae die and decompose, oxygen in the water is depleted, wiping out populations of aquatic organisms.

checkpoint How is the phosphorus cycle different from the water, carbon, and nitrogen cycles?

FIGURE 3-20

Phosphorus Cycle Different chemical forms of phosphorus (mostly phosphates) circulate among land, water, and organisms in this simplified model of the phosphorus cycle. (Yellow box sizes do not represent relative reservoir sizes.)

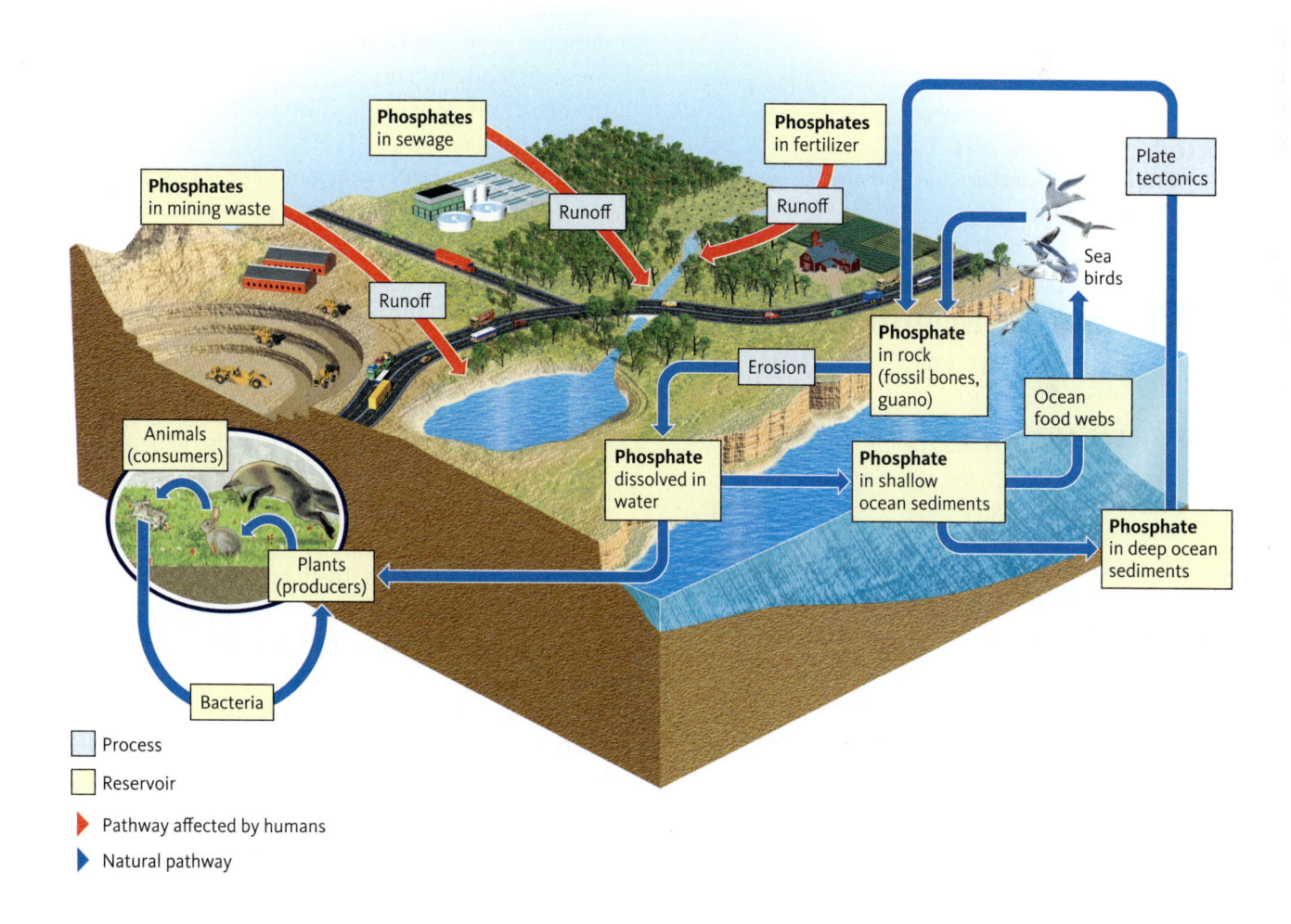

LESSON 3.4   91

3.4 Assessment

1. **Identify Main Ideas** What is the role of the hydrologic cycle in relation to the carbon, nitrogen, and phosphorus cycles?
2. **Summarize** Describe at least one way humans impact each of the following cycles: hydrologic, carbon, nitrogen, and phosphorus.

SCIENCE AND ENGINEERING PRACTICES

3. **Using Math** There are about 11,200,000 billion metric tons of O_2 in the atmosphere. Based on an estimated photosynthesis rate of 600 billion metric tons per year, how many years might it have taken to reach current O_2 levels? (Note: Assume for this exercise that the rate of O_2 production is constant even though it has changed over time.)

SCIENCE AND ENGINEERING PRACTICES

4. **Developing Models** Create a simplified model to illustrate the role of photosynthesis and cellular respiration in the cycling of carbon. Include how carbon cycles through the biosphere, atmosphere, hydrosphere, and geosphere.

CROSSCUTTING CONCEPTS

5. **Energy and Matter** How is matter conserved in the nutrient cycles?

3.5 How Do Scientists Study Ecosystems?

CORE IDEAS AND SKILLS

- Compare and contrast the advantages and disadvantages between field research and laboratory research.
- Explain why mathematical models are an important tool for studying natural systems.
- Describe how a better scientific understanding of planetary boundaries can help measure the health of ecosystems.

KEY TERMS

Holocene   Anthropocene

Ecologists Study Ecosystems Directly

Scientists such as ecologists use several approaches to increasing the scientific understanding of ecosystems. Field research involves going into a natural setting to study one or more features of an ecosystem. Ecologists have numerous methods of field research, including taking water or soil samples, identifying the species in an area, doing population counts, observing feeding behaviors, and using GPS to track animals' movements. In this lesson you will read about the pioneering field research of one scientist, National Geographic Explorer Thomas E. Lovejoy. Much of what people know about ecosystems comes from data obtained through fieldwork.

Scientists who study tropical rain forests, for example, use a variety of methods to study those ecosystems. Most animals in a tropical forest live in the canopy, which can be as high as 28 meters (92 feet) above the ground. Scientists often use ropes and pulleys to access a canopy (Figure 3-21). They may also install rope walkways, spiral staircases, and temporary platforms. Some long-term scientific projects have built construction cranes that tower over the surrounding trees. These "canopy cranes" can bring researchers and heavy equipment to many different points within the forest canopy. All of these methods help researchers identify and observe the rich diversity of species living or feeding in treetop habitats.

Ecologists may carry out controlled experiments in the field by isolating and changing a variable within a defined area. They then compare the results with unchanged areas nearby. You learned about a classic example of this in the Case Study in Chapter 2.

Advances in technology allow ecologists to obtain data in new ways. Satellites and aircraft equipped with sophisticated cameras can scan and collect data about Earth's surface. Scientists use geographic information system (GIS) software to capture, store, analyze, and display this data. For example, a GIS can convert digital satellite images into global, regional, and local maps. The maps can show variations in vegetation, gross primary productivity, air pollution emissions, and other variables. More recently, scientists have used drones to photograph, document, and monitor rates of deforestation.

Some researchers attach small radio transmitters to animals and use global positioning systems (GPS) to track where and how far the animals travel. This technology is an important tool for studying endangered species, which you will learn about in Chapter 7.

checkpoint What are some of the methods ecologists use in field research?

FIGURE 3-21
National Geographic Explorer and tropical ecologist Greg Goldsmith pauses to take notes a mere 30 meters (98 feet) above the ground. Some of Greg's fieldwork involves hanging out with wildlife in the upper canopy of this montane cloud forest in Costa Rica.

Ecologists Study Ecosystems Indirectly

Most ecologists supplement their field research by conducting research in laboratories. In a lab, scientists can set up, observe, and make measurements of model ecosystems and populations. They can create simplified systems in culture tubes, bottles, aquariums, and greenhouses, and in indoor and outdoor chambers.

By isolating biological systems, scientists can control variables such as temperature, light, CO_2, and humidity. Scientists must consider how well their observations and measurements in laboratory conditions reflect what actually takes place in the more complex and often changing conditions found in nature. Although they provide only part of the picture, controlled experiments (in the field or a lab) offer the best means by which to identify cause-and-effect relationships.

Scientific knowledge advances when multiple lines of evidence support the same explanation, so ecologists often use a combination of indirect and direct observation.

Ecologists also use mathematical modeling. Mathematical models can simulate large and complex systems with many variables and large data sets. The models usually require so many variables and so much data that they can only be run on a high-speed supercomputer. Whole systems such as lakes, oceans, forests, and Earth's climate cannot be observed in their entirety or modeled physically. The scope of these systems is too large, and the timescales may be too long, to allow for direct study. Mathematical modeling, however, is ideally suited to these large-scale natural systems.

checkpoint What are some of the methods ecologists use in laboratory research?

NATIONAL GEOGRAPHIC | EXPLORERS AT WORK
Thomas E. Lovejoy Tropical and Conservation Biologist

Conservation biologist and National Geographic Fellow Thomas E. Lovejoy (1941–2021) is credited with coining the term *biological diversity*. Throughout his career, Dr. Lovejoy played a major role in educating people about the need to protect biodiversity. His conclusions about biodiversity were supported by data he and others had collected over many years of field research.

Lovejoy began carrying out field research in the Amazon forests of Brazil in 1965. His work there led him to help start the world's largest and longest-running study of habitat fragmentation. The study is called the Biological Dynamics of Forest Fragments Project (BDFFP). Since 1979, the project has measured the impacts of fragmentation across a 1,000 square-kilometer (386 square-mile) area of the central Amazon.

A goal of the BDFFP is to define the minimum amount of land area necessary for sustaining biodiversity in a tropical forest. As forests become increasingly fragmented, a better understanding of how fragmentation affects biodiversity is more important than ever before.

Threats to tropical forest ecosystems—such as deforestation, poaching, and pollution—are a matter of public interest today largely through Lovejoy's efforts to increase public awareness.

Lovejoy founded the popular and widely acclaimed public television series *Nature*. He also wrote numerous articles and books on issues related to the conservation of biodiversity.

In addition to teaching environmental science and policy at George Mason University, Lovejoy held several important posts over the years. He served as director of the World Wildlife Fund's conservation program and was president of the Society for Conservation Biology. He was also executive director of the UN Environment Programme (UNEP). In 2012, he was awarded the Blue Planet Prize in recognition of his efforts to understand and sustain Earth's biodiversity.

FIGURE 3-22

Planetary Boundaries The table describes planetary boundaries for nine major systems that help sustain life. A team of scientists estimated that human activities may have exceeded the boundary limits of four of these systems (the first four, shown in red).

Nitrogen and Phosphorus Cycles	Disruption of the nitrogen and phosphorus cycles caused by greatly increased use of fertilizers
Climate Change	Altering the carbon cycle, mostly by overloading it with CO_2 produced by burning fossil fuels
Biodiversity Loss	Replacing biologically diverse forests and grasslands with fields of single crops
Land Use	Land system change from agriculture and urban development
Freshwater Use	Global consumption of water per year
Ocean Acidification	Acidification of the ocean caused by increased carbonate concentration
Ozone Depletion	Stratospheric ozone concentration
Atmospheric Aerosols	Atmospheric aerosol loading (microscopic particles in the atmosphere that affect climate and living organisms)
Chemical Pollution	Levels of toxic heavy metals and endocrine disruptors Introduction of novel entities (e.g., organic pollutants, radioactive wastes, nanomaterials, and microplastics)

SCIENCE FOCUS 3.4

TESTING PLANETARY BOUNDARIES

For most of the past 10,000 years, humans have lived in an epoch called the **Holocene**—a period of relatively stable climate and other environmental conditions. This general stability has allowed humans to develop agriculture and expand the human population around the world.

Although most geologists argue that we are still living in the Holocene, a growing number of other scientists think we are living in a new epoch, which they call the **Anthropocene**. According to their argument, the Anthropocene began around 1750 with the Industrial Revolution. Since that time, people have been consuming a much greater share of Earth's resources and have become the dominant cause of changes to the planet's major systems that sustain life.

In 2009, an international group of 28 scientists, led by Will Steffen and Johan Rockström of the Stockholm Resilience Centre, identified the boundaries, or ecological tipping points, of nine major planetary systems that play a key role in supporting life. (Recall that an ecological tipping point is like a system's "point of no return," resulting in severe degradation or collapse.) Subsequent research published by the Stockholm group indicated four of these boundaries have now been exceeded (Figure 3-22).

These scientists warn if humans exceed too many boundaries, we could trigger abrupt and long-lasting environmental changes that will degrade the planet's ability to support life. They argue that there is an urgent need for more research to better define the boundary limits of these planetary systems, which are still not exact. They also say we need to learn more about what exceeding the boundaries will do to the health of humans and other species.

If this is the Anthropocene, how will it end for humans and other species? What will Earth's next epoch look like?

Thinking Critically
Infer Which boundaries are most affected by urban development? Which ones are most affected by agriculture?

LESSON 3.5   95

3.5 Assessment

1. **Compare and Contrast** Compare and contrast the advantages and disadvantages between field research and laboratory research.
2. **Summarize** Why are mathematical models an important tool for studying complex natural systems?
3. **Draw Conclusions** How can a better scientific understanding of planetary boundaries help scientists determine the health of ecosystems?

CROSSCUTTING CONCEPTS

4. **Stability and Change** Use the text and your own research to argue for or against the view that we are living in a new epoch called the Anthropocene.

TYING IT ALL TOGETHER STEM
The Energy Cost of Cutting Rain Forests

The Case Study at the beginning of this chapter discusses the importance of the world's incredibly diverse tropical rain forests. Producers in rain forests rely on solar energy to produce a vast amount of biomass through photosynthesis. A huge variety of forest species take part in and depend on the flow of the sun's energy and cycling of nutrients through the ecosystem. In other words, rain forests are highly complex and highly productive.

Recall that net primary productivity (NPP) is the rate at which producers can make the chemical energy that is potentially available to the rest of the organisms in an ecosystem. Figure 3-23 provides an estimated annual average NPP for Earth's major types of ecosystems.

Look through the table in light of all you have learned about how energy moves through ecosystems. What meaning can be drawn from these numbers? In this exercise, you will analyze these data and synthesize ideas from this chapter with your own thinking.

Use the table to help you answer the questions that follow.

1. Highly complex ecosystems are relatively resilient, meaning they are likely to recover over time from "small" changes, such as the loss of a species. Explain this in terms of food webs.
2. Even highly complex ecosystems cannot recover from extreme changes. Give an example of a change from which a rain forest would likely not recover.
3. Look at the units for NPP in Figure 3-23. Kilocalorie (kcal) is a unit of energy. Develop a formula to determine the impact of clear-cutting one square meter per year on the energy produced by an ecosystem.
4. An estimated 5,000 m² of Amazon rain forest is cleared annually. Use your formula to calculate the impact on NPP.
5. Compare your answer from number 4 to the impact of clear-cutting an equal area of desert scrub.
6. Work with a partner. Use your answers from numbers 1–4 to help you evaluate the following statement. Revise the statement and add a paragraph that explains your reasoning. Support your reasons with data.

"Damage to rain forests results in less harm to Earth's life-support system than damage to simpler ecosystems because complex ecosystems are more resilient to change."

FIGURE 3-23

Ecosystem/Life Zone	Average Net Primary Productivity (kcal/m²/year)
Terrestrial Ecosystems	
Swamp and marsh	9,000
Tropical rain forest	9,000
Temperate forest	5,800
Taiga	3,400
Savanna	3,000
Woodland and shrubland	2,600
Temperate grassland	2,200
Tundra	600
Desert scrub	200
Aquatic Ecosystems	
Estuary	9,000
Lake and stream	2,200
Continental shelf	1,500
Open ocean	1,100

Source: R. H. Whittaker, *Communities and Ecosystems*, 2nd ed., New York: Macmillan, 1975.

CHAPTER 3 SUMMARY

3.1 What are Earth's major spheres, and how do they support life?

- Earth's capacity to support life depends on the proper functioning of, and interaction among, four major planetary spheres. The major planetary spheres are the atmosphere (air), hydrosphere (water), geosphere (land), and biosphere (living things).

3.2 What are the major ecosystem components?

- Ecology is the study of how organisms interact with one another and their nonliving environment of matter and energy. Ecologists focus on one or more levels of organization.
- Ecosystems are composed of abiotic and biotic factors. Species are classified into trophic levels based on how they obtain food. Producers make their own food, while consumers feed on other organisms or the wastes and remains of other organisms.
- Consumers are herbivores, carnivores, omnivores, or decomposers. Decomposers recycle nutrients back to producers by decomposing the wastes and remains of other organisms.

3.3 What happens to energy in an ecosystem?

- Life is sustained by the one-way flow of energy, mainly from the sun, through the biosphere. Energy flows through ecosystems in food chains and food webs. The amount of energy available to organisms decreases at each successive trophic level.
- Gross primary productivity (GPP) is the rate at which plants convert solar energy into chemical energy. The net primary productivity (NPP) is the rate at which producers use photosynthesis to produce and store chemical energy minus the rate at which they release some of this stored energy through aerobic respiration.

3.4 What happens to matter in an ecosystem?

- The flow of energy drives the cycling of matter within Earth's biosphere. Matter in an ecosystem travels in the form of nutrients. These nutrients cycle in and among ecosystems and the biosphere, geosphere, hydrosphere, and atmosphere.
- Four key nutrient cycles are the water cycle, carbon cycle, nitrogen cycle, and phosphorus cycle. Human activities affect all four cycles.

3.5 How do scientists study ecosystems?

- Scientists use field and laboratory research, in addition to mathematical and other models, to learn about ecosystems.
- Research indicates that four out of nine planetary boundaries, or tipping points, have likely been exceeded. Scientists need to generate more data to determine the current state of these and other possible planetary tipping points.

MindTap If you have been provided with access to a MindTap course, additional resources are available at login.cengage.com.

CHAPTER 3 ASSESSMENT

Review Key Terms

Select the term that best fits the definition. Not all terms will be used.

aerobic respiration	geosphere	phosphorus cycle
anaerobic respiration	greenhouse effect	photosynthesis
Anthropocene	gross primary productivity	primary consumer
aquifer	groundwater	producer
atmosphere	herbivore	secondary consumer
biosphere	Holocene	stratosphere
carbon cycle	hydrologic cycle	surface runoff
carnivore	hydrosphere	tertiary consumer
consumer	net primary productivity	trophic level
decomposer	nitrogen cycle	troposphere
detritivore	nutrient cycle	
food chain	omnivore	
food web		

1. Process in which solar energy interacts with carbon dioxide, water vapor, and other gases in the air; it warms the troposphere
2. Organism that cannot produce its own nutrients and gets them by feeding on other organisms or their wastes and remains
3. Animal that eats both plants and animals
4. A complex network of interconnected food chains
5. Component of Earth's life-support system that consists of a thin spherical envelope of gases surrounding the planet's surface
6. Includes all the water on or near Earth's surface
7. Component of Earth's life-support system where life is found
8. Uses oxygen to convert glucose and other organic molecules back into carbon dioxide and water
9. An organism's feeding level within an ecosystem
10. Precipitation that sinks through soil into underground layers of rock, sand, and gravel
11. The rate at which producers convert radiant energy into chemical energy
12. Continuous movement of elements and compounds that make up nutrients through air, water, soil, rock, and living organisms

Review Key Concepts

13. Name and describe four large-scale systems, or spheres, that sustain life on Earth.
14. Define organism, population, communities, ecosystems, and the biosphere.
15. Which level(s) of organization are the main focus of Vitor Becker's work?
16. Give two examples each of abiotic factors and biotic factors found in a tropical rain forest.
17. What are decomposers? What purpose do they serve in an ecosystem?
18. Summarize the process of photosynthesis.
19. Summarize the process of aerobic respiration.
20. Define and distinguish between a food chain and a food web.
21. Distinguish between gross primary productivity (GPP) and net primary productivity (NPP).
22. Explain how nutrient cycles connect past, present, and future life.
23. How are humans affecting the water, carbon, nitrogen, and phosphorus cycles?
24. Explain why we need more research about the structure and condition of the world's ecosystems.

Think Critically

25. How would you explain the importance of tropical rain forests to people who think such forests have no connection to their lives?
26. Explain why energy from the sun is essential for the cycling of nutrients.
27. Explain the interaction of energy, nutrients, and organisms in an ecosystem.
28. Make a list of the food you ate today. Trace each food item back to a producer species. Diagram the sequence of trophic levels that led to your consumption of that food item.
29. Why are there so few top predators in an ecosystem when compared with the number of primary consumers?

30. What would happen to an ecosystem if
    a. all decomposers and detritus feeders were eliminated?
    b. all producers were eliminated?
    c. all insects were eliminated?
    d. only producers and decomposers existed?

31. Explain the importance of the roles of microbes in the biosphere.

32. Research indicates that we may have exceeded four planetary boundaries. Describe how exceeding each of these boundaries might affect you, your children, or your grandchildren.

Chapter Activities

A. Develop Models: Greenhouse Effect STEM

The greenhouse effect provides the warmth needed to sustain life in Earth's troposphere. The greenhouse effect is produced by an interaction between the sun and Earth's atmosphere (Lesson 3.1). Look at the materials below. Design a model of the greenhouse effect using these or other materials approved by your teacher.

Materials

plastic bottles   thermometers   heat lamps
glass jars

1. Describe how your model will work using a diagram to show the flow of energy through your system. Build and test your model.

2. How did your model demonstrate the greenhouse effect?

3. What are some limitations of your model? How could your model be improved?

4. How could your model be used to test the effect of increased carbon dioxide on atmospheric temperature? Write a hypothesis.

B. Citizen Science

National Geographic Learning Framework

Attitudes | Curiosity
Skills | Collaboration
Knowledge | Critical Species

Plankton are microscopic producers and the primary consumers that form the basis of Earth's ocean food webs (Figure 3-15). Although they may be small, their role in the biosphere is enormous. Earth's ocean ecosystems collectively produce more biomass per year than any other ecosystem.

Yet, plankton populations are declining. Because they are so small and their habitats are so vast and inaccessible, scientists are relying on citizens to help them gather and analyze the data they need.

Join a citizen science project asking volunteers to help collect data on plankton or analyze video footage. Then gather in small groups to discuss your answers to the questions.

1. What was the name of the project, website, or app that you used? What institution is running the project?

2. Describe your findings and contributions.

3. What was your experience like working on the project? Do you think you will continue?

4. Discuss any ideas you have to improve upon or expand the project.

STEM ENGINEERING PROJECT 1
DESIGN A METHOD FOR TREATING CONTAMINATED SOIL

Eating food from your own garden has its rewards. Chances are, it is fresher, and tastes better, than any produce you can buy in a store. Growing your food has many other benefits, including reducing your ecological footprint, preserving green space, and improving your health. However, if your garden soil is polluted, the food grown in it may contain unhealthy contaminants. Plants absorb contaminants through their roots and may pass them along to you when you eat the plants.

Building a raised garden using fresh soil reduces the likelihood that you are growing food in polluted soil. But what about the soil you are adding? Even soil that is completely free of pesticides and chemical fertilizers contains trace amounts of heavy metals such as lead and arsenic. In fact, all soil contains these metals in very small amounts because they are natural elements in the environment. As rock weathers, for example, arsenic is released naturally into the biosphere. Heavy metals become pollutants when they reach toxic (harmful) levels. Toxic levels of heavy metals damage the cardiovascular and central nervous systems.

Mining and other human activities increase the release of heavy metals from rock into the biosphere. For instance, coal sludge is liquid waste produced from washing coal. Heavy-metal toxins such as arsenic and lead can spread from coal sludge into the surrounding soil and water. Once heavy metals become part of the soil, they are difficult to remove. How to detoxify soil of heavy metals is an unsolved problem.

One solution gaining attention is *phytoremediation*. *Phyto-* means "plant," and *-remediation* means "providing a remedy." Recall that plants absorb contaminants through their roots. Some plants, such as sunflowers, are especially good at absorbing heavy metals. Phytoremediation is the use of plants as a natural technology to clean up pollution. That sounds great, but how well does phytoremediation really work? Can sunflowers really reduce the amount of heavy metals in soil? What would it look like to implement a "sunflower solution"? Much more data are needed.

In this challenge, you will find answers to some of these and other questions. You will work with a team to design a pollution-cleanup solution and test your solution. Engineering usually involves many rounds of testing and retesting, so the process of design is often represented as a cycle rather than a series of steps. Engineers use this cycle to translate ideas into practical solutions. Armed with these practices and your own ingenuity, your team may find answers that no one has found before.

Engineering DESIGN CYCLE
- Defining problems
- Developing and using models
- Planning and carrying out investigations
- Analyzing data and using math
- Designing solutions
- Forming arguments from evidence
- Obtaining, evaluating, and communicating information

Defining Problems

1. Define the problem you want to solve.
   - Describe the problems associated with heavy metals in the soil and their cleanup.
2. Underlying every problem is a need. What do people need that is not currently available in terms of soil?
3. What would a soil cleanup solution need to do? List the criteria.

Developing and Using Models

4. Brainstorm ideas for ways people could use sunflowers to reduce soil contamination by heavy metals.
5. Select one idea as a team.
6. Develop a simplified model of your solution that can be created within the time and material constraints given by your teacher.

Planning and Carrying Out Investigations

7. Plan a controlled experiment to test your solution.
8. Consider any limitations of your design plan and revise it if needed. Make sure your team is controlling for all known variables.
9. Refine your experiment to make it as safe as possible.
10. Create your model and carry out your experiment. Record your data in a table.

Analyzing Data and Using Math

11. Analyze your data. Explain your findings.
12. How well does your solution meet each of the criteria you identified in step 3?

Designing Solutions

13. What are the strengths of your solution? What are its weaknesses?
14. Suggest an improvement to your solution based on your findings.

Forming Arguments from Evidence

15. What claims can you make about your solution? Support your claims with evidence.
16. Apply your solution to an outdoor garden.
    - When and where would sunflowers be planted?
    - How should sunflowers be disposed of?

Obtaining, Evaluating, and Communicating Information

17. Present your solution to the class. Summarize your model, experiment, and results.
18. Provide respectful and specific feedback to other teams.
19. Write a final report. Include recommendations for further changes, testing, and scientific research.

UNIT 2
Biodiversity

A flock of flamingos, known as a *flamboyance*, gathers in Punta Gallinas, La Guajira, Colombia.

Flamingos are known for their distinctive, bright-pink feathers and long, stilt-like legs. Their coloration is the result of the beta-carotene obtained through their diet of brine shrimp, mollusks, crustaceans, algae, small insects, and larvae. Well-nourished flamingoes have an even brighter pink hue, making them more attractive to potential mates. A variety of flamingo species can be found around the globe in the Americas and the Caribbean, Africa, Asia, and the Mediterranean. The birds are social animals, living in colonies that can consist of as many as 1,000 adults and chicks.

Los Flamencos Sanctuary is a wetland and forest reserve located in the Guajira Peninsula of northern Colombia along the Caribbean Sea. The wildlife reserve was established in 1977 and is home to a large protected population of American flamingos (*Phoenicopterus ruber*). The park is considered one of the most diverse protected areas, providing a home to a large variety of aquatic species and migratory birds.

In this unit, you will investigate the richness of life and its many forms. You will explore how the diversity of habitats and ecosystems has influenced the abundance of species on the planet.

CHAPTER 4
BIODIVERSITY AND EVOLUTION

CHAPTER 5
SPECIES INTERACTIONS, ECOLOGICAL SUCCESSION, AND POPULATION CONTROL

CHAPTER 6
ECOSYSTEMS AND CLIMATES

ENGINEERING PROJECT 2: Design a System to Assess a Local Species

CHAPTER 4
BIODIVERSITY AND EVOLUTION

NEW GUINEA and its surrounding islands are known for their diversity of life. The third-largest block of tropical rain forest is found on New Guinea and at least 5% of the world's species live there. Many are found nowhere else. Take for example the more than three dozen species of birds-of-paradise that live in the forest canopy. With their striking colors and fancy plumage, these birds demonstrate Earth's biodiversity in dramatic fashion.

KEY QUESTIONS

4.1 What is biodiversity and why is it important?

4.2 What roles do species play in ecosystems?

4.3 How does life on Earth change over time?

4.4 What factors affect biodiversity?

A male greater bird-of-paradise sits high in the canopy of Badigaki Forest on Wokam Island, off the coast of New Guinea.

NATIONAL GEOGRAPHIC | EXPLORERS AT WORK

Fighting a Deadly Frog Disease

with National Geographic Explorer Anna Savage

Some of Anna Savage's earliest memories are of chasing frogs in a forest pond near her family's home in New Hampshire. She's not surprised her friends generally don't share her fascination with frogs. As a scientist, Savage knows well the importance of maintaining and conserving the planet's biodiversity. But once she points out that without amphibians, some species would starve, insects and diseases would thrive, and some ecosystems would collapse, her friends tend to pay attention.

These days, Dr. Savage teaches and leads research on evolutionary genetics at the University of Central Florida. Her research picks up on her early interest in frogs, and there's a good reason: Frogs are in trouble.

A devastating disease known as chytrid has been linked to the decline of hundreds of amphibian species in many different environments worldwide. Chytrid is caused by a fungal infection and leads to thickening of the skin in amphibians. This symptom can be deadly, as it reduces an amphibian's ability to absorb water, air, and necessary salts through its skin. That can lead to dehydration and cardiac arrest.

In many ways, chytrid is a puzzle. Where did it come from? How is it able to spread so quickly? And why does it affect each frog species and each individual frog differently?

Savage's approach to the puzzle can be tricky to follow. She began by examining frogs' immune system genes to see if there were differences that account for why some are able to fight off the disease while others are not. Though researchers have found some variations in genes that are associated with high survival rates, selectively breeding frogs that possess the "good" genes is problematic. What if, by breeding frogs with the genes that resist chytrid, scientists inadvertently destroy adaptations that provide resistance to other diseases?

Instead, Savage and others are researching three variables they believe affect a frog's immune response to chytrid. The first is an individual frog's genetic makeup. The second is a frog's environment. The third is the genetic composition of the specific organism causing the disease. It's hoped a database can be developed to track genetic and environmental factors that may influence susceptibility to chytrid. Savage notes, "Eventually we'll be able to say the specific things that you should do as a population manager or conservationist to promote the right environmental conditions to help your specific populations fight off the disease."

In the meantime, Savage believes it's important to protect frog habitats so that successful breeding can increase the level of genetic diversity that's needed to fight off disease. Her efforts may help save frog species everywhere.

Thinking Critically
Infer How might Anna Savage respond to someone who suggests "letting nature take its course" when it comes to frogs and chytrid?

Anna Savage hopes her genetics research with frogs in Arizona helps shed light on how to conserve frog species around the world that are threatened by chytrid disease.

CASE STUDY
Amphibians at Risk

Amphibians are a class of animals that includes frogs, toads, and salamanders. Amphibians were among the first vertebrates (animals with backbones) to leave Earth's waters and live on land. These cold-blooded animals make use of both aquatic and land environments at various points in their life cycle. Today, more than 6,700 amphibian species inhabit Earth. They are found on every continent except Antarctica.

Despite their diversity, amphibians face a very uncertain future. Since 1980, populations of hundreds of species have declined or vanished (Figure 4-1). According to the International Union for Conservation of Nature, about 41% of known amphibian species face extinction, which is the death of an entire species.

No single cause can account for this trend. However, scientists believe the decline of amphibians is linked to a number of human activities. For example, pesticides and other chemicals can contaminate water, soil, and the tissues of living organisms. These pollutants may affect amphibians at various stages of their life cycle. Frog eggs lack shells to protect embryos inside from exposure to water pollutants. Adult frogs are exposed to insecticides contained in the insects they eat.

Why should people care if some amphibian species become extinct? Scientists claim these species are critical. They serve as sensitive biological indicators of changes to environmental conditions. These changes include habitat loss, pollution, increased levels of ultraviolet (UV) radiation, and a warming climate.

Amphibians also play important roles in biological communities. Adult amphibians eat more insects than do many species of birds. In turn, amphibians are a food source for a variety of animals. They are prey for aquatic insects, reptiles, birds, fish, mammals, and other amphibians. In some habitats, the loss of amphibian species could trigger the loss of those species as well.

Amphibians even play a role in human health. Compounds extracted from amphibians' skin secretions have been developed into a variety of pharmaceutical products. Examples include painkillers, antibiotics, and treatments for burns and heart disease. If amphibians vanish, so do these potential medicinal benefits.

The threat to amphibians is part of a greater threat to Earth's biodiversity. In this chapter, you will learn about biodiversity, how it arose on Earth, why it is important, and how it is threatened. You will also consider possible solutions to these threats.

As You Read Think about the biodiversity of different habitats within your community. Do some habitats support more biodiversity than others? For example, how might the biodiversity of a mowed lawn compare to that of a weedy field?

◀ FIGURE 4-1
Vanishing Amphibians These specimens represent some of the more than 100 amphibian species that have gone extinct since the 1970s.

4.1 What Is Biodiversity and Why Is It Important?

CORE IDEAS AND SKILLS
- Describe the four components of biodiversity.
- Explain how biodiversity leads to more resilient ecosystems.
- Understand how biodiversity relates to natural capital.

KEY TERMS

biodiversity
species diversity
genetic diversity
ecosystem diversity
biome
functional diversity
insurance hypothesis

Biodiversity Is the Variety of Life

In one form or another, life has taken hold virtually everywhere on Earth, from seemingly barren glaciers and deserts to colorful coral reefs and lush rain forests. In the Chapter 3 Case Study, you read about tropical rain forests and the incredible variety of life they support. Tropical rain forests cover only about 7% of Earth's land surface but contain half of the plant and animal species found on land. To put this in perspective, consider the diverse plant and animal life in Serra Bonita Reserve, established by National Geographic Explorer Vitor Becker. The reserve covers an area roughly one-third the size of Manhattan. Yet 458 species of trees grow within its borders. More than 350 species of birds take shelter in the reserve. So do a whopping 5,000 species of moths and butterflies—more moth and butterfly species than are found in all of North America. From flea-sized frogs to powerful pumas, Serra Bonita Reserve is packed with plants and animals of all shapes and sizes. As such, it is an area rich in biodiversity.

Biodiversity, or biological diversity, is the variety of life on Earth. It is the variety of Earth's species, the genetic material they contain, and the ecosystems in which they live. It is also the variety of ecosystem processes such as energy flow and nutrient cycling. Recall that biodiversity is one of three scientific factors of sustainability. It has four components.

Species Diversity The most common way to study biodiversity is to study **species diversity**. This component of biodiversity describes both the variety of species present in a specific ecosystem and their abundance within that ecosystem (Figure 4-2). To date, scientists have identified about 2 million species of organisms on Earth. They estimate the actual number of species most likely ranges from 7–10 million and may be as many as 100 million.

FIGURE 4-2
Species diversity describes the number of species in an ecosystem and their abundance there. How many different species of animals do you see at this watering hole in Kenya?

LESSON 4.1 109

FIGURE 4-3
Genetic diversity is demonstrated by the variations in color and banding patterns among members of the species *Polymita picta*, also known as the Cuban land snail.

Genetic Diversity The second component of biodiversity is **genetic diversity**, which is the variety of genes found in a population or in a species. Genes contain instructions, or codes, called genetic information, that give rise to specific traits, or characteristics. Those traits are then passed on to offspring through reproduction. Variations in traits contribute to genetic diversity (Figure 4-3). Species have a better chance of surviving and adapting to environmental changes if they have greater genetic diversity. There is a greater chance that individuals will possess favorable traits needed to survive, such as resistance to disease.

Ecosystem Diversity The third component of biodiversity is **ecosystem diversity**. This term refers to Earth's diversity of biological communities such as deserts, grasslands, forests, lakes, rivers, and wetlands. A number of ecosystems may be represented within larger geographic regions called **biomes**. Biomes are characterized by a distinct climate and certain species (particularly vegetation) that are able to survive there. Figure 4-4 shows the major biomes found across the midsection of North America. You will further study biomes in Chapter 6.

Functional Diversity The fourth component of biodiversity is **functional diversity**—the variety of processes such as energy flow and matter cycling that occur within ecosystems (Figure 3-11, Lesson 3.2). This component of biodiversity generally concerns the variety of ecological roles organisms play in their communities and the impact these roles have on the overall ecosystem. Examples may include where an organism fits into a food chain or web, what its feeding strategy is, and how its behavior affects the chemical and physical conditions of its environment.

Multiple species may share the same function, such as when two animals that are grazers share the same feeding strategy. If an ecosystem contains many species that share functional traits, that ecosystem will be better able to withstand some species loss without losing functionality. However, if an ecosystem has few species, each with different functional traits, species loss will have a greater impact on functionality of that ecosystem.

How does biodiversity relate to ecosystem productivity and stability? More biologically diverse ecosystems are more productive. With a greater variety of producer species, an ecosystem will produce more plant biomass, which in turn will support a greater variety of consumer species. More biologically diverse ecosystems are also more stable.

An ecological concept known as the **insurance hypothesis** states that "biodiversity ensures ecosystems against a decline in their functioning because many species provide greater guarantees of functioning even if others fail." Ecosystems that are biologically diverse are more likely to include species with traits that enable them to adapt to changes in the environment, such as disease or drought. Such species could help an ecosystem maintain its resilience even if other species are lost.

checkpoint What is the significance of the insurance hypothesis?

Biodiversity Builds Natural Capital

Biodiversity is vital to maintaining the natural capital that helps keep humans alive and supports their economies. A 2020 report by the World Economic Forum noted that more than half the world's gross domestic product is "moderately or highly dependent on nature." Biodiversity supports global food security and provides raw materials for other resources such as medicine, building materials, and fuel.

Biodiversity enhances natural ecosystem services such as air and water purification, soil regeneration, decomposition of wastes, and pollination. It also helps offset natural threats such as flooding, climate change, and the spread of disease. It even has cultural value—certain species are associated with spiritual practices or national identity, for example. Finally, Earth's variety of genetic information, species, and ecosystems provides raw materials for the evolution of new species and ecosystem services, as species respond to changing environmental conditions.

checkpoint How do humans make use of biodiversity resources?

FIGURE 4-4 ▼

North America's Biomes The midsection of North America contains a variety of biomes, each with its own characteristic climate and mix of species.

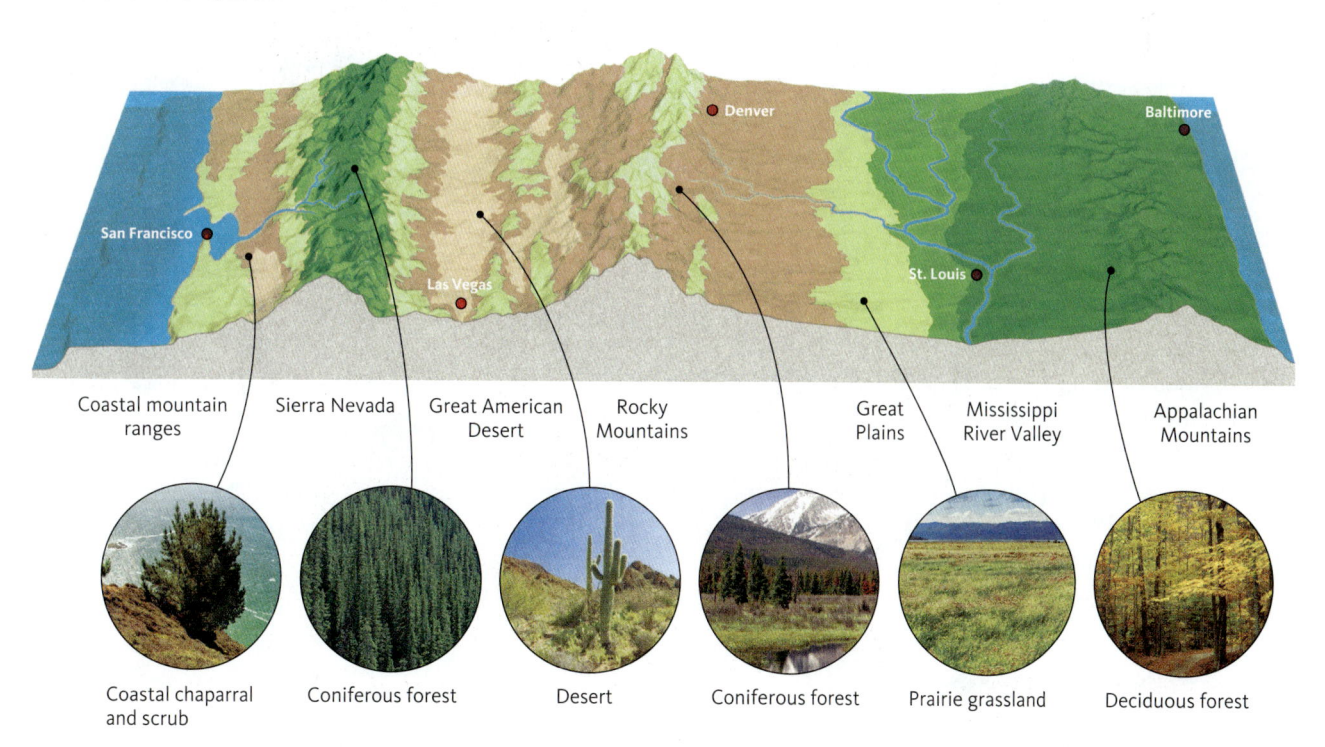

Coastal mountain ranges	Sierra Nevada	Great American Desert	Rocky Mountains	Great Plains	Mississippi River Valley	Appalachian Mountains
Coastal chaparral and scrub	Coniferous forest	Desert	Coniferous forest	Prairie grassland		Deciduous forest

LESSON 4.1   111

112　CHAPTER 4　BIODIVERSITY AND EVOLUTION

FIGURE 4-5
ON ASSIGNMENT National Geographic photographer David Liittschwager found a creative way to measure species diversity in five different terrestrial and aquatic ecosystems around the world. Liittschwager built a one-cubic-foot metal frame, which he left in place for several weeks at each location. With help from biologists, Liittschwager identified and photographed all organisms he observed passing through the cube during that time—even those barely more than a millimeter in size. The species shown here were photographed at a coral reef in French Polynesia. What does this collage of species suggest about the biodiversity of this ecosystem?

LESSON 4.1

FIGURE 4-6
These dung beetles perform an important ecosystem service. They aid in the decomposition of animal waste, which releases nutrients back into the soil.

SCIENCE FOCUS 4.1

INSECTS AND ECOSYSTEM SERVICES

Many people tend to consider insects as pests based on their experience with them. Insects may compete with us for food. They may bite or sting us or even infect us with diseases such as malaria. They may invade our lawns, gardens, and houses.

But insects also play a number of vital roles in sustaining life on Earth. They are a source of food for myriad animals—as well as people in some parts of the world. Insects also provide a number of ecosystem services:

- Pollination is a crucial environmental service that allows flowering plants to reproduce. When pollen grains from the flower of one plant are transported to the receptive part of another plant of the same species, reproduction occurs. Some 80% of flowering plant species depend on animals for pollination, and the vast majority of these pollinators are insects.
- Insects may improve soil health. Insects that are decomposers or detritivores help return organic matter and nutrients to soil (Figure 4-6). Insect activities may also loosen topsoil. This helps aerate the soil and improves water drainage.
- Insects that eat other insects provide natural pest control. They limit populations of at least half the insect species considered to be pests.

Certain environmental changes around the world are threatening insect populations and their ecosystem services. Human activities are causing some of the changes. Entomologists—scientists who study insects—are seeking to better understand impacts from these environmental threats.

One major threat that concerns entomologists is colony collapse disorder, a disease that is plaguing honeybees in the United States and elsewhere. Honeybees are extremely important pollinators. Their decline threatens to disrupt whole ecosystems that depend on bees for pollination. It will also disrupt the production of crops important to our food supply. This serious environmental problem is discussed more fully in Chapter 7.

Thinking Critically

Analyze How are insects both a threat and an aid to a sustainable food supply for humans?

4.1 Assessment

1. **Generalize** How does biodiversity help sustain life on Earth?
2. **Analyze** How might a reduction in species diversity affect the other three components of biodiversity?
3. **Synthesize** In your own words, explain why scientists believe that threats to amphibians present a warning about a number of environmental threats to Earth's biodiversity. Refer to the Case Study.

SCIENCE AND ENGINEERING PRACTICES

4. **Asking Questions** Suppose researchers are studying biodiversity in tide pool ecosystems. The researchers study two tide pools and identify 10 tide pool species at each site. Develop a set of questions researchers could investigate to compare species diversity, genetic diversity, and functional diversity for both sites.

4.2 What Roles Do Species Play in Ecosystems?

CORE IDEAS AND SKILLS

- Describe how a species fills an ecological niche within its ecosystem.
- Explain how ecologists classify native, nonnative, keystone, and indicator species.
- Understand how threats to keystone species and indicator species in turn threaten the ecosystems they inhabit.

KEY TERMS

ecological niche
habitat
generalist species
specialist species
native species
nonnative species
keystone species
indicator species

Each Species Plays a Role in Its Ecosystem

Species and Niches Each species has a specific role within the ecosystem it inhabits. Ecologists describe the role that a species plays in an ecosystem as its **ecological niche**, or simply its niche. A niche is a species' way of life in its community and includes all factors that affect its survival and reproduction. Those factors might be how much water and sunlight a species needs, how much space it requires, what it feeds on, what feeds on it, and the temperature and other conditions it can tolerate. A species' niche should not be confused with its **habitat**. A habitat is the place, or type of ecosystem, in which a species lives and obtains what it needs to survive.

Ecologists can use a species' ecological niche to classify it as being mainly a generalist or specialist. A **generalist species**, such as the raccoon, has a broad niche (Figure 4-7, right curve). Generalists can live in many different places, eat a variety of foods, and often tolerate a wide range of environmental conditions. Examples of other generalist species include American crows, coyotes, house sparrows, and white-tailed deer.

In contrast, a **specialist species**, such as the giant panda, occupies a narrow niche (Figure 4-7, left curve). Specialists may be able to live in only one type of habitat, eat only one or a few types of food, or tolerate a narrow range of environmental conditions. Some shorebirds, such as the American avocet, are also specialist species. Not only are they selective about the food items they eat, they are also selective about the areas where they forage along beaches and their adjoining coastal wetlands (Figure 4-8).

Because of their narrow niches, specialists are more prone to becoming endangered or extinct when environmental conditions change. For example, China's giant panda is highly endangered due to a combination of habitat loss, low birth rate, and a specialized diet that mainly consists of bamboo.

FIGURE 4-7
Ecological Niches Specialist species, such as the giant panda, have a narrow niche (left curve). Generalist species, such as the raccoon, have a broad niche (right curve).

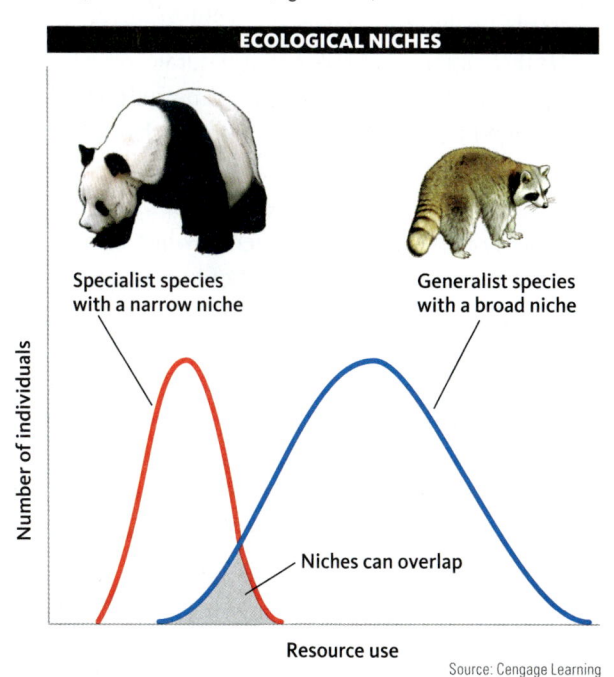

ECOLOGICAL NICHES

Specialist species with a narrow niche

Generalist species with a broad niche

Niches can overlap

Number of individuals

Resource use

Source: Cengage Learning

LESSON 4.2

FIGURE 4-8 ▼
Feeding Niches Various bird species in a coastal wetland occupy specialized feeding niches. This specialization reduces competition and allows sharing of limited resources.

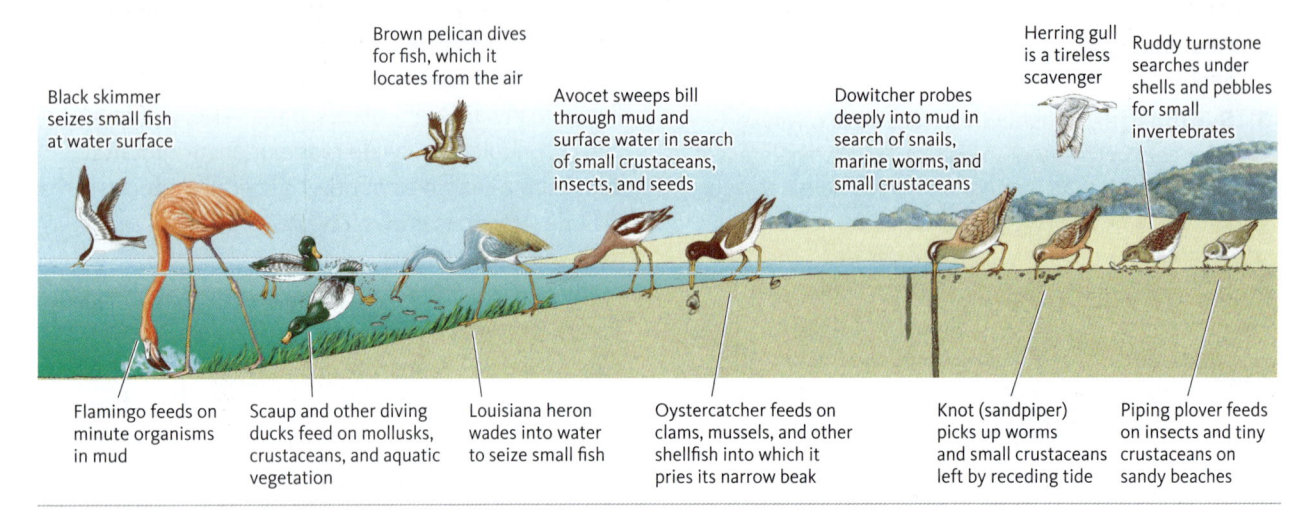

Brown pelican dives for fish, which it locates from the air

Black skimmer seizes small fish at water surface

Avocet sweeps bill through mud and surface water in search of small crustaceans, insects, and seeds

Dowitcher probes deeply into mud in search of snails, marine worms, and small crustaceans

Herring gull is a tireless scavenger

Ruddy turnstone searches under shells and pebbles for small invertebrates

Flamingo feeds on minute organisms in mud

Scaup and other diving ducks feed on mollusks, crustaceans, and aquatic vegetation

Louisiana heron wades into water to seize small fish

Oystercatcher feeds on clams, mussels, and other shellfish into which it pries its narrow beak

Knot (sandpiper) picks up worms and small crustaceans left by receding tide

Piping plover feeds on insects and tiny crustaceans on sandy beaches

Is it better to be a generalist or a specialist species? It depends. When environmental conditions are fairly constant, specialists have an advantage because they have fewer competitors. Under rapidly changing environmental conditions, the more adaptable generalist usually is better off.

Species and Ecosystems Ecologists use three metrics to further classify organisms. The first is their history within ecosystems (native and nonnative species). The second is the impact their behavior has on maintaining healthy ecosystems (keystone species). The third is what their presence may indicate about environmental conditions within ecosystems (indicator species).

Native species are those that naturally originated in a given ecosystem and have become suited to the environmental conditions there. **Nonnative species** are those that are introduced to an ecosystem in which they have not occurred previously. This introduction may be natural, such as when a species migrates. Or, it may be the result of human actions, either accidental or deliberate. Nonnative species are also referred to as alien species, exotic species, or invasive species. Invasive species are those that disrupt ecosystems and replace native species.

People may assume nonnative species always harm their new environments. This is incorrect. For example, tomatoes, chickens, cattle, and fish, were all introduced to the United States. As food sources, these nonnative species certainly benefit humans.

However, nonnative species can compete with and disrupt an ecosystem's native species, often causing unintended or unexpected consequences. They are then considered invasive species. In 1956, Brazil imported wild African honeybees in order to breed them with native honeybees. The hope was to create a honeybee that could better survive Brazil's warm climate, which would in turn increase honey production. The opposite occurred. The more aggressive African bees displaced some of Brazil's native honeybee populations, which led to a reduced honey supply. African honeybees have since spread across South and Central America and into Mexico and the southern United States. They have killed thousands of domesticated animals and approximately 1,000 people in the Western Hemisphere, many of them allergic to bee stings.

This example demonstrates how nonnative species can spread rapidly if they find a new location with favorable conditions. In their new niches, some of these species may not face the predators and diseases they faced in their native niches. They may outcompete some native species in their new locations. You will learn more about invasive species in Chapter 8.

A **keystone species** is an organism that plays a unique role in the way an ecosystem functions. Without keystone species, the ecosystem would be dramatically different or cease to exist.

Keystone species play several critical roles in helping to sustain ecosystems. For example, mountain lions and wolves help control the distribution and population of prey animals such as deer and rabbits. They influence where these species feed and raise their young. Sea otters help sustain kelp forests by feeding on sea urchins, which destroy kelp plants as they graze on them. Without keystone predators to limit the size of certain prey populations, those populations can explode. The entire ecosystem can then collapse. (See the Chapter 5 Case Study for more on the sea otter as a keystone species.)

The loss of a keystone species can lead to population crashes and extinctions of other species in a community that depends on them for certain ecosystem services. This is why it is important for scientists to identify and protect keystone species.

American Alligator as a Keystone Species

The American alligator is a keystone species in subtropical wetland ecosystems in the southeastern United States. American alligators play several important ecological roles. They dig deep depressions, also known as gator holes. These depressions hold fresh water during dry periods and serve as refuges for aquatic life. The depressions supply fresh water and food for fish, insects, snakes, turtles, birds, and other animals. The large nesting mounds alligators build also provide nesting and feeding sites for herons and egrets. Red-bellied turtles lay their eggs in old alligator nests as well.

When alligators excavate holes and build nesting mounds, these activities also prevent vegetation from invading shorelines and open-water areas. Without this ecosystem service, freshwater ponds and coastal wetlands fill in with shrubs and trees. This transition to heavy vegetation would cause dozens of species to disappear from these ecosystems. Likewise, alligators help insure the presence of game fish such as bass and bream by eating large numbers of gar, a predatory fish that hunts these species.

In the 1930s, hunters began killing large numbers of American alligators for their exotic meat and soft belly skin. People used the skin to make expensive shoes, belts, and purses. Other people hunted the alligators for sport. By the 1960s, hunters and poachers had wiped out 90% of the American alligators in Louisiana. The American alligator population in the Florida Everglades also was near extinction.

In 1967, the U.S. government placed the American alligator on the endangered species list. By 1987, because it was protected, its populations had made a strong comeback and the American alligator was removed from the endangered species list. Today, there are more than 1 million of these alligators in Florida. The state now allows property owners to kill alligators that stray onto their land.

To conservation biologists, the comeback of the American alligator is an important success story in wildlife conservation. Recently, however, large and rapidly reproducing Burmese and African pythons have invaded the Florida Everglades. These snakes, released deliberately or accidentally by humans, are hurting this ecosystem. The nonnative invaders feed on young alligators, threatening the long-term survival of this keystone species of the Everglades. Threats posed by the spread of Burmese pythons are further discussed in Chapter 7.

Sharks as Keystone Species

As keystone species, certain sharks play crucial roles in keeping their ecosystems functioning. Sharks feeding at or near the tops of their food webs remove injured and sick animals. Without this ecosystem service, the oceans would teem with dead and dying fish and marine mammals. Shark activity also influences the distribution and feeding habits of other species, which helps maintain balance in marine ecosystems.

Media reports on shark attacks greatly exaggerate the dangers that sharks pose to humans. Every year, members of a few species, such as great white, bull, and tiger sharks, injure 60 to 80 people. In a typical year, sharks kill 6 to 10 people worldwide. However, for every shark that injures or kills one person, people kill about 1.2 million sharks. Sharks are harvested for their liver oil, meat, hides, jaws, and fins. Humans also kill sharks out of fear.

Shark fin soup is considered a delicacy in some countries and demand for fins drives the majority of shark fishing. As many as 73 million sharks are killed each year for their fins, a practice called shark finning. Fishermen capture sharks, cut off their fins, and then throw the sharks back into the ocean. These sharks either bleed to death or drown because they can no longer swim.

As of 2021, 31% of the world's shark species were reported to be threatened, primarily because of overfishing. Because some sharks are keystone species, their extinction can threaten the ecosystems they inhabit and their associated ecosystem services. Sharks are especially vulnerable to population declines because they grow slowly, mature late, and have only a few offspring per generation. Today, sharks are among the world's most vulnerable and least-protected animals.

Fortunately, campaigns to increase public awareness and to encourage bans on the transport and sale of shark fins have led to declines in consumer demand, including in Hong Kong, the world's largest supplier of shark fins (Figure 4-9).

Ecologists identify **indicator species** as those whose presence or absence indicates the quality or characteristics of certain environmental conditions. For example, certain species may be indicators for levels of water or air pollution. Indicator species are sensitive to environmental changes and are affected almost immediately by damage to their ecosystems. As such, they can provide an early warning that environmental conditions are deteriorating.

Frogs as Indicator Species In this chapter's Case Study, you learned that some frogs and other amphibians are classified as indicator species. Scientists are trying to determine why certain frog populations are declining and how factors affecting amphibians could affect other species. They have identified several natural and human-related factors that are likely acting in combination to cause the decline and disappearance of these species.

Viral and fungal diseases are natural factors that impact frogs and other amphibians. Chytrid, the disease that Dr. Anna Savage is researching, is one example. Such diseases can spread easily because adults of many amphibian species gather in large numbers to breed.

Parasites are also natural factors that affect the survival of frogs and other amphibians. Parasites live on or inside other organisms. *Ribeiroia ondatrae*, a type of parasitic flatworm, burrows into tadpoles during part of its life cycle. This kills the tadpoles or causes them to become malformed later in their development (Figure 4-10).

Research has shown that human activities are also impacting frogs. For example, frog habitats have been fragmented, mainly as a result of deforestation and the destruction of freshwater wetlands for

CONSIDER THIS

A discovery made by middle school students spurred research on malformed frogs in Minnesota and beyond.

In the fall of 1995, students from Minnesota New Country School accompanied their teacher on a field trip to study pond biology at a nearby nature center. Students quickly noticed there was something wrong with the leopard frogs they were catching. The frogs were missing hind limbs. Students collected 22 frogs that day, and 11 of them had the same deformity. Their teacher decided to contact the Minnesota Pollution Control Agency, and its biologists did a separate survey with similar results. When the students' discovery made the news, reports of malformed frogs began pouring in from around the state. At the same time, biologists began reporting similar trends in frog populations in other states and other countries. Today, research continues to determine what is causing an apparent rise in malformations of frogs.

FIGURE 4-9
Shark fins are laid out to dry atop a building in Hong Kong. Growing public backlash to shark finning has forced merchants to work in such out-of-sight places.

FIGURE 4-10
Parasitic flatworms caused this American bullfrog's malformation.

farming and urban development. Increased exposure to UV radiation is another human-caused problem affecting amphibians. High levels of UV radiation can harm frog embryos in shallow ponds as well as adults basking in the sun for warmth. Historically, ozone in the stratosphere screened UV radiation. Chemical pollutants released from human sources (e.g., spray cans of aerosol paint, hairspray, and furniture polish) have entered the stratosphere and destroyed some of that protective ozone. International action has been taken to limit the release of these chemicals, but it will take about 50 years for ozone levels to recover to levels observed in the 1960s.

Climate change is also a concern. Amphibians are sensitive to even slight changes in temperature and moisture patterns. These environmental "cues" influence the timing of life cycle events such as breeding, metamorphosis, and hibernation. Warming temperatures are causing some amphibian species to breed earlier in spring. Tadpoles hatch earlier, too, and are more vulnerable to cold weather events that can kill them. Extended dry periods also can lead to a decline in amphibian populations.

Amphibians' responses to habitat loss, pollution, and climate change can serve as a biological "smoke alarm," alerting scientists to changes in ecosystems that may be impacting amphibians as well as other species found there.

checkpoint How could the disappearance of a keystone species lead to the collapse of an ecosystem?

4.2 Assessment

1. **Evaluate** What is meant by the statement "a niche is a species' way of life?"
2. **Compare and Contrast** How is the loss of a keystone species from an ecosystem similar to and different from the loss of an indicator species?
3. **Classify** Would you describe yourself as belonging to a specialist species or a generalist species? Give reasons for your choice.

SCIENCE AND ENGINEERING PRACTICES

4. **Engaging in Argument** Some species of sharks are on the verge of extinction because of shark finning. Washington, California, Guam, and Hawai'i prohibit selling, trading, and possessing shark fins. Do you believe sharks should be protected? Should the United States and other countries support this ban? Why or why not?

LESSON 4.2

4.3 How Does Life on Earth Change Over Time?

CORE IDEAS AND SKILLS

- Explain the scientific theory of biological evolution.
- Describe genetic variability and natural selection as mechanisms for evolution.
- Understand that natural selection has limits.

KEY TERMS

biological evolution
natural selection
fossil
genetic variability
mutation
adaptation

Evolution Explains How Organisms Change Over Time

How did such an amazing diversity of species come about on Earth? The scientific answer is **biological evolution**, or simply evolution. Evolution is the process by which species genetically change over time. These changes occur within the genes of populations of organisms from generation to generation. According to this scientific theory, species have evolved from earlier, ancestral species through **natural selection**. Through this process, individuals with certain genetic traits are more likely to survive and reproduce under a specific set of environmental conditions. These individuals then pass these traits on to their offspring.

A large body of evidence supports this idea. As a result, biological evolution through natural selection is the most widely accepted scientific theory that explains how life has changed over the past 3.8 billion years, giving rise to today's diversity of life.

Most of what we know about the history of life on Earth comes from **fossils** (Figure 4-11). Fossils are the preserved remains or traces of prehistoric organisms. Fossils may be mineralized or petrified skeletons, bones, teeth, shells, leaves, and seeds. They may be the impressions in rock that such structures leave behind. Or they may be impressions left behind by animal activity such as tracks, trails, and burrows. Scientists have discovered fossil evidence in successive layers of sedimentary rock such as limestone and sandstone.

In addition to analyzing fossils excavated from sedimentary rocks, scientists have also studied evidence of ancient life contained in ice core samples drilled from glacial ice at Earth's Poles

FIGURE 4-11 ▼

Ancient Fossil This fossil shows the mineralized remains of an early ancestor of the present-day horse. It lived on Earth more than 35 million years ago. What other fossil do you see here?

and on mountaintops. And they have studied ancient organisms trapped in amber, or fossilized tree resin.

The total body of fossil evidence, known as the fossil record, is uneven and incomplete. Many past forms of life left no fossils and some fossils have decomposed. Scientists estimate the fossils found so far represent only 1% of all species that have ever lived. Many questions still remain about the details of evolution by natural selection, and research continues in this area.

checkpoint What evidence do scientists examine to better understand evolution?

Evolution Depends on Genetic Variability and Natural Selection

The idea that organisms change over time and are descended from a single common ancestor is not a new one. In fact, it has been discussed since the time of the early Greek philosophers. The first

NATIONAL GEOGRAPHIC | EXPLORERS AT WORK
Edward O. Wilson — Biologist and Author

As a boy growing up in the southeast, Edward O. Wilson (1929–2021) became interested in insects at the age of 9. He once said, "Every kid has a bug period. I never grew out of mine."

Before entering college, Wilson decided he would specialize in the study of ants. He became one of the world's experts on ants, unlocking secrets to their methods of communication and social behaviors.

Over time, Dr. Wilson steadily widened his focus to include the entire biosphere. Today, he is considered one of the greatest naturalists of his time and a leading authority on biodiversity.

One of Wilson's landmark works is *The Diversity of Life*, published in 1992. In that book, he presented the principles and practical issues of biodiversity more completely than anyone had to that point. Wilson spent the rest of his life writing and lecturing about the need for global conservation efforts. He also worked on Harvard University's *Encyclopedia of Life*, an online database for Earth's known and named species.

In 2011, Wilson became involved in efforts to restore Gorongosa National Park in Mozambique, Africa, which he called "ecologically the most diverse park in the world." Wilson led numerous expeditions to catalog the park's biodiversity.

During his career, Wilson won more than 100 national and international awards and wrote 28 books, two of which won the Pulitzer Prize for General Nonfiction. In 2013, at age 84, he received the National Geographic Society's highest award, The Hubbard Medal. About the importance of biodiversity, he wrote, "How can we save Earth's life forms from extinction if we don't even know what most of them are?. . .I like to call Earth a little known planet."

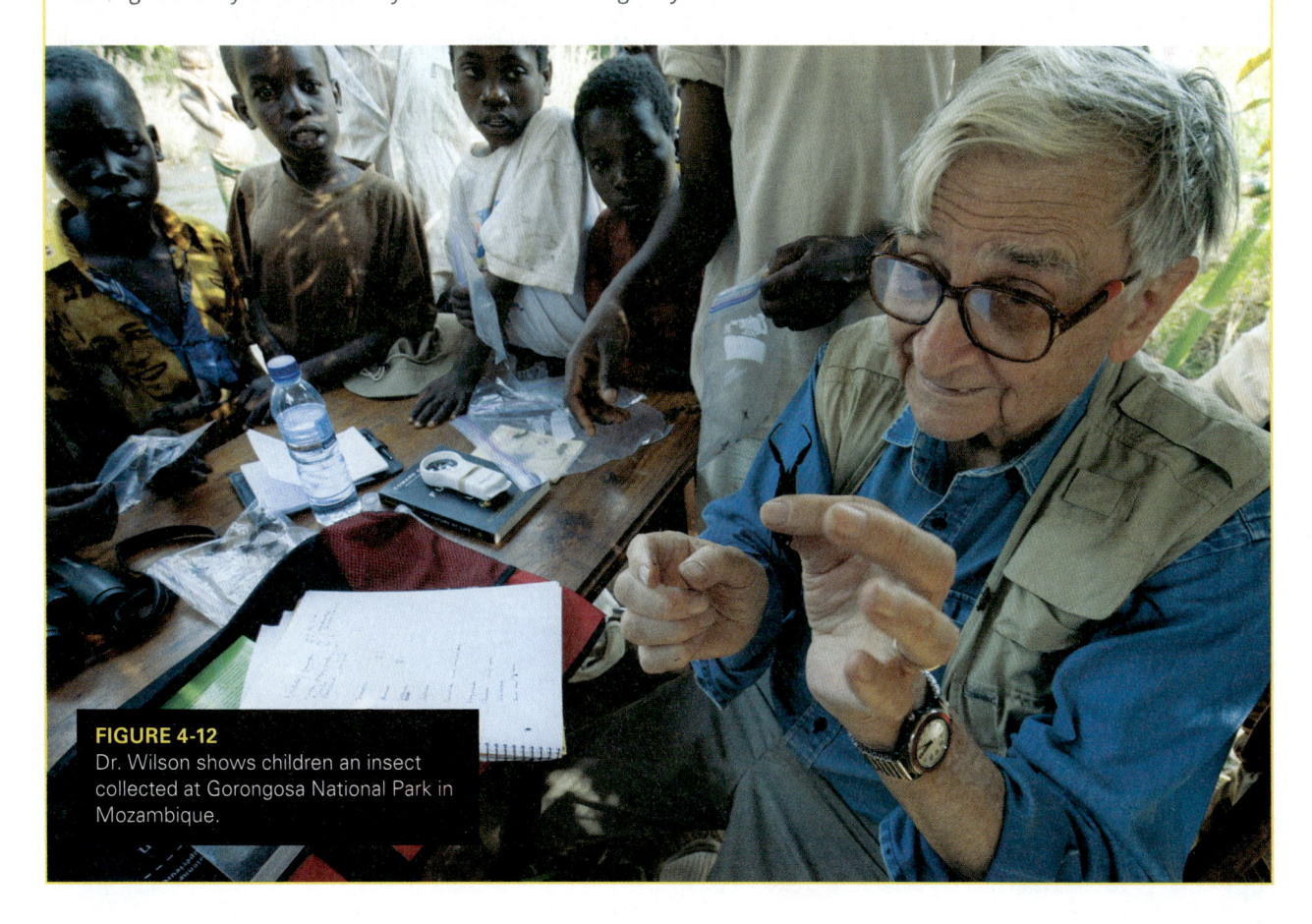

FIGURE 4-12
Dr. Wilson shows children an insect collected at Gorongosa National Park in Mozambique.

convincing explanation for evolution was put forward in 1858. Naturalists Charles Darwin (1809–1882) and Alfred Russel Wallace (1823–1913) independently proposed the concept of natural selection as a mechanism for biological evolution. Darwin gathered evidence for this idea and published it in his 1859 book *On the Origin of Species by Means of Natural Selection*.

Biological evolution by natural selection involves changes in a population's genetic makeup through successive generations. Recall that a population is a group of individuals of the same species living in a particular space. It is important to note that evolution does not apply to individual organisms, nor does it take place over the course of a single life cycle. Evolution occurs over the course of generations.

Genetic Variability The development of genetic variability is one factor necessary for evolution to occur. **Genetic variability** refers to variety in the genetic makeup of individuals in a population. It primarily results from mutations. A **mutation** is a permanent change in the DNA sequence within a gene in any cell inherited.

During an organism's lifetime, the DNA in its cells is copied each time one of its cells divides and whenever it reproduces—millions of times, in all. Most mutations result from random changes in the DNA's coded genetic instructions and occur in only a tiny fraction of these millions of divisions. Some mutations result from exposure to external agents such as radioactivity, UV radiation, and certain natural and human-made chemicals called mutagens.

Mutations can occur in any cell, but only those that take place in the genes of reproductive cells may be passed on to offspring. Sometimes a mutation can result in a new genetic trait, called a heritable trait, which can be passed from one generation to the next. In this way, populations develop genetic differences among their individuals. Some mutations are harmful to offspring, some are beneficial, and some have little or no harmful or beneficial effect at all.

Natural Selection The scientific concept of natural selection explains how populations evolve in response to changes in environmental conditions by changing their overall genetic makeup. Through natural selection, environmental conditions favor increased survival and reproduction of individuals that are better suited to those conditions. Such individuals possess an **adaptation**, or adaptive trait—any heritable trait that provides an individual with some advantage over the other individuals in a given population.

An example of natural selection at work is genetic resistance. Genetic resistance results when one or more organisms in a population are genetically predisposed to tolerate exposure to some condition that normally would be fatal. Genetic resistance can develop quickly in populations of organisms such as bacteria and insects that produce large numbers of offspring over short periods of time.

For example, certain bacteria have developed genetic resistance to widely used antibiotics (Figure 4-13). Often, when such drugs are used, a few bacteria that are genetically resistant to them survive. They rapidly produce more offspring than

FIGURE 4-13 ▼
Evolution by Natural Selection (a) A population of bacteria is exposed to an antibiotic, which (b) kills all individuals except those possessing a trait that makes them resistant to the drug. (c) The resistant bacteria survive and reproduce with other bacteria, eventually (d) replacing all or most of the nonresistant bacteria.

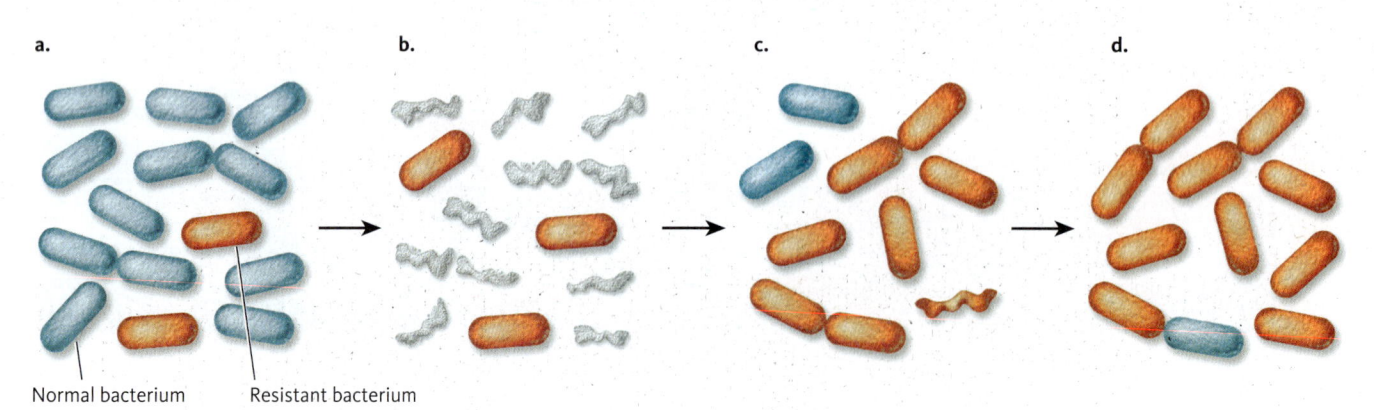

the bacteria that were killed by the drug could have produced. The genetically resistant bacteria keep reproducing while those that are susceptible to the drug die off. As a result, the antibiotic eventually loses its effectiveness.

Through natural selection, humans have evolved traits that have enabled them to survive in many environments and reproduce successfully. In fact, in the evolutionary blink of an eye, humans have developed powerful technologies and taken over much of Earth's net primary productivity.

Evolutionary biologists attribute humans' evolutionary success to three major adaptations:

- Strong opposable thumbs allowed humans to grip and use tools better than the few other animals that have thumbs.
- The ability to walk upright gave humans agility and freed up their hands for many uses.
- A complex brain allowed humans to develop many skills, including the ability to talk, read, and write in order to transmit complex ideas.

To summarize the process of biological evolution by natural selection: Genes mutate, certain individuals are selected, and populations evolve that are better adapted to survive and reproduce under existing environmental conditions.

Limits of Natural Selection In the not-too-distant future, will adaptations to new environmental conditions through natural selection protect us from harm? Will adaptations allow our descendants' skin to become more resistant to the harmful effects of UV radiation? Could future generations' lungs cope with air pollutants, and will their livers be able to detoxify pollutants in their bodies?

Scientists in this field say it is not likely because of two limitations on adaptation through natural selection. First, a change in environmental conditions can lead to such an adaptation only if genetic traits are already present in a population's gene pool or if they arise from mutations—which occur randomly.

Second, even if a heritable trait is present in a population, the pace at which a population can adapt may be limited by its reproductive capacity. Populations of genetically diverse species that reproduce quickly often adapt to a change in environmental conditions in a matter of days to years. Examples include organisms such as dandelions, mosquitoes, rats, bacteria, and cockroaches.

But some species cannot rapidly produce large numbers of offspring. Elephants, sharks, orangutans,

CONSIDER THIS

Are humans still evolving?

The answer is yes...and in some cases very rapidly by evolutionary standards. One example is the evolution of lactose tolerance. Lactose is a form of sugar found in milk. All humans are able to digest lactose as infants, when their diet includes breast milk. But until humans began keeping cattle some 10,000 years ago, weaned children did not drink milk. As a result, they would stop making the enzyme lactase, which breaks down lactose into simple sugars, and become "lactose intolerant" as a result.

After humans began herding cattle, being able to digest milk as an adult became very useful. This advantage led dairy farming populations in Europe, the Middle East, and Africa to evolve lactose tolerance independently through natural selection. Today, more than 90% of northern Europeans are lactose tolerant. Cultures not dependent on cattle, such as the Chinese and Inuit, remain lactose intolerant.

and humans are a few such species. It takes them thousands to millions of years to adapt through natural selection.

checkpoint What three adaptations have helped humans to be successful according to evolutionary biologists?

4.3 Assessment

1. **Summarize** How does evolution explain the diversity of life on Earth?
2. **Explain** How do mutation and natural selection influence the passing of heritable traits from one generation to the next?
3. **Recall** What are the two limits of natural selection?
4. **Evaluate** Charles Darwin said, "It is not the strongest of the species that survives, nor the most intelligent, but the one most responsive to *change.*" Explain what Darwin meant by this statement.

CROSSCUTTING CONCEPTS

5. **Cause and Effect** The Centers for Disease Control and Prevention claims overuse of antibiotics leads to antibiotic resistance in bacteria. How does this happen and what might be the consequences to human health?

4.4 What Factors Affect Biodiversity?

CORE IDEAS AND SKILLS

- Explain how speciation and extinction determine Earth's biodiversity.
- Understand how artificial selection and genetic engineering allow humans to select species' traits.

KEY TERMS

speciation
geographic isolation
reproductive isolation
biological extinction
endemic species
background extinction rate
mass extinction
artificial selection
genetic engineering
synthetic biology

Speciation Gives Rise to New Species

Under certain circumstances, natural selection can lead to an entirely new species. Through this process, called **speciation**, one species evolves into two or more different species. Speciation most commonly occurs in species that reproduce sexually. It can begin when distant migration or a physical barrier separates two or more populations of a given species. Genes are no longer exchanged between members of the two groups. Each population may begin to evolve in distinct ways, as it undergoes random mutations and adapts to the selective pressures associated with conditions of its particular environment. True speciation occurs when, even if members from the two populations come back into contact, they are unable to interbreed and produce fertile offspring.

Geographic Isolation When two different groups of the same species become physically isolated from one another, **geographic isolation** results. For example, part of a population may migrate in search of food and then begin living as a separate population in an area with different environmental conditions. Populations may be separated by a physical barrier, such as a mountain range or a valley. Or, populations may be separated due to a natural event, such as a hurricane, earthquake, or volcanic eruption. Even blowing winds or flowing water may carry a few individuals far away, where they establish a new population. Human activities, such as construction of dams or the clearing of forests, can also create physical barriers for certain species.

Reproductive Isolation Once populations have been geographically separated, the exchange of genes between them stops. This leads to **reproductive isolation**. Random mutations and changes in response to natural selection begin to operate independently within the respective gene pools of geographically isolated populations. If this process continues long enough, members of isolated populations can become very different in their genetic makeup. In fact, these populations may no longer interbreed and produce live, fertile offspring if the populations come together again. One species has become two (Figure 4-14).

checkpoint What evidence do scientists use to conclude that speciation has truly occurred?

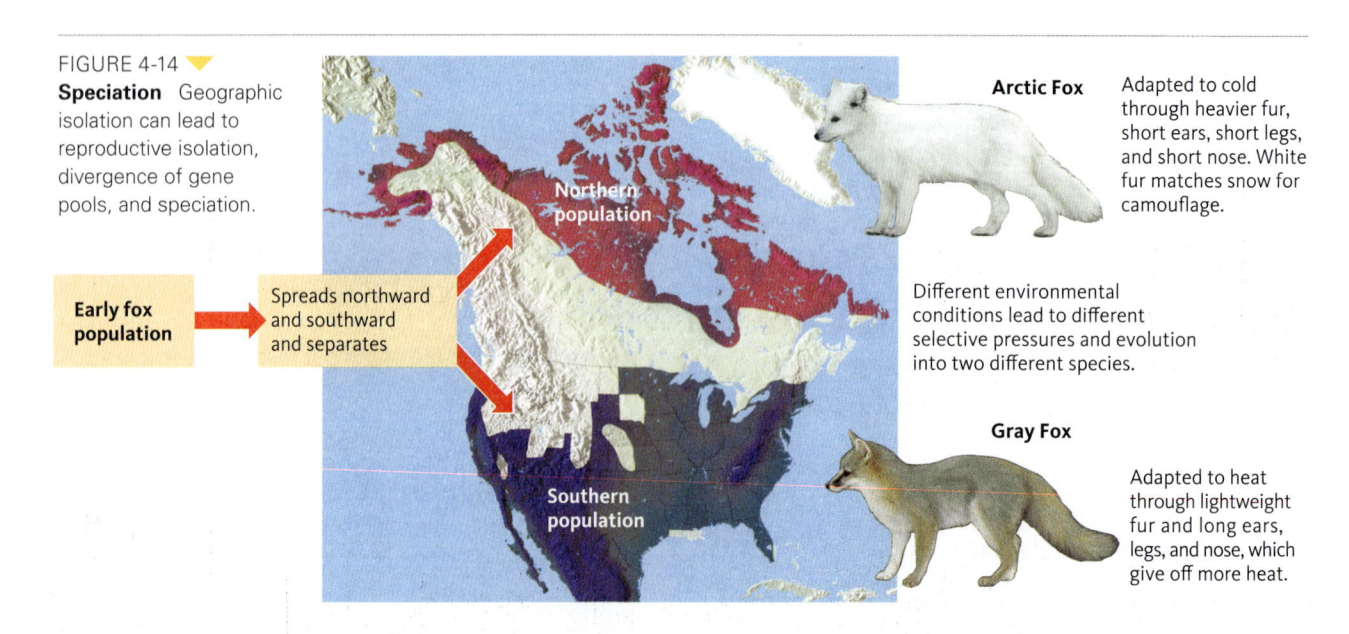

FIGURE 4-14 ▼
Speciation Geographic isolation can lead to reproductive isolation, divergence of gene pools, and speciation.

Early fox population → Spreads northward and southward and separates

Northern population

Southern population

Arctic Fox Adapted to cold through heavier fur, short ears, short legs, and short nose. White fur matches snow for camouflage.

Different environmental conditions lead to different selective pressures and evolution into two different species.

Gray Fox Adapted to heat through lightweight fur and long ears, legs, and nose, which give off more heat.

124   CHAPTER 4   BIODIVERSITY AND EVOLUTION

Extinction Eliminates Species

Another factor affecting Earth's biodiversity is biological extinction. **Biological extinction**, or simply extinction, occurs when an entire species ceases to exist (Figure 4-15). When environmental conditions change, a species must adapt to survive and reproduce in response. Or, it must move to a new environment with more favorable conditions. Otherwise, it will eventually become extinct.

Endemic species are species that are found in only one area and that have highly specialized roles. These organisms may be found on islands or other isolated areas and in other unique areas such as tropical rain forests. Endemic species are especially vulnerable to extinction. They are unlikely to be able to migrate or adapt in the face of rapidly changing environmental conditions. For example, the golden toad became extinct in 1989 even though it lived in the well-protected Monteverde Cloud Forest Reserve in the mountains of Costa Rica. Scientists believe the most likely reason for this species' extinction is that it could not adapt to the severe drought conditions that affected its habitat.

Fossils and other scientific evidence indicate that all species eventually become extinct. In fact, evidence indicates that 99.9% of all the species that have existed on Earth are now extinct. Throughout most of Earth's long history, species have disappeared at a naturally low rate, called the **background extinction rate**. Based on the fossil record and analysis of ice cores, biologists estimate that the average annual background extinction rate has been less than one species lost annually for every 1 million species on Earth.

Evidence indicates that life on Earth has been sharply reduced by periods of mass extinction. A **mass extinction** involves a significant rise in extinction rates well above the background rate. Fossil and geological evidence indicate Earth has experienced five mass extinctions during the past 500 million years. Each of these events wiped out approximately 50–90% or more of existing species.

The causes of mass extinctions are unknown. Possible events include volcanic eruptions and collisions with giant meteors and asteroids. Such events would trigger drastic environmental changes on a global scale. For example, massive amounts of debris might be released into the atmosphere that would block sunlight for an extended period of time. Toxic or heat-trapping gases might also be released, the latter of which could spur rapid global warming.

A mass extinction provides an opportunity for the evolution of new species that can fill unoccupied ecological niches or newly created ones. Scientific evidence indicates that each past mass extinction was followed by an increase in species diversity. This happened over the course of several million years.

The balance over time between speciation and extinction determines Earth's biodiversity. The existence of millions of species today means speciation, on average, has kept ahead of extinction.

FIGURE 4-15
The dusky seaside sparrow became extinct in 1987, when this male—the last of his kind—died in captivity.

However, evidence indicates the global extinction rate is rising dramatically. Many scientists argue that we are actually entering a sixth mass extinction. There is also considerable evidence that human activity is the primary cause, especially activity that degrades or destroys habitats.

checkpoint What is the background extinction rate?

Humans Select Species' Traits

Humans have developed methods to selectively manipulate and combine genes of species in order to enhance genetic traits that better meet a particular human need. In this way, humans have learned to bypass natural selection. And while the outcome of natural selection is the emergence of traits that allow

SCIENCE FOCUS 4.2

GEOLOGICAL PROCESSES AND BIODIVERSITY

Earth's surface has changed dramatically over its long history. Scientists have discovered that huge flows of molten rock within the planet's interior have broken its surface into a number of gigantic solid plates. Known as tectonic plates, these rigid slabs of solid rock underlay Earth's oceans and continents. For hundreds of millions of years, these plates have been moving very slowly, gliding over the planet's mantle.

Rock and fossil evidence indicates that 200–250 million years ago, all of Earth's present-day continents were connected. They formed a supercontinent called Pangaea (Figure 4-16, left). About 135 million years ago, Pangaea began to slowly split apart as Earth's tectonic plates moved. Ultimately, that splitting led to the present-day locations of the continents (Figure 4-16, right).

The movement of tectonic plates has had two important effects on the evolution and distribution of life on Earth. First, the locations of the continents and oceanic basins have greatly influenced Earth's climate. This, in turn, has helped to determine where plants and animals can live. Second, as tectonic plate movement has caused landmasses to break apart or join together throughout Earth's history, species have been able to move, adapt to new environments, and form new species through speciation.

Along boundaries where they meet, tectonic plates may pull away from, collide into, or slide past each other. Tremendous forces produced by these interactions along plate boundaries can lead to earthquakes and volcanic eruptions. These geological activities can also affect biological evolution. Earthquakes cause fissures in Earth's crust that, on rare occasions, can separate and isolate populations of species. That can lead to speciation. Volcanic eruptions can also affect extinction and speciation by destroying habitats and reducing, isolating, or wiping out populations of species. These processes are further discussed in Chapter 11.

Thinking Critically
Analyze If Earth's tectonic plates reversed their movements and started returning to a supercontinent, how might this affect the planet's current biodiversity? Explain your conclusion.

FIGURE 4-16
Shifting Continents Over millions of years, Earth's continents have moved very slowly on several gigantic tectonic plates.

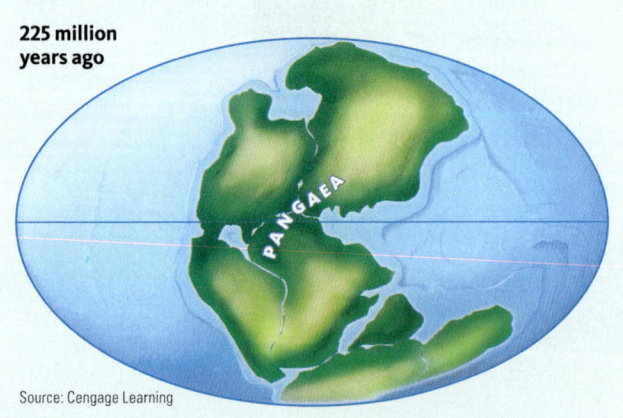

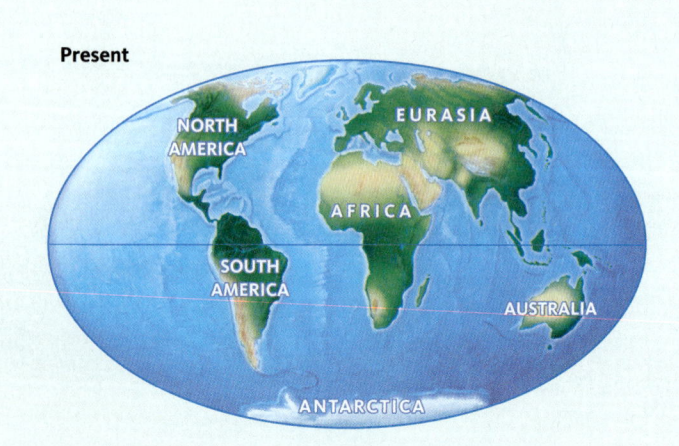

225 million years ago — PANGAEA

Present — NORTH AMERICA, EURASIA, AFRICA, SOUTH AMERICA, AUSTRALIA, ANTARCTICA

Source: Cengage Learning

for better survival of a species, this is not necessarily the goal of humans' selective methods.

Artificial Selection Through **artificial selection**, humans change the genetic characteristics of populations with similar genes. First, they select one or more desirable genetic traits that already exist in the population of a plant or animal. Then they control which members of a population have the opportunity to reproduce. They use selective breeding, or crossbreeding, to increase the numbers of individuals in a population with these desired traits.

Crossbreeding is a practice that dates back some 10,000 years, to the time when humans began to herd animals and cultivate crops. Most of the grains, fruits, and vegetables we eat have been produced by artificial selection. Artificial selection also has given rise to food crops with higher yields and cows that produce more milk.

As mentioned earlier, artificial selection may be at odds with natural selection. It may favor traits that would not naturally be expressed, as they do not contribute to a species' survival. Artificial selection accounts for the fact that a breed of cat may be hairless or a breed of horse may be miniature in size. It is the mechanism by which a single species of wild mustard plant has given rise to Brussels sprouts, cabbage, cauliflower, broccoli, and kale.

Artificial selection is limited to crossbreeding between genetic varieties of the same species or between species that are genetically similar. Artificial selection is also limited by the time it takes animals to grow and reproduce. It may be a slow process, taking years or even thousands of years.

Genetic Engineering Now scientists also use **genetic engineering** to rapidly manipulate genes in order to select desirable traits or eliminate undesirable ones. Scientists are able to alter an organism's genetic material by adding, deleting, or changing segments of its DNA. This process is called gene splicing. Genetic engineering also enables scientists to transfer genes between different species that would not interbreed in nature.

Genetic engineering techniques have been applied to a variety of industries, most notably food and medicine. For example, corn and soybeans have been genetically engineered to better resist pests and disease. Commercial species of fish have been genetically engineered to grow faster and bigger. Bacteria have even been genetically engineered to be able to digest toxic metals.

New organisms created by genetic engineering are called genetically modified organisms (GMOs). According to the U.S. Department of Agriculture (USDA), at least 80% of food products on supermarket shelves contain some form of genetically engineered ingredients. The USDA also states that GMOs pose no health safety threat, although this conclusion is controversial. Consumer concern remains regarding the presence of GMOs in food, and many shoppers actively look for products labeled as "non-GMO" or "100% organic." To further aid consumers, a law was passed in 2018 called the National Bioengineered Food Disclosure Standard (NBFDS). It requires a "bioengineered" label for products containing GMOs and applies to most national food manufacturers and importers. They must be in full compliance as of 2022.

A new and rapidly growing form of genetic engineering is **synthetic biology**. Whereas many genetically modified crops today contain a single engineered gene, synthetic biology makes it easier to generate larger clusters of genes and gene parts. This technology enables scientists to make new sequences of DNA. They use genetic information to design and create cells, tissues, organisms, and devices, and redesign existing natural biological systems. Much of the activity in synthetic biology is in the fields of pharmaceuticals, diagnostic tools, chemistry, and energy products such as biofuel (Chapter 13).

checkpoint How does artificial selection differ from natural selection?

4.4 Assessment

1. **Explain** What are three possible outcomes for species when rapid environmental changes occur?
2. **Compare and Contrast** How are the results of artificial selection and genetic engineering alike and different?

SCIENCE AND ENGINEERING PRACTICES

3. **Engaging in Argument** How do farmers benefit from growing GMO crops? Do consumers benefit in any way? Do these benefits warrant the use of GMO crops or not? Use the text and your own research to develop and support your argument.

CROSSCUTTING CONCEPTS

4. **Stability and Change** Describe how the balance between speciation and extinction determines Earth's biodiversity and what happens when extinction outpaces speciation.

TYING IT ALL TOGETHER STEM
Piecing Together the Puzzle of Chytrid Disease

In this chapter, you learned about chytrid, a disease that threatens many amphibian species worldwide. You also learned that genetic diversity may allow species to better respond to disease by enabling them to adapt tolerance for or resistance to disease over time.

Dr. Anna Savage is studying lowland leopard frogs to see if some populations may have adapted, through the process of evolution by natural selection, to have genetic disease resistance to chytrid. She has focused on immune system genes in these frogs, which control their ability to identify a disease agent and produce an immune response to it.

Specifically, Savage is studying the different *alleles* of the genes. Alleles are different versions of a gene with slightly different DNA sequences, which are expressed in slightly different ways. (Think hair color or eye color, for example.) Savage wants to know if frogs with particular alleles of immune system genes are better or worse at fighting off chytrid.

In one study, Savage collected skin swabs from different populations of lowland leopard frogs in Arizona. She analyzed the skin swabs to see if the fungus that causes chytrid, *Batrachochytrium dendrobatidis* (*Bd*), was present in these populations, either as infection or mortality (death) from infection (Figure 4-17). She also performed DNA analysis on tissue samples from members of the same populations to determine their genetic makeup. Savage studied "allelic richness," or variability, of the immune system genes (Figure 4-18). **Use the figures to answer the questions that follow.**

1. Anna Savage is studying adaptation through natural selection. Define this process and explain its importance to evolution.
2. Examine the colored bars in Figure 4-17. Savage categorized each frog population she sampled as susceptible (black), tolerant (gray), or uninfected (white). Explain how Savage decided to assign each population to its given category based on her results.
3. If the "AS" and "HS" populations are uninfected, what can be said about the *Bd* fungus in these populations?
4. Examine Figure 4-18. What relationship does the data suggest exists between allelic richness for immune system genes and tolerance for chytrid?
5. Lab experiments have enabled Anna Savage to identify a specific allele that appears responsible for tolerance to chytrid. What method described in this chapter might be used to intentionally increase the presence of this gene in the populations she studies?

FIGURE 4-17 ▼
Infection and Mortality This bar graph shows the prevalence of infection by the *Bd* fungus in different populations of lowland leopard frogs (abbreviations on *x*-axis). *Prevalence* refers to the proportion of each population infected with the disease (*y*-axis). Data is broken down to show infection prevalence and mortality prevalence. Sample size for each population is shown above each bar.

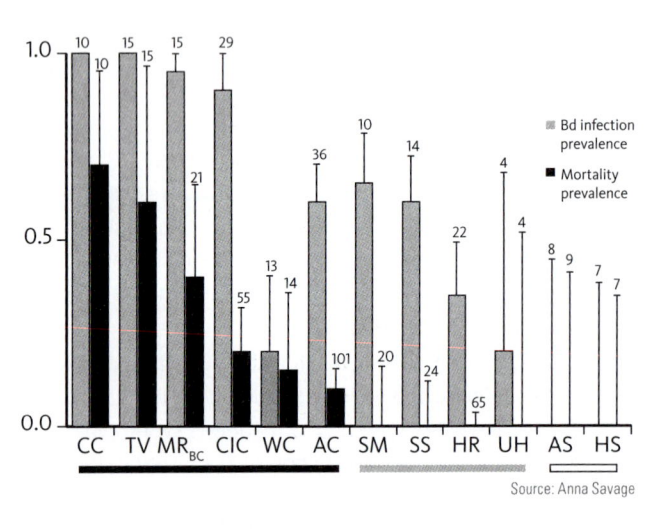

FIGURE 4-18 ▼
Genetic Variability This bar graph shows the results of one aspect of DNA analysis performed on different populations of lowland leopard frogs (abbreviations on *x*-axis). *Allelic richness* (*y*-axis) indicates variation in the immune system gene. Sample size for each population is shown above each bar.

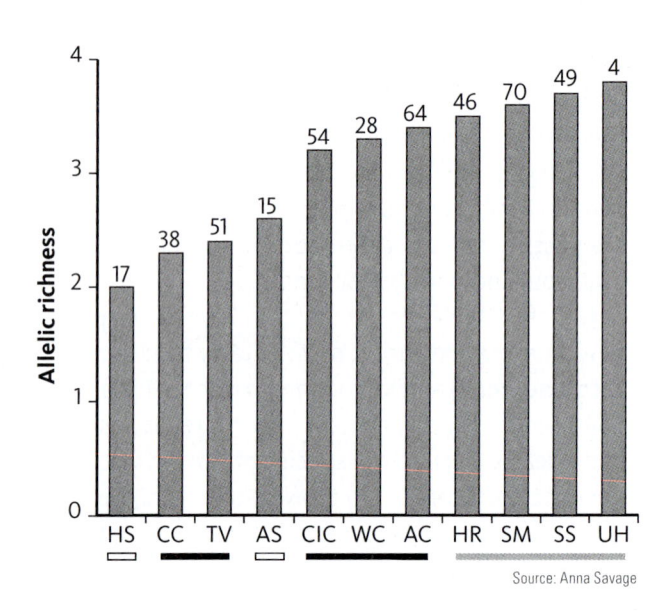

CHAPTER 4 SUMMARY

4.1 What is biodiversity and why is it important?

- Biodiversity is the variety of life on Earth. The four components of biodiversity are species diversity, genetic diversity, ecosystem diversity, and functional diversity.
- According to the inheritance hypothesis, biodiversity makes ecosystems more resilient by increasing the odds that species will be present that can respond to changes there.
- Biodiversity provides natural resources and ecosystem services that help build natural capital.

4.2 What roles do species play in ecosystems?

- Each species plays a specific ecological role called its niche. Within their niches, species are classified as generalists or specialists. Within ecosystems, species may be classified as native, nonnative, keystone, or indicator species.
- Native species are those that naturally originated in a given ecosystem and have become adapted to the environmental conditions there. Nonnative species are those that are introduced or migrate to an ecosystem in which they have not occurred previously.
- Keystone species play critical roles in sustaining ecosystems. Their loss can lead to population crashes and extinctions of other species. Indicator species are those whose presence or absence indicates the quality of certain environmental conditions.

4.3 How does life on Earth change over time?

- Biological evolution through natural selection is the widely accepted scientific theory that explains how life has changed over the past 3.8 billion years and has led to the diversity of life today. Most evidence for the history of life on Earth comes from the study of fossils, which are the preserved remains or traces of prehistoric organisms.
- Evolution depends on genetic variability, which refers to variety in the genetic makeup of individuals in a population as well as the likelihood of genes changing in response to environmental or genetic factors. Genetic variability primarily results from mutations.
- Evolution also depends on natural selection, a process whereby organisms that are better adapted to their environment are more likely to survive, reproduce, and pass their genetic traits on to offspring.

4.4 What factors affect biodiversity?

- Speciation occurs when one species evolves into two or more different species that can no longer interbreed. For sexually reproducing organisms, this occurs after two populations of a species become geographically and reproductively isolated from each other over time.
- Extinction occurs when an entire species ceases to exist. Throughout most of Earth's history, species have disappeared at a naturally low background rate.
- Earth has experienced five mass extinctions during the past 500 million years, most likely due to catastrophic events such as volcanic eruptions or impacts from meteors or asteroids.
- Artificial selection involves selectively crossbreeding plants or animals to increase the presence of a desired genetic trait in a population. Genetic engineering involves altering segments of an organism's DNA. The altered gene can be transferred to another species.

CHAPTER 4 ASSESSMENT

Review Key Terms

Select the term that best fits the definition. Some terms will not be used.

adaptation	functional diversity	mass extinction
artificial selection	generalist species	mutation
background extinction rate	genetic diversity	native species
biodiversity	genetic engineering	natural selection
biological evolution	genetic variability	nonnative species
biological extinction	geographic isolation	reproductive isolation
biome	habitat	specialist species
ecological niche	indicator species	speciation
ecosystem diversity	insurance hypothesis	species diversity
endemic species	keystone species	synthetic biology
fossil		

1. The variety of Earth's species
2. A species that naturally lives and thrives in a particular ecosystem
3. The place or type of ecosystem in which a species lives and obtains what it needs to survive
4. Process by which Earth's life forms change over time
5. Large geographic region characterized by a distinct climate and mix of species
6. Method used to alter an organism's genetic material by adding, deleting, or changing segments of its DNA
7. A species that has a large effect on the types and abundance of other species in an ecosystem
8. The abundance and distribution of species within an ecosystem
9. Permanent change in the DNA sequence within a gene
10. Species whose presence, absence, or abundance is linked to a specific environmental condition and as a result is sensitive to environmental change
11. Occurs when an entire species ceases to exist
12. Low rate at which species ceased to exist throughout much of Earth's history
13. Species that can live in many different places, eat a variety of foods, and often tolerate a wide range of environmental conditions

Review Key Concepts

14. How does the loss of amphibian species impact the ecosystems in which they live? How does their loss impact humans?
15. Define biodiversity and describe its four major components.
16. Describe three ecosystem services that insects provide.
17. Compare and contrast a niche and a habitat.
18. Distinguish between generalist species and specialist species and give one example of each.
19. Describe native, nonnative, keystone, and indicator species and give one example of each.
20. Describe the role of the American alligator as a keystone species.
21. Provide a convincing argument for protecting sharks. Use the terms *keystone species*, *population decline*, and *human activities* in your argument.
22. What are fossils, and what information do they provide to scientists?
23. Distinguish between geographic isolation and reproductive isolation. Explain how these factors can lead to the formation of a new species.

Think Critically

24. What would be the consequence of losing the ecosystem services provided by Earth's biodiversity?
25. Is the human species a keystone species? Explain. If humans became extinct, what are three species that might also become extinct, and what are three species whose populations would probably grow?
26. If you had to choose between saving the giant panda from extinction and saving a shark species, which would you choose and why?
27. How would you respond to someone who tells you we should not worry about air pollution because natural selection will enable humans to develop lungs that can detoxify pollutants?
28. Why might people be uncomfortable with the idea of consuming genetically modified foods, while products created through artificial selection do not seem to cause them concern?

Chapter Activities

A. Investigate: Species Diversity STEM

A *quadrat study* is one method for measuring species diversity. This method involves studying several small plots within a habitat to assess the presence and distribution of species there. Plan and conduct an investigation that uses this method to study species diversity within a habitat on or near your school campus. You will need to research the terms *quadrat sampling, random sampling*, and *systematic sampling* as you plan your investigation. Also consider how to use the suggested materials and/or any others your teacher approves to conduct the investigation. Work with a partner or small group.

Materials

- rope
- map
- hula hoop
- ruler or meter stick
- notebook or tablet
- stakes
- smart phone
- camera
- colored craft sticks
- field identification guide(s)

1. Develop a list of steps that outline the procedure for your investigation. Create maps and diagrams as needed to illustrate materials and methods used.
2. Conduct your investigation. How did you record and display your data?
3. Examine the data you collected at each of your plots. Consider which species you observed and where. What patterns do you observe and what relationships might they suggest?
4. What are possible sources of error for your data? How could your procedure be improved?
5. Compare your data with that of a group that sampled a different habitat. Discuss how your data is similar or different and suggest reasons why.
6. Based on the data you collected, identify one question you could investigate that relates to biodiversity. Write a hypothesis.

B. Citizen Science

National Geographic Learning Framework
Attitudes | Curiosity
Skills | Observation
Knowledge | New Frontiers

The task of cataloging Earth's species diversity is huge! That is why a number of projects are depending on citizen scientists to submit data about species found in their area. This data is logged into a larger biodiversity database. Some projects focus on a specific location or a specific class of animals, such as birds. Others accept information about any plant or animal species from any location. Often, individuals participate by simply snapping a picture of a local species on their smart phone, recording some information about it, and uploading the photo to a website where it is shared with others.

Join a citizen science project asking volunteers to share photographs of species in their area. You might try the search term "citizen science photography for biodiversity" to locate a suitable project. Form a small group to take photographs of local species and submit them to the project. Then, gather to discuss the questions that follow.

1. What is the name of the project you joined? Who is the project's sponsor?
2. How did you use technology to collect and share data?
3. What data did you collect and for which species?
4. Did the project enable you to learn more about the species you photographed and if so, how?
5. What was your experience like working on the project? Do you think you will continue?
6. Discuss any ideas you have to improve upon or expand the project.

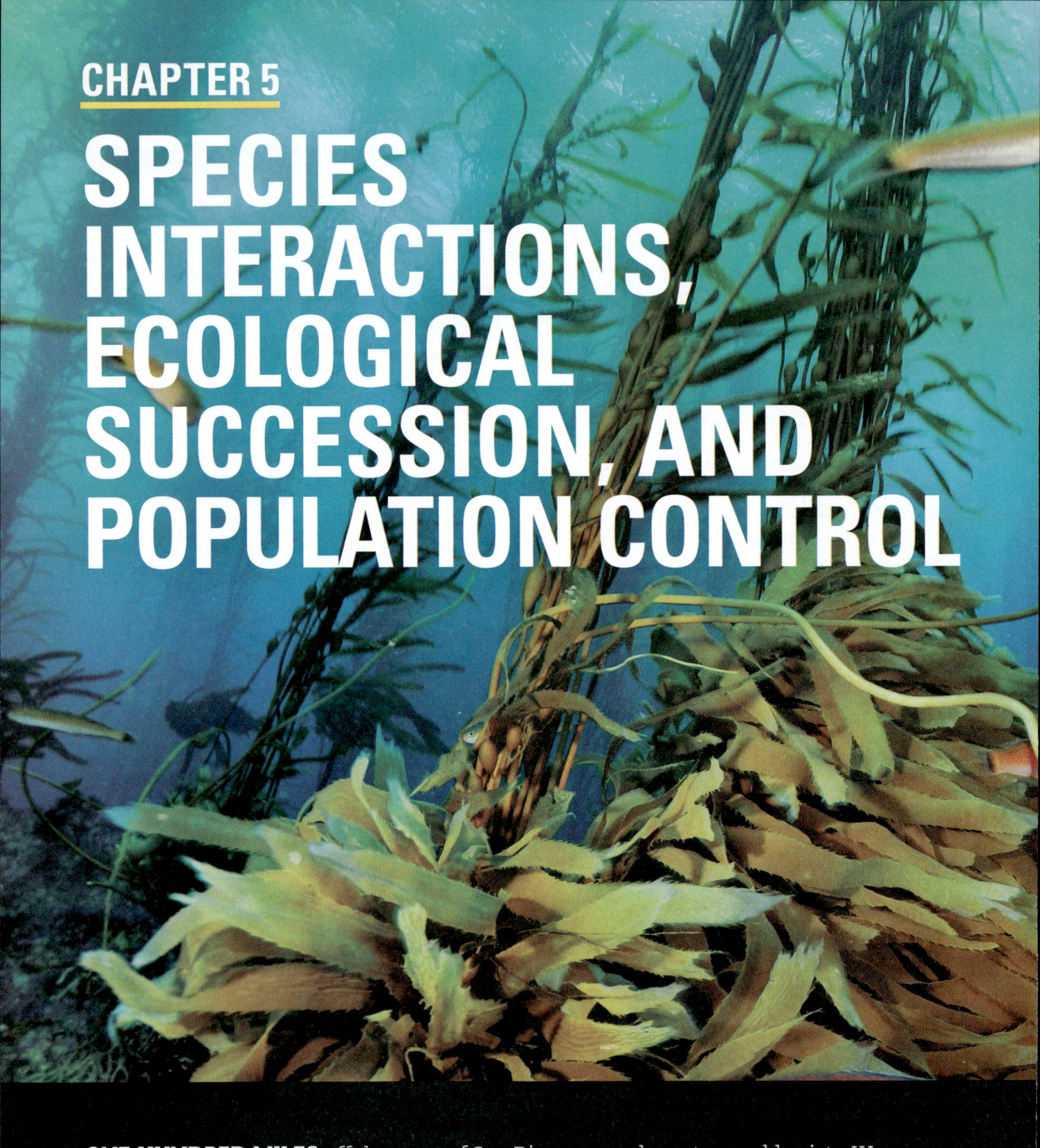

CHAPTER 5
SPECIES INTERACTIONS, ECOLOGICAL SUCCESSION, AND POPULATION CONTROL

ONE HUNDRED MILES off the coast of San Diego, an underwater world exists. Wrasses and senorita fish glide over corals and dart among clusters of giant kelp. Here on Cortes Bank, populations of fish, plants, and algae live together, each affecting the other and each in constant flux. Human populations enter into the system as fishing boats and surfers skim the surface waters above.

KEY QUESTIONS

5.1 How do species interact?

5.2 How do ecosystems respond to changing conditions?

5.3 What limits the growth of populations?

NATIONAL GEOGRAPHIC | EXPLORERS AT WORK

Sea Otters Step Up
with National Geographic Explorer Jim Estes

They may be cute and fun to watch, but sea otters do much more than entertain. They play a key role in protecting kelp forests from the destruction that comes when sea urchins experience unchecked growth. No one knows this better than National Geographic Explorer Jim Estes. Dr. Estes is a Distinguished Professor of Ecology and Evolutionary Biology at the University of California, Santa Cruz.

Estes and his research partners have studied sea otter populations for decades. Sea otters are marine mammals native to the coasts of the northeastern Pacific Ocean. Once hunted widely for their fur, sea otters were in danger of extinction during the 20th century. Conservation efforts and hunting regulations combined to improve the size of sea otter populations. Even so, the sea otter is still considered endangered. (See this chapter's Case Study for more about this.) The sea otter's endangered status is significant, given that it is a good example of a keystone species. As you read in Chapter 4, keystone species are critical to the populations of many other species in an ecosystem.

In one study, Estes and his partners synthesized 40 years of data to examine the relationship between sea otters and local ocean kelp forests. Kelp forests are critical because they provide food and shelter for diverse marine species. Kelp also absorb dissolved carbon dioxide (CO_2) from the atmosphere through photosynthesis. Protecting kelp forests could help slow climate change, as they are important participants in the carbon cycle. Research by Estes and other scientists indicates that an otter-assisted kelp forest is usually in much better shape than a kelp forest in an environment with no sea otters.

Here's why. Sea otters could be described as marine-based "foodies." When spotted in the water, a sea otter is often seen floating on its back, smacking a sea urchin against a rock on its chest. This maneuver helps to expose the tender meat inside. Without sea otters to feed on them, those same sea urchins would graze relentlessly on the kelp. In fact, they would basically raze the kelp to the ocean floor, leaving bare branches in their wake.

The good news, according to Jim Estes and his colleagues, is that sea otters can make a difference by keeping those urchin populations from expanding. It's pretty clear that by protecting sea otters, we also protect kelp forests. In turn, those kelp forests can do their job of absorbing CO_2 and providing habitat for vital marine biodiversity. Protecting sea otters does not solve global environmental problems in and of itself. But it is a great example of how managing predator-prey populations can affect a local ecosystem—and how that ecosystem can contribute to better health for the planet.

In evaluating how much a species can potentially benefit ecosystems, Estes concludes, "[How] species are linked to the carbon cycle is going to be very important." Estes and his current research team have projects underway in the Aleutian Islands, central California, the Channel Islands, and New Zealand.

Thinking Critically
Synthesize In your own words, explain the effect of sea otters on kelp forests, and what Estes's findings suggest about managing predator-prey relationships in local environments.

Jim Estes steers a boat through the waters surrounding Kiska Island, off the coast of Alaska. Estes has been exploring the Pacific Ocean for decades as part of his research on sea otters.

CASE STUDY
The Southern Sea Otter: A Species in Recovery

Southern sea otters live in giant kelp forests in shallow waters along parts of the Pacific coast of North America. Most of the remaining members of this endangered species are found off the California coast between Santa Cruz and Santa Barbara.

Fast and agile swimmers, the otters dive to the ocean bottom looking for shellfish and other food. Sea otters are air-breathing mammals. They swim on their backs on the ocean's surface and use their bellies as a table (Figure 5-1). Each day, a sea otter consumes 20–30% of its weight in clams, mussels, crabs, sea urchins, abalone, oysters, and other species of bottom-dwelling organisms.

Estimates show that around 16,000 southern sea otters once lived in California's coastal waters. By the early 1900s, they had been hunted almost to extinction by fur traders, who killed them for their luxurious fur. Commercial fishermen also killed otters, as they competed with them for valuable abalone and other shellfish.

The southern sea otter population has grown from a low of 50 in 1938 to an estimated 2,962 in 2019. An important step in this partial recovery took place in 1977. That year, the U.S. Fish and Wildlife Service declared the species threatened in most of its range, with a population of only 1,850 individuals. Even so, the sea otter population must expand further to justify removing it from the threatened species list.

Why should we care about the southern sea otters of California? One reason is ethical: Many people believe it is wrong to allow human activities to cause the extinction of a species. Another reason is that people love to look at these appealing and highly intelligent animals as they play in the water. As a result, otters help generate money in the form of tourism dollars. A third reason—and a key reason in our study of environmental science—is that biologists classify them as a keystone species. Scientists hypothesize that in the absence of southern sea otters, sea urchins and other kelp-eating species would probably destroy the Pacific coast kelp forests and much of the rich biodiversity they support. Biodiversity is an important part of Earth's natural capital and a scientific factor of sustainability.

In this chapter, you will look at how species interact and help control one another's population sizes. You will also explore how communities, ecosystems, and populations of species respond to changes in environmental conditions.

As You Read Think about how humans interact with other species and how these interactions affect the whole ecosystem. What examples can you think of that demonstrate positive or negative interactions with other species?

FIGURE 5-1
A southern sea otter feeds on a sea urchin in Monterey Bay, California.

5.1 How Do Species Interact?

CORE IDEAS AND SKILLS

- Explain how species compete with one another for certain resources.
- Recognize feeding relationships as a major category of interaction among species.
- Understand how interactions between predator and prey species can drive each other's evolution.
- Differentiate among parasitism, mutualism, and commensalism.

KEY TERMS

interspecific competition
resource partitioning
predation
predator
prey
predator-prey relationship
coevolution
parasitism
mutualism
commensalism

Species Compete for Resources

Ecologists have identified several basic types of interactions among species as they vie for limited resources such as food, shelter, and space. Interactions include: competition, predation, parasitism, mutualism, and commensalism. Species interactions greatly affect population sizes and the use of resources in an ecosystem. Competition is the most common interaction among species.

Types of Competition Species compete for limited resources, such as food, water, light, and space. When competition for resources occurs within a single species, it is called intraspecific competition. Competition among different species is called **interspecific competition**. Most interspecific competition involves one species becoming more efficient than others at obtaining resources.

When two species compete for the same resources, their niches overlap. The greater the overlap, the more the species compete for key resources. What happens if one species takes over the largest share of a key resource? Each of the other competing species must move to another area, adapt, suffer a population decline, or become extinct in that area. Humans compete with many other species for space, food, and other resources. As our ecological footprints grow and spread, we take over or degrade the habitats of many of those species.

Resource Partitioning If given enough time for natural selection to occur, populations can develop adaptations that enable them to reduce or avoid competition with other species. **Resource partitioning** occurs when species competing for similar scarce resources evolve specialized traits that allow them to "share" the same resources. Sharing resources can mean using parts of the resources or using resources at different times or in different ways.

FIGURE 5-2 ▼
Spreading the Wealth This illustration shows resource partitioning among five species of insect-eating warblers in the spruce forests of Maine. Each species spends at least half its feeding time in its associated highlighted area, as shown on these spruce trees.

Blackburnian warbler   Black-throated green warbler   Cape May warbler   Bay-breasted warbler   Yellow-rumped warbler

Source: R.H. MacArthur, "Population Ecology of Some Warblers in Northeastern Coniferous Forests," *Ecology* 36:533–536, 1958.

FIGURE 5-3 A cheetah cub learns to chase down its prey in the African grasslands.

Figure 5-2 shows resource partitioning by some insect-eating bird species. The birds' adaptations allow them to reduce competition by using different portions of spruce trees and by feeding on different insects.

checkpoint Explain what occurs if the niches of two species overlap and they share the same resources.

Species Prey on Other Species

In **predation**, a member of one species is the **predator** (hunter) and feeds directly on all or part of a living organism as part of a food web. The **prey** is the species that is hunted. The cheetah and the Thomson's gazelle are examples of species that engage in a **predator-prey relationship** (Figure 5-3). Interactions between predators and prey have strong effects on population sizes as well as other factors in many ecosystems.

For example, populations within the giant kelp forest ecosystem hinge on the relationship between predators such as southern sea otters and sea urchins. Sea urchins feed on kelp, a type of seaweed. As a keystone species, southern sea otters prey on sea urchins and prevent them from destroying the kelp forests.

Predators have a variety of methods to help them capture prey. Some predators, such as cheetahs, catch prey by chasing it down. Other predators, such as the American bald eagle, can fly and have keen eyesight to spot prey. Still others hunt in groups, coordinating their actions to capture prey. Female African lions use this method to run down large or fast-running animals of the savanna.

Other predators use camouflage to hide in plain sight and ambush their prey. Praying mantises sit on flowers or plants of a color similar to their own to ambush visiting insects. Ermines (a type of weasel), snowy owls, and Arctic foxes hunt their prey in snow-covered areas. Their white coloration keeps them well hidden as they hunt. Some predators use chemicals to attack their prey. For example, some spiders and snakes use venom to paralyze prey. They can also use their venom to defend against their own predators.

Prey species have evolved many ways to avoid predators too (Figure 5-4). Some prey species have highly developed senses of sight, hearing, or smell that alert them to the presence of predators. Other adaptations include protective body coverings such as shells (armadillos and turtles) and spines (porcupines and sea urchins).

Other prey species use camouflage; they are colored or shaped to blend in with their surroundings. For example, some insect species resemble twigs

(Figure 5-4a) or bird droppings on leaves. A leaf insect blends well with its background (Figure 5-4b), as does an Arctic hare in its white winter fur.

Chemical defenses are another common strategy used by prey species. For example, their bodies may contain or emit chemicals that are irritating (bombardier beetles, Figure 5-4c) or foul smelling (skunks and stinkbugs). Chemical defenses may confer a bad taste (monarch butterflies, Figure 5-4d) or even make prey poisonous (poison dart frogs, Figure 5-4e). When attacked, some species of squid and octopus emit clouds of black ink. This confuses predators long enough to allow them to escape.

Many bad-tasting, bad-smelling, toxic, or stinging prey species have also evolved warning coloration. Their bright colors and bold patterns make them easily recognizable to predators. These predators quicky learn to avoid them once they taste how unpleasant they are to eat. Examples are the brilliantly colored, foul-tasting monarch butterfly (Figure 5-4d) and poison dart frog (Figure 5-4e). When a bird eats a monarch butterfly, it usually vomits. It later avoids monarchs as a result. Some butterfly species gain protection by looking and acting like other, more dangerous species, a device called mimicry. The nonpoisonous viceroy butterfly (Figure 5-4f) mimics the monarch's warning coloration.

Other prey species have evolved behaviors to avoid predation. For example, some attempt to fool predators by appearing larger. They puff up (blowfish) or spread their wings (peacocks). Some moths, such as the Io moth (Figure 5-4g), have wing spots that look like the eyes of much larger animals. Some prey may mimic a predator's features (snake caterpillars, Figure 5-4h). Other prey species gain some protection by living in large groups, such as schools of fish and herds of antelope.

checkpoint What is the difference between camouflage and mimicry?

Predator-Prey Interactions Can Drive Evolution

At the individual level, predation benefits those members of a species that are predators. By contrast, predation harms those members of a species that become prey. At the population level, predation plays a role in natural selection. Predators tend to kill the sick, weak, aged, and least-fit members of a prey population because they are the easiest to catch. Individuals with better defenses against predation tend to survive longer. They produce more offspring with adaptations for avoiding predation. Over time, a prey species develops traits that make it harder to catch. In turn, its predators face selection pressures that favor traits improving their ability to catch their prey. Then, the prey species must further adapt to elude the more effective predators.

FIGURE 5-4
Prey Species: Tricks of the Trade Each of these prey species demonstrates a strategy to avoid predation: (a, b) camouflage, (c, d, e) toxic chemicals, (d, e, f) warning coloration, (f) mimicry, (g) deceptive looks, and (h) deceptive behavior.

(a) Span worm

(b) Wandering leaf insect

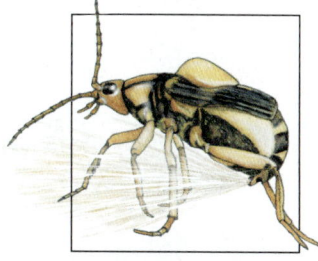

(c) Bombardier beetle

(d) Foul-tasting monarch butterfly

(e) Poison dart frog

(f) Viceroy butterfly mimics monarch butterfly

(g) Hind wings of Io moth resemble eyes of a much larger animal

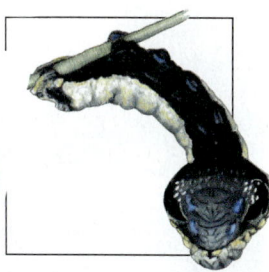

(h) When touched, snake caterpillar changes shape to look like head of snake

FIGURE 5-5
ON ASSIGNMENT Photographer Tim Laman caught the moment a moray eel opened wide to let a cleaner shrimp feed on parasites in its mouth. Which type of interspecies relationship does this image illustrate?

Coevolution is a natural selection process in which changes in the gene pool of one species lead to changes in the gene pool of another species. When populations of two species interact as predator and prey over a long period of time, genetic changes can occur in both species in response to this interaction. Such genetic changes help both species adapt to their biotic environment.

Coevolution can be observed between bats and certain species of moths they feed on. Bats hunt at night using echolocation. They emit pulses of high-frequency sound that bounce off objects. The bats then capture the returning echoes, which tell them where their prey is located. Over time, certain moth species evolved hearing that is especially sensitive to the sound frequencies bats use to find them. When moths hear these frequencies, they drop to the ground or fly evasively. Some bat species evolved ways to counter this defense by changing the frequency of their sound pulses. In response, some moths developed their own high-frequency clicks to jam the bats' echolocation systems. Some bat species then adapted by silencing their echolocation systems and using the moths' own clicks to find them.

Thus, the complex predator-prey relationship plays an important role in controlling population growth of predator and prey species. It enables them to contribute to important ecosystem services such as pollination. Predator-prey relationships can be disrupted when nonnative predator species are introduced, either accidentally or deliberately.

checkpoint What is coevolution and how does it work?

Some Species Form Close Relationships

Parasitism When one organism (a parasite) lives in or on another organism (a host) and benefits at the host's expense, their relationship is called **parasitism**. A parasite weakens its host but rarely kills it. Doing so would eliminate the source of its benefits. Parasites may be plants, animals, or microorganisms. Tapeworms are parasites that live part of their life cycle inside their hosts. Mistletoe plants and blood-sucking sea lampreys attach themselves to the outsides of their hosts and suck nutrients from them. Some parasites move from one host to another (fleas and ticks), while others (protozoa, tuberculosis) spend their adult stages within a single host. Parasites harm individual hosts, but help to keep the populations of their hosts in check.

Mutualism In **mutualism**, two species behave in ways that benefit both species by providing each with food, shelter, or some other resource. Pollination of flowering plants is one example of mutualism. Pollinator species include honeybees, hummingbirds, and butterflies that feed on the nectar of flowers. Mutualism also occurs between certain birds and large animals such as elephants, rhinoceroses, and impalas. The birds ride on the animals' heads or backs. They remove and eat parasites and other pests from the animals' bodies. The birds often make noises warning the larger animals when predators approach.

In gut inhabitant mutualism, armies of bacteria in the digestive systems of animals break down the animals' food. In turn, the hosts supply bacteria with food and habitat. Trillions of bacteria in your gut secrete enzymes that help you digest your food. Mutualism might appear as a form of cooperation between species. However, each species is concerned only for its own survival. The benefits received from mutualism are helpful, but unintentional.

Commensalism In **commensalism**, one species benefits but the interaction has little, if any, beneficial or harmful effect on the other species. One example involves plants called epiphytes (air plants). Epiphytes attach themselves to the trunks or branches of trees in tropical and subtropical forests. They benefit by having a solid base on which to grow in an elevated location. The location gives them better access to sunlight, water from the humid air and rain, and nutrients falling from the tree's upper leaves and limbs. Epiphytes do not directly use their host plants for nutrients, and their presence apparently does not harm host trees. The same is true for birds that use tree branches for nesting.

checkpoint How is parasitism different from mutualism?

5.1 Assessment

1. **Identify** Define resource partitioning and give an example.
2. **Explain** Explain the predator-prey relationship.
3. **Describe** If you could choose how you interact with another species, would you choose parasitism, mutualism, or commensalism? Explain.
4. **Relate** How could coevolution lead to mimicry?

SCIENCE AND ENGINEERING PRACTICES

5. **Communicating Information** Draw a diagram to describe the coevolution between bats and moths.

5.2 How Do Ecosystems Respond to Changing Conditions?

CORE IDEAS AND SKILLS

- Understand how the species composition of a community or ecosystem can change.
- Recognize that living systems are sustained through constant change.

KEY TERMS

ecological succession
primary ecological succession
secondary ecological succession
inertia
resilience

Ecosystems Experience Succession

The types and numbers of species in biological communities and ecosystems change in response to changing environmental conditions. Changes may be caused by events such as fires, volcanic eruptions, climate change, and clearing forests to plant crops. The normally gradual change in species composition in a given area is called **ecological succession**.

Ecologists recognize two major types of ecological succession. The classification depends on the conditions present at the beginning of the process. **Primary ecological succession** involves the gradual establishment of communities of different species in mostly lifeless areas. Primary ecological succession begins where there is no soil in a terrestrial ecosystem or no bottom sediment in an aquatic one. Bare rock exposed by a retreating glacier (Figure 5-6) allows for this type of succession. So does newly cooled lava from a volcanic eruption, an abandoned highway or parking lot, or a newly created shallow pond or reservoir. Primary succession can span hundreds to thousands of years because it takes a long time to build up fertile soil or aquatic sediments. These substrates and their nutrients must be present to establish a community of producers.

The other, more common type of ecological succession is **secondary ecological succession**. This type of succession occurs when communities or ecosystems develop in an area where an ecosystem has been disturbed or destroyed but some soil or bottom sediment remains. Candidates for secondary ecological succession include abandoned farmland (Figure 5-7), burned or cut forests, heavily polluted streams, and flooded land. Because some soil or sediment is present, new vegetation can begin to grow, usually within a few weeks. Growth starts with the germination of seeds present in the soil as well as those transported in by wind or in animal droppings.

Ecological succession is an important ecosystem service because it tends to enrich the biodiversity of communities and ecosystems. It does so by increasing the diversity of species and the interactions among them. Such interactions strengthen an ecosystem's sustainability by promoting population control and increasing the complexity of food webs. That, in turn, enhances energy flow and nutrient cycling. As part of Earth's natural capital, succession is an example of natural ecological restoration.

The traditional view of ecological succession described it as proceeding in a predictable, orderly sequence before reaching a climax community. This view defined a climax community as one dominated by a few long-lived plant species, often within a mature forest (see the "final" stages of Figure 5-6 and Figure 5-7). This equilibrium model of succession is what people once meant when they talked about the balance of nature.

Over the past several decades, a different picture of succession has begun to emerge. Ecological succession doesn't always follow a predictable path. It does tend to lead to more complex, diverse, and presumably sustainable ecosystems. However, a close look at almost any terrestrial community or ecosystem reveals continual change. Forest ecosystems are more like ever-changing patchworks of vegetation in different stages of succession.

checkpoint What is the difference between primary and secondary ecological succession?

> **CONSIDER THIS**
>
> **Life regenerates after a volcanic eruption.**
>
> One of the world's most active volcanoes is Kilauea on the island of Hawai'i. When a volcano erupts, the lava flow destroys everything in its path. Regeneration of vegetation depends on rainfall, the type of lava or ash, and how close the flow is to the volcano's opening. Where conditions are favorable, lichens are the first to appear, growing directly on cooled lava.

FIGURE 5-6

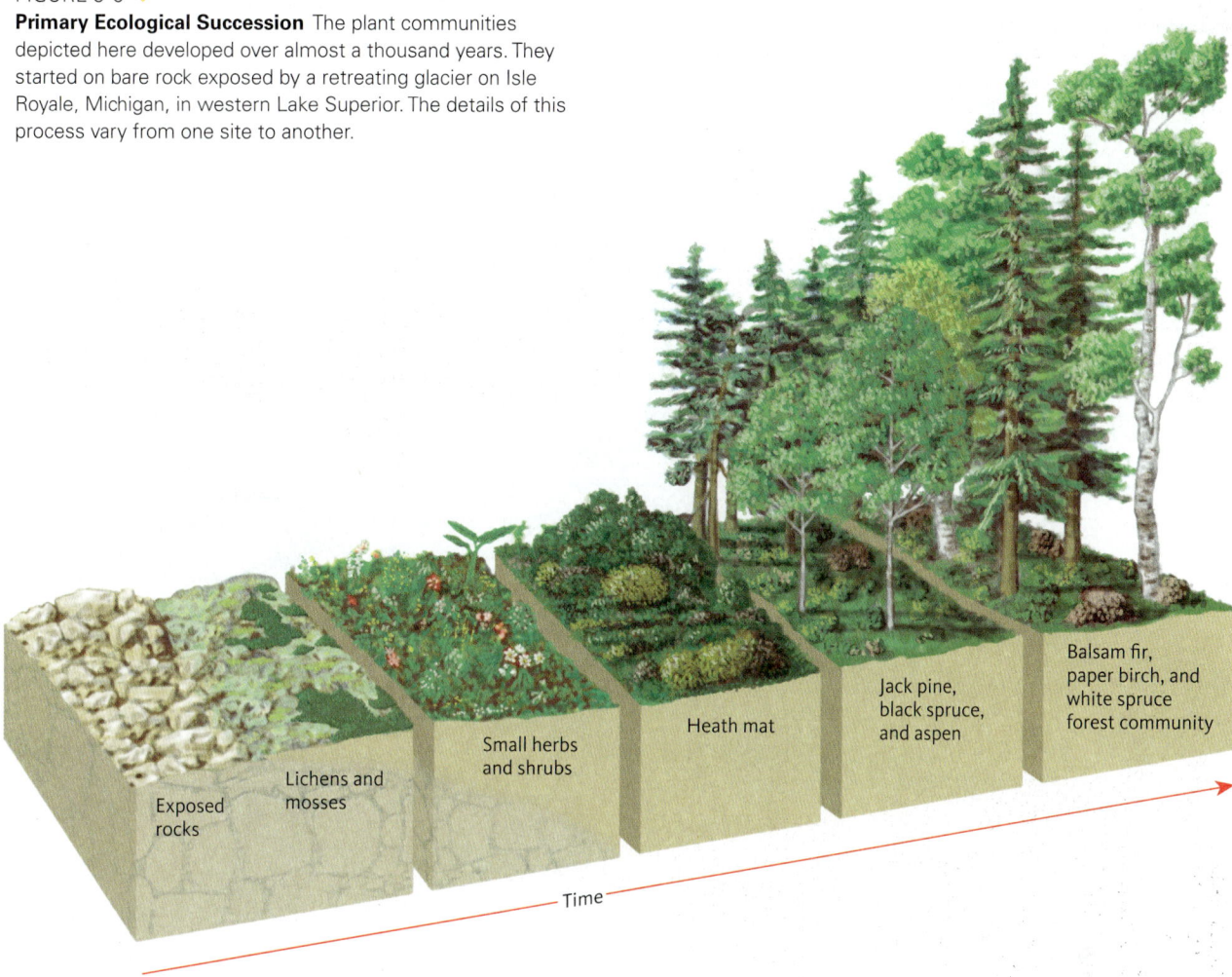

Primary Ecological Succession The plant communities depicted here developed over almost a thousand years. They started on bare rock exposed by a retreating glacier on Isle Royale, Michigan, in western Lake Superior. The details of this process vary from one site to another.

Exposed rocks → Lichens and mosses → Small herbs and shrubs → Heath mat → Jack pine, black spruce, and aspen → Balsam fir, paper birch, and white spruce forest community

Time →

Living Systems Arise from Constant Change

Living systems arise out of complex processes that result in some degree of stability over time, or sustainability. For example, in a mature tropical rain forest, some trees die and others take their places. Despite this activity, unless the forest is cut, burned, or otherwise destroyed, you would still recognize it as a tropical rain forest 50 or 100 years later.

It is useful to distinguish between two aspects of stability in ecosystems. One is inertia, or persistence. Ecological **inertia** is the ability of an ecosystem to survive moderate disturbances. A second factor is resilience. Ecological **resilience** is the ability of an ecosystem to be restored through secondary ecological succession after a severe disturbance.

Evidence suggests that some ecosystems have one of these properties but not the other. Tropical rain forests have high species diversity and high inertia. Thus, they are resistant to lower levels of change or damage. But once a large tract of rain forest is cleared or severely damaged, the resilience of the degraded forest ecosystem may be so low, it reaches an ecological tipping point. Once this point is reached, the forest might not be restored by secondary ecological succession. Most of the nutrients in a typical rain forest are stored in its vegetation, not in the topsoil. Once the nutrient-rich vegetation is gone, daily rains on a large, cleared area of land will wash away the remaining soil nutrients. That will prevent the rain forest from returning.

By contrast, grasslands are much less diverse than most forests when studied on a large scale. Therefore they have low inertia and can burn easily. But most of their plant matter is stored in underground roots. These root systems can survive fire to produce new grasses. Thus, grassland ecosystems have high resilience and can recover quickly.

checkpoint How is stability maintained in a living system?

LESSON 5.2

FIGURE 5-7

Secondary Ecological Succession Disturbed land on an abandoned farm field in North Carolina was restored naturally. It took 150–200 years after the farmland was abandoned for the area to become covered with a mature oak and hickory forest.

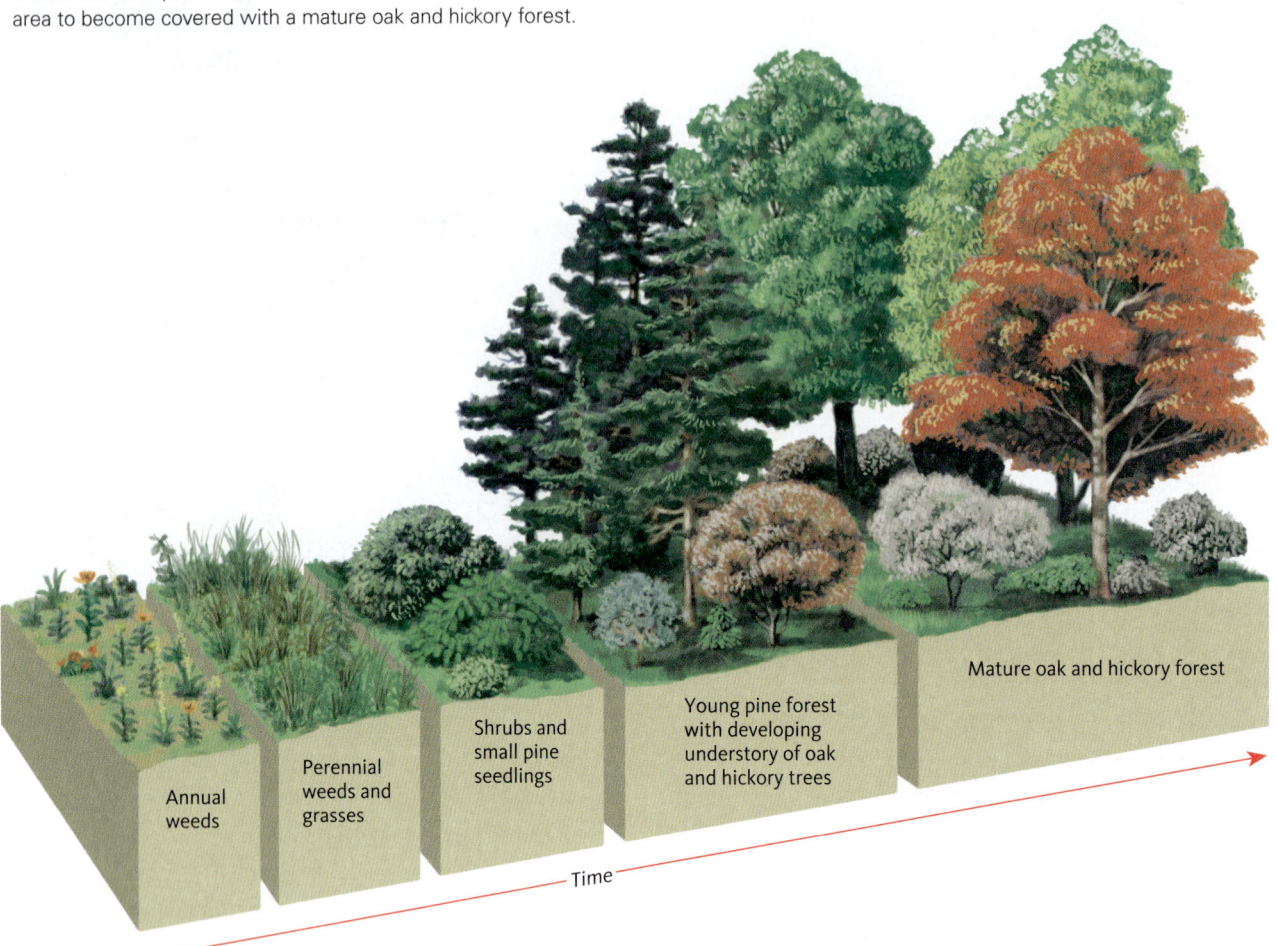

Annual weeds → Perennial weeds and grasses → Shrubs and small pine seedlings → Young pine forest with developing understory of oak and hickory trees → Mature oak and hickory forest

Time

5.2 Assessment

1. **Recall** What is the gradual change in species composition in a given area called?
2. **Identify** Which type of ecological succession occurs in an abandoned parking lot?
3. **Define** Define and give an example of resilience and inertia in an ecosystem.
4. **Explain** How can grasslands regenerate quickly after being burned to the ground?

CROSSCUTTING CONCEPTS

5. **Stability and Change** Describe how a rain forest can reach a point when it cannot be restored by secondary ecological succession.

5.3 What Limits the Growth of Populations?

CORE IDEAS AND SKILLS
- Identify the variables that govern changes in population size and the factors that limit population size.
- Explain reproductive and survivorship patterns of populations.

KEY TERMS

population
age structure
range of tolerance
limiting factor
population density
environmental resistance
carrying capacity
population crash
r-selected species
K-selected species
survivorship curve

Populations Can Grow, Shrink, or Remain Stable

A **population** is a group of interbreeding individuals of the same species. Most populations live together in groups, such as packs of wolves, schools of fish, and flocks of birds. Living in groups allows them to cluster where resources are available. It also provides some protection from predators and gives predator species a better chance of getting a meal.

Four variables—birth, death, immigration, and emigration—govern changes in population size. A population increases through birth and immigration (the arrival of individuals from outside the population). A population decreases through death and emigration (the departure of individuals from the population).

> population change = (births + immigration) − (deaths + emigration)

A population's **age structure** describes the distribution of individuals among various age groups. Age structure can greatly impact how rapidly a population grows or declines. Age groups are usually described in terms of organisms not mature enough to reproduce (pre-reproductive stage), those capable of reproduction (reproductive stage), and those too old to reproduce (post-reproductive stage).

Each population in an ecosystem has a **range of tolerance**—a range of variations in its physical environment under which it can survive (Figure 5-8). Individuals within a population may have slightly different tolerance ranges for temperature, chemical factors, or other physical factors. These occur because of small differences in their genetic makeup, health, and age. A trout population may do best within a narrow band of temperatures (the optimum level or range), but a few individuals can survive above and below that band. If, however, the water temperature greatly exceeds that narrow band or range, none of the trout can survive.

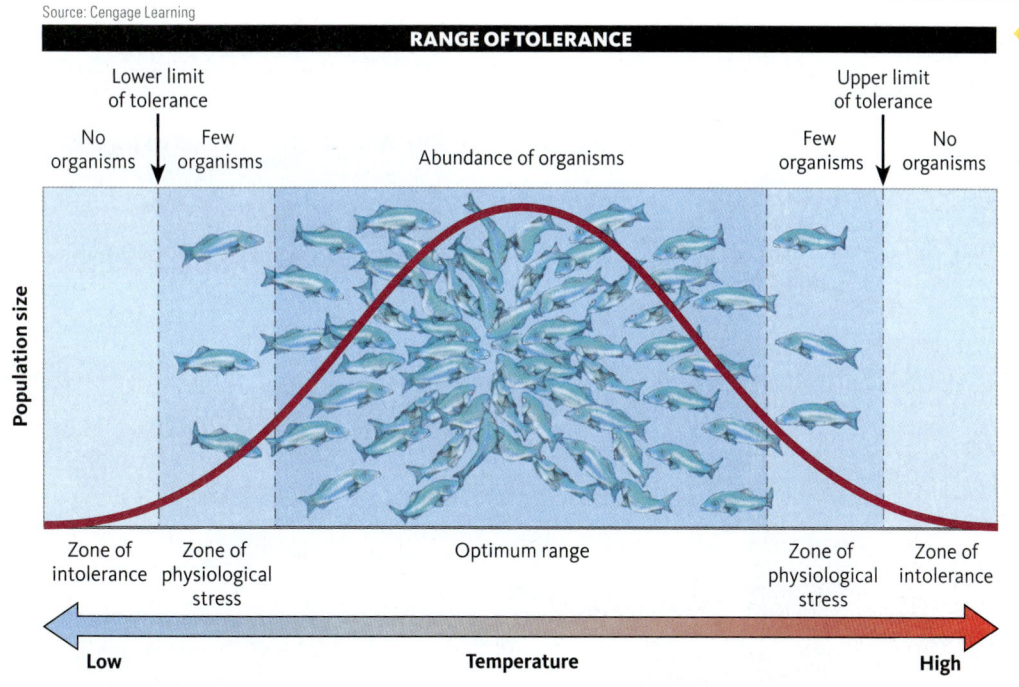

FIGURE 5-8
Range of Tolerance A trout population thrives within an optimum range of temperatures.

NATIONAL GEOGRAPHIC | EXPLORERS AT WORK
Zeb Hogan Ecologist and Photographer

Freshwater ecosystems around the world are home to a great diversity of fish species. They also support the livelihoods of hundreds of millions of people in the fishing industry. However, these systems are some of Earth's most fragile ecosystems. Freshwater species of fish are rapidly slipping away from existence. The largest fish—the megafish—are often the first to go.

But not if National Geographic Fellow Zeb Hogan can help it. Dr. Hogan studies freshwater ecosystems around the planet. He works to protect endangered species that depend on these ecosystems. He has come face to face with some of the largest freshwater megafish species in the world, some more than 3 meters (10 feet) long.

Hogan leads the National Geographic Society's Megafish Project. The Megafish Project covers six continents and aims to document, understand, and protect Earth's giant freshwater fish species before they are gone. Hogan also hosts a NatGeo Wild TV show, *Monster Fish*. He uses photography to help connect people to amazing fish they may have never known existed.

The global population of megafish has declined by 94% (Figure 5-9). What makes megafish species' survival so precarious? Research by Hogan and others suggests that multiple factors put the freshwater giants at greater risk of extinction than their smaller counterparts.

The population structures of fish play a role in their ability to withstand environmental degradation. Age structure is one factor. Large fish must survive a greater length of time before they reach reproductive age. In fished waters, they get caught and removed from the population before they can have any offspring.

Large fish also tend to have low population densities and higher demands for habitat size. These factors make them more vulnerable to the effects of habitat loss and degradation.

Another factor putting megafish at risk is that humans often target the largest animals to fish. Finally, these fish tend to live out of sight, and thus out of the minds of many conservationists. But that is changing, thanks to Zeb Hogan.

FIGURE 5-9
Zeb Hogan presents a Eurasian giant trout, or taimen, one of many megafish species in danger of disappearing.

Limits to Population Growth Chemical and physical factors can determine the number of organisms in a population. Sometimes one or more factors, known as **limiting factors**, are more important than other factors in regulating population growth. On land, precipitation is often the limiting factor. Low precipitation levels in desert ecosystems can limit desert plant growth. Conversely, too much water or fertilizer can kill plants.

Important limiting physical factors for populations in aquatic systems include water temperature, depth, and clarity (allowing for more or less sunlight). Nutrient availability, acidity, salinity, and the level of oxygen gas in the water (dissolved oxygen content) are also important factors. Too much acidity in an aquatic environment can harm some of its organisms.

Density is another factor that can affect the size of populations. **Population density** is the number of individuals in a population found within a defined area (square foot, square mile, square inch) or volume (cubic feet, cubic yards, cubic inches, or gallons). Density-dependent factors become more important as a population's density increases. In a dense population, parasites and diseases can spread more easily. That leads to higher death rates. On the other hand, a higher population density helps sexually reproducing individuals find mates more easily.

Other factors such as drought and climate change are considered density-independent. They can affect population sizes regardless of density.

The populations of some species have an incredible ability to increase their numbers exponentially. Plotting these numbers against time yields a J-shaped curve of exponential growth (Figure 5-10, left, and Figure 5-11). Members of such populations typically reproduce at an early age and have many offspring each time they reproduce. They reproduce many times, with short intervals between generations.

However, research reveals that a rapidly growing population of any species eventually reaches some size limit imposed by limiting factors. These factors include sunlight, water, temperature, space, nutrients, or exposure to predators or infectious diseases. **Environmental resistance** is the sum of all such factors in any habitat. Limiting factors largely determine any area's **carrying capacity**, the maximum population of a species that a habitat can sustain indefinitely. As a population approaches the carrying capacity of its habitat, the J-shaped curve of its exponential growth becomes an S-shaped curve. This shape indicates logistic growth, or growth that fluctuates around a certain level (Figure 5-10, right).

FIGURE 5-10 ▼

Patterns of Population Growth When resources are plentiful, populations of species can undergo exponential growth, which is represented by a J-shaped curve (left). As resources become limited, a population approaches carrying capacity and undergoes logistic growth. This growth pattern is represented by a more S-shaped curve (right).

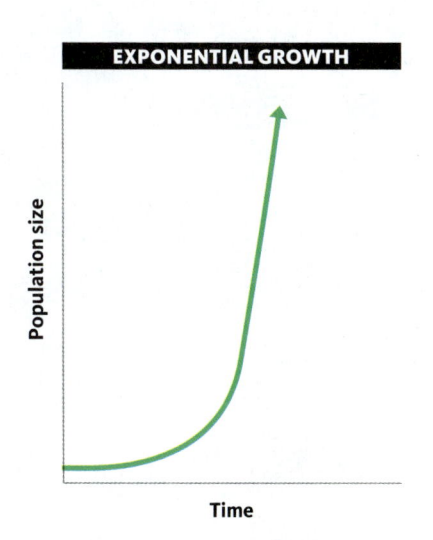

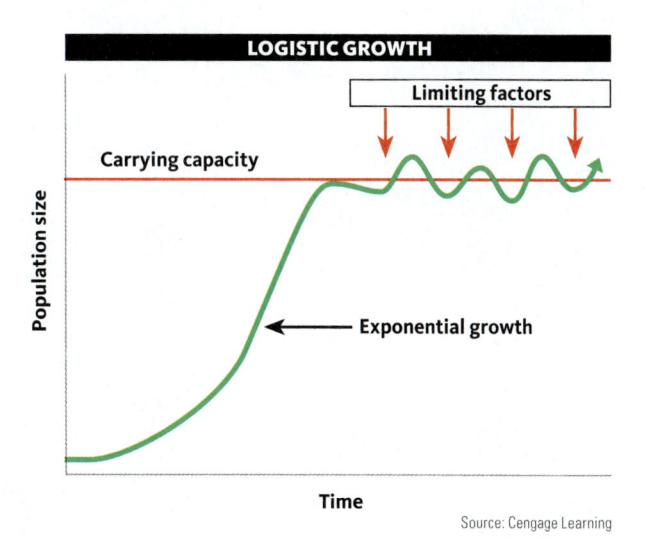

Source: Cengage Learning

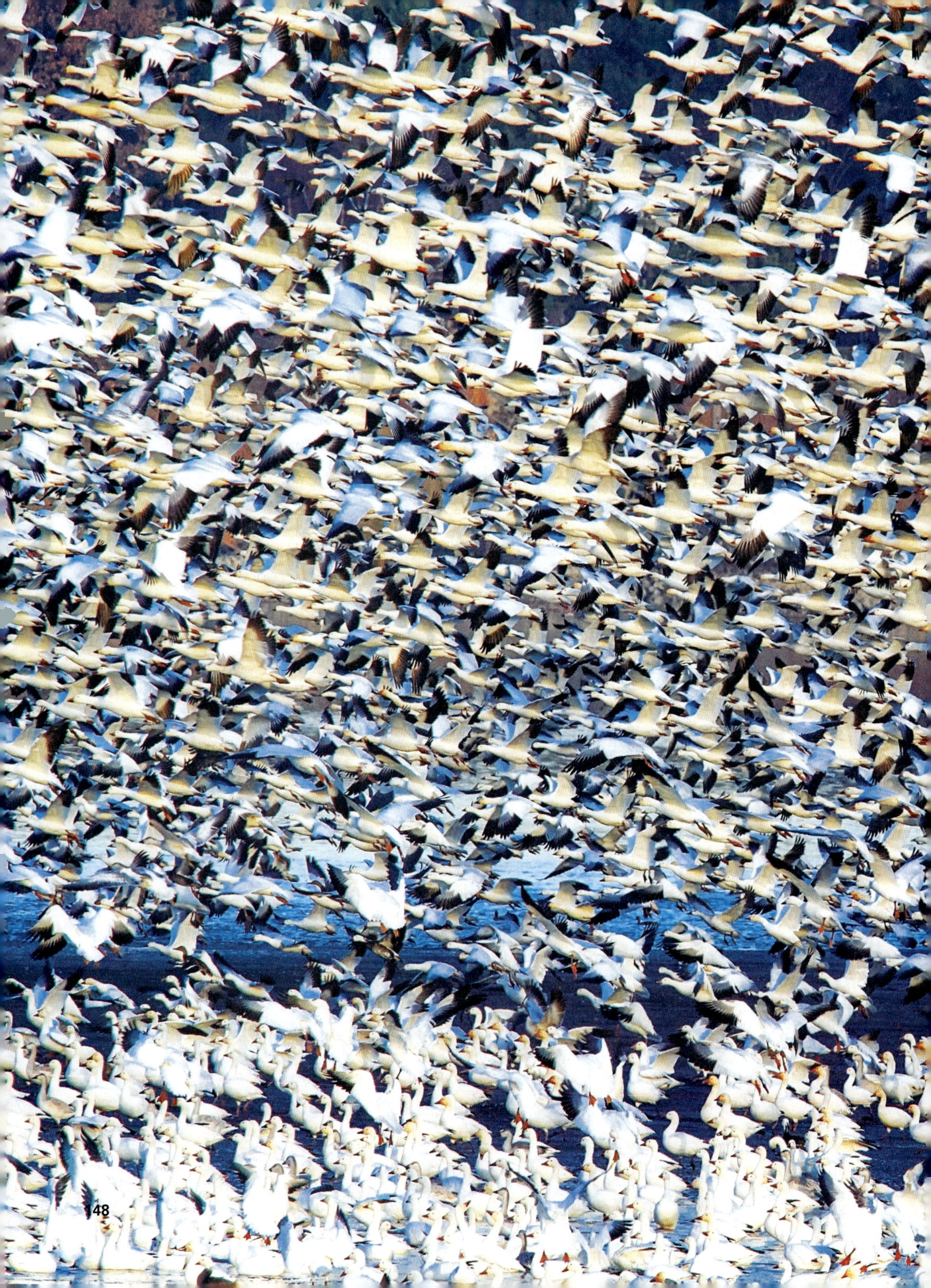

FIGURE 5-11
ON ASSIGNMENT National Geographic photographer David Doubilet captured this dizzying image of migrating snow geese at a fall stopover in Quebec. North America's snow goose population exploded from roughly 2 million in the 1970s to about 15 million in 2015 owing to an increased food supply. But no population can increase indefinitely. Snow geese population growth is declining as the birds destroy their own habitat.

Some populations do not make a smooth transition from exponential growth to logistic growth. Instead, they use up their resource supplies and temporarily overshoot, or exceed, the carrying capacity of their environment. In such cases, the population suffers a sharp decline, called a dieback or **population crash**, unless part of the population can switch to new resources or move to an area that has more resources. Such a population crash occurred when reindeer were introduced onto a small island in the Bering Sea in the early 1900s (Figure 5-12).

checkpoint What is a population's range of tolerance?

Different Species Have Different Reproductive Patterns

Species vary in their reproductive patterns. Species with a capacity for a high rate of population increase (r) are called **r-selected species**. These species tend to have short life spans and produce many offspring, which are usually small in size. Examples include algae, bacteria, and most insects. The parents provide little or no care or protection of their offspring. To overcome massive losses in their offspring, r-selected species produce huge numbers of descendants so a few will likely survive. Each new generation repeats this reproductive pattern.

As a rule, r-selected species are opportunists. They reproduce and disperse rapidly when conditions are favorable. Disturbances such as fire or clear-cutting may also allow them to invade a new habitat or niche. Once established, their populations may crash due to unfavorable changes in environmental conditions or invasion by more competitive species. Most opportunist species undergo irregular and unstable boom-and-bust cycles in their population sizes.

K-selected species are at the other end of the spectrum. **K-selected species** tend to reproduce later in life, have few offspring, and have long life spans. Typically, the offspring of K-selected mammal species develop inside their mothers (where they are safe) and are born fairly large. After birth, they mature slowly and are cared for and protected by one or both parents. In some cases they live in herds or groups until they reach reproductive age and begin the cycle again. Most organisms have reproductive patterns between the extremes of r-selected and K-selected species.

K-selected species do well in competitive conditions when their populations are near the carrying capacity (K) of their environments. Most large mammals (such as elephants, whales, and humans) are K-selected. So are birds of prey and large, long-lived plants (such as saguaro cactus and most rain forest trees). Many of these species are

FIGURE 5-12 ▼
Population Crash After a population overshoots its habitat's carrying capacity, it can experience a population crash. This crash occurred in a population of reindeer that were introduced onto the small Bering Sea island of St. Paul in 1911.

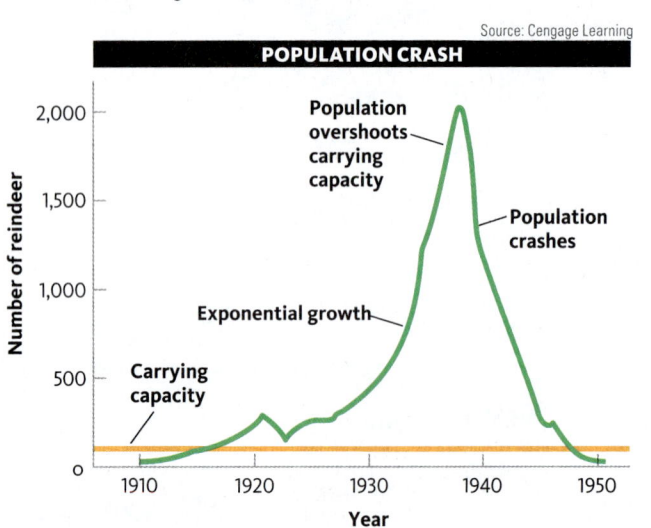

FIGURE 5-13 ▼
Survivorship Curves This graph shows three different general survivorship curves for populations of different species.

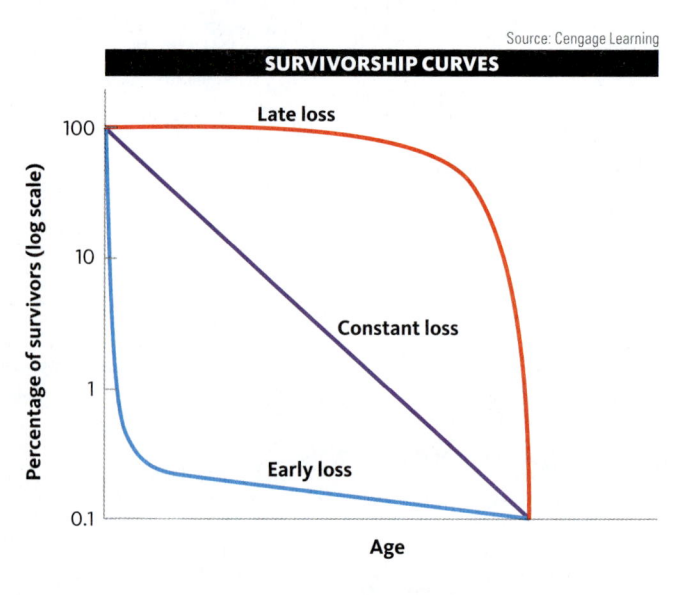

vulnerable to extinction—especially those with low reproductive rates. Among them are elephants, sharks, giant redwood trees, and California's southern sea otters.

Different reproductive strategies are associated with different life expectancies. This can be illustrated by a **survivorship curve**, which shows the percentages of the members of a population surviving at different ages. There are three general types of survivorship curves: late loss, early loss, and constant loss (Figure 5-13). A late-loss population typically has high survivorship to a certain age and then experiences high mortality (Figure 5-14). A constant-loss population (common among songbirds) shows a fairly constant death rate at all ages. For an early-loss population (common among annual plants and many bony fish species), survivorship is low early in life. These generalized survivorship curves only approximate the realities of nature.

checkpoint Describe the reproductive patterns of *r*-selected species.

5.3 Assessment

1. **Identify** What four variables govern changes in population growth?
2. **Compare and Contrast** Compare and contrast *K*-selected species and *r*-selected species.
3. **Recall** Species with which type of reproductive pattern are most vulnerable to extinction? Explain your answer.
4. **Define** What does a survivorship curve show, and what are the three types?
5. **Apply** Which type of survivorship curve applies to humans? Explain.

CROSSCUTTING CONCEPTS

6. **Cause and Effect** Describe the types of data needed to provide causal evidence of space as a limiting factor on the size of a fish population.

FIGURE 5-14
African elephants are a *K*-selected, late-loss species. Young elephants mature slowly under the care of their older female relatives.

TYING IT ALL TOGETHER STEM
Southern Sea Otters Face an Uncertain Future

Along the western coast of North America, sea otters are part of a complex kelp forest ecosystem. The population size of southern sea otters has fluctuated in response to changes in environmental conditions. One change influencing this species' populations has been an increase in populations of the orcas (killer whales) that feed on them. Scientists hypothesize that orcas started feeding on southern sea otters when populations of their normal prey, sea lions and seals, began declining.

Also, between 2010 and 2012, the number of sea otters killed or injured by sharks increased from an average of 6% during the 1980s to more than 50%. Scientists are trying to determine why this increase is happening.

Another factor affecting sea otters may be parasites that breed in the intestines of cats. Scientists hypothesize that sea otter deaths are linked to pet owners flushing feces-laden cat litter down their toilets or dumping it in storm drains that empty into coastal waters. The feces contain parasites that may be taken up by mussels and other shelled animals, which in turn are eaten by otters.

Otters are also threatened by toxic algae blooms. The algae are fed by urea, a key ingredient in fertilizer that washes into coastal waters. Other pollutants released by human activities are fat-soluble toxic chemicals such as PCBs. These chemicals accumulate to high levels in the tissues of the shellfish that otters eat. Because southern sea otters feed at a high trophic level and live close to the shore, they are vulnerable to these and other pollutants in coastal waters.

The impacts from all these factors are made worse by the fact sea otters have fairly low reproductive rates and rising mortality rates. Thus, the ability of this endangered species to rebuild its population has been hindered (Figure 5-15). In 2012, the National Geographic Society funded a project to learn why juvenile sea otters were suffering a high mortality rate. The study was led by Nicole Thometz, a biologist at the University of California, Santa Cruz. The aim of this study was to track changes in the development of the otters and understand how physiological variables affected their foraging ability and success.

Such information could be used to help biologists fine-tune recovery plans for the otter. According to the U.S. Geological Survey, the California southern sea otter population would have to reach at least 3,090 animals for three consecutive years before it could be considered for removal from the endangered species list.

Use the graph to help you answer the questions that follow.

1. Has the southern sea otter population ever met the criteria needed for this species to be considered for removal from the endangered species list? What data support your answer? Do you think this species will be delisted? Why or why not?
2. Are otter reproductive patterns more like *r*-selected species or *K*-selected species? What are the implications of this reproductive pattern?
3. What shape would you expect the otter's survivorship curve to take? Consider their high juvenile mortality rate.
4. Design an experiment to test a hypothesis about the effect of one of the factors described on sea otter population size.

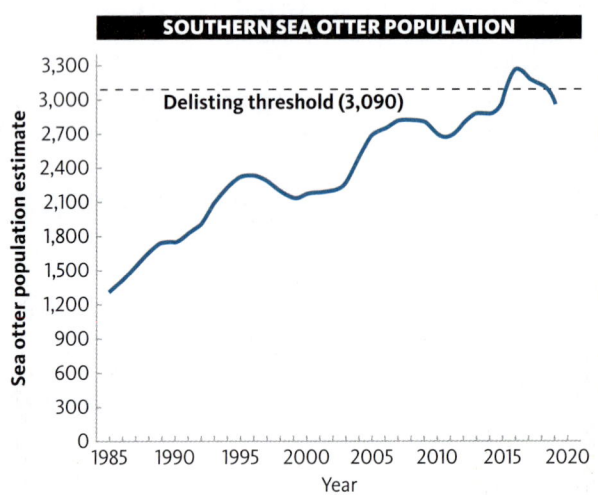

Source: U.S. Geological Survey

SOUTHERN SEA OTTER POPULATION

152 CHAPTER 5 SPECIES INTERACTIONS, ECOLOGICAL SUCCESSION, AND POPULATION CONTROL

CHAPTER 5 SUMMARY

5.1 How do species interact?

- Most species compete with one another for certain resources. Intraspecific competition occurs among members of the same species. Interspecific competition occurs among members of different species. Resource partitioning occurs when species competing for similar limited resources evolve specialized traits that allow them to "share" those resources.
- Predators are hunters, and prey species are the hunted. Predator-prey relationships affect population sizes. Some predator species use camouflage to ambush their prey, while some prey species use camouflage to hide from predators. Many predator and prey species use chemical defenses. The chemicals can be poisonous, irritating, foul smelling, or bad tasting.
- Coevolution happens when interaction between species drives adaptive changes in both species.
- Parasitism occurs when one organism (the parasite) lives in or on another organism (the host). In mutualism, both species benefit from their relationship. In commensalism, one species benefits but does not significantly help or harm the other species.

5.2 How do ecosystems respond to changing conditions?

- Ecological succession occurs when the composition of a community or ecosystem changes in response to changing environmental conditions. Ecological succession can be either primary or secondary, depending on whether or not there is soil or bottom sediment at the site of succession.
- Inertia, or persistence, is the ability of an ecosystem to survive moderate disturbances in the environment. Resilience is the ability of an ecosystem to be restored through secondary ecological succession after a severe disturbance.

5.3 What limits the growth of populations?

- A population is a group of interbreeding individuals of the same species. Most populations live together in groups (packs of wolves, flocks of birds).
- Variables that govern changes in population size are birth, death, immigration, and emigration. Other factors that affect population size are age structure, range of tolerance, limiting factors, and population density.
- A rapidly growing population of any species eventually reaches some size limit imposed by limiting factors such as sunlight, water, temperature, space, nutrients, or exposure to predators or infectious diseases.
- *r*-selected species, with short life spans, little parental care, and many offspring, have a capacity for a high rate of population increase. *K*-selected species reproduce later in life, with greater parental care, fewer offspring, and longer life spans.
- A survivorship curve illustrates the different life expectancies of species, showing the percentages of the members of a population surviving at different ages.

CHAPTER 5 ASSESSMENT

Review Key Terms

Select the key term that best fits each definition. Not all terms will be used.

age structure
carrying capacity
coevolution
commensalism
ecological succession
environmental resistance
inertia
interspecific competition
K-selected species
limiting factor
mutualism
parasitism
population
population crash
population density
predation
predator
predator-prey relationship
prey
primary ecological succession
range of tolerance
resilience
resource partitioning
r-selected species
secondary ecological succession
survivorship curve

1. A sharp decline in a population
2. The number of individuals in a population found within a defined area or volume
3. All the population-limiting factors in a habitat
4. The ability of an ecosystem to survive moderate disturbances
5. A member of a species that feeds directly on all or part of a living organism
6. The distribution of individuals among various age groups in a population
7. Chemical or physical factor that determines the number of organisms in a population
8. A species interaction that has a strong effect on population sizes and other factors in many ecosystems
9. An organism that is eaten by a predator
10. The maximum population of a species that a habitat can sustain indefinitely
11. Occurs when species competing for similar scarce resources evolve specialized traits that allow them to share resources
12. The ability of an ecosystem to be restored through secondary ecological succession after a severe disturbance
13. Species that has a capacity for a high rate of population increase
14. A range of variations in a population's physical environment under which it can survive
15. A natural selection process in which changes in the gene pool of one species lead to changes in the gene pool of another species
16. A graph that shows the percentages of the members of a population surviving at different ages
17. Occurs when one organism feeds on another organism, usually by living on or inside the host
18. Species that tends to reproduce later in life, has a small number of offspring, and has a long life span
19. An interaction between two species that benefits one species but has little, if any, beneficial or harmful effect on the other

Review Key Concepts

20. Define and give an example of interspecific competition.
21. Define and give an example of resource partitioning and explain how it can increase species diversity.
22. Explain why people should preserve kelp forests.
23. Describe three ways in which predators can increase their chances of feeding on their prey and three ways in which prey species can avoid their predators.
24. Define and give an example of coevolution.
25. Define parasitism, mutualism, and commensalism and give an example of each.
26. Distinguish between primary ecological succession and secondary ecological succession.
27. List and explain the four variables that govern changes in population size.
28. What is a population's age structure, and what are the three major age groups called?

Think Critically

29. What difference would it make if the southern sea otter (Case Study) became extinct?
30. How would you reply to someone who argues that people should not worry about the effects that human activities have on natural systems because ecological succession will repair the damage?

31. Which reproductive strategy do most species of insect pests and harmful bacteria use? Why does this strategy make it difficult for people to control their populations?

32. Are human populations exempt from population crashes? Explain your answer.

Chapter Activities

A. Investigate: Populations STEM

Predator and prey populations exert strong influences on each other. This activity will demonstrate some possible outcomes for an American alligator population given changes in resource availability. One-third of the class will represent alligators and the remaining two-thirds will represent resources necessary for survival.

Materials
paper and writing utensils to record
cones or tape for lines

1. Stand with your group behind one of two lines placed on opposite ends of a large open space. Facing away from each other, members of the alligator group select which resource they will look for. Members of the resources group decide which resource they will become. (Food = hands over stomach, Water = hands over mouth, Shelter = hands together over head)

2. On the count of three, turn to face the other group. Alligators locate your matching resource. (Only the alligators should move.)

3. The alligators whose needs are met are able to "survive." Both alligators and their matching natural resources return to the alligator line to become new alligators for the next round. (Only one alligator can take each resource.) If an alligator does not meet its need, then it "dies" and becomes part of the habitat, joining the resources line.

4. Repeat this for 10 rounds. Record the number of alligators in each round.

5. Using the data collected during the activity, create a graph to show the change in alligator population over the 10 rounds (representing 10 years).

6. What information can you obtain from this graph?

7. What patterns do you notice in your data? How can you explain them?

8. Using your graph of alligator population changes, predict how the population of a predator of alligators, such as a Burmese python, might change over the same 10 years. Add the predator population to your graph, making sure to identify each population. Explain the reasoning behind your prediction.

B. Citizen Science

National Geographic Learning Framework
Attitudes | Responsibility
Skills | Observation
Knowledge | Critical Species

Scientists are concerned about how environmental changes are impacting penguin populations in Antarctica. Populations of several species in this part of the world are declining from the combined effects of climate change, commercial fishing activities, and human disturbance. Researchers are working to collect data about penguin populations to better understand and predict population trends.

Since 2009, the Penguin Watch project at the University of Oxford has used cameras to monitor populations of various penguin species throughout the Southern Ocean and along the Antarctic Peninsula. The project invites citizen scientists to count and identify penguins, chicks, and eggs in the images they collect. Work with a group to analyze images collected by this organization. Then gather in small groups to describe your experience with your classmates.

CHAPTER 6

ECOSYSTEMS AND CLIMATE

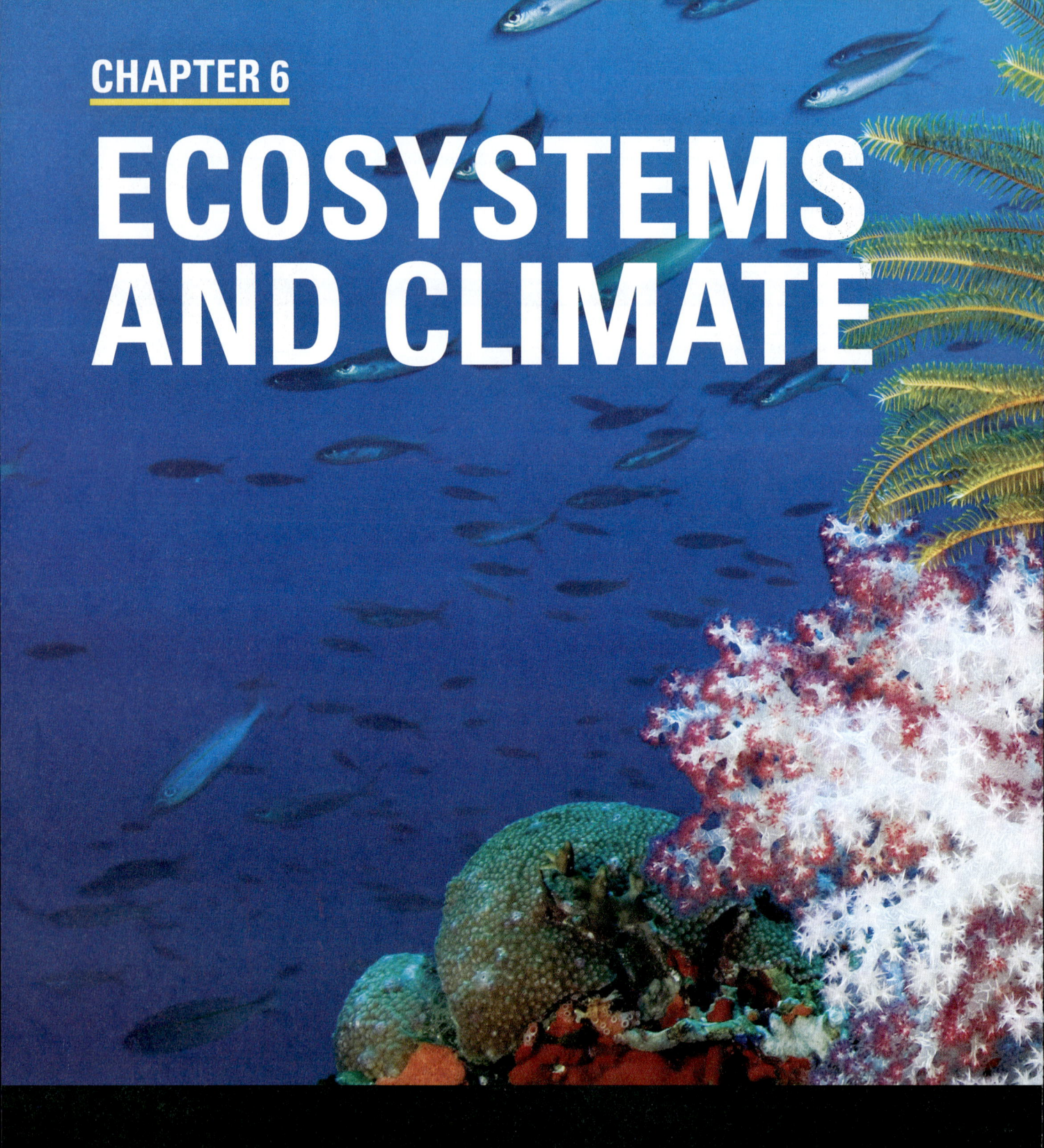

CORAL REEF ECOSYSTEMS are extremely rich and complex centers of biodiversity, providing habitat for roughly one-fifth of the world's marine fish species. Corals themselves are ancient organisms whose ancestors have been found in the fossil record from more than 400 million years ago. These well-established communities depend on a stable climate, as corals have a limited tolerance for changes in temperature.

KEY QUESTIONS

6.1 What factors influence climate?

6.2 What are the major types of terrestrial ecosystems?

6.3 What are the major types of marine ecosystems?

6.4 What are the major types of freshwater systems?

NATIONAL GEOGRAPHIC | EXPLORERS AT WORK

Saving the Last Wild Places in the Ocean

with National Geographic Explorer Enric Sala

When Enric Sala was a young boy, he was glued to the television screen watching famed undersea explorer Jacques Cousteau document life in the ocean's depths. Excited to see vast blue seas teeming with fascinating fish and marine creatures, Sala looked forward to snorkeling off the Mediterranean coast—but what did he find when he got there? Just some seaweed and a few fish the size of his hand. Later he learned that over many decades the Mediterranean Sea had been overfished.

Disappointed by what he saw (and what he didn't see), Sala decided that he wanted to become a marine biologist so that he could study the impacts of fishing on ocean life. He eventually became a professor at the Scripps Institution of Oceanography in La Jolla, California. As all of the ocean places he loved so much continued to decline, Dr. Sala says he found himself "writing the obituary of nature with increasing precision."

"I felt like a doctor telling the patient how she is going to die, with excruciating detail. If I were that patient, I would have fired myself and looked for a doctor who would look for a solution," says Sala.

Now a National Geographic Explorer-in-Residence, Sala is working to "revive his patient" through conservation efforts aimed at protecting the last pristine marine ecosystems in the world. Sala has helped protect multiple marine areas, including the Pacific Remote Islands Marine National Monument near Hawai'i, the Motu Motiro Hiva Marine Park off the coast of Chile, and the Seamounts Marine Managed Area of Costa Rica.

In 2008, Sala launched the Pristine Seas project to identify, survey, protect, and restore the last truly wild places in the ocean. The Pristine Seas project is one of the National Geographic Society's key initiatives dedicated to marine preservation. Some of the problems the Pristine Seas project is trying to address are:

Marine Pollution: Pollutants such as trash and toxic chemicals are dumped into the ocean at an alarming rate every single day, threatening marine species and their habitats.

Overfishing: Disrupting the balance of life in the ocean through overfishing has significant consequences for the millions of people who rely on fish as their primary protein source.

Sea Temperature Rise: Warmer oceans threaten coral reefs and fish species, and they cause sea levels to rise.

Acidification: As the ocean absorbs carbon emissions, its surface pH level drops, destabilizing marine environments.

Invasive Species: Aquatic invasive species can be harmful to the balance of native marine ecosystems. Especially near shorelines, nonnative species are wiping out native communities and interfering with human-built structures.

Enric Sala observes healthy corals in the South Pacific Ocean.

Today, Enric Sala is a familiar presence in the halls of National Geographic's headquarters in Washington, D.C. From there, he collaborates with an international community of marine biologists, ecologists, conservationists, and other scientists as part of Pristine Seas. Marine reserves like those Sala is supporting help protect critical ocean areas and restore the health of the ocean. Scientists are able to study intact marine environments in a reserve and understand how the ocean appeared and functioned before extensive human impact.

Since it began, Pristine Seas has helped secure 2.2 million square kilometers (850,000 square miles) of ocean protection worldwide. The commitment of individuals like Enric Sala is making a difference that will carry far into the future.

Thinking Critically
Evaluate Should the oceans be important to people who don't live on or near one, or who don't directly rely on one from day to day? Why or why not? What can people who don't live near an ocean do to help protect them?

CASE STUDY
The Importance of Coral Reefs

Shallow coral reefs form in clear, warm, coastal waters in tropical areas. These rich natural wonders are among the world's oldest, most diverse, and most productive ecosystems.

Massive colonies of tiny animals called polyps (close relatives of jellyfish) form coral reefs. Polyps build reefs by secreting a protective crust of limestone (calcium carbonate) around their soft bodies. When the polyps die, their empty crusts remain as part of a platform for more reef growth. The resulting elaborate network of calcium carbonate serves as shelter for a variety of marine animals. This buildup of calcium carbonate into a coral reef can take up to 10,000 years.

Coral reefs form out of a mutualism between polyps and tiny single-celled algae called zooxanthellae (zoh-ZAN-thel-ee) that live in the tissues of the polyps. The algae provide the polyps with food and oxygen through photosynthesis. The algae help corals produce calcium carbonate. Algae also give the reefs their stunning coloration. The polyps, in turn, provide the algae with a well-protected home and some of their nutrients.

Although shallow and deep-water coral reefs occupy only 0.2% of the ocean floor, they provide important ecosystem and economic services. Reefs act as natural barriers that protect 15% of the world's coastlines from flooding and erosion caused by battering waves and storms. They also provide habitats, food, and spawning grounds for one-fourth to one-third of the organisms living in the ocean. Reefs produce one-tenth of the global fish catch. Through tourism and fishing they provide goods and services worth about $40 billion a year.

Coral reefs are vulnerable to damage because they grow slowly and are disrupted easily. Runoff of soil can cloud the water and block the sunlight algae need for photosynthesis. The water where shallow reefs form must have a temperature of 18–30°C (64–86°F) and cannot be too acidic. This explains why the two major threats to coral reefs are climate change and ocean acidification.

Climate change could raise water temperatures above tolerable limits, killing the corals' algae. Acidification reduces the amount of carbonate available to polyps.

One result of such stressors is coral bleaching. First, the algae that corals depend on for food die off. Then the polyps themselves may die. When coral polyps die, they leave behind a skeleton of calcium carbonate (Figure 6-1).

Studies by the Global Coral Reef Monitoring Network and others estimate that since the 1950s, 45–53% of the world's shallow coral reefs have been destroyed or degraded and that another 25–33% could be lost within 20 to 40 years. These centers of biodiversity are by far the most threatened marine ecosystems. This chapter will explore the factors that determine climate and will tour each of Earth's major terrestrial (land-dwelling) and aquatic (water-dwelling) ecosystems.

As You Read Think about the ecosystem where you live as well as those far from you. How do your activities affect both nearby and faraway ecosystems?

FIGURE 6-1
A sea turtle swims over a seafloor covered with dead corals.

6.1 What Factors Influence Climate?

CORE IDEAS AND SKILLS

- Understand the difference between weather and climate.
- Relate ocean currents and air circulation to Earth's climate zones.
- Explain how greenhouse gases enter the atmosphere and how these gases affect Earth and its atmosphere.
- Define the rain shadow effect and explain how it affects climate.

KEY TERMS

weather
climate
ocean current
greenhouse gas
rain shadow

Climate Differs from Weather

The biodiversity of a region is defined in part by its climate. Climate differs from weather. **Weather** is a set of physical conditions of the lower atmosphere that includes temperature, precipitation, humidity, wind speed, cloud cover, and other factors. Weather occurs in a given area over a period of hours or days and often fluctuates daily and seasonally.

Weather differs from **climate**, which is the pattern of atmospheric conditions in an area over periods ranging from at least three decades to thousands of years. You can think of climate as typical or average weather. The climate of Florida is warm and humid whereas the weather one day may be cold and rainy. The key factors that influence an area's climate are incoming solar energy, Earth's rotation, global patterns of air and water movement, gases in the atmosphere, and Earth's surface features.

Climate changes occur more slowly than weather changes because climate is the average of weather conditions over a span of at least 30 years. Some climate changes are caused by natural events such as a change in Earth's orbit, the amount of energy from the sun, volcanic eruptions, and changes in the ocean. Human activities, such as changes to atmospheric gases, can lead to relatively rapid climate changes such as those that are occurring in northern North America (Figure 6-2). You will learn more about climate change in Chapter 16.

checkpoint What factors determine a region's climate?

FIGURE 6-2
The cold climate in Canada's northern MacKenzie River Valley is slowly becoming warmer. The average temperature has increased by about 2°C (3.6°F) since the 1970s.

FIGURE 6-3 ▼

Ocean Currents and Climate This generalized map of Earth's current climate zones shows the major ocean currents and upwelling areas (where currents bring nutrients from the ocean bottom to the surface). Based on this map, what is the general type of climate where you live?

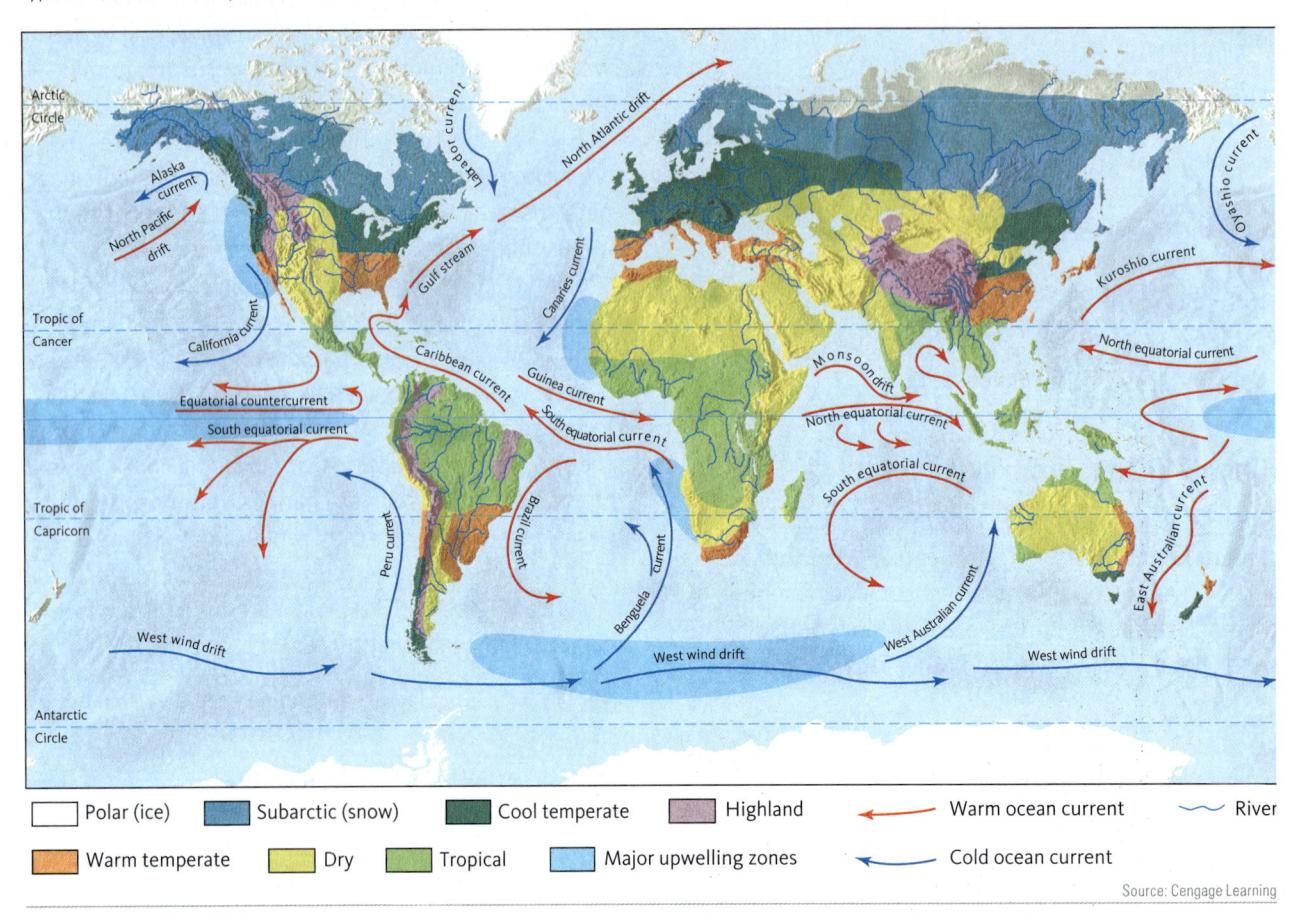

☐ Polar (ice)	☐ Subarctic (snow)	☐ Cool temperate	☐ Highland	⟵ Warm ocean current	~ River
☐ Warm temperate	☐ Dry	☐ Tropical	☐ Major upwelling zones	⟵ Cold ocean current	

Source: Cengage Learning

Climate Is Influenced by Water and Air Circulation

Some of Earth's general climate zones are shown in Figure 6-3 as colored regions on the map. The distribution of climate zones is just what you might expect in some ways. For example, climates near the Poles are cold and climates near the Equator are warm. But as you can see, there is more to the story. Climate varies among Earth's regions primarily because, over long periods of time, winds and ocean currents distribute heat and precipitation unevenly across the planet through the process of convection.

Ocean currents have a major effect on climate. **Ocean currents** are mass movements of surface water driven by winds and shaped by landforms. Figure 6-3 illustrates the general movement of Earth's ocean currents in addition to showing climate zones. Warm currents are shown in red and cold currents are shown in blue. Three major factors affect the way in which water circulates on Earth. One factor is the uneven heating of Earth by the sun. Another factor is Earth's rotation on its axis. A third factor that affects ocean currents, and thus climate, is the physical properties of air, water, and land.

Uneven Heating of Earth's Surface Air is heated more at the Equator, where the sun's rays strike directly, than at the Poles, where sunlight strikes at a lower angle and spreads out over a much greater area (Figure 6-4). These differences in the input and radiation of solar energy into the atmosphere help explain why tropical regions near the Equator are hot and polar regions are cold. Temperate regions between these two areas generally have both warm and cool temperatures. The intense input of solar radiation in tropical regions leads to greatly increased evaporation of moisture from forests, grasslands, and bodies of water. As a result, tropical regions normally receive more precipitation than do other areas.

CHAPTER 6 ECOSYSTEMS AND CLIMATE

FIGURE 6-4

Global Air Circulation As air rises and falls in Hadley cells (right), it is also deflected to the east or west (left), depending on where the cell is located. The resulting global winds help to distribute heat and moisture in the atmosphere. How do Hadley cells relate to the areas of forest, desert, and grassland?

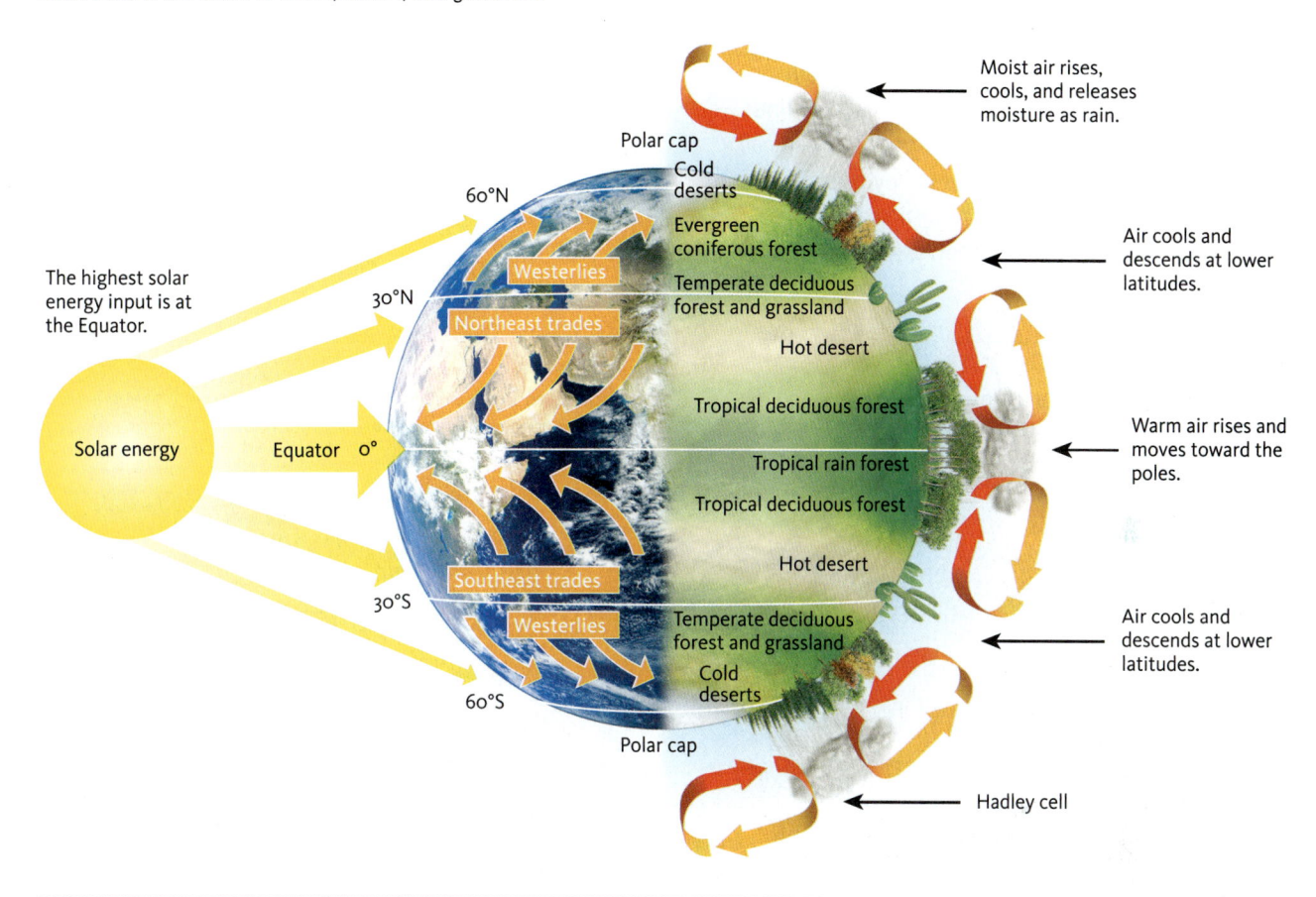

Earth's Rotation As Earth rotates on its axis, Earth's surface near the Equator spins faster than the regions to the north and south. Air masses that move north or south appear to turn, or deflect, relative to the land below them. The deflection of a mass as it moves over a spinning surface is called the Coriolis effect. Winds that move north or south toward the Equator curl to the west because they are moving slower than the land below them. This movement produces the Northeast and Southeast trade winds (Figure 6-4). Air masses that move away from the Equator deflect eastward because they are moving faster than the land below them. These winds "from the west" are called westerlies. Surface air that moves toward the Poles causes easterly winds, which blow from the east. Trade winds, westerly winds, and easterly winds are prevailing winds. Prevailing winds blow almost continuously. They distribute heat and moisture over Earth's surface and drive surface ocean currents.

> **CONSIDER THIS**
>
> **Trade winds can change.**
>
> Normal prevailing winds blowing east to west cause upwellings of cold, nutrient-rich bottom water in the tropical Pacific Ocean. Every few years, a shift in trade winds, known as the El Niño–Southern Oscillation (ENSO), disrupts this pattern for 1–2 years.

Properties of Air, Water, and Land Heat from the sun evaporates ocean water and transfers heat from the oceans to the atmosphere, especially near the Equator. This evaporation of water creates giant cyclical Hadley cells that move air, heat, and moisture both vertically and from place to place in the atmosphere. Hadley cells are depicted in Figure 6-4 by looping red and orange arrows.

LESSON 6.1

Hadley cells transfer energy by convection (Figure 6-5). Convection is the physical process by which warm, wet air rises into the atmosphere, then cools and releases heat and moisture as precipitation. The cooler, denser, dry air then sinks back to the surface, warms up, and absorbs moisture as it flows across Earth's surface to begin the cycle again.

Earth's air circulation patterns, prevailing winds, and the configuration of continents and oceans all play a part in the formation of Hadley cells. Together, these factors lead to the irregular distribution of climates and the resulting deserts, grasslands, and forests.

The ocean currents you have read about so far are surface water currents. Water also moves vertically in the oceans as denser water sinks while less dense water rises. This creates a connected loop of deep and shallow ocean currents (Figure 6-6). This loop acts like a giant conveyor belt as it moves heat from the surface to the deep sea and transfers warm and cold water between the tropics and the Poles. This global conveyor belt takes about 1,000 years to circulate, but plays an important role in regulating weather and climate and cycling nutrients.

checkpoint What is the Coriolis effect and what causes it?

FIGURE 6-5
Convection The same phenomenon that moves heated water in a pot moves air in the atmosphere—convection. Cool, dense, dry air warms as it sinks and absorbs moisture and heat from Earth's surface. The resulting warm air cools as it rises, releasing moisture as precipitation.

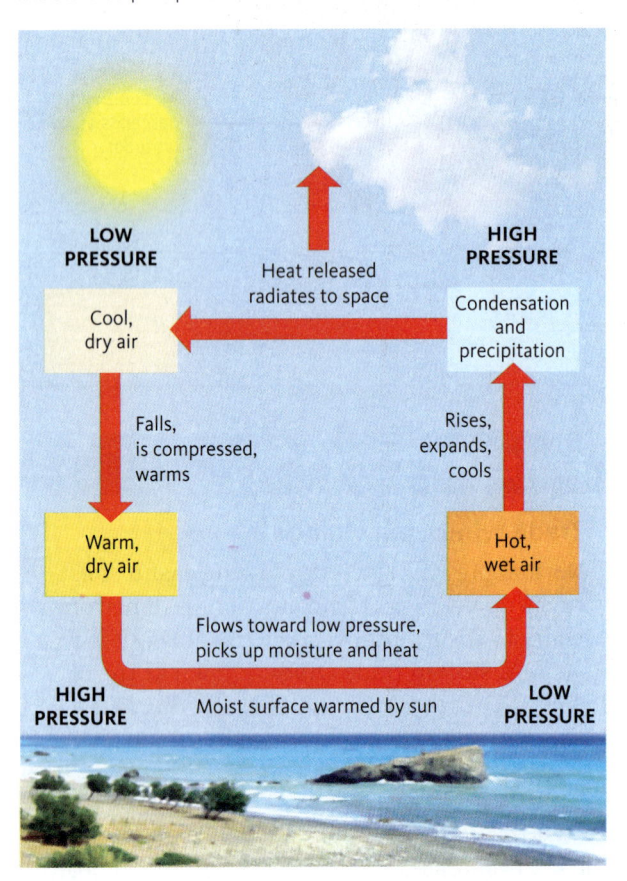

FIGURE 6-6
Connected Loop of Currents A "conveyer belt" of deep and shallow ocean currents moves warm and cool water to various parts of Earth.

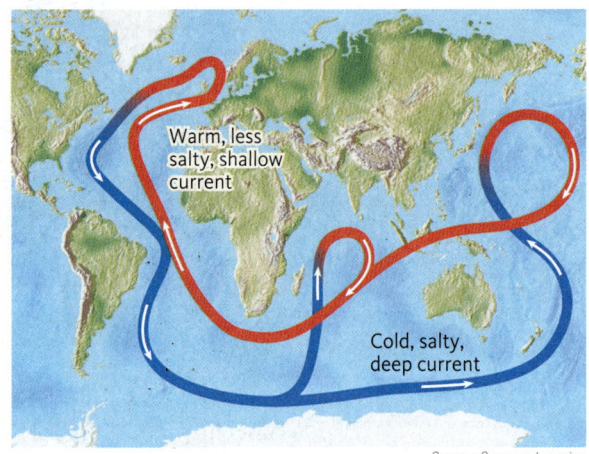

Source: Cengage Learning

164 CHAPTER 6 ECOSYSTEMS AND CLIMATE

FIGURE 6-7
Positive Feedback The greenhouse effect is intensified by a positive feedback loop with melting Arctic sea ice. As temperatures rise, more ice melts, reducing the surface area of reflective land. Less reflective land means more energy is absorbed by the ocean. The ocean thus warms and the cycle accelerates.

Temperatures rise. → Ice melts. → More heat is absorbed. →

SCIENCE FOCUS 6.1

GREENHOUSE GASES AND CLIMATE

Gases in the lower atmosphere affect Earth's climates. As energy flows from the sun to Earth's surface, some energy is reflected back into the atmosphere. Molecules of certain gases in the atmosphere, including water vapor (H_2O), carbon dioxide (CO_2), methane (CH_4), and nitrous oxide (N_2O), absorb some of this solar energy and release a portion of it as infrared radiation, which warms the lower atmosphere and Earth's surface. These gases that warm the lower atmosphere are called **greenhouse gases**. Greenhouse gases play a role in determining the lower atmosphere's average temperatures and, therefore, Earth's climates. The natural warming of the lower atmosphere by greenhouse gases is called the greenhouse effect (Lesson 3.1). Without this natural warming effect, Earth would be a cold and mostly lifeless planet.

Human activities, such as producing and burning fossil fuels, clearing forests, and growing crops, release large amounts of carbon dioxide and methane into the atmosphere. An enormous body of scientific evidence, combined with climate model projections, indicate that humans are releasing greenhouse gases into the atmosphere faster than they can be removed by natural processes such as the carbon and nitrogen cycles. These emissions intensify Earth's natural greenhouse effect (Figure 6-7) and alter global climates. The complex problem of climate change and how it can be addressed will be discussed in depth in Chapter 16.

Thinking Critically
Evaluate Is the greenhouse effect harmful or beneficial?

LESSON 6.1 165

FIGURE 6-8
Rain Shadow Landforms can affect climate. Winds lose most of their moisture as rain and snow fall on the windward slopes of a mountain range. This leads to semiarid and arid conditions on the leeward side of the mountain range and on the land beyond.

Prevailing winds pick up moisture from an ocean.

On the windward side of a mountain range, air rises, cools, and releases moisture.

On the leeward side of the mountain range, air descends, warms, and releases little moisture, causing a rain shadow effect.

Earth's Surface Features Affect Local Climate

Another factor that influences a region's climate in addition to ocean currents, winds, and the angle of the sun's rays is Earth's topography. Topography refers to the shape of Earth's surface and its features. Mountains interrupt the flow of prevailing winds and the movement of storms. When moist air from an ocean blows inland and reaches a mountain range, the air mass is forced upward. As the air mass rises, it cools, expands, and loses most of its moisture as rain and snow that fall on the windward slope of the mountain.

As the drier air mass passes over mountaintops, it flows down the leeward slopes (facing away from the wind) and warms. This warmer air can hold more moisture, but it typically does not release much of it. Instead, the air tends to dry out plants and soil. This effect is called a **rain shadow**. Over many decades, the rain shadow effect results in semiarid or arid conditions on the leeward side of a high mountain range (Figure 6-8). On the West Coast of the United States, warm, moist winds from the Pacific Ocean encounter the Sierra Nevada mountain range, forming the Mojave Desert in California, Nevada, Utah, and Arizona on the leeward side.

Cities can also affect local climate by creating distinct microclimates. Bricks, concrete, asphalt, and other building materials conduct heat differently than natural surface coverings. Buildings block wind flow. Motor vehicles and the heating and cooling systems of buildings release large quantities of heat and pollutants. As a result, cities tend to have more haze, smog, higher temperatures, and lower wind speeds than the surrounding countryside. These factors make cities "heat islands" in what is known as the urban heat island effect.

checkpoint How do mountains affect the climate of a region?

6.1 Assessment

1. **Contrast** How does climate differ from weather?
2. **Explain** How does Earth's rotation on its axis affect climate?
3. **Summarize** Summarize the greenhouse effect in a series of steps.
4. **Analyze** What is the role of convection in producing the rain shadow effect?
5. **Predict** Explain how urban expansion might affect a region's climate.

CROSSCUTTING CONCEPTS

6. **Patterns** How can studying climate at the global level provide information about the climate where you live?

166 CHAPTER 6 ECOSYSTEMS AND CLIMATE

6.2 What Are the Major Types of Terrestrial Ecosystems?

CORE IDEAS AND SKILLS

- Describe how climate and vegetation vary with latitude and elevation.
- Identify the types of deserts, grasslands, and forests.
- Define the ecological roles of mountains and their importance in ecosystem services.
- Describe some ways in which humans alter terrestrial ecosystems.

KEY TERMS

edge effect   permafrost

Climate Helps Determine Where Terrestrial Organisms Can Live

In general, the wildlife in North America is more like that in Europe than in Mexico or Central America. Even though Mexico and Central America are geographically closer to North America, Europe has a similar temperate climate owing to its latitude. The latitude of a region is its distance from the Equator. Climate and vegetation tend to vary with latitude. Combinations of average annual precipitation and temperatures, along with global air circulation patterns and ocean currents, lead to the formation of tropical (hot), temperate (moderate), and polar (cold) areas. Each of these areas includes deserts, grasslands, and forests, as summarized in Figure 6-9.

Climate and vegetation also vary according to elevation. The elevation of a region is its height above sea level. If you travel from the Equator to Earth's northern polar region (latitude) you can observe changes in plant life similar to those you would encounter in climbing a tall mountain from its base to its summit (elevation). Latitude and elevation, then, influence a region's dominant vegetation. The dominant plant life, in turn, helps determine the other types of plants and the types of animals and decomposers that live there.

FIGURE 6-9 ▼
Climate and Dominant Plant Life Average precipitation and temperature help to determine the type of desert, grassland, or forest in any particular area.

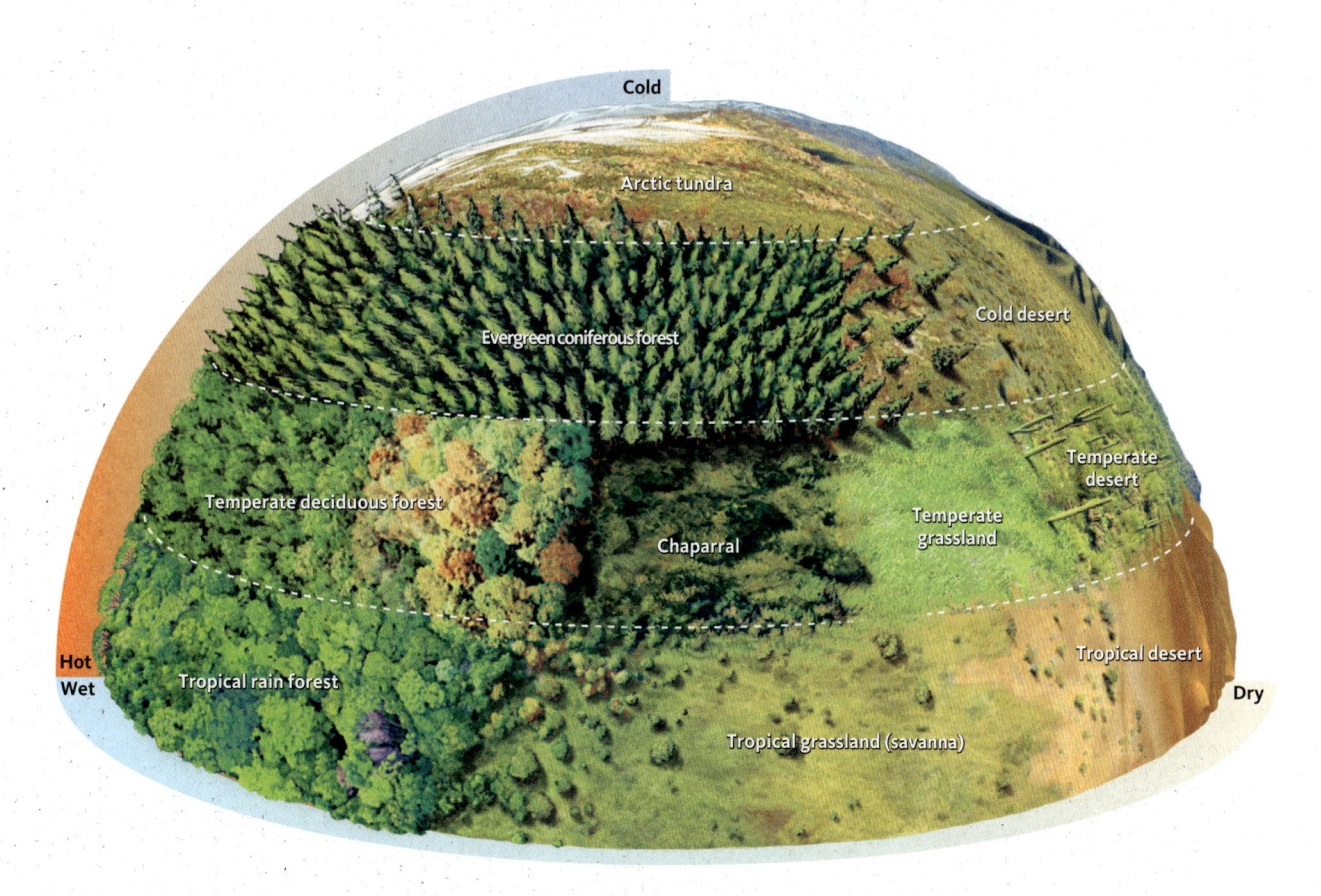

LESSON 6.2 **167**

FIGURE 6-10

Earth's Major Biomes A biome is a broad classification of land based on the climate and dominant life there. Each biome contains many ecosystems.

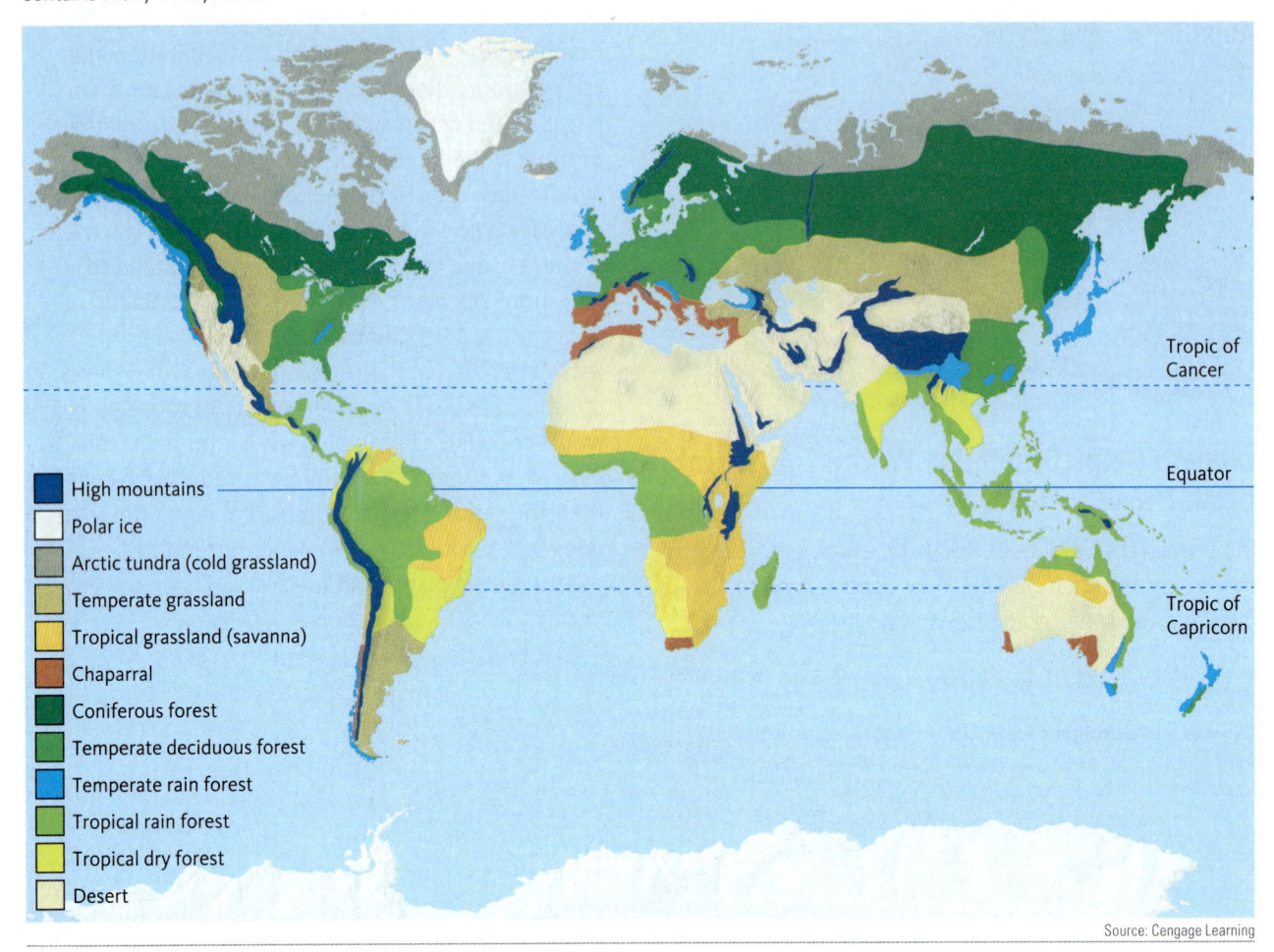

Legend:
- High mountains
- Polar ice
- Arctic tundra (cold grassland)
- Temperate grassland
- Tropical grassland (savanna)
- Chaparral
- Coniferous forest
- Temperate deciduous forest
- Temperate rain forest
- Tropical rain forest
- Tropical dry forest
- Desert

Source: Cengage Learning

Figure 6-10 shows one way in which scientists have divided the world into large terrestrial regions called biomes. You may recall from Chapter 4 that a biome is characterized by a specific type of climate and certain combinations of dominant plant life. The variety of terrestrial biomes and aquatic systems produce biodiversity—a vital part of Earth's natural capital and one of the three scientific factors of sustainability. Scientists have different ways of defining Earth's biomes. Some scientists recognize as few as five biomes while others classify dozens.

Maps show major biomes with sharp boundaries and uniform vegetation. But in reality, biomes are not uniform. They consist of a patchwork of areas, each with somewhat different biological communities, yet with similarities typical of the biome. These patches occur because of the irregular distribution of resources needed by plants and animals. Also, human activities have removed or altered the natural vegetation in many areas.

There are also differences along the transition zone (called the ecotone) between two different ecosystems or biomes. The ecotone contains habitats common to both ecosystems along with other habitats that are unique to the transition zone. The **edge effect** is the tendency for a transition zone to have greater species diversity and a higher density of organisms than found in either of the individual ecosystems.

checkpoint What two elements characterize a biome?

FIGURE 6-11
Thick, fleshy stems, widely spreading roots, and thin spines in place of leaves help cacti conserve water and survive in dry climates, such as the Uyuni salt flat in Bolivia.

SCIENCE FOCUS 6.2

DESERT ADAPTATIONS

Adaptations for survival in the desert have two main themes: beat the heat and every drop of water counts. Desert plants have evolved a number of adaptations based on such strategies.

During long hot-and-dry spells, some desert plants drop their leaves to survive in a dormant state. Succulent (fleshy) plants such as the cacti shown in Figure 6-11 have no leaves, which means they don't lose water to the atmosphere through transpiration. They reduce water loss by opening their pores only at night to take up carbon dioxide (CO_2). Succulents also store water and produce nutrients in their expandable, fleshy tissue. The spines of these and many other desert plants guard them from being eaten by herbivores seeking the precious water they hold.

Some desert plants use deep roots to tap into groundwater. Others such as the prickly pear and saguaro cacti use widely spread shallow roots to collect water after brief showers and store it in their spongy tissues. Still other desert plants have wax-coated leaves that reduce water loss. Annual wildflowers and grasses store much of their biomass in seeds that remain inactive, sometimes for years, until they receive enough water to germinate. Shortly after a rain, these seeds sprout, grow, and carpet deserts with dazzling arrays of colorful flowers that last several weeks.

Most desert animals are small. Some beat the heat by hiding in cool burrows or rocky crevices by day and coming out at night or in the early morning when it's cooler. Others become dormant during periods of extreme heat or drought. Larger animals such as camels drink massive amounts of water when it is available and store it in their fat for use as needed. Camels do not sweat, which reduces water loss through evaporation. Kangaroo rats seldom drink water. They get the water they need by breaking down fats in seeds they consume.

Humans also have adaptations for surviving the heat. The human body maintains homeostasis, or stability, through sweating. When the body begins to overheat, sweat is released from glands in the skin. Evaporating sweat cools blood that moves by the surface of the skin. Tiny vessels bring cooled blood from the head and face straight to the human's large brain. Sweating helps humans beat the heat, but not conserve water. A human can lose as much as 12 liters of water (about 3 gallons) during one day in the desert heat.

Thinking Critically
Explain What methods or devices could you use to stay cool in a desert?

LESSON 6.2

Types of Deserts

In a desert, annual precipitation is low and often scattered unevenly throughout the year. During the day, the sun warms the ground and evaporates water from plant leaves and the soil. At night, most of the heat stored in the ground radiates quickly into the atmosphere. This explains why you may be hot during the day but shiver at night when you are in a desert. A combination of low rainfall and varying average temperatures creates a variety of desert types—tropical, temperate, and cold.

Tropical Desert *Tropical* does not always mean hot and wet. Tropical deserts such as the Sahara (Figure 6-13) and the Namib of Africa are hot but dry most of the year. They have few plants and a hard, windblown surface strewn with rocks and sand. The top graph in Figure 6-12 shows the average annual temperature and precipitation for a typical tropical desert.

Temperate Desert In temperate deserts, daytime temperatures are high in summer and low in winter. Temperate deserts receive more precipitation than tropical deserts. Their sparse vegetation is mostly widely dispersed, drought-resistant shrubs and cacti or other succulents adapted to the dry conditions and temperature variations. The middle graph in Figure 6-12 shows the typical climate for a temperate desert.

Cold Desert In cold deserts such as the Gobi Desert in Mongolia, vegetation is sparse. Winters are cold, summers are warm or hot, and precipitation is low in cold deserts (Figure 6-12, bottom).

Desert ecosystems are fragile. The lack of vegetation, especially in tropical and polar deserts, makes them vulnerable to heavy wind erosion. They can be easily damaged or destroyed because they have slow plant growth, low species diversity, slow nutrient cycling (due to low bacterial activity in the soils), and little water. It takes decades to centuries for desert soils to recover from disturbances such as off-road vehicle traffic. Traffic can also destroy the habitats for a variety of animal species that live underground.

checkpoint What characteristics must a biome have to be classified as desert?

FIGURE 6-12 ▼

Desert Climates These climate graphs display typical variations in annual temperature (red) and precipitation (blue) in tropical, temperate, and cold deserts in the Northern Hemisphere. How are the climates alike and different?

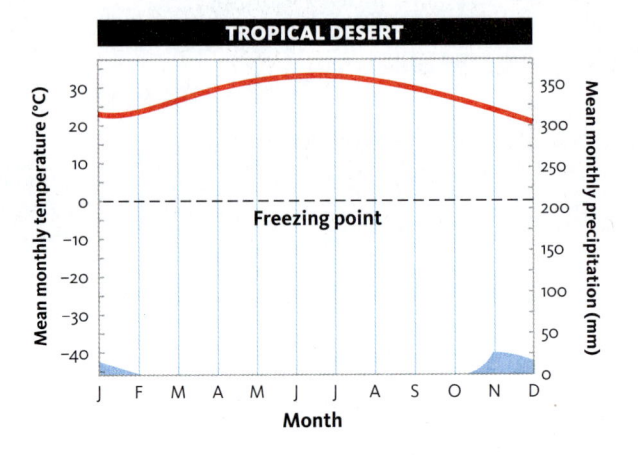

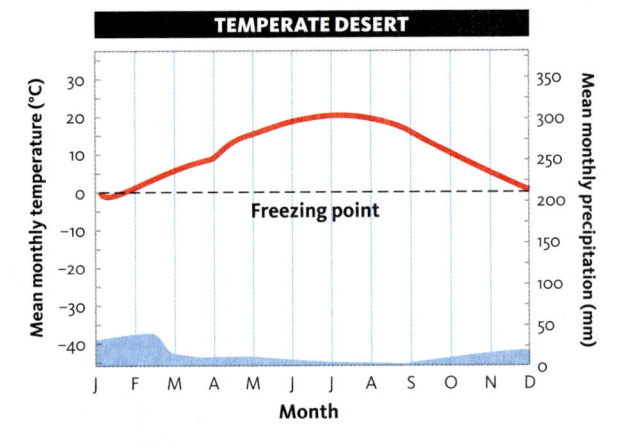

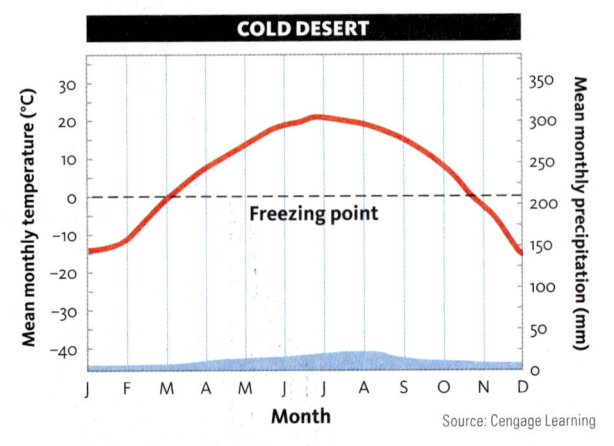

CONSIDER THIS

Is the legend true that camels store water in their humps?

The answer is yes *and* no. To survive in the dry desert climate, camels store fat in their humps. The fat can later be converted into energy and water, allowing camels to endure long periods without drinking. Camel humps are sometimes covered with thick fur, which seems to go against all reason in the desert. But the air spaces in the camel's fur actually serve to insulate it against the outside heat.

Source: Cengage Learning

FIGURE 6-13
The Sahara in northern Africa is the world's largest tropical desert. Temperatures here are high and rainfall is unreliable. Some plants and animals, such as camels, have adapted to the harsh climate. Camels provide a viable means of transport across this vast, roadless landscape.

LESSON 6.2   171

Types of Grasslands

Grasslands occur primarily in the interiors of continents in areas that are too moist for deserts to form and too dry for forests to grow. Grasslands are dominated by grass species but typically contain many other types of flowering plants as well. Different types of grasslands arise from climatic differences in temperature and precipitation (Figure 6-14).

Tropical Grassland The largest tropical grassland is in Africa, but there are also large areas in Australia, South America, and India. The main type of tropical grassland is the savanna, which contains widely scattered clumps of trees. The savanna has warm temperatures year-round and alternating dry and wet seasons. Tropical savannas in Africa (Figure 6-15) are home to grazing (grass-eating) and browsing (twig- and leaf-nibbling) hoofed animals such as wildebeests, gazelles, zebras, giraffes, and antelopes. Lions and hyenas are predators of the savanna.

Temperate Grassland Temperate grassland winters can be bitterly cold, and the summers are hot and dry. Annual precipitation is fairly sparse and falls unevenly throughout the year. Because the above-ground parts of most of the grasses die and decompose each year, organic matter accumulates to produce deep, fertile topsoil. This topsoil is held in place by a thick network of the intertwined grass roots. If the topsoil is plowed, soil can be lost to high winds. Periodic droughts and fires burn the above-ground parts of plants but leave the roots unharmed. Many of the world's temperate grasslands have been converted to farmland because their fertile soils are useful for growing crops and grazing cattle.

Cold Grassland Cold grasslands or Arctic tundra lie south of the Arctic polar ice cap. During most of the year, these treeless plains are bitterly cold, swept by frigid winds, and covered with ice and snow. Winters are long with few hours of daylight, and the scant precipitation falls primarily as snow. A thick, spongy mat of low-growing plants lies under the snow. Trees and tall plants cannot survive in the cold and windy tundra. One outcome of the extreme cold is the formation of permafrost. **Permafrost** is underground soil where captured water can stay frozen for more than two consecutive years. The permafrost layer prevents melted snow and ice from draining into the ground during the brief summer. Hordes of mosquitoes, black flies, and other insects thrive in the shallow surface pools that result. The insects serve as food for large colonies of birds that migrate from the south to nest and breed in the tundra's summer bogs and ponds.

checkpoint Describe three types of grasslands.

FIGURE 6-14 ▼

Grassland Climates These climate graphs display typical variations in annual temperature (red) and precipitation (blue) in tropical, temperate, and cold grasslands in the Northern Hemisphere. How are the climates alike and different?

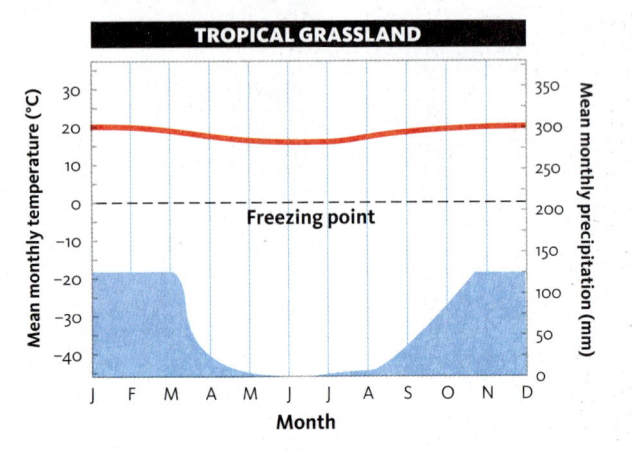

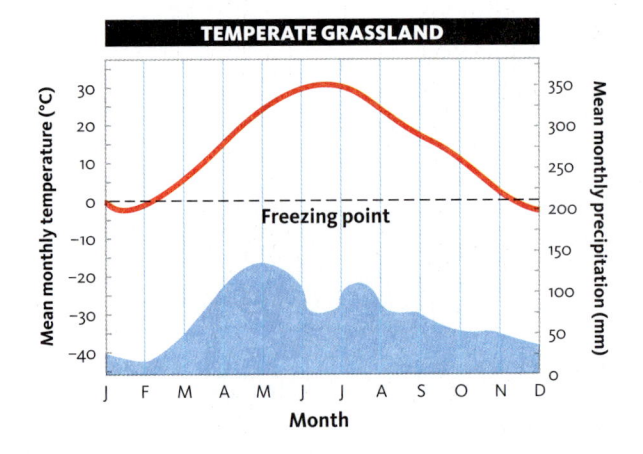

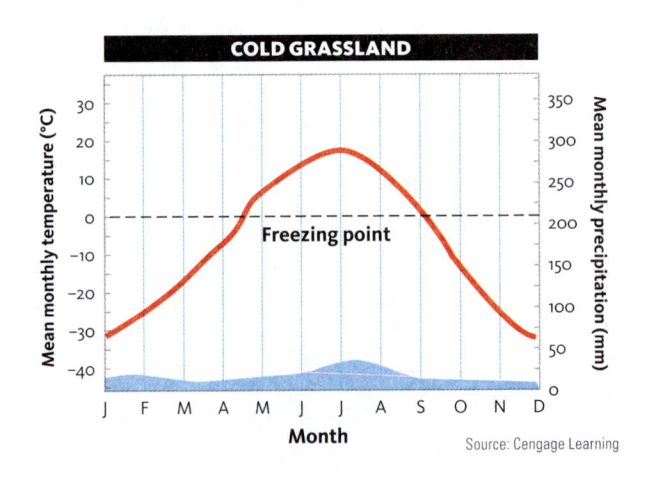

Source: Cengage Learning

FIGURE 6-15
The African savanna is a complex, tropical grassland ecosystem characterized by large grassy plains, dispersed trees, and an abundance of animals. Large herds of zebras, for example, prune the grasses and fertilize the soil as they graze.

Types of Forests

Forests are lands that are dominated by trees. The different types of forest result from combinations of varying precipitation levels and varying average temperatures (Figure 6-16).

Tropical Rain Forests Tropical rain forests are found near the Equator where hot, moisture-laden air rises and releases its moisture. These lush forests have year-round warm temperatures, high humidity, and almost daily heavy rainfall. This fairly constant warm, wet climate is ideal for a wide variety of plants and animals (Figure 6-18). Tropical rain forests are dominated by broadleaf evergreen plants that keep most of their leaves year-round. The tops of the trees form a dense canopy that blocks most of the sunlight from reaching the forest floor. Many of the plants living at the ground level have enormous leaves to capture what little sunlight filters down to them. Some trees are draped with vines (called lianas) that reach the treetops to gain access to sunlight. In the canopy, the vines grow from one tree to another, providing walkways for many species living there. When a large tree is cut down, its network of lianas can pull down other trees.

Rain forest species occupy a variety of specialized niches in distinct layers that contribute to their high species diversity. Vegetation layers are structured according to the plants' needs for sunlight. Much of the animal life, particularly insects, bats, and birds, lives in the sunny canopy layer. The canopy provides abundant shelter with supplies of leaves, flowers, and fruits. Dropped leaves, fallen trees, and dead animals decompose quickly because of the warm, moist conditions and the hordes of decomposers.

Temperate Forest Temperate forests can be deciduous or coniferous, based on the dominant type of trees. Temperate deciduous forests typically experience warm summers, cold winters, and abundant precipitation—rain in summer and snow in winter months. These areas are dominated by a few species of broadleaf deciduous trees such as oak, hickory, maple, aspen, and birch. Animal species living in these forests include predators such as wolves, foxes, and wildcats. They feed on herbivores such as white-tailed deer, squirrels, rabbits, and mice. Warblers, robins, and other bird species live in these forests during the spring and summer. A thick layer of slowly decaying leaf litter on the forest floor provides a storehouse of nutrients.

Another type of temperate forest, the temperate rain forest, is found scattered within coastal temperate areas with ample rainfall and moisture from dense ocean fog. Thick stands of these forests with large conifers like the Sitka spruce, Douglas fir, giant sequoia, and redwoods once dominated areas along the west coast of North America.

FIGURE 6-16

Forest Climates These climate graphs display typical variations in annual temperature (red) and precipitation (blue) in tropical, temperate, and cold forests in the Northern Hemisphere. How are the climates alike and different?

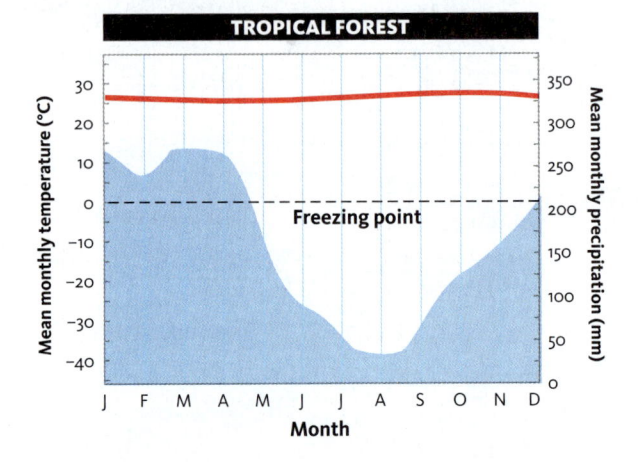

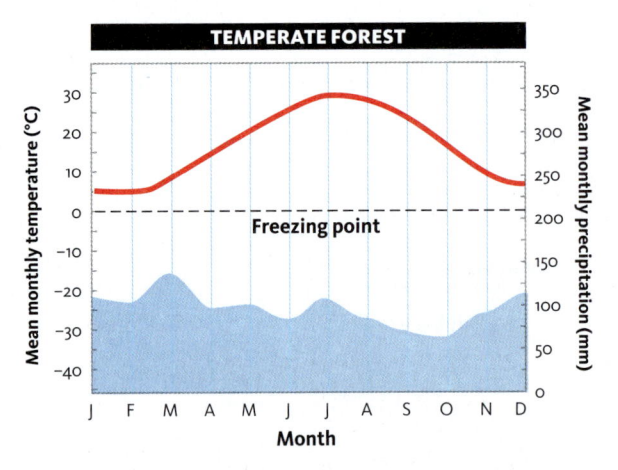

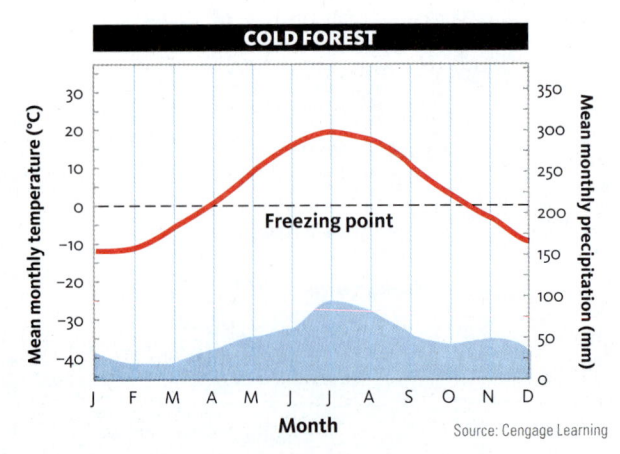

Source: Cengage Learning

Boreal Forests Cold, or northern coniferous forests, also called boreal forests or taigas (TIE-guhs), are found south of the Arctic tundra. In this subarctic, cold, and moist climate, winters are long and extremely cold. Most boreal forests are dominated by a few species of coniferous evergreen trees (or conifers) such as spruce, fir, cedar, hemlock, and pine. Plant diversity is low because few species can survive the winters when soil moisture is frozen.

Beneath the stands of trees in boreal forests is a deep layer of partially decomposed needles. The slowly decomposing conifer needles make the nutrient-poor topsoil acidic and prevent most other plants from growing on the forest floor. Year-round wildlife includes bears, wolves, moose, lynx, and many burrowing rodent species. During the brief summer, warblers and other birds feed on insects.

checkpoint Name and describe three types of forests.

NATIONAL GEOGRAPHIC | EXPLORERS AT WORK
Alizé Carrère Cultural Ecologist

National Geographic Explorer Alizé Carrère is an explorer in the true sense of the word. Raised in a quasi-treehouse on a lake in upstate New York, Carrère moved to Montreal to study environmental science and international development at McGill University, where she also completed her master's degree in bioresource engineering.

Carrère travels the planet, searching out solutions for climate change, learning about different cultures, and sharing her adventures through writing, photography, and speaking.

Carrère's broad perspective and creativity enable her to learn from innovative solutions she finds around the globe. For example, severe deforestation in Madagascar has led to the formation of massive holes in the ground, known locally as "lavakas." Carrère was inspired when she heard one of her professors, Dr. Jon Unruh, speak about lavakas. Unruh discussed lavakas not as just another example of environmental degradation, but as an opportunity. Water and nutrients concentrate at the base of the lavakas, forming rich, fertile soil. Agricultural communities have formed in these areas, contributing to local food security and environmental management. Carrère now collaborates with Dr. Unruh on researching such adaptive systems as part of a growing field of study known as cultural ecology.

FIGURE 6-17
A lemur accompanies Alizé Carrère as she explores a Malagasy rain forest.

Harpy eagle
Harpia harpyja

Blue-and-yellow macaw
Ara ararauna

Brown-throated three-toed sloth
Bradypus variegatus

Emerald tree boa
Corallus caninus

EMERGENT LAYER

160

140

CANOPY

120

100

Brazil nut tree

Red-and-green macaw

Brown-throated three-toed sloth

Liana

Harpy eagle

176

Blue morpho
Morpho menelaus

Green iguana
Iguana iguana

Jaguar
Panthera onca

Brazilian tapir
Tapirus terrestris

UNDERSTORY

FOREST FLOOR

0 ft
20
40
60
80

Jaguar

Bromeliad

Brazilian tapir

White-lipped peccary

NATIONAL GEOGRAPHIC
Rain Forest Strata

FIGURE 6-18
In a tropical rain forest, specialized plant and animal niches are stratified, or arranged roughly in layers. Filling such specialized niches enables many species to avoid or minimize competition. The result is a great wealth of species. (Representative animals are not shown to scale.)

177

Mountains Play Important Ecological Roles

About one-fourth of Earth's land surface is steep, high-elevation land. In mountains, dramatic changes in altitude, slope, climate, soil, and vegetation occur over short distances. Of the world's population, more than a billion people live in mountain ranges or their foothills. More than half of all people depend on mountain systems for all or some of their water. The soil on the steep slopes of mountains erodes easily when the vegetation holding it in place is disturbed. Natural disturbances like landslides and avalanches, or human activities such as timber cutting and agriculture, increase soil erosion. Many mountains are "islands of biodiversity" surrounded by lower-elevation landscapes that have been transformed by human activities.

Mountains contain the majority of the world's forests, which are habitats for much of the planet's terrestrial biodiversity. Montane (mountain) forests often are habitats for endemic species—species found nowhere else on Earth. Mountains serve as sanctuaries for animals capable of migrating to higher altitudes and surviving in such environments. Every year, more of these animals are driven from lowland areas to mountain habitats by human activities and a warming climate.

Mountains also play a critical role in the hydrologic cycle by serving as major storehouses of water. During winter, precipitation is stored as ice and snow. In the warmer weather of spring and summer, much of this snow and ice melts. The melted water is released to streams and used by wildlife and humans for drinking and irrigating crops. Because the atmosphere has warmed over the last 40 years, mountaintop snowpacks and glaciers are melting earlier in the spring each year. This can lead to lower food production in certain areas. Much of the water needed throughout the summer for irrigation is released too quickly and too early in the season.

Scientific measurements and climate models indicate that a large number of the world's mountaintop glaciers may disappear during this century if the atmosphere continues to warm. Many people may be forced to move from their homelands in search of new water supplies and places to grow crops. Despite the ecological, economic, and cultural importance of mountain ecosystems, methods for protecting these areas have eluded many governments and environmental organizations.

checkpoint What role do mountains play in the hydrologic cycle?

Human Impacts on Terrestrial Ecosystems

Humans have not treaded lightly on terrestrial ecosystems, as summarized in Figure 6-19. According to Johan Rockström of the Stockholm Resilience Center (Science Focus 3.4), the spread of unsustainable agriculture is one of the greatest threats to terrestrial ecosystems. In 2015, the Stockholm group estimated the ecological tipping point for land use change to be surpassed. That means that recovering terrestrial ecosystems and their associated aquatic systems may be exceedingly difficult. Over a third of Earth's ice-free land is presently devoted to agriculture. As the human population continues to grow, it will require more land for agriculture. Innovative solutions must be applied to make existing farmland more productive and natural capital securely protected. Land is also cleared to expand urban development in more densely populated regions.

FIGURE 6-19 ▼

Major Human Impacts on Terrestrial Ecosystems			
Deserts	**Grasslands**	**Forests**	**Mountains**
Large desert cities	Conversion to cropland	Clearing for agriculture, livestock grazing, timber, and urban development	Agriculture
Destruction of soil and underground habitat by off-road vehicles	Release of CO_2 to atmosphere from burning grassland	Conversion of diverse forests to tree plantations	Hydroelectric dams and reservoirs
Depletion of groundwater	Overgrazing by livestock	Damage from off-road vehicles	Air pollution blowing in from urban areas and power plants
Land disturbance and pollution from mineral extraction	Oil production and off-road vehicles in Arctic tundra	Pollution of forest streams	Soil damage from off-road vehicles

FIGURE 6-20
Areas of the Amazon rain forest are burned to create pasture for cattle.

Human activities such as agriculture and urban development have destroyed or disturbed at least half of all tropical rain forests (Figure 6-20). Tropical rain forest ecosystems are brimming with natural capital. Although they cover only about 2% of Earth's total surface, ecologists estimate that they contain at least 50% of the known terrestrial plant and animal species. A single tree in these forests may support several thousand insect species. Plants from tropical rain forests are a source of several important natural chemicals, many of which act as blueprints for making most of the world's prescription drugs.

Because decomposition occurs rapidly in rain forests, about 90% of the nutrients released are quickly taken up and stored by trees, vines, and other plants. Nutrients not taken up are soon leached from the thin topsoil by the frequent rainfall. As a result, little plant litter builds up on the ground. The lack of fertile soil explains why rain forests are not good places to clear and grow crops or graze cattle sustainably. Ecologists warn that without strong protective measures, most of these forests could be gone by the end of this century and replaced by grassland. Their unique biodiversity and valuable ecosystem services will disappear with them.

checkpoint Why is agriculture on cleared rain forest soil unsustainable?

6.2 Assessment

1. **Apply** Select a region of the world and explain how its latitude, elevation, geographical features, and relationship to winds and ocean currents determine its climate.
2. **Predict** Where in a forest are you likely to find the greatest biodiversity?
3. **Describe** What ecosystem services do mountains provide?

CROSSCUTTING CONCEPTS

4. **Stability and Change** Identify and describe a feedback loop that affects your region's climate.

SCIENCE AND ENGINEERING PRACTICES

5. **Communicate** Using the climate graphs shown in this lesson as a guide, draw a generalized climate graph for a temperate rain forest ecosystem. You do not need to show specific values for temperature or precipitation.

LESSON 6.2   179

6.3 What Are the Major Types of Marine Ecosystems?

CORE IDEAS AND SKILLS

- Define aquatic life zones and explain the difference between marine and freshwater life zones.
- Discuss the difference between the euphotic zone and the bathyal zone of the ocean.
- Explain the causes of ocean acidification.

KEY TERMS

aquatic life zone	estuary	open sea
marine life zone	brackish water	ocean
coastal zone	coastal wetland	acidification

Oceans Provide Vital Ecosystem Services

About 71% of Earth's surface is covered with ocean water. Although the global ocean is a single and continuous body of salt water, the International Hydrographic Organization defines four oceans—the Arctic, the Atlantic, the Indian, and the Pacific. Together, the oceans hold almost 98% of Earth's water. Oceans provide vital functions for all ecosystems. Among other ecosystem services, oceans provide nutrient cycling, including carbon dioxide absorption and oxygen production. Oceans also provide economic services such as food, tidal and wave energy, transportation routes, minerals, recreation, and tourism.

The aquatic equivalents of terrestrial biomes are called **aquatic life zones**. These are the saltwater and freshwater portions of the biosphere that can support life. The distribution of many aquatic organisms is determined largely by the water's salinity. Salinity is the amount of various salts (such as sodium chloride) dissolved in a given volume of water. Saltwater environments can be divided into marine life zones. **Marine life zones** include oceans and their bays, estuaries, coastal wetlands, shorelines, coral reefs, and mangrove forests. Open oceans can be further divided into euphotic, bathyal, and abyssal zones.

In most aquatic systems, the key factors that determine the types and numbers of organisms found at various depths are water temperature, dissolved oxygen content, availability of food, and availability of light and nutrients required for photosynthesis, such as carbon (as dissolved CO_2 gas), nitrogen (as NO_3^-), and phosphorus (mostly as PO_4^{3-}).

Coastal Zone The **coastal zone** is the warm, nutrient-rich, shallow water extending from the high-tide mark on land to the gently sloping, shallow edge of the continental shelf (the submerged part of a continent). It makes up less than 10% of the world's ocean area, but contains 90% of all marine species. It is also the site of most large commercial marine fisheries. This zone's aquatic systems include estuaries, coastal marshes, mangrove forests, and coral reefs (Figure 6-22).

An **estuary** is where a river meets the sea. It is a partially enclosed body of water where salt water mixes with the river's fresh water, as well as nutrients and pollutants in runoff from the land. In the "in-between" environments of estuaries and river deltas, water can form layers based on differences in density. **Brackish water** also forms. Brackish water is water that has a salinity between that of salt water and fresh water.

Estuaries are associated with **coastal wetlands**—coastal land areas covered with water all or part of the year. These wetlands are some of Earth's most productive ecosystems and include coastal marshes and mangrove forests. Seagrass beds are ecosystems located in shallow coastal waters (Figure 6-21). These areas host as many as 60 species of underwater flowering grasses that form dense beds, some of which are large enough to be seen from space. Seagrass beds provide critical habitat for a wide diversity of fish and invertebrates and store as much or more carbon than terrestrial forests of the same size.

These coastal aquatic systems provide important ecosystem and economic services. They maintain water quality in tropical coastal zones by filtering toxic pollutants, excess plant nutrients, and sediments, and by absorbing other pollutants. They provide food, habitats, and nursery sites for a variety of aquatic and terrestrial species. Coastal wetlands also reduce storm damage and coastal erosion by absorbing waves and storing excess water produced by storms and tsunamis.

checkpoint What key factors determine the types and numbers of organisms found at various depths in aquatic ecosystems?

Zones of the Open Sea

A sharp increase in water depth at the edge of the continental shelf separates the coastal zone from the vast volume of the ocean, or the **open sea**. This

FIGURE 6-21
An octopus hides in seagrass while waiting for unsuspecting prey. Seagrass beds are complex and productive ecosystems located along most continental coastlines.

aquatic life zone is divided into three vertical zones based on the degree of penetration of sunlight (Figure 6-22). Temperatures also vary with ocean depth. Scientists use temperature to define zones of species diversity in these layers.

Euphotic Zone The euphotic zone is the brightly lit upper zone that contains drifting phytoplankton. These organisms perform about 40% of the world's photosynthetic activity. Large, fast-swimming predatory fish like swordfish, sharks, and bluefin tuna also populate this zone.

Bathyal Zone The bathyal zone is the dimly lit middle zone that receives little sunlight and does not contain photosynthesizing producers. Zooplankton and smaller fish, many that migrate to feed on the surface at night, are found in this zone.

Abyssal Zone The deepest zone, called the abyssal zone, is the dark and cold region near the ocean floor. There is no sunlight to support photosynthesis, and this zone contains little dissolved oxygen. Nevertheless, the deep ocean floor flows with life because it contains enough nutrients to support many species (Figure 6-24). Most organisms in the deep waters and on the ocean floor get food from showers of dead and decaying organisms—called marine snow—that drift down from upper layers.

checkpoint How do organisms obtain enough food to survive in the abyssal zone?

LESSON 6.3   181

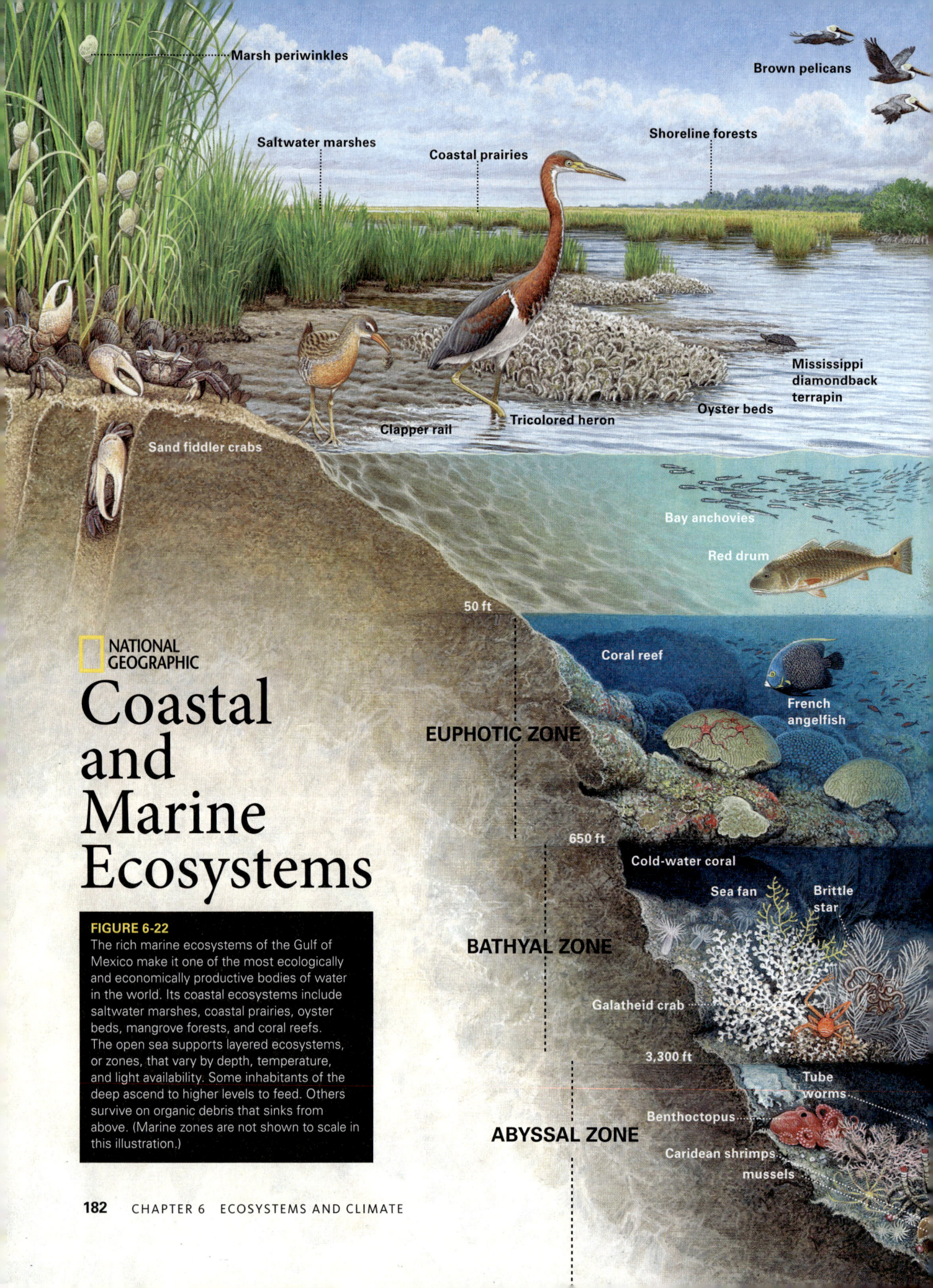

NATIONAL GEOGRAPHIC
Coastal and Marine Ecosystems

FIGURE 6-22
The rich marine ecosystems of the Gulf of Mexico make it one of the most ecologically and economically productive bodies of water in the world. Its coastal ecosystems include saltwater marshes, coastal prairies, oyster beds, mangrove forests, and coral reefs. The open sea supports layered ecosystems, or zones, that vary by depth, temperature, and light availability. Some inhabitants of the deep ascend to higher levels to feed. Others survive on organic debris that sinks from above. (Marine zones are not shown to scale in this illustration.)

Marsh periwinkles
Brown pelicans
Saltwater marshes
Coastal prairies
Shoreline forests
Sand fiddler crabs
Clapper rail
Tricolored heron
Oyster beds
Mississippi diamondback terrapin
Bay anchovies
Red drum
50 ft
Coral reef
French angelfish
EUPHOTIC ZONE
650 ft
Cold-water coral
Sea fan
Brittle star
BATHYAL ZONE
Galatheid crab
3,300 ft
Tube worms
Benthoctopus
ABYSSAL ZONE
Caridean shrimps
mussels

182 CHAPTER 6 ECOSYSTEMS AND CLIMATE

Royal tern

Frigatebird

Mangrove forests

Wilson's storm petrels

Bottlenose dolphins

Sargassum seas seaweed

Mahimahi

Juvenile loggerhead turtle

Whale shark

Human Impacts In 2010, BP's *Deepwater Horizon* oil rig exploded, spewing millions of barrels of oil into the Gulf. The costs charged to BP totaled more than $50 billion. Other threats to the Gulf include overfishing, loss of wetlands, and nutrient pollution.

Red snapper

Atlantic bluefin tuna

Leatherback turtle

Brown shrimps

Sperm whale

Cutlass fish

Giant isopod

Crevalle jacks

Chain cat shark

Bioluminescent bamboo coral

Giant squid

Dragon fish

Elbowed squid

Midshipman

Deep-sea jellyfish

LESSON 6.3 **183**

Human Impacts on Marine Ecosystems

Certain human activities disrupt and degrade many of the ecosystem and economic services provided by marine aquatic systems. Areas most affected are coastal marshes, shorelines, mangrove forests, and coral reefs. Possible ways to manage the harmful effects of human activities on marine ecosystems will be discussed in Chapter 10.

As noted in the Case Study, coral reefs are some of the world's oldest, most diverse, and most productive ecosystems. These centers of aquatic biodiversity are the marine equivalents of tropical rain forests. They have complex interactions among diverse populations of species.

Worldwide, human activities are destroying coral reefs at an alarming rate. One growing threat is **ocean acidification**—the rising levels of acidity in ocean waters. This occurs because the oceans absorb at least 25% of the CO_2 emitted into the atmosphere by human activities, especially the burning of carbon-based fossil fuels. The CO_2 reacts with ocean water to form a weak acid (carbonic acid, H_2CO_3). This reaction decreases the levels of carbonate ions (CO_3^{2-}) necessary for formation of coral reefs. Carbonate ions form the shells and skeletons of marine organisms such as crabs, oysters, mussels, and some phytoplankton. The lower levels of carbonate ions make it harder for these species to thrive and reproduce. At some point, rising acidity could slowly dissolve corals and the shells and skeletons of these marine species. Entire productive ecosystems are at risk.

checkpoint How does ocean acidification threaten coral reefs?

FIGURE 6-23

Major Human Impacts on Marine Ecosystems	
Marine Ecosystems	**Coral Reefs**
Half of all coastal wetlands lost to agriculture and urban development	Ocean warming
	Rising ocean acidity
More than one-fifth of mangrove forests lost to agriculture, aquaculture, and development	Rising sea levels
	Soil erosion
	Algae growth from fertilizer runoff
Beaches eroding due to development and rising sea levels	Bleaching
	Increased UV exposure
Ocean-bottom habitats degraded by dredging and trawler fishing	Damage from anchors and from fishing and diving
	Overfishing
At least 2% of coral reefs severely damaged and 25–30% more threatened	

6.3 Assessment

1. **Compare and Contrast** How are the zones of an ocean similar to and different from the layers of a rain forest?

2. **Contrast** What is the difference between the euphotic zone and the bathyal zone of the ocean?

3. **Predict** Scientists project that by 2100, ocean acidification may increase by 170% over levels measured in the 1800s. How will this affect humans, and how can this be prevented?

4. **Evaluate** Which two threats to marine ecosystems do you think are the most serious? Why?

SCIENCE AND ENGINEERING PRACTICES

5. **Planning Investigations** Design an experiment to test the cause-and-effect relationship between atmospheric CO_2 and ocean acidification. Include a hypothesis in the form of an "if, then" statement.

FIGURE 6-24
Deep-sea fish have some unusual adaptations. This anglerfish uses a bioluminescent lure to attract its prey.

6.4 What Are the Major Types of Freshwater Systems?

CORE IDEAS AND SKILLS

- Understand that a river typically flows through three zones.
- Describe the seven ecosystem and economic services that inland wetlands provide.
- Describe human activities that are degrading freshwater systems.

KEY TERMS

surface water
freshwater life zone
runoff
watershed
eutrophication
delta
inland wetland

Lakes, Streams, Rivers, and Wetlands

As you learned in Chapter 3, Earth has a fixed supply of water, and the hydrologic cycle collects, purifies, and distributes this supply. Precipitation that does not sink into the ground or evaporate becomes **surface water**. This is fresh water that flows or is stored in bodies of water on Earth's surface. Fresh water is water that contains no more than 1,000 ppm salt. **Freshwater life zones** include standing bodies of water such as lakes, ponds, and inland wetlands.

Flowing freshwater systems include rivers and streams. Surface water that flows into such bodies of water is called **runoff**.

A **watershed**, or drainage basin, is the land area that delivers runoff, sediment, and dissolved substances to streams, lakes, or wetlands. Although freshwater systems cover about 1% of Earth's surface, they provide many ecosystem services. Freshwater systems are sources of climate moderation, nutrient cycling, waste treatment, flood control, groundwater recharge, and habitat. They provide economic services including food, drinking water, irrigation water, hydroelectricity, transportation corridors, and recreation.

Lake Systems Lakes are large natural bodies of standing fresh water formed when precipitation, runoff, streams, rivers, and groundwater seepage fill depressions in Earth's surface. Causes of such depressions include movement of glaciers, displacement of Earth's crust, and volcanic activity. A lake's watershed supplies it with water from rainfall, melting snow, and streams. Freshwater lakes vary greatly in size, depth, and nutrient content. Deep lakes normally consist of distinct life zones, defined by their depth and distance from shore. Figure 6-25 illustrates the life zones of a typical deep lake and the biodiversity they can support.

FIGURE 6-25
Lake Zones A typical deep lake has distinct zones of life. How are deep lakes like rain forests?

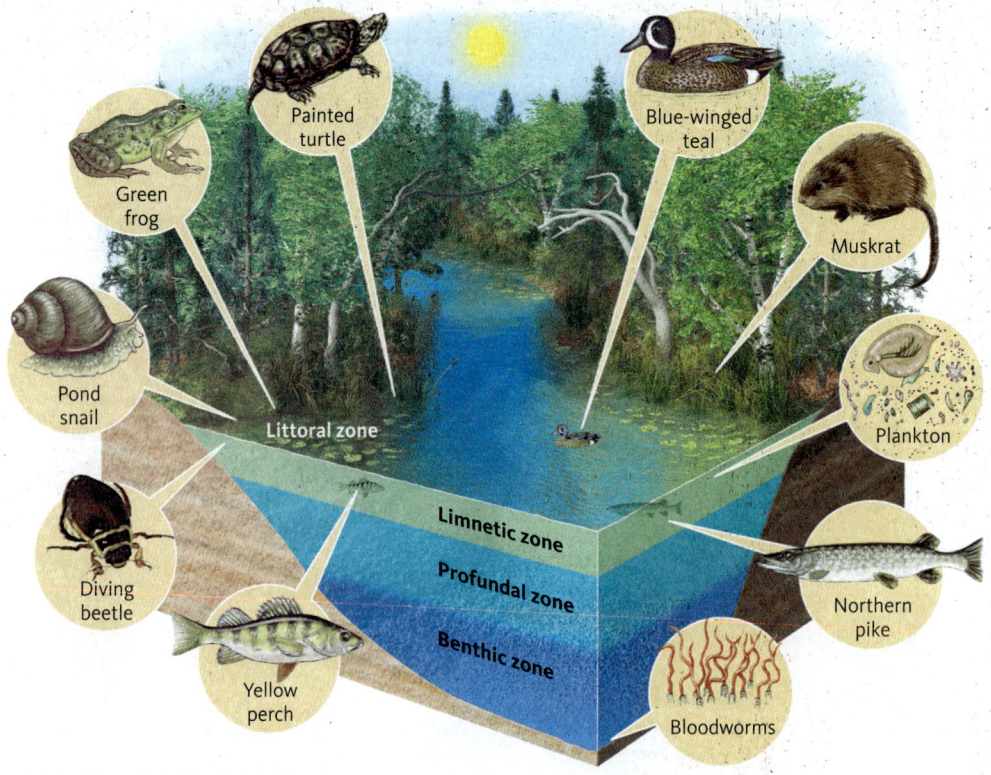

FIGURE 6-26 ▼

Flowing Freshwater Zones Water flowing downhill forms three zones—the source zone, transition zone, and floodplain zone. Over time, the friction of moving water and sediment shapes the land, cutting valleys and canyons. Sand, gravel, and soil carried by streams and rivers are deposited in low-lying areas. A delta may form at the river's mouth.

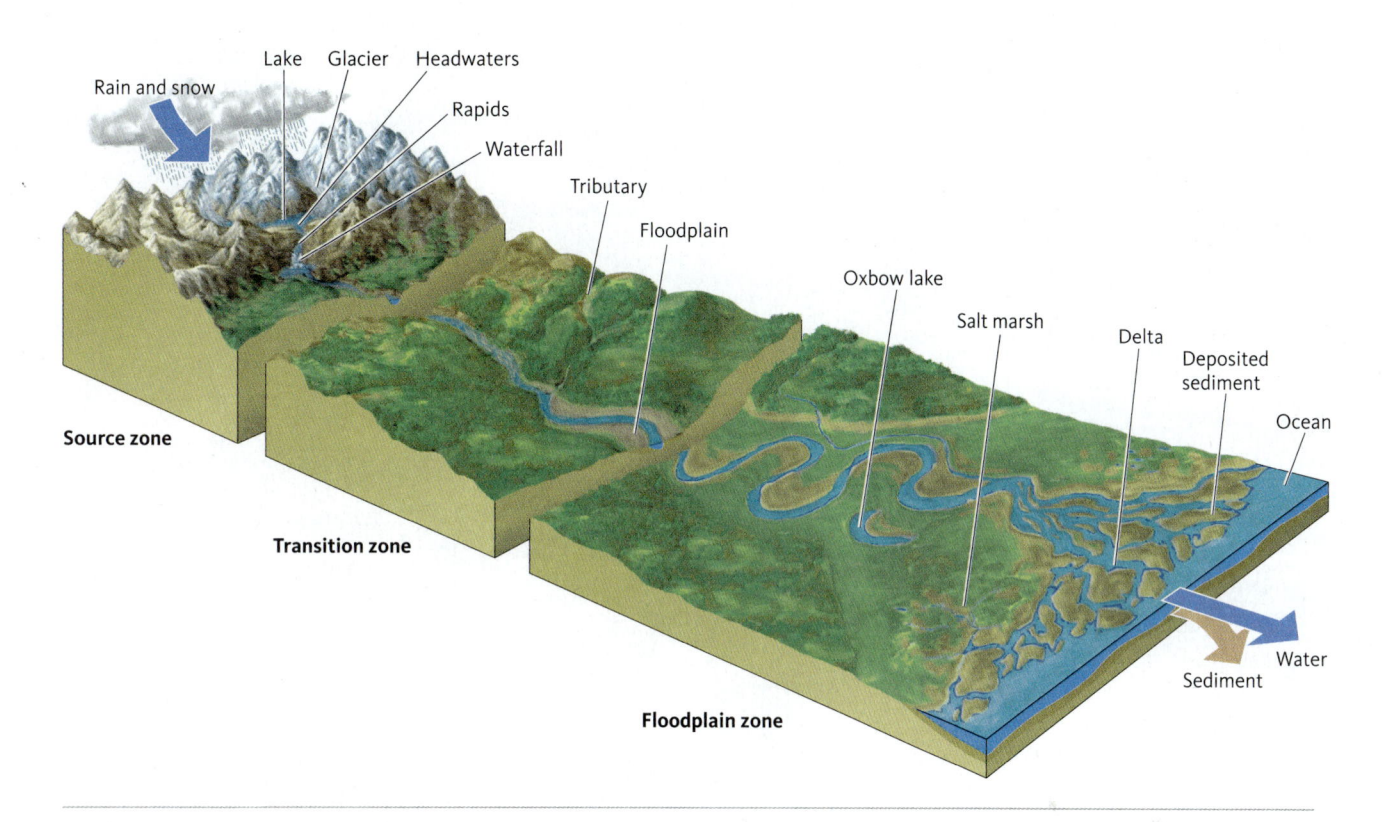

Ecologists classify lakes according to their nutrient content and primary productivity. A lake that has a small supply of plant nutrients is called an oligotrophic lake. This type of lake is often deep and has steep banks. Glaciers and mountain streams supply water to many of these lakes. They usually have crystal-clear water and small populations of phytoplankton and fish such as smallmouth bass and trout. Because of their low levels of nutrients, these lakes have a low net primary productivity.

Over time, sediments, organic material, and inorganic nutrients wash into most oligotrophic lakes. Plants grow and decompose to form bottom sediments. The process by which lakes gain nutrients is called **eutrophication**. Eutrophic lakes are typically shallow with murky brown or green water. With their high levels of nutrients, these lakes have a high net primary productivity. Water pollution can accelerate this natural process of eutrophication and disrupt lake ecosystems (Lesson 10.4).

checkpoint Give examples of standing and flowing bodies of fresh water.

Streams and Rivers In watersheds, water accumulates in small streams that join to form rivers. Collectively, rivers carry huge amounts of water from highlands to lakes and oceans. A typical river system flows through three zones (Figure 6-26). The source zone contains headwater streams found in highlands and mountains. The transition zone has wider, lower-elevation streams. In the floodplain zone, rivers empty into larger rivers or the ocean.

As streams flow downhill, they shape the land. At its mouth, a river may divide into many channels as it flows through its **delta**—an area at the mouth of a river built up by deposited sediment that often contains estuaries and coastal wetlands. These important forms of natural capital absorb and slow the approach of floodwaters from coastal storms and hurricanes and provide habitats for a wide variety of marine life.

Inland Wetlands Located away from coastal areas, **inland wetlands** are lands that are covered with fresh water all or part of the time.

LESSON 6.4   187

FIGURE 6-27
Swamps are inland forested wetland ecosystems that contain water year-round and are dominated by trees. At Caddo Lake on the Texas-Louisiana border, Spanish moss hangs from towering bald cypress trees and aquatic plants float on the water's surface. Swamps like this one provide critical sheltered habitat for nesting birds, fish, amphibians, and reptiles.

Inland wetlands include swamps (Figure 6-27), marshes, small ponds, and prairie potholes. Prairie potholes are depressions carved out by ancient glaciers. Other examples are floodplains, which receive excess water from streams or rivers during heavy rains and floods. Some wetlands are covered with water year-round. Others remain under water for only a short time each year, such as prairie potholes, floodplain wetlands, and the Arctic tundra.

Inland wetlands provide a number of ecosystem and economic services. They:

- filter and degrade toxic wastes and pollutants;
- reduce flooding and erosion by absorbing storm water and releasing it slowly, and by absorbing overflows from streams and lakes;
- sustain stream flows during dry periods;
- recharge groundwater aquifers;
- maintain biodiversity by providing habitats for a variety of species;
- supply valuable products such as fish and shellfish, blueberries, cranberries, and wild rice; and
- provide recreation for birdwatchers, nature photographers, boaters, anglers, and waterfowl hunters.

checkpoint Describe three river zones.

Human Impacts on Freshwater Ecosystems

In addition to overfishing (Chapter 8), human activities disrupt and degrade many of the ecosystem and economic services provided by freshwater rivers, lakes, and wetlands in four major ways:

1. Dams and canals restrict the flow of about 40% of the world's 237 largest rivers. This alters or destroys terrestrial and aquatic wildlife habitats along these rivers, coastal deltas, and estuaries. Dams and canals reduce water and the flow of sediments to river deltas.

2. Flood control levees and dikes built along rivers disconnect the rivers from their floodplains and destroy aquatic habitats by altering or degrading the functions of adjoining wetlands.

3. Cities and farms add pollutants and excess plant nutrients to nearby streams, rivers, and lakes. For example, runoff of nutrients into a lake (eutrophication) causes explosions in the populations of algae and cyanobacteria that deplete the lake's dissolved oxygen. This may cause fish and other species to die off, initiating a major loss in biodiversity.

4. Many inland wetlands are drained or filled to grow crops. Some have been covered with concrete, asphalt, and buildings. More than half of the inland wetlands estimated to exist in the continental United States during the 1600s no longer exist. Eighty percent of the lost wetlands were drained to grow crops. This loss of natural capital has been an important factor behind increased flood damage in parts of the United States. Many other countries have suffered similar losses. In Germany and France, 80% of all inland wetlands have been destroyed.

In spite of the many challenges of achieving a more sustainable use of Earth's freshwater ecosystems, there is reason to be hopeful. A large number of scientists and other individuals are devoting their lives to learning how humans can use freshwater resources more sustainably. You will learn more about what you and others can do in Chapter 10.

checkpoint How do human activities degrade freshwater systems?

6.4 Assessment

1. **Relate** How do rivers affect the land through which they flow?
2. **Recall** Describe the characteristics of the floodplain zone.
3. **Identify** Identify and describe a unique service provided by inland wetlands.
4. **Evaluate** How can flood control levies and dikes built along rivers degrade freshwater systems?

SCIENCE AND ENGINEERING PRACTICES

5. **Obtaining and Evaluating Information** Gather, read, and evaluate 2–3 sources of information on a local wetland or other local ecosystem. Evaluate the validity of each source based on the credentials of the author(s) and their sources of information. Discuss the reliability of each source based on your ability to verify its data.

SCIENCE AND ENGINEERING PRACTICES

6. **Communicating Information** If a public meeting was held to drain and fill a nearby wetland for a shopping mall, what comments would you present at the meeting?

TYING IT ALL TOGETHER STEM
Coral Reef Recovery

In this chapter's Case Study, you read about the ecological and economic importance of the world's coral reefs. Coral reefs are living examples of three factors of sustainability in action. They thrive on solar energy, play key roles in the cycling of carbon and other nutrients, and sustain a great deal of aquatic biodiversity. About half of the world's shallow coral reefs have been destroyed or severely damaged. A number of factors have played a role in this loss: ocean warming, sediment from coastal soil erosion, excessive algal growth from fertilizer runoff, coral bleaching, rising sea levels, ocean acidification, overfishing, and damage from hurricanes.

Scientists have identified a connection between coral reef recovery and fishing. In recent studies, scientists predict that damaged coral reefs around the world can recover in 35–50 years if unfished. However, creating marine reserves is not always feasible because many people's livelihoods depend on fishing. The question in many regions is: Can coral reefs be managed as sustainable fisheries?

Work with a group to complete the following steps.

1. Select one of the following questions to research:
   - Which types of fishing are the most damaging? Include blast fishing and cyanide spraying in your research.
   - Are there any species of fish that can be singled out and protected to aid in coral reef recovery?
   - How can seasonal or spatial closures help coral reefs recover?
   - Should catch limits be imposed, and if so, how?
   - What structures can be built to encourage coral reef growth?
2. Locate three sources of information on your topic. Critically analyze the sources. Who are the authors and what are their credentials?
3. Reorganize into groups with members representing each research question. In your new group, create a list of best practices for coral reef recovery. At the end of your list, provide recommendations for further research.
4. Create a Reef Management Strategies report. Cite evidence to support your claims whenever possible. Document your sources.
5. Did any conflicting information or ideas arise? If so, did your group reach agreement?
6. Read another group's report. How does it differ from yours?

CHAPTER 6 SUMMARY

6.1 What factors influence climate?

- Key factors that influence an area's climate are incoming solar energy, Earth's rotation, global patterns of air and water movement, gases in the atmosphere, and Earth's surface features.
- Weather is a set of physical conditions of the lower atmosphere, including temperature, precipitation, humidity, wind speed, cloud cover, and other factors. Climate is the general pattern of atmospheric conditions in a given area over periods ranging from three decades to thousands of years.
- Major greenhouse gases are water vapor, carbon dioxide, methane, and nitrous oxide. The greenhouse effect is the natural warming of the lower atmosphere.
- The rain shadow effect is a reduction of rainfall and loss of moisture as air moves up and over a mountain, which can lead to a desert.

6.2 What are the major types of terrestrial ecosystems?

- Earth's terrestrial regions can be classified into biomes. Each biome is characterized by its climate and dominant plant life.
- Desert, grassland, and forest biomes are tropical, temperate, or cold depending upon their locations.
- Mountain biomes are important ecologically because they store water used in the hydrologic cycle and provide habitats for rare species.
- Human activities are disrupting ecosystems and economic services provided by many of Earth's deserts, grasslands, forests, and mountains.

6.3 What are the major types of marine ecosystems?

- Oceans make up most of Earth's surface and provide vital ecosystem and economic services that are being disrupted by human activities.
- The open sea is divided into three vertical zones based on the degree of sunlight each receives and its temperature.
- Coral reefs contain large amounts of diverse and productive ecosystems that are threatened by ocean acidification and increasing temperatures.

6.4 What are the major types of freshwater systems?

- Fresh water is classified as either standing or flowing.
- Lakes are standing water systems. Lakes with low levels of nutrients have clear water. Lakes with high levels of nutrients from natural and human sources have green or brown water.
- Freshwater streams and rivers carry large volumes of water through three zones.
- Inland wetlands include marshes, swamps, ponds, and prairie potholes. These wetlands provide numerous ecosystem and economic services.
- Human activities disrupt and degrade freshwater rivers, lakes, and wetlands.

MindTap   If you have been provided with access to a MindTap course, additional resources are available at login.cengage.com.

CHAPTER 6 ASSESSMENT

Review Key Terms

Select the key term that best fits each definition. Not all terms will be used.

> aquatic life zone
> brackish water
> climate
> coastal wetland
> coastal zone
> delta
> edge effect
> estuary
> eutrophication
> freshwater life zone
> greenhouse gas
> inland wetland
> marine life zone
> ocean acidification
> ocean current
> open sea
> permafrost
> rain shadow
> runoff
> surface water
> watershed
> weather

1. The tendency for a transition zone to have greater species diversity and a higher density of organisms than found in either of the individual ecosystems
2. A set of physical conditions of the lower atmosphere that includes temperature, precipitation, humidity, wind speed, cloud cover, and other factors
3. Aquatic life zone such as oceans and their bays, estuaries, coastal wetlands, shorelines, coral reefs, and mangrove forests
4. Aquatic life zone such as lakes, rivers, streams, and inland wetlands
5. The general pattern of atmospheric conditions in a given area over periods ranging from three decades to thousands of years
6. The land area that delivers runoff, sediments, and dissolved substances to a stream, lake, or wetland
7. An area at the mouth of a river built up by deposited sediment that often contains estuaries and coastal wetlands
8. Land located away from coastal areas that holds fresh water all or part of the time; includes marshes, swamps, ponds, and prairie potholes
9. Fresh water, originating from precipitation, that flows or is stored in bodies of water on Earth's surface
10. The warm, nutrient-rich, shallow water extending from the high-tide mark on land to the gently sloping, shallow edge of the continental shelf
11. Saltwater and freshwater portion of the biosphere that can support life

Review Key Concepts

12. Define and give three examples of a greenhouse gas.
13. What is coral bleaching and how can it be prevented?
14. Explain the rain shadow effect and how it leads to the formation of deserts.
15. Why do cities tend to have more haze and smog, higher temperatures, and lower wind speeds than the surrounding countryside?
16. Describe how climate and vegetation vary with latitude and elevation.
17. How do desert plants and animals survive?
18. Why is there so much biodiversity in tropical rain forests?
19. Why does a thick layer of decaying litter typically cover the floors of temperate deciduous forests?
20. How do most species of coniferous evergreen trees survive the cold winters in boreal forests?
21. What are the three major life zones in an ocean?
22. Explain the ecological and economic importance of coastal marshes, mangrove forests, and seagrass beds.
23. List five human activities that pose major threats to marine systems.

Think Critically

24. You are a defense attorney arguing in court for sparing a tropical rain forest from being cut down. Give your three best arguments for the defense of this ecosystem. Do the same for sparing a threatened coral reef.
25. Suppose you have a friend who owns property that includes a freshwater wetland. Your friend tells you she is planning to fill the wetland to make more room for her lawn and garden. What would you say to this friend?

26. What are three steps that governments and private interests could take to protect the world's remaining coral reefs?

27. How might the distribution of the world's forests, grasslands, and deserts shown in Figure 6-9 differ if the prevailing winds shown in Figure 6-4 did not exist?

Chapter Activities

A. Investigate: Leaf Characteristics STEM

Each of Earth's biomes is characterized by vegetation adapted to its climate. The dominant vegetation drives the types of animals and other plants found in that biome.

Materials
a variety of leaves   hand lens

1. Carefully observe the different types of leaves using the hand lens.

2. Choose three of the leaves and complete a triple Venn diagram to describe their similarities and differences. Look carefully at each leaf's properties such as the texture, thickness, covering, color, and other features.

3. Make a claim about the probable climate conditions that each plant is adapted to. Support your claims with observational evidence.

4. Infer which type of biome each plant thrives within and explain your reasoning.

5. What additional plant specimens would help further support your claims?

6. Choose one biome and describe all the characteristics that a plant would need to thrive in that environment. Explain why each characteristic would be useful to the plant's survival.

7. How might climate change impact the biome you chose? How would this affect the traits needed for the survival of a plant?

B. Citizen Science

National Geographic Learning Framework

Attitudes | Curiosity
Skills | Observation
Knowledge | Our Human Story

Join a citizen science project, such as Old Weather, to catalog historical weather data. Old Weather enlists citizens, students, and researchers to explore, mark, and transcribe real climate-related entries from nearly 300 Royal Navy ships of the World War I era. In addition to climate data, the logs contain observations of political and social significance.

The data provided through historical weather projects help climate scientists build better models. In the case of Old Weather, data can reveal how sea ice has changed since old whaling ships navigated the Arctic.

The project also helps maritime historians piece together voyages and events that happened on the ships. After your experience, gather in small groups to discuss the questions below.

1. What was the name of the project you joined?

2. Describe your findings or contributions. Were there any surprises?

3. What was your experience like working on the project? Do you think you will continue?

4. Discuss any ideas you have to improve upon or expand the project.

STEM ENGINEERING PROJECT 2
DESIGN A SYSTEM TO ASSESS A LOCAL SPECIES

In this challenge, you will get to use your own ingenuity and problem-solving skills. You will work with a team to design a solution to a real environmental problem, build a model of your system, and test it. The problem in this case is related to a single species in your ecosystem. As you well know, the parts of an ecosystem are interconnected: A change in one species' population can affect the entire ecosystem.

Have you heard about a plant or animal species in your area that may be a cause for concern? For example, maybe there is an invasive insect or weed that is affecting local businesses. Perhaps a top predator is disappearing or recently moved into your area. Maybe an important pollinating insect or bird is on the decline and no one knows why. Or maybe a migrating species has changed its patterns. Whether it is an invasive species on the rise, a pollinating species on the decline, or another issue, you will select an organism in your area that needs better monitoring and come up with a new method to capture data about it.

The process of testing an engineering design involves a series of practices similar to the practices used by scientists to test hypotheses. Engineering usually involves many rounds of testing and retesting, so the process of design is often depicted as a cycle rather than a series of steps. Engineers use this cycle to translate ideas into practical solutions.

Engineering DESIGN CYCLE
- Defining problems
- Developing and using models
- Planning and carrying out investigations
- Analyzing data and using math
- Designing solutions
- Forming arguments from evidence
- Obtaining, evaluating, and communicating information

Defining Problems

1. Look at the list of local species provided by your teacher. Select one species. Use the Internet to learn about the problem associated with your species. Nature centers and university extension offices are also great resources for research.
   - What species will your team focus on? Include the scientific name and the common name.
   - What is the problem associated with the species?
2. Identify a need that underlies the problem.
   - What data might help solve the problem?
   - How could a system of cell phones be used to gather that data?
3. What would a successful solution be able to achieve? List the criteria.
4. Consider a possible population assessment system that uses cell phones. List the design constraints, such as time, equipment, or people.

Developing and Using Models

5. Brainstorm ideas for a solution with your team. Consider the factors provided by your teacher.
6. Discuss two different ideas in more depth with your team. Then choose one idea to carry out.
7. Describe the design of your system both in words and with a diagram or flow chart.
8. Develop a simplified version of your system that your team can use to test your ideas. Keep in mind that you will evaluate your model using the criteria from step 3.

Planning and Carrying Out Investigations

9. Plan a test of your model with your group. Write out your plan in steps.
10. How reliable do you think your data will be? Consider limitations and refine your design accordingly.

11. What social considerations can you think of? Refine your design to make it as safe and ethical as possible.

12. Create your model and carry out your test. Record your data in a table.

Analyzing Data and Using Math

13. Analyze your data. What patterns do you notice?

14. What limitations exist in your data?

15. How well does your system meet each of the criteria from step 3?

Designing Solutions

16. What were the strengths and weaknesses of your design?

17. How can your design be improved? Revise your description and diagram accordingly.

Forming Arguments from Evidence

18. Describe how you revised your design.

19. What evidence justified the change you made?

20. Test your new design and compare the results to your original design.

21. What claims can you make about your solution? Use evidence to support your claims.

Obtaining, Evaluating, and Communicating Information

22. Present your team's solution to the class. Describe how your system works and how it was optimized to meet the criteria. Summarize your testing process and the results.

23. Offer thoughtful and specific criticism of other teams' designs.

24. Conduct follow-up research. Do your data and observations match up with those in the studies?

25. Write a scientific question that could be asked using your proposed design solution.

26. Write a final report. Include recommendations for further changes, testing, and scientific research.

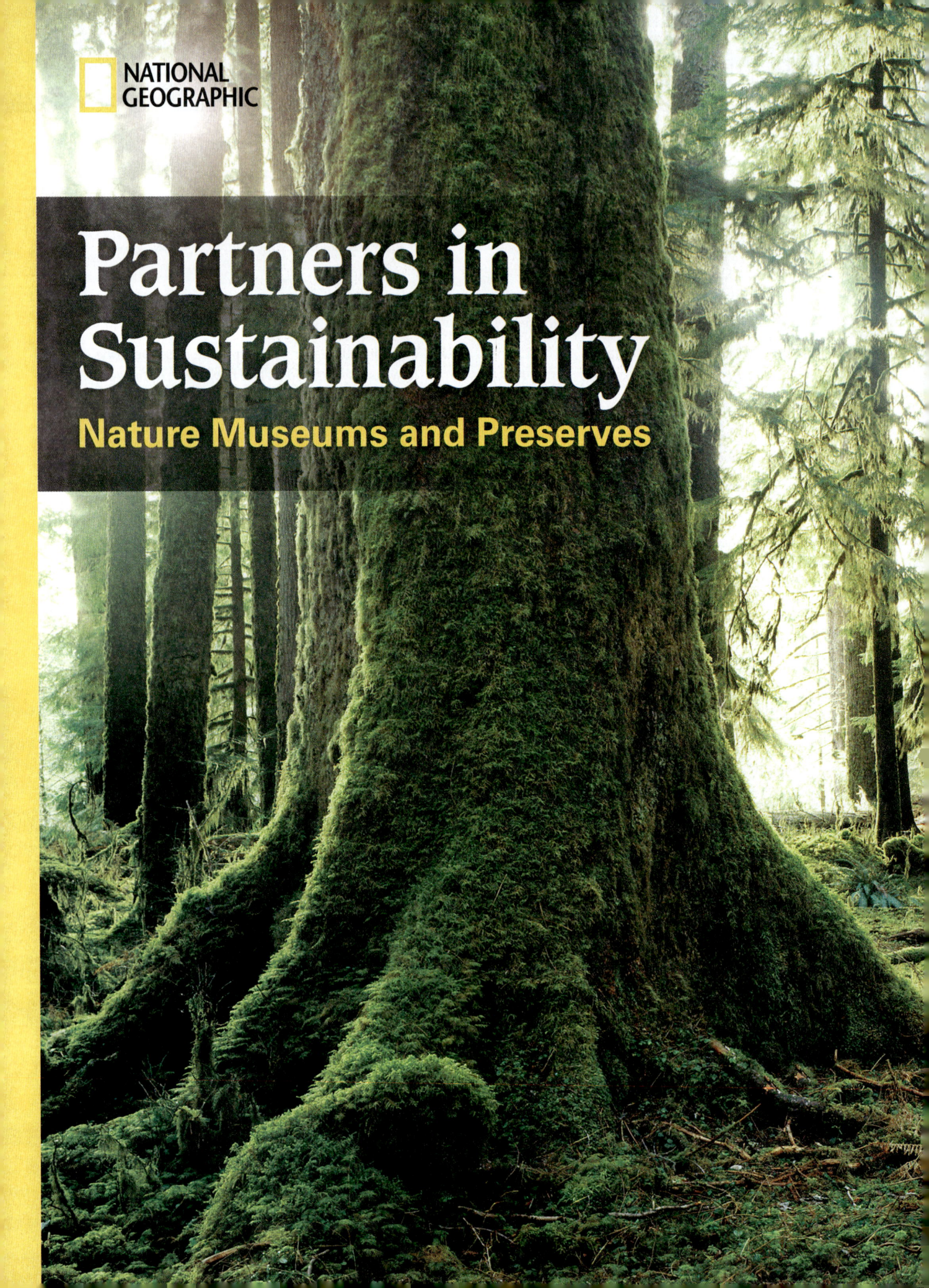

NATIONAL GEOGRAPHIC

Partners in Sustainability

Nature Museums and Preserves

"A lifetime can be spent in a Magellanic voyage around the trunk of a single tree."

—Edward O. Wilson, biologist, author, and National Geographic Hubbard Award recipient

Natural History

When was the last time you walked on soil, grass, sand, or rocks? If you're like most people and you live in an urban area of almost any size, it's easy to go an entire week without setting foot on the natural surface of the planet. Likewise, it's easy to drown out the sound of birds with your favorite music, or to ignore insects and weather by staying indoors, endlessly entertained by the Internet, which is always at your fingertips. All of the wonderful conveniences of modern life make it blissfully easy to forget that none of it would be possible without a healthy planet.

Yet no matter where you live, you can still connect with the natural world simply by going outside.

That's part of what nature museums, parks, botanical gardens, zoos, wildlife preserves, and conservation organizations are here to help you do. They're here to open your nature-deprived eyes, ears, hearts, and minds to the wonders of the natural world—which is also right at your fingertips.

Ever hear of natural history? Natural history involves the scientific study of plants and animals through observation rather than experimentation. When natural history museums first began to appear in the 19th century, they mostly exhibited models of plants and animals—often in large dioramas. Over time, however, natural history centers began to champion conservation and other environmental concerns. They established research programs and really great citizen science initiatives. If you've ever identified a bird or a wild plant, hiked somewhere, or studied an animal in its habitat—or if you think that's something you could maybe do—then you're ready to learn natural history.

Hikers explore the Hoh Rain Forest in Olympic National Park, one of a dozen natural UNESCO World Heritage Sites in the United States.

OLYMPIC NATIONAL PARK

America's national parks have been called "the best idea [the country] ever had. Absolutely American, absolutely democratic, they reflect us at our best rather than our worst." The idea, realized about 100 years ago, was to preserve and protect the country's natural wonders and the wildlife that makes these places their home. National parks are democratic because they are open for the benefit and enjoyment of all people.

Olympic National Park in the state of Washington contains a wide variety of landscapes and ecosystems, from the glaciers of Mount Olympus to teeming tide pools along the shoreline. Vast, diverse, and ancient forests rise between these two landscapes, some with trees more than 1,000 years old. Among the forest communities in the park is one type you may not expect: rain forests. But these aren't the steamy, tropical ones you would find in places such as South America. Olympic's rain forests are temperate, which means they thrive in climates with mild winters, cool summers, and plenty of rain. The park's Hoh Rain Forest is one of the most spectacular temperate rain forests in the world. Many species of moss grow on the forest's trees, some of which are endemic to the area.

The best way to learn about Olympic and its many ecosystems is to drop in to one of the park's many visitor centers. You can see exhibits on the park's natural and cultural history and then borrow a Discovery Backpack for the day. The backpack's worksheets, field guides, and games help you and your family discover the purpose and meaning of the park's natural wonders. You can also explore Olympic's flora and fauna on a field trip with your class, using a "traveling trunk." Trunks include plant and animal profiles, guides, and activities. With resources like these, you can hit the trails and really investigate Olympic's living laboratory.

> The Sol Duc River carves through Olympic National Park, which receives as much as 4.3 meters (14 feet) of rain every year.

EVERGLADES NATIONAL PARK

Not all national parks contain towering mountains, deep valleys, and lush forests. Everglades National Park in south Florida is mostly flat and located on the edge of the Atlantic Ocean and the Gulf of Mexico. For thousands of years, the Everglades existed as wetlands, with its marshes, palm trees, tropical plants, and alligators. This watery landscape changed in the early 1900s, when developers began to transform the land into farmland and communities. The park was created in 1947 to preserve the remaining 25% of the Everglades' ecosystem and protect the endangered species that live there, including the manatee, American crocodile, and Florida panther.

Do you know the difference between an alligator and a crocodile? The easiest way to tell them apart is by examining their snouts. An alligator's snout, or mouth and nose, is wide and U-shaped, while that of a crocodile is narrow and pointy. You can get close enough to safely tell the difference as you walk the trails and raised boardwalks along the park. You can also slog through the shallow waters of the Everglades on a ranger-led expedition to study the wildlife there. (But don't worry! The tour doesn't include an encounter with alligators or crocodiles.)

Teacher-led field trips are encouraged as well, transforming the park into a classroom. Teachers can lead their students on self-guided trips along the boardwalks or on boat tours through the Florida Bay or Gulf Coast. Conversely, educators can bring the Everglades into the classroom. The park offers lesson plans and activities on topics including the role non-native species play in the Everglades' ecosystems. Interested in a career in the national park system? A ranger can come to your school to attend a career day or just bring an Everglades experience to you. Who knows? Maybe one day you'll take part in the important preservation and restoration work going on at the park.

An American crocodile smiles for the camera. The United Nations has classified the Everglades ecosystem, another World Heritage Site, as endangered.

PEGGY NOTEBAERT NATURE MUSEUM

Want to explore animal homes, admire artwork painted by monkeys, or walk through a butterfly haven? (Be honest with yourself. Who wouldn't?!) You can do this at the Peggy Notebaert Nature Museum. Located in the heart of Chicago, this museum allows visitors to experience nature's sights, sounds, and smells. Educators from the museum even come to Chicago schools to teach hands-on science lessons, often using specimens from the museum's collection of animals, plants, fossils, and artifacts.

Another neat thing about the Peggy Notebaert is the chance it offers the public to work with biologists and see science in action. The Blanding's Turtle Restoration Project is a good example. Since 2008, the museum has exhibited the Blanding's, a yellow-throated species of turtle that is native to the Great Lakes region. When numbers of the turtles sharply declined, biologists at Peggy Notebaert stepped in to try to save the endangered species. The biologists take care of the turtles during the first few years of their lives—beginning at the egg stage—and then release them into the wild with tracking devices to study their range. What's especially cool, though, is that the museum allows anyone to purchase one of the tracking devices and become part of the team. Participants can learn about the Blanding's and help the biologists track a turtle's activity.

The museum doesn't stop with turtles. Other initiatives invite the public to monitor butterflies, observe squirrels, or adopt an animal from Peggy Notebaert's living collection. These programs, and many others, encourage people to connect with nature—even in the city. They also help the museum in its mission to protect species and raise interest and awareness in environmental issues. If there's a nature museum near you, go check it out for similar opportunities.

The museum's Teenagers Exploring and Explaining Nature and Science (TEENS) program gives high schoolers opportunities to participate in hands-on scientific research. Participants learn about college and career paths in the natural sciences, conservation, and education.

A gray hawk at the museum hunts for prey. Gray hawks nearly disappeared from Arizona. Recent conservation efforts have succeeded in restoring their population.

ARIZONA-SONORA DESERT MUSEUM

The Arizona-Sonora Desert Museum in Tucson is not your typical botanical garden. In fact, it is part botanical garden, zoo, art gallery, natural history museum, and aquarium. And even though it's called a museum, most of the experience takes place outdoors. Visitors tour desert habitats that demonstrate the interdependence of the native plants and animals that live and grow there, as well as the culture of the region. An important part of the Desert Museum is honoring the region's tie to Sonora, the state of Mexico south of Arizona.

Before the museum opened in 1952, many of the local residents of Tucson had shown little appreciation for the desert. Much of what they knew about the environment came from visiting roadside snake farms where the animals were kept in cages. The founders of the museum wanted to change all that. They sought to educate the public on the diversity of the Sonoran Desert and to live in harmony with it. To that end, the museum instituted research and education programs on such issues as habitat conservation and water usage in the desert.

Outreach initiatives for adults include the Desert Ark Community Programs, in which museum educators bring artifacts and live animals from the Sonoran Desert to groups in southern Arizona for lectures and demonstrations. The museum also offers classes, trips, and knowledgeable docents, or guides, who answer questions posed by visitors on the grounds and in the community.

Many docents are adults, but high school students can become docents, too, through the Teen Conservation Leadership Corps. In the corps' two-tier program, teens begin by becoming Earth Ambassadors and learning about the Sonoran Desert through exploration, study, and hands-on research under the guidance of museum experts. Once teens spend a year as Earth Ambassadors, they can undergo further training to become Junior Docents and share what they've learned with museum visitors. Students can also take guided visits with their teachers or participate in museum-sponsored classes and labs. Science really does come alive in this wonderful outdoor classroom.

The National Zoo maintains an "Orangutan Transit System," called the "O Line" for short, that allows its orangutans to choose which of two buildings they want to spend time in.

MONTEREY BAY AQUARIUM

When you visit the Monterey Bay Aquarium in Monterey, California, you can learn about the ability of the stumpy cuttlefish to camouflage itself before attacking unsuspecting prey and the tactics bigfin reef squid use to appear bigger when confronted with their predators. But when you dive deeper into the aquarium, you'll find that it's about much more than watching fish in a tank. It's really about keeping the planet—and the people who live on it—healthy.

The oceans are an indicator of global ecosystem health. Oceans not only provide food for the world, but tiny plants called phytoplankton also supply much of the oxygen people breathe. That's why the Monterey Aquarium is committed to keeping the oceans healthy. Scientists at the aquarium are working to protect California's marine ecosystems, ensure the survival of such species as sea otters, sharks, and bluefin tuna, and promote sustainable seafood and fisheries by combating overfishing and illegal fishing practices and by setting standards for wild-capture fisheries.

To help in this effort, the aquarium is grooming the next generation of marine conservationists. The Teen Conservation Leader program is designed to train young volunteers to help out at the aquarium. After a two-week long instruction in marine biology and conservation, teens are ready to answer visitors' questions and take on leadership roles at the aquarium's summer camp programs. Lucky students in the Watsonville, California, area (a community about 50 kilometers, or 30 miles, from Monterey) can take part in a year-long program, learning about the ecosystem and helping to restore local habitats. At the end of the program, the students can also conduct their own field research project. Teens can also join an oceanography club, and young women can take advantage of a program designed just for them to learn about science, the ocean, and conservation. As you can see, the Monterey Bay Aquarium offers many ways to explore what's below the surface. Do you live near an ocean or aquarium? Have you taken the plunge? If not, then it's time to get your feet wet.

SMITHSONIAN'S NATIONAL ZOO

When you think of Smithsonian's National Zoo in Washington, D.C., you may picture the giant pandas that have been housed here since 1972. If you visit the zoo, you'll notice the animals' spacious enclosures in natural settings. Zoos used to keep animals in small cages. The enforced inactivity caused some animals to pace or exhibit other symptoms of boredom and depression. Today, many zoos—including the National Zoo—try to re-create an animal's habitat and provide stimulating activities that mimic what the animal might do in the wild.

The zoo attracts visitors with its collection of animals, but it's also in the business of conserving species and training future conservationists. Did you know that more than 100 amphibian species have become extinct since 1980? This rate far surpasses that of birds and mammals during the same period. In response, scientists at the National Zoo helped identify a deadly skin disease responsible for the decline of many amphibians and provided a home for a rescued population of Panamanian golden frogs that had been wiped out by the disease. The zoo also encourages the public to take part in conservation and biodiversity efforts through such programs as Neighborhood Nestwatch. Participants in this citizen science program identify and observe migratory bird nests in their backyards and report their findings to zoo scientists.

Divers hand-feed the fish twice a day in the aquarium's Kelp Forest exhibit. The surrounding kelp forest is 8.5 meters (28 feet) tall.

INTERNATIONAL CRANE FOUNDATION

Many of the institutions you've been introduced to in these pages focus on protecting local plants, animals, and habitats. Conservation organizations, on the other hand, work to protect wildlife around the world. As you might guess, the International Crane Foundation concentrates on protecting the 15 species that belong to the crane family. Eleven of the species face extinction. In its headquarters in Baraboo, Wisconsin, the foundation maintains birds of all 15 species in an on-site zoo to help ensure their survival. The wild cranes belonging to these species range over five continents, and the foundation works to sustain them in the places where they are most threatened.

Specifically, the crane foundation tries to secure the birds' ecosystems, watersheds, and flyways. For example, the foundation works with their partners in East Asia to conserve wetlands along the flyway routes of migrating cranes and develop policies that protect their habitats. These measures not only help safeguard the cranes but also improve the environment for the people who live there.

To get its message out, the foundation seeks to educate the public on the plight of the cranes. Local students can take field trips to the Wisconsin headquarters with their teachers and use activity packets to enhance their visit. Using their packets, high school students learn about crane conservation while studying the birds' behavior and biology. In addition, classrooms can adopt a crane and provide food and medical care for the bird for a year. In their teaching, the foundation also stresses the cultural significance of cranes in different parts of the world. In Japan, for instance, many people practice the art of paper folding, or origami. According to legend, anyone who makes 1,000 paper cranes will be granted a wish. The staff at the International Crane Foundation would probably use theirs to save the cranes.

> Armed with a hand puppet, a volunteer teaches a black-necked crane chick to hunt for food. The hand puppet mimics the movements of an adult crane, better preparing the bird for life in the wild.

LAGUNA ATASCOSA NATIONAL WILDLIFE REFUGE

Imagine a place where birds, mammals, reptiles, amphibians, butterflies, and other wildlife live together in safety. When you do, you'll picture a place much like Laguna Atascosa Wildlife Refuge. Located in southern Texas along the border between the United States and Mexico, Laguna Atascosa is a sanctuary for wildlife and habitats, spreading out over more than 39,000 hectares (97,000 acres). Hundreds of species of migratory birds stop over in the refuge, and the endangered ocelot makes its home there. A wide variety of habitats, from coastal prairie to thorn forest to wetlands, provide homes and resting places for the abundant wildlife. The Laguna Atascosa, for which the refuge is named, means "muddy lagoon" and is the winter home of an estimated 80% of North American redhead ducks.

Unfortunately, even a refuge can't cut itself off completely from the world. Climate change, with its unusual hot spells and sudden storms, is having a negative impact on Laguna Atascosa. And its effects create a chain reaction. For instance, rising temperatures may cause some insects to hatch earlier or later, leaving migratory birds less to eat. Warmer weather may also alter vegetation that certain species depend on for survival. The scientists at Laguna Atascosa are trying to anticipate changes to the refuge's ecosystems so that they can minimize the impact on wildlife. But they also ask people in local communities to help by planting native trees and shrubs and reducing their carbon footprint.

Of course, education is key to protecting the refuge. Local students can take field trips to the refuge with their class. In addition to exploring the wildlife and their habitats, students can take notes or make sketches in nature journals of all they see, hear, and smell. Scientists from Laguna Atascosa will also come to local schools to tell the stories of the wildlife of South Texas and explain what students can do to help protect those species that are endangered. Not bad for a place that's largely for the birds.

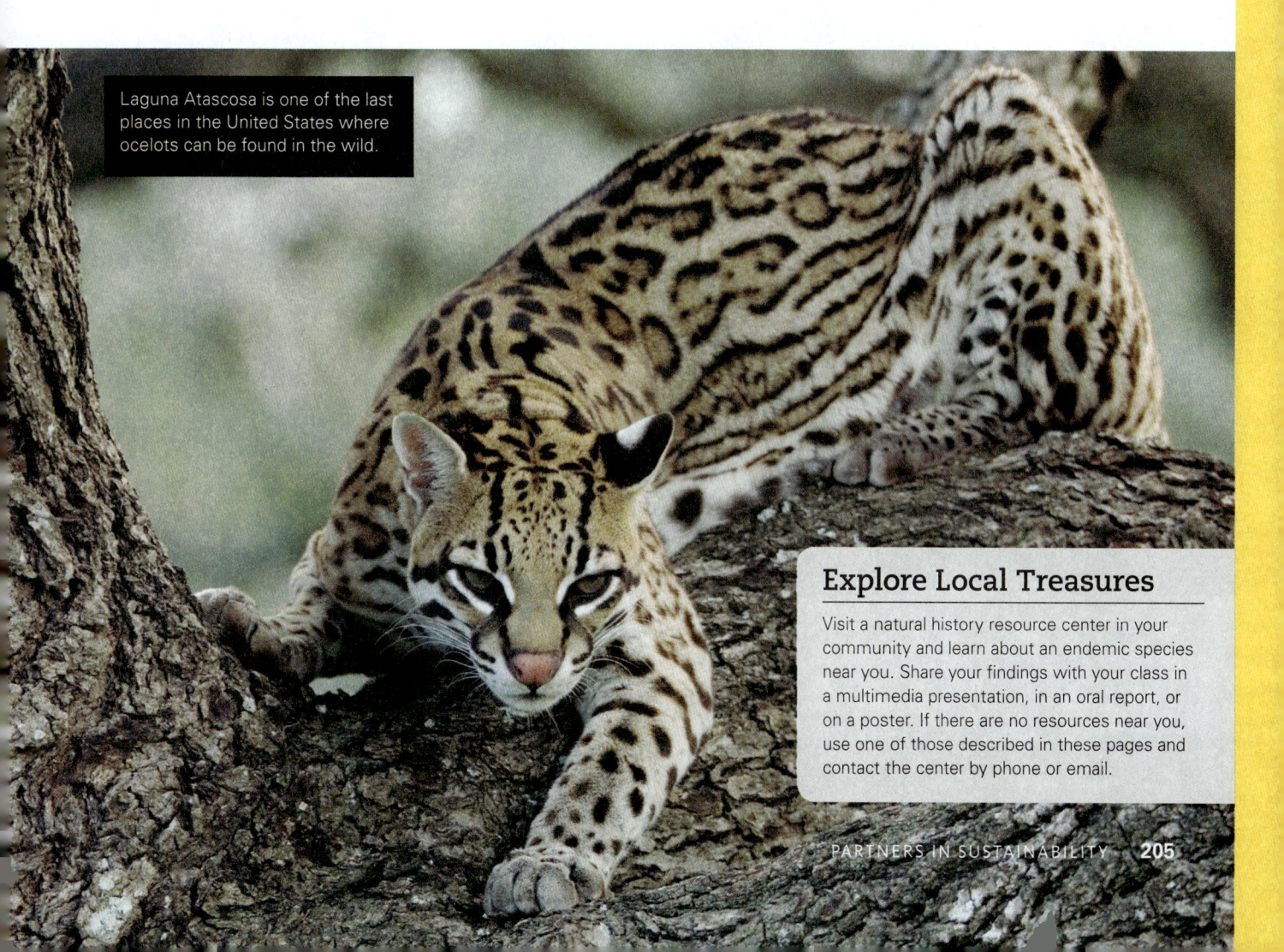

Laguna Atascosa is one of the last places in the United States where ocelots can be found in the wild.

Explore Local Treasures

Visit a natural history resource center in your community and learn about an endemic species near you. Share your findings with your class in a multimedia presentation, in an oral report, or on a poster. If there are no resources near you, use one of those described in these pages and contact the center by phone or email.

UNIT 3
Sustaining Biodiversity

Bluebonnets bloom in a Texas roadside field.

The Texas bluebonnet or Texas lupine (*Lupinus texensis*) is a wildflower found in Texas and northern Mexico. The species was named the official flower of Texas in 1901. The flowers are often found in open spaces along highways, where they are planted to beautify the environment as well as conserve water, control erosion, and provide habitat for wildlife.

The Texas Department of Transportation (TxDOT) started its Wildflower Program in 1932 and today it buys and sows about 30,000 pounds of wildflower seed each year. According to TxDOT, the program "not only helps our highways look good but also reduces the cost of maintenance and labor by encouraging the growth of native species that need less mowing and care." The bluebonnets also provide an essential source of nectar and pollen for bees and other pollinators. Several species of butterfly rely on the plants as larval hosts.

In this unit, you will learn about multiple approaches to conserving ecosystems and sustaining biodiversity, and why this is important. You will use evidence to develop arguments for the benefits of ecosystem management solutions.

CHAPTER 7
SAVING SPECIES AND
ECOSYSTEM SERVICES

CHAPTER 8
SUSTAINING BIODIVERSITY:
AN ECOSYSTEM APPROACH

ENGINEERING PROJECT 3: Design a Solar Cooker

CHAPTER 7
SAVING SPECIES AND ECOSYSTEM SERVICES

PROTECTING EARTH'S BIODIVERSITY requires careful management of ecosystems and their natural resources. Today, habitat loss and other impacts from human activities pose a growing threat to wild species such as snow leopards. In order to save these big cats and other endangered species, individuals and governments will need to cooperate at local, national, and international levels.

KEY QUESTIONS

7.1 What causes extinction and what are its impacts?

7.2 What role do humans play in the loss of species and ecosystem services?

7.3 How can we sustain wild species and ecosystem services?

An endangered snow leopard triggers a remote-controlled camera while prowling through India's Hemis National Park.

NATIONAL GEOGRAPHIC | EXPLORERS AT WORK

A Focus on Saving Honeybees

with National Geographic Explorer Anand Varma

Honeybees play an important role in sustaining our environment—and it's not to produce the delicious honey so many people enjoy. No, the primary role of honeybees in nature is to pollinate trees and other plants that depend on insects for reproduction. Many of these plants are important to our food supply.

Over the last 10 years, scientists have become very concerned about the global decline of honeybee populations. A plague of issues, including pests such as the varroa mite, diseases, habitat loss, and chemicals used in agriculture, have made life difficult for honeybees. They are also thought to have triggered an increase in "colony collapse disorder," a mysterious disease that destroys entire bee colonies.

Anand Varma is a National Geographic Explorer and photographer. When Varma got an assignment to photograph a story on honeybees, he wanted to get up close and personal. He chose to observe bees in their home—the hive. Varma knew a local beekeeper in his hometown of Berkeley, California. She let him move one of her hives into a shed in his backyard so he could study them intimately to see how they live.

"Being able to actually watch them in the hive, you get to observe the things you read about or are told about," Varma says. "I got a lot more excited about bees because I was able to watch them make their living."

Varma became particularly interested in finding a way to photograph the full development of an egg into an adult worker bee. This process takes 21 days from start to finish. So, Varma needed to safely remove the cap from a brood cell and watch the developing bee inside without killing it. Varma solved this problem by using an incubator inside a lab at the University of California, Davis. He set the incubator to the same temperature and humidity found within the hive. Then he brought in brood cells and programmed his camera to photograph a single brood cell for a week at a time.

It took Varma dozens of tries over six months to capture all stages of development. He was finally able to create a time-lapse video of a honeybee's growth from start to finish. "Once it started to work, I was like, this is going to be the coolest thing ever if I can show every part of this process," he recalls.

Varma collaborated with a variety of scientists throughout the project to make sure he didn't harm the bees' development as he did his work. "The people who research bees all tend to be very passionate about bees," he says. "Because I had gotten excited and learned about bees myself, I think that was helpful. I think having gone through that experience helped me relate to the scientists better."

In the same way, Anand Varma's unique photos can help people better appreciate honeybees. Perhaps they can also help scientists in their search for ways to keep these important insects from disappearing.

Thinking Critically
Infer How might Anand Varma's photography of honeybees help scientists who are researching colony collapse disorder?

Clockwise from top: Anand Varma uses his camera to capture a honeybee in flight; a varroa mite attacks a developing honeybee; an adult honeybee emerges from its brood cell.

CASE STUDY
A Honeybee Mystery

In meadows, forests, farm fields, and gardens around the world, hard-working honeybees buzz from one flowering plant to another (Figure 7-1). The bees collect liquid nectar to bring back to their hives. Nectar is the main ingredient in honey, which adult honeybees feed on year-round. The bees also collect pollen grains, which stick to stiff hairs on their legs. Protein-rich pollen is fed to young honeybees.

As a bee travels from flower to flower, some of the pollen that clings to its body inevitably brushes off. In this way, pollen is spread from male to female reproductive organs of the same flower or among different flowers. Fertilization can then occur. This process is called pollination. It is a vital ecosystem service provided by honeybees and other pollinators. Pollination leads to the production of fruits and seeds to grow new plants.

Honeybees pollinate many plant species, including important food crops. In fact, bee pollination in the United States contributes to more than $15 billion in crops annually. Large-scale farms rent bees from commercial beekeepers to make sure plants are pollinated during the weeks they are in bloom. The beekeepers truck about 2.7 million hives to farms across the country. They use European honeybees to pollinate various crops, from grapefruit to almonds. As a result, this single bee species pollinates about three-quarters of the fruit and vegetable species that provide 90% of the world's food and one-third of the U.S. food supply.

Unfortunately, European honeybees have been in decline since the 1980s. That's when several problems began to plague them, including new parasites, viruses, and fungal diseases. In the last decade, another threat has also emerged. Massive numbers of bees in the United States and other countries have been disappearing from their colonies during winter. This phenomenon is named colony collapse disorder (CCD). It affects 30–45% of U.S. honeybee colonies each winter.

Why are European honeybees abandoning their hives and simply vanishing? Research has targeted a range of causes. Diseases and pests may be to blame. So might exposure to toxic chemicals. Habitat loss, poor nutrition, or stress from being transported cross-country are other possibilities. Most likely, all these factors interact to set the stage for CCD. Most certainly, human activities add to the problem.

The honeybee crisis is a classic case of how the decline of a species can threaten vital ecosystem and economic services. It also shows how human activities are increasingly causing biodiversity loss. This chapter examines how and why humans are impacting biodiversity and ecosystem services and how those impacts might be reduced.

As You Read Think about how ecosystem services impact your quality of life. How might the loss of these services affect you? What responsibility do you have to conserve ecosystem services?

FIGURE 7-1
A European honeybee aids in flower pollination as it works to collect nectar and pollen.

7.1 What Causes Extinction and What Are Its Impacts?

CORE IDEAS AND SKILLS

- Describe the causes of extinction and its part in Earth's history.
- Discuss Earth's past mass extinctions and their impacts, and explain why Earth might be entering a sixth mass extinction.
- Understand how species may be classified as endangered or threatened.
- Describe how extinction threatens Earth's natural capital.

KEY TERMS

endangered species
threatened species
bioprospector

Extinction Is Part of Earth's History

Some organisms—such as sharks—have existed on Earth for millions of years. Other organisms—such as dinosaurs—have come and gone and are only found in the fossil record. You learned in Chapter 4 that when a species is no longer found anywhere on Earth, it has experienced biological extinction. How does a species become extinct? A species may not be able to adapt and successfully reproduce under new environmental conditions. Or all of its members may die out due to a catastrophic environmental event.

Extinction is ongoing, but it does not happen at a constant rate. For example, there have been times throughout Earth's history when mass extinctions have occurred. Recall that when a mass extinction takes place, many species become extinct in a relatively short period of geologic time—tens of thousands of years or even as little as 5,000 years.

Geologic, fossil, and other records indicate that Earth has experienced five such mass extinctions. During each of these events, 50–90% of the species present at the time went extinct (Figure 7-2). The causes of these past mass extinctions probably involved global changes in environmental conditions. Scientists believe most were spurred by extreme volcanic eruptions or the impacts of large asteroids or comets. Both types of events would have had global effects, such as months of darkness caused by clouds of ash in the atmosphere, global warming or cooling, changes in sea levels, and changes in ocean water acidity and oxygen levels.

The largest mass extinction took place some 250 million years ago. The Permian-Triassic extinction wiped out about 90% of the world's existing species, including 96% of all marine species. Scientists hypothesize that massive volcanic eruptions over a long time period may have been responsible. They would have released huge amounts of the greenhouse gases carbon dioxide (CO_2) and methane (CH_4) into the atmosphere. Earth's climate would have been altered as a result.

Mass extinctions obviously devastated life on Earth. But they also provided opportunities for new life forms to emerge and diversify as surviving species adapted to fill empty ecological niches. For example, dinosaurs appeared after the Permian-Triassic extinction about 250 million years ago. Then the dinosaurs themselves disappeared during Earth's next mass extinction, about 65 million years ago. In turn, that time marked the period in Earth's history when mammals rapidly diversified.

Scientific evidence indicates that after each mass extinction, Earth's overall biodiversity eventually returned to equal or higher levels. However, each recovery required several million years. The existence of millions of species today means speciation, on average, has kept ahead of extinction. It also demonstrates the importance of biodiversity as a factor of sustainability.

checkpoint Which natural events likely caused Earth's past mass extinctions?

Extinction Rates Are Rising

With the exception of mass extinction events, species extinctions have occurred at a naturally low rate throughout most of Earth's history. Recall from Chapter 4 that this low rate is called the background extinction rate. It is estimated to be fewer than a single species lost per year (0.1) for every 1 million species living on Earth.

However, rates of extinction began rising after modern humans evolved some 200,000 years ago. Since then, periods of human migration have been followed by increases in species extinctions. Human populations have grown and occupied increasingly larger ecological footprints. Extinction rates started rising significantly after the Industrial Revolution, particularly in the last century (Figure 7-3).

Scientists now estimate that the annual rate at which wild species are becoming extinct is at least 1,000 times the natural background extinction rate.

LESSON 7.1   213

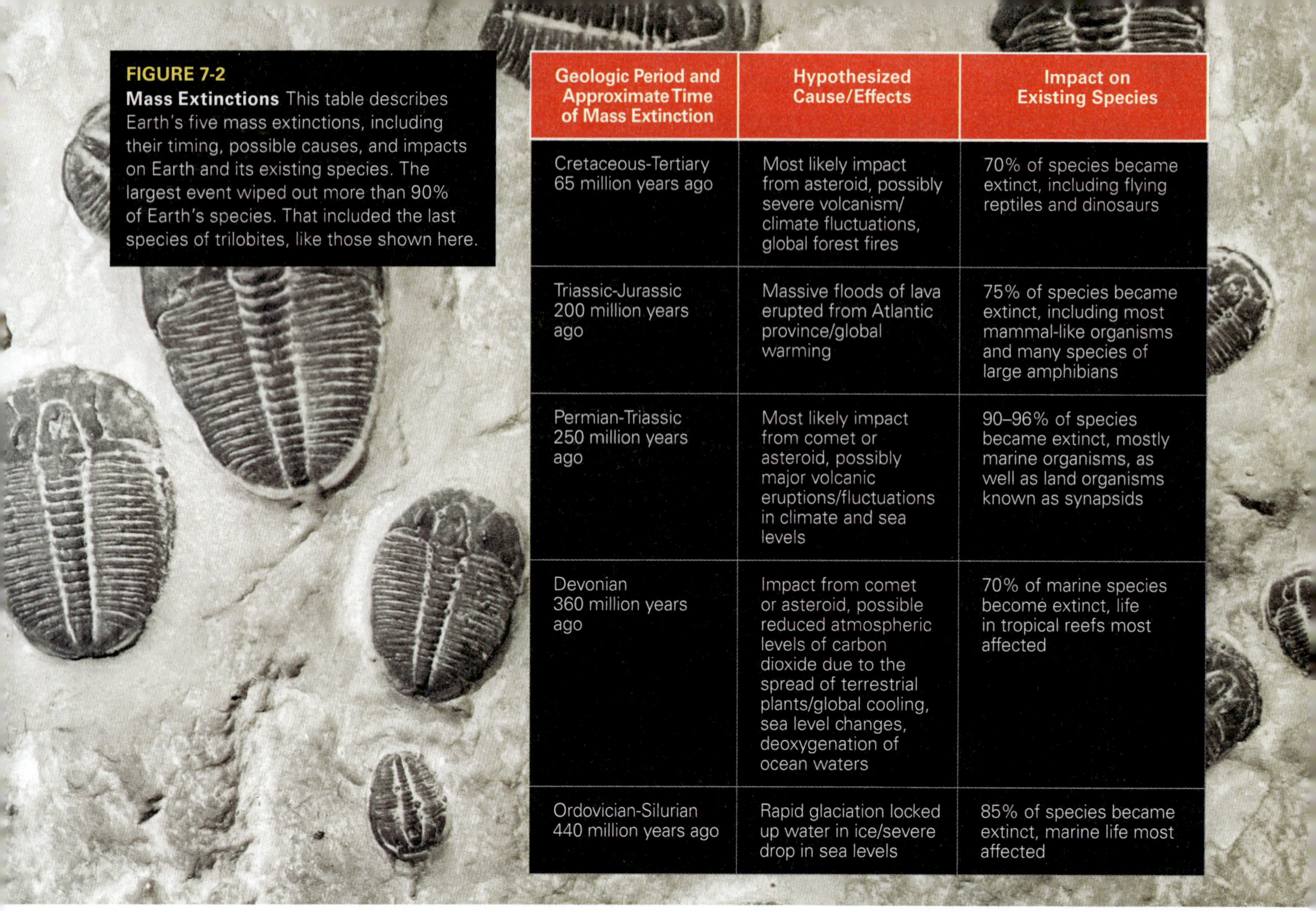

FIGURE 7-2

Mass Extinctions This table describes Earth's five mass extinctions, including their timing, possible causes, and impacts on Earth and its existing species. The largest event wiped out more than 90% of Earth's species. That included the last species of trilobites, like those shown here.

Geologic Period and Approximate Time of Mass Extinction	Hypothesized Cause/Effects	Impact on Existing Species
Cretaceous-Tertiary 65 million years ago	Most likely impact from asteroid, possibly severe volcanism/climate fluctuations, global forest fires	70% of species became extinct, including flying reptiles and dinosaurs
Triassic-Jurassic 200 million years ago	Massive floods of lava erupted from Atlantic province/global warming	75% of species became extinct, including most mammal-like organisms and many species of large amphibians
Permian-Triassic 250 million years ago	Most likely impact from comet or asteroid, possibly major volcanic eruptions/fluctuations in climate and sea levels	90–96% of species became extinct, mostly marine organisms, as well as land organisms known as synapsids
Devonian 360 million years ago	Impact from comet or asteroid, possible reduced atmospheric levels of carbon dioxide due to the spread of terrestrial plants/global cooling, sea level changes, deoxygenation of ocean waters	70% of marine species become extinct, life in tropical reefs most affected
Ordovician-Silurian 440 million years ago	Rapid glaciation locked up water in ice/severe drop in sea levels	85% of species became extinct, marine life most affected

In other words, extinctions are happening more on the order of 100 extinctions per million species annually. Some scientists predict the extinction rate could rise to 10,000 times the background extinction rate during this century—amounting to a loss of 10,000 species annually if Earth has 10 million species.

If these estimates are correct, Earth is entering a *sixth* mass extinction. And it will be driven mainly by human activities. According to this scenario, most species of big cats—including snow leopards, cheetahs, tigers, and lions—will probably exist only in zoos and other sanctuaries by the end of the century. Most elephants, rhinoceroses, gorillas, chimpanzees, and orangutans will also likely disappear from the wild. In fact, one estimate predicts between 25% and 50% of Earth's identified plant and animal species will become extinct during this timeframe. This estimate does not include any species yet to be identified.

Some experts actually consider a projected extinction rate of 10,000 times the background extinction rate to be low. They point to the fact that the human population will continue to grow, as will humans' demand for resources. This will cause the rate of extinction and resulting threats to ecosystem services to rise sharply in the next 50 to 100 years. In addition, current and projected extinction rates in the world's biodiversity hotspots—areas that are

> **CONSIDER THIS**
>
> **Are we living in the "Age of Man"?**
>
> The geologic time period we live in today is referred to as the Holocene epoch. It is considered to have begun 11,700 years ago. However, some scientists are asserting that a new epoch has begun due to impacts modern humans have had on the planet, especially since the Industrial Revolution. Scientists call this new epoch *Anthropocene*—from *anthropo-*, meaning "man," and *-cene*, meaning "new." It suggests the role humans now play in altering Earth's ecosystems. Human activities are causing species extinctions, changing patterns of land use, and even altering the composition of Earth's atmosphere.

CHAPTER 7   SAVING SPECIES AND ECOSYSTEM SERVICES

FIGURE 7-3 ▼

Plant Species Extinctions This graph shows extinctions among plant species since 1500 as confirmed by the IUCN. Extinction rates began rising significantly after the rise of the Industrial Revolution.

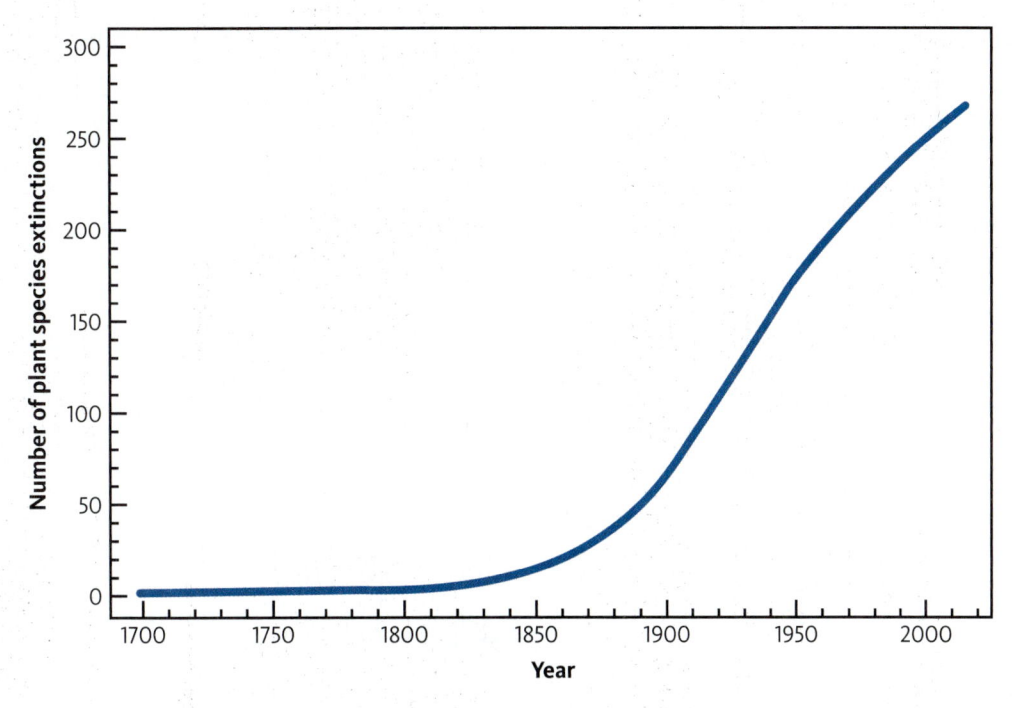

Source: Roux, Johannes J. Le, Cang Hui, Maria L. Castillo, José María Iriondo, Jan-Hendrik Keet, Anatoliy A. Khapugin, Frédéric Médail, Marcel Rejmánek, Genevieve Theron, Florencia A. Yannelli and Heidi Hirsch. "Recent Anthropogenic Plant Extinctions Differ in Biodiversity Hotspots and Coldspots." *Current Biology* 29 (2019): 2912-2918.e2.

highly endangered centers of biodiversity—are much higher than the global average.

Finally, human activities are negatively impacting many biologically diverse environments that serve as potential sites for the emergence of new species. Examples include tropical forests, coral reefs, wetlands, and estuaries. So in addition to greatly increasing the rate of extinction, humans may be limiting the long-term recovery of biodiversity by eliminating the places where new species can evolve.

Unlike previous mass extinctions, which stemmed from catastrophic but natural events, this mass extinction is stemming from human activities. And, unlike previous mass extinctions, which played out over many thousands of years, this mass extinction could unfold over the course of a human lifetime. One thing does still hold true, however: recovery from such a mass extinction would take millions of years.

checkpoint What evidence suggests Earth is entering a sixth mass extinction?

Certain Species Are More Vulnerable to Extinction

Some species have characteristics that increase their chances of becoming extinct (Figure 7-4). As a result, scientists monitor and track changes in species' populations to identify those that may become extinct without efforts to protect them. Biologists classify species heading toward biological extinction as either endangered or threatened. An **endangered species** has so few individual survivors that the species could soon become extinct. A **threatened species** has enough remaining individuals to survive in the short term, but because of declining numbers, it is likely to become endangered in the near future.

The International Union for Conservation of Nature (IUCN) is an organization that has taken on the responsibility of monitoring the status of species worldwide. For nearly 60 years, the IUCN has published its Red List, which identifies species that are critically endangered, endangered, or vulnerable to extinction. The IUCN bases its list on criteria such as population trends, geographic range, and habitat needs for each species surveyed.

The 2021 IUCN Red List identified 38,500 species as threatened with extinction. The actual number of species in trouble is very likely much higher.

checkpoint What criteria does the IUCN use to develop its Red List?

LESSON 7.1

Extinction Impacts Biodiversity and Ecosystem Services

More than half of the world's primates—human beings' closest living relatives—are threatened with extinction. Among them are the orangutans. As of 2021, the World Wildlife Fund (WWF) estimated that roughly 100,000 Bornean, 13,800 Sumatran, and 800 Tapanuli orangutans remained in the wild. Orangutans are the largest arboreal mammals on Earth. They live among the trees in dense tropical rain forests. The greatest threat to these species' survival is habitat loss. Much of the orangutans' habitat is being cleared for agriculture, particularly oil palm plantations. Palm oil is the world's most widely used vegetable oil. It is an ingredient in numerous products, from cookies to cosmetics to biofuels. Road building, mining, and logging also destroy orangutan habitat. Poaching and smuggling impact orangutans as well. An illegally smuggled, live orangutan sells on the black market for $20,000 or more.

The effects of these activities are made worse by the fact that orangutans have the lowest birth rate of all mammals. They give birth to one young about once every eight years. So, even small changes in numbers can be devastating to these primates.

Without urgent protective action, the endangered orangutan may disappear in the wild within the next two decades. Does it matter that the orangutan—or any species, for that matter—faces extinction largely due to human activities? After all, the argument could be made that new species eventually evolve to take the places of those species lost through mass extinctions. Why should we care if we greatly speed up the extinction rate over the next 50 to 100 years?

One compelling reason is the loss of ecosystem services due to extinction of species. As discussed in earlier chapters, biodiversity helps sustain life on Earth by providing free ecosystem services. Some services are of obvious importance to us. For example, pollination is one ecosystem service that is vital to the production of many agricultural crops. Other ecosystem services may not be so obvious. Aquatic species that live in streams can help purify flowing water. Trees produce oxygen that organisms need to survive. Earthworms aerate topsoil, which helps improve soil health.

Orangutans also provide an important ecosystem service. They are considered keystone species in the ecosystems they inhabit. Recall that a keystone species plays a unique role in the way an ecosystem

FIGURE 7-4

Vulnerable to Extinction Certain characteristics can place a species at greater risk of becoming extinct.

	Characteristic	Examples
	Low reproductive rate (*K*-strategist)	Blue whale, giant panda, rhinoceros
	Specialized niche	Blue whale, giant panda, Everglades kite
	Narrow distribution	Elephant seal, desert pupfish
	Feeds at high trophic level	Bengal tiger, bald eagle, grizzly bear
	Fixed migratory patterns	Blue whale, whooping crane, sea turtle
	Rare	African violet, some orchids
	Commercially valuable	Snow leopard, tiger, elephant, rhinoceros, rare plants and birds
	Require large territories	California condor, grizzly bear, Florida panther

functions. Without keystone species, the ecosystem would be dramatically different or cease to exist altogether. Orangutans are herbivores that eat a wide range of foods. More than 500 plant species have been identified in their diet. More than half that diet is fruit. When orangutans eat fruit and subsequently pass along the seeds that fruits contain, they aid in seed dispersal. Ecosystem services such as seed dispersal help keep rain forests healthy. If orangutans disappear, many rain forest plants—and the other animals that consume them—may be threatened as well.

People could also lose economic services due to extinction of species. Various plant species provide economic value as food crops, wood for fuel, lumber for construction, pulp for paper, and substances for medicines. **Bioprospectors** search tropical forests and other ecosystems for plants and animals that may serve as biological sources for medicinal drugs. To date, less than half of 1% of known plant species have been studied for their medicinal properties.

NATIONAL GEOGRAPHIC | EXPLORERS AT WORK
Joel Sartore Photographer

National Geographic Fellow Joel Sartore has a passion for documenting endangered species and landscapes in order to show a world worth saving. He is well known for his work as a photographer, speaker, author, teacher, and conservationist.

As a contributor to *National Geographic* magazine for more than 30 years, Sartore has turned his lens on conservation issues across every continent.

Sartore says his latest mission, Photo Ark, was born out of desperation to halt, or at least slow, the loss of global biodiversity. It's a trend he has seen firsthand, having witnessed the effects of overconsumption and habitat loss around the globe. "Frankly, I didn't know what else to do," he says.

Photo Ark is Sartore's 20-year project to document as many of the world's 15,000 captive species as possible. He is more than halfway through and has photographed more than 12,000 species at responsible zoos and aquariums worldwide. Sartore features each animal on a black or white background, to allow the viewer "to look every animal in the eye, the window to the soul" (Figure 7-5 and Figure 7-6).

Sartore hopes these up-close portraits inspire people to change how we treat Earth and its wildlife. "It is folly to think that we can destroy one species and ecosystem after another and not affect humanity," he says. "When we save species, we're actually saving ourselves."

FIGURE 7-5
Joel Sartore zooms in with his camera lens to snap a shot of a false water cobra for his Photo Ark project.

218

FIGURE 7-6
ON ASSIGNMENT This composite image features portraits of many primate species that National Geographic photographer Joel Sartore has created for Photo Ark. Numerous primate species are endangered or threatened, including orangutans. Can you identify the orangutans shown here or any other species you know?

219

FIGURE 7-7
Medicinal Properties of Plants These are just a few examples of plant species used as sources of medicine.

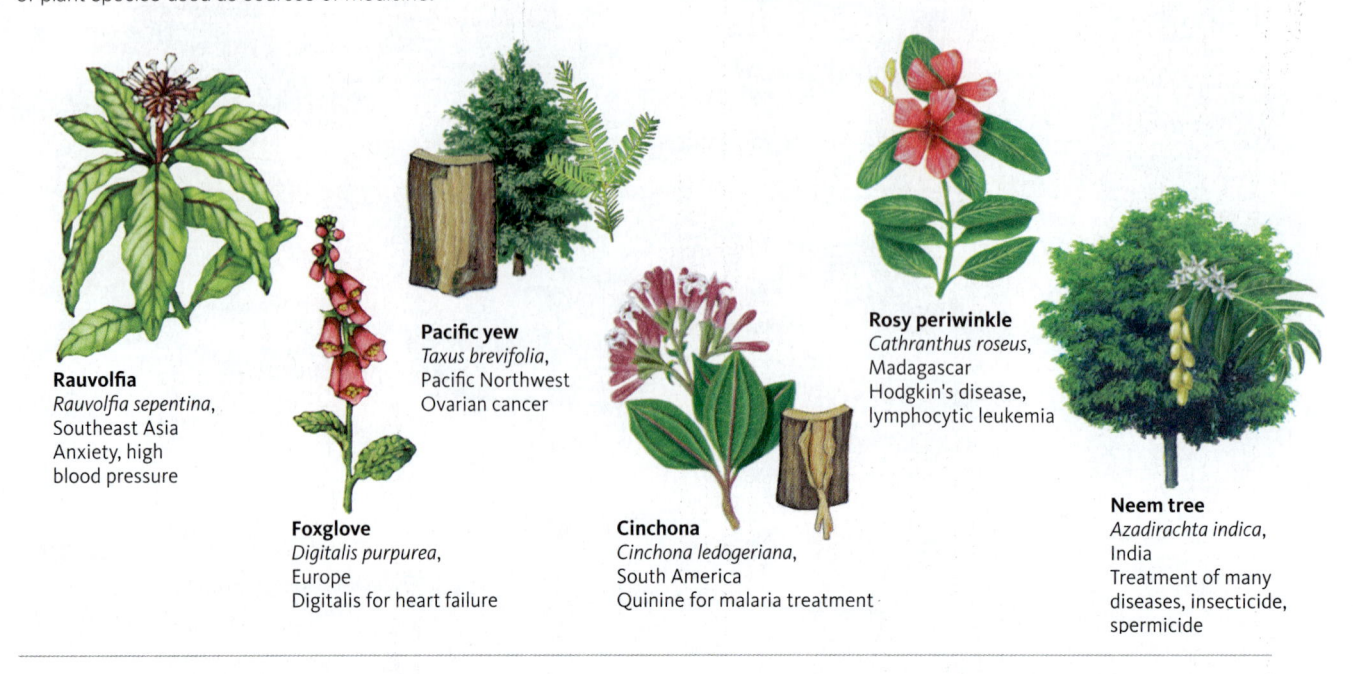

Rauvolfia
Rauvolfia sepentina, Southeast Asia
Anxiety, high blood pressure

Pacific yew
Taxus brevifolia, Pacific Northwest
Ovarian cancer

Foxglove
Digitalis purpurea, Europe
Digitalis for heart failure

Cinchona
Cinchona ledogeriana, South America
Quinine for malaria treatment

Rosy periwinkle
Cathranthus roseus, Madagascar
Hodgkin's disease, lymphocytic leukemia

Neem tree
Azadirachta indica, India
Treatment of many diseases, insecticide, spermicide

Yet it is estimated that 62% of all cancer drugs have been derived from the discoveries of bioprospectors. In addition, of the more than 10,000 known phytochemicals—certain chemicals that naturally occur in plants—many are believed to have the potential to slow aging, reduce pain, prevent various cancers, and control diseases such as diabetes (Figure 7-7). Consider once again the extinction of orangutans or other species. Their disappearance may threaten plant species whose medicinal benefits have yet to be discovered.

Preserving species and their habitats also provides economic benefits in the form of ecotourism. A rapidly growing industry, ecotourism specializes in responsible travel to natural areas. Ecotourism also promotes conservation, low environmental impact, respect for local cultures, and support for local economies. Travelers who sign up for ecotours have the chance to see endangered species such as orangutans in the wild. Some tours also allow travelers to help with data collection or other scientific work. Revenues from ecotourism generate more than $496 million per day in tourist expenditures worldwide.

Finally, many people believe that wild species such as orangutans have a right to exist, regardless of their usefulness to us. According to this worldview, we have an ethical responsibility to limit the impact of our ecological footprint and protect Earth's species from extinction. Also at stake is the well being of future generations. Research indicates it will take 5–10 million years for natural speciation to rebuild the biodiversity that is likely to be lost during this century due to extinction.

checkpoint What is the main reason orangutans are endangered?

7.1 Assessment

1. **Explain** How does a species become extinct?
2. **Recall** What types of natural events caused mass extinctions in the past?
3. **Define** What is the background extinction rate? Why is it rising?
4. **Compare and Contrast** How would a sixth mass extinction be similar to and different from those that have previously taken place on Earth?
5. **Form Opinions** Do you agree that wild species such as orangutans have the right to exist whether or not they are useful to humans? Explain.

CROSSCUTTING CONCEPTS

6. **Cause and Effect** Use the text and your own research to create a cause-and-effect diagram that illustrates how a rain forest ecosystem would be affected by the extinction of orangutans. Use it to predict how food chains would be altered and how other species' populations would be affected.

7.2 What Role Do Humans Play in the Loss of Species and Ecosystem Services?

CORE IDEAS AND SKILLS

- Discuss how human population growth and activities lead to habitat fragmentation and increase wild species extinctions.
- Understand how invasive species disrupt ecosystems.
- Discuss examples of pollution and their effects.
- Explain how overexploitation threatens wild species.

KEY TERMS

habitat fragmentation biomagnification
bioaccumulation

Habitat Destruction Poses the Greatest Threat

Certain human activities are increasingly responsible for driving species to extinction and degrading natural capital. Biodiversity researchers use the acronym HIPPCO to classify these activities as follows: **H**abitat loss, degradation, and fragmentation; the spread of **I**nvasive species; **P**opulation growth and increasing resource use; **P**ollution; **C**limate change; and **O**verexploitation. This chapter provides an overview of impacts from these activities. Most are also more closely examined in later chapters. Figure 7-8 illustrates how HIPPCO impacts Earth's terrestrial biomes.

The greatest threat to wild species is habitat destruction (Figure 7-9). Activities that negatively impact habitat include agriculture, industrial development, aquaculture, oil and gas exploration, deforestation, and water diversion. Habitats may be destroyed outright, such as when forests are clear-cut or when wetlands are filled in. They may be degraded by factors that affect their function, such as invasive species or pollution.

Habitats may also become divided, such as when large, intact areas of habitat are broken up by roads, logging operations, farm fields, and urban development. This process is called **habitat fragmentation** and it results in smaller, isolated patches or habitat islands. Habitat fragmentation can create barriers and isolate populations of a species. It causes them to be more vulnerable to predators, competitor species, disease, and natural events such as storms and fires. It also causes difficulties with migration, dispersal, colonizing areas, locating adequate food supplies, and finding mates.

Loss of habitat in tropical rain forests results in the greatest threat to wild species and to the ecosystem services they provide. At least half of these ecosystems have been destroyed or disturbed by human development. Other threats to habitat include the destruction and degradation of coastal wetlands and coral reefs, conversion of grasslands to agricultural crops, and the pollution of streams, lakes, and oceans.

Island species—many of them found nowhere else on Earth—are especially vulnerable to extinction

FIGURE 7-8
Threats to Biodiversity This map illustrates the fact that 58% of Earth's land surface is significantly impacted by ongoing threats of habitat loss, degradation, and fragmentation; invasive species; population growth; pollution; climate change; and overexploitation.

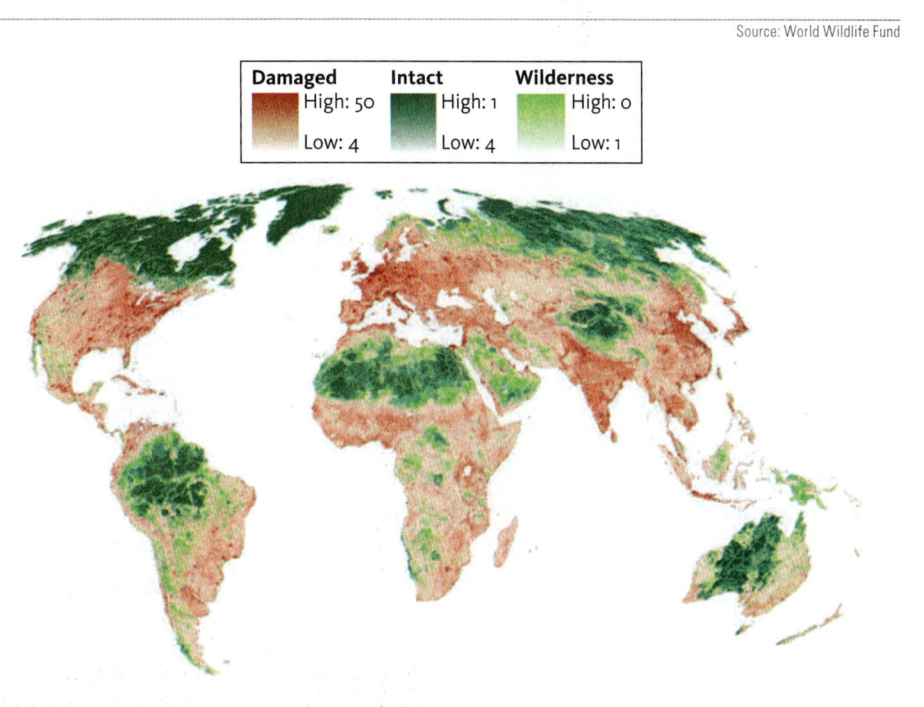

Source: World Wildlife Fund

LESSON 7.2 221

FIGURE 7-9

Shrinking Ranges These maps show the greatly reduced ranges of four wild species, mainly due to habitat destruction and illegal hunting.

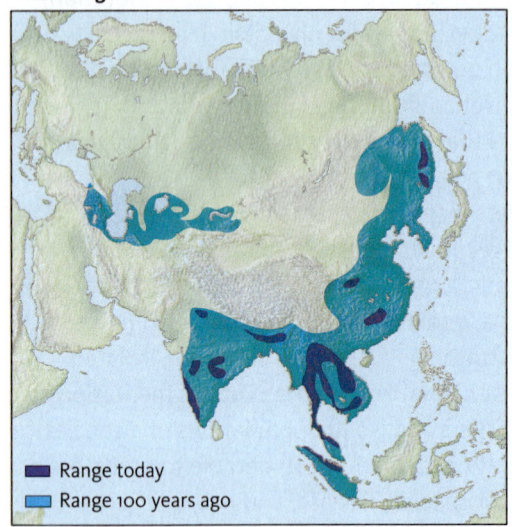

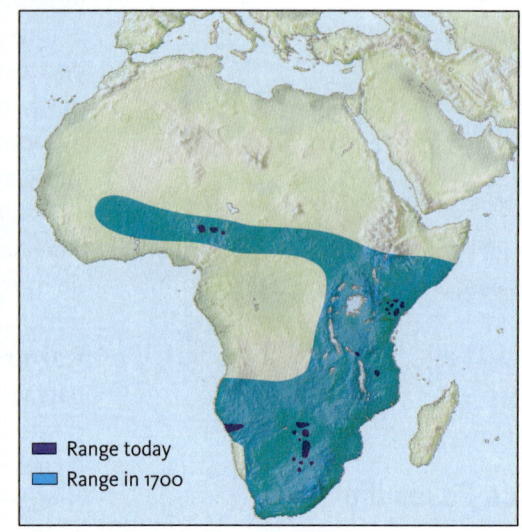

Indian Tiger
- Range today
- Range 100 years ago

Black Rhino
- Range today
- Range in 1700

African Elephant
- Range today
- Probable range in 1600

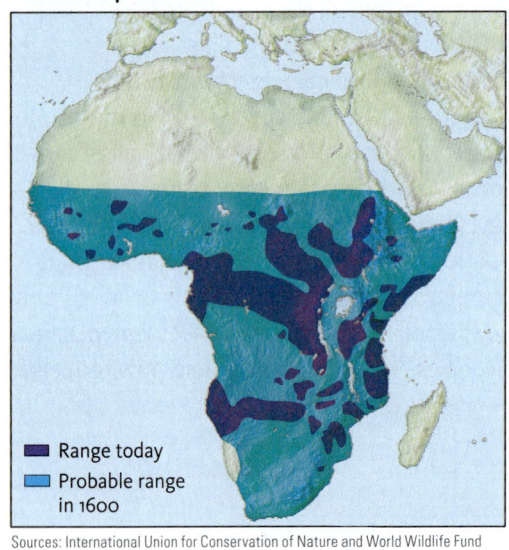

Asian or Indian Elephant
- Range today
- Former range

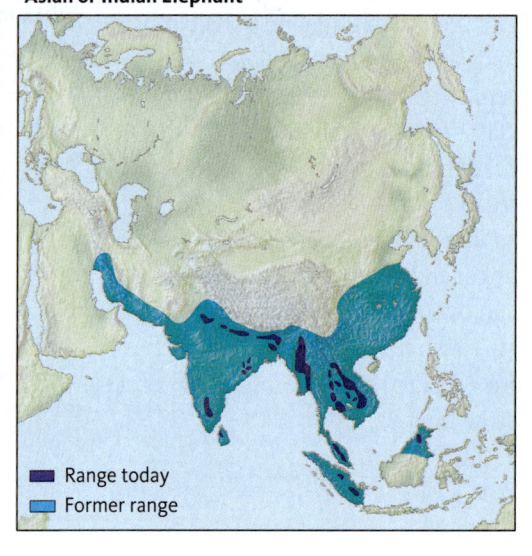

Sources: International Union for Conservation of Nature and World Wildlife Fund

---

after habitat loss occurs. They have nowhere else to go. In fact, the Hawaiian Islands are sometimes called America's "extinction capital," as they contain more than one-third of U.S. species known to be at risk.

checkpoint How does habitat fragmentation threaten wild species?

Invasive Species Disrupt Ecosystems

After habitat destruction, the spread of invasive species represents the second largest cause of extinction and loss of ecosystem services. Recall that an invasive species is one that negatively impacts an ecosystem where it has been introduced. For example, an invasive species may prey upon or parasitize native species. It may compete with them for food. It may transmit new diseases or alter the structure or function of the environment. An invasive species may also negatively impact human health or economic activities.

A newly introduced organism often does not face the same threats as it did in its native habitat. For example, it may not encounter the natural predators, competitors, parasites, viruses, bacteria, or fungi that controlled its numbers in its native habitat. As a result, some invasive species are able to spread rapidly. They crowd out populations of many native species and disrupt ecosystem services. According to the U.S. Fish and Wildlife Service (USFWS), 42% of the species considered endangered in the United

FIGURE 7-10

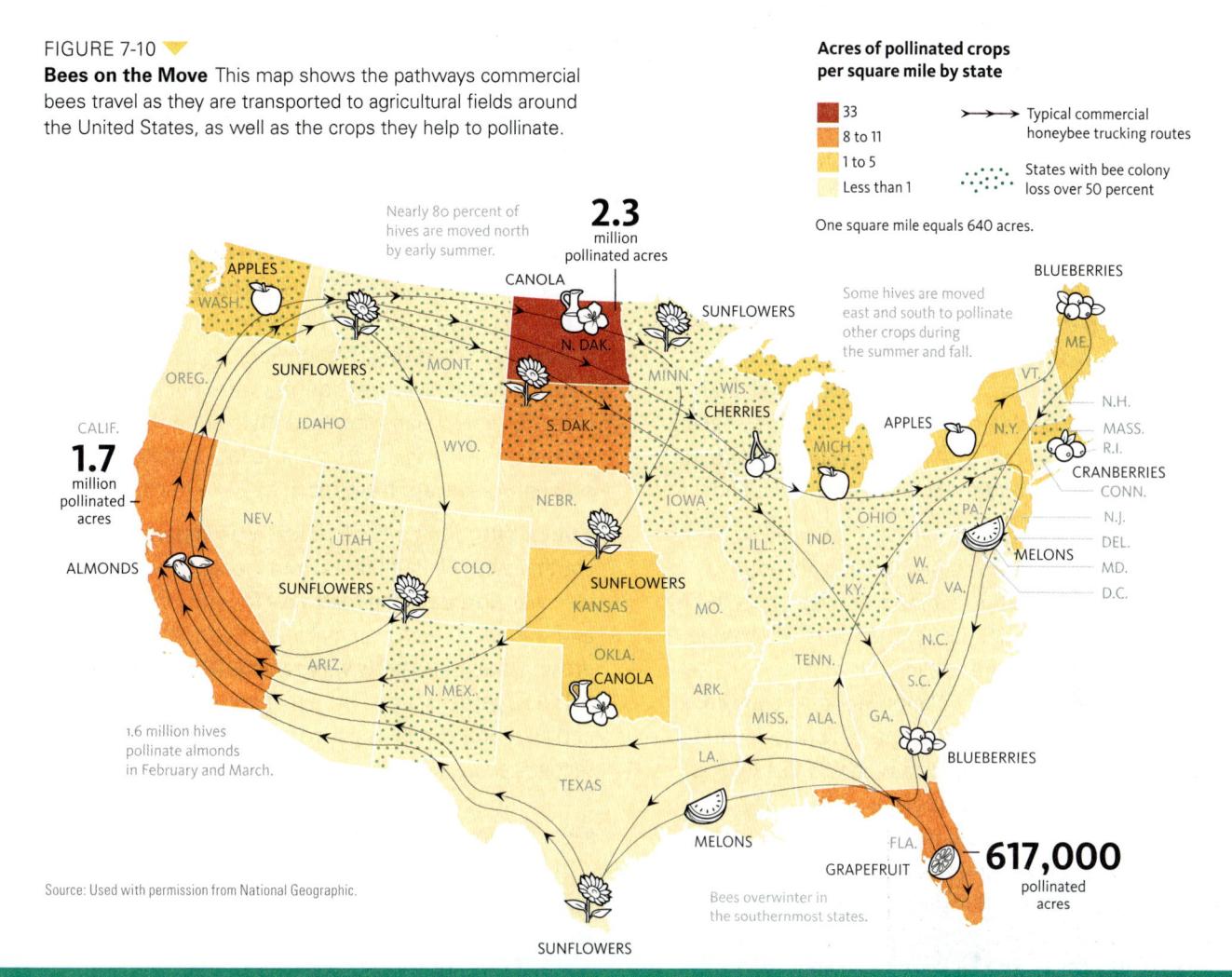

Bees on the Move This map shows the pathways commercial bees travel as they are transported to agricultural fields around the United States, as well as the crops they help to pollinate.

Acres of pollinated crops per square mile by state
- 33
- 8 to 11
- 1 to 5
- Less than 1
- → Typical commercial honeybee trucking routes
- ⋯ States with bee colony loss over 50 percent

One square mile equals 640 acres.

Nearly 80 percent of hives are moved north by early summer.

2.3 million pollinated acres — CANOLA, N. DAK.

APPLES — WASH.
SUNFLOWERS — OREG., MONT., MINN.
CHERRIES — WIS., MICH.
BLUEBERRIES — ME.
APPLES — N.Y.
CRANBERRIES
MELONS
SUNFLOWERS — S. DAK.
SUNFLOWERS — KANSAS
CANOLA — OKLA.
MELONS
BLUEBERRIES — GA.
GRAPEFRUIT — FLA.
617,000 pollinated acres
SUNFLOWERS

CALIF. **1.7 million pollinated acres** ALMONDS
1.6 million hives pollinate almonds in February and March.

Some hives are moved east and south to pollinate other crops during the summer and fall.

Bees overwinter in the southernmost states.

Source: Used with permission from National Geographic.

SCIENCE FOCUS 7.1

HONEYBEE LOSSES

What causes colony collapse disorder (CCD)? Research has focused on several factors, which likely act together to cause CCD.

Pests: The varroa mite feeds on the blood of both adult honeybees and their larvae, weakening their immune systems and shortening their lives. Varroa mites have killed millions of honeybees since the late 1980s. Scientists suspect honeybees are now even more vulnerable to varroa mites because their immune systems are weakened by other factors, described below.

Viruses: A recently discovered virus, called the tobacco ringspot virus, may infect honeybees that feed on pollen where it is present. The virus is thought to attack the bees' nervous systems. The virus has also been detected in varroa mites, which may help spread the virus as they feed on honeybees.

Pesticides: As honeybees forage for nectar, they can come in contact with insecticides sprayed on crops. Some research indicates a class of pesticides called neonicotinoids may play a role in CCD. Exposure to trace amounts of neonicotinoids appears to harm honeybees' nervous and immune systems. It can affect their ability to navigate back to their hives.

Stress: Honeybees that are transported across the United States travel long distances and face a longer-than-usual pollination season (Figure 7-10). Stress from transport and overcrowding can weaken their immune systems.

Diet: In natural ecosystems, honeybees feed on a variety of flowering plants. But commercial honeybees feed mostly on one or just a few crops and may lack the nutrients they need.

Thinking Critically

Apply How do practices used in large-scale agriculture contribute to the factors associated with CCD?

States and 95% of those in Hawai'i are at risk mainly due to threats from invasive species.

While some invasive species are introduced deliberately, most accidentally reach new environments due to human activities. For example, invasive species may be transported unintentionally. They may travel in ballast water inside boats, on fruits and vegetables, among foreign packing materials, even in people's luggage. Exotic pets have also escaped or have been released into the wild to become invasive species. Global trade, travel, and transport of goods all help to spread invasive species. Following are examples of invasive species found in the United States and the impacts they have had.

Asian Carp The term "Asian carp" actually covers several carp species including silver, bighead, and black carp. These filter-feeding fish were imported to the southern United States in the 1970s. They were used to help keep water clean in fish farm ponds and at wastewater treatment plants. Flooding allowed Asian carp to escape into the Mississippi River and later to the connecting Missouri and Illinois Rivers.

As these invasive species have spread, they have negatively impacted native fish. Asian carp reproduce in great numbers, grow quickly, and can become very large. Adults have no natural predators. They are aggressive and have outcompeted native species for food and habitat. For example, Asian carp have greatly reduced populations of plankton, tiny organisms that form the basis of many aquatic food webs. They also pose a threat to humans. When startled, these fish leap from the water (Figure 7-11). The sound of a motor can cause a flurry of flying fish that can collide with people and boats.

Scientists worry that Asian carp may be on their way to entering the Great Lakes. The Illinois River is connected to the Great Lakes by a system of waterways called the Chicago Sanitary and Ship Canal. Several electrical barriers were built in the canal to repel Asian carp. Scientists hope these barriers, as well as others such as locks and dams, will prevent Asian carp from reaching Lake Michigan.

Kudzu Vine In the 1930s, kudzu was imported to the southeastern United States from Japan. The ornamental vine was planted along roadways for erosion control and was also used for animal feed. Kudzu did control erosion, but it also grew very rapidly. It quickly engulfed hillsides, gardens, trees, stream banks, cars, houses, and anything else in its path. By the 1950s, kudzu was recognized as a weed and was no longer considered suitable for planting. In 1998, it was classified as a federally noxious (harmful and destructive) plant.

Nicknamed "the vine that ate the South," kudzu has now spread throughout much of the southeastern United States. It could spread to the north if the climate gets warmer as scientists predict.

FIGURE 7-11
Boaters on the Illinois River try to navigate through a swarm of Asian carp leaping from the waters around them. The fish are reacting to the sound of the boat motor.

Red Imported Fire Ant In the 1930s, the extremely aggressive red imported fire ant was accidentally introduced into the United States. It is thought to have arrived among shiploads of cargo imported from South America. Fire ants have no natural predators in the southern United States. They spread rapidly by land and also by water because they can float. When these ants invade an area, they can displace up to 90% of native ant populations. Native ants provide important ecosystem services. They help enrich topsoil, disperse plant seeds, and control pest species such as flies, bedbugs, and cockroaches.

When disturbed, as many as 100,000 fire ants might swarm out of their nest to attack. The ants deliver painful, burning stings that can actually be fatal. Fire ants have killed deer fawns, ground-nesting birds, baby sea turtles, newborn calves, pets, and at least 80 people who were allergic to their venom.

Widespread pesticide application in the 1950s and 1960s temporarily reduced fire ant populations. However, the pesticide use reduced the populations of many native ant species as well. Less competition allowed the fire ants to invade even more quickly. Pesticide use also promoted the development of genetic resistance to pesticides in the fire ants through natural selection. The introduction of certain parasitic fly species and one type of fungus have shown some success in reducing fire ant populations. However, more research is needed to understand the long-term impacts of these biological remedies.

Zebra Mussel The zebra mussel is a thumbnail-sized mollusk that is native to eastern Europe and western Asia. This invasive species was first observed in 1988 in Lake St. Clair, between Lake Huron and Lake Erie. It is thought to have arrived in the ballast tanks of ocean-going ships. Today, zebra mussels have spread to all the Great Lakes as well as a number of major rivers. They have been reported in nearly three dozen states.

Zebra mussels reproduce rapidly and have no known natural enemies. They attach themselves to hard surfaces, forming dense blankets of up to 100,000 individuals per square meter. Zebra mussels have displaced other mussel species and depleted populations of plankton, a key link in many species' food webs. Their filter feeding behavior has also drastically changed water clarity, which has led to algae outbreaks. Decaying algae that washes up on shore contaminates beaches and harbors harmful bacteria. Zebra mussels also cause massive economic damage. They clog irrigation pipes and shut down cooling water intake pipes for power plants and city water supplies. They also jam ship rudders and form large masses on boat hulls, piers, and other exposed aquatic surfaces.

Burmese Python The Burmese python is an example of what can happen when nonnative species escape or are released into the wild and become invasive species. These snakes are imported from Asia for sale as pets. Some buyers, after learning these reptiles make poor pets, simply let them go in the wild. Burmese pythons have invaded wetlands of the Florida Everglades and are breeding in Everglades National Park. In fact, more than 13,000 Burmese pythons have been removed from the park and surrounding areas since 2000 (Figure 7-12).

The Burmese python can live 20 to 25 years, growing as long as 5 meters (16 feet). These snakes have huge appetites. They feed at night, eating a variety of birds and mammals such as rabbits, foxes, raccoons, and white-tailed deer. Occasionally the pythons will eat other reptiles, including the American alligator—a keystone species in the Everglades ecosystem. Pythons have also been known to eat cats and dogs, small farm animals, and geese.

Research indicates that pythons are proving more than a nuisance to local residents. Their feeding habits are actually altering the complex food webs and ecosystem services of the Everglades. Pythons are hard to find and kill and they reproduce rapidly. Trapping and moving the snakes from one area to another has proven ineffective. They are able to return to the areas where they were captured.

> ### CONSIDER THIS
>
> **You can help fight the spread of invasive species.**
>
> Take these steps to protect native species.
> - Do not buy wild plants and animals or remove them from natural areas.
> - Do not release wild pets in natural areas.
> - Do not dump aquarium contents or unused fishing bait into waterways or storm drains.
> - When camping, use only local firewood.
> - Brush or clean pets, shoes, mountain bikes, watercraft, motors, fishing tackle, and other gear before entering or leaving wild areas.

Control of Invasive Species Once a harmful invasive species becomes established, removing it is almost impossible. Americans pay $116 million annually on average to eradicate or control an increasing number of invasive species—without much success. Clearly, the best way to limit the harmful impacts of these organisms is to prevent them from being introduced into ecosystems.

How might the spread of invasive species be prevented or controlled? First, more research is needed to identify the major characteristics of successful invaders as well as ecosystems where they have the greatest impact. Research is also needed to identify natural predators, parasites, bacteria, and viruses that could be used to control populations of established invaders. Additional ground surveys and satellite observations could help to track invasive plant and animal species. Better computer models are needed for predicting how a species spreads and what harmful effects it could have.

Countries can also work to create international treaties banning the transfer of invasive species from one country to another, as is now done for endangered species. They can also step up inspection of imported goods to enforce such bans. Finally, it is important to educate the public about the effects of releasing exotic plants and pets into the environment near where they live.

checkpoint How does international shipping of goods and cargo aid the spread of invasive species?

Human Population Growth Strains Resources

Fewer than 1 billion people lived on Earth in the year 1800. However, improvements in people's standards of living during the Industrial Revolution enabled the human population to increase dramatically. In fact, it doubled by 1930. By 1975, the human population had reached 4 billion, and today it stands at just over 7.9 billion. The United Nations estimates the human population will reach 9.8 billion by 2050. (Human population growth is the focus of Chapter 14.)

Growth in the human population has been accompanied by a growth in resource use per person, which has greatly expanded the human ecological footprint (Figure 1-10). People have eliminated, degraded, and fragmented vast areas of wildlife habitat as they have spread over the planet. They have consumed resources at increasing rates. The

FIGURE 7-12
This Burmese python was captured in the Florida Everglades. Some researchers fear this invasive species will spread to other wetlands in the southern half of the United States.

FIGURE 7-13 ▼

Human Population Growth This graph shows the dramatic rise in the human population since the Industrial Revolution (left). This population increase has been accompanied by a dramatic increase in extinction rates (right). Scientists worry that Earth is now entering a sixth mass extinction.

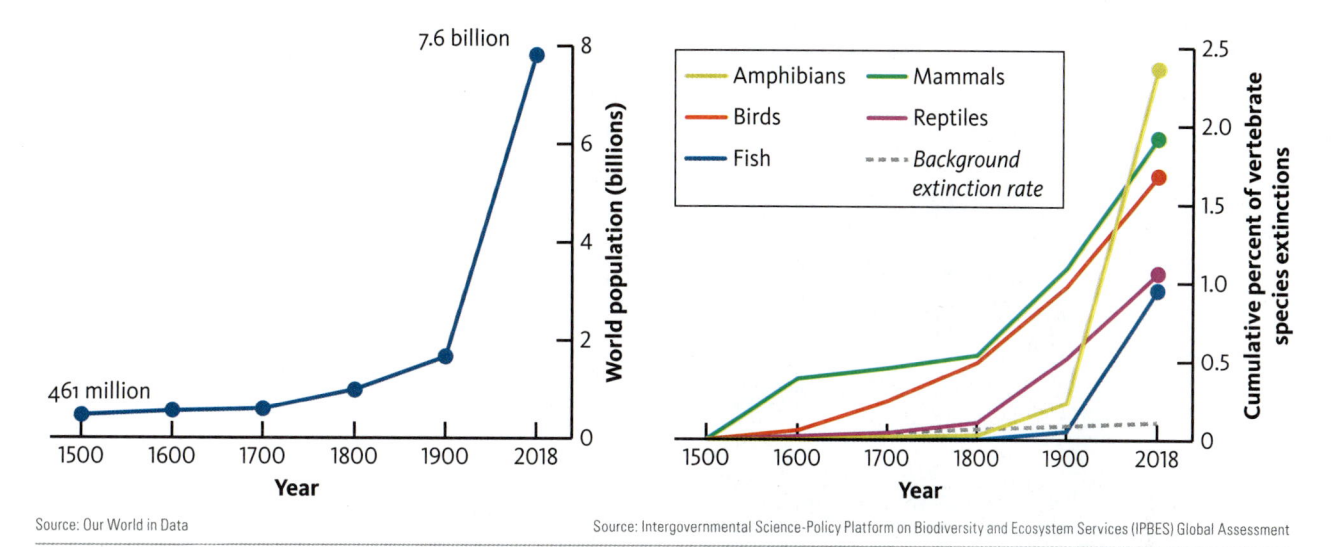

Source: Our World in Data

Source: Intergovernmental Science-Policy Platform on Biodiversity and Ecosystem Services (IPBES) Global Assessment

dramatic increase in the human population has been accompanied by a dramatic increase in wild species extinctions (Figure 7-13).

checkpoint How much is the human population predicted to increase between now and 2050?

Pollution Threatens Organisms and Ecosystems

Pollution from human activities has impacted ecosystems worldwide. In fact, ecosystems may contain pollutants from sources hundreds or thousands of miles away. Pollutants may contaminate soil, water, and air. They may kill organisms outright or impact their ability to reproduce or perform other functions. Two human activities that cause significant pollution are the use of agricultural chemicals and the emission of greenhouse gases. These and other pollutants are discussed in later chapters.

Agricultural Chemicals The use of pesticides and fertilizers has increased significantly in the last 50 years. These products have made possible huge increases in crop production. However, their use has also had harmful effects on ecosystems and organisms. Pesticides and fertilizers may wash into waterways or contaminate soil and groundwater. Pesticides may kill nontarget organisms, including beneficial insects. Each year pesticides kill about one-fifth of the European honeybee colonies that pollinate almost one-third of all U.S. food crops (Case Study). A 2021 EPA study estimated 93% of the country's endangered and threatened species are harmed by just one single pesticide, glyphosate.

During the 1950s and 1960s, populations of fish-eating birds such as ospreys, brown pelicans, and bald eagles plummeted. The cause for these population declines was tied to a pesticide called DDT. After DDT was sprayed to kill mosquitoes, it remained in the environment and was taken up by organisms. Instead of being metabolized, a chemical derived from the DDT accumulated in the organisms' fatty tissues, a process called **bioaccumulation**.

The chemical then became more concentrated as it moved up the food chain, a process called **biomagnification**. Eventually, predatory birds were affected (Figure 7-14). They had difficulty producing calcium during egg laying, and their eggshells were very thin as a result. Many eggs cracked before hatching, and the birds could not reproduce successfully. Other predatory birds hit hard in those years were the prairie falcon, sparrow hawk, and peregrine falcon. All these birds of prey are important to controlling populations of rabbits, ground squirrels, and other crop eaters. Since the United States banned DDT in 1972, most of these bird species have made a comeback.

LESSON 7.2

FIGURE 7-14 ▼

Biomagnification This illustration models DDT biomagnification in a food chain featuring species found in New York state. DDT levels are measured in parts per million (ppm) and are represented by the gold bands (not to scale).

DDT in fish-eating birds (ospreys) 25 ppm

DDT in large fish (needlefish) 2 ppm

DDT in small fish (minnows) 0.5 ppm

DDT in zooplankton 0.04 ppm

DDT in water 0.000003 ppm, or 3 ppt

Greenhouse Gas Emissions Greenhouse gases play a role in determining the lower atmosphere's average temperatures and thus Earth's climates. Since the 1800s, human activities have led to significant increases in the emission of greenhouse gases, specifically carbon dioxide (CO_2) and methane (CH_4). These activities have mainly included burning fossil fuels, deforestation, and agriculture. Most scientists now agree that Earth's atmosphere is warming at a rate that is likely to lead to significant climate change. (Climate change is discussed in Chapter 16.)

According to a study by Conservation International, projected climate change could drive one-quarter to one-half of all land animals and plants to extinction by the end of this century. The effects of climate change include increased melting of polar ice and glaciers, a rise in sea levels, increased ocean acidification, and more extreme weather. The habitat loss associated with climate change effects will impact biodiversity and ecosystem services on every continent.

The hardest-hit species will be those that have limited ranges or low tolerance for temperature changes. For example, studies indicate that polar bears are threatened because of higher temperatures and melting sea ice in their polar habitat. The shrinkage of floating ice is making it harder for polar bears to find seals, their favorite prey. They use the ice as a platform for diving to find food and for giving birth to their pups. A recent study has suggested that smaller areas of sea ice during the mating season are reducing mating opportunities and long-range breeding dispersal. As a result, polar bears in some areas are inbreeding, which reduces genetic diversity.

checkpoint What is the difference between bioaccumulation and biomagnification?

Overexploitation Threatens Biodiversity

Humans have learned to exploit wild species for food, clothing, medicine, pets, and many other purposes. Advances in technology and increasing demands of a growing human population are leading to overexplotation of many wild species. Overexploitation occurs when so many members of a species' population are removed, it can no longer sustain itself without intervention. It is in danger of becoming extinct.

Overharvesting Overharvesting is one example of overexploitation. Overharvesting occurs when too many individuals of a given species are taken by means of hunting, gathering, or fishing.

For centuries, indigenous people in much of West and Central Africa have sustainably hunted wildlife as a source of food. This source of meat is known as "bushmeat." In the last three decades, bushmeat hunting in some areas has skyrocketed. Logging roads in once-inaccessible forests have made hunting bushmeat much easier. Rapidly growing populations have fueled demand for bushmeat, as has interest in serving exotic meats at restaurants in major cities.

Bushmeat hunting has driven at least one species—Miss Waldron's red colobus monkey— to complete extinction. It is also a factor in the decline of some populations of orangutans, gorillas, chimpanzees, elephants, and hippopotamuses.

The U.S. Agency for International Development (USAID) is trying to reduce unsustainable hunting for bushmeat in some areas of Africa by introducing alternative sources of food, including farmed fish. They are also showing villagers how to breed large rodents such as cane rats as a source of protein.

Exotic Pet Trade The legal and illegal trade in wild species for use as pets is a huge and profitable business worldwide and can lead to overexploitation.

Many owners of exotic pets don't realize that for every live animal captured and sold in the pet market, many others are killed or die in transit. According to the IUCN, more than 260 bird species, mostly parrots, are endangered or threatened because of the wild bird trade (Figure 7-15). In response, the United States passed the Wild Bird Conservation Act in 1992. The Act makes it illegal to import parrots into the United States. Any parrot purchased today must be from a domestic breeder.

Wild species such as parrots that are allowed to remain in the wild can prove valuable in other ways. For example, collectors of exotic birds might pay $10,000 or more for an endangered hyacinth macaw smuggled out of Brazil. However, during its lifetime a single hyacinth macaw left in the wild could attract an estimated $165,000 in ecotourism revenues.

Buyers of wild animals might also be unaware that some imported exotic animals carry diseases. For example, they may be infected with hantavirus, Ebola virus, Asian bird flu, herpes B virus (carried by most adult macaques), or Salmonella bacteria (from pets such as hamsters, turtles, and iguanas). These diseases can spread easily from pets to their owners and then to other people.

Other wild species threatened by the pet trade include many amphibians, various reptiles, and tropical fishes taken mostly from the coral reefs near Indonesia and the Philippines. Some divers catch tropical fish by using plastic squeeze bottles of poisonous cyanide to stun them. For each fish caught alive, many more die. To make matters worse, the cyanide solution kills coral polyps, the tiny animals that create the reefs.

Some exotic plants, especially orchids and cacti, are endangered because they are gathered and sold, often illegally. A mature crested saguaro cactus can earn a cactus rustler as much as $15,000. An orchid collector might pay $5,000 for a single rare orchid.

Illegal Trade Illegal hunting and trade can also contribute to overexploitation of wild species. Illegal hunting is also known as poaching. Illegal trade is the sale of protected species or their parts for use as pets, medicine, food, spiritual items, and souvenirs. Globally, illegal trade in wildlife brings in as much as $23 billion per year. At least two-thirds of all live animals smuggled around the world die in transit. Few of the smugglers are caught or punished.

For smugglers and poachers of endangered species, the financial rewards of their illegal activities may seem worth the risk of getting caught. Consider the black market prices they are able to charge for one of each of the following:

- live eastern mountain gorilla (roughly 1,000 remain in the wild): $150,000
- pelt of a giant panda (roughly 1,800 remain in the wild): $100,000
- coat made from tiger fur (roughly 3,900 remain in the wild): up to $100,000
- horn from black rhinoceros (roughly 5,500 remain in the wild): up to $500,000

Four of the five rhinoceros species are critically endangered, mostly because so many have been illegally killed for their horns (Figure 7-16). A rhino's horn is composed of keratin, the same protein that makes up fingernails and hair. Powdered rhino horn is still sought for use in traditional Asian medicine, which is why it sells for so much on the black market.

Poaching of elephants for their valuable ivory tusks continues to take place despite an international ban on ivory trade that began in 1990. A set of ivory tusks from an adult male elephant can command $375,000 on the black market. In 2020, according to the World Wildlife Fund, 20,000 elephants were killed for their ivory. After implementing a ban on domestic sales of ivory in 2017, demand for it in China has declined, but China still remains the largest market for illegal ivory.

FIGURE 7-15

Illegal Trade These caged scarlet macaws were captured for sale as part of the illegal pet trade in Guyana, South America.

FIGURE 7-16
This female southern white rhinoceros, named Thandi, was lucky to survive after losing her horn to poachers. Thandi lives in Kariega Game Reserve in Eastern Cape Province, South Africa.

In 1900, there were an estimated 100,000 tigers in the wilds of Asia in a rapidly shrinking range. Today there are only 3,900 and half of them are in India. Tigers have lost 90% of their habitat due to rapid human population growth. They have also been targets of poaching, much of which is motivated by poverty. The World Wildlife Fund reports that tiger populations appear to be stable or increasing in India, Nepal, Bhutan, Russia, and China but cautions much work is needed to secure tigers' future protection.

Small "anti-poaching" aerial drones are now being used worldwide to help protect rhinos, elephants, tigers, and other endangered species from poachers. These drones use infrared and thermal imaging to locate illegal poachers working at night. The drones then stream back live video to park rangers.

checkpoint What factors have led to overharvesting of bushmeat?

7.2 Assessment

1. **Recall** What are the five factors related to human activities that threaten wild species with extinction? Which factor poses the greatest threat?
2. **Apply** A new species is introduced to an ecosystem. What evidence would suggest it has become an invasive species?
3. **Explain** How can chemicals sprayed at one location end up harming organisms and ecosystems elsewhere?
4. **Summarize** What factors contribute to overexploitation of wild species?

CROSSCUTTING CONCEPTS

5. **Cause and Effect** Draw a flowchart that illustrates how greenhouse gas emissions could eventually cause the extinction of polar bears.

7.3 How Can We Sustain Wild Species and Ecosystem Services?

CORE IDEAS AND SKILLS

- Understand how international treaties and national laws protect species and ecosystems.
- Discuss the benefits of wildlife refuges and other protected lands for wild species and ecosystems.
- Explain methods used to preserve the seeds of plant species.
- Discuss how captive breeding programs preserve endangered animal species.

KEY TERMS

Convention on International Trade in Endangered Species of Wild Fauna and Flora (CITES)
Convention on Biological Diversity (CBD)
Endangered Species Act (ESA)
seed bank
botanical garden
arboretum
egg pulling
captive breeding

Treaties and Laws Can Protect Endangered Species

Governments are working to reduce species extinction and sustain ecosystem services by establishing and enforcing international treaties and conventions as well as national environmental laws.

Convention on International Trade in Endangered Species of Wild Fauna and Flora Several international treaties and conventions help protect endangered and threatened wild species. The farthest-reaching international agreements are embodied in the 1975 **Convention on International Trade in Endangered Species of Wild Fauna and Flora (CITES)**. CITES bans the hunting, capturing, and selling of threatened or endangered species. It covers the sale of live specimens as well as any products made from parts of threatened or endangered species.

Signed by 183 countries, CITES grants varying degrees of protection for approximately 38,000 species of plants and animals. The treaty currently restricts commercial trade of 931 species because they are in danger of extinction. It also controls the trade of roughly 5,900 animal species and 32,000 plant species at risk of becoming threatened.

CITES has helped reduce the international trade of many threatened animals, including rhinoceroses, elephants, crocodiles, cheetahs, and chimpanzees.

LESSON 7.3 231

The treaty has also raised public awareness about illegal wildlife trade and poaching. And it has inspired campaigns against these practices (Figure 7-17).

However, CITES is limited in some other ways. Enforcement varies among countries, and convicted violators often pay only small fines. Member countries can exempt themselves from protecting any listed species. Also, much of the highly profitable illegal trade in wildlife and wildlife products goes on in countries that have not signed the treaty.

Convention on Biological Diversity Another key international treaty is the **Convention on Biological Diversity (CBD)**, which legally commits participating governments to reduce the global rate of biodiversity loss and to share benefits from use of the world's genetic resources. It also aims to prevent or control the spread of invasive species. The CBD has been ratified (formally approved) by 196 countries.

The CBD is a landmark in international law because it focuses on ecosystems rather than on individual species. It also links biodiversity protection to issues such as the traditional rights of indigenous peoples. However, because some key countries have not ratified it (including the United States as of 2021), implementation has been slow. Also, the law lacks severe penalties or other enforcement mechanisms.

The U.S. Endangered Species Act The United States enacted its own national environmental law called the Endangered Species Act in 1973 (and amended it in 1978, 1982, 1988, and 2004). The **Endangered Species Act (ESA)** is designed to identify and protect endangered and threatened species in the United States and abroad. The ESA creates recovery programs for the species it lists. The goal of the ESA is to help each species' numbers recover to levels where legal protection is no longer needed. In that case, a species can be taken off the list, or delisted.

The ESA is probably the most successful and far-reaching environmental law adopted by any nation. Between 1973 and November 2021, the number of U.S. species on the official endangered and threatened species lists increased from 92 to 1,667. Almost three-quarters of listed species had active recovery plans. As of 2021, 99% of the listed species had been saved from extinction. Nearly half of all listed species had received designations of critical habitat as well. Successful recoveries include the American alligator, gray wolf, peregrine falcon, bald eagle, humpback whale, and brown pelican.

Under the ESA, the National Marine Fisheries Service (NMFS) is responsible for identifying and listing endangered and threatened ocean species. The USFWS identifies and lists all other endangered and threatened species. Any decision by either agency to list or delist a species must be based on biological factors alone, without consideration of economic or political factors. However, the two agencies can use economic factors in deciding whether and how to protect endangered habitat and in developing recovery plans for listed species.

The ESA forbids federal agencies (except the Defense Department) to carry out, fund, or authorize projects that would jeopardize any endangered or threatened species or destroy or modify its critical habitat. If endangered or threatened species inhabit private lands, the ESA prohibits landowners from modifying habitat in any way that might harm these species. The ESA also makes it illegal for Americans to sell or buy any product made from an endangered or threatened species. Furthermore, they cannot hunt, kill, collect, or injure such species in the United States.

For offenses committed on private lands, fines as high as $100,000 and one year in prison can be imposed. This provision has rarely been used. However, it is potentially very significant. More than half of all species listed as endangered or threatened spend at least part of their life cycles on private lands. Since 1982, the ESA has been amended to give private landowners various economic incentives to save endangered species living on their lands.

The ESA also requires all commercial shipments of wildlife and wildlife products to enter or leave the country through one of 18 designated airports and ocean ports. The 113 full-time USFWS inspectors can inspect only a small fraction of the more than 200 million wild animals brought legally into the United States annually. Each year, tens of millions of wild animals are also brought in illegally, but few illegal shipments of endangered or threatened animals or plants are confiscated.

Since 1995, there have been numerous efforts to weaken the ESA and reduce its annual budget. Opponents of the ESA say it places the rights of endangered plants and animals above people's rights. Some critics call the ESA an expensive failure because as of late 2021, only 54 species had recovered enough to be removed from the endangered list.

FIGURE 7-17
National Geographic Explorer Asher Jay describes herself as a creative conservationist. She uses her artistic talents to raise public awareness about issues such as poaching and illegal wildlife trade. Jay calls this artwork "Broken Record." What message does it convey? How does Jay's work apply the ethics factor of sustainability?

LESSON 7.3   233

In 2021, the USFWS proposed removing 23 species from the endangered species list due to extinction, which would add to the 11 species formally removed previously.

A national poll conducted by the Center for Biological Diversity found that two out of three Americans want the ESA strengthened or left alone. Despite this sentiment, the ESA has been challenged in recent the years. In 2019, a number of rule changes were implemented that served to weaken the ESA. They included changes in how habitat and critical habitat are defined, less stringent rules for protecting threatened species, and more freedom to consider economic impacts when making conservation decisions. Fortunately, a plan to reverse these actions was put forward by the USFWS in 2021. In addition, for fiscal year 2021, the U.S. Congress increased funding for the ESA program by 3.7%, reversing several years of budget cuts.

Supporters of the ESA claim that it is still one of the most successful and suitable models for protecting wild species. They argue that because species are listed only when they face serious danger of extinction, some are bound to be beyond the help of the ESA. Not every species listed will fully recover. They also argue that it takes many decades for a species to reach the point where it is in danger of extinction. Therefore, it takes many decades to bring a species back to the point where it can be delisted. ESA supporters also cite federal data that suggests the conditions of more than half the listed species are stable or improving.

According to a study by The Nature Conservancy, 33% of the country's species are at risk of extinction, and 15% of all species are at high risk, which is far more than the current number listed under the ESA. Therefore, this law has the potential to help many more wild species. It also has the potential to be made more effective. Suggested improvements include increased funding, an emphasis on developing recovery plans more quickly, and better technical and financial assistance to landowners. In addition, when a species is first listed, more effort could also be focused on identifying the core of its habitat as critical for its survival. That area could then be given maximum protection.

checkpoint What are the goals of the Endangered Species Act?

Protected Lands Can Help Sustain Ecosystems and Wild Species

One key to protecting wild species is protecting the areas where they live. According to the IUCN, a protected area is "a clearly defined geographical space, recognized, dedicated, and managed, through legal or other effective means, to achieve the long term conservation of nature with associated ecosystem services and cultural values." Protected areas may range in size from community-managed conservation areas to privately or publicly owned reserves, to wilderness areas, national parks, and wildlife refuges.

In the United States, the USFWS oversees the National Wildlife Refuge System. This network of protected areas covers both terrestrial and aquatic habitats. It is the only system of federal lands in the nation that is dedicated to wildlife conservation.

In 1903, President Theodore Roosevelt established the first U.S. federal wildlife refuge at Pelican Island, Florida, to protect the brown pelican and other birds from extinction. In 2009, the brown pelican was removed from the U.S. Endangered Species list, thanks to the early protection Roosevelt's administration awarded to the bird. By 2021, there were 567 refuges in the National Wildlife Refuge System (Figure 7-19). Each year, more than 60 million Americans visit these refuges to hunt, fish, hike, and watch birds and other wildlife.

A majority of the refuges serve as wetland sanctuaries that are vital for protecting migratory waterfowl. At least one-fourth of all U.S. endangered and threatened species have habitats in the refuge system, and some refuges have been set aside specifically for certain endangered species. Such areas have aided the recovery of populations of brown pelicans, trumpeter swans, and Key deer (found only in Florida).

The refuge system falls short in some ways, however. Activities that can be harmful to wildlife are also legally allowed on some refuge lands. Such activities include mining, oil drilling, and use of off-road vehicles. Biodiversity researchers also feel the U.S. government needs to set aside more refuges and to increase funding for the system.

checkpoint What activities are allowed in a wildlife refuge?

FIGURE 7-18
These specimens are part of a huge collection of animal parts housed at the USWFS National Forensics Lab. Scientists use them for reference when trying to identify evidence sent to the lab.

SCIENCE FOCUS 7.2

WILDLIFE FORENSICS

The U.S. Fish and Wildlife Service runs one of the world's only forensic labs that is dedicated to solving wildlife crimes. The lab operates in much the same way as a regular forensics lab. It receives physical evidence collected by wildlife agents and inspectors. These professionals work to enforce treaties and laws such as CITES and the ESA. They might deliver evidence to the lab in the form of animal parts, such as bones, tissue, teeth, hair, feathers, claws, and blood. The lab can compare these items to those in its own reference collection for identification (Figure 7-18).

Evidence may also be delivered in the form of weapons used to kill an animal or products that have been made from an animal.

Forensic specialists analyze evidence in an effort to answer such questions as:
- What species of animal did it come from?
- Where did the animal live?
- What was the cause of the animal's death?
- Was the animal protected by law?
- Who killed the animal?

To carry out their analysis, forensics experts may run a series of very sophisticated tests. They might analyze DNA, look for traces of poison, test fingerprints left behind by perpetrators, and examine molecules of blood.

Experts use what they learn about physical, genetic, and chemical characteristics of the evidence to theorize what happened and if there is cause for legal action. In addition to laboratory analysis, forensics experts may visit actual crime scenes or testify in court. These specially trained scientists provide unique help in the fight against illegal hunting and trade.

Thinking Critically
Infer How do you think forensic experts are able to determine the cause of death from evidence such as blood or bones?

Seed Storage and Cultivation Can Preserve Plant Species

Recent research suggests that between 60,000 and 100,000 species of plants are in danger of extinction—roughly one-quarter of all known plant species. Because the seeds and spores of plants can survive long periods of dormancy, conservation biologists are now working to safely store them away so they are available for the future.

Seed banks are refrigerated, low-humidity storage environments used to preserve genetic information and endangered plant species. About 1,750 seed banks around the world collectively hold millions of samples. The most ambitious of these is the Millennium Seed Bank in England, which as of 2020 held in storage more than 2.4 billion seeds collected from 97 countries.

Seed banks can vary in quality, be expensive to operate, and be vulnerable to fire or other mishaps. However, the Svalbard Global Seed Vault has been designed to offer the safest long-term storage of seeds (Figure 7-20). It has been built underground, beneath the permafrost on a remote island in the Arctic. It is expected to withstand natural and human-made disasters. As of 2021, this facility's holdings included more than 5,000 plant species.

The world's 1,800 **botanical gardens** contain living plants that represent almost one-third of the world's known plant species. They contain only a small percentage of the world's rare and threatened plant species and have limited space and funding to preserve most of those species. An **arboretum** is a facility for the scientific study and public display of various species of trees and shrubs.

checkpoint What are the limitations of seed banks?

Captive Breeding Programs Can Increase Populations of Endangered Animals

Zoos, aquariums, game parks, and animal research centers often house collections of animals that are threatened or endangered. These facilities may manage breeding programs to help increase the populations of such animals. Their goal is to maintain populations that are large enough to be self-sustaining and genetically healthy. They may also have the long-term goal of reintroducing animals from their collections into protected wild habitats.

Two techniques for preserving endangered terrestrial species are egg pulling and captive breeding. **Egg pulling** involves collecting eggs laid in the wild by critically endangered bird species and then hatching them in zoos or research centers. In **captive breeding**, some or all of the wild individuals of a critically endangered species are collected for breeding in captivity, with the aim of reintroducing their offspring into the wild.

FIGURE 7-19
Caribou gather on the slopes of the Brooks Range foothills in Alaska's Arctic National Wildlife Refuge.

FIGURE 7-20
Cary Fowler stands with seed samples in front of the Svalbard Global Seed Vault. Fowler helped establish the vault, which is also called the "doomsday seed vault."

Breeding programs may also use a procedure called artificial insemination, which involves direct insertion of semen into a female's reproductive system. Another technique is embryo transfer, the surgical implantation of fertilized eggs of one species into a surrogate mother of another species. After birth, young can be kept in incubators or raised by parents of a similar species, a method called cross fostering. Scientists also match individuals for mating by using DNA analysis along with computer databases that hold information on family lineages of endangered zoo animals. Keeping records of breeding helps scientists to increase genetic diversity.

The ultimate goal of some captive breeding programs is to build populations to a level where they can be reintroduced into the wild. Examples of successful reintroductions include the black-footed ferret, the golden lion tamarin (a highly endangered monkey species), the Arabian oryx, and the California condor. For reintroduction to be successful, individuals raised in captivity must be able to survive in the wild. They must have suitable habitat. And they must be protected from overhunting, poaching, pollution, or other environmental hazards. All these challenges can make reintroduction difficult.

One problem for captive breeding programs is that a captive population of an endangered species must typically contain 100 to 500 individuals. Otherwise, it is still in danger of extinction due to accidental deaths, diseases, or the loss of genetic diversity through inbreeding. Recent genetic research indicates that 10,000 or more individuals are needed for an endangered species to maintain its capacity for biological evolution. Zoos and research centers do not have the funding or the space required to house such large populations.

Public aquariums that exhibit unusual and attractive species of fish and marine animals such as seals and dolphins help to educate the public about the need to protect such species. They also invest in research and take part in conservation projects in the field. However, mostly due to limited funds, public aquariums have not served as effective gene banks for endangered marine species, especially marine mammals that need large volumes of water.

In some cases, farm facilities have been set up to breed and raise endangered species for a different purpose: commercial sale. The goal is to reduce pressures on wild populations caused by overharvesting or poaching. For example, in Florida, alligators are raised on farms for their meat and hides. Various endangered species of fish and shellfish, including sturgeon and abalone, are successfully being farm-raised for sale as well.

checkpoint What are challenges to captive breeding?

Protecting Wild Species Raises Difficult Questions

Efforts to prevent the extinction of wild species and the accompanying losses of ecosystem services require the use of limited financial and human resources. This raises some challenging questions:

- Should we focus on protecting species or should we focus more on protecting ecosystems and the ecosystem services they provide?
- How do we allocate limited resources between these two priorities?
- How do we decide which species should get the most attention in our efforts to protect as many species as possible? For example, should we focus on protecting the most threatened species or on protecting keystone species?
- Protecting species that are appealing to humans, such as panda bears and orangutans, can increase public awareness of the need for wildlife conservation. Is this more important than focusing on the ecological importance of species when deciding which ones to protect?
- How do we determine which habitat areas are the most critical to protect? How do we allocate limited resources among these areas?

Conservation biologists struggle with these questions all the time. They must decide which species will get priority, because resources are finite.

A growing trend in deciding on conservation actions involves calculating benefits, costs, and likelihood of success. There is also a growing emphasis on protecting entire ecosystems to benefit as many species as possible. (This approach is discussed in Chapter 8.)

checkpoint How can a species' "appeal" impact decisions to conserve it?

7.3 Assessment

1. **Recall** What is CITES and what is its purpose?
2. **Explain** What legal protections are given to species under the Endangered Species Act?
3. **Infer** Why is it important to preserve seeds of plant species so they are available in the future?
4. **Evaluate** Should zoos have the goal of reintroducing animals to the wild or should they keep animals safe in captivity?

SCIENCE AND ENGINEERING PRACTICES

5. **Engaging in Argument** People tend to donate to conservation programs that target "popular" endangered species, such as pandas and tigers. There are arguments for and against the value of this approach to conservation. Work with a partner to research both sides of this argument. Construct an argument in support of the position with which you agree.

TYING IT ALL TOGETHER STEM
Colony Collapse Disorder

This chapter's Case Study discussed the decline of European honeybee populations in the United States and elsewhere. Colony collapse disorder was identified as the cause of this decline. The Case Study described the impact CCD is having on European honeybees and the agriculture industry that relies on their pollination services. Science Focus 7.1 described possible causes for CCD that researchers are investigating.

Work with a group to complete the following steps.
1. Choose one of the possible factors for CCD listed in Science Focus 7.1. Decide which questions you will investigate to learn more about this factor as it relates to CCD.
2. Use the data and information you collect to construct an explanation for how this factor contributes to CCD.
3. Next, investigate a possible solution that might reduce this factor's impact on honeybees.

If necessary, make a model to show how this solution might be implemented. Your solution should take into consideration technical, social, economic, and environmental impacts.

4. Present your solution to the class. Be prepared to defend the merits of your solution based on your evidence.

CHAPTER 7 SUMMARY

7.1 What causes extinction and what are its impacts?

- Biological extinction occurs when a species can no longer be found on Earth. Extinction is a natural process but natural events and human activities can alter the extinction rate.
- Five mass extinctions have occurred on Earth. These extinctions probably resulted from natural events such as asteroid impacts, volcanic eruptions, and prolonged cooling or warming periods.
- For most of Earth's history, extinction has occurred at a low rate, called the background extinction rate. Today, scientists estimate annual extinction rates are at least 1,000 times the background extinction rate.
- Earth may be entering a sixth mass extinction caused largely by human activities.
- Biologists classify species heading toward biological extinction as threatened or endangered.
- It is important to sustain wild species and the ecosystem and economic services they provide. Many people feel wild species have a right to inhabit Earth alongside humans.

7.2 What role do humans play in the loss of species and ecosystem services?

- Human-related activities that hasten extinction include habitat destruction, degradation, and fragmentation; spread of invasive species; population growth; pollution; climate change; and overexploitation.
- Invasive species can threaten native species and alter their new ecosystems. They may also threaten human health or economic activities.
- Significant growth in Earth's human population has led to increased demand for natural resources and increased species extinctions.
- Agricultural chemicals can pollute ecosystems and organisms and increase extinctions. Greenhouse gas emissions contribute to climate change that can increase species extinctions.
- Species can become endangered when they are overexploited due to overharvesting, legal and illegal hunting, and legal and illegal trade.

7.3 How can we sustain wild species and ecosystem services?

- One key to protecting wild species is protecting areas where they live. Protected lands include conservation areas, wilderness areas, national parks, and wildlife refuges.
- The Convention on International Trade in Endangered Species of Wild Fauna and Flora (CITES) and Convention on Biological Diversity (CBD) are international treaties created to protect endangered species and biodiversity.
- The U.S. Endangered Species Act (ESA) identifies species that are endangered or threatened. Species on the list receive legal protection and are managed for recovery with the goal of being delisted.
- Seed banks, botanical gardens, and arboretums are facilities that store, preserve, and sometimes display plant species.
- Captive breeding programs at zoos, aquariums, and commercial farms can increase populations of endangered animal species.
- Conservation biologists must make difficult decisions about which endangered species will get priority because resources are finite.

CHAPTER 7 ASSESSMENT

Review Key Terms

Select the key term that best fits each definition. Not all terms will be used.

> arboretum
> bioaccumulation
> biomagnification
> bioprospector
> botanical garden
> captive breeding
> Convention on Biological Diversity
> Convention on International Trade in Endangered Species of Wild Fauna and Flora
> egg pulling
> endangered species
> Endangered Species Act
> habitat fragmentation
> seed bank
> threatened species

1. Occurs when a large, intact area of habitat is divided into smaller, isolated patches
2. Person who searches tropical forests and other ecosystems to find plants and animals that can be used to make medicinal drugs
3. The increase in concentration of a chemical in an organism's tissues
4. Species that has so few individual survivors that the species could soon become extinct
5. Place for the scientific study and public display of various species of trees and shrubs
6. Refrigerated, low-humidity storage environment used to preserve genetic information and endangered plant species
7. Program whereby wild individuals of a critically endangered species are collected for breeding in captivity, with the aim of reintroducing their offspring into the wild
8. Method that involves collecting eggs laid in the wild by critically endangered bird species and then hatching them in zoos or research centers
9. International treaty that bans hunting, capturing, and selling of threatened or endangered species and products made from such species
10. Species that has enough remaining individuals to survive in the short term, but is likely to become endangered in the near future

Review Key Concepts

11. Explain how rates of wild species extinction have changed since the appearance of modern humans.
12. What percentage of the world's species is likely to become extinct during this century, largely as a result of human activities?
13. Explain how protecting wild species can improve human health.
14. What is ecotourism and how does it differ from traditional tourism?
15. Why can populations of invasive species increase rapidly in new environments?
16. What methods can be used to increase captive populations of endangered species?
17. Give an example of an invasive species in the United States and describe its ecological impact.
18. What do countries that participate in the Convention on Biological Diversity agree to do?

Think Critically

19. Identify three aspects of your lifestyle that might directly or indirectly contribute to declines in European honeybee populations. What changes could you make to limit your impact?
20. Give your response to the following statement: "Eventually, all species become extinct. So it does not really matter that the world's remaining tigers or orangutans disappear because of what humans are doing." Be honest about your reaction, and give arguments to support your position.
21. Did your reading of this chapter make you feel more hopeful, less hopeful, or indifferent about the extinction of wild species that is happening this century? Why do you think you feel this way?
22. As an individual, what personal changes would you be willing to make to protect wild species and ecosystems? What changes would be unacceptable to you and why?

Chapter Activities

A. Develop Models: Endangered Species Data STEM

Figure 7-22 presents data from the 2021 U.S. Endangered Species list. In this activity, you will use the data and the materials listed to design a model that shows the proportion of endangered and threatened species each category represents.

Materials

calculator
sheets of colored paper
 (8.5 x 11 or larger)
protractor
ruler
tape
scissors

1. Work with a partner. Calculate what percentage of the total number of endangered and threatened species each category of animals represents. Round your results to the nearest tenth of a percent.

2. Use the materials listed or others approved by your teacher to create a model that visually illustrates your calculations. Your completed model should include:
   a. name of each animal category
   b. percentage of total species listed that are from each category

3. Consider which category of animals represents the greatest percentage of those listed as endangered or threatened. What thoughts do you have about why this might be the case? What is one question you can investigate to learn more about the data presented in your model?

FIGURE 7-22

U.S. Species of Animals on 2021 Endangered Species List

Group	Endangered	Threatened	Total Listings
Corals	0	7	7
Snails	39	12	51
Amphibians	22	16	38
Reptiles	16	29	45
Mammals	68	28	96
Clams	76	16	92
Fishes	93	77	170
Arachnids	12	0	12
Insects	75	14	89
Crustaceans	25	4	29
Birds	76	23	99
Animal totals	**502**	**226**	**728**

Source: U.S. Fish and Wildlife Service

B. Take Action

National Geographic Learning Framework

Attitudes | Curiosity
Skills | Collaboration
Knowledge | Critical Species

Consider creating a honeybee habitat in a suitable location on your school campus. Pick a sunny location where flowers will grow and where the presence of bees will not bother anyone. Start by identifying flowers that are native to your area. Contact a local garden supply store, garden club, or university extension office for help. Choose a variety of flowers in different colors. Also choose flowers that bloom in shifts over the spring, summer, and fall seasons. This will make pollen and nectar available to bees for as long as possible. Plant flowers close together, and keep flowers of the same variety together too. You might also include a water source for bees, such as a shallow pan filled with water. Place a few large rocks in the pan to weigh it down and to provide a landing spot for bees that come to drink.

CHAPTER 8
SUSTAINING BIODIVERSITY: AN ECOSYSTEM APPROACH

NATIONAL PARKS play an important role in sustaining biodiversity. These protected public lands are managed to provide large, healthy ecosystems for wild species. Denali National Park and Preserve in Alaska features a mix of subarctic ecosystems. Tundra, spruce forests, meadows, glaciers, and mountains form the landscape. Some of North America's largest mammals, including grizzly bears, live within Denali's borders.

KEY QUESTIONS

8.1 How can forests be better managed?

8.2 How can grasslands be better managed?

8.3 How can protected lands be better managed?

8.4 How does the ecosystem approach help protect terrestrial biodiversity?

8.5 How does the ecosystem approach help protect aquatic biodiversity?

NATIONAL GEOGRAPHIC | EXPLORERS AT WORK

Protecting Biodiversity One Bird at a Time

with National Geographic Explorer Çağan Şekercioğlu

It doesn't surprise those who know Çağan Şekercioğlu (cha-HAN shay-KER-jeoh-loo) that his first name means "hawk." This scientist and wildlife photographer has been documenting vanishing bird species for years in his home country of Turkey and beyond, and with good reason. As he explains, "Many pressures that will ultimately affect other animals, and even people, are happening to birds first."

Scientists predict that by the end of this century, up to one-quarter of the world's bird species may be extinct. Dr. Şekercioğlu works to estimate bird extinctions by combining data from fieldwork and scientific literature with atmospheric warming and habitat loss scenarios. In the course of his work, he's observed 55% of Earth's bird species across 70 countries and on every continent.

The data Şekercioğlu collects on bird ecology, conservation status, and migration eventually become part of a global bird database. Ornithologists (scientists who study birds) use this tool to track changing bird populations around the globe. Recently, Şekercioğlu began focusing his research on Costa Rica, to assess the risk to tropical mountain birds there. Data for these species are lacking.

Şekercioğlu monitors birds in Costa Rica's forests and agricultural areas to study the effects of habitat loss and farmland development. As he points out, "Protected areas alone aren't enough. We must work with local people to improve the biodiversity capacity of agricultural land and increase connectivity of protected areas." In other words, protecting forest habitat only works if it isn't fragmented by farmland, which makes it difficult for birds to make use of it. Şekercioğlu's work is extensive. In one ongoing project, more than 50,000 birds and more than 250 species have been captured and banded at stations he put in place.

Şekercioğlu notes that climate change is also driving birds into narrowing ranges higher in the mountains to find the right conditions for survival. He calls it "the escalator to extinction" for these birds.

Şekercioğlu also works hard to address conservation challenges closer to home. As director of an environmental organization in Turkey, he has established village-based biocultural tourism that brings bird-loving visitors to a lake where more than half the country's bird species are known to gather. "It's become a crucial source of local pride, income, and incentive for protecting the lake," he says.

Çağan Şekercioğlu's passion for photography helps bring the problems birds face to the forefront. "Facts and figures are often not enough," he observes. "People need to feel the personal, emotional connection photography can provide." The world's bird populations couldn't have a stronger advocate.

Thinking Critically
Explain How does Dr. Şekercioğlu get local people involved with his conservation projects? How do you think this involvement has helped his projects' success?

Çağan Şekercioğlu says he has always been drawn to nature. Here, he admires a songbird for a moment before releasing it.

CASE STUDY
Costa Rica—A Global Conservation Leader

Costa Rica is a tiny nation located in Central America. It is smaller in area than the U.S. state of West Virginia and is about one-tenth the size of France. Costa Rica was once blanketed by tropical forest. But much of that forest disappeared between 1963 and 1983 as politically powerful ranching families cleared it in order to graze their cattle.

Despite such widespread forest loss, Costa Rica remains a superpower of biodiversity. The nation is home to an estimated 500,000 plant and animal species, including monkeys, jaguars, and unusual snakes, spiders, and frogs. A single park in Costa Rica is home to more bird species than are found in all of North America.

Costa Rica's biodiversity mainly is a product of its tropical location. The country is situated between two oceans. It has both coastal and mountainous regions that provide a variety of microclimates and habitats for wildlife. The government has helped protect this biodiversity by maintaining a strong commitment to conservation. As a result, Costa Rica is a small but mighty model for others to follow.

Costa Rica established a system of nature reserves and national parks in the mid 1970s (Figure 8-1). By 2012, this system included more than 25% of the nation's land, with 6% of it reserved for indigenous peoples. In fact, Costa Rica has devoted a larger proportion of its land to biodiversity conservation than any other country.

To reduce deforestation, or the widespread removal of forests, the government eliminated subsidies for converting forestlands to grazing lands. Instead, it pays landowners to maintain or restore tree cover. The strategy has worked—Costa Rica has gone from having one of the world's highest deforestation rates to having one of the lowest.

Ecologists warn that human population growth, economic development, and poverty are placing increasing pressure on Earth's ecosystems and the ecosystem services they provide. Earth's biodiversity is also threatened. This chapter describes some of the challenges and solutions to managing terrestrial and aquatic ecosystems in ways that are sustainable.

As You Read What factors do you think Costa Rica's government must weigh when deciding how to serve society's interests while also meeting its conservation goals? Keep these factors in mind as you read about other ecosystems and their management in this chapter.

FIGURE 8-1
Mist rises over the Rio Sirena in Corcovado National Park, Costa Rica.

8.1 How Can Forests Be Better Managed?

CORE IDEAS AND SKILLS

- Identify examples of ecosystem and economic services forests provide.
- Describe ways scientists classify forests based on their age and structure.
- Identify various methods of harvesting timber and their impacts on forests, and explain how deforestation impacts forests.
- Describe management solutions that help reduce impacts of timber harvesting, fires, and deforestation.

KEY TERMS

old-growth forest
second-growth forest
tree plantation
deforestation
prescribed burn
debt-for-nature swap
conservation concession

Forests Help Support Life on Earth

Forests cover one-third of Earth's landmass. While forests may look very different from region to region, these ecosystems all provide vital ecosystem and economic services. For example, forests remove carbon dioxide (CO_2) from the atmosphere through photosynthesis and store it in organic compounds. By performing this ecosystem service, forests help stabilize average atmospheric temperatures and climate conditions.

Forests also provide habitats for about two-thirds of Earth's terrestrial species. And they are home to more than 300 million people. About 1 billion people living in extreme poverty depend on forests for their survival. Forests also play a role in maintaining human health. Traditional medicines, used by 80% of the world's people, are mostly made from plant species native to forests. Chemicals found in a number of tropical forest plants serve as the basis for making most prescription drugs.

Forests have tremendous economic value. They provide raw materials for numerous products, from paper to fuel. Examples of important ecosystem and economic services that forests provide are listed in Figure 8-2. Scientists and economists can use various methods to estimate the monetary value of major ecosystem services provided by ecosystems such as forests (Science Focus 8.1). The importance of forests to human economies and life on Earth means they must be managed responsibly.

checkpoint What role do forests play in the carbon cycle?

Forests Vary in Age and Structure

One way scientists classify forests is based on their age and structure. Forest structure refers to the distribution of vegetation, both horizontally and vertically, as well as the type, size, and shape of that vegetation. An **old-growth forest**, or primary forest, is an uncut or regenerated (regrown) forest that has not been seriously disturbed by human activities or natural disasters for 200 years or more. Old-growth forests are reservoirs of biodiversity because they are able to provide ecological niches for a multitude of wildlife species.

A **second-growth forest** is a stand of trees resulting from secondary ecological succession. Recall from Chapter 5 that ecological succession is a gradual change in species composition in a given area. Second-growth forests develop after the trees in an area have been removed by natural forces, such as wildfires and hurricanes, or by human activities, such as clear-cutting for timber or conversion to farmland.

A **tree plantation** is a managed forest made up of only one or two species of trees that are all the same age. It is also known as a tree farm or commercial forest. Tree plantations are often established by first clearing old-growth or second-growth forest. Trees grown on plantations are usually harvested by clear-cutting as soon as they become commercially valuable. The land is then replanted and clear-cut again in a regular cycle (Figure 8-3).

FIGURE 8-2 ▼

Examples of Ecosystem and Economic Services of Forests	
Ecosystem Services	**Economic Services**
Store atmospheric carbon	Wood for fuel
Support energy flow and nutrient cycling	Lumber
Produce oxygen	Wood products
Reduce soil erosion	Pulp to make paper
Absorb and release water	Mining
Purify water and air	Livestock grazing
Influence local and regional climates	Recreation
Provide numerous wildlife habitats	Jobs

SCIENCE FOCUS 8.1

WHAT ARE ECOSYSTEM SERVICES WORTH?

The value of forests and other ecosystems is currently estimated based on their economic services. However, ecologists and ecological economists have also tried to calculate the monetary value of the ecosystem services that forests provide. Ecological economists consider the value of nature's goods and services and how they contribute to human economies. Taking this value into account is a way to apply the full-cost pricing factor of sustainability.

In 2014, a team of ecologists, economists, and geographers, led by ecological economist Robert Costanza, estimated the monetary value of 17 ecosystem services. Their cautious estimate was that the global monetary value of these services is at least $125 trillion per year—much more than the $76 trillion that the entire world spent on goods and services in 2014. This means that as of 2014, Earth provided everyone in the world with ecosystem services worth at least $17,123 on average.

The researchers also determined that the world has been losing a value of ecosystem services greater than the gross national product of the United States over that time.

According to the team's research, the world's forests alone provide ecosystem services worth at least $9.1 trillion per year—hundreds of times more than their economic value in terms of lumber, paper, and other wood products. The researchers pointed out their estimates were very conservative.

Costanza and other researchers estimate that preserving ecosystems in a global network of nature reserves occupying 15% of Earth's land surface and 30% of its oceans would provide $4.4–$5.2 trillion of ecosystem services. That amounts to 100 times the value of converting those systems for human uses.

Relying on the accuracy of these estimates, we can draw three important conclusions from the study. First, Earth's ecosystem services are essential for all humans and their economies. Second, the economic value of these services is huge. And third, these ecosystem services will provide an ongoing source of ecological income as long as they are used sustainably.

Thinking Critically

Evaluate Why might it be harder to convince people of the value of a forest's ecosystem services as opposed to the value of its economic services?

FIGURE 8-3 ▼
Harvest Cycle Trees in a pine tree plantation such as the one in the photo typically are harvested according to a 20- to 30-year cycle, as shown below.

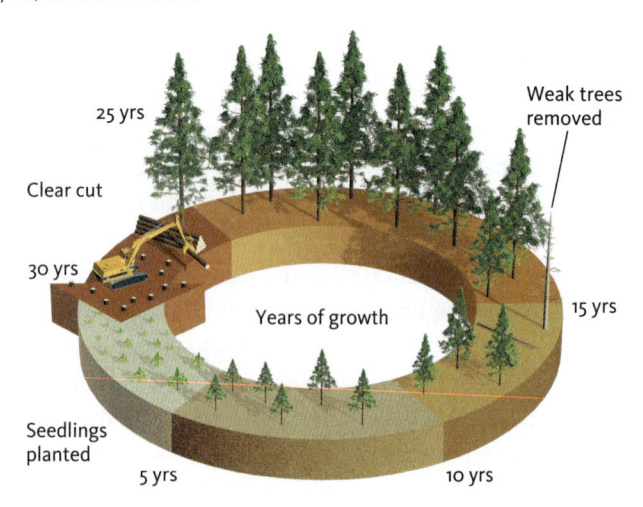

CONSIDER THIS

Cleared forests can grow back.

Today, forests cover about 30% of the United States—which is more than they did roughly a century ago. The primary reason for this is that many of the old-growth forests cleared or partially cleared between 1620 and 1920 have grown back naturally through secondary ecological succession. Tree plantations have also added forest cover. There are now diverse second-growth (and in some cases third-growth) forests in every region of the United States except in much of the West. Environmental writer Bill McKibben called this forest regrowth "the great environmental success story of the United States, and in some ways, the whole world."

When managed carefully, tree plantations can produce wood rapidly and supply most of the wood used for industrial purposes, including papermaking and construction. This protects old-growth and second-growth forests, as long as they are not cleared to make room for the tree plantations.

In 2011, the Food and Agriculture Organization (FAO) of the United Nations reported that the net total forest cover in several countries changed very little or even increased between 2000 and 2010. The United States was one of those countries. While some of the increases in tree cover resulted from natural secondary ecological succession, the spread of commercial tree plantations was also a factor. Many tree plantations were established as part of a program sponsored by the United Nations Environment Programme (UNEP) to plant billions of trees throughout much of the world.

The downside of tree plantations is that they contain only one or two tree species. As a result, they are less biologically diverse and sustainable than old-growth and second-growth forests. Repeated cutting and replanting can also eventually deplete the nutrients in the forest's topsoil. Any type of forest is less likely to regrow on such land.

Fires Impact Forest Structure Two types of fires can affect the structure of forest ecosystems. Surface fires usually burn only undergrowth and leaf litter on the forest floor. They kill seedlings and small trees, but spare most mature trees and allow most wild animals to escape.

Occasional surface fires can be beneficial to a forest. Such fires can:
- burn away flammable material such as dry brush and prevent more destructive fires;
- free valuable plant nutrients trapped in slowly decomposing litter and undergrowth;
- release seeds from the cones of tree species such as lodgepole pines and stimulate the germination of other seeds such as those of the giant sequoia; and
- help control the presence of destructive insects and tree diseases.

Another type of fire, called a crown fire, is an extremely hot fire that leaps from treetop to treetop, burning whole trees. Crown fires usually start in forests that have not experienced surface fires for several decades. The absence of fire allows dead wood, leaves, and other flammable ground litter to build up. When a surface fire breaks out, these burning materials create a "ladder effect." The surface fire is able to jump to the canopy and become a crown fire. Wind conditions can also aid the spread of crown fires. Crown fires can destroy most vegetation, kill wildlife, increase topsoil erosion, and burn or damage buildings and homes.

checkpoint What is the difference between an old-growth forest and a second-growth forest?

Poor Management Negatively Impacts Forests

As described in Figure 8-2, forests provide a wide variety of economic services and ecosystem services. Forests are often managed with the goal of maximizing one or more of these services. It is important to consider how management practices also impact the health, biodiversity, and productivity of forest ecosystems.

Impacts of Timber Harvest Trees have immense economic value, and harvesting wood from forests is one of the world's major industries. The first step in harvesting wood is to build roads for access and timber removal. Even carefully designed logging roads can have harmful impacts. Building and using these roads can increase topsoil erosion, sediment runoff into waterways, habitat fragmentation, and biodiversity loss. Logging roads also increase chances of invasion by disease-causing organisms and nonnative pests, as well as disturbances from human activities like farming and ranching.

LESSON 8.1 249

Loggers use a variety of methods to harvest trees. When selectively cutting trees, loggers will cut intermediate-aged or mature trees singly or in small groups (Figure 8-4A). However, loggers may choose to remove all the trees from an area in what is called a clear-cut (Figure 8-4B). Clear-cutting is the most efficient and sometimes the most cost-effective method for harvesting trees. While clear-cutting may be an easier method for landowners and timber companies to use, it can be harmful to forest ecosystems. Wholesale removal of trees leads to increased soil erosion, increased sediment pollution of nearby waterways, and decreased biodiversity. (See the Chapter 2 Case Study.)

A variation of clear-cutting that produces a more sustainable timber yield without widespread destruction is strip cutting (Figure 8-4C). Strip cutting involves clear-cutting a strip of trees along the contour of the land within a corridor narrow enough to allow natural forest to grow back within a few years. After one strip grows back, loggers cut another strip next to it, and so on.

Impacts of Deforestation The temporary or permanent removal of large expanses of forest for agriculture, settlements, or other uses is known as **deforestation**. Figure 8-5 lists some of the negative impacts of this process. Surveys by the World Resources Institute (WRI) reveal that during the past 8,000 years, deforestation has eliminated almost half of Earth's old-growth forest cover. Most of this loss has taken place in the last 60 years. In addition, the WRI has predicted that at current deforestation rates, 40% of the world's remaining intact forests will be logged or converted to other uses within two decades, if not sooner.

The impacts of deforestation have been especially significant in tropical forests. Today, tropical forests cover 6% of Earth's landmass—roughly the area of the continental United States. Climate and biological data indicate mature tropical forests once covered at least twice the area they do now. Most loss of the world's tropical forests has taken place since 1950. Between 2000 and 2013, the world lost the equivalent of more than 50 soccer fields of tropical forest every minute, mainly due to deforestation.

FIGURE 8-4 ▼
Harvesting Methods The three major ways to harvest trees are (A) selective cutting, (B) clear-cutting, and (C) strip cutting.

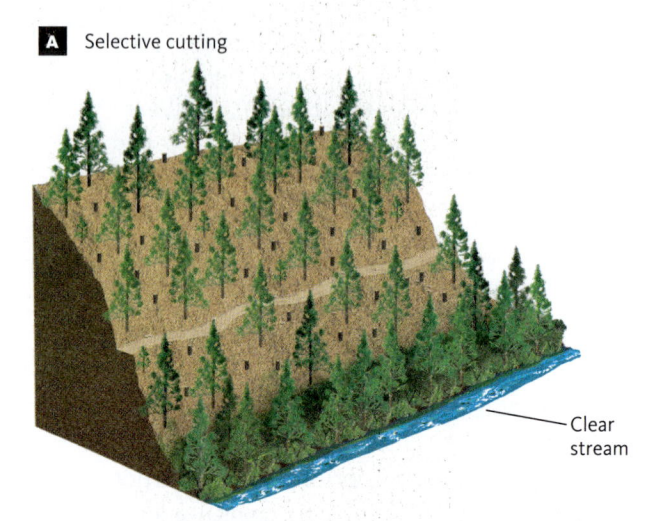

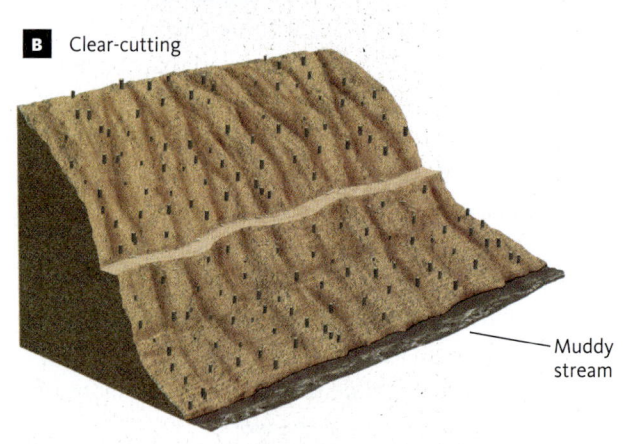

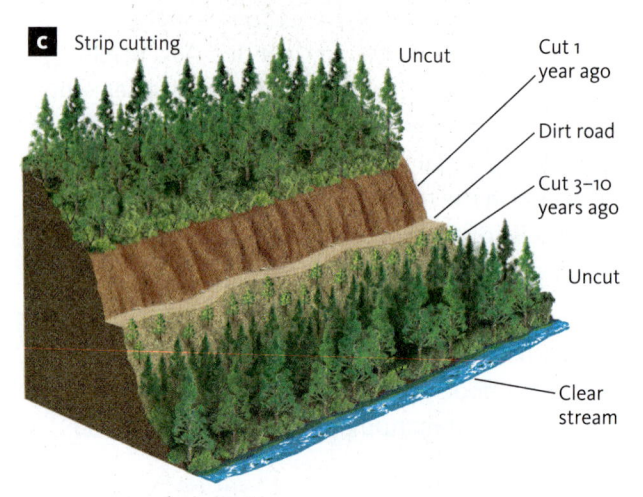

A Selective cutting

B Clear-cutting

C Strip cutting

FIGURE 8-5 ▼

Harmful Effects of Deforestation
Water pollution and soil degradation from erosion
Acceleration of flooding
Local extinction of specialist species
Habitat loss for native and migrating species
Release of CO_2 and loss of CO_2 absorption

A number of indirect and direct factors contribute to tropical deforestation. Pressures from population growth and poverty push the landless poor and subsistence farmers (farmers who grow only enough crops to feed themselves and their families) into tropical forests. People resort to cutting or burning trees for firewood or clearing land to grow any food they can to survive (Figure 8-6).

The major direct causes of deforestation in tropical forests vary by area. Tropical forests in Brazil and other South American countries are cleared or burned primarily for cattle grazing and large soybean plantations. In Indonesia, Malaysia, and other areas of Southeast Asia, large oil palm plantations are replacing tropical forests. Palm oil is used to make products such as biodiesel fuel and cosmetics. Government subsidies can accelerate other direct causes such as large-scale logging and ranching by reducing their costs of doing business.

Tropical deforestation follows a predictable cycle in most cases. First, access roads are built deep in the forest, often by international corporations. Then loggers selectively cut down the largest and best trees. After trees are harvested, the land is usually sold to ranchers, who burn any remaining timber to clear the land for cattle grazing. Within a few years, the land typically is overgrazed. The ranchers move on, selling the degraded land to farmers or settlers for agriculture. After a few more years, the land becomes useless, as the soil is eroded and is stripped of its limited nutrients. The farmers and settlers move on and the deforestation cycle continues.

Deforestation destroys biodiversity in tropical forests. Studies indicate that half the world's known species of terrestrial plants, animals, and insects live in these ecosystems. These species often have specialized niches. They are vulnerable to extinction when their forest habitats are destroyed or degraded.

Deforestation can also impact rainfall in tropical forests. Water evaporating from trees and vegetation in a tropical rain forest plays a major role in determining the amount of rainfall there. Removing large numbers of trees can also lead to drier conditions that dehydrate the topsoil by exposing it to sunlight and allowing it to be blown away. It becomes difficult for a forest to grow back in the area, which is often replaced instead by tropical grassland or savanna.

Deforestation has global as well as local consequences. Currently, tropical forests absorb and store one-third of the world's terrestrial carbon emissions as part of the carbon cycle (Figure 3-18).

FIGURE 8-6
Tropical deforestation is a serious problem in the island nation of Madagascar. Slash-and-burn agriculture (shown here), logging, and fuelwood and charcoal production all contribute to the problem.

> **CONSIDER THIS**
>
> **Harvesting one tree often means taking down many others.**
>
> When loggers cut down one big tree in a tropical forest, many other trees often fall with it. A network of vines called lianas grows throughout the forest canopy and weaves trees together. So, one large tree may be connected to many others. As it falls, it tugs on the other trees around it. They are easily pulled down because the soil in a tropical forest is thin and trees there have shallow roots.

Eliminating these forests reduces their carbon absorption and contributes to climate change. Burning and clearing tropical forests also adds carbon to the atmosphere, accounting for 10–15% of global greenhouse gas emissions.

checkpoint How does poverty contribute to tropical deforestation?

People Can Harvest Timber More Sustainably

Biodiversity researchers and a growing number of foresters have called for more sustainable forestry practices (Figure 8-7). These practices recognize that sustaining a forest's *ecosystem* services is key to maintaining its *economic* services over time.

To preserve greater amounts of natural forest, loggers can employ a variety of sustainable measures in their operations. For example, they can use selective cutting and strip cutting to harvest tropical trees for lumber instead of clear-cutting the forests. To reduce damage to neighboring trees when cutting and removing individual trees in tropical forests, loggers can first cut the lianas (canopy vines) that connect them.

Governments can also begin making a shift to more sustainable forest management. They can phase out subsidies and tax breaks that encourage forest degradation and deforestation and replace them with forest-sustaining economic rewards. Doing so would likely lead to higher prices on unsustainably produced timber and wood products, in keeping with the full-cost pricing factor of sustainability. Costa Rica is taking a lead in using this approach (Case Study). Governments can also encourage tree-planting programs to help restore degraded forests.

Certification The nonprofit Forest Stewardship Council (FSC) oversees the certification of forestry operations that meet specific sustainable forestry standards. According to the FSC, these standards promote "environmentally sound, socially beneficial, and economically prosperous management." To become certified, operators must demonstrate that they do not cut trees at a rate that exceeds long-term forest regeneration in a given area. They must maintain roads and use harvesting systems in ways that limit ecological damage. They must also prevent unreasonable damage to topsoil and leave downed wood and standing dead trees to provide wildlife habitat (Figure 8-8).

The FSC reported that by 2021, 10% of the world's forest area in 89 countries had been certified according to FSC standards. The FSC also certified 5,400 manufacturers and distributors of wood products. In this way, certification helps consumers play a role in sustainable forestry. They can look for products labeled with the FSC-certified logo.

checkpoint Why might consumers choose to buy products with the FSC-certified logo?

People Can Better Manage Forest Fires

The U.S. Forest Service and the National Advertising Council launched the Smokey Bear campaign in the 1940s to educate the public about the danger of forest fires. The campaign is credited with preventing numerous forest fires, saving many lives, and preventing billions of dollars in losses of trees, wildlife, and human-built structures. However, the

FIGURE 8-7 ▼

Sustainable Forestry Practices
Include ecosystem services of forests in estimates of their economic value.
Identify and protect highly diverse forest areas.
Stop logging in old-growth forests.
Stop clear-cutting on steep slopes.
Reduce road building in forests and rely more on selective cutting and strip cutting.
Leave most standing dead trees and larger fallen trees for wildlife habitat and nutrient cycling.
Grow tree plantations only on deforested and degraded land.
Certify timber grown using sustainable methods.

FIGURE 8-8
When left in place, a downed tree can serve as a "nurse log" by supporting the growth of tree seeds that take root on top if it. The growing trees use nutrients released by the log as it decays.

campaign also convinced much of the public that all forest fires are bad and should be prevented or put out. This is not the case.

In fact, ecologists warn that trying to prevent all forest fires can actually make forests more vulnerable to fires in the long run. It increases the likelihood of destructive crown fires due to the accumulation of highly flammable underbrush in some forests. Ecologists and forest fire experts recommend several strategies for limiting the harmful effects of forest fires instead:

- Use carefully planned and controlled fires, called **prescribed burns**, to remove flammable small trees and underbrush within the highest-risk forest areas.
- Allow some fires on public lands to burn underbrush and smaller trees, as long as the fires do not threaten human-built structures or human lives.
- Protect houses and other buildings in fire-prone areas by thinning trees and other vegetation in a zone around them and eliminating the use of highly flammable construction materials such as wood shingles.
- Use solar-powered microdrones, equipped with infrared sensors, to detect forest fires and monitor progress in fighting them.

checkpoint What is an unintended side effect of preventing or putting out all forest fires?

People Can Reduce Demand for Harvested Trees

According to the Worldwatch Institute and forestry analysts, up to 60% of the wood consumed in the United States is wasted unnecessarily. This waste results from inefficient use of construction materials, excessive packaging, overuse of junk mail, inadequate paper recycling, and the failure to reuse or find substitutes for wooden shipping containers. One simple step individuals can take to reduce such waste is to choose reusable plates, cups, napkins, and bags over throwaway paper products made from trees.

Another way to reduce demand for harvested trees is to produce tree-free paper. Pulp from trees is typically used to make paper, but paper manufacturers can also use fibers from non-tree sources. For example, China uses rice straw and other agricultural residues to make much of its paper.

Of the small volume of tree-free paper produced in the United States, the majority is made from the fibers of a rapidly growing woody annual plant called kenaf (kuh-NAHF). Kenaf and other non-tree fiber sources such as hemp yield more pulp per area of land than tree farms do and can be grown using fewer pesticides and herbicides.

More than 2 billion people in less-developed countries use fuelwood and charcoal (made from wood) for heating and cooking purposes. Most of these less-developed countries are experiencing fuelwood shortages because people have been harvesting trees for fuelwood and forest products 10–20 times faster than new trees are being planted.

FIGURE 8-9 ▼

Sustainable Forestry Practices in Tropical Forests
Prevention
Protect the most diverse and endangered areas.
Educate settlers about sustainable agriculture and forestry.
Subsidize only sustainable forest use.
Protect forests through debt-for-nature swaps and conservation concessions.
Certify sustainably grown timber.
Reduce poverty and slow population growth.
Restoration
Encourage regrowth through secondary succession.
Rehabilitate degraded areas.
Concentrate farming and ranching in already-cleared areas.

In Guatemala, fuelwood harvesting is estimated to account for more than 55% of deforestation annually. Establishing small plantations of fast-growing trees and shrubs around farms and in community woodlots can help ease the severity of the fuelwood crisis.

checkpoint What are the advantages of using kenaf to make paper?

People Can Reduce Tropical Deforestation

Analysts have suggested various ways to protect tropical forests and use them more sustainably (Figure 8-9). At the international level, debt-for-nature swaps can make it financially attractive for countries to protect their tropical forests. Under the terms of **debt-for-nature swaps**, participating countries act as custodians of protected forest reserves in return for foreign aid or debt relief. For example, the United States forgave Guatemala $24 million in debt in 2006 as part of a debt-for-nature swap. In return, Guatemala agreed to instead use the funds to protect its tropical forests.

Governments or private conservation organizations may also use a similar strategy, called **conservation concessions**, whereby they *pay* governments or landowners in other nations to preserve their land's natural resources for a set amount of time. Conservation concessions are described as a way to "rent biodiversity."

National governments can also take important steps to reduce deforestation. Between 2005 and 2013, Brazil cut its deforestation rate by 80% by enforcing logging laws and setting aside a large conservation reserve in the Amazon Basin. Governments can also subsidize sustainable forestry and tree planting projects.

Consumers can help reduce the demand for unsustainable and illegal logging in tropical forests as well. They can choose to buy only wood and wood products that have been certified as sustainably produced by the FSC and other organizations, including the Rainforest Alliance and the Sustainability Action Network.

checkpoint What would motivate a country to participate in a debt-for-nature swap or make conservation concessions?

CONSIDER THIS

Individuals can make a big difference in forest conservation.

The late Wangari Maathai, a Nobel Peace Prize winner, promoted tree planting in her native country of Kenya and worldwide in what became the Green Belt Movement. Her efforts inspired the United Nations Environment Programme to launch a global effort to plant at least 1 billion trees a year beginning in 2006. By 2012, the year Maathai died, about 12.6 billion trees had been planted in 193 countries.

8.1 Assessment

1. **Describe** How can fire both benefit and threaten forest ecosystems?
2. **Summarize** What cycle of events typically takes place in the process of tropical deforestation?
3. **Explain** How can consumers help reduce demand for harvested trees? Give two examples.

SCIENCE AND ENGINEERING PRACTICES

4. **Engaging in Argument** Use the text and your own research to investigate the success of debt-for-nature swaps and develop an argument for or against this conservation approach.

8.2 How Can Grasslands Be Better Managed?

CORE IDEAS AND SKILLS
- Identify ecosystem services provided by grasslands.
- Define rangelands and pastures, and explain how overgrazing impacts these areas.
- Describe methods for managing rangelands that limit impacts of overgrazing and help preserve grassland habitat.

KEY TERMS

rangeland   overgrazing
pasture

Grasslands Are Overgrazed

Grasslands cover about one-quarter of Earth's landmass. They provide important ecosystem services, including soil formation, erosion control, nutrient cycling, storage of atmospheric CO_2 in biomass, and maintenance of biodiversity.

Other than forests, grasslands are the most widely used and altered ecosystems. **Rangelands** are unfenced grasslands in temperate and tropical climates that supply forage, or vegetation for grazing (grass-eating) and browsing (shrub-eating) animals. Livestock also graze in **pastures**, which are managed grasslands or fenced meadows that are often planted with domesticated grasses or other forage crops such as alfalfa and clover (Figure 8-10).

Cattle, sheep, and goats graze on about 42% of the world's grasslands. The 2005 UN Millennium Ecosystem Assessment—a four-year study by 1,360 experts from 95 countries—estimated that number could increase to 70% by 2050.

Blades of rangeland grass grow from the base, not at the tip as broadleaf plants do. As long as only the upper portion of the blade is eaten and the lower portion remains in the ground, rangeland grass is a renewable resource that can be grazed again and again. Moderate levels of grazing are healthy for grasslands because removal of mature vegetation stimulates rapid regrowth and encourages greater plant diversity.

Overgrazing occurs when too many animals graze an area for too long, damaging grasses and their roots and exceeding the area's carrying capacity. Overgrazing reduces grass cover, exposes the topsoil to erosion by water and wind, and compacts the soil, which reduces its capacity to hold water.

Overgrazing also promotes the invasion of species such as sagebrush, mesquite, cactus, and cheatgrass, which cattle will not eat. The FAO has estimated that overgrazing by livestock has reduced productivity on as much as 20% of the world's rangelands.

checkpoint How does overgrazing impact soil health?

Rangelands Can Be Protected

Managing rangelands more sustainably typically involves controlling several factors. For example, rangeland managers can control the timing of grazing. They can avoid grazing an area at the same stage of plant growth year after year. They can also avoid grazing an area too frequently during the growing season. Managers should consider which livestock have diets best suited to the vegetation in a given area. Finally, they should limit how many animals are allowed to graze in a given area, in order to avoid exceeding the area's carrying capacity.

One way to prevent overgrazing is to use rotational grazing, in which cattle are confined by portable fencing to one area for a few days and then moved to a new location. Cattle prefer to graze around ponds and other natural water sources, especially streams and rivers lined by strips of vegetation known as riparian zones. Overgrazing can destroy the vegetation in riparian zones, but rotational grazing helps prevent this. Ranchers can also fence off damaged areas, which allows for eventual restoration through natural ecological succession.

A more expensive and less common method of rangeland management involves suppressing the growth of undesirable plants by using herbicides, mechanical removal, or controlled burning. A cheaper way to discourage unwanted vegetation in some areas is to allow controlled, short-term trampling by large numbers of livestock such as sheep, goats, and cattle. Trampling destroys the root systems of the undesirable plants.

Since 1980, millions of people have moved to the southwestern United States, into areas typically devoted to ranching. Developers have been buying ranchlands to build housing developments, condominiums, and small "ranchettes." This change in land use has concerned some ranchers, ecologists, and environmentalists in the region. They have joined together to preserve a number of cattle ranches in an effort to sustain important remaining grasslands and the habitats they provide for native species.

FIGURE 8-10
A fence separates a grazed pasture from an ungrazed area. What methods could be used to prevent overgrazing in this pasture?

Land trust groups have also gotten involved. They pay ranchers to establish conservation easements—legally binding restrictions that ensure future owners will not develop the land. These groups are also pushing for local governments to zone the land in order to prevent large-scale development in ecologically fragile rangeland areas.

checkpoint How can ranchers protect riparian zones?

8.2 Assessment

1. **Compare and Contrast** What is the difference between a pasture and a rangeland?
2. **Explain** Why is rangeland grass considered a renewable resource?
3. **Describe** What is rotational grazing and what are its advantages?

CROSSCUTTING CONCEPTS

4. **Stability and Change** Describe how impacts on grassland conditions due to livestock grazing can change how these ecosystems function.

8.3 How Can Protected Lands Be Better Managed?

CORE IDEAS AND SKILLS

- Identify challenges of managing national parks and other protected lands, especially in less-developed countries.
- Explain factors that place stress on U.S. national parks.
- Understand how buffer zones can make it easier to sustainably manage reserves.
- Explain how wilderness areas are managed to protect wild species and ecosystems.

KEY TERMS

buffer zone   biosphere reserve   wilderness area

The World's Parks Face Many Challenges

According to the International Union for Conservation of Nature (IUCN), there are now more than 6,600 major national parks located in more than 120 countries. However, most of these parks are too small to sustain many large animal species. The invasion of harmful nonnative species has also impacted many parks, as they have outcompeted native species for resources and reduced those species' populations. And some national parks are so popular that large numbers of visitors are degrading the very features that make the parks attractive.

Parks in less-developed countries have the greatest biodiversity of all the world's parks, but only 1% of them are protected. In countries struggling with poverty, local people often enter parks illegally in search of wood, game animals, and other natural products they need to survive. Loggers and miners also operate illegally in many of these parks, as do wildlife poachers. Park services in most less-developed countries have too little money and too few personnel to combat these problems, either through enforcement or through education.

checkpoint What illegal activities take place in some national parks?

U.S. National Parks Are Challenged

The U.S. National Park System includes 59 major national parks, along with 339 monuments and historic sites. States, counties, and cities also operate public parks.

Popularity is one of the biggest problems for many U.S. national parks. Between 1970 and 2020, the number of recreational visitors to these protected areas increased from about 7.72 million to about 237 million. In some U.S. national parks and on other public lands, noisy and polluting dirt bikes, dune buggies, jet skis, snowmobiles, and other off-road vehicles have become a problem. These recreational vehicles destroy or damage fragile vegetation and disturb wildlife. They also negatively affect the park experience for other visitors.

Many U.S. national parks have become islands of biodiversity surrounded by commercial development. Their wildlife and recreational value is threatened by nearby activities such as mining, logging, livestock grazing, water diversion, operation of coal-fired power plants, and urban development. The National Park Service reports that air pollution, mainly caused by coal-fired power plants and dense vehicle traffic, impairs scenic views more than 90% of the time in many of its parks.

A number of parks also suffer damage from the migration or deliberate introduction of nonnative species. European wild boars, imported into the state of North Carolina in 1912 for hunting, now threaten vegetation in parts of the Great Smoky Mountains National Park. Nonnative mountain goats in Washington State's Olympic National Park trample and destroy the root systems of native vegetation and accelerate soil erosion.

Native species—some of which are threatened or endangered—are hunted or illegally removed from almost half of all U.S. national parks. Some have disappeared altogether from the ecosystems they once inhabited. Their absence can have a number of ecological consequences, especially if they are keystone species in their ecosystems.

This was illustrated by the case of gray wolves in Yellowstone National Park. Wolves disappeared from the park in the mid 1920s due to overhunting. With no wolves to prey upon them, deer, elk, and moose became overpopulated in the park. They overbrowsed vegetation along rivers, which in turn led to increased soil erosion. Other animals that fed on the same vegetation declined, as did the species that preyed on them. For example, overbrowsing of willows led to a scarcity of beavers. Scavenger species, such as eagles, grizzly bears, and foxes, were impacted because they depended on partially eaten carcasses the wolves left behind. Populations of coyotes increased dramatically because they no longer had to compete with wolves for prey.

Wolves were reintroduced to Yellowstone in the mid 1990s (Figure 8-11). As they re-established themselves, the park changed once again. For example, the wolves reduced the elk herds, and surviving elk became more spread out across the park. Overbrowsed vegetation recovered to some extent as a result. Scavenger species became more common, too, because they had access to leftover wolf kills once again. Coyote populations fell to normal levels, allowing populations of their prey species to recover as well.

checkpoint What is one way Yellowstone National Park's ecosystem changed after gray wolves disappeared from the park?

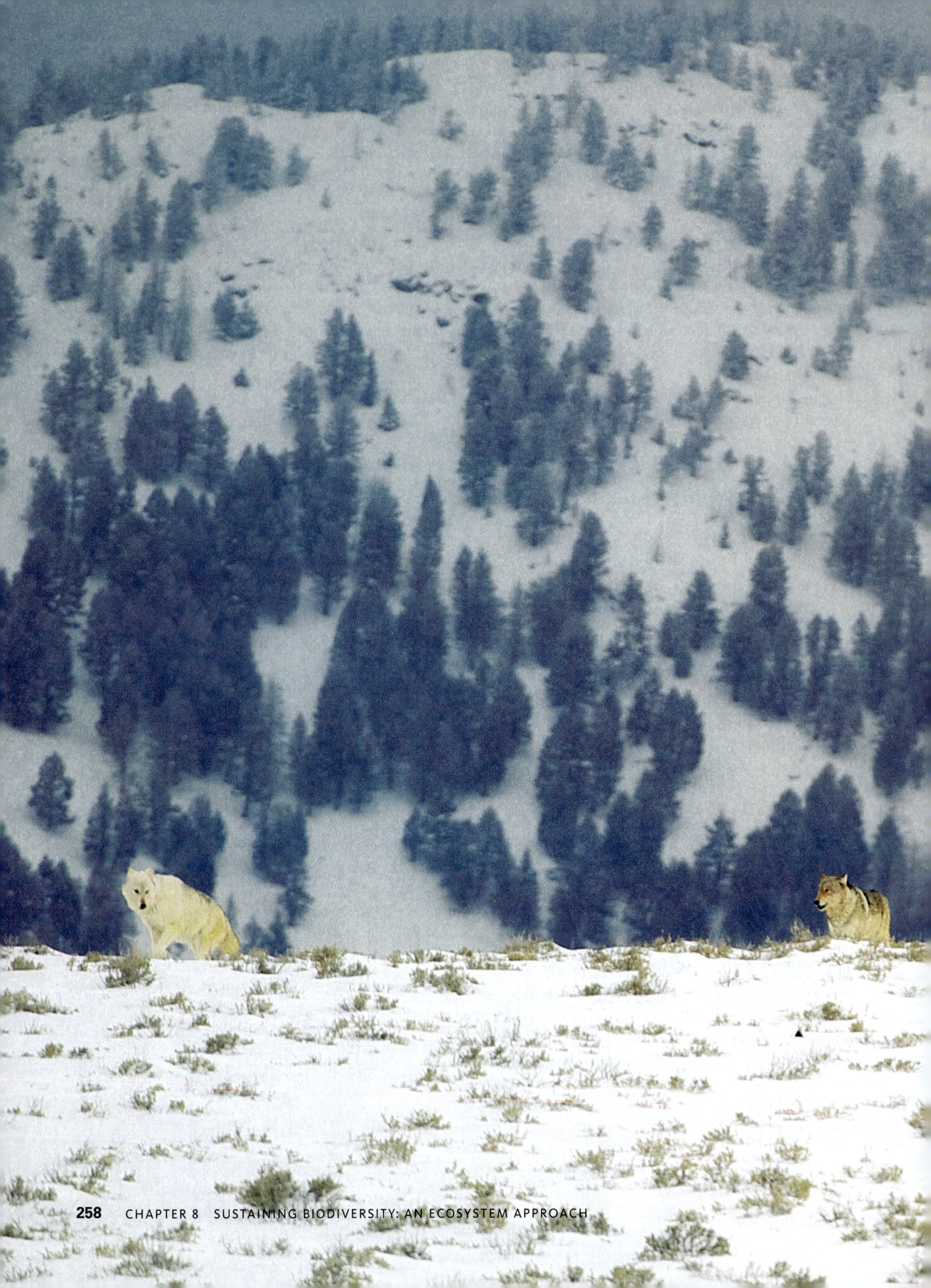

FIGURE 8-11
ON ASSIGNMENT National Geographic photographer Barrett Hedges captured the moment a pack of wolves appeared on a ridge in Yellowstone National Park. This species was reintroduced to the park in 1995. A dozen or so packs live throughout the park today.

People Can Set Aside More Areas for Protection

Most ecologists and conservation biologists argue that the best way to preserve biodiversity is to create a worldwide network of protected areas. In 2021, less than 15% of Earth's land area (excluding Antarctica) was protected in more than 177,000 wildlife refuges, nature reserves, parks, and wilderness areas. However, no more than 6% of Earth's land is strictly protected from potentially harmful human activities.

Conservation biologists call for fully protecting at least 30% of Earth's land area in a global system of biodiversity reserves. The system would include multiple sites representing each of Earth's biomes. In 2020, the United Nations Convention on Biological Diversity estimated the cost of achieving this vision at roughly $140 billion a year, compared to the $24 billion spent on such protection today.

Buffer Zones Conservation biologists suggest using the buffer zone approach whenever possible to design and manage nature reserves. Establishing a **buffer zone** means strictly protecting the inner core of a reserve as well as establishing another, outer zone in which local people can extract resources sustainably. By 2015, the United Nations had used the buffer zone concept to create a global network of 651 **biosphere reserves** in 120 countries (Figure 8-12).

Costa Rica's government used the buffer zone concept to consolidate the country's parks and reserves into several large conservation areas, or megareserves (Figure 8-13). Each megareserve contains a protected inner core surrounded by two buffer zones that local and indigenous people can use for sustainable logging, crop farming, cattle grazing, hunting, fishing, and ecotourism. Instead of shutting people out of reserve areas, this approach enlists local people as partners in protecting reserves from activities such as illegal logging and poaching. It is an application of the biodiversity and win-win factors of sustainability.

Costa Rica's megareserves are intended to sustain about 80% of the country's biodiversity. In addition to their ecological benefits, megareserves have paid off financially. Today, Costa Rica's largest source of income is its travel and tourism industry. Ecotourism accounts for a majority of that industry.

FIGURE 8-12
Big Bend National Park in Texas is part of the Chihuahuan Desert Biosphere Reserve. Big Bend serves as the inner core of the reserve. All of its natural and cultural resources are fully protected.

FIGURE 8-13

Megareserves Costa Rica has created several megareserves. Green areas are protected natural parklands. Yellow areas are surrounding buffer zones.

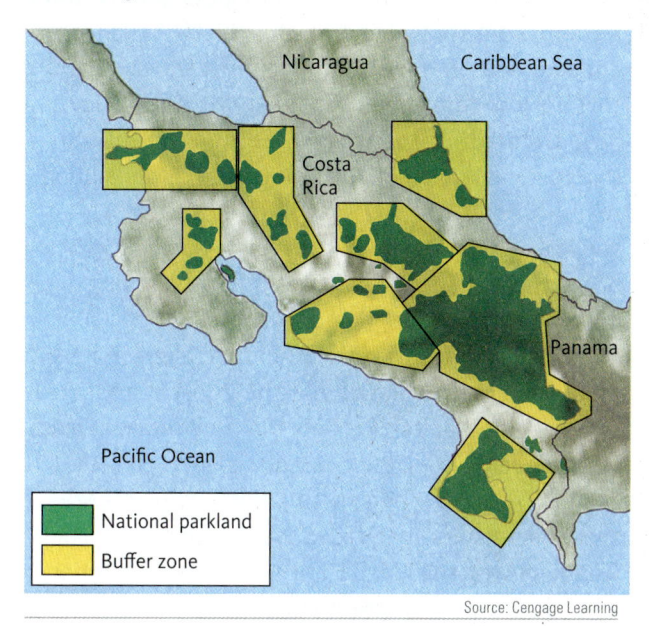

Source: Cengage Learning

Wilderness Areas One way to protect undeveloped lands from human exploitation is to designate them as wilderness areas. Designated **wilderness areas** are essentially undisturbed by humans, and they are protected by law from activities that would alter their ecology. For example, forestry, road and trail development, mining, and building construction are not allowed.

In 1964, the U.S. Congress passed the Wilderness Act, which allowed the government to protect undeveloped tracts of U.S. public land from development as part of the National Wilderness Preservation System. By 2012, the area of protected wilderness was nearly 12 times as large as in 1964. Even so, only 5% of all U.S. land is protected as wilderness—more than 54% of it in Alaska. Only 2.7% of land area in the lower 48 states is protected as wilderness, most of it in the West.

Most developers and resource extractors oppose protecting even 13% of Earth's land in any form. They argue that these areas might contain valuable resources that would provide short-term economic benefits. Ecologists and conservation biologists take a longer view. They argue that such areas provide a long-term "ecological insurance policy" for humans and other species. They are islands of biodiversity and ecosystem services that will be necessary to support life and human economies in the future. In addition, undeveloped lands are needed as centers for future evolution in response to changes in environmental conditions.

> **CONSIDER THIS**
>
> **Fewer young people are exploring America's wild spaces.**
>
> National parks and other protected lands may still be popular in the United States, but their appeal to young people has been declining since the 1980s. Several factors are responsible for this trend. More people live in urban areas, far from national parks and other wildlands. Young people are now more likely to opt for indoor activities involving electronic devices. They are becoming more plugged into technology but more disconnected from the natural world. Some scientists and other analysts warn that, if this trend continues, there will be less citizen pressure for protecting wilderness and for setting aside nature reserves.

checkpoint What human activities are not allowed in wilderness areas?

8.3 Assessment

1. **Summarize** What are the challenges to managing parks in less-developed countries?
2. **Define** What is the buffer zone concept? Give an example of how it is used.
3. **Infer** Why do you think more than 54% of U.S. wilderness areas are located in Alaska and only 2.7% of land area in the lower 48 states is protected as wilderness, most of it in the West? Explain your thinking.

SCIENCE AND ENGINEERING PRACTICES

4. **Designing Solutions** Popularity is one of the most significant problems for U.S. national parks. Use the text and your own research to describe this problem and to develop a possible solution.

8.4 How Does the Ecosystem Approach Help Protect Terrestrial Biodiversity?

CORE IDEAS AND SKILLS

- Identify the steps to sustaining terrestrial biodiversity using an ecosystem approach.
- Define biodiversity hotspots and explain their importance.
- Understand the importance of protecting ecosystem services and human communities that depend on these services.
- Describe methods used to repair damaged ecosystems, and understand how reconciliation ecology enables humans to share ecosystems with other species.

KEY TERMS

biodiversity hotspot ecological restoration reconciliation ecology

The Ecosystem Approach: A Five-Point Plan

Most wildlife biologists and conservationists believe that an ecosystem approach is the most effective way to slow species extinctions due to human activities. This approach focuses on protecting threatened habitats and their ecosystem services. To protect terrestrial biodiversity, the ecosystem approach generally employs a five-point plan as outlined in Figure 8-14.

With the ecosystem approach, land, water, and living resources are all managed to promote conservation and sustainable use in an equitable way. This approach recognizes that humans are a component of many ecosystems.

checkpoint What is biodiversity-friendly development?

Protecting Biodiversity Hotspots

The ecosystem approach calls for identifying and protecting Earth's biodiversity hotspots. **Biodiversity hotspots** are areas that are rich in highly endangered endemic species—species found nowhere else—and that are also threatened by human activities. They have suffered serious ecological disruption, mainly due to rapid human population growth and resulting pressure on natural resources and ecosystem services. Biologists feel that saving biodiversity hotspots will have a significant impact on saving Earth's biodiversity.

To qualify as a biodiversity hotspot, a region must meet two criteria. First, it must have lost at least 70% of its original habitat. Second, it must contain at least 1,500 endemic species of vascular plants (greater than 0.5% of the world's total). Vascular plants are those that have a specialized system for transporting water, mineral salts, and sugars. Ferns, conifer trees, and flowering plants are examples of vascular plants.

Figure 8-15 shows 34 terrestrial biodiversity hotspots biologists have identified. According to the IUCN, these areas cover little more than 2% of Earth's land surface but contain an estimated 50% of the world's flowering plant species. They also hold 42% of all terrestrial vertebrates (mammals, birds, reptiles, and amphibians).

A large majority of the world's endangered or critically endangered species live in biodiversity hotspots. One-sixth of the world's population—about 1.2 billion people—lives there as well. However, only 5% of the total area of these hotspots is truly protected through sufficient government funding and law enforcement.

checkpoint What criteria are used to identify biodiversity hotspots?

FIGURE 8-14

Ecosystem Approach to Sustaining Terrestrial Biodiversity	
Goal	**Approach**
1. Map and inventory ecosystems	Map the world's terrestrial ecosystems and create an inventory of the species contained in each of them, along with the ecosystem services they provide.
2. Identify resilient and fragile ecosystems	Identify terrestrial ecosystems that are resilient and can recover if not overwhelmed by harmful human activities, along with ecosystems that are fragile and need protection.
3. Protect endangered plants and ecosystem services	Protect the most endangered terrestrial ecosystems and species, with emphasis on protecting plant biodiversity and ecosystem services.
4. Restore degraded ecosystems	Seek to restore as many degraded ecosystems as possible.
5. Make development biodiversity-friendly	Provide significant financial incentives (such as tax breaks and subsidies) and technical help to private landowners who agree to help protect endangered ecosystems.

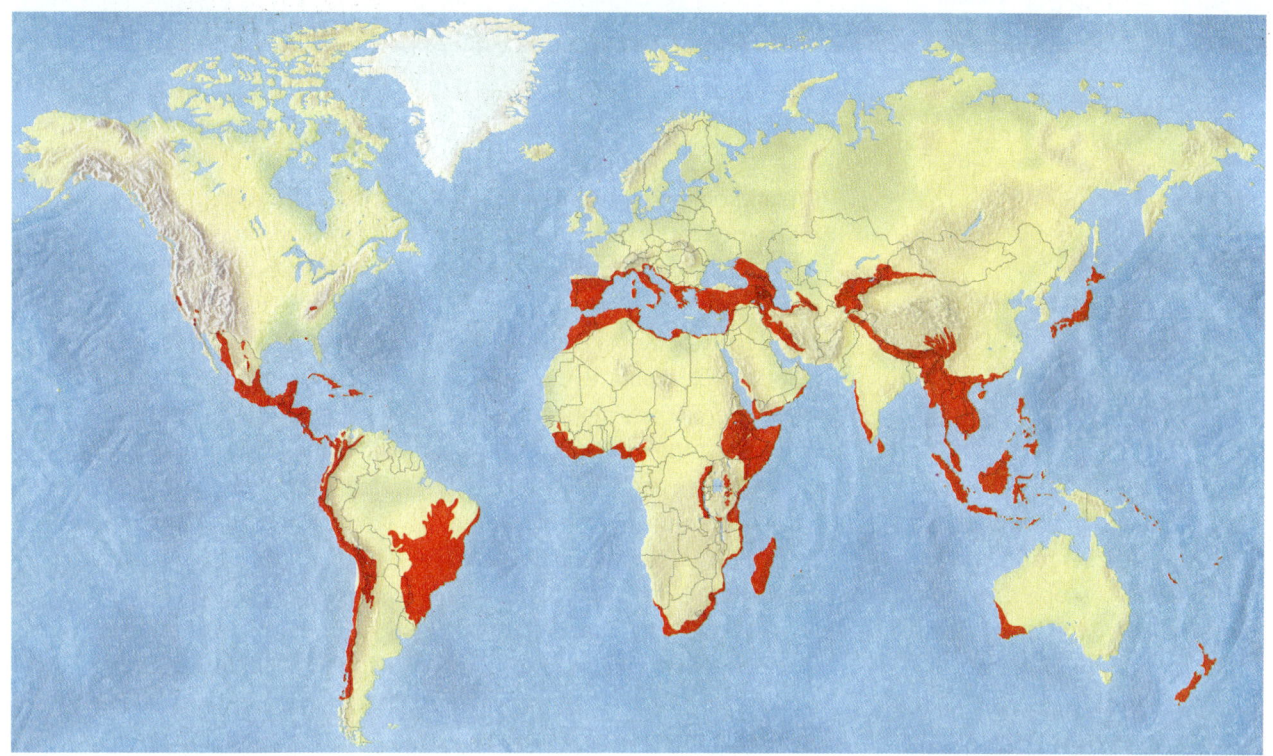

FIGURE 8-15
Biodiversity Hotspots Biologists have identified these 34 biodiversity hotspots.

Source: Center for Applied Biodiversity Science at Conservation International

Protecting Ecosystem Services

One study done for the UN Millennium Ecosystem Assessment suggested that humans degrade or overuse 60% of the ecosystem services provided by various ecosystems worldwide. When the quality of ecosystem services declines, species biodiversity is threatened and so is the well-being of human populations.

For this reason, the ecosystem approach calls for identifying and protecting areas in which vital ecosystem services are at risk. This approach goes beyond simply setting aside land for protection. It also calls for protecting the human communities that exist in these areas. Without addressing such issues as poverty, population growth, urbanization, and resource use, ecosystem services will continue to decline.

Scientists suggest identifying highly stressed "life raft" ecosystems. These are generally areas where poverty is very high and most people heavily depend on ecosystem services for survival. This dependence has degraded ecosystems severely enough to threaten humans and other species.

In life raft ecosystems, residents, public officials, and conservation scientists can work together to protect human communities along with the natural biodiversity and ecosystem services that support all life and economies. Instead of pitting people against nature, this approach offers a way to apply the win-win factor of sustainability.

checkpoint Why is protecting human communities important to protecting ecosystem services?

Restoring Damaged Ecosystems

Almost every natural place on Earth has been impacted to some degree by human activities, most often in negative ways. However, it is possible to partially reverse much of the damage to ecosystems caused by human activities. The approach to repairing damaged ecosystems varies, as described in Figure 8-16. Each approach helps speed the natural recovery that results from ecological succession.

For example, the process of **ecological restoration** can help return a damaged habitat or ecosystem to a state very close to its original one. Examples of ecological restoration include replanting forests, reintroducing native species, removing harmful invasive species, freeing river flows by removing dams, and restoring grasslands, coral reefs, wetlands, and stream banks.

LESSON 8.4

NATIONAL GEOGRAPHIC | EXPLORERS AT WORK
Kamaljit S. Bawa Conservation Biologist

Kamaljit S. Bawa is an expert on global change and biodiversity. Dr. Bawa's career has certainly been global in scope. Currently, Bawa is a Distinguished Professor of Biology at the University of Massachusetts at Boston. He is also the president of the Ashoka Trust for Research in Ecology and the Environment in Bangalore, India. And he is involved with numerous international organizations and committees, including National Geographic's Committee for Research and Exploration, which awards grants for science research worldwide.

Bawa has spent time researching India's four biodiversity hotspots and has written books about two of them: the Western Ghats mountain range along the southwestern coast of India, and the Eastern Himalaya region, which spans parts of India, Nepal, and Bhutan. The Eastern Himalaya is home to species ranging from orchids to snow leopards, tigers, red pandas, and rhinoceroses. New species continue to be discovered there today.

Bawa trekked across parts of the Eastern Himalaya to study biodiversity there as well as impacts from climate change, deforestation, pollution, dam building, and hunting of rare species. He visited 18 villages to talk with people about changes they see happening in the region, such as changing weather patterns, and how to address them. Bawa published his experiences in a coffee table book called *Himalayas: Mountains of Life*. The book features photography by Sandesh Kadur, who is also a National Geographic Explorer.

Bawa hopes the Eastern Himalaya and India's other biodiversity hotspots receive the attention they deserve for conservation. "Life on Earth is spectacular and very fragile. And I think we are blessed in India to have so much diversity," he says.

Researchers have suggested the following four-step strategy for carrying out most forms of ecological restoration and rehabilitation:

1. Identify the causes of the degradation, such as pollution, farming, overgrazing, mining, or invasive species.
2. Stop the degradation by eliminating or sharply reducing these factors.
3. Reintroduce key species to restore natural ecological processes, as was done with gray wolves in the Yellowstone ecosystem.
4. Protect the area from further degradation to allow natural recovery.

checkpoint How is ecological restoration related to ecological succession?

Sharing Ecosystems with Other Species

There are simply not enough ecosystems left on Earth where wild species can exist free of the effects of human activities. And it is not realistic to think that vanishing species can simply be closed off in areas set aside for them. For these reasons, protecting terrestrial biodiversity must focus on finding ways for humans to share some of the spaces they dominate with other species. Ecologist Michael L. Rosenzweig calls this approach *reconciliation ecology*.

Reconciliation ecology involves establishing and maintaining new habitats to conserve species diversity in places where people live, work, or play. It calls for redesigning ways in which people use their communities so that they can also support wild species living among them. Ecotourism is one example of reconciliation ecology. Ecotourism allows people to protect local wildlife and ecosystems while providing economic resources for their communities.

Reconciliation ecology is also a way to protect vital ecosystem services. For example, people are learning how to protect insect pollinators in their communities, such as butterflies and honeybees, which are vulnerable to pesticides and habitat

FIGURE 8-16

How to Repair Damaged Ecosystems	
Goal	Approach
Restoration	Return a degraded habitat or ecosystem to a condition as similar as possible to its natural state in cases where this is possible.
Rehabilitation	Turn a degraded ecosystem into a functional or useful ecosystem without trying to restore it to its original condition. Examples include removing pollutants from abandoned mining or industrial sites and replanting trees to reduce soil erosion in clear-cut forests.
Replacement	Replace a degraded ecosystem with another type of ecosystem. For example, a degraded forest may be replaced by a productive pasture or tree plantation.
Creation of artificial ecosystems	Intentionally establish environments that function for the most part like their natural counterparts. For example, artificial wetlands have been created in some areas to reduce flooding and to treat sewage.

CONSIDER THIS

You can help sustain terrestrial biodiversity.

You can take the following steps to conserve natural resources and protect terrestrial biodiversity where you live:

- Plant trees and take care of them.
- Recycle paper and buy recycled paper products.
- Buy sustainably produced wood and wood products and wood substitutes such as recycled plastic furniture and decking.
- Landscape your yard with a diversity of native plants.

loss. Neighborhoods and municipal governments are doing so by reducing or eliminating the use of pesticides on their lawns, fields, golf courses, and parks. Neighbors can also work together to plant gardens of flowering plants as a source of food for bees and other pollinators.

People have also taken steps to protect bluebirds within human-dominated habitats where most of the bluebirds' nesting trees have been cut down and bluebird populations have declined. Specially designed boxes have provided artificial nesting spots for bluebirds. Their widespread use has allowed populations of this species to grow. These and many other examples of people working together on projects to restore degraded ecosystems all involve applications of the biodiversity and win-win factors of sustainability.

checkpoint Why must there be more emphasis on conserving species diversity in areas where humans are dominant?

8.4 Assessment

1. **Explain** What are biodiversity hotspots, and why are they given priority under the ecosystem approach?
2. **Recall** What are four approaches to repairing damaged ecosystems?
3. **Draw Conclusions** How do issues of poverty impact decisions about protecting ecosystems?
4. **Analyze** Ecologist Michael L. Rosenzweig is credited with coining the term *reconciliation ecology*. Consider the definition for *reconciliation* and explain how it relates to this approach.

CROSSCUTTING CONCEPTS

5. **Patterns** Examine the map of biodiversity hotspots shown in Figure 8-15. Then examine the map showing the human ecological footprint in Figure 1-10. What patterns do you notice when you compare these maps? What might these patterns suggest?

LESSON 8.4

8.5 How Does the Ecosystem Approach Help Protect Aquatic Biodiversity?

CORE IDEAS AND SKILLS

- Describe how human activities are destroying aquatic biodiversity, and explain the impacts of overfishing, ocean acidification, and dead zones.
- Explain the role of marine protected areas and marine reserves in sustaining marine biodiversity.
- Identify the steps in the ecosystem approach to sustaining aquatic biodiversity.

KEY TERMS

fishery   fishprint

Human Activities Are Threatening Aquatic Biodiversity

Human activities have destroyed or degraded much of the world's coastal wetlands, coral reefs, and mangroves—and even the ocean floor. They also have disrupted many freshwater ecosystems. Threats to aquatic ecosystems range from pollution and overfishing to habitat disruption and the spread of invasive species. For example, the USFWS claims invasive species have caused two-thirds of all fish extinctions in the United States since 1900.

In freshwater aquatic zones, human activities are destroying aquatic habitats, degrading water flows, and disrupting biodiversity. Such activities include dam building and excessive water withdrawal from rivers for irrigation and urban water supplies.

Human activities are also taking a toll on oceans. Sea-bottom habitats are threatened by impacts from dredging operations, which involve scooping shellfish off the sea floor, and by fishing trawlers. Like giant submerged bulldozers, trawlers drag huge nets weighted down with chains and steel plates over the ocean floor to harvest a few species of bottom fish and shellfish (Figure 8-18).

Each year, thousands of trawlers scrape and disturb an area of ocean floor many times larger than the annual global total area of forests that are clear-cut. Some marine scientists call bottom trawling the largest human-caused disturbances to the biosphere.

Changing sea levels also threaten saltwater ecosystems. Scientists expect that sea levels will rise during this century, mainly due to projected climate change. Warming temperatures are expected to cause melting of glaciers and ice sheets. As sea levels rise, they are likely to destroy many coral reefs and flood some low-lying islands, along with their protective coastal mangrove forests.

Ocean waters themselves are growing warmer, mostly due to heat absorbed from a warmer atmosphere. This change is already having negative impacts on coral reef ecosystems. (See the Chapter 6 Case Study.) A 2011 WRI study estimated 75% of the world's shallow coral reefs are at risk of being destroyed, mainly by warmer waters. Overfishing, increasing water acidity, and pollution are also contributing factors.

Overfishing Fish and fish products provide 20% of the world's animal protein and are consumed by billions of people. A **fishery** is a concentration of an aquatic species suitable for commercial harvesting in an ocean area or inland body of water. Today, 4.4 million fishing boats harvest fish from oceans.

Industrial fishing fleets use a variety of methods to harvest marine fishes and shellfish. They also use global satellite positioning technology, sonar fish-finding devices, huge nets, long fishing lines, spotter planes, and refrigerated factory ships to process and freeze their catches. These highly efficient fleets supply the growing demand for seafood, but critics say they are vacuuming the seas. As a result, they are reducing marine biodiversity and degrading important marine ecosystem services. Figure 8-17 shows the major methods used in commercial harvesting of marine fishes and shellfish.

One result of the increasingly efficient global hunt for fish is that larger individuals of commercially valuable wild species—including cod, marlin, swordfish, and tuna—are becoming scarce. A study conducted by scientists in Canada found that 90% or more of these and other large, predatory, open-ocean fishes disappeared between 1950 and 2006. One side effect of this trend: When larger predatory species start to dwindle, rapidly reproducing invasive species such as jellyfish can easily take over and disrupt ocean food webs.

The decline in commercially valuable large species also has meant the fishing industry has begun working its way down marine food webs by shifting to smaller marine species known as forage fish. Examples include anchovies, herring, sardines, and shrimp-like krill. Ninety percent of this catch is converted to fishmeal and fish oil and fed to farmed fish. Scientists warn this reduces the food supply

FIGURE 8-17

Methods of Harvest This diagram illustrates major commercial fishing methods, as well as methods used to raise fish through aquaculture.

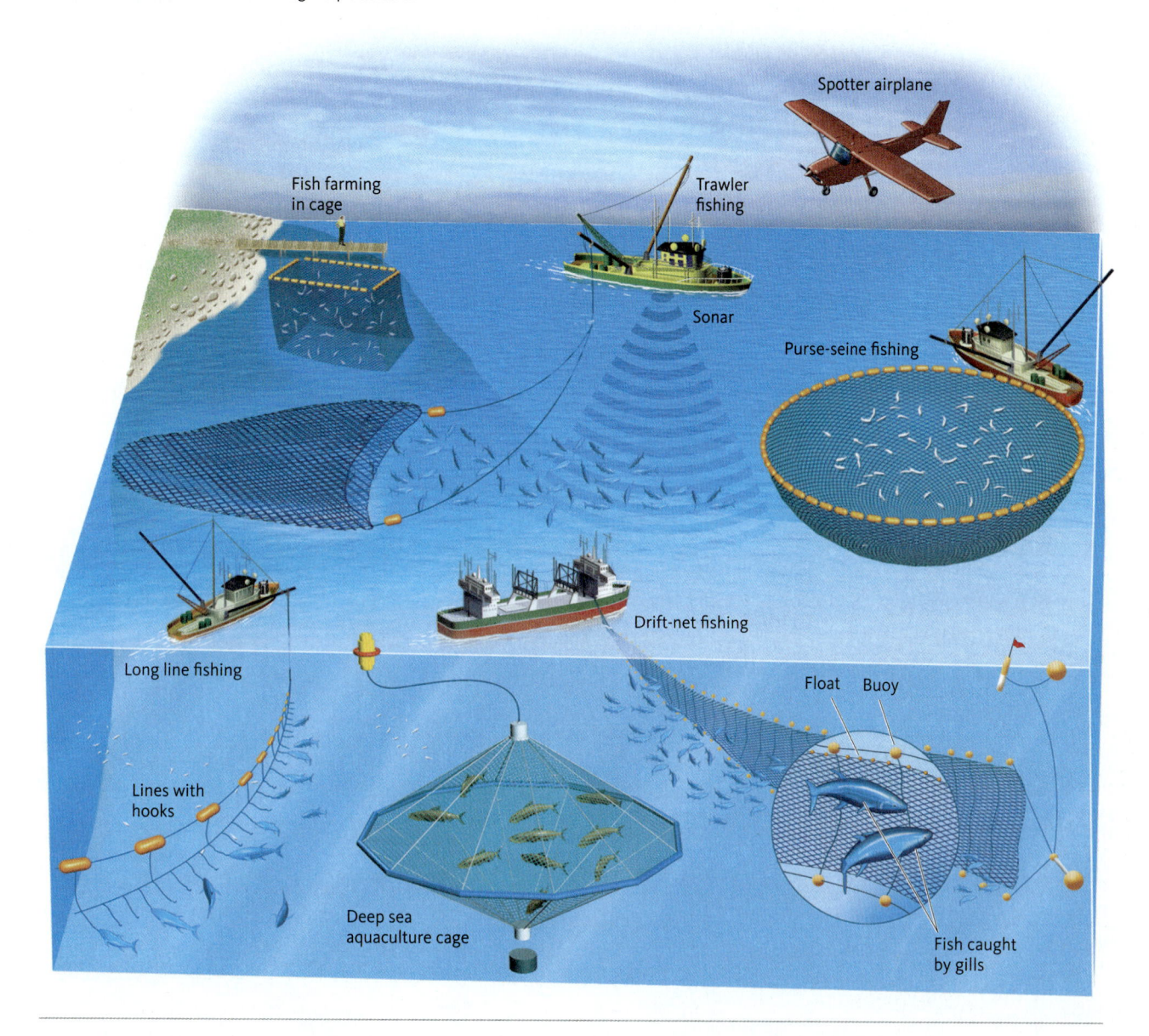

for larger species that will likely have a harder time rebounding from overfishing. The end result will be further disruption of marine ecosystems and their ecosystem services.

If current fishing patterns continue, all major commercial fish species likely will suffer population collapse by 2048. People are simply harvesting more wild fish than can be restocked over time. A fishprint can help assess the pressures that fishing activities place on fisheries. A **fishprint** is defined as the area of ocean needed to sustain the fish consumption of an average person, nation, or the world, based on the weight of fish they consume annually. The term is based on the concept of an ecological footprint.

Ocean Acidification Today, coral reefs are exposed to the warmest, most acidic ocean waters of the past 400,000 years. These changes are tied to an increase in carbon dioxide (CO_2) in the atmosphere. By burning increasingly larger amounts of carbon-containing fossil fuels, especially since 1950, people are adding CO_2 to the atmosphere faster than it can be removed by the carbon cycle.

According to 97% of the world's climate scientists, human activities are causing increased levels of CO_2 in the atmosphere. Increased CO_2 levels are in turn warming the lower atmosphere and are contributing to climate change. Earth's oceans have now absorbed one-fourth of this excess CO_2 from the atmosphere.

LESSON 8.5

Deep-Sea Trawling

FIGURE 8-18
This diagram illustrates methods for deep-sea trawling, which badly damages marine ecosystems. Trawlers drag huge fishing nets across the ocean bottom. Heavy doors keep the mouth of the net open and positioned along the sea floor. Floats lift the net open, and rollers carve out a path below the net as it drags along.

Trawler

Door

Floats

Rollers

Door

Due to excess CO_2, the chemistry of ocean waters is being changed through a process called ocean acidification (Figure 8-19).

When absorbed CO_2 combines with water, it forms carbonic acid (H_2CO_3), a weak acid also found in carbonated drinks. This process increases the level of hydrogen ions (H^+) in the water and makes the water less basic (lower pH). It also reduces the level of carbonate ions (CO_3^{2-}) in the water because the ions react with hydrogen ions (H^+) to form bicarbonate ions (HCO_3^-).

The problem is that many aquatic species, including phytoplankton, corals, sea snails, crabs, and oysters, use carbonate ions to produce calcium carbonate ($CaCO_3$), the main component of their shells and bones. As a result, in less basic waters with lower concentrations of carbonate ions, shell-building species and coral reefs grow more slowly. Furthermore, when the hydrogen ion concentration in surrounding seawater reaches high enough levels, the calcium carbonate in these organisms' shells and bones begins to dissolve.

According to most marine scientists, slowing these changes requires a quick and significant reduction in the use of fossil fuels. It also requires a sharp reduction in energy waste and a shift to a mix of low-carbon renewable energy resources such as solar power and wind power. These changes would cut the massive inputs of CO_2 into the air and, subsequently, into the ocean. Protecting and restoring mangrove forests, sea grass beds, and coastal wetlands can also help combat ocean acidification. These aquatic habitats take up and store some of the atmospheric CO_2 that is at the heart of the problem.

Dead Zones Pollution from excess soil nutrients sometimes causes oxygen-depleted areas in the ocean called dead zones. In the spring when farmers plant fields, they add fertilizers to the soil to enhance the growth of crops. Some of the fertilizers are washed off during spring rains and the runoff enters streams, rivers, and the ocean. This nutrient pollution causes overgrowth of algae in ocean waters.

The algae eventually die and sink to the bottom, where bacteria begin to decompose them. Because decomposition requires oxygen, levels of this dissolved gas in the water become depleted. Marine organisms either suffocate due to lack of dissolved oxygen in the water or they are forced to leave. The area becomes lifeless. Each year, excess fertilizer runoff creates a dead zone in the northern Gulf of Mexico, threatening that ecosystem's rich biodiversity. (See Figure 6-22.)

checkpoint How does acidification affect pH levels in ocean waters?

Protecting Marine Biodiversity Is Difficult

Protecting marine biodiversity is difficult for several reasons. For example, the human ecological footprint and fishprint are expanding so rapidly it is difficult to monitor their impacts on marine biodiversity. Many people incorrectly view oceans as an inexhaustible resource that can absorb an almost infinite amount of waste and pollution and still produce seafood and other products in demand. They are unaware of much of the damage to oceans and other bodies of water because it is not easily visible.

Most of the world's ocean area also lies outside the legal jurisdiction of any country. Thus, much of it is an open-access resource, subject to overexploitation. This is a classic example of the tragedy of the commons (Lesson 1.2).

Since 1986, the IUCN has helped establish a global system of marine protected areas (MPAs)—areas

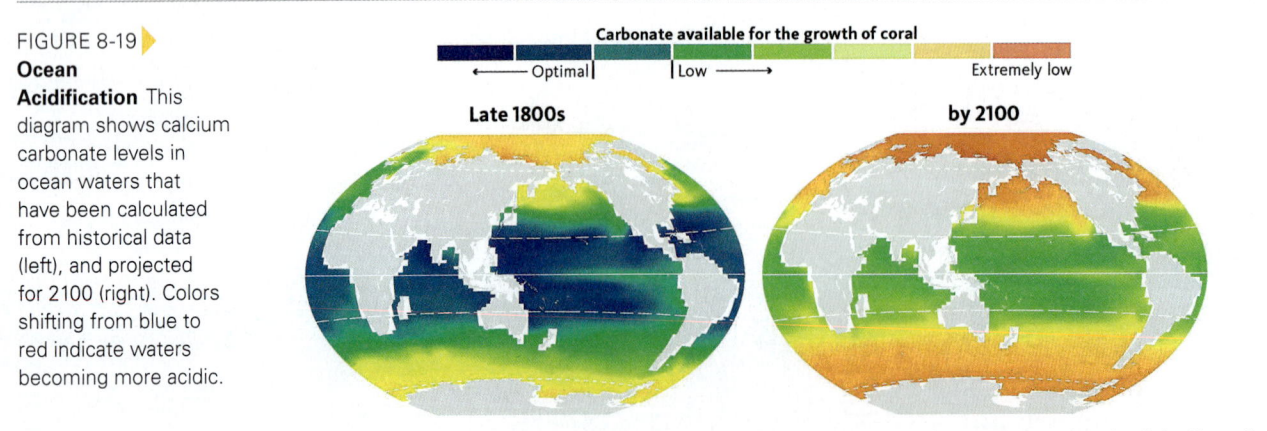

FIGURE 8-19
Ocean Acidification This diagram shows calcium carbonate levels in ocean waters that have been calculated from historical data (left), and projected for 2100 (right). Colors shifting from blue to red indicate waters becoming more acidic.

Carbonate available for the growth of coral
←Optimal | Low→ | Extremely low

Late 1800s | by 2100

Source: Used with permission from National Geographic.

FIGURE 8-20

Ecosystem Approach to Sustaining Aquatic Biodiversity	
Goal	Approach
1. Map and inventory ecosystems	Complete the mapping of the world's aquatic biodiversity, identifying and locating as many plant and animal species as possible.
2. Identify biodiversity hotspots and fragile ecosystems	Identify and preserve the world's aquatic biodiversity hotspots and areas where deteriorating ecosystem services threaten people and other forms of life.
3. Establish protected marine reserves for recovery of aquatic ecosystems	Create large and fully protected marine reserves to allow damaged marine ecosystems to recover and to allow fish stocks to be replenished.
4. Restore degraded ecosystems	Protect and restore the world's lakes and river systems, which are the most threatened ecosystems.
5. Initiate restoration projects	Initiate worldwide ecological restoration projects in systems such as coral reefs and inland and coastal wetlands.
6. Initiate reconciliation projects	Find ways to increase the incomes of people who live in or near protected lands and waters so that they can become partners in the protection and sustainable use of ecosystems.

of ocean partially protected from human activities. According to the U.S. National Ocean Service, there are more than 5,800 MPAs worldwide (more than 1,600 in U.S. waters). 41% of U.S. waters are protected in some manner, including the Great Lakes. Activities such as dredging, trawler fishing, and other types of resource extraction are allowed in MPAs. As a result, they only offer partial protection. And many are too small to be effective in protecting larger species of animals.

On the other hand, marine reserves can offer full protection for marine species. These reserves, some of which already exist, encompass entire ecosystems. Unlike MPAs, marine reserves are declared off-limits to commercial fishing, dredging, mining, and waste disposal. Studies show that in fully protected marine reserves, commercially valuable fish populations double, fish size increases by almost one-third, fish reproduction triples, and species diversity increases by almost one-fourth. These improvements can be seen two to four years after strict protection begins.

Even so, marine reserves face challenges in funding, management, and monitoring. The fact remains that 98.8% of Earth's oceans are not effectively protected from harmful human activities.

checkpoint How do marine reserves benefit fish populations?

The Ecosystem Approach Can Help

Biodiversity expert Edward O. Wilson has adapted the ecosystem approach to protecting terrestrial biodiversity in order to apply it to aquatic ecosystems. The steps involved in applying this adapted ecosystem approach are summarized in Figure 8-20.

Protecting Earth's vital biodiversity and limiting humans' harmful environmental impact will not be implemented without bottom-up political pressure on elected officials from individual citizens and groups. It also will require cooperation among scientists, engineers, and key people in government and the private sector. People can "vote with their wallets" by purchasing only products and services that do not harm terrestrial and aquatic biodiversity.

checkpoint How are the ecosystem approaches to aquatic and terrestrial diversity similar to each other?

CONSIDER THIS

Damage to ecosystems can still be reversed.

There is growing evidence that many of the harmful impacts of human activities on both terrestrial and aquatic biodiversity and ecosystem services could be reversed over the next two decades using an ecosystem approach. According to Edward O. Wilson, such a conservation strategy would cost $30 billion a year—an amount that could be provided by a tax of one penny per cup of coffee consumed in the world each year.

8.5 Assessment

1. **Summarize** What factors make it difficult to protect marine biodiversity?
2. **Explain** What effect do rising carbon dioxide (CO_2) emissions have on aquatic species whose shells and bones are composed of calcium carbonate ($CaCO_3$)?
3. **Apply** How does overexploitation of oceans illustrate the tragedy of the commons?
4. **Evaluate** The Food and Agriculture Organization claims "the state of world fisheries presents us with pressing ecological, economic, social, and political challenges with significant ethical implications." Use details from the text to evaluate this statement.

SCIENCE AND ENGINEERING PRACTICES

5. **Communicating Information** Draw a diagram to illustrate the chain of events that leads to formation of a dead zone.

TYING IT ALL TOGETHER STEM
Rates of Tropical Forest Loss

In this chapter, you learned how human activities are destroying or degrading much of the world's terrestrial and aquatic biodiversity. For example, you learned about deforestation and its impact on forests, particularly tropical forests. The causes of deforestation are complex and involve a combination of social and economic factors. Solutions to this problem are equally complex, and often require cooperation among individuals, groups, and government agencies.

The Case Study provided a strong example of what can be achieved when this kind of cooperation takes place. Costa Rica has become a leader in conserving tropical forests. Unfortunately, tropical forest loss continues to take place at a high rate worldwide. Figure 8-21 illustrates the impact of tropical forest loss for a number of hypothetical countries.

Use the table to help you answer the questions that follow.

1. What is the annual rate of tropical forest loss, as a percentage of total forest area, in each of the five countries?
2. What is the annual rate of tropical deforestation collectively in all of the countries represented in the table?
3. According to the table, and assuming the rates of deforestation remain constant, which country's tropical forest will be totally destroyed first?
4. Assuming the rate of deforestation in country C remains constant, how many years will it take for all of its tropical forests to be destroyed?
5. If you were a government representative in one of these countries, what measures would you recommend to reduce the rates of tropical deforestation in your country and why?

FIGURE 8-21 ▼

Country	Area of Tropical Rain Forest (square kilometers)	Area of Deforestation per Year (square kilometers)
A	1,800,000	50,000
B	55,000	3,000
C	22,000	6,000
D	530,000	12,000
E	80,000	700

CHAPTER 8 SUMMARY

8.1 How can forests be better managed?

- Forests provide a variety of valuable ecosystem and economic services.
- Scientists classify forests as old-growth forests, second-growth forests, or tree plantations based on their age and structure.
- Unsustainable timber harvest and deforestation are the chief threats to forest ecosystems.
- More sustainable forestry practices include less clear-cutting, less logging of old-growth forests, and intentionally leaving standing dead trees and larger fallen trees in place.
- Measures to promote sustainable forestry include certification of environmentally responsible forestry operations, reducing waste, debt-for-nature swap and conservation concessions programs, and better fire management.

8.2 How can grasslands be better managed?

- Grasslands provide ecosystem services including soil formation, erosion control, nutrient cycling, storage of atmospheric CO_2 in biomass, and maintenance of biodiversity.
- Rangelands and pastures are overgrazed when too many animals graze one area for too long. Rotational grazing and limiting the amount of livestock use can prevent overgrazing.

8.3 How can protected lands be better managed?

- Managed parks face many threats, including illegal poaching and logging, high volumes of tourists, the spread of invasive species, and the decline of native species.
- Sustaining biodiversity requires more protection of existing parks and nature reserves, as well as remaining undisturbed land area. Creating buffer zones can protect reserves while allowing local people to meet their daily living needs.

8.4 How does the ecosystem approach help protect terrestrial biodiversity?

- People can sustain terrestrial biodiversity and limit humans' harmful environmental impact by following an ecosystem approach, identifying and protecting biodiversity hotspots, and employing restoration ecology and reconciliation ecology.
- People can share resources with other species by establishing and maintaining new habitats to conserve species diversity in places where people live, work, or play.

8.5 How does the ecosystem approach help protect aquatic biodiversity?

- Human activities such as dam building and excessive water withdrawal impact the function of freshwater ecosystems. Dredging and trawler fishing degrade ocean environments. Human activities also lead to ocean acidification and development of dead zones.
- Most commercial fisheries are projected to collapse within the next several decades. Adding marine protected areas and marine reserves can protect marine ecosystems.
- The ecosystem approach to protecting terrestrial biodiversity can be applied to protecting aquatic biodiversity as well.

CHAPTER 8 ASSESSMENT

Review Key Terms

Select the key term that best fits each definition. Not all terms will be used.

biodiversity hotspot	deforestation	prescribed burn
biosphere reserve	ecological restoration	rangeland
buffer zone	fishery	reconciliation ecology
conservation concession	fishprint	second-growth forest
debt-for-nature swap	old-growth forest	tree plantation
	overgrazing	wilderness area
	pasture	

1. Uncut or regenerated forest that has not been seriously disturbed by human activities or natural disasters for 200 years or more

2. Area of ocean needed to sustain the fish consumption of an average person, nation, or the world, based on the weight of fish they consume annually

3. Deliberately set, controlled small fire that removes flammable small trees and underbrush in the highest-risk forest areas

4. Temporary or permanent removal of large expanses of forest for agriculture, settlements, or other uses

5. Arrangement in which a government or private conservation organization pays a government or private landowner in another nation for agreeing to preserve natural resources

6. Unfenced grassland in temperate and tropical climates that supplies forage for grazing and browsing animals

7. Managed forest containing only one or two species of trees that are all the same age

8. Scientific approach that focuses on establishing and maintaining new habitats to conserve species diversity in places where people live, work, or play

Review Key Concepts

9. What major ecological and economic benefits do forests provide?

10. What are some ecological benefits of occasional surface fires?

11. Explain how increased reliance on tree plantations can reduce overall forest biodiversity and degrade forest topsoil.

12. What is certified sustainably grown timber?

13. What ecosystem services do grasslands provide?

14. Explain how trawling negatively impacts habitats along the ocean floor.

15. How does the way in which people view oceans make it difficult to protect marine biodiversity?

16. What happens when larger species of commercial fish disappear from fisheries?

Think Critically

17. China uses 80 billion disposable wood chopsticks each year. To meet that demand, China cuts down 20 million trees to make the chopsticks. What recommendation(s) would you give to prevent cutting down 20 million trees?

18. You are a defense attorney arguing in court for preserving an old-growth forest that developers want to clear for a suburban development. Give your three strongest arguments for preserving this ecosystem. How would you counter the argument that preserving the forest would harm the economy by causing a loss of jobs in the timber industry?

19. Many of Earth's biodiversity hotspots are also home to communities of people who are living at the subsistence level—able to obtain just what they need for daily survival. Is it realistic to think conservation plans for such regions can be designed to serve local people as well? Use what you have learned in this chapter to explain why or why not.

20. What opportunities might there be to practice reconciliation ecology in your community? Which individuals or groups in your community would likely support the idea of reconciliation ecology and why? Which would likely oppose it and why? How might you convince those opponents of the value and feasibility of reconciliation ecology projects?

Chapter Activities

A. Investigate: Ocean Acidification STEM

As you read in this chapter, Earth's oceans absorb about one-quarter of the CO_2 humans release into the atmosphere every year. When absorbed CO_2 chemically reacts with seawater, the water's acidity changes. During this experiment, you will model the process by which ocean water absorbs CO_2 to answer the question, "How do increased levels of absorbed CO_2 in seawater affect its acidity?"

Materials

artificial seawater	glass	pH scale
graduated cylinder	eye dropper	straw
	pH test solution	timer
	spoon	

1. Use a graduated cylinder to measure 10 mL of the seawater and pour it into a glass.

2. Use an eye dropper to add 10 drops of pH test solution to the glass. Gently mix the solution with a spoon. Let the solution sit for one minute. Compare the color of the seawater with the pH scale your teacher provides.

3. Consider how you can use the suggested materials or others your teacher approves to create a source of CO_2 that can be added to seawater. (Hint: Consider your breath.) What real-world sources of CO_2 will you be modeling using your source of CO_2?

4. Write a statement that predicts what will happen when you add CO_2 to your seawater.

5. Use the method and materials you chose to add CO_2 to the seawater for 30 seconds. Let the solution sit for one minute. Then compare the color of the solution to the pH scale. What does the color change indicate? Was your prediction correct?

B. Citizen Science

National Geographic Learning Framework

Attitudes | Curiosity
Skills | Observation
Knowledge | Our Living Planet

One way researchers are trying to better understand the impacts of deforestation is to identify areas where deforestation has occurred and determine how land use in these areas has changed. Has the land been converted to ranchland or farmland, for example? Satellite images can be very useful because they capture aerial views of forest ecosystems over time. They can provide a "before" and "after" look at a particular area, which can help researchers determine if deforestation has taken place and, if it has, why it might have happened.

A number of organizations are using citizen scientists to analyze satellite images for signs of deforestation. Work with a group to identify one of these organizations. Use the search term "citizen science and deforestation." When you finish your analysis, discuss as a group how easy or difficult the work was, what surprised you most, and how your contributions will help researchers better understand deforestation.

CHAPTER 8 ASSESSMENT   275

STEM ENGINEERING PROJECT 3
DESIGN A SOLAR COOKER

Technology does not have to be complex to make a big difference in the world. Solar cookers are a good example. A solar cooker is basically a container designed to collect and trap solar energy to heat water and cook food. Solar cookers don't require electricity or fuel such as wood or charcoal. As a result, these simple ovens offer a low-cost, environmentally friendly way to prepare food in places where sunlight is plentiful.

Currently about 2 billion people depend on wood-fueled ovens to sanitize their water and cook their food. Using fuelwood to meet people's basic needs may have once been a sustainable way of life, but most regional populations are much too large for that today. The reliance on fuelwood now poses multiple problems. Harvesting local wood from forests degrades or destroys this source of natural capital. As forests dwindle, people must travel farther, often on foot, to access fuelwood. Additionally, the smoke produced from burning wood harms people's health.

On the island nation of Madagascar, household air pollution from ovens is the second leading cause of death, and the majority of its victims are children. Ovens there use wood or charcoal made from wood. The wood is harvested from local forests full of rare and endangered plant and animal species. Much of these forests have already been destroyed to clear pasture for cattle. Since the early 2000s, a Swiss organization called Association pour le Développement de l'Energie Solaire (ADES) has been working in Madagascar to introduce solar cookers. Since then, tens of thousands of people have made the switch to solar cooking.

ADES and other organizations share solar cooker designs, ideas, and information on the Internet. In this challenge, your team will research these designs, create your own solar cooker, and test it. You will use the engineering design cycle to translate your ideas into a practical solution.

Engineering DESIGN CYCLE
- Defining problems
- Developing and using models
- Planning and carrying out investigations
- Analyzing data and using math
- Designing solutions
- Forming arguments from evidence
- Obtaining, evaluating, and communicating information

Defining Problems

1. Define the problem.
   - Describe the problems associated with wood- and charcoal-fueled ovens.
2. Underlying every problem is a need.
   - What do people living in a remote mountain community in Madagascar need in an oven?
3. A practical, successful solution must meet the following criteria. What else can you add to this list?
   - The solar cooker must reach and maintain a temperature of at least 150°C (300°F) for 1 hour.
   - The cooker must be able to cook 1 cup of white rice.
   - The cooker must be safe and easy to use.
4. What constraints will your team be working within in addition to the budget constraint given below? Make a list.
   - The total cost of the materials used to build the solar cooker cannot be more than $5.

Developing and Using Models

5. Research existing solar cooker designs. What is the difference between a parabolic and box cooker?
6. Brainstorm your own ideas as a team.
7. Weigh the pros and cons of each design. Then select one solution for your team to pursue further.
8. Draw a labeled diagram of your solar cooker design. Does it meet all the design constraints from step 4?

Planning and Carrying Out Investigations

9. Plan a test of your solution using water.
10. Refine your solar cooker design to make sure it is safe to operate.
11. Gather your materials and build your solar cooker.
12. Plan and conduct a test of your design. Take periodic temperature measurements of the water and record your data in a table.

Analyzing Data and Using Math

13. Plot your data on a temperature versus time graph.
14. Analyze your data. Explain your findings.
15. What limitations exist in your data?
16. How well does your solution meet each of the criteria you listed in step 3?

Designing Solutions

17. Make an improvement to your design based on your findings. Record your changes.
18. Plan a test of your design using rice.
19. Carry out your test and record your findings.

Forming Arguments from Evidence

20. What are the strengths of your design? What are its weaknesses?
21. Optimize your design based on your findings and draw a new diagram.
22. Make a claim about the effectiveness of your solar cooker. Use evidence to support your claim.

Obtaining, Evaluating, and Communicating Information

23. Present your solar cooker to the class. Summarize your testing process, findings, and revisions.
24. Provide respectful and specific feedback to other teams.
25. Write a final report. Include a technical explanation of how wave interactions convert solar energy into thermal energy in your design. Consult a physical science resource if needed.

UNIT 4
Environmental Quality

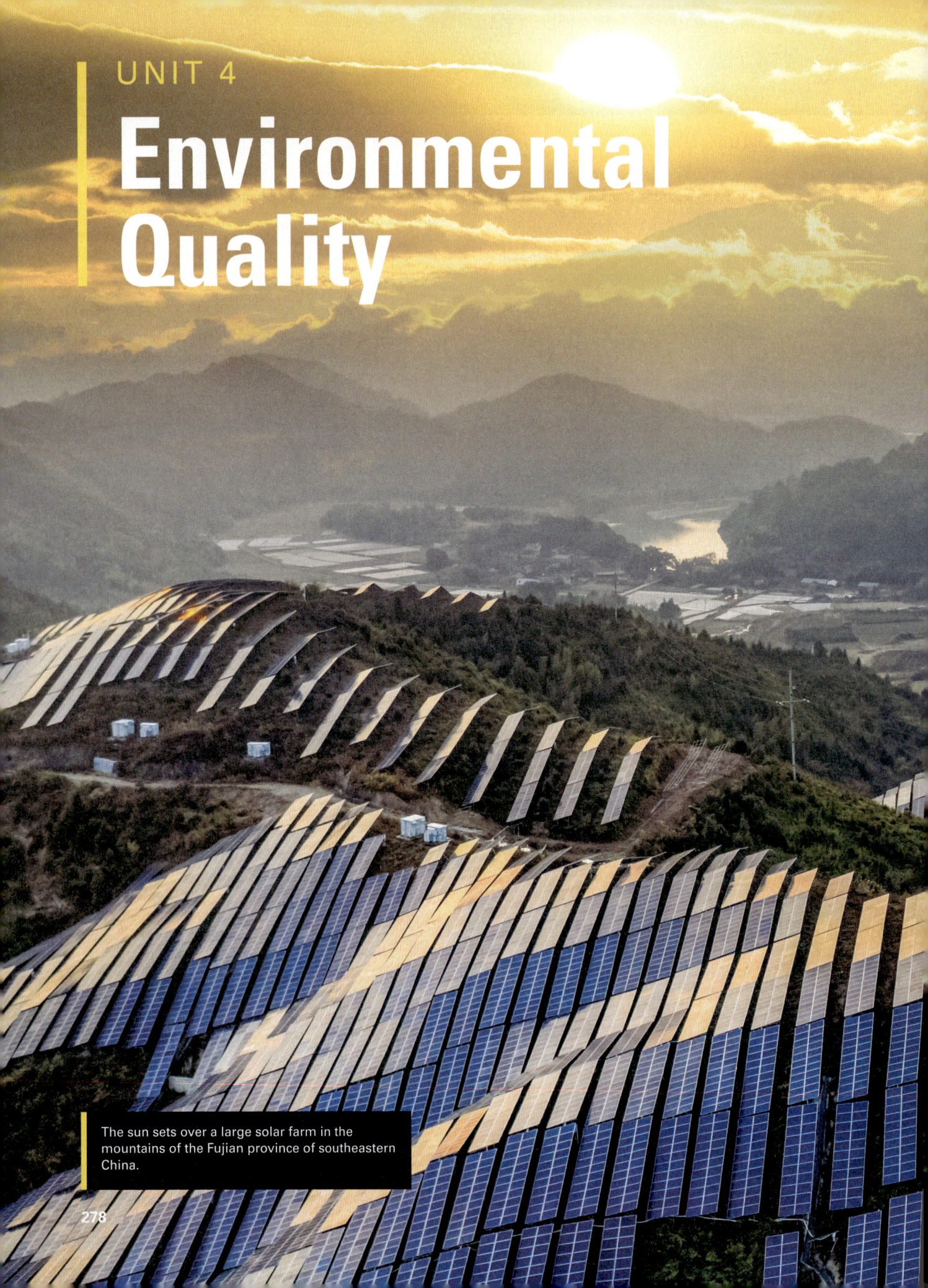

The sun sets over a large solar farm in the mountains of the Fujian province of southeastern China.

A major global shift has taken place in the 21st century as people begin to assess our relationship to and responsibility for the planet. While human activity has contributed to habitat loss and climate change, citizens and governments are increasingly on the search for alternative solutions to resource management and energy production.

In this unit, you will learn about the vast range of solutions being explored and deployed to preserve vital Earth systems. You will assess the benefits and costs of different proposals, and how realistically they may be employed at different scales.

CHAPTER 9
FOOD, SOIL, AND PEST MANAGEMENT

CHAPTER 10
WATER RESOURCES AND WATER POLLUTION

CHAPTER 11
GEOLOGY AND NONRENEWABLE MINERAL RESOURCES

CHAPTER 12
NONRENEWABLE ENERGY RESOURCES

CHAPTER 13
RENEWABLE ENERGY RESOURCES

ENGINEERING PROJECT 4: Design a Wind-Powered Generator

CHAPTER 9
FOOD, SOIL, AND PEST MANAGEMENT

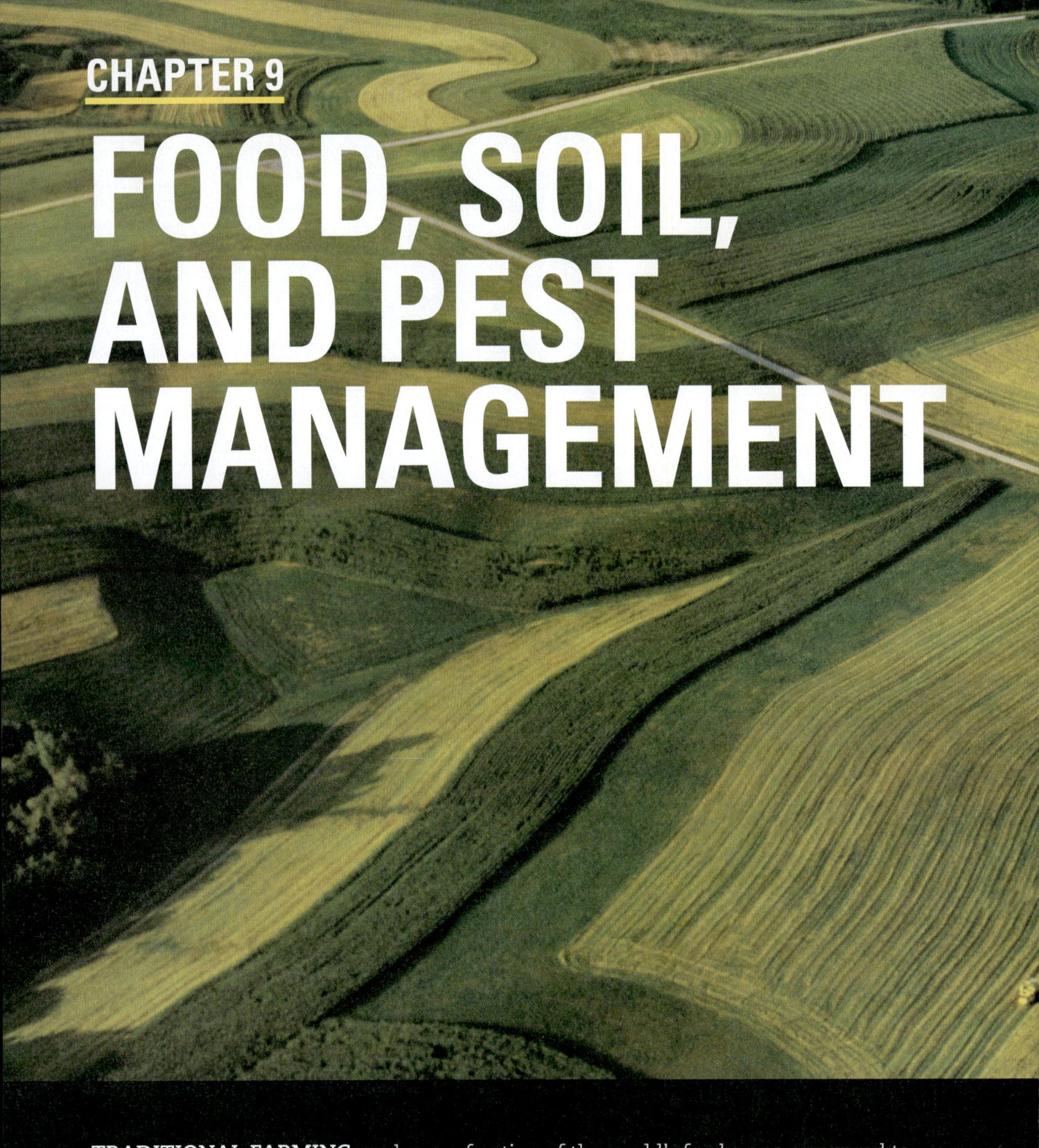

TRADITIONAL FARMING produces a fraction of the world's food crops compared to large, industrialized "agribusinesses," while using nearly three times as much of Earth's farmable land. However, industrialized farms require large inputs of resources and often produce large amounts of pollution. To sustainably feed the world's growing population, people will need to explore other solutions.

KEY QUESTIONS

9.1 What is food security?

9.2 How is food produced?

9.3 How are environmental issues and food production connected?

9.4 How can society manage agricultural pests more sustainably?

9.5 What are sustainable solutions for food production?

A farm in Wisconsin uses contour planting and strip cropping to reduce soil erosion and prevent irrigation waste.

NATIONAL GEOGRAPHIC | EXPLORERS AT WORK

Sustainable Agriculture: Is It Possible?

with National Geographic Explorer Jennifer Burney

Jennifer Burney has done the math and she doesn't like how it adds up. "We are a world of plenty, yet almost a billion people don't have enough to eat," she notes. "At the same time, the way we produce, distribute, and cook food contributes tremendously to climate change. Finding a way to lift the world out of perpetual hunger without creating an even worse climate crisis is an intricate, interwoven problem."

Dr. Burney is an assistant professor at the University of California, San Diego. Her fieldwork takes her to parched sub-Saharan farmland where farmers grow rain-fed crops on small plots of land and struggle to live on about a dollar a day. Crops, income, and nutrition are all at the mercy of the rain. Dry seasons are hungry seasons, lasting half the year or more.

In a sun-baked land, solar energy offers answers. Solar irrigation systems enable Africa's farmers to grow more fruits and vegetables, boosting nutrition and family incomes. The electricity that powers traditional motorized irrigation systems is too costly and the supply is too unreliable. Burney works with the Solar Electric Light Fund (SELF) in the country of Benin to implement and evaluate solar solutions.

"I design and evaluate village-scale, solar-powered, drip irrigation systems. These systems weaken rainfall dependence by boosting dry-season production of fruits and vegetables. Systems have now been installed in 10 villages, impacting thousands of lives," says Burney.

These systems combine the efficiency of drip irrigation with the reliability of solar-powered water pumps. Solar panels speed pumping on sunny days and slow it on cloudy days so crops always get the water they need. Farmers are now able to grow fruits and vegetables on a larger scale, improving food security as well as overall income. Families can now support access to safe drinking water and build schools and small businesses.

Finding ecologically sound ways to cook food is another project of Burney's. Traditional wood-burning cook stoves used in many less-developed countries create black carbon (soot). The sooty air inside homes causes respiratory infections. Outside, soot can have the same impact as greenhouse gases. In contrast, eco-stoves offer improved ventilation and efficiency by significantly limiting emissions and cutting fuel use. Burney hopes that the data she's collected will help encourage governments and development groups to invest in this green technology.

Thinking Critically
Evaluate Could solar energy applications be effective anywhere in the world? Why or why not?

Jennifer Burney works with farmers in Africa to implement solar irrigation systems while helping the farmers run their businesses.

CASE STUDY
Growing Power—An Urban Food Oasis

The USDA estimates that more than 38 million Americans, including 12 million children, are food insecure. They live in food deserts—urban neighborhoods or rural areas without easy access to nutritious food. In some instances, residents must rely on convenience stores and fast-food restaurants for their meals. A diet of such high-calorie, highly processed foods can lead to significant health issues.

Will Allen (Figure 9-1) was one of six children of a sharecropper and grew up on a farm in Maryland. He left farm life for college and professional basketball, followed by a successful corporate marketing career. In 1993, Allen decided to return to his roots. He bought the last working farm within the city limits of Milwaukee, Wisconsin, and created a food oasis in a food desert.

On this small urban plot, Allen developed Growing Power, Inc. As an ecologically based farm, it showcases forms of agriculture that apply all three scientific factors of sustainability. Allen's farm is powered partly by solar electricity and solar hot water systems and makes use of several greenhouses to capture solar energy for growing food year-round. The farm produces an amazing diversity of crops with about 150 varieties of organic produce. It also produces organically raised chickens, turkeys, goats, fish, and honeybees. At the farm, nutrients are recycled in creative ways. For example, wastes from the farmed fish provide nutrients for some of the crops.

The farm's products are sold locally at Growing Power farm stands throughout the region as well as to restaurants. Allen worked with the City of Milwaukee to establish the Farm-to-City Market Basket program through which people can sign up for weekly deliveries of organic produce at modest prices.

Growing Power's efforts are far-reaching. Schoolchildren visit the farm and learn about where food comes from. Allen also trains about 1,000 people every year who want to learn organic farming methods. The farm partnered with the City of Milwaukee to create 150 new green jobs for unemployed and low-income workers, such as building greenhouses and growing food organically. Growing Power has expanded, opening an urban farm in a neighborhood of Chicago, Illinois, and setting up training sites in five other states.

This chapter explores several conventional methods for producing food, the environmental effects of food production, and how people can produce food more sustainably. It begins by examining the importance of making healthy food available for everyone.

As You Read Think about the foods you eat and where they come from. Consider what impact producing those foods might have on the environment.

◀ FIGURE 9-1
Will Allen has won several prestigious awards for his efforts through Growing Power, Inc. He is most proud of the fact that his urban farm helps feed more than 10,000 people every year and puts people to work raising good food.

9.1 What Is Food Security?

CORE IDEAS AND SKILLS
- Explain the concepts of food security and insecurity.
- Identify the main causes of food insecurity.
- Discuss the consequences of food insecurity on human health.
- Explore ways to improve food security.

KEY TERMS

food security, food insecurity, chronic malnutrition, overnutrition, subsidy

Poverty Is the Root of Food Insecurity

Over a third of Earth's ice-free land is devoted to growing food crops. More than 1 billion people work in agriculture to cultivate these resources. The resulting crops produce more than enough food to meet the basic nutritional needs of every person on the planet. This food surplus should guarantee **food security**, or access to enough safe, nutritious foods for a healthy and active lifestyle. Despite the surplus of production, one out of every nine people in the world faces **food insecurity**, or a lack of access to nutritious food.

Most agricultural experts agree that the root cause of food insecurity is poverty, which prevents poor people from growing, buying, or accessing enough food to meet their needs. This is not surprising given that three out of ten people in the world are trying to survive on the equivalent of less than $3.10 a day.

Food security is also tied to environmental, political, and economic issues. Obstacles to food security include war, corruption, political upheaval, climate change, and bad weather (such as prolonged drought, flooding, and heat waves). Another reason for food insecurity is that not all calories from grain production end up as food for people. About 33% of the calories are used to produce meat that many poor people cannot afford, 5% are used to produce biofuels for cars, and another 33% are wasted.

Also, as the world population continues to grow, many people live in the major cities of less-developed countries where they are isolated from food supplies. Humanity faces the challenge of feeding a potential 9.7 billion people by 2050 without causing serious harm to the environment.

checkpoint What causes food insecurity?

Many People Suffer from Lasting Hunger and Malnutrition

To maintain good health and resist disease, people need certain amounts of nutrients. People need relatively large amounts of macronutrients, such as carbohydrates, proteins, and fats. People need smaller amounts of micronutrients—vitamins, such as A, B, C, and E, and minerals such as iron, iodine, and calcium.

Hunger and Malnutrition People who cannot grow or buy enough food to meet their basic energy needs suffer from chronic undernutrition, or hunger. Most of the world's hungry people live in low-income, less-developed countries. They typically can afford only a low-protein, high-carbohydrate, vegetarian diet consisting mainly of grains such as wheat, rice, and corn. In more-developed countries, people living in food deserts (Figure 9-2) have a similar problem. Their diet is heavy on cheap food loaded with fats, sugar, and salt but lacking in nutrition (Figure 9-3). In both cases, people often suffer from **chronic malnutrition**, a condition in which they do not get enough protein and other key nutrients. Chronic malnutrition can make people more vulnerable to diseases and hinder the normal physical and mental development of children. About one of every eight people on the planet is undernourished or malnourished.

Lack of Vitamins and Minerals About 2 billion people, most of them in less-developed countries, suffer from a deficiency of one or more vitamins and minerals—usually vitamin A, iron, and iodine. Micronutrient deficiencies can lead to health issues such as blindness, intellectual disability, and stunted growth. Fortunately, many of these problems have potential solutions, given enough logistical, political, and financial support.

checkpoint How do food deserts impact the health of Americans?

Health Problems Arise from Overeating

Overnutrition occurs when food energy intake exceeds energy use and causes excess body fat. Too many calories, too little exercise, or both can cause overnutrition. Diets high in calories and processed foods can lead to higher rates of obesity. In the United States, about 73% of adults over age 20 and 33% of all children are overweight or obese.

FIGURE 9-2 ▼

Food Deserts People living in food deserts in urban areas like Houston, Texas, do not have access to nutritious food.

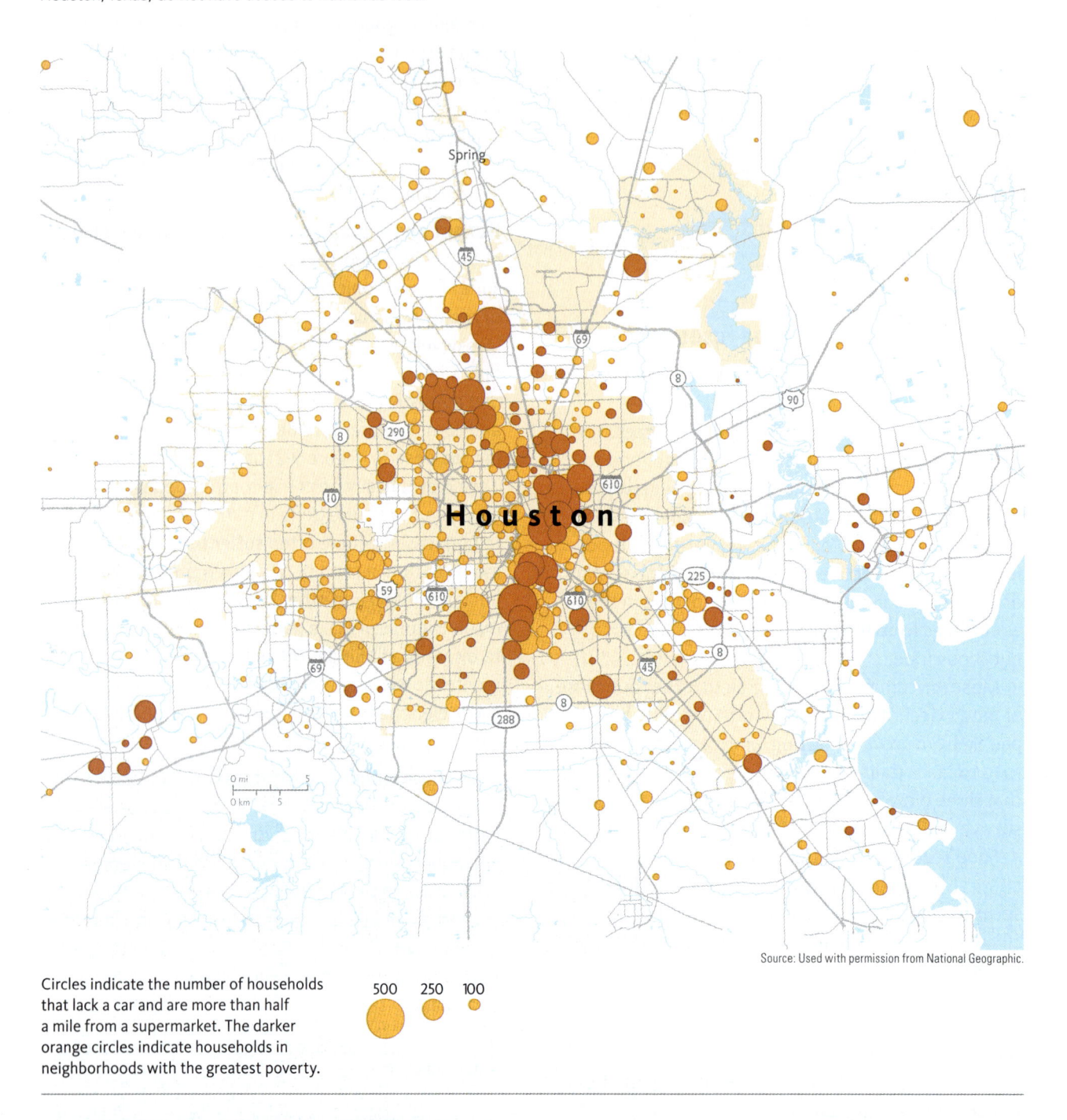

Circles indicate the number of households that lack a car and are more than half a mile from a supermarket. The darker orange circles indicate households in neighborhoods with the greatest poverty.

500 250 100

Source: Used with permission from National Geographic.

People who are underfed and underweight and those who are overfed and overweight face similar health problems: lower life expectancy, greater susceptibility to disease and illness, and lower productivity and quality of life. In addition to the 795 million people who face health problems because they are undernourished, another 2.1 billion have health problems because they eat too much.

Overnutrition plays a role in heart disease, stroke, type 2 diabetes, and some forms of cancer. The World Health Organization (WHO) estimates that obesity contributes to 2.8 million deaths every year.

checkpoint How can a person be overweight and malnourished at the same time?

286   CHAPTER 9   FOOD, SOIL, AND PEST MANAGEMENT

FIGURE 9-3

Calorie Sources People living in food deserts may get their calories from sugary drinks, snacks, and other foods lacking key macro- and micronutrients.

Top ten sources of calories for low-income individuals
Age two and older, per person per day

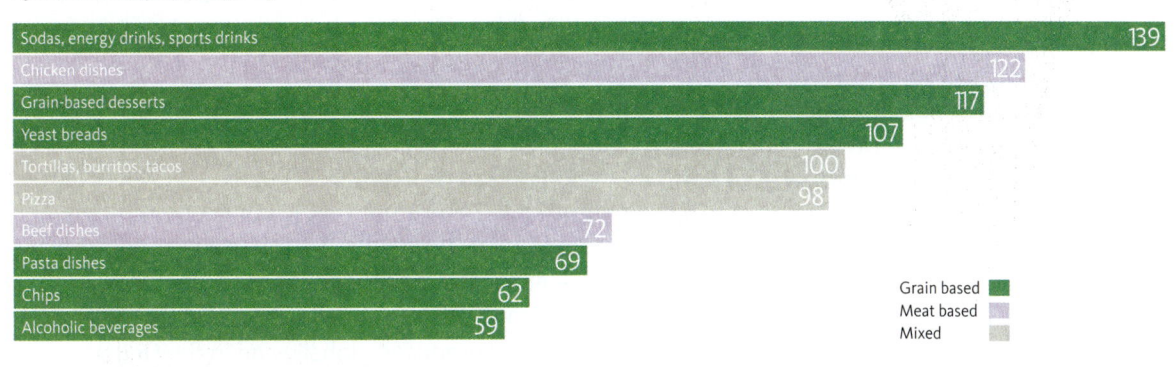

Sodas, energy drinks, sports drinks	139
Chicken dishes	122
Grain-based desserts	117
Yeast breads	107
Tortillas, burritos, tacos	100
Pizza	98
Beef dishes	72
Pasta dishes	69
Chips	62
Alcoholic beverages	59

Grain based ■
Meat based ■
Mixed ■

Source: Used with permission from National Geographic.

Combined Efforts Improve Food Security

Multiple combined efforts can lead to improved food security worldwide. Government policies can support and improve food production. Public and private programs can work to reduce poverty and malnutrition. Education can lead to healthier eating habits and more sustainable food production. Individuals can take action and make choices that support global food security.

Government Subsidies Agriculture is a financially risky business. Whether farmers have a good or bad year depends on factors over which they have little control, including weather, crop prices, pests and diseases, interest rates on loans, and global food markets.

Governments use two main approaches to influence food production. One approach is to place a legally mandated upper limit on food prices in order to keep them artificially low. This helps consumers but farmers may find it harder to earn a living.

Governments also can subsidize farming. **Subsidies** are forms of government support intended to help businesses—like farms—survive. Farm subsidies can be in the form of price supports, tax breaks, and other financial support that encourages farmers to increase food production (Figure 9-4). If, however, government subsidies are too generous and farmers face a year with few challenges, they may produce more food than they can sell.

Some opponents of subsidies call for ending such programs. Some traditional subsidies encourage environmental degradation, such as overfishing and soil erosion. Other analysts would like to see a shift to subsidies that promote more sustainable food production.

Other Efforts Many organizations help individuals, communities, and nations produce food more sustainably. One effort is the transfer of technologies to less-developed nations. National Geographic Explorer Jennifer Burney (Explorers at Work) is helping farmers in Africa implement more sustainable forms of irrigation that rely on renewable energy rather than fossil fuels. Others believe that some researchers should help less-developed nations develop simple, sustainable, local food production and distribution systems. Growing Power's Will Allen (Case Study) believes this type of solution gives farmers more control over their food security.

Still other groups are addressing nutrition problems (Figure 9-5). Some health problems from malnutrition could be avoided through simple, inexpensive programs. For example, one study indicates that more than half of nutrition-related childhood deaths could be prevented at an average annual cost of $5–$10 per child. This involves simple measures such as immunizing more children against childhood diseases, preventing dehydration due to diarrhea, and combating blindness by giving children an inexpensive vitamin A capsule twice a year. Another approach involves research and engineering. One bowl of a genetically modified strain of rice, called "golden rice," could provide 60% of the daily requirement of vitamin A.

LESSON 9.1

The Food and Agriculture Organization (FAO) and the WHO estimate that adding traces of iodine to salt would cost the equivalent of only two to three cents per year for every person in the world and could potentially eliminate goiter (a thyroid condition that can lead to deafness).

Individual Actions A growing number of consumers are becoming locavores. That means they buy as much of their food as possible from local and regional producers. Many people also participate in community-supported agriculture (CSA) programs. In these programs, people buy shares of a local farmer's crop and receive a box of fresh, seasonal produce on a regular basis. In some communities, mobile grocers-on-wheels businesses bring fresh and healthy food to residents who live in food deserts.

An increase in the demand for locally grown food results in more small, diversified farms and minimally processed foods. Buying locally grown food reduces the fossil fuel energy costs for food producers, as well as the greenhouse gas emissions from storing and transporting food products over long distances.

Approximately 15% of the world's food is grown in urban areas today, and this percentage could easily be doubled. More people are planting gardens and raising small livestock, like chickens and goats, in urban and suburban backyards. Others are establishing community gardens in vacant lots and building raised garden beds in parking lots—a practice known as asphalt gardening. People also grow small gardens in containers on rooftops, balconies, and patios at home and at work. One report suggests that converting 10% of American lawns into food-producing gardens would supply one-third of the country's fresh produce.

A major strategy to reduce food insecurity is to cut food waste. Studies have found that 33% of all food produced globally and in the United States is lost during production or thrown away. Educating farmers, retailers, and consumers about how to reduce food waste should be the first step in addressing this issue.

checkpoint Name three strategies to address food insecurity.

FIGURE 9-4

Farm Subsidies This figure shows the breakdown of federal subsidies for farm products in the United States.

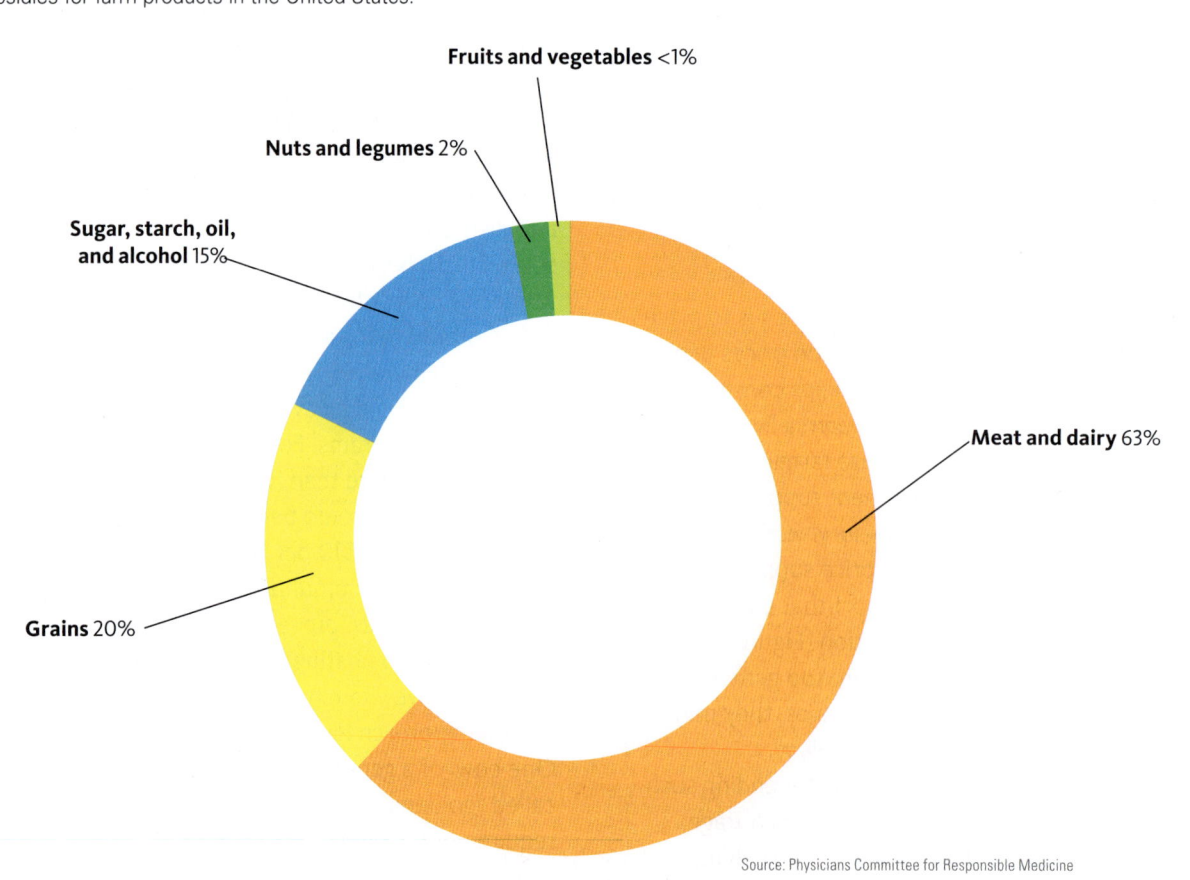

- Fruits and vegetables <1%
- Nuts and legumes 2%
- Sugar, starch, oil, and alcohol 15%
- Grains 20%
- Meat and dairy 63%

Source: Physicians Committee for Responsible Medicine

288 CHAPTER 9 FOOD, SOIL, AND PEST MANAGEMENT

FIGURE 9-5
Public and private programs, like this soup kitchen in the U.S. Midwest, may be the only source of nutritious food for some families.

9.1 Assessment

1. **Recall** What are three health consequences of malnutrition?

2. **Compare and Contrast** How do you think your food sources compare to the breakdown shown in Figure 9-3?

3. **Explain** How does grain used for meat production and biofuel contribute to food insecurity?

4. **Draw Conclusions** Do you think the challenges society faces with regard to feeding the world's growing population is an example of the concept of ecological carrying capacity? Why or why not?

CROSSCUTTING CONCEPTS

5. **Cause and Effect** How could political, environmental, and economic issues impact food security?

9.2 How Is Food Produced?

CORE IDEAS AND SKILLS
- Identify three major systems used to produce food.
- Describe the types of agriculture that use croplands.
- Understand the pros and cons of crossbreeding and genetic engineering in food production.

KEY TERMS

industrialized agriculture
monoculture
traditional agriculture
polyculture
organic agriculture
green revolution
aquaculture

Food Production Has Increased Dramatically

Three systems supply most of the world's food. Croplands produce grains such as rice, wheat, and corn. Croplands account for about 77% of the world's food. A second system produces meat and meat products. This system includes rangelands, pastures, and feedlots. The third system supplies fish and shellfish and accounts for most of the remaining food supply. Since 1960, there has been a staggering increase in global food production from all three of the major food production systems.

These three systems depend on a small number of plant and animal species. Of the estimated 50,000 edible plant species, only 14 supply most of the world's food calories. About two-thirds of the world's population survive primarily by eating three grain crops—rice, wheat, and corn—because they cannot afford meat and do not have access to fresh produce. Only a few species of mammals and fish provide most of the world's meat and seafood.

Such food specialization makes humans vulnerable. If any of the small number of crop strains, livestock breeds, and fish and shellfish species that people depend on were to become depleted, the consequences would be severe. This could occur because of disease, environmental degradation, or climate change. Food specialization violates the biodiversity factor of sustainability, which calls for depending on a variety of food sources as an ecological insurance policy against changing environmental conditions.

checkpoint What categories of food do the world's three food systems produce?

Land-Based Agriculture

Farming on croplands is the world's major food-producing system. Together, industrialized and conventional agriculture make up for the vast majority of the world's food production. Organic agriculture produces a small percentage of the world's food but is growing as an important component of sustainable agriculture.

Industrialized Agriculture **Industrialized agriculture**, or high-input agriculture, uses motorized equipment along with large amounts of financial capital, fossil fuels, water, commercial inorganic fertilizers, and pesticides (Figure 9-6). Industrialized agriculture produces a single crop at a time, a practice known as **monoculture**. The major goal of industrialized agriculture is to steadily increase each crop's yield—the amount of food produced per unit of land. Industrialized agriculture is practiced on 25% of all cropland, mostly in more-developed countries, and produces about 80% of the world's food.

In the United States, industrialized farming has evolved into agribusinesses. A few giant, multinational corporations increasingly control the growing, processing, distribution, and sale of food at home and abroad. Since 1960, U.S. industrialized agriculture has more than doubled the yields of key cash crops (crops grown for profit, such as wheat, corn, and soybeans) without cultivating more land. Increased yields have saved large areas of U.S. forests, grasslands, and wetlands from being converted to farmland.

Plantation agriculture is a form of industrialized agriculture used primarily in less-developed tropical countries. It involves growing cash crops such as bananas, coffee, soybeans (mostly to feed livestock), sugarcane (for sugar and ethanol fuel), and palm oil (for cooking oil and biodiesel fuel). These crops are grown on large monoculture plantations, mostly for export to more-developed countries.

Traditional Agriculture **Traditional agriculture**, or conventional agriculture, is low-input compared with industrialized agriculture. Traditional agriculture provides about 20% of the world's food crops on about 75% of its cultivated land, mostly in less-developed countries (Figure 9-7). Traditional subsistence agriculture combines energy from the sun with the labor of humans and draft animals to produce enough crops for a farm family's survival, with little left over to sell or store as a reserve.

In traditional intensive agriculture, farmers try to obtain higher crop yields by increasing their inputs of human and draft-animal labor, animal manure for fertilizer, and water. If weather cooperates, farmers can produce enough food to feed their families and have some left over to sell for income.

Some traditional farmers focus on cultivating a single crop, but many grow several crops on the same plot simultaneously. This practice is known as **polyculture**. Polyculture relies on solar energy and natural fertilizers such as animal manure. Because the various crops mature at different times, polyculture provides food year-round and keeps the topsoil covered, reducing erosion from wind and water. This method also lessens the need for fertilizer and water because root systems at different depths in the soil capture nutrients and moisture efficiently.

Polyculture is an application of the biodiversity factor of sustainability. Crop diversity helps protect and replenish the soil and reduces the chance of losing most or all of the year's food supply to pests, bad weather, and other misfortunes. Research shows that low-input polyculture produces higher average yields than high-input industrialized monoculture, while using less energy and fewer resources. Low-input, high-yield polyculture generally leads to greater food security for small landowners. The Growing Power farm (Case Study) practices polyculture by growing a variety of crops in inexpensive greenhouses—applying the solar energy *and* biodiversity factors of sustainability.

Organic Agriculture A fast-growing sector of U.S. and world food production is **organic agriculture**. Organic crops are grown without the use of synthetic pesticides, synthetic inorganic fertilizers, and genetically engineered varieties. Animals are raised on 100% organic feed without the use of antibiotics or growth hormones.

In the United States, by law, a label of 100% organic (or USDA certified organic) means that a food product is produced only by organic methods, contains all organic ingredients, and has undergone a certification process (Figure 9-8). About 13,000 of the 2.2 million farms in the United States are USDA certified organic. Products labeled "organic" must contain at least 95% organic ingredients. Those labeled "made with organic ingredients" must contain at least 70% organic ingredients. The food label "natural" is not associated with any certification.

FIGURE 9-6
This industrial farm relies on expensive equipment and uses large amounts of seed, manufactured inorganic fertilizer, pesticides, and fossil fuels to produce crops.

FIGURE 9-7
ON ASSIGNMENT National Geographic photographer Robin Hammond captured this view of banana plantations on a corporate farm bordering small traditional farms in Mozambique, Africa.

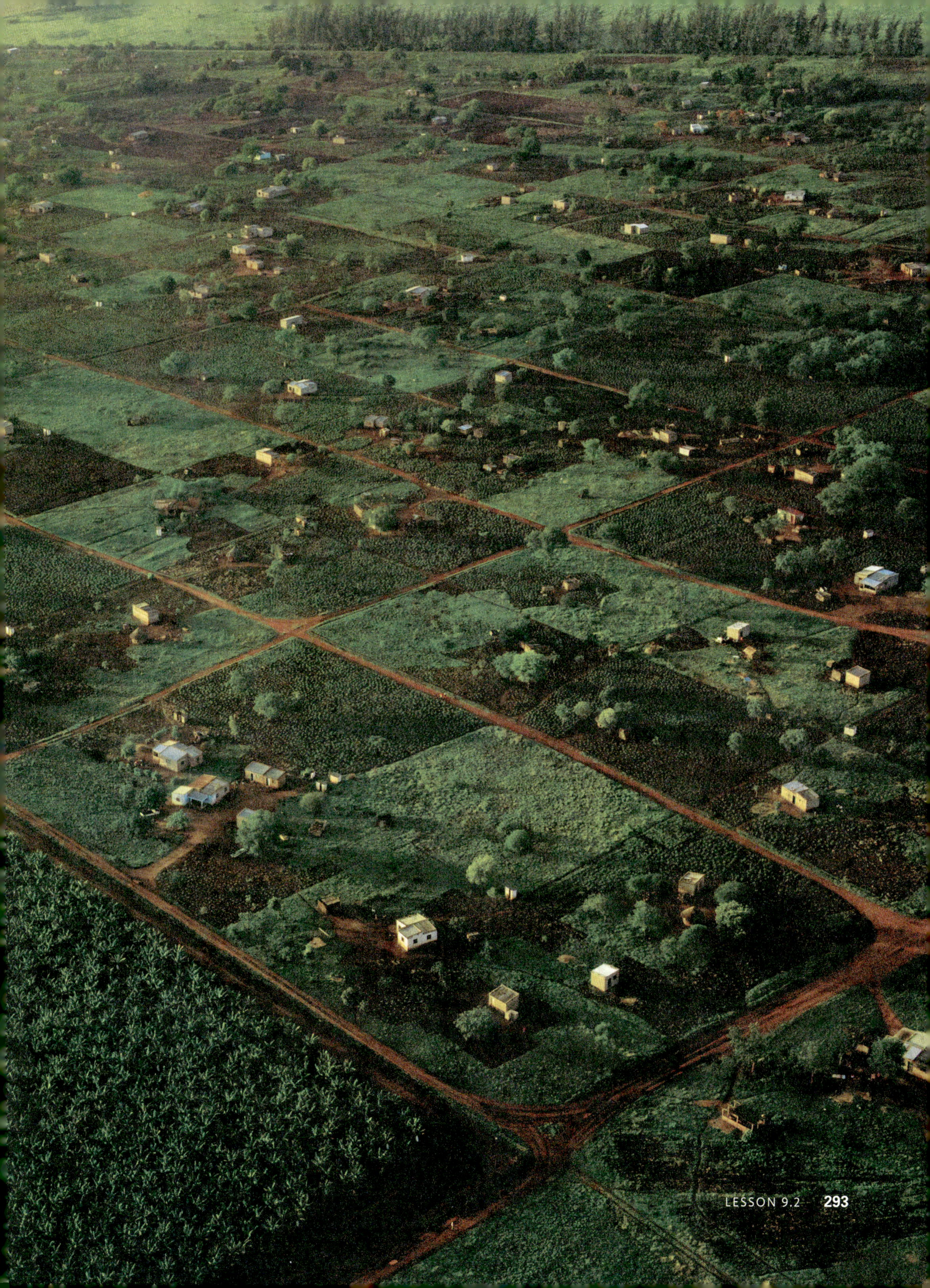

Research shows that organic farming has a number of environmental advantages over industrialized farming. One advantage is that eating only certified organic foods reduces exposure to pesticide residues and to bacteria-resistant antibiotics that can be found in some nonorganic foods.

On the other hand, conventional agriculture can usually produce higher yields from monoculture crops on smaller areas of land than organic agriculture can. Another drawback is that most organically grown food costs more than conventionally produced food, primarily because organic farming is more labor intensive. However, some farms, like Growing Power (Case Study), are able to use sustainable farming methods to get higher yields of a variety of organic crops at affordable prices.

checkpoint What are the three types of land-based agriculture?

Green Revolutions Increase Yields

Farmers have two ways to produce more food: farm more land or increase yields from existing cropland. Since 1950, most of the increase in global grain production has been the result of increased yields from high-input agriculture. The **green revolution** describes the process of increasing crop yields through industrialized agriculture. The first step in this process is to develop and plant selectively bred or genetically engineered varieties of key grain crops such as rice, wheat, and corn. Next, farmers use large inputs of water, synthetic inorganic fertilizers, and pesticides to increase yields from the planted crops. Finally, farmers increase the number of crops grown per year on a plot of land through multiple cropping.

In the first green revolution, which occurred between 1950 and 1970, this high-input approach dramatically raised crop yields in most of the world's more-developed countries, especially the United States. In the second green revolution, which began in 1967, fast-growing varieties of rice and wheat, specially bred for tropical and subtropical climates, were introduced into middle-income, less-developed countries, including India, Brazil, and China (Figure 9-9).

Largely because of the two green revolutions, world grain production increased by 312% and per capita grain production grew by 37% between 1950 and 2014. The world's three largest grain-producing countries—China, India, and the United States—produce almost half of the world's grains.

FIGURE 9-8
Certified organic labels empower consumers to "vote with their fork" by purchasing more sustainable foods.

Since 1960 there has been a staggering increase in global food production. This was made possible by technological advances such as greater use of tractors and other farm machinery and high-tech fishing equipment. Another major advancement has been improved irrigation. Other technological developments include increased use of manufactured inorganic fertilizers and pesticides.

CONSIDER THIS

Consider pesticides when choosing food.
According to the Environmental Working Group (EWG), a research organization that helped pass the Food Quality Protection Act, individuals could reduce pesticide intake by up to 90% by not eating 12 types of conventionally grown fruits and vegetables. These 12 foods, which EWG has nicknamed the "dirty dozen," tend to have the highest pesticide residues. In 2022, these foods were strawberries, apples, nectarines, peaches, celery, grapes, cherries, spinach, tomatoes, bell and hot peppers, pears, and kale, collard, and mustard greens.

FIGURE 9-9
Improved varieties of rice have led to increased yields, helping rice production keep up with population growth thus far.

Limits to the Expansion of Green Revolutions

So far, several factors have limited the success of the green revolutions and may limit them even more in the future. For example, without large inputs of water and synthetic inorganic fertilizers and pesticides, most green-revolution and genetically engineered crop varieties produce yields that are no higher (and are sometimes lower) than those from traditional strains. Climate change and increasing world population also limit the success of green revolutions, as does cost. The high inputs that define green revolutions cost too much for most subsistence farmers in less-developed countries.

Scientists point out that where such inputs do increase yields, there comes a point where yields stop growing because of the inability of crop plants to take up nutrients from additional fertilizer and irrigation water. This helps to explain the slowdown in the rate of growth in global yields for most major crops since 1990.

Since 1978, the amount of irrigated land per person has been declining and it is projected to fall much more by 2050. One reason for this per-person decline is population growth. Other factors that contribute to this decline are the limited availability of irrigation water, buildup of salts in soil, and limited financial means of most farmers to irrigate crops. In addition, projected climate change during this century is likely to melt some of the mountain glaciers that provide irrigation and drinking water for many millions of people in China, India, and South America.

Farming already uses about 38% of Earth's ice-free land surface for croplands and pastures. Clearing tropical rain forests and irrigating arid land could more than double the area of the world's cropland. However, massive clearing of forests would contribute significantly to climate change, topsoil erosion, and biodiversity loss. In addition, many of the forests and arid lands that could be used for farming have poor soil fertility, steep slopes, or both. Cultivating such land would be expensive and probably ecologically unsustainable.

During the 21st century, fertile croplands in coastal areas are likely to be flooded by rising sea levels resulting from projected climate change. These areas include many of the major rice-growing floodplains and river deltas in Asia that contribute a significant portion of calories to the world's poorest people. Food production could also drop sharply in some major food-producing areas because of the longer and more intense droughts and heat waves that are very likely to result from climate change.

checkpoint What factors could limit the expansion of green revolutions?

Industrialized Food Production

People directly consume about half of the world's grain production (Figure 9-10). Most of the rest of the grain is fed to livestock and thus is indirectly consumed by people who can afford to eat meat products and purchase biofuels such as ethanol and biodiesel. About 40% of the U.S. corn crop is currently used to produce biofuels (mostly energy-inefficient ethanol). Low-income people in less-developed countries typically spend most of their income on food. By contrast, Americans spend only an average of 9% of their disposable income on food.

There are a number of hidden costs related to food production and consumption. Most American consumers are not aware that their actual food costs are much higher than the market prices they pay. Such hidden costs include the costs of pollution, environmental degradation, and higher health insurance bills related to the harmful environmental and health effects of industrialized agriculture. Other hidden costs of food include the taxes to pay for farm subsidies (Figure 9-4).

checkpoint How are global crops consumed and used?

FIGURE 9-10

Crop Destinies About half of the world's croplands produce food that goes directly to people's plates (green). The other half produces feed for livestock and fuel for vehicles (purple).

Where the calories are produced
FEED and FUEL for animals and industry — FOOD for people
100% calories — 50% — 100%

How global crop calories are used

45% FEED (36%) and FUEL* (9%)   **55%** FOOD

Source: Used with permission from National Geographic.

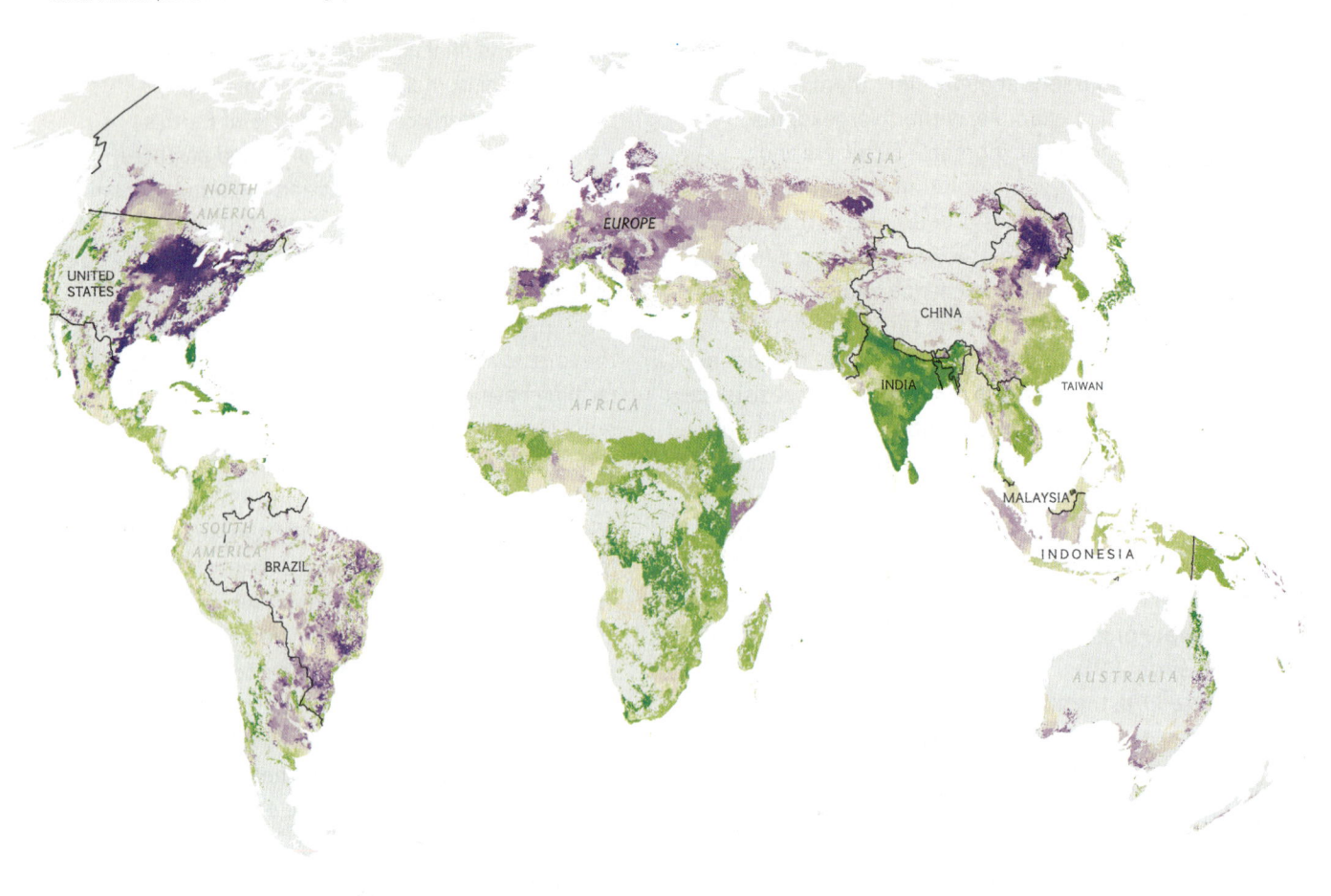

Scientists Produce New Varieties of Food

For centuries, farmers and scientists have used crossbreeding to produce new varieties of food. Through artificial selection, farmers develop genetically improved varieties of crops and livestock animals. Such selective breeding in this first "gene revolution" has yielded amazing results. For example, ancient ears of corn were about the size of a finger and wild tomatoes were once the size of grapes. Most of the large varieties used now were selectively bred for specific desirable traits.

Traditional crossbreeding is a slow process. It can take 15 years or more to produce a commercially valuable new crop. Crossbreeding can only combine traits from species that are genetically similar. Typically, resulting varieties remain useful for only five to ten years before pests and diseases reduce their yields. Important advances are still being made in this field.

Today a second gene revolution is taking place. Scientists and engineers are developing genetically modified strains of crops and livestock animals. They use a process called gene splicing to add, delete, or change segments of an organism's DNA and thus alter its genetic material. The goal of genetic modification is to add desirable traits or eliminate undesirable ones by transferring genes between species that would not normally interbreed in nature. The resulting organisms are called genetically modified organisms, or GMOs (Figure 9-11).

Compared with traditional crossbreeding, developing a new crop variety through gene splicing takes about half as long and usually costs less. It allows for the insertion of genes from almost any organism into crop or animal cells. According to the U.S. Department of Agriculture (USDA), at least 80% of the food products on U.S. supermarket shelves contain some form of genetically engineered food or ingredients, and that percentage is growing.

FIGURE 9-11
Scientists test new, genetically modified strains of rice for their tolerance to salt water. This rice variety was engineered with rising sea levels, a result of climate change, in mind.

FIGURE 9-12
ON ASSIGNMENT Feedlots like this cattle farm require high inputs of energy and resources and produce large amounts of waste. National Geographic photographer Brian Finke captured this image of the nearly 50,000 cows present at this farm in Tulia, Texas.

> **CONSIDER THIS**
>
> **GM crops influence organic food prices.**
> The spread of GM (genetically modified) crop genes by windborne pollen from field to field threatens the production of certified organic crops. Organic crops must be grown without genetically modified genes to be officially certified. Organic farmers have to perform expensive tests to detect GMOs. They may have to take costly planting measures to prevent the spread of GMOs to their fields from nearby farms. This results in higher production costs that must be passed on to consumers.

On January 1, 2022, the United States became the sixty-fifth country to require the labeling of food with genetically modified ingredients. Polling has shown that such laws are very popular with consumers.

Bioengineers hope to develop new GMO varieties of crops that are resistant to heat, cold, drought, insect pests, parasites, viral diseases, herbicides, and salty or acidic soil. They also hope to develop crop plants that can grow faster and survive with little or no irrigation and with less use of fertilizer and pesticides. Some scientists think that such innovations could improve global food security.

Controversy over GMOs Some critics of GMOs argue that GMOs would allow a small number of seed companies to patent genetically modified crops and control most of the world's food production, and therefore food prices. Other critics point out that most of the GM crops developed so far have provided very few benefits. In some cases, controversy exists because some herbicide-resistant genetically engineered crops have led to increased herbicide use and to herbicide-resistant superweeds.

checkpoint What is a GMO and why are GM foods controversial?

Meat Consumption Has Grown Steadily

Meat and animal products such as eggs and milk are sources of high-quality protein and represent the world's second major food-producing system. Between 1950 and 2014, according to the FAO, production of these foods—mostly beef, pork, poultry, mutton, milk, cheese, and eggs—grew more than sixfold. Meanwhile, the average per-person consumption of these foods more than doubled. Global meat consumption is likely to more than double again by 2050 as incomes rise in rapidly developing countries.

About half of the world's meat comes from livestock grazing on grass in unfenced rangelands and enclosed pastures. The other half is produced through an industrialized factory farm system. This involves raising large numbers of animals bred to gain weight quickly, mostly in crowded feedlots or in crowded pens and cages in huge buildings. These operations are called concentrated animal feeding operations, or CAFOs (Figure 9-12).

In CAFOs, the animals are fed grain, soybeans, fishmeal, or fish oil, and some of this feed contains growth hormones and antibiotics to accelerate livestock growth. Because of the crowding and runoff of animal wastes, feedlots and CAFOs can have serious impacts on the air and water. These impacts are examined later in this chapter.

checkpoint Why are animals in CAFO facilities fed growth hormones and antibiotics?

Fish and Shellfish Production Has Risen

The world's third major food-producing system consists of fisheries and aquaculture. A fishery is a concentration of particular aquatic species suitable for commercial harvesting in a given ocean area or inland body of water. Industrial fishing fleets use a variety of methods to harvest most of the world's marine catch

> **CONSIDER THIS**
>
> **Meat production leads to dead zones.**
> Huge amounts of synthetic inorganic fertilizers are used in the midwestern United States to produce corn for animal feed and ethanol fuel for cars. Much of this fertilizer washes from croplands into the Mississippi River. The added nitrate and phosphate nutrients over-fertilize coastal waters in the Gulf of Mexico, where the river flows into the ocean. Each year this creates an oxygen-depleted "dead zone" that is often larger than the U.S. state of Massachusetts. Dead zones threaten one-fifth of the nation's seafood yield and negatively impact marine ecosystems. Dead zones are an example of how agricultural practices in one part of the world can have far-reaching impacts on the environment.

FIGURE 9-13
ON ASSIGNMENT National Geographic photographer Brian Skerry captured this image of a diver at an underwater, sustainable scallop farm. This operation is sustainable in part because scallops are filter feeders and require very little input of food.

of wild fish.

Fish and shellfish are also produced through aquaculture. **Aquaculture** is the practice of raising fish and shellfish for human consumption and is also known as fish farming. Aquaculture is the world's fastest-growing type of food production (Figure 9-14). In 2014, aquaculture accounted for 42% of the world's fish and shellfish production. The World Bank estimates that by 2030 aquaculture could produce 62% of all seafood. Aquaculture operations are built in freshwater ponds, lakes, reservoirs, and rice paddies, and even in underwater cages in coastal and deeper ocean waters (Figure 9-13).

Most of the world's aquaculture involves raising species that feed on algae or other plants—mainly carp in China and India, catfish in the United States, tilapia in several countries, and shellfish in a number of coastal countries. However, the farming of meat-eating species such as shrimp and salmon is growing rapidly, especially in more-developed countries. Such species are often fed fishmeal and fish oil produced from other fish and their wastes.

checkpoint How has the production of aquaculture compared to wild fisheries changed since 1950?

9.2 Assessment

1. **Compare and Contrast** What are some similarities and differences between monoculture and polyculture agriculture?
2. **Explain** What are GMOs and how are they produced?
3. **Infer** How does biodiversity in food species contribute to food security?

CROSSCUTTING CONCEPTS

4. **Energy and Matter** Recall from Chapter 3 how energy transfers between trophic levels. Why is more energy required to produce meat and fish products than to produce grains, vegetables, and other plant crops?

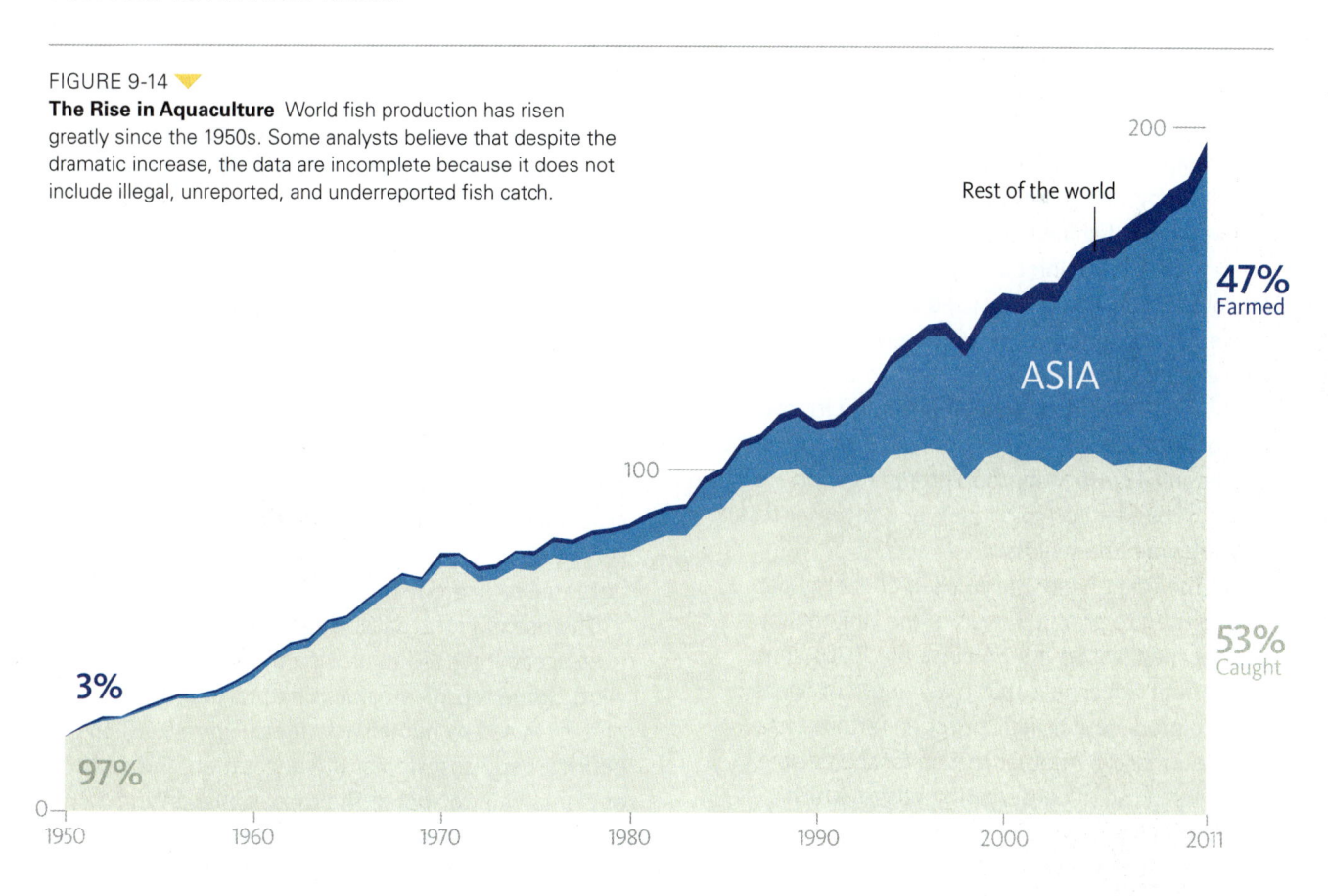

FIGURE 9-14 ▼
The Rise in Aquaculture World fish production has risen greatly since the 1950s. Some analysts believe that despite the dramatic increase, the data are incomplete because it does not include illegal, unreported, and underreported fish catch.

Source: Used with permission from National Geographic.

9.3 How Are Environmental Issues and Food Production Connected?

CORE IDEAS AND SKILLS
- Describe how food production contributes to environmental issues.
- Understand how environmental issues impact food production and, therefore, food security.

KEY TERMS
topsoil
desertification
soil salinization
agrobiodiversity

Food Production Requires Energy

The industrialization of food production has been made possible by using fossil fuels. Energy—mostly from oil and natural gas—is required to run farm machinery and fishing vessels, to pump irrigation water for crops, and to produce synthetic pesticides and synthetic inorganic fertilizers. All together, agriculture accounts for about 17% of all energy use in the United States—more than any other industry. Food processing and transport contribute significantly to that statistic. In the United States, for example, food items travel an average of 2,400 kilometers (1,300 miles) from farm to plate.

The world's fishing fleets use about 12.5 units of energy for every 1 unit of food energy from seafood. By comparison, the energy used to grow, store, process, package, transport, refrigerate, and cook all plant and animal food produced on U.S. farms takes about 10 units of fossil fuel energy for every 1 unit of food energy. Today's systems for producing, processing, transporting, and preparing food are highly dependent on fossil fuels and together result in a large net energy loss.

On the other hand, the amount of energy per calorie used to produce crops in the United States has declined by about 50% since the 1970s. One factor in this decline is that the amount of energy used to produce synthetic nitrogen fertilizer has dropped sharply. Another reason for the decline is the rising use of conservation tillage, which substantially reduces energy use and the harmful environmental effects of plowing.

checkpoint Explain why industrialized food production results in a net energy loss.

Agriculture Affects the Environment

Industrialized agriculture has allowed farmers to use less land to produce more food. Using less land for farms has many environmental benefits. However, many analysts point out that industrialized agriculture has greater overall harmful environmental impacts than any other human activity. These impacts may limit future food production.

According to the United Nations Environment Programme (UNEP), agriculture uses massive amounts of the world's resources and pollutes the air and water. It uses about 70% of the world's fresh water, produces about 60% of all water pollution, degrades and erodes topsoil, emits about 25% of the world's greenhouse gases, and utilizes about 38% of the world's ice-free land. As a result, many analysts view today's industrialized agriculture as environmentally and economically unsustainable. However, proponents of industrialized agriculture argue that its benefits outweigh its harmful effects.

checkpoint What are some of the drawbacks of industrialized agriculture?

Agriculture Impacts Soil and Vice Versa

Topsoil, on which all terrestrial life directly or indirectly depends, is one of the most important components of Earth's natural capital (Figure 9-15). Topsoil stores and purifies water and supplies most of the nutrients needed for plant growth. It recycles these nutrients endlessly as long as they are not removed faster than natural processes replenish them. Organisms living in topsoil remove and store carbon dioxide from the atmosphere, thereby helping to control Earth's climate as part of the carbon cycle. Topsoil underlies all forests, grasslands, and croplands. Agriculture relies on the ecosystem services that topsoil provides but also has the potential to degrade these services.

The rise of industrialized agriculture has exposed irreplaceable topsoil to erosion by water and wind. Some topsoil erosion is natural, but much of it is caused by human activities (Figure 9-16). In undisturbed ecosystems with vegetation, the roots of plants help to anchor the topsoil, preventing some erosion. However, topsoil can erode when soil-holding grasses, trees, and other vegetation are removed through activities such as farming, deforestation, and overgrazing.

FIGURE 9-15

The Anatomy of Soil Different-aged soils have different characteristics. As the soil "matures," nutrients become more available to plants growing on the surface.

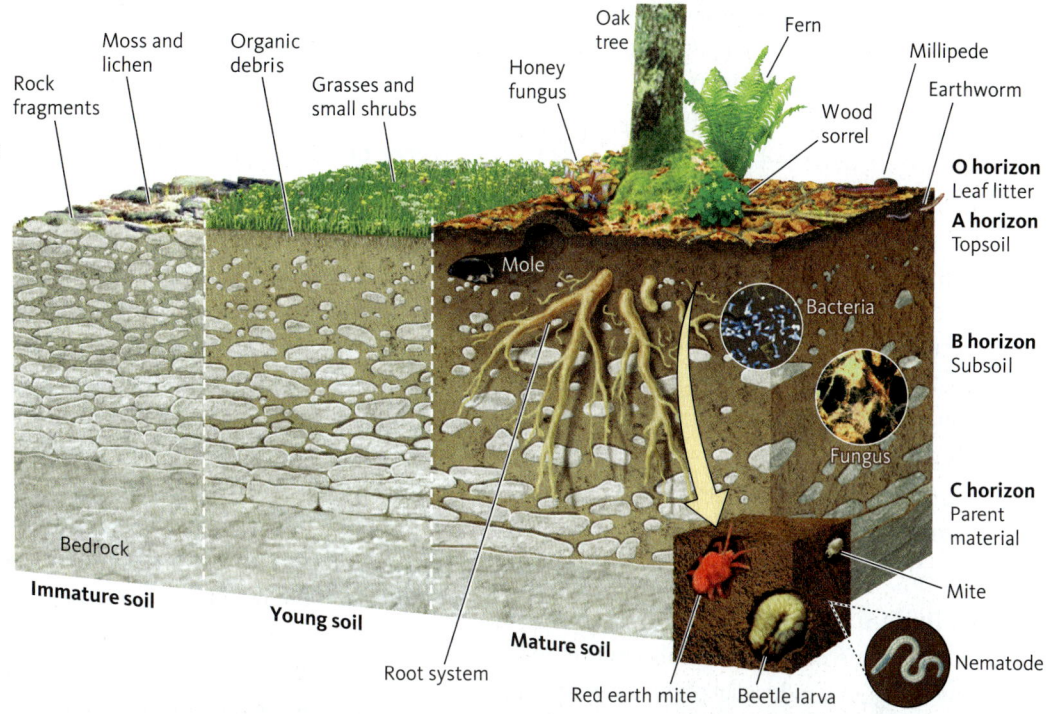

SCIENCE FOCUS 9.1

SOIL SCIENCE

Soil is a complex mixture of eroded rock, mineral nutrients, decaying organic matter, water, air, and billions of living organisms, most of them microscopic decomposers. Soil formation begins when bedrock is slowly broken down into fragments and particles by physical, chemical, and biological processes called weathering.

Each layer, or horizon, of mature soil (Figure 9-15) has a distinct texture and composition that varies with different types of soils. Most of a soil's organic matter is concentrated in its two upper layers, the O horizon of leaf litter and the A horizon of topsoil. The roots of most plants grow in these two upper layers.

In healthy soils, the two top layers teem with decomposers—bacteria, fungi, earthworms, small insects—all interacting in complex food webs. Decomposers break down some of the soil's complex organic compounds into a mixture of partially decomposed plants and animals, called humus, and inorganic materials such as clay, silt, and sand. Soil moisture carrying these dissolved nutrients is drawn up by the roots of plants, which transport the nutrients through the stems and into the leaves. This process is a key component of nutrient cycling, one of the factors of sustainability.

The B horizon (subsoil) and the C horizon (parent material) contain most of a soil's inorganic material, such as broken-down rock consisting of varying mixtures of sand, silt, clay, and gravel. Much of the inorganic material is transported by water from the A horizon. The C horizon lies on a base of parent material, which is often bedrock.

The spaces, or pores, between the solid organic and inorganic particles in the upper and lower soil layers contain varying amounts of air (mostly nitrogen and oxygen gas) and water. Plant roots use the oxygen for cellular respiration. As long as the O and A horizons are anchored by vegetation, the soil layers store water and nutrients, and release them slowly and steadily as plants need them.

Although topsoil is a renewable resource, it is renewed very slowly, which means it can be depleted. Just 1 centimeter (0.4 inch) of topsoil can take more than 100 years to form. It can be washed or blown away in a matter of weeks or months when farming practices leave topsoil unprotected.

Thinking Critically

Draw Conclusions How does soil contribute to each of the four components of biodiversity described in Chapter 4?

Topsoil is eroding faster than it forms on about one-third of the world's cropland. Erosion of topsoil has three major harmful effects.

- Soil fertility is lost as plant nutrients in the topsoil are depleted.
- Eroded topsoil ends up as sediment in surface waters. This sediment can kill fish and shellfish and clog irrigation ditches, boat channels, reservoirs, and lakes. Additional water pollution occurs when the eroded sediment contains pesticide residues that can be ingested by aquatic organisms and biomagnified within food webs.
- Carbon that was stored in the soil by vegetation is released into the air and water, impacting the carbon cycle and atmospheric levels of CO_2.

Soil Pollution Industrial pollution often affects agriculture. Some of the chemicals emitted into the atmosphere by industrial plants, power plants, and motor vehicles can pollute soil and irrigation water. For example, a study by China's environment ministry found that 19% of China's farmable land is contaminated, especially with toxic metals such as cadmium, arsenic, and nickel. The ministry also estimated that 2.5% of the country's cropland is too contaminated to grow food safely.

The agricultural industry can also pollute soil. While some farmers intentionally add synthetic pesticides and fertilizers, too much or careless application can pollute the soil locally. It can also spread to and contaminate nearby organic farms.

Desertification Some farming practices, combined with drought and climate change, can cause **desertification**. In this process, the productive potential of topsoil falls by 10% or more because of human activities such as overgrazing, deforestation, and excessive plowing. These activities can expose topsoil to erosion (Figure 9-17). The severity of desertification can vary. In extreme cases, it leads to desert conditions. Desertification is especially problematic in the world's water-short drylands, which are home to about 2 billion people.

Over thousands of years, Earth's deserts have expanded and contracted primarily because of natural variances in climate. However, human uses of the land, especially agriculture, have increased desertification in some parts of the world.

checkpoint How does topsoil replenish itself?

FIGURE 9-16 ▼
Farming practices that do not protect soil can increase erosion. Topsoil, stripped by wind, clouds the air at this farm in Iowa.

FIGURE 9-17
Severe soil erosion due to overgrazing by cattle has led to desertification.

NATIONAL GEOGRAPHIC | EXPLORERS AT WORK
Jerry Glover Agroecologist

Jerry Glover is a soil scientist and agroecologist—a scientist who uses the principles of ecology to study agricultural systems.

A pit cut into a field reveals to Dr. Glover what is happening below the surface. On one side of the test site, traditional wheat grows five feet above solid soil, roots barely noticeable. On the other side, perennial wheat grows to a similar height, but the roots reach deep into the soil and create a visible mesh of fibers. According to Glover, perennials are the key to feeding the world.

Most farmers plant annuals. Their seeds have to be replanted every year. When farmers harvest their crops, they remove all the nutrients the plants absorbed from the soil. Any remaining nutrients from applied fertilizers are often washed or blown away when the field is bare after harvest. The annuals have shallow root systems that do little to hold soil in place.

Perennial crops, which live for several years, help farmers copy nature by better using and conserving natural resources—sunlight, soil, and water. Since there is no need to till the soil and replant each year, topsoil erosion and water pollution are hardly an issue. Perennials also reduce the need for irrigation because their deep roots retain more water than do the shorter roots of annuals. A reduced need for chemical fertilizers and pesticides results in less pollution. Perennials also remove and store more carbon from the atmosphere, and growing them requires less energy input.

So why don't all farmers use perennials? As humans selected and developed improved seed varieties, more and more annuals were introduced to farmers. Now, the traits that farmers and consumers desire in crops are often only found in annuals, not in perennials. The goal of soil scientists like Glover is to research and develop perennial varieties that are just as good as the annuals. This, according to Glover, could take at least 15 years.

Excessive Irrigation Pollutes Soil

The use of irrigation is a major reason farmers have had success in boosting productivity. Irrigation accounts for about 70% of the water that humanity uses and contributes to the production of nearly half of the world's food. However, irrigation has a downside. Most irrigation water is a dilute solution of various salts, such as sodium chloride, that are picked up as the water flows over or through soil and rocks. Irrigation water that is not absorbed into the topsoil evaporates and leaves behind a thin crust of dissolved mineral salts in the topsoil. Repeated application of irrigation water in dry climates leads to the gradual accumulation of salts in the upper soil layers. This process is called **soil salinization** and degrades the soil (Figure 9-18). Soil salinization stunts crop growth, lowers crop yields, and can eventually kill plants and ruin the land.

CONSIDER THIS

Corn, ethanol, and soil conservation compete.

The United States started a successful program in the 1980s to conserve topsoil by "retiring" erodible land. But in recent years, some U.S. farmers have taken erodible land out of the reserve in order to receive more generous government subsidies for planting corn. (Corn removes nitrogen from the soil and reduces the ability of soil to remove and store carbon from the atmosphere.) The farmers used the corn to make ethanol for use as a motor vehicle fuel. This has led to mounting political pressure to cut back on the nation's topsoil conservation efforts.

FIGURE 9-18
White alkaline salts have displaced crops that once grew on this heavily irrigated land in Colorado.

The FAO estimates that severe soil salinization affects 20 to 50% of the world's irrigated cropland. Salinization affects almost one-fourth of irrigated cropland in the United States, especially in western states.

Another problem with irrigation is waterlogging, in which water accumulates underground and gradually raises the water table. Waterlogging is especially prevalent when farmers apply large amounts of irrigation water in an effort to reduce salinization by leaching salts deeper into the soil. Waterlogging lowers the productivity of crop plants and kills them after prolonged exposure. Plants are eventually deprived of the oxygen they need to survive. At least 10% of the world's irrigated land suffers from this worsening problem, according to the FAO.

checkpoint How can agriculture cause soil pollution?

Agriculture Contributes to Air Pollution and Climate Change

Agricultural activities create a great deal of air pollution. Clearing and burning of forests to raise crops or livestock adds dust and smoke into the air. Plowing fields also releases dust. Applying fertilizer and pesticides spreads particles and various harmful chemicals into the air. Ammonia from livestock can lead to acid rain and can alter the pH of water- and land-based ecosystems.

Besides harmful particles in the air, agriculture also accounts for more than a quarter of all human-generated emissions of carbon dioxide (CO_2). The emissions of this and other greenhouse gases warm the atmosphere and contribute to climate change, which can threaten crop production and food security. Other greenhouse gases include methane (CH_4) and nitrous oxide (N_2O). Methane has about 25 times more warming potential per molecule than CO_2. It is emitted into the atmosphere through the digestive processes of livestock such as cows and from manure stored in waste lagoons. Nitrous oxide has about 300 times the warming potential of CO_2. It is emitted into the atmosphere from livestock manure and synthetic inorganic fertilizers.

checkpoint How can agriculture contribute to air pollution?

Food and Biofuel Production Reduce Biodiversity

Natural biodiversity and some ecosystem services are threatened when forests are cleared and when croplands replace grasslands to produce food and biofuels. One of the fastest-growing threats to the world's biodiversity is happening in Brazil. Large areas of tropical rain forest in the Amazon Basin are being lost. In addition, a huge tropical grassland region south of the Amazon Basin called *cerrado* is being lost. These lands are being cleared for cattle ranches, large plantations of soybeans grown for cattle feed, and sugarcane used for making ethanol fuel.

Another example of agricultural land replacing biodiveristy-rich land is happening in Indonesia. There, tropical forests are burned to plant oil palm trees. The trees produce palm oil, which is increasingly used to produce biodiesel fuel. Biodiversity is threatened in these and many other areas because tropical forests and grasslands host more varieties of organisms than agricultural land.

A related problem is the increasing loss of **agrobiodiversity**—the genetic variety of animal and plant species used on farms to produce food. Scientists estimate that since 1900, the world has lost three-quarters of the genetic diversity of agricultural crops. For example, India once planted 30,000 varieties of rice. Now most of its rice production comes from only ten varieties. Soon, almost all of its production might come from just one or two varieties.

LESSON 9.3

FIGURE 9-19
The Svalbard Global Seed Vault looks like a regular warehouse at a glance. On closer inspection, its frozen, underground location becomes apparent. The cold climate and remote location help ensure longevity of the seeds and offer security.

In the United States, about 97% of the food plant varieties available to farmers in the 1940s no longer exist, except perhaps in small amounts in seed banks and in occasional home gardens.

In losing agrobiodiversity, ecologists warn that farming practices are rapidly shrinking the world's genetic "library" of plant varieties, which are critical for increasing food yields. This failure to preserve agrobiodiversity is a serious violation of the biodiversity factor of sustainability.

Scientists are making efforts to save endangered varieties of crops and wild plant species important to the world's food supply. About 1,400 refrigerated seed banks store individual plants and seeds. They are also stored in agricultural research centers and botanical gardens scattered around the world.

The world's most secure seed bank is the underground Svalbard Global Seed Vault, also called the "doomsday seed vault" (Figure 9-19). The Global Seed Vault was carved into the Arctic permafrost on a frozen Norwegian Arctic island. It is being stocked with duplicates of much of the world's seed collections. This provides insurance against irreversible losses of stored seeds at other sites from possibilities like power failures, fires, storms, or wars.

However, the seeds of many plants cannot be stored successfully in seed banks. Even seeds that store well do not remain alive indefinitely. They must be planted and germinated periodically and new seeds must be regularly collected.

checkpoint How do seed banks support agrobiodiversity?

Meat Production Requires High Input

Meat is a major food source for many people. Proponents of industrialized meat production point out that it has increased meat supplies, reduced overgrazing, and kept food prices down. But feedlots and CAFOs produce harmful health and environmental effects. Some analysts point out that meat produced by industrialized agriculture is artificially cheap because most of its harmful environmental and health costs are not included in the market prices of meat. This is a violation of the full-cost pricing factor of sustainability. Figure 9-20 summarizes some of the advantages and disadvantages of industrialized meat production.

FIGURE 9-20

Animal Feedlots and CAFOs	
Advantages	**Disadvantages**
Increased meat production	Animals unnaturally confined and crowded
Higher profits	Large inputs of grain, fishmeal, water, and fossil fuels
Less land use	Greenhouse gas (CO_2 and CH_4) emissions
Reduced overgrazing	Concentration of animal wastes that can contaminate land and water
Reduced soil erosion	Use of antibiotics can increase genetic resistance in microbes in humans
Protection of biodiversity	Reduction of genetic diversity in livestock

The FAO reports that overgrazing and erosion by livestock has degraded about 20% of the world's grasslands and pastures. The same report estimated that rangeland grazing and industrialized livestock production caused more than half of all topsoil erosion and sediment pollution, and one-third of the water pollution from runoff of nitrogen and phosphorus from synthetic fertilizers.

Feedlots and CAFOs use large amounts of water to irrigate feed crops for livestock and wash away animal waste. One estimate suggests that producing a pound of beef requires more water than an American uses in a year's worth of showering. Livestock production also results in water pollution from animal waste, hormones and antibiotics, and synthetic pesticides and fertilizers.

Industrialized meat production also uses large amounts of energy (mostly from oil), which helps to make it a major source of air and water pollution and greenhouse gas emissions. The Environmental Working Group (EWG) has estimated that production of meat and meat products generates 10–20 times more greenhouse gases per unit of weight than does production of common vegetables and grains.

Another growing problem is the use of antibiotics in industrialized livestock production facilities. In 2011, the U.S. Food and Drug Administration (FDA) estimated that about 80% of all antibiotics sold in the United States (and 50% of those in the world) are added to animal feed. This is done to reduce the spread of diseases in crowded feedlots and CAFOs and to promote the growth of the animals before they are slaughtered. According to FDA data and several studies, this plays a role in the rise of genetic resistance among many disease-causing bacteria. Such resistance can reduce the effectiveness of some antibiotics used to treat humans for infections. It can also promote the development of new, more genetically resistant infectious disease organisms.

Finally, according to the U.S. Department of Agriculture (USDA), animal waste produced by the American meat industry amounts to about 130 times the amount of waste produced by the country's human population. Globally, only about half of all manure is returned to the land as nutrient-rich fertilizer—a violation of the nutrient cycling factor of sustainability. Much of the other half ends up polluting aquatic systems, producing foul odors, and emitting large quantities of climate-changing greenhouse gases into the atmosphere.

checkpoint What are some pros and cons of meat production?

Fish and Shellfish Production Has Environmental Drawbacks

While aquaculture offers several advantages, it also poses several environmental problems (Figure 9-21). Some analysts warn that the harmful environmental effects of aquaculture could limit its future production potential. About a third of the catch from marine fisheries is converted to fish oil and fishmeal and fed to farmed fish. The added pressure on marine fisheries in order to produce feed reduces wild fish populations and impacts marine food webs—a serious threat to marine biodiversity and ecosystem services.

Another problem with aquaculture is that some fishmeal and fish oil fed to farm-raised fish is contaminated with long-lived toxins such as PCBs that are picked up from the ocean floor. Aquaculture producers contend that the concentrations of these chemicals are not high enough to threaten human health, but some scientists disagree.

Fish farms, especially those that raise carnivorous fish such as salmon and tuna, also produce large amounts of wastes, including pesticides and antibiotics used on fish farms. Yet another problem is that farmed fish can escape their pens and mix with wild fish and possibly disrupt the gene pools of wild populations or become invasive species.

FIGURE 9-21 ▼

Aquaculture	
Advantages	**Disadvantages**
High efficiency	The use of fish oil and fishmeal on fish farms depletes wild fisheries.
High yield	Large waste output
Reduces overharvesting of wild fisheries	Loss of mangrove forests and estuaries
Jobs and profits	Dense populations vulnerable to disease

FIGURE 9-22 ▼
Bycatch, a result of overfishing and inefficient fishing methods, results in unwanted fish being discarded. Shrimp fisheries are particularly bad, producing multiple pounds of bycatch, such as these fish and rays, for only one pound of shrimp.

Major seed companies are now pushing to use patented, genetically modified soybeans as the primary feed for farm-raised fish and shellfish. This could increase water pollution because fish that are fed soy tend to produce more waste than those that are not fed soy. It would also give a small number of seed companies control over much of the world's seafood production, along with seafood prices. Additionally, this shift could worsen the problems associated with growing industrialized monocultures for animal feed.

Wild-caught fisheries also have environmental issues (Figure 9-22). According to the FAO, about 87% of the world's commercial ocean fisheries are either being harvested at full capacity or are being overfished. By targeting large fish, such as tuna, the fishing industry has "fished down" the food web and removed the genetic stock of the largest fish. Increasingly, the industry is left with smaller species on lower trophic levels such as some jellies and squid. This practice alters food webs and impacts entire ocean ecosystems. Some fishing practices and fishing gear can also degrade ocean habitat, especially those that target species living on the ocean floor.

checkpoint What is "fishing down the food web"?

9.3 Assessment

1. **Synthesize** Name one way that each of the three food production systems (crops, meat, and fish/shellfish) can result in environmental problems.
2. **List** What are two ways that pollution impacts food security?
3. **Classify** What is the difference between the environmental impacts of particles, such as dust and smoke, released from livestock production and greenhouse gases released from livestock production?

CROSSCUTTING CONCEPTS

4. **Systems** Describe two ways the industrialized meat production system interacts with Earth's hydropshere.

9.4 How Can Society Manage Agricultural Pests More Sustainably?

CORE IDEAS AND SKILLS

- Understand why pests need to be managed to ensure food security.
- Discuss the trade-offs of synthetic pesticides.
- Explore alternative, more sustainable ways to manage agricultural pests.

KEY TERMS

pest
synthetic pesticide
integrated pest management (IPM)

Nature Helps Controls Many Pests

A **pest** is any species that interferes with human welfare by eating society's food; invading homes, lawns, or gardens; destroying building materials; spreading disease; invading ecosystems; or simply being a nuisance. Worldwide, only about 100 species of plants (weeds), animals (mostly insects), fungi, and microbes cause most of the damage to crops. However, the damage and potential damage to crops by pests is significant and must be combated to ensure food security. For example, the FAO estimates that 20% of global crops are lost to plant-eating pests.

In natural ecosystems and in many polyculture crop fields, natural enemies (predators, parasites, and disease organisms) control the populations of most potential pest species. This free ecosystem service is an important part of Earth's natural capital. Biologists estimate that about 40,000 known species of spiders kill far more crop-eating insects every year than humans do by using insecticides. Most spiders, including the wolf spider (Figure 9-23), do not harm humans.

Many of these natural population checks and balances are upset by clearing forests and grasslands, planting monoculture crops, and applying chemicals to fields. Society must then devise and pay for other ways to protect monoculture crops, tree plantations, lawns, and golf courses from insects, weeds, and other pests that nature helps to control at no charge.

checkpoint Why do pests need to be managed?

LESSON 9.4

FIGURE 9-23 ▼
This ferocious-looking wolf spider with a grasshopper in its mouth is one of many important insect predators that is harmless to humans but can be affected by synthetic pesticides.

Synthetic Pesticides Are an Option

Scientists and engineers have developed a variety of **synthetic pesticides**—chemicals used to kill or control populations of organisms considered undesirable. Common types of synthetic pesticides include insecticides (insect killers), herbicides (weed killers), fungicides (fungus killers), and rodenticides (rat and mouse killers). Since 1950 the use of synthetic pesticides has increased more than 50-fold.

Selective, or narrow-spectrum, synthetic pesticides are effective against a narrowly defined group of organisms. Examples of selective pesticides are fungicides, compounds that inhibit the molting process of insects, and hormones that disrupt the life cycle of insects. Some synthetic pesticides, called broad-spectrum agents, are toxic to beneficial species as well as to pests. Some, such as the broad-spectrum agent DDT, remain in the environment for years and can be biomagnified in food chains (Figure 7-14). Others, such as organophosphates, are active for days or weeks and are not biologically magnified but can be highly toxic to humans.

In 1962, Rachel Carson's book, *Silent Spring*, was published and created an uproar of concern from scientists, politicians, doctors, and the general public about the use of DDT (Chapter 1). As a result, DDT was banned for agricultural use in the United States in 1972. However, DDT is still being used in tropical areas such as Asia, South America, Africa, and Malaysia. It is mainly used for controlling populations of mosquitoes that can transmit diseases such as malaria.

The use of synthetic pesticides has advantages and disadvantages. Proponents say the benefits of synthetic pesticides outweigh their harmful effects. They argue that if used as directed, synthetic pesticides do not remain in the environment at levels high enough to cause serious environmental or health problems. Opponents say the harmful effects of these chemicals outweigh their benefits. Figure 9-24 summarizes the advantages and disadvantages of using synthetic pesticides.

FIGURE 9-24 ▼

Synthetic Pesticides	
Advantages	**Disadvantages**
Save human lives	Promote genetic resistance to pesticides
Expand food supplies	Kill pests' natural enemies
Work fast	Can harm wildlife and people
Raise profits	Pollute air, water, and land
Safe if used properly	Expensive for farms

checkpoint What is the difference between broad-spectrum and narrow-spectrum pesticides?

Laws Regulate Synthetic Pesticides

When the Federal Insecticide, Fungicide, and Rodenticide Act (FIFRA) was passed in 1947, it established procedures for registering synthetic pesticides with the United States Department of Agriculture (USDA) and established labeling provisions. The law has been amended numerous times and in its current form, FIFRA mandates that the Environmental Protection Agency (EPA) regulate the use and sale of synthetic pesticides to protect human health and preserve the environment. However, according to studies by the National

Academy of Sciences, the federal laws that regulate pesticide use are generally inadequate and poorly enforced.

Despite national regulations, pesticide use in different areas of the world can still affect people and the environment. If one country bans synthetic pesticides but continues to export them, residues of these pesticides can come back on imported food. Winds can also carry persistent pesticides from one country to another. International treaties now exist to reduce the harmful effects of synthetic pesticides, but they will have limited effectiveness until all countries sign on.

checkpoint How are countries vulnerable to the harmful effects of synthetic pesticide use in other countries?

There Are Alternatives to Synthetic Pesticides

Scientists encourage the use of biological, ecological, and other alternative methods for controlling pests and diseases that affect crops and human health. Here are some of these alternatives:

- Fool the pest. A variety of cultivation practices can be used to minimize pest problems. Examples include rotating the types of crops planted in a field each year and adjusting planting times so that major insect pests either starve or get eaten by their natural predators.
- Provide homes for pest enemies. Farmers can increase the use of polyculture, which uses plant diversity to provide habitats for the predators of pest species.
- Implant genetic resistance. Use genetic engineering to speed up the development of pest- and disease-resistant crop strains.
- Bring in natural enemies. Use biological control by importing natural predators such as the wolf spider, wasps, parasites, and disease-causing bacteria and viruses to help regulate pest populations (Figure 9-25). This approach is nontoxic to other species and is usually less costly than applying synthetic pesticides.

FIGURE 9-25 ▶
In this example of biological pest control, a wasp is parasitizing a gypsy moth caterpillar.

LESSON 9.4 **313**

- Use insect scents. Trace amounts of sex attractants (called pheromones) can be used to lure pests into traps or to attract their predators into crop fields.
- Use insect hormones. Hormones control developmental processes at different stages of life. Scientists have learned how to identify and use hormones that disrupt an insect's normal life cycle, thereby preventing it from reaching maturity and reproducing.
- Use natural methods to control weeds. Organic farmers control weeds by methods such as crop rotation, mechanical cultivation, hand weeding, and the use of cover crops and mulches.

checkpoint Name two forms of biological pest control.

Integrated Pest Management Is One Sustainable Solution

Many pest control experts and farmers believe the best way to control crop pests is through **integrated pest management (IPM)**. IPM is a carefully designed program that evaluates each crop and its pests as parts of an ecosystem.

The overall aim of IPM is to reduce crop damage to an economically tolerable level with minimal use of synthetic pesticides. Farmers using IPM start with biological methods (natural predators, parasites, and disease organisms) and cultivation controls (such as altering planting times and growing different crops on fields from year to year to disrupt pests). They apply small amounts of synthetic pesticides only when pest populations reach a threshold where the potential cost of pest damage to crops outweighs the cost of applying the pesticide.

According to the U.S. National Academy of Sciences, a well-designed IPM program can reduce synthetic pesticide use and pest control costs by 50–65%, without reducing crop yields and food quality. IPM can also reduce inputs of fertilizer and irrigation water and slow the development of genetic resistance. Pests are treated less often and with lower doses of pesticides. IPM is an important form of pollution prevention that reduces risks to wildlife and human health and applies the biodiversity factor of sustainability.

Many countries have seen reductions in synthetic pesticide use through the adoption of IPM. In the United States, scientists are urging the USDA to promote the practice. Despite its potential, IPM still has some drawbacks. It requires expert knowledge about each pest situation and takes more time than relying strictly on synthetic pesticides. Methods developed for a crop in one area might not apply to areas with even slightly different growing conditions. Initial costs of IPM may be higher, although long-term costs are typically lower than the use of synthetic pesticides alone. Widespread use of IPM has been hindered in the United States by government subsidies that support the use of synthetic pesticides, opposition from pesticide manufacturers, and a shortage of IPM experts.

checkpoint Define integrated pest management.

9.4 Assessment

1. **Explain** How is integrated pest management a more sustainable way to manage agricultural pests than using synthetic pesticides?
2. **Apply** Imagine you have planted a garden and caterpillars are eating your plants. What steps would you take to manage the issue? Explain your answer.

CROSSCUTTING CONCEPTS

3. **Stability and Change** Under what circumstances, if any, might the benefits of synthetic pesticide use outweigh the risks?

9.5 What Are Sustainable Solutions for Food Production?

CORE IDEAS AND SKILLS
- Recognize that sustainable solutions to agriculture must conserve soil, water, and energy.
- Identify a variety of solutions to produce crops and animal products sustainably.

KEY TERMS

soil conservation
organic fertilizer
synthetic inorganic fertilizer
aquaponics

There Are Many Ways to Protect Soil

Past human innovation led to higher-yield food production methods and supported a human population explosion. Now, people are increasingly looking for practices that avoid or minimize the environmental impacts of food production so that it can be sustained over the long run. For example, reducing topsoil erosion is a key component of more sustainable agriculture. **Soil conservation** involves using a variety of methods to reduce topsoil erosion and restore soil fertility. Terracing, as shown in Figure 9-26, involves converting steeply sloped land into a series of broad, nearly level terraces that run across the land's contours.

On less steeply sloped land, contour planting can be used. It involves plowing and planting crops in rows across the slope of the land rather than up and down. Each row acts as a small dam to help hold topsoil by slowing runoff. Similarly, strip-cropping helps to reduce erosion and to restore soil fertility with alternating strips of a row crop (such as corn or cotton) and a row of cover crop that completely covers the soil (such as alfalfa, clover, oats, or rye). The cover crop traps topsoil that erodes from the row crop and catches and reduces water runoff.

Alley cropping, or agroforestry, is another way to slow erosion and maintain soil fertility. One or more crops, usually legumes or other crops that add nitrogen to the soil, are planted together in alleys between orchard trees or fruit-bearing shrubs, which provide shade. This reduces water loss by evaporation and helps regulate soil moisture.

Farmers can also establish windbreaks, or shelterbelts, of trees around fields to reduce wind erosion. The trees also retain soil moisture, supply wood for fuel, and provide habitats for birds and insects that help with pest control and pollination.

Another way to greatly reduce topsoil erosion is to eliminate or minimize the plowing and tilling of topsoil and to leave crop residues on the ground. Such conservation-tillage farming uses specialized machines that inject seeds and fertilizer directly through crop residues into minimally disturbed topsoil. Weeds are controlled with herbicides. This type of farming increases crop yields and greatly reduces water pollution from sediment and fertilizer runoff. However, one drawback is that the greater use of herbicides promotes the growth of herbicide-resistant weeds that force farmers to use larger doses of weed killers or, in some cases, to return to plowing.

One sustainable solution to conserve topsoil is to retire the estimated one-tenth of the world's highly erodible cropland. The goal would be to identify erosion hotspots, stop cultivation, and plant them with grasses or trees.

Another way to protect soil is to restore some of the lost plant nutrients that have been washed, blown, or leached out of topsoil, or that have been removed by repeated crop harvesting.

FIGURE 9-26
Each level on this terraced rice farm in Bali, Indonesia, retains water for crops and reduces topsoil erosion by controlling runoff.

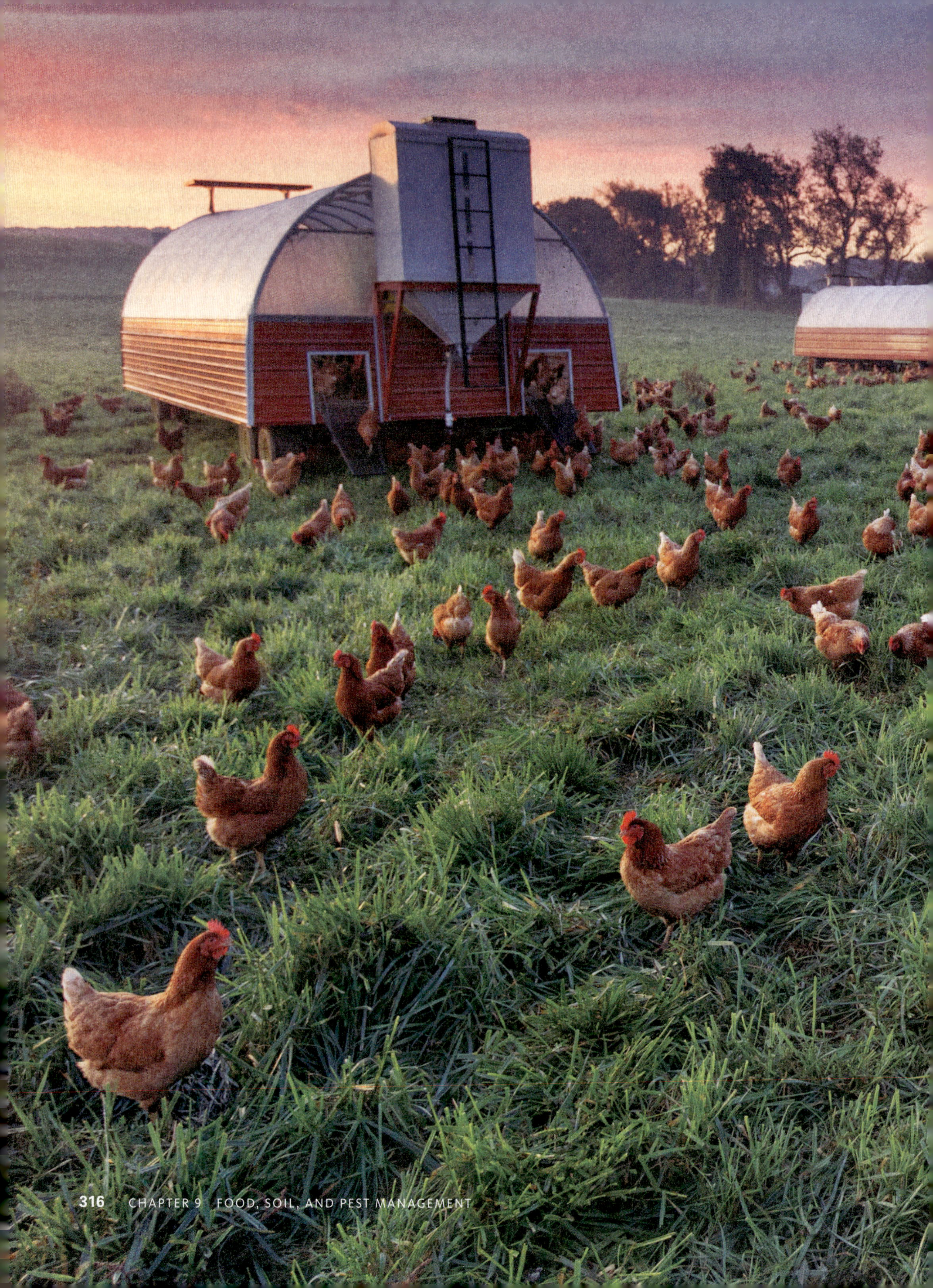

FIGURE 9-27
ON ASSIGNMENT National Geographic photographer Peter Essick walked among the chickens that fertilized the soil on this Pennsylvania farm. Mobile chicken coops are moved daily to distribute the manure more evenly and reduce runoff of fertilizer.

LESSON 9.5

To restore the soil, farmers can use **organic fertilizer** derived from plant and animal materials or synthetic fertilizer. **Synthetic inorganic fertilizer** is made of inorganic compounds that contain nitrogen, phosphorus, and potassium along with trace amounts of other plant nutrients. There are several types of organic fertilizers. One is animal manure—the dung and urine of cattle, horses, poultry (Figure 9-27), and other farm animals. Manure improves topsoil structure, adds organic nitrogen, and stimulates the growth of beneficial soil bacteria and fungi. Another type, called green manure, consists of freshly cut or live green vegetation that is plowed into the topsoil to increase the organic matter and humus available to the next crop. A third type is compost, produced when microorganisms break down organic matter such as leaves, crop residues, food wastes, paper, and wood in the presence of oxygen.

Practicing crop rotation replenishes soil nutrients that plants use to grow. A farmer plants an area with a nutrient-depleting crop one year, and the next year, plants the same area with legumes, whose root nodules add nitrogen to the soil. This method also reduces erosion by keeping the topsoil covered with vegetation.

Many farmers, especially those in more-developed countries, rely on synthetic inorganic fertilizers. The use of these products has grown more than ninefold since 1950, and now accounts for about 25% of the world's crop yield. While these fertilizers can replace depleted inorganic nutrients, they do not replace organic matter. Restoring topsoil nutrients requires both inorganic and organic fertilizers.

Reducing Salinization and Desertification

While soil salinization solutions exist, they are often costly. Farmers can reduce irrigation or use more efficient irrigation systems to reduce salinization. To clean up damaged soils, farmers can flush the soil, install underground drainage systems, or stop growing crops for a number of years. All of these solutions would be costly to implement.

Farmers do not have control over weather and other external factors that cause desertification. However, people can reduce population growth, overgrazing, deforestation, and destructive forms of planting and irrigation in dryland areas, which have left much land vulnerable to desertification. Society can also work to decrease the human contribution to climate change, which is projected to increase the severity of droughts in some areas.

checkpoint How can farmers protect topsoil?

Crops Can Be More Sustainable

Modern industrialized food production yields large amounts of food at affordable prices. However, the yield and the processes are not sustainable because they violate the three scientific factors of sustainability. Industrialized food production relies heavily on nonrenewable fossil fuels and thus adds greenhouse gases and other air pollutants to the atmosphere and contributes to climate change. Industrialized agriculture also reduces biodiversity and interferes with the cycling of plant nutrients. These harmful effects are hidden from consumers. Most of the harmful environmental and health costs of food production are not included in the market prices of food. This violates the full-cost pricing factor of sustainability.

A more sustainable food production system would have several different components and require less input. Compared to high-input farming, low-input farming produces similar yields with less energy input and lower greenhouse gas emissions. Low-input farms better preserve topsoil fertility and reduce erosion.

One component of sustainable food production is organic agriculture. Many experts support a shift to organic farming because it sharply reduces the harmful environmental and health effects of industrialized farming, improves the condition of topsoil, and reduces pollution of air and water.

Another important component of more sustainable agriculture would be to rely less on conventional monoculture and more on organic polyculture. Of particular interest to some scientists is the idea of using polyculture to grow perennial crops—crops that grow back year after year on their own (Explorers At Work).

Another key aspect of sustainable agriculture is to shift from using fossil fuels to relying more on renewable energy for food production—an important application of the solar energy factor of sustainability. Farmers can make greater use of renewable solar energy, wind, flowing water, and biofuels produced from farm wastes.

Analysts suggest five major strategies to help farmers and consumers make the transition to more sustainable agriculture over the next 50 years.

1. Increase research on organic farming, perennial polyculture, and improving human nutrition.
2. Establish sustainable agriculture education and training programs for students, farmers, and government agricultural officials.

FIGURE 9-28

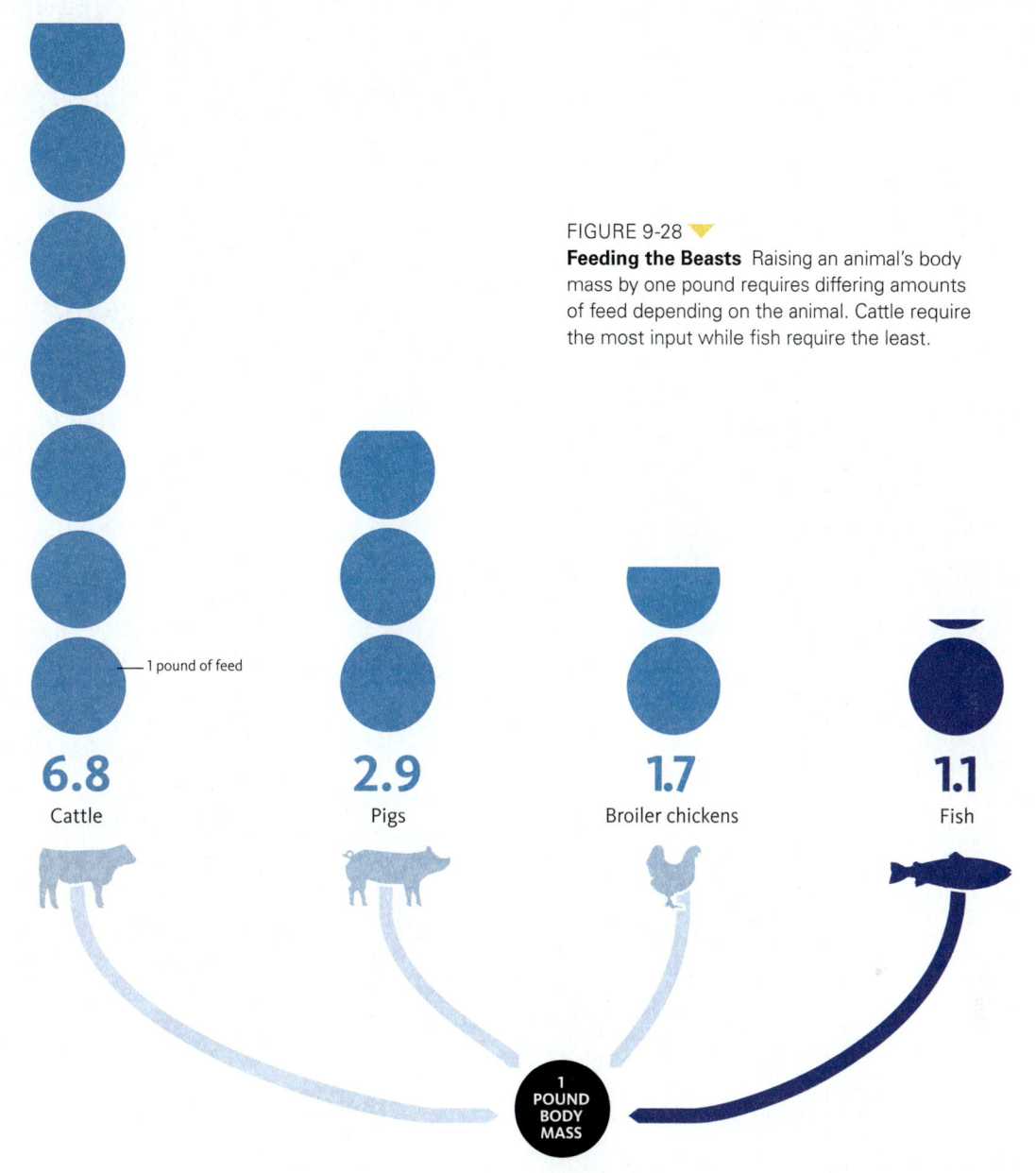

Feeding the Beasts Raising an animal's body mass by one pound requires differing amounts of feed depending on the animal. Cattle require the most input while fish require the least.

6.8 — Cattle
2.9 — Pigs
1.7 — Broiler chickens
1.1 — Fish

1 pound of feed → 1 POUND BODY MASS

Source: Used with permission from National Geographic.

3. Set up an international fund to give farmers in poor countries access to various types of sustainable agriculture.
4. Replace government subsidies for environmentally harmful forms of industrialized food production with subsidies that encourage more sustainable food production.
5. Educate consumers about where their food comes from, how it is produced, and the harmful environmental and health effects of industrialized food production.

checkpoint Describe the components of a sustainable farm.

Animal Products Can Be Sustainable

The production of meat, dairy, fish, and shellfish has a huge environmental impact. Meat consumption is the largest factor in the growing ecological footprints of individuals in affluent nations.

Meat Production An example of a more sustainable form of meat production is a free-range chicken farm that uses chicken manure to fertilize the soil (Figure 9-27). Buying chicken from such farms is an alternative to buying chicken from CAFOs, which produce large amounts of manure that is not returned to the soil.

FIGURE 9-29
ON ASSIGNMENT National Geographic photographer Jim Richardson visited TableTop Farm in Iowa and discovered several sustainable aspects of the farm. These include IPM with organic chemicals, buffers and cover crops to prevent erosion and retain soil moisture, alternatives to tillage, and organic polyculture crops. Sally Gran and her farmers sell their produce locally and regionally and offer CSA in the community.

Another way to move toward sustainable meat production and consumption involves shifting from less grain-efficient forms of animal protein to more grain-efficient forms (Figure 9-28). Beef, pork, and carnivorous fish produced by aquaculture require large inputs of feed compared with poultry and plant-eating farmed fish.

A growing number of people have one or two meatless days per week. Others go further and eliminate most or all meat from their diets. They replace the meat with fruits, vegetables, and protein-rich foods such as peas, beans, and lentils. According to one estimate, if all Americans picked one day per week to have no meat, the reduction in greenhouse gas emissions would be equivalent to taking 30 to 40 million cars off the road for a year.

Fish and Shellfish Production Various organizations have established guidelines, standards, and certifications to encourage more sustainable fishing and aquaculture practices. The Aquaculture Stewardship Council (ASC), for example, has developed aquaculture sustainability standards, although it has certified only about 4.6% of the world's aquaculture production operations. The Marine Stewardship Council (MSC) performs a similar program for wild-caught fisheries. Like the organic certification, the programs have associated labels for food products that empower consumers to choose to purchase a more sustainable option.

Scientists and engineers are working on ways to make aquaculture more sustainable. One approach is open-ocean aquaculture, which involves raising large carnivorous fish in underwater pens. Some are located as far as 300 kilometers (190 miles) offshore where currents can wash away and dilute fish wastes.

> **CONSIDER THIS**
>
> **Sustainable fish guides are available.**
> The MSC, NRDC, Monterey Bay Aquarium, and other organizations offer easy-to-access information on their websites that can help guide your fish choices. Fish that often top the lists of sustainability include wild Alaskan salmon, U.S. farmed rainbow trout, Alaskan or Canadian sablefish (black cod), and farmed shellfish. Fish to avoid include Atlantic cod, Atlantic wild halibut, Atlantic salmon, swordfish, and bluefin tuna.

Other fish farmers are reducing coastal damage from aquaculture by raising shrimp and fish species in inland facilities using zero-discharge freshwater ponds and tanks. In such recirculating systems, the water used to raise the fish is continually recycled.

Making aquaculture more sustainable will require some fundamental changes for producers and consumers. One change is for more consumers to choose fish species such as tilapia and catfish that eat algae and other vegetation rather than fish oil and fishmeal produced from other fish. Raising carnivorous fishes such as salmon, trout, tuna, grouper, and cod contributes to overfishing and population crashes of species used to feed these carnivores. Aquaculture producers can avoid this problem by raising plant-eating fishes such as carp, tilapia, and catfish, as long as they do not try to increase yields by feeding fishmeal to such species.

Fish farmers can also emphasize polyaquaculture, which has been part of aquaculture for centuries, especially in Southeast Asia. This practice applies the nutrient cycling and biodiversity factors of sustainability. Polyaquaculture operations raise fish or shrimp along with algae, seaweeds, and shellfish in coastal lagoons, ponds, and tanks. The wastes of the fish or shrimp feed the other species. In another operation called **aquaponics**, fish waste provides fertilizer for food crops that are grown in water. The plants purify the water and provide oxygen and nutrients that support the fish.

checkpoint Why are vegetarian diets more sustainable than meat-based diets, for both humans and livestock?

9.5 Assessment

1. **Recall** What are the barriers to cleaning up soil that has been polluted from salinization?
2. **Contrast** What is the difference between aquaponics and polyaquaculture?
3. **Apply** What, if any, programs exist that allow your community to purchase local food products?
4. **Infer** How has the availability of natural resources, such as clean water and fertile soil, and natural hazards, such as drought and erosion, driven more sustainable food production processes?

SCIENCE AND ENGINEERING PRACTICES

5. **Developing Models** How could you build a garden where you live? What would you grow and where would you grow it? How would you ensure you had enough nutrients, fertile soil, and water?

TYING IT ALL TOGETHER STEM
Sustainability Starts with You

Throughout this chapter you explored different ways that food is produced and brought to your plate. Individuals have the power to influence the sustainability of food systems by "voting with their forks." At the grocery store, customers can purchase foods that are lower on the food chain, are produced organically, are in season or grown locally, or are produced using other sustainable practices.

How can the consumer know? This chapter explored some consumer information, like certified organic labels, GMO labeling, and identifying local produce. But the responsibility ultimately lies with consumers to educate themselves. At the store, consumers can ask questions of the grocer, the butcher, the fish monger, and other salespeople about where the food came from and how it was produced. At restaurants, diners can ask the server, chef, or purchaser where an item came from and how it was produced. The employees don't always know the answer but by asking, the consumer puts pressure on the employees and owners to provide sustainable food options.

The internet can help you choose more sustainable foods. Perform an internet search for "sustainable seafood" and "dirty dozen foods." **Then answer the questions below.**

1. For each phrase, visit the websites of three different organizations that provide guidance. For each website, record:
   a. The name of the organization.
   b. The type of tool, guide, or information offered to help consumers.
   c. How easy it is to understand the information that is presented.
2. For each website, how likely would you be to follow their recommendations? Explain your reasoning.
3. In a group, develop and design a new buying guide for a sustainable food category (e.g., beef, eggs, or grain). Use the findings from your internet search to make the guide easy to use and informational for the consumer. Use information from the text and other sources to determine the sustainability of the foods.
4. What challenges did your group experience when considering information to include in the guide?
5. As a group, present your consumer guide to the rest of the class.

CHAPTER 9 SUMMARY

9.1 What is food security?

- Obstacles to providing enough nutritious food for the world's growing population include lack of nutrition information, poverty, war, bad weather, climate change, and the harmful effects of industrialized food production.
- Food security can be improved by reducing poverty and chronic malnutrition, producing food more sustainably, relying more on locally grown food, and cutting food waste.

9.2 How is food produced?

- The three agricultural systems that provide most of the world's food are (1) croplands, (2) rangelands, pastures, and feedlots, and (3) fisheries and aquaculture.
- Increased livestock and fish production are a result of using feedlots, concentrated animal feeding operations (CAFOs), ocean fishing, and aquaculture.
- Genetically improved varieties of crops and livestock animals are developed by crossbreeding through artificial selection and genetically modified organisms (GMOs).

9.3 How are environmental issues and food production connected?

- Environmental problems from food production include soil erosion, biodiversity loss, air pollution, water pollution, and threats to human health.

9.4 How can society manage agricultural pests more sustainably?

- To create and maintain food security, society must manage pests. This includes the use of natural and synthetic pesticides and managing the environmental and health impacts of using such chemicals.
- More sustainable food production involves sharply cutting pesticide use without decreasing crop yields by using a mix of cultivation techniques, biological pest controls, and small amounts of selected synthetic pesticides as a last resort. This can be achieved with integrated pest management (IPM).

9.5 What are sustainable solutions for food production?

- To make food production more sustainable, society must reduce erosion, increase soil fertility, reduce salinization and desertification, and improve the production of animal products.
- Other measures are to train farmers in sustainable practices, establish enforceable laws and guidelines to govern food production, and educate consumers about how food is produced and how it can be produced more sustainably.

CHAPTER 9 ASSESSMENT

Review Key Terms

Select the key term that best fits each definition. Some terms will not be used.

agrobiodiversity	integrated pest management	subsidy
aquaculture	monoculture	synthetic inorganic fertilizer
aquaponics	organic agriculture	synthetic pesticides
chronic malnutrition	organic fertilizer	topsoil
desertification	overnutrition	traditional agriculture
food insecurity	pest	
food security	polyculture	
green revolution	soil conservation	
industrialized agriculture	soil salinization	

1. Access at all times to enough nutritious food for an active, healthy life
2. Natural capital that stores the water and nutrients needed by plants
3. Practice that uses heavy equipment along with large amounts of financial capital, fossil fuels, water, synthetic fertilizers, and synthetic pesticides to produce monocultures
4. Lack of access to nutritious food
5. Government support intended to help businesses survive and thrive
6. Raising fish in freshwater ponds, lakes, reservoirs, and rice paddies, and in underwater cages in coastal and deeper ocean waters
7. Process in which the productive potential of topsoil falls by 10% or more because of prolonged drought or human activities that expose topsoil to erosion
8. Repeated irrigation in dry climates that leads to the accumulation of salts in the topsoil
9. The genetic variety of animal and plant species used on farms to produce food

Review Key Concepts

10. Define food security and food insecurity.
11. What three systems supply most of the world's food?
12. What is traditional subsistence agriculture?
13. Distinguish between crossbreeding through artificial selection and genetic engineering.
14. What are the health effects of eating meat from animals raised by organic methods versus eating meat from those raised in CAFOs?
15. How does compost contribute to the sustainability of food production?
16. Define soil and describe its formation along with the major layers in mature soils.
17. What are the harmful effects of desertification?
18. Compare the advantages and disadvantages of using GMOs in food production.
19. How have industrialized food production systems caused losses in biodiversity?
20. How have some governments used subsidies to influence food production, and what have been some of the effects?
21. What are three important ways in which individual consumers can help promote more sustainable food production?

Think Critically

22. Imagine you got a job with Growing Power (Case Study) and were assigned to turn an abandoned suburban shopping center and its large parking lot into an organic farm. Write a plan for accomplishing this and include a model of the proposed site.
23. Do you support or oppose increasing the use of (a) genetically modified food production, and (b) organic perennial polyculture? Explain your reasoning.
24. What might happen to industrialized food production if oil prices rose sharply? How might this affect your life? How might it affect the next generation? List two ways you would deal with these changes.
25. If mosquitoes in your area were proven to carry malaria or a dangerous virus, how would you combat them? Explain your reasoning.

Chapter Activities

A. Develop Models: Fish Consumption STEM

Figure 9-30 gives the world's fish harvest and population data. By analyzing data like these, scientists can determine how much food is available and used for human consumption.

Materials

graph paper
cork tile
scissors
glue
12 push pins
colored string

1. Use the world fish harvest and population data to calculate the per capita fish consumption for each year from 2000 to 2012. (Assume all fish caught were consumed.)
2. Cut the graph paper to match the size of the cork tile and glue it on to the tile.
3. Using the materials available, create a 3D graph of either total fish harvest or per capita fish consumption.

Questions

1. Compare your graph to a classmate's graph that shows the other set of data. How do the graphs compare?
2. How would you describe the trends of fish harvest and per capita fish consumption?
3. In what year(s) did per capita fish consumption decrease?
4. What do you think might have caused the trends you identified?

FIGURE 9-30

Years	Fish Harvest (billion kilograms)	Human Population (billions)
2000	131.0	6.07
2001	130.6	6.15
2002	133.0	6.22
2003	132.5	6.31
2004	140.5	6.39
2005	142.7	6.46
2006	143.7	6.54
2007	142.2	6.61
2008	142.2	6.69
2009	145.7	6.82
2010	148.0	6.90
2011	156.2	7.00
2012	156.7	7.05

Source: UN Food and Agriculture Organization and Earth Policy Institute

---

B. Take Action

National Geographic Learning Framework

Attitudes | Responsibility
Skills | Problem-Solving
Knowledge | Our Living Planet

For one week, weigh the food that is purchased in your home and the food that is thrown out. Keep track of the types of food you eat from day to day, using categories like fruits, vegetables, meats, dairy, and even more specific categories if you wish. For instance, you might weigh packaging and separate your data into "landfill" and "recyclable" data.

1. Record and compare your data.
2. Develop a plan to increase your food sustainability, for example, by cutting your household food waste in half.
3. Develop a similar study for your school cafeteria and report the results and your recommendations to school officials.

CHAPTER 10
WATER RESOURCES AND WATER POLLUTION

WATER IS ABUNDANT ON EARTH but much of it is held in oceans, locked in ice, or stored underground. Only a tiny percentage is available as liquid fresh water—a resource vital to human health and economies. Getting water to where it is needed and protecting water quality is not always easy. Managing water resources raises social, economic, political, and environmental concerns. One thing is certain: The world needs water.

KEY QUESTIONS

10.1 Why is fresh water in short supply?

10.2 How can people increase freshwater supplies?

10.3 How can people use fresh water more sustainably?

10.4 How can people reduce water pollution?

Hoover Dam converts energy from moving water into electricity that is delivered to California, Nevada, and Arizona. Its reservoir, Lake Mead, provides fresh water to more than 20 million people in these states as well.

NATIONAL GEOGRAPHIC | EXPLORERS AT WORK

Rescuing the Colorado River Delta

with National Geographic Explorer Osvel Hinojosa Huerta

Once upon a time the Colorado River surged through the Grand Canyon on its journey to the Sea of Cortez. Back then, Mexico's Colorado River delta flourished with willow, cottonwood, and mesquite forests. Thousands of herons and other birds lined its shores. Fast-forward a century, and millions of Americans and Mexicans rely on the Colorado River for drinking, bathing, irrigation, and industry. The river that was once powerful enough to sculpt the Grand Canyon has been reduced to a trickle.

The delta that once teemed with life now stretches for miles as desert, bare soil, and salty mudflats. Invasive species choke out the native trees and fish. Migratory bird populations that depend on wetlands for wintering or stopover habitats have experienced devastating declines. Even more troubling, the river no longer reaches the sea during most of the year.

National Geographic Explorer and wetlands conservationist Osvel Hinojosa Huerta is determined to rescue wetlands on the Colorado River delta from the effects of the damming that has brought the river to a halt. As Dr. Hinojosa Huerta explains, "When dams were planned and constructed, all water rights were allocated to cities, agriculture, and industry. No one thought about the needs of nature. . . . Today about 80% of those wetland areas have been lost."

Hinojosa Huerta works with environmental coalitions, governments, businesses, and citizen groups along the Colorado River's drainage basin to save the delta. He and his team use mapping and remote sensing to monitor wetland loss. He believes restoring just 1–3% of the river's flow would allow it to reconnect with the sea as well as renew about 80,900 hectares (200,000 acres) of delta wetland.

One successful experiment in 2014 used a "pulse flow," a controlled release of water from behind dams into the Colorado River. The pulse flow dramatically changed the river's ecosystem. Copepods (microscopic crustaceans) that had been dormant in the sand for more than a decade suddenly appeared along the river's edge, feeding on algae. Dragonflies fed on the copepods, and in turn they attracted carp from upstream. In 2021, another pulse flow was released, delivering much-needed water to restoration and recreational areas. More are planned for the future.

The goal of restoration efforts, Hinojosa Huerta says, is to create habitat connectivity. That is key to ensuring that birds and other wildlife can find enough places in the delta to feed, breed, rest, and nest.

Hinojosa Huerta encourages students to be informed about the water they use. "We turn on our tap, but we don't know where the water comes from," he says. "So learn about your watershed . . . and the conservation concerns of that watershed."

Thinking Critically
Analyze What ecological factors do you think might come into play when deciding how to time a pulse flow?

Osvel Hinojosa Huerta observes evidence of water in a previously dry channel, the result of a "pulse flow" of water released into the Colorado River delta. With him is National Geographic Freshwater Fellow Sandra Postel (featured later in this chapter).

CASE STUDY
The Colorado River Story

The Colorado River, the major river of the arid Southwest, flows 2,300 kilometers (1,400 miles) through seven states to the Gulf of California (Figure 10-1). Most of its water comes from snowmelt in the Rocky Mountains. During the past 100 years, the Colorado has been tamed by 15 major dams, reservoirs, and canals. The canals reduce flooding and supply water to farms, ranches, and cities.

Major dams along the river have hydroelectric plants that supply electricity to roughly 40 million people in seven states. The river's water is used to produce 15% of the nation's crops and livestock. The system also supplies water to some of the nation's driest, hottest cities. Without it, Las Vegas and Phoenix would become uninhabitable desert. San Diego and Los Angeles could not support their present populations. In the Imperial Valley, California's most productive agricultural region, crops would wither and die.

So much water is withdrawn from the Colorado that little of it reaches the sea, and the delta has dried up. Since 1999, the river's watershed system has also experienced severe drought, a prolonged period in which precipitation is lower and evaporation is higher than normal. Managing this shared resource will become more challenging as local populations and economies grow.

The Colorado River is part of a larger story. In fact, emerging shortages of water for drinking and irrigation in several parts of the world represent one of the major environmental challenges of the 21st century.

In this chapter, you will learn about the freshwater resources available to people and why they are important. You will learn about current and projected shortages in freshwater supplies and discover steps people can take toward more sustainable water use.

As You Read Think about your daily water use. Where could you conserve? Do you know where your fresh water comes from?

FIGURE 10-1

The Colorado River Basin The area drained by this large river basin covers much of the southwestern United States. This map shows 6 of the river's 15 dams. The dams allow little water to reach the delta, which is in Mexico.

330 CHAPTER 10 WATER RESOURCES AND WATER POLLUTION

10.1 Why Is Fresh Water in Short Supply?

CORE IDEAS AND SKILLS
- Explain the importance of fresh water.
- Differentiate between surface water and groundwater and describe reliable surface runoff.
- Explain the causes and effects of freshwater scarcity.

KEY TERMS

zone of saturation
water table
reliable surface runoff
virtual water
water footprint

People Often Manage Fresh Water Poorly

You live on a planet that is unique because of a precious layer of water—most of it salt water—covering about 71% of its surface. Water is an unusual compound with properties that keep humans and every other species alive. You could survive for several weeks without food, but only a few days without fresh water. There is no substitute for this vital form of natural capital.

Access to clean fresh water is important to human health and economies as well as to national and global security and the environment. First, it is a global health issue. The World Health Organization (WHO) estimates an average of 4,400 people die each day from waterborne infectious diseases because they lack access to safe drinking water. Second, access to fresh water is an economic issue because water is vital for producing food and energy as well as reducing poverty. Third, access to fresh water is a national and global security issue because of increasing tensions within and between nations over the limited freshwater resources they share.

Finally, water is an environmental issue. Excessive withdrawal of water from rivers and aquifers has resulted in dwindling rivers, shrinking lakes, and disappearing wetlands. This, in combination with water pollution in many areas of the world, has degraded water quality. It has also reduced fish populations, hastened the extinction of some aquatic species, and degraded aquatic ecosystem services.

checkpoint Explain how access to fresh water affects human health, economies, security, and the environment.

Most of Earth's Fresh Water Is Not Available

Only about 0.024% of the planet's water supply is available to people as liquid fresh water. This water is found in lakes, rivers, streams, and accessible underground deposits. Surface runoff is another source. Fortunately, fresh water is continually recycled, purified, and distributed in Earth's hydrologic cycle (Figure 3-16). This irreplaceable water recycling and purification works well on its own. The system begins to fail when the cycle is altered, overloaded with pollutants, or overdrawn.

Earth's fresh water is not distributed evenly. Canada, with only 0.5% of the world's population, has 20% of the world's fresh water. China, with 19% of the world's people, has only 6.5% of the freshwater supply. Changes to the hydrologic cycle make this uneven distribution more pronounced. Global atmospheric warming alters the water cycle by evaporating more water into the atmosphere. As a result, wet places see more frequent and heavier flooding. Dry places experience more intense drought. Storms can be more forceful.

Groundwater Much of Earth's fresh water is stored in the ground. Precipitation soaks into the ground and sinks through spaces in soil, gravel, and rock until an impenetrable layer of rock or clay stops it. The fresh water in these underground spaces is groundwater, and it is a key component of Earth's natural capital.

The spaces in soil and rock close to Earth's surface hold little moisture. However, below a certain depth, in the **zone of saturation**, these spaces are completely filled with fresh water. The top of the groundwater zone is the **water table**. The water table rises in wet weather. The water table falls in dry weather or when people remove groundwater faster than nature can replenish it.

Deeper down lie geological layers called aquifers. Aquifers are underground caverns and porous layers of sand, gravel, or rock. Some caverns have rivers of groundwater flowing through them. The layers of sand, gravel, or rock in most aquifers are like large sponges. Groundwater seeps through these layers—typically moving only a meter or so (about 3 feet) per year and rarely more than 0.3 meter (1 foot) per day. Watertight layers of rock or clay below aquifers keep fresh water from sinking deeper into Earth.

LESSON 10.1

Pumps bring groundwater to the surface for irrigating crops, supplying households, and meeting the needs of industries. Most aquifers replenish naturally by precipitation that sinks through exposed soil and rock. Others recharge from nearby lakes, rivers, and streams. However, aquifers recharge slowly. And in urban areas, so much of the landscape has been developed that fresh water can no longer penetrate the ground to recharge aquifers as needed.

Surface Water Groundwater is one critical freshwater resource. Surface water is another. Surface water is the fresh water from rain and melted snow that flows or is stored in lakes, reservoirs, wetlands, streams, and rivers. Precipitation that does not infiltrate the ground or return to the atmosphere by evaporation is called surface runoff. Recall that the land from which surface runoff drains into a particular body of water is called a watershed or drainage basin. The drainage basin of the Colorado River is shown in yellow and green on the map in the Case Study. Two-thirds of the annual surface runoff into rivers and streams is lost in seasonal floods and is not available for human use. The remaining one-third is called **reliable surface runoff** and is regarded as a stable source of fresh water from year to year.

The human population more than tripled in the last century. Global water use has increased sixfold, and per capita withdrawals have quadrupled. Some water experts predict that people are likely to remove up to 90% of reliable runoff by 2025. Factors behind this prediction include population growth, rising rates of water use per person, longer dry periods in some areas, and failure to reduce unnecessary water losses.

Worldwide, people use 70% of the fresh water taken each year from rivers, lakes, and aquifers to irrigate cropland and raise livestock. In arid regions, 90% of all water is used for food production. Industry uses roughly another 20% of the water withdrawn globally each year. Cities and residences use the remaining 10% (Figure 10-2).

You use most water indirectly, as shown in Figure 10-3. This water is called virtual water. **Virtual water** is fresh water that is used to produce food and other products. Indirect water use makes up a large part of a person's water footprint, especially in more-developed countries. Your **water footprint** is a rough measure of the volume of fresh water you use directly and indirectly to support your lifestyle. Agriculture accounts for 92% of humanity's water footprint. Producing and delivering a typical quarter-pound hamburger, for example, takes about 2,400 liters (about 630 gallons) of fresh water. That is enough to fill more than 16 bathtubs. Most of that water is used to grow grain to feed cattle.

checkpoint What is the difference between direct and indirect use of water?

FIGURE 10-2

Daily Water Use According to the American Water Works Association, the average American directly uses about 261 liters (about 69 gallons) of fresh water each day. That's enough to fill almost two bathtubs.

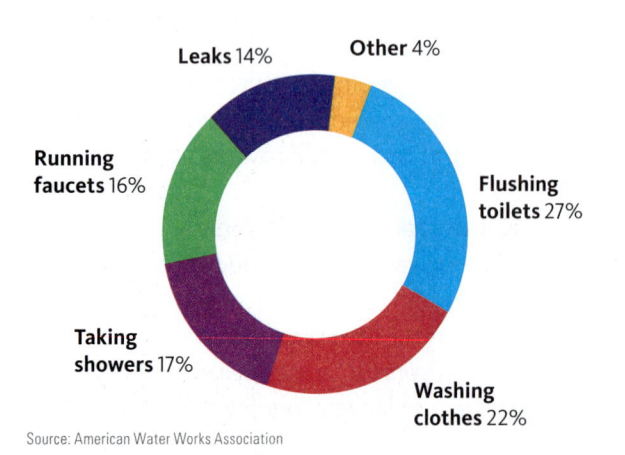

- Leaks 14%
- Other 4%
- Running faucets 16%
- Flushing toilets 27%
- Taking showers 17%
- Washing clothes 22%

Source: American Water Works Association

FIGURE 10-3

Hidden Water Use Most of the water people use each day is virtual water, which is water used to produce the products that they consume. How does the amount of water used to produce and deliver one quarter-pound hamburger compare to the amount of water the average American uses directly in a day?

These items...	take this many bathtubs full of water to produce.
Cup of coffee	1
2-liter soda	3
Loaf of bread	4
1 dozen eggs	14
1 pound of cheese	15
Hamburger	16
T-shirt	17
Blue jeans	72
Small car	2,600
Medium-sized house	16,000

Sources: UN Food and Agriculture Organization, UNESCO-IHE Institute for Water Education, World Water Council, and Water Footprint Network

Freshwater Shortages Will Grow

The main causes of freshwater scarcity in a region are a dry climate, drought, too many people using a water supply, and inefficient use of water. Freshwater scarcity stress is a comparison of the amount of fresh water available with the amount used by humans. More than 30 countries now experience stress from freshwater scarcity. By 2050, as many as 60 countries—most of them in Asia—are likely to suffer from freshwater scarcity stress. The United States has more than enough renewable fresh water to meet its needs. However, this water is unevenly distributed. Much of it is contaminated by agricultural and industrial practices. Figure 10-4 identifies varying levels of risk to water supplies in the western half of the United States.

Currently, 30% of Earth's land experiences severe drought. That is an area roughly five times the size of the United States. By 2059, as much as 45% of Earth's land could see extreme drought from a combination of natural drought cycles and projected climate change. In 276 of the world's water basins, two or more countries share the available freshwater supplies. However, countries in only 118 of those basins have water-sharing agreements. At the same time, populations are growing, demand for water is increasing, and freshwater supplies are shrinking. Conflicts over shared freshwater resources are likely to happen more frequently as a result. The *politics of scarcity* is a term used to describe the view that depletion of essential resources such as water leads to political and military conflicts.

In 2020, the United Nations and the World Health Organization reported that roughly 1 in 4 people lacked safely managed drinking water in their homes. In addition, nearly half the world's population lacked safely managed sanitation. Most of those people were impoverished. The report did note some progress: between 2016 and 2020, the global population with safely managed drinking water at home increased from 70% to 74%. Still, many analysts view the likelihood of expanding water shortages in many parts of the world as one of the most serious environmental, health, and economic problems that society faces.

checkpoint What is freshwater scarcity stress?

10.1 Assessment

1. **List** List the ways in which water is an environmental issue.
2. **Describe** What is reliable surface runoff?
3. **Summarize** What are the main causes of freshwater scarcity?
4. **Explain** Explain the concept of virtual water.

CROSSCUTTING CONCEPTS

5. **Stability and Change** Consider how lifestyle affects the use of and demand for fresh water. How are use and demand related to each other? Can you think of a feedback loop involved in this system?

FIGURE 10-4 ▼

Water Scarcity Risk Areas marked with darker colors represent those with higher risk to water supplies. Which of these areas are found in the Colorado River basin?

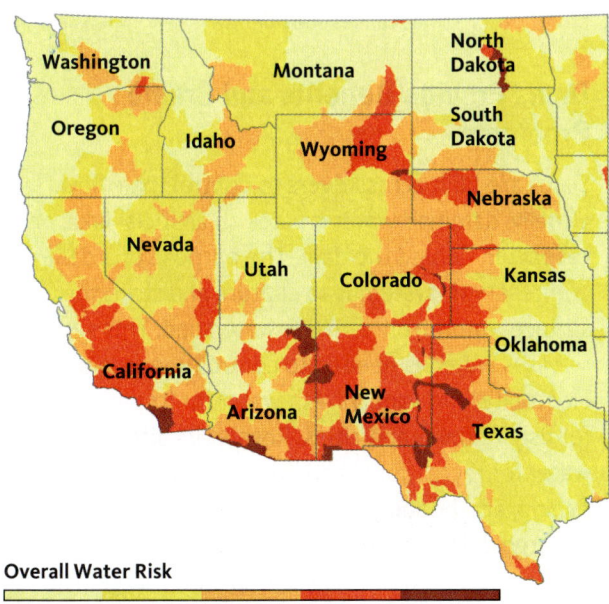

Overall Water Risk

Low	Low - Medium	Medium - high	High	Extremely high
(0-1)	(1-2)	(2-3)	(3-4)	(4-5)

Sources: World Resources Institute

10.2 How Can People Increase Freshwater Supplies?

CORE IDEAS AND SKILLS
- Explain the harmful effects of aquifer depletion.
- Understand the benefits and problems associated with dams and reservoirs.
- Discuss the problems associated with water transfer and desalination.

KEY TERMS

subsidence   aqueduct   desalination
reservoir   snowpack

Aquifers Are Being Depleted

Aquifers provide drinking water for nearly half of the world's people. In the United States, aquifers supply almost all drinking water in rural areas (20% in urban areas) and 37% of irrigation water. However, test wells and other data indicate that water tables are falling in many areas of the world. Aquifers are being pumped faster than they can recharge. Most water is drawn from aquifers to irrigate crops. According to the Consultative Group on International Agricultural Research, an estimated 14–17% of food produced with use of groundwater relies on unsustainable withdrawal of groundwater resources. This practice is referred to as *overpumping*.

Overpumping aquifers contributes to limits on food production, rising food prices, and widening economic gaps in some areas. As water tables drop, the energy and costs of pumping water rise sharply. Farmers must drill deeper wells, buy larger pumps, and use more electricity to run the pumps. The farmers who cannot afford to do this often lose their land. Struggling farmers must work for wealthier farmers or migrate to cities that are already overcrowded with people.

> **CONSIDER THIS**
>
> **Eating less meat conserves water.**
> One of the easiest ways to cut back on water consumption is to eat less meat. Growing corn and other feed crops for livestock uses a tremendous quantity of water. People who don't consume any meat or dairy products use nearly 2,270 fewer liters (600 fewer gallons) of water per day than the average American.

FIGURE 10-5 ▼

Groundwater Depletion	
Prevention	**Control**
Use water more efficiently	Raise price of water to discourage waste
Subsidize water conservation	Tax water pumped from wells near surface water
Limit number of wells	Build rain gardens in urban areas
Stop growing water-intensive crops in dry areas	Use permeable paving material on streets, sidewalks, and driveways

Withdrawing large amounts of groundwater can sometimes cause land to collapse. **Subsidence** occurs when the land above an aquifer *subsides*, or sinks. Once an aquifer becomes compressed by subsidence, recharge is impossible. Subsidence can also damage roadways, water and sewer lines, and the foundations beneath buildings.

Overpumping groundwater in coastal areas, where many of the world's largest cities are found, can pull salt water into freshwater aquifers. The resulting contaminated groundwater is undrinkable and can't be used for irrigation. This problem is especially serious in the U.S. states of California, Texas, Florida, Georgia, South Carolina, and New Jersey. It is also a prevalent in Turkey, Thailand, and the Philippines. Figure 10-5 outlines strategies for preventing and controlling depletion of groundwater supplies.

checkpoint What causes subsidence?

Dams Provide Benefits and Create Problems

Dams are built across rivers to control their flow. Usually, dammed water creates an artificial lake, or **reservoir**, behind the dam. The purpose of a dam-and-reservoir system is to capture and store the surface runoff from a river's watershed. This system releases water as needed to control floods, generate electricity, and supply fresh water for irrigation and for use in towns and cities. Reservoirs also provide water for recreational activities such as swimming, fishing, and boating.

Large dams and reservoirs provide several important benefits. The world's 45,000 large dams capture and store 14% of Earth's surface runoff. They provide water for almost half of all irrigated cropland.

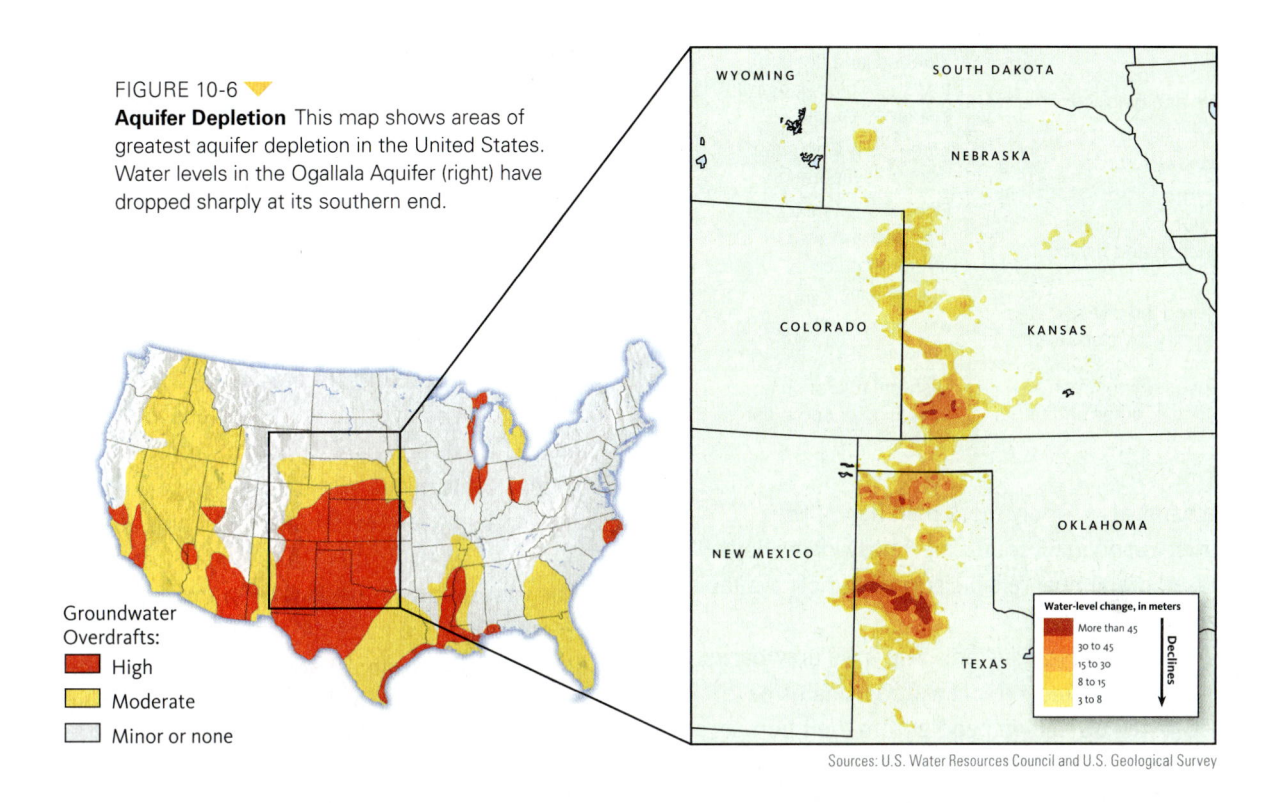

FIGURE 10-6

Aquifer Depletion This map shows areas of greatest aquifer depletion in the United States. Water levels in the Ogallala Aquifer (right) have dropped sharply at its southern end.

Groundwater Overdrafts:
- High
- Moderate
- Minor or none

Water-level change, in meters
- More than 45
- 30 to 45
- 15 to 30
- 8 to 15
- 3 to 8

Declines

Sources: U.S. Water Resources Council and U.S. Geological Survey

SCIENCE FOCUS 10.1

AQUIFER DEPLETION IN THE UNITED STATES

In the United States, groundwater is being withdrawn from aquifers four times faster, on average, than it is being replenished. One of the most serious overdrafts of groundwater is evident in the lower half of the Ogallala Aquifer. The Ogallala Aquifer is one of the world's largest known aquifers. It stretches beneath eight states from South Dakota to Texas (Figure 10-6).

The Ogallala Aquifer supplies one-third of all the groundwater used for irrigation in the United States. This aquifer turned the Great Plains into one of world's most productive irrigated agricultural regions. The Ogallala is essentially a deposit of liquid natural capital with a slow rate of recharge.

In parts of the southern half of the Ogallala, groundwater is being pumped out 10–40 times faster than the natural recharge rate. This has lowered water tables and raised pumping costs, especially in Texas. Overpumping, along with urban development and restricted access to Colorado River water, has led to reduced cropland in Texas, Arizona, Colorado, and California. It has also increased competition for water among farmers, ranchers, and growing urban areas.

Government subsidies—payments or tax breaks designed to increase crop production—have encouraged farmers to continue growing thirsty crops in dry areas. This has accelerated depletion of the Ogallala Aquifer. In particular, corn has been planted widely on fields watered by the Ogallala. Serious aquifer depletion is also taking place in California's semiarid Central Valley. This region supplies half of the country's fruits and vegetables.

Thinking Critically
Analyze What economic effects might result from the depletion of groundwater supplies from the Ogallala Aquifer?

LESSON 10.2

FIGURE 10-7

Using Dam-and-Reservoir Systems	
Advantages	Disadvantages
Provides irrigation water above and below dam	Flooded land destroys forests or cropland and displaces people
Provides water for drinking	Large losses of water through evaporation
Reservoir useful for recreation and fishing	Deprives downstream cropland and estuaries of nutrient-rich silt
Can produce cheap electricity (hydropower)	Risk of failure and devastating downstream flooding
Reduces downstream flooding of cities and farms	Disrupts migration and spawning of some fish

Large dams also supply more than half of the electricity used in 65 countries. Finally, dams have increased the annual reliable surface runoff available to people by nearly 33%.

Dam-and-reservoir systems also have drawbacks. For example, the reservoirs created by dams have displaced 40–80 million people from their homes. They have flooded large areas of mostly productive land. Dams have also impaired some of the important ecosystem services that rivers provide. As a result of dams, excessive water withdrawals, and prolonged drought in some areas, only 21 of the planet's 177 longest rivers consistently reach the sea. This fact explains in part why estimated extinction rates for freshwater life are four to six times higher than for marine or terrestrial species. Figure 10-7 summarizes the advantages and disadvantages of using dams.

Reservoirs have limited life spans. Within 50 years, reservoirs typically fill up with sediments (mud and silt). This makes them useless for storing water or producing electricity. In the Colorado River system, the equivalent of roughly 20,000 dump-truck loads of silt are deposited on the bottoms of Lake Powell and Lake Mead every day. Sometime this century, these two reservoirs will be too full of silt to function as designed. About 85% of U.S. dam-and-reservoir systems will be 50 years old or older by 2025. Some of those aging dams are already being removed because their reservoirs have filled with silt.

If climate change occurs as predicted during this century, water shortages will only intensify. For example, billions of people in South America and Asia depend on river flows fed by mountain glaciers. These glaciers serve as freshwater savings accounts.

They tend to store precipitation as ice and snow in wet periods and melt slowly in dry periods. As of 2020, many of these mountain glaciers had been shrinking for more than 25 consecutive years. If the glaciers eventually disappear, nearly 3 billion people will likely be short of fresh water for all purposes, including food production.

checkpoint What is the goal of a dam-and-reservoir system?

Water Transfers Can Be Harmful

Canals and pipelines are used to transfer large volumes of water from water-rich areas to water-poor areas. When you consume lettuce in the United States, chances are it was grown in the arid Central Valley of California. It was irrigated in part with water transferred in from the mountains of northeastern California. The California State Water Project is one of the world's largest freshwater transfer projects. This project uses a maze of giant dams, pumps, and **aqueducts** (lined canals), to transfer fresh water to the vast farms and heavily populated cities of water-poor central and southern California.

The California State Water Project has yielded many benefits. The heavily irrigated Central Valley supplies half the nation's fruits and vegetables. However, the project has also reduced the flow of the Sacramento River. This in turn has threatened fisheries and reduced the flushing action that helps cleanse San Francisco Bay of pollutants. The flow of fresh water to the Bay's coastal marshes and other ecosystems has dropped as well. These factors have placed stress on the species that depend on the Bay's many ecosystems.

Subsidies provided to the California State Water Project promote inefficient uses of water, such as irrigating thirsty crops like lettuce, alfalfa, and almonds in desert-like areas. In central California, agriculture consumes three-fourths of the water that is transferred. Much of that is lost through inefficient irrigation systems. Studies show that making irrigation just 10% more efficient would provide all the water necessary for domestic and industrial uses in southern California. However, taxpayer subsidies lower the price of water for farmers and urban residents. Consequently, they have little incentive to invest in more efficient irrigation or in water-saving toilets, showerheads, and other devices.

Projected climate change will make matters worse in California by reducing surface water availability.

FIGURE 10-8
The Carlsbad Desalination Project opened in Carlsbad, California, in late 2015. The plant uses reverse osmosis to produce 189 million liters (50 million gallons) of drinkable water each day. It is the largest desalination plant in the Western Hemisphere.

In northern California, many people depend on mountain snowpacks in the Sierra Nevada for more than 60% of their fresh water during the hot, dry summer. **Snowpacks** are bodies of densely packed, slowly melting snow. Projected climate change could shrink the snowpacks by as much as 40% by 2050 and by as much as 90% by the end of this century. Shrinking snowpacks will sharply reduce the amount of fresh water available for northern California residents and ecosystems in the region.

checkpoint What impacts has the California Water Project had on the Sacramento River?

Removing Salt from Seawater Is Costly and Harmful

Saltwater processing could potentially provide a source of fresh water. **Desalination** is the process of removing dissolved salt from ocean water or from brackish (slightly salty) water in aquifers or lakes (Figure 10-8). Currently, the two most widely used methods for desalinating water are distillation and reverse osmosis. Distillation involves heating salt water until it evaporates, leaving behind salts in solid form. The water vapor is then cooled and condenses as fresh water. Reverse osmosis uses high pressure to force salt water through a membrane filter with pores small enough to remove the salt and other impurities. After passing through the membrane, fresh water is collected and waste materials are flushed away.

Desalination supplies just 1% of the world's demand for fresh water and only 0.4% of demand in the United States. Three main problems prevent desalination from being more widely used.

1. Desalination is costly because removing salt from seawater requires a lot of energy.
2. Pumping large volumes of seawater through pipes requires chemicals to sterilize the water and prevent algae growth. This kills many marine organisms.
3. Desalination produces huge quantities of salt waste that require proper disposal. Dumping salt waste into coastal ocean waters increases the salinity of those waters. Increased salinity threatens food resources and aquatic life, especially if salt waste is dumped near coral reefs, marshes, or mangrove forests. Disposing of salt waste on land can contaminate groundwater and surface water.

Currently, desalination is practical only for water-short countries and cities that can afford its high cost. However, scientists and engineers are working to develop better and more affordable desalination technologies.

checkpoint What is the difference between distillation and reverse osmosis methods of desalination?

10.2 Assessment

1. **Explain** What harmful effects can occur from overpumping aquifers?
2. **Restate** What are the benefits and drawbacks of damming a river?
3. **Summarize** What are the three main problems associated with desalination?
4. **Evaluate** Look again at the advantages and disadvantages of dam-and-reservoir systems presented in Figure 10-7. Which advantage and which disadvantage do you think are most important? Explain your answers.

SCIENCE AND ENGINEERING PRACTICES

5. **Defining Problems** Identify a problem mentioned in this lesson that could be solved or lessened through an improved engineering design. What criteria would an ideal solution meet?

10.3 How Can People Use Fresh Water More Sustainably?

CORE IDEAS AND SKILLS

- Explain how low-cost water leads to waste of fresh water in irrigation, industries, and homes.
- Describe ways to reduce freshwater losses.

KEY TERMS

flood irrigation   drip irrigation   gray water

Reducing Freshwater Losses

Experts say that the world's freshwater needs can be met if supplies are used wisely. Improving irrigation practices would achieve the largest reduction in water loss. Other reductions would result from reducing government subsidies, repairing infrastructure in cities, increasing water reuse, and educating the public about water conservation.

338 CHAPTER 10 WATER RESOURCES AND WATER POLLUTION

FIGURE 10-9

Irrigation Methods Traditional irrigation methods rely on gravity and flowing water (left). Newer systems, such as drip irrigation (middle) and center pivot irrigation with sprinklers (right), are far more efficient.

Gravity flow
(efficiency 60% and 80% with surge valves)

Water usually comes from an aqueduct system or a nearby river.

Drip irrigation
(efficiency 90–95%)

Above- or below-ground pipes or tubes deliver water to individual plant roots.

Center pivot
(efficiency 80–95% depending on sprinkler type)

Water is usually pumped from underground and sprayed from a mobile boom with sprinklers.

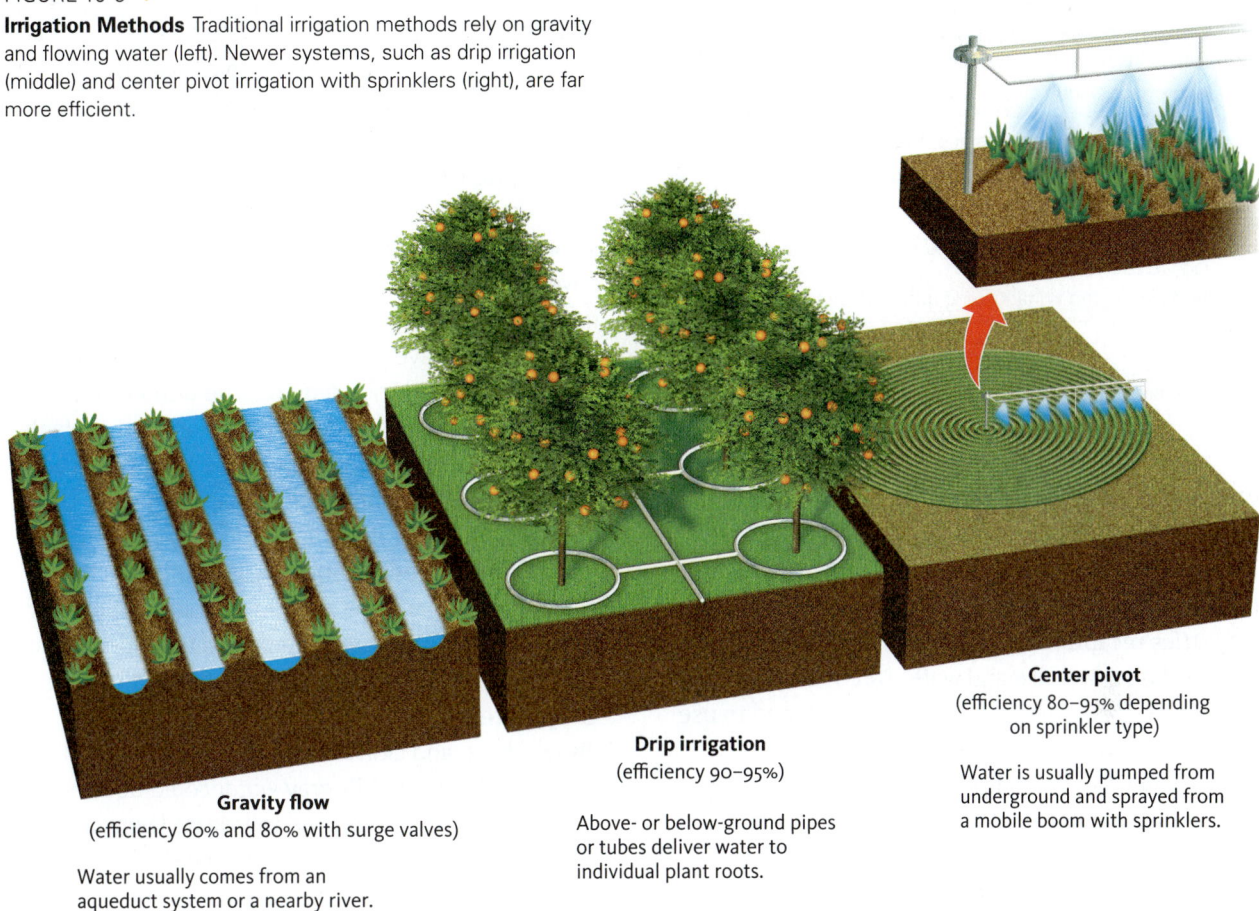

Government subsidies help to create the artificially low prices that most people pay to use fresh water. Governments provide irrigation water or the electricity and diesel fuel that farmers use to pump fresh water from rivers and aquifers at below-market prices. Americans and Chinese—two of the world's largest water users—pay only 2.9% and 0.5%, respectively, of their disposable income for water. Failing to apply full-cost pricing gives users little or no reason to invest in water-saving technologies. In addition, there is a lack of subsidies for improving the efficiency of water use. Replacing subsidies that encourage wasting water with subsidies that encourage saving it would reduce water losses.

Irrigation Most irrigation systems are highly inefficient. In fact, only 40–60% of the world's irrigation water reaches any crops. Flood irrigation is the most inefficient irrigation system and is commonly used in less-developed countries. In **flood irrigation**, water is pumped from a groundwater or surface water source into unlined ditches and flows by gravity to crops (Figure 10-9, left). This method delivers far more water than is needed for crops to grow. About 45% of the water is lost through evaporation, seepage, and runoff. With existing irrigation technologies (Figure 10-9, middle and right), this loss could be reduced to 5–10%.

Drip irrigation is a more efficient system for delivering water to crops. **Drip irrigation** involves using pipes or tubes to deliver water directly to plant roots. The global area of cropland where drip irrigation is used has increased sixfold since the early 1990s. However, drip irrigation is used on less than 5% of the world's irrigated crop fields because these systems are costly. If fresh water were priced closer to the value of the ecosystem services it provides and if government subsidies leading to inefficient water use were eliminated, farmers would have more incentive to use drip irrigation.

According to the United Nations, reducing the global withdrawal of water for irrigation by just 10% would save enough water to grow crops and meet the estimated additional water demands of Earth's cities and industries through 2025. Figure 10-10 lists several ways to reduce water losses in irrigation.

LESSON 10.3

FIGURE 10-10

Reducing Irrigation Water Losses
Avoid growing thirsty crops in dry areas.
Import water-intensive crops and meat.
Encourage organic farming and polyculture to retain soil moisture.
Monitor soil moisture to add water only when necessary.
Expand use of drip irrigation and other efficient methods.
Irrigate at night to reduce evaporation.
Line canals that bring water to irrigation ditches.

Industries and Homes Producers of chemicals, paper, oil, coal, primary metals, and processed foods consume almost 90% of the fresh water used by industries in the United States. Some of these industries recapture, purify, and recycle water to reduce their water use and water treatment costs. Most industrial processes could be redesigned to use much less water.

Flushing toilets with fresh water is the single largest use of domestic fresh water in the United States. It accounts for one-fourth of home water use. Since 1992, U.S. government standards have required that new toilets use no more than 6.1 liters (1.6 gallons) of water per flush. Even at this rate, just two flushes use more than the total daily amount of water available to many of the world's poor who live in arid regions.

Other water-saving appliances are widely available. Low-flow showerheads save large amounts of water by cutting the flow of a shower in half. Front-loading clothes washers use 30% less water than top-loading machines. According to an estimate by the American Water Works Association, the typical American household could cut its daily water use nearly one-third by using low-flow appliances and stopping leaks.

Many homeowners and businesses in water-short areas use drip irrigation to cut water losses. Some use smart sprinkler systems with moisture sensors that have cut water use on lawns up to 40%. Others copy nature by replacing lawns with a mix of native plants. Native plants have adapted to the area's climate, weather, and soil. They require little or no watering or spraying with pesticides. In addition, native plants provide shelter and food for wildlife and pollinators, thus encouraging biodiversity. Despite these benefits, a number of communities have passed local laws requiring green lawns. The laws prohibit the planting of native vegetation in their place. Some of these communities are located in water-short areas.

Water used in homes can be reused and recycled. **Gray water** is used water from showers, sinks, dishwashers, and clothes washers. About 50–75% of a typical household's gray water could be stored in a holding tank and reused to irrigate lawns and nonedible plants, flush toilets, and wash cars. Such efforts mimic the way nature recycles water, and thus follow the nutrient cycling factor of sustainability.

The relatively low cost of water in most communities is a major cause of excessive water use and waste. About one-fifth of U.S. public water systems do not use water meters, which help track water use and can reveal water leaks. These public water systems charge a single, low annual rate for almost unlimited use of high-quality water.

Large amounts of drinkable fresh water are used to flush away industrial, animal, and household wastes. If current growth trends in population and water use continue, within 40 years the world's entire reliable flow of river water will be needed just to dilute and transport wastes. Much of this fresh water could be saved by recycling and reusing gray water from homes and businesses and wastewater from sewage treatment plants. These alternative water supplies could be reused for purposes such as cleaning equipment, watering lawns, and watering crops.

checkpoint What are the benefits of using native plants instead of green grass in arid climates?

> **CONSIDER THIS**
>
> **Given enough time, small leaks can fill swimming pools.**
>
> Even small leaks waste more water than many homeowners realize. A faucet leaking one drop per second can drip about 11,500 liters (about 3,000 gallons) of water down the drain in a year—enough to fill about 80 bathtubs. Toilets are another common source of water leakage. Anyone can easily identify whether a toilet has an undetected leak simply by adding a few drops of food coloring to the toilet tank and waiting five minutes. If the color shows up in the bowl, there is a leak.

NATIONAL GEOGRAPHIC | EXPLORERS AT WORK
Sandra Postel Freshwater Conservationist

Sandra Postel is one of the world's most respected authorities on water issues. In 1994, she founded the Global Water Policy Project, a research and education organization that promotes more sustainable use of Earth's finite, or limited, freshwater supply. Dr. Postel has authored or co-authored several influential books and written dozens of articles about using water more sustainably.

In her quest to educate people about water supply issues, Postel has appeared on several television news shows. She has taken part in many environmental documentary films (including the BBC's *Planet Earth*), and addressed the European Parliament. From 2010 to 2015, she also served as Freshwater Fellow for the National Geographic Society.

Postel also is co-director of *Change the Course*, a national freshwater conservation and restoration campaign being piloted in the Colorado River basin. In 2017, *Change the Course* won the US Water Prize, which honors innovative and sustainable approaches to water management.

FIGURE 10-11
Sandra Postel and her colleagues paddle along the Colorado River. Postel is hard at work helping to restore this important freshwater ecosystem.

LESSON 10.3

10.3 Assessment

1. **Identify** What is the major cause of excessive water use in homes and industries?
2. **Contrast** Explain the difference between flood irrigation and drip irrigation. Which method uses less water?
3. **Define** What does the word *loss* mean with regard to fresh water?
4. **Make Judgments** The text discusses several human practices that contribute to freshwater losses. Choose one of these practices and suggest an alternative, more sustainable practice that would serve the same purpose.

SCIENCE AND ENGINEERING PRACTICES

5. **Evaluate Information** Many claim that freshwater subsidies enable the farming of unproductive land, stimulate local economies, and help keep food and electricity prices low. Do you think this is reason enough for governments to continue providing subsidies to farmers and cities to keep the price of fresh water low? Why or why not?

10.4 How Can People Reduce Water Pollution?

CORE IDEAS AND SKILLS

- Compare and contrast point and nonpoint sources of water pollution.
- Explain why streams and rivers can recover from pollution more easily than lakes and reservoirs and why groundwater contamination should be prevented.
- Identify the sources of ocean pollution and their impacts on ocean ecosystems.
- Discuss sustainable ways to reduce or eliminate water pollution.

KEY TERMS

water pollution   wastewater   cultural eutrophication

Water Pollution Comes from Point and Nonpoint Sources

Water pollution is any change in water quality that can harm living organisms or make water unfit for human uses such as drinking, irrigation, and recreation. Water pollution can come from single (point) sources or from larger and dispersed (nonpoint) sources.

Sources of Water Pollution *Point sources* discharge pollutants into bodies of surface water at specific locations, often through ditches, sewer lines, or drain pipes (Figure 10-12). Examples include factories, sewage treatment plants (which remove some, but not all, pollutants), underground mines, oil wells, and oil tankers. Point sources exist in specific places. As a result, they are relatively easy to identify, monitor, and regulate. Most more-developed countries have laws that control point-source discharges of harmful chemicals into aquatic systems. In most less-developed countries, adequate control of such discharges is lacking.

Nonpoint sources are broad and diffuse areas where runoff from rainfall or snowmelt washes pollutants off the land into bodies of surface water. Examples of pollutants include eroded soil and chemicals such as fertilizers and pesticides. Runoff transports these pollutants from cropland, feedlots, logged forests, city streets, parking lots, lawns, and golf courses. It is difficult and expensive to identify and control discharges from so many diffuse sources. According to the U.S. Environmental Protection Agency, nonpoint-source pollution is the main reason why 40% of U.S. rivers, lakes, and estuaries are still not clean enough for fishing and swimming, despite major water pollution control laws. (See Lesson 1.2 for more about point and nonpoint sources of pollution.)

Agricultural activities are by far the leading source of water pollution. Sediment eroded from croplands is the most common pollutant. Other major agricultural pollutants include fertilizers

---

CONSIDER THIS

Plastics are a major source of pollution.
Scientists have discovered that chemicals leach from plastic containers into water. Bisphenol A (BPA) has received the most headlines. BPA has been linked to cancer, neurological difficulties, early puberty in girls, and reduced fertility in women. One study determined that 96% of American women have BPA in their bodies. BPA can be found in food storage containers and cans, contact lenses, the lining of tin cans, and dental sealants. One way to protect yourself from exposure to BPA is to use glass or stainless steel bottles and containers when possible.

FIGURE 10-12
Wastewater from a power plant pours from a drain pipe and is an example of point-source pollution. How might this polluted water affect the aquatic ecosystem into which it drains?

and pesticides, bacteria from livestock and food-processing wastes, and excess salts from irrigated soil. Point sources from industrial facilities that emit a variety of harmful chemicals are the second largest source of water pollution. Mining is the third largest source. Surface mining disturbs the land, which in turn leads to major erosion of sediments and runoff of toxic chemicals.

Another form of water pollution originates from the widespread use of human-made materials such as plastics to make millions of products. People throw away many of these products, and they can end up in rivers, lakes, and oceans. Consider plastic water bottles. The United States is the world's largest consumer market for bottled water. Americans drink 1,500 bottles of water every second but recycle only 38% of the 50 billion empty plastic water bottles they generate each year. This wasted plastic is worth at least $1 billion. Two million tons of single-use water bottles are dumped in landfills every year instead of being recycled. Discarded plastic water bottles take more than 1,000 years to biodegrade.

Flooding and Water Pollution Floodwater can be polluted with agricultural and industrial chemicals as well as untreated human and animal wastes. Three major human activities contribute to flooding and associated water pollution.

The first activity is the installation of infrastructure to reduce the threat of flooding on floodplains, which are land areas adjacent to streams and rivers. Some rivers are narrowed, straightened, and lined with protective levees (long mounds of earth along their banks). In the long run, however, such measures can lead to greatly increased flooding and surface water pollution when heavy snowmelt or prolonged rains cause levees to break.

A second activity that makes flooding more likely is the removal of water-absorbing vegetation, especially on hills. Trees on a hillside may be cut for timber, fuel, livestock grazing, or farming. Water from precipitation then rushes down the naked slope, and precious topsoil is carried away. This practice increases flooding and pollution in local streams.

A third human activity that increases the severity of flooding is the draining of wetlands that naturally absorb floodwaters. These areas often end up being covered with pavement and buildings, which leads to greatly increased runoff. Increased runoff contributes to flooding and pollution of surface water.

Many scientists argue that people can reduce flooding and water pollution by relying less on engineered devices and more on nature's systems. For example, wetlands absorb floodwaters and can filter pollutants from water passing through them.

LESSON 10.4

FIGURE 10-13 ▼

Stream Pollution This diagram shows how bacteria break down biodegradable pollutants where they enter a stream. The process requires a significant amount of oxygen (red line). As a result, levels of dissolved oxygen drop sharply immediately downstream of the pollution (blue line). The amount of dissolved oxygen impacts what types of organisms are present.

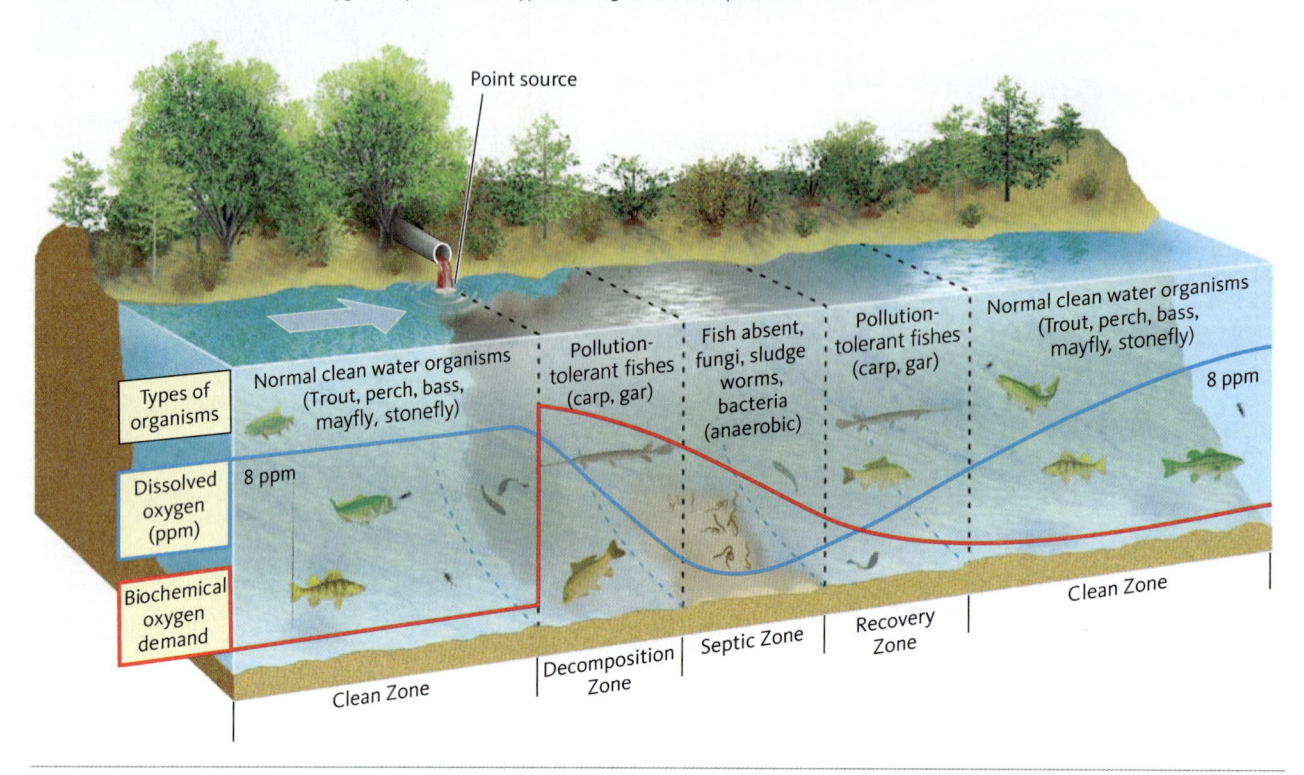

Preserving existing wetlands and restoring degraded wetlands that lie in floodplains takes advantage of the natural ecosystem services they provide.

checkpoint Identify three major causes of water pollution.

Pollution of Rivers, Lakes, and Reservoirs

Because they are flowing, rivers and streams can recover from pollutants more easily than can stationary lakes and reservoirs.

River and Stream Pollution Flowing rivers and streams can recover rapidly from moderate levels of biodegradable wastes. The wastes are diluted by flowing water and are also broken down by bacteria (Figure 10-13). However, this process does not work when a stream is overloaded with biodegradable pollutants. It does not work when drought, damming, or water diversion reduce flow. Also, this process does not eliminate slowly biodegradable and non-biodegradable pollutants. As bacteria break down biodegradable wastes they also deplete dissolved oxygen in the stream, which affects other organisms.

Federal and state laws enacted in the 1970s to control water pollution led to a greatly increased number of facilities that treat wastewater.

Wastewater is any water that contains sewage and other wastes from homes and industries. Wastewater can be treated to remove or reduce pollutants. Wastewater treatment is common in most more-developed countries. Environmental laws also require industries to reduce or eliminate their point-source discharges of harmful chemicals into surface water.

In most less-developed countries, stream pollution from discharges of untreated sewage, industrial wastes, and discarded trash is a serious and growing threat. Most cities in less-developed countries dump 80–90% of their untreated sewage directly into rivers, streams, and lakes. These bodies of water are often used for drinking, bathing, and washing clothes.

According to the World Commission on Water for the 21st Century, half of the world's 500 major rivers are heavily polluted. Most of these polluted waterways run through less-developed countries. Most of these countries cannot afford to build waste treatment plants and do not have, or do not enforce, laws to control water pollution.

Lake and Reservoir Pollution Lakes and reservoirs are generally less effective at diluting pollutants than streams. First, lakes and reservoirs often contain

344 CHAPTER 10 WATER RESOURCES AND WATER POLLUTION

layers of different densities that don't mix together much. Second, lakes and reservoirs have low flow rates or no flow at all. The flushing and changing of water in lakes and large artificial reservoirs can take from 1 year to 100 years. The same process takes just several days to several weeks for streams. As a result, lakes and reservoirs are more vulnerable than streams to contamination. Sources include runoff or discharges of plant nutrients, oil, pesticides, and non-biodegradable toxic substances like lead, mercury, and arsenic. Many toxic chemicals also enter lakes and reservoirs from the atmosphere.

Eutrophication is the natural nutrient enrichment of a shallow lake, a coastal area at the mouth of a river, or a slow-moving stream. It is a natural process that occurs in lakes and ponds (Lesson 6.4). However, lakes become more eutrophic as plant nutrients such as nitrates and phosphates are added to a watershed. Human activities greatly accelerate the input of nutrients near urban or agricultural areas. Sources include farmland, feedlots, city streets, parking lots, chemically fertilized lawns, mining sites, and sewage treatment plants. This process of accelerated eutrophication is called **cultural eutrophication**.

During hot weather, cultural eutrophication can produce dense growths or "blooms" of organisms such as algae and cyanobacteria (Figure 10-14). When the algae die, they are decomposed by swelling populations of oxygen-consuming bacteria. The bacteria use up the dissolved oxygen in the water. The lack of oxygen kills fish and other aerobic aquatic animals. If excess nutrients continue to flow into a lake, anaerobic bacteria take over. The bacteria, which can survive without oxygen, produce gaseous products such as highly toxic hydrogen sulfide and flammable methane.

FIGURE 10-14
Cultural eutrophication produced an algal bloom in this river-and-lake system in China. What human activities contribute to cultural eutrophication?

Cultural eutrophication can be prevented. Advanced (but expensive) waste treatment processes can remove nitrates and phosphates from wastewater before it enters a body of water. Another approach is to mimic Earth's natural cycles by recycling nutrients into the soil instead of dumping them into waterways. Banning or limiting the use of phosphates in household cleaners could reduce nutrient inputs. Employing soil conservation and land-use control could reduce nutrient runoff.

Lakes suffering from cultural eutrophication can be cleaned up by removing weeds, adding herbicides, and pumping air into them. However, these methods are expensive and energy intensive. Most lakes can eventually recover from cultural eutrophication if excessive inputs of plant nutrients are stopped.

Groundwater Pollution Common pollutants such as fertilizers, pesticides, gasoline, and organic solvents can seep into groundwater from numerous sources (Figure 10-15). People who dump or spill gasoline, oil, and organic solvents like paint thinners onto the ground also risk contaminating groundwater. Hydraulic fracturing, or fracking, presents a new and

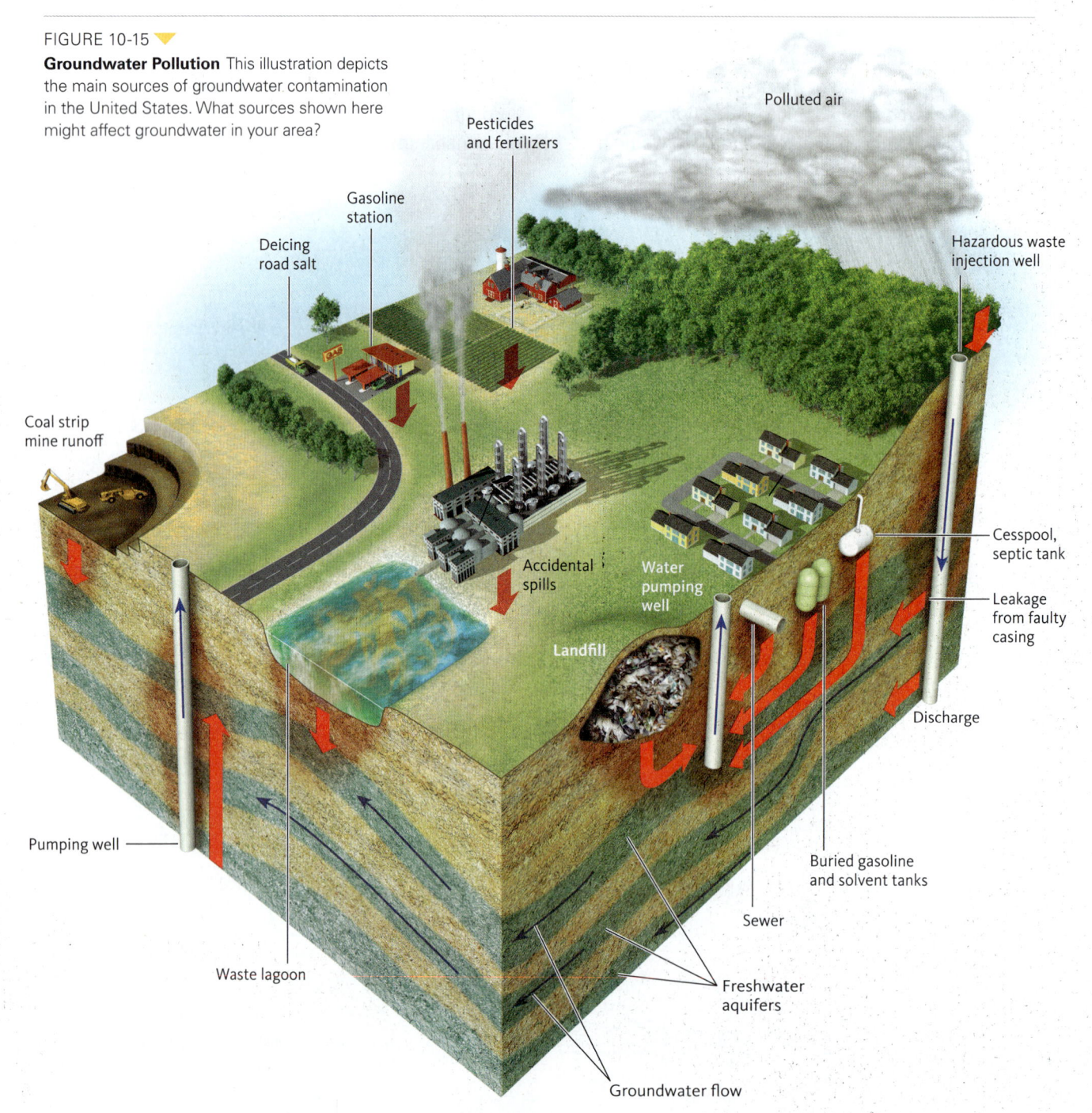

FIGURE 10-15 ▼

Groundwater Pollution This illustration depicts the main sources of groundwater contamination in the United States. What sources shown here might affect groundwater in your area?

346 CHAPTER 10 WATER RESOURCES AND WATER POLLUTION

growing potential threat to groundwater. Fracking involves drilling thousands of oil and natural gas wells in parts of the United States. (This practice is further discussed in Chapter 12.) Groundwater contamination can result from leaky pipes and pipe fittings in gas wells and from contaminated wastewater brought to the surface during fracking operations.

Groundwater pollution is a serious hidden threat to human health globally. Contaminated groundwater threatens water supplies used for drinking and irrigation (Figure 10-16).

When groundwater becomes contaminated, it cannot cleanse itself of degradable wastes as quickly as flowing surface water. Groundwater flows so slowly that contaminants are not diluted and dispersed effectively. In addition, groundwater usually has much lower concentrations of dissolved oxygen (which decomposes many contaminants) and smaller populations of decomposing bacteria. The typically cold temperatures of groundwater slow down chemical reactions that decompose wastes.

When groundwater becomes polluted, it can take tens to thousands of years to become purified of slowly degradable wastes (such as DDT). On a human timescale, non-biodegradable wastes (such as lead and arsenic) remain in the water permanently. If toxic chemicals reach an aquifer, effective cleanup is often not possible or is too costly. Although there are ways to clean up contaminated groundwater, such methods are very expensive. Cleaning up a single contaminated aquifer can cost anywhere from $10 million to $10 billion. Thus, preventing groundwater contamination is the only effective way to deal with this serious water pollution problem.

Freshwater Purification In more-developed countries, drinking water from surface water sources is typically stored in a reservoir for several days. This improves clarity and taste by increasing dissolved oxygen content and allowing suspended matter to settle. The water is then pumped to a purification plant and treated. In areas with pure groundwater, little treatment is needed to meet government drinking water standards. Many less-developed countries lack such laws, or they are not enforced.

Technology exists that can convert sewer water into pure drinking water. One process begins with microfiltration to remove bacteria and suspended solids. The wastewater then undergoes reverse osmosis to remove minerals, viruses, and various organic compounds. Finally, hydrogen peroxide and ultraviolet light remove additional organic compounds. In a world where people will face increasing shortages of drinking water, the business of wastewater purification is likely to experience major growth. However, the reclamation of wastewater still faces opposition from citizens and health officials who are unaware of the advances in this technology.

checkpoint Why is prevention the only feasible way to deal with groundwater pollution?

Ocean Pollution Is a Growing Problem

Why should people care about ocean pollution? Oceans foster life. They provide and recycle the planet's fresh water through the water cycle. Oceans affect weather and climate and help regulate Earth's temperature. They also absorb some of the massive amounts of carbon dioxide human activities emit into the atmosphere.

Coastal areas such as wetlands, estuaries, coral reefs, and mangrove swamps bear the brunt of ocean pollution. In coastal areas of less-developed countries, 80–90% of city sewage is dumped into oceans without being treated. This often overwhelms the natural ability of the coastal waters to degrade the wastes and upsets the marine ecosystem. Adding excessive nitrates and phosphates to the ocean instead of recycling these vital plant nutrients to the soil alters the nitrogen and phosphorus cycles (Lesson 3.4). The dumping of wastes in coastal waters represents a serious violation of the nutrient cycling factor of sustainability.

In deeper waters, the oceans can dilute, disperse, and degrade large amounts of raw sewage and other types of degradable pollutants. However, recent studies of some U.S. coastal waters found huge colonies of viruses thriving in raw sewage and other wastes from sewage treatment plants and leaking septic tanks. It's been estimated that one-fourth of the people using coastal beaches in the United States become ill after swimming in seawater containing infectious viruses and bacteria. They develop ear infections, sore throats, eye irritations, respiratory diseases, or gastrointestinal diseases.

Runoff of sewage and agricultural wastes into coastal waters introduces large quantities of nitrate (NO_3^-) and phosphate (PO_4^{3-}), plant nutrients that cause explosive growths of algae. These harmful algal blooms—known as red, brown, or green toxic tides—lead to the release of waterborne and airborne toxins.

FIGURE 10-16
ON ASSIGNMENT National Geographic photographer Joel Sartore captured visual evidence of threats to groundwater supplies in Arizona. The brown water in this basin is filled with tailings—wastes left over from copper mining. Chemicals in the wastewater could seep into and contaminate groundwater. This in turn could contaminate drinking water supplied to residents in nearby homes.

349

FIGURE 10-17

Ocean Pollution Residential areas, factories, and farms all contribute to the pollution of coastal waters.

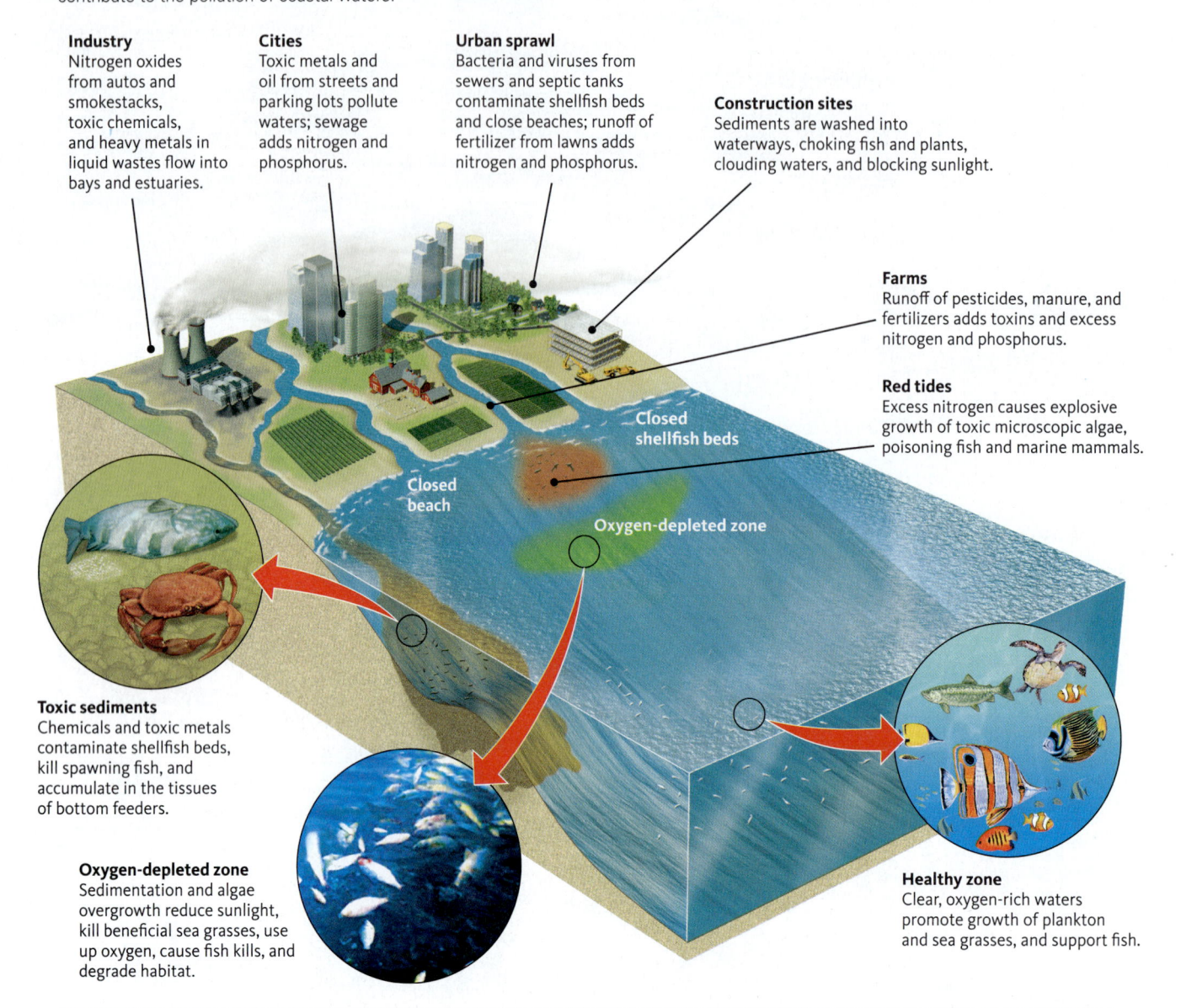

Industry
Nitrogen oxides from autos and smokestacks, toxic chemicals, and heavy metals in liquid wastes flow into bays and estuaries.

Cities
Toxic metals and oil from streets and parking lots pollute waters; sewage adds nitrogen and phosphorus.

Urban sprawl
Bacteria and viruses from sewers and septic tanks contaminate shellfish beds and close beaches; runoff of fertilizer from lawns adds nitrogen and phosphorus.

Construction sites
Sediments are washed into waterways, choking fish and plants, clouding waters, and blocking sunlight.

Farms
Runoff of pesticides, manure, and fertilizers adds toxins and excess nitrogen and phosphorus.

Red tides
Excess nitrogen causes explosive growth of toxic microscopic algae, poisoning fish and marine mammals.

Toxic sediments
Chemicals and toxic metals contaminate shellfish beds, kill spawning fish, and accumulate in the tissues of bottom feeders.

Oxygen-depleted zone
Sedimentation and algae overgrowth reduce sunlight, kill beneficial sea grasses, use up oxygen, cause fish kills, and degrade habitat.

Healthy zone
Clear, oxygen-rich waters promote growth of plankton and sea grasses, and support fish.

Toxins produced by algal blooms poison seafood, damage fisheries, kill some fish-eating birds, and reduce tourism. Each year, algal blooms also lead to the poisoning of about 60,000 Americans who eat shellfish contaminated by the algae.

Harmful algal blooms occur annually in several hundred *oxygen-depleted zones* (dead zones) around the world. They form mainly in temperate coastal waters and in large bodies of water with restricted outflows, such as the Baltic and Black Seas (Lesson 8.5). The largest such zone in U.S. coastal waters forms every year in the northern Gulf of Mexico.

As discussed earlier in this lesson, much of the plastic that is improperly discarded finds its way into the oceans. There are currently six huge ocean garbage patches (Lesson 8.5). Plastic that has been broken down into tiny pieces that can harm fish, birds, and other marine creatures makes up 80% of the material in these patches. The North Pacific Gyre garbage patch is estimated to be larger than the size of Texas.

You can help prevent these patches from growing larger. For example, you can avoid using plastic water bottles as much as possible, and always dispose of the bottles responsibly. Figure 10-17 summarizes the major sources of ocean pollution as well as their impact on marine wildlife and ecosystems.

Ocean Pollution from Oil Crude petroleum (oil) and petroleum that has been refined into fuel oil, diesel, gasoline, and other processed products are highly disruptive ocean pollutants. In 2010, the BP *Deepwater Horizon* oil rig exploded, burned, and sank in the Gulf of Mexico. Eleven crew members were killed. A total of 507 million liters (134 million gallons) of crude oil poured into the Gulf before the leaking well was capped three months later. BP has spent more than $70 billion for cleanup, fines, and other costs related to the disaster. Figure 6-22 depicts the marine ecosystems affected by this tragedy.

Although less dramatic than an oil spill or blowout, the largest source of ocean oil pollution is urban and industrial runoff from land. Oil leaks from pipelines, refineries, and other oil-handling and storage facilities are the major sources of polluted runoff. An estimated one-third to one-half of ocean oil pollution comes from oil and oil products intentionally dumped or accidentally leaked onto the land or into sewers by homeowners and industries.

Chemicals in oil called volatile organic hydrocarbons kill many aquatic organisms immediately on contact. Animals in their larval forms are especially vulnerable. Other chemicals in oil form floating, tarlike globs that coat the feathers of seabirds and the fur of marine mammals. The oil destroys the animals' buoyancy and natural heat insulation. Many of them subsequently drown or die from loss of body heat. Heavy oil components sink to the ocean floor or wash into estuaries and coastal wetlands. They can smother bottom-dwelling organisms such as crabs, oysters, mussels, and clams. They can also make these organisms unsafe for people to eat. Some oil spills have killed coral reefs.

Oil spills that are not too large can be partially cleaned up by mechanical means. Methods include floating booms, skimmer boats, absorbent devices such as giant pillows filled with feathers or hair, and even cleaning by hand (Figure 10-18). But scientists estimate that current cleanup methods can recover typically no more than 15% of oil from a major spill.

In the long run, preventing oil pollution is the most effective and least costly approach. One of the best ways to prevent oil tanker spills is to use tankers with double hulls. This design employs two watertight surfaces along the bottom and sides. Stricter safety standards and inspections could reduce oil well blowouts at sea. Most importantly, businesses, institutions, and citizens living in coastal areas must take care to prevent leaks or spills of even the smallest amounts of oil.

checkpoint Name three major sources of ocean pollution.

Ways to Reduce Water Pollution

There are a number of ways to reduce water pollution from both nonpoint and point sources, as well as ways to reduce water pollution resulting from sewage.

Reducing Nonpoint-Source Pollution Most nonpoint sources of pollution come from agricultural practices. To reduce this type of pollution, farmers can keep cropland covered with vegetation and use other methods to reduce soil erosion. They can minimize the amount of fertilizer that runs off into surface water by using slow-release fertilizer. They can also opt not to use fertilizer on steeply sloped land and can plant buffer zones of vegetation between farmland and nearby lakes and streams.

Organic farming and other forms of sustainable food production use few if any synthetic fertilizers. Thus, they can prevent water pollution caused by nutrient overload. Farmers can reduce pesticide runoff by applying pesticides only when needed and relying more on integrated pest management (Chapter 9). In addition, they can control runoff by planting buffer zones and locating feedlots, pastures, and animal waste storage sites away from steeply sloped land, surface water, and flood zones.

Reducing Pollution from Point Sources The Federal Water Pollution Control Act of 1972 (renamed the Clean Water Act when it was amended in 1977) and the 1987 Water Quality Act form the basis of U.S. efforts to control pollution of the country's surface water. The Clean Water Act sets standards for allowable levels of 100 key pollutants. The Act requires companies that use certain industrial processes to obtain a permit that sets limits on emissions and establishes record keeping and reporting requirements. Some scientists call for strengthening the Clean Water Act. Suggested improvements include:

- Shifting the focus of the law from end-of-pipe removal of specific pollutants to water pollution prevention.
- Greatly increased monitoring for violations of the law, with much larger mandatory fines for violators.
- Regulating irrigation water quality.

FIGURE 10-18
Workers clean up an oil spill that has contaminated a stretch of beach and nearby waters. The work is painstaking and may take weeks, months, or even years to complete.

> **CONSIDER THIS**
>
> **Cruise ships are an underreported source of ocean pollution.**
> Cruise ships are leaving waste in their wakes. A cruise liner can carry as many as 6,300 passengers and 2,300 crew members. They can generate as much waste (toxic chemicals, garbage, sewage, and waste oil) as a small city. Many cruise ships dump these wastes at sea. In U.S. waters, such dumping is illegal, but some ships continue dumping secretively, usually at night.

Another suggestion is to expand the rights of citizens to bring lawsuits to ensure water pollution laws are enforced. Still another suggestion is to rewrite the Clean Water Act to clarify that it covers all waterways. This was the original intent of the law, but subsequent court decisions have weakened it. Some polluters have taken advantage of the resulting confusion to continue polluting in many areas.

Many people oppose these proposals, contending that the Clean Water Act's regulations are already too restrictive and costly. Some state and local officials argue that in many communities, it is unnecessary and too expensive to test all the water for pollutants as required by federal law.

Treating Sewage In urban areas in the United States and other more-developed countries, most waterborne wastes from homes, businesses, and storm runoff flow through sewer pipes to a sewage treatment plant. Raw sewage reaching a treatment plant typically undergoes one or two levels of treatment. The first is *primary sewage treatment*, a physical process that removes large floating objects and allows solids such as sand and rock to settle out. Then the waste stream flows into a primary settling tank where suspended solids settle out as sludge.

The second level is *secondary sewage treatment*. During this process, aerobic bacteria are used to remove as much as 90% of dissolved and suspended organic materials and nutrients. A combination of primary and secondary treatment removes 95–97% of the suspended solids and organic wastes and 70% of most toxic metal compounds and synthetic organic chemicals. In addition, 70% of the phosphorus and 50% of the nitrogen are removed. However, the treatment process removes only a tiny fraction of potentially toxic organic substances found in some pesticides and in discarded medicines. In addition, it does not kill viruses, bacteria, and other disease-causing agents.

Before discharge, water from sewage treatment plants usually undergoes bleaching to remove water coloration. It also undergoes disinfection to kill disease-carrying bacteria and some (but not all) viruses. The usual method for accomplishing disinfection is chlorination. But chlorine can react with organic materials in water to form small amounts of chlorinated hydrocarbons. Some of these chemicals cause cancers in test animals. They also can increase the risk of miscarriages and can damage the human nervous, immune, and endocrine systems. Use of other disinfectants, such as ozone and ultraviolet light, is increasing. But they are more costly and have shorter-lasting effects.

Some scientists call for redesigning the sewage treatment system to prevent toxic and hazardous chemicals from reaching sewage treatment plants. One option is to require businesses to remove toxic and hazardous wastes from water before sending it to sewage treatment plants. That would implement the full-cost pricing factor of sustainability by increasing the cost of creating waste and pollution. By reducing or eliminating the use of toxic chemicals, the expense of complying with control laws would be reduced.

Another option is to require more buildings to eliminate sewage outputs by switching to waterless, odorless composting toilet systems. These systems were pioneered several decades ago in Sweden. They convert nutrient-rich human fecal matter into soil-like humus that can be used as a fertilizer supplement. This process returns plant nutrients in human waste to the soil, thereby mimicking the nutrient cycling factor of sustainability.

Unlike a conventional toilet, a compost toilet does not use water, and waste is not flushed away. Instead, waste is mixed with materials such as sawdust inside a composter that is part of the system. Naturally occurring aerobic bacteria break down the waste with the aid of air and heat. Composting toilets do not require the use of any chemicals.

On a larger scale, composting toilet systems would be cheaper to install and maintain than current sewage systems. They do not require vast networks of underground pipes connected to centralized sewage treatment plants. They save large amounts of water and help lower water bills. They also reduce the amount of energy used to pump and purify water.

FIGURE 10-19
The Solar Sewage Treatment Plant in Providence, Rhode Island, is an ecological wastewater purification system, also called the Eco-Machine. Biologist John Todd invented this ecological process for purifying wastewater by using the sun and a series of tanks containing living organisms.

ENGINEERING FOCUS 10.2

WORKING WITH NATURE TO TREAT SEWAGE

Some communities and individuals are seeking better ways to purify sewage by working with nature. Biologist John Todd has developed an ecological approach to treating sewage, called the Eco-Machine (Figure 10-19).

This purification process begins when sewage flows into a passive solar greenhouse or outdoor site containing rows of large open tanks. The tanks are populated by an increasingly complex series of organisms. In the first set of tanks, algae and microorganisms decompose organic wastes. Sunlight helps to speed up the process. Water hyacinths, cattails, bulrushes, and other aquatic plants growing in the tanks take up the nutrients that are released.

After flowing though several of these natural purification tanks, the water passes into an artificial marsh. The marsh is made of sand, gravel, and bulrushes that filter out algae and remaining organic waste. Some of the plants absorb toxic metals like lead and mercury. Some secrete natural antibiotic compounds that kill pathogens.

Next, the water flows into aquarium tanks. There, snails and zooplankton consume microorganisms and in turn are consumed by crayfish, tilapia, and other fish that can be eaten or sold as bait. After 10 days, the clear water flows into a second artificial marsh for final filtering and cleansing. The water can be made pure enough to drink by treating it with ultraviolet light or by passing the water through an ozone generator, usually placed under water and out of sight as part of an attractive pond or wetland habitat.

Operating costs are about the same as those of a conventional sewage treatment plant. These systems are widely used on a small scale. However, they have been difficult to maintain on a scale large enough to handle the typical variety of chemicals in the sewage wastes from more-developed urban areas.

More than 800 cities and towns worldwide (150 in the United States) use natural or artificial wetlands to treat sewage as a lower-cost alternative to expensive waste treatment plants. Both this approach and the Eco-Machine system apply all three scientific factors of sustainability: using solar energy, employing natural processes to remove and recycle nutrients and other chemicals, and relying on a diversity of organisms and natural processes.

Thinking Critically

Make Judgments Can you think of any disadvantages of using a nature-based system instead of a conventional sewage treatment plant? Do you think the disadvantages outweigh the advantages? Why or why not?

FIGURE 10-20

Water Pollution Solutions
Prevent groundwater contamination.
Reduce nonpoint runoff.
Work with nature to treat sewage and reuse treated wastewater.
Find substitutes for toxic pollutants.
Practice the four Rs of resource use (refuse, reduce, reuse, recycle).
Reduce air pollution.
Reduce poverty.
Slow population growth.

In rural and suburban areas with suitable soils, sewage from each house is discharged into a septic tank and leach field. Sewage is pumped into the septic tank. Grease and oil rise to the top, and solids settle to the bottom to be decomposed by bacteria. The partially treated wastewater that results is discharged into a *leach field*. A leach field is a large drainage field with small holes in perforated pipes embedded in porous gravel or crushed stone just below the soil's surface. As wastes drain from the pipes and sink downward, the soil filters potential pollutants and soil bacteria decompose biodegradable materials.

About one-fourth of all homes in the United States are served by septic tanks. They work well as long as they are not overloaded. In addition, any solid wastes that are not decomposed by bacteria must be regularly pumped out of the septic tank. Otherwise it will overflow.

Eliminating Water Pollution Many environmental and health scientists are asking the question: How can we avoid producing water pollutants in the first place? Figure 10-20 lists ways to achieve this goal over the next several decades. The shift to pollution prevention will not take place unless citizens put political pressure on elected officials.

You can take personal steps to prevent water pollution. If you use fertilizer on a yard or garden, choose compost over commercial inorganic fertilizer. Limit the use of pesticides, especially near bodies of water. Keep yard wastes from entering storm drains.

Don't use water fresheners in toilets and never flush medicines down the toilet. Finally, never pour pesticides, paints, solvents, oil, antifreeze, or other harmful chemicals down the drain or onto the ground. They will eventually find their way into the water supply.

checkpoint How can water pollution from nonpoint and point sources be reduced?

10.4 Assessment

1. **Contrast** What is the difference between point and nonpoint sources of water pollution?
2. **Identify** How can flood damage be reduced?
3. **Summarize** Summarize the process of cultural eutrophication in a series of steps.
4. **Explain** Explain why groundwater is more difficult to clean than surface water.
5. **Restate** Describe how a septic tank and leach field treat sewage and wastewater.
6. **Evaluate** Two major sources of pollution in our oceans are plastic wastes and oil spills from deep-water drilling. Suppose you are the director of the EPA. Choose which problem to address first and describe what measures you would take to reduce that particular problem.

SCIENCE AND ENGINEERING PRACTICES

7. **Engaging In Argument** Do you think the Clean Water Act is important? Should the act be strengthened or weakened? If so, in what way?

TYING IT ALL TOGETHER STEM
Saving the Colorado River

The Case Study at the beginning of this chapter discussed the problems and tensions that can occur when several states in a water-short region share a limited river water resource. Roughly 40 million people in the United States depend on the Colorado River for water, electricity, and other ecosystem services. The rules governing the use of the Colorado River are currently a complex combination of federal (national), state, and local laws. These laws are known collectively as the "Law of the River."

Some districts in the Colorado River basin and elsewhere are turning to *tiered pricing* to help manage water supplies. Tiered pricing is a rate structure in which the price to the end user increases based on use. The more water a customer uses, the higher its price per unit. Tiered pricing is meant to discourage the wasteful use of fresh water.

Work with a group to answer the questions that follow.

1. Decision makers must take many factors into account when changing water laws. The following list of factors illustrates one district's priorities. Would you prioritize these factors in the same way? If not, how would your priorities be different?
   - Meeting water utility revenue requirements
   - Conservation of water supply
   - Saving extra water in case of drought
   - Enabling economic development
   - Affordability to the average customer
   - Simplicity to customers
   - Ease of implementation

2. Suppose water costs $5.00 per 1,000 gallons for the first 10,000 gallons. For every 10,000 additional gallons the price increases $0.15 more per gallon. What would it cost to use 10,000 gallons of water? What would it cost to use 20,000 gallons?

3. What types of businesses do you think would be most affected by tiered pricing?

4. Should customers using untreated water, such as farmers, be charged the same amount as customers using treated water? Explain.

5. How does tiered pricing differ from taxation?

6. Some state laws prohibit utilities from charging more money for water than the cost of producing and delivering it. A 2015 ruling in California called one city's tiered pricing structure illegal because of state laws. How does the definition of "cost" affect such rulings?

7. At which level—federal, state, or local—do you think most decisions regarding fresh water use should be made? Why?

8. How might tiered pricing disproportionately affect low-income customers?

9. How can gray water setups help residences and businesses save money in a tiered-pricing system?

CHAPTER 10 SUMMARY

10.1 Why is fresh water in short supply?

- Humans and other species need fresh water to survive. The availability of fresh water is not only an environmental issue; it also affects global health, economics, and security.
- Groundwater is fresh water that sinks through spaces in soil, gravel, and rock, and collects above a layer of rock or clay in the ground. Fresh water from rain and melted snow that flows or is stored above ground is surface water. Precipitation that does not sink into the ground or evaporate is surface runoff. About one-third of runoff is reliable surface runoff, a stable source of fresh water from year to year.
- Supplies of fresh water are not evenly distributed, and many people lack access to clean water. Even where large freshwater supplies exist, they are often poorly managed.

10.2 How can people increase freshwater supplies?

- Aquifers are renewable sources of fresh water if not overpumped or contaminated. Overpumping leads to shortages of water needed for irrigation, which may reduce food production, resulting in higher food prices and hunger. Overpumping may also cause the land above the aquifer to subside, or sink.
- Dam-and-reservoir systems provide fresh water to cities, generate electricity, provide recreation areas, and supply fresh water for irrigation. In a drought, reservoirs can shrink, agriculture production can drop, and ecosystems can disappear.
- Government subsidies of water transfers lead to inefficient use of water because they reduce incentives for farmers and residents to conserve it.
- Desalination of ocean water is expensive, requires the use of chemicals that kill marine life, and produces salty wastewater that is difficult to dispose of properly.

10.3 How can people use fresh water more sustainably?

- Government subsidies that encourage efficient water use would reduce water losses.
- Improving irrigation efficiency and cutting water losses in industries and homes would also lead to sustainable water use.

10.4 How can people reduce water pollution?

- Point sources of pollution discharge pollutants at specific locations. Nonpoint sources are broad and diffuse areas where rain or snowmelt washes pollutants off the land into bodies of surface water.
- Because they flow, streams and rivers can recover from pollutants more easily than stationary lakes and reservoirs.
- Because groundwater takes so long to cleanse itself and is so costly to clean, prevention is the only feasible way to deal with groundwater contamination.
- Sources of ocean pollution include sewage from coastal areas, wastes from cruise ships, runoff of agricultural wastes, plastics, and oil from spills and industrial runoff.
- Sustainable ways to reduce water pollution include reducing soil erosion from farms and using organic farming methods, government regulation of polluters and policies that promote pollution prevention or reduction, sewage treatment, and individual efforts.

CHAPTER 10 ASSESSMENT

Review Key Terms

Select the key term that best fits the definition. Not all terms will be used.

aqueduct	reliable surface runoff	water footprint
cultural eutrophication	reservoir	water pollution
desalination	snowpack	water table
drip irrigation	subsidence	zone of saturation
flood irrigation	virtual water	
gray water	wastewater	

1. Area under ground in which all pores are filled with fresh water
2. Overnourishment of aquatic ecosystems with plant nutrients such as nitrates and phosphates
3. The top of the zone of saturation
4. Fresh water not directly consumed but used to produce food and other products
5. A rough measure of the volume of fresh water used directly and indirectly to support one's lifestyle
6. The process of removing salts from ocean water or brackish water
7. Lined canal used to transfer fresh water to farms and cities
8. Occurs when land above an aquifer sinks
9. Irrigation method in which water is delivered directly to plant roots through pipes or tubes
10. Method in which water is pumped from a source through unlined ditches and flows by gravity to crops
11. Used water from bathtubs, showers, sinks, dishwashers, and clothes washers
12. Any change in water quality that can harm living organisms or make the water unfit for human uses
13. Water containing sewage and other wastes from homes and industries

Review Key Concepts

14. Describe the dilemma of disposing of sludge produced by sewage treatment plants.
15. Explain how human activities interfere with the hydrologic cycle.
16. Define a water footprint and virtual water and give two examples of each.
17. Give an advantage and a disadvantage of withdrawing groundwater.
18. List two problems that result from overpumping aquifers.
19. Define desalination and describe its drawbacks.
20. Describe the contribution of plastic to water pollution and list ways to reduce plastic pollution.

Think Critically

21. Are you for or against government subsidies that provide farmers with water at low cost? Would you be in favor of government subsidies for improving irrigation efficiency? Explain.
22. How might you be contributing directly or indirectly to groundwater pollution? What are three things you could do to reduce your contribution?
23. Research the environmental and economic costs that have resulted from the BP *Deepwater Horizon* oil rig disaster. Do any areas affected by the spill still face recovery challenges today and if so, what are they? Summarize your findings in three paragraphs.
24. What role does population growth play in water supply problems? Relate this to the water supply problems of the Colorado River basin.
25. List three ways you could use water more efficiently. Which, if any, of these measures do you already take?
26. Should the same sewage treatment laws that are applied to point sources be applied to mobile sources (like cruise ships)? Explain.

Chapter Activities

A. Develop Models: Oil Spill Cleanup STEM

Oil spills that are not too large can be partially cleaned using mechanical (physical) means. Oil is hydrophobic, meaning that it naturally separates from water. Oil is also less dense than water, causing it to form a layer on the water's surface. Despite these physical properties, oil cleanup is a daunting and dirty job.

Scientists estimate that only about 15% of the oil from a spill can be removed even using the best methods. Typically, a floating "wall" called a boom surrounds the oil until a boat called a skimmer can come by and pick it up. Oil-absorbing materials are sometimes used to help gather the oil. In this activity you will model a solution for cleaning up an oil spill from ocean waters.

Materials

"oil spill"	dish soap	string
paper towels	foil	cloth
plastic wrap	craft sticks	feathers
cotton balls	paper cups	straw

1. Look over the materials provided by your teacher and develop a plan to clean up the oil spill. Write out the steps of your plan.
2. Develop a way to estimate the percentage of oil removed.
3. Carry out your plan and record the percentage of oil removed.
4. Evaluate your results and revise your plan. Carry out your revised plan and compare your results.
5. Which aspects of your original plan were most and least successful?
6. Did your revisions result in an improved solution? Explain.
7. How did you estimate the percentage of oil removed?
8. Share your results with the class. Which methods were most effective?
9. Could a similar solution be employed on a larger scale? Explain.

B. Take Action

National Geographic Learning Framework

Attitudes | Responsibility
Skills | Problem Solving
Knowledge | Our Living Planet

Use a water footprint calculator, such as the one provided online by the GRACE Communications Foundation. Then analyze your household's water use. Look for ways to improve your water footprint and take care of your water supply. Use the following steps as a guide.

1. Take an inventory of appliances that use water in your residence (toilets, kitchen and bathroom faucets, baths/showers, outside faucets, dishwasher, washing machine). List ways to reduce water use and waste from these appliances.
2. Refer to the graph of daily water use (Figure 10-2) and the table of hidden water use (Figure 10-3). Which change in your habits could make the biggest impact?
3. Use the internet to identify the source of fresh water for your residence. Have there been issues with pollution or shortages in the past 5–10 years? What was done about the issue(s)?

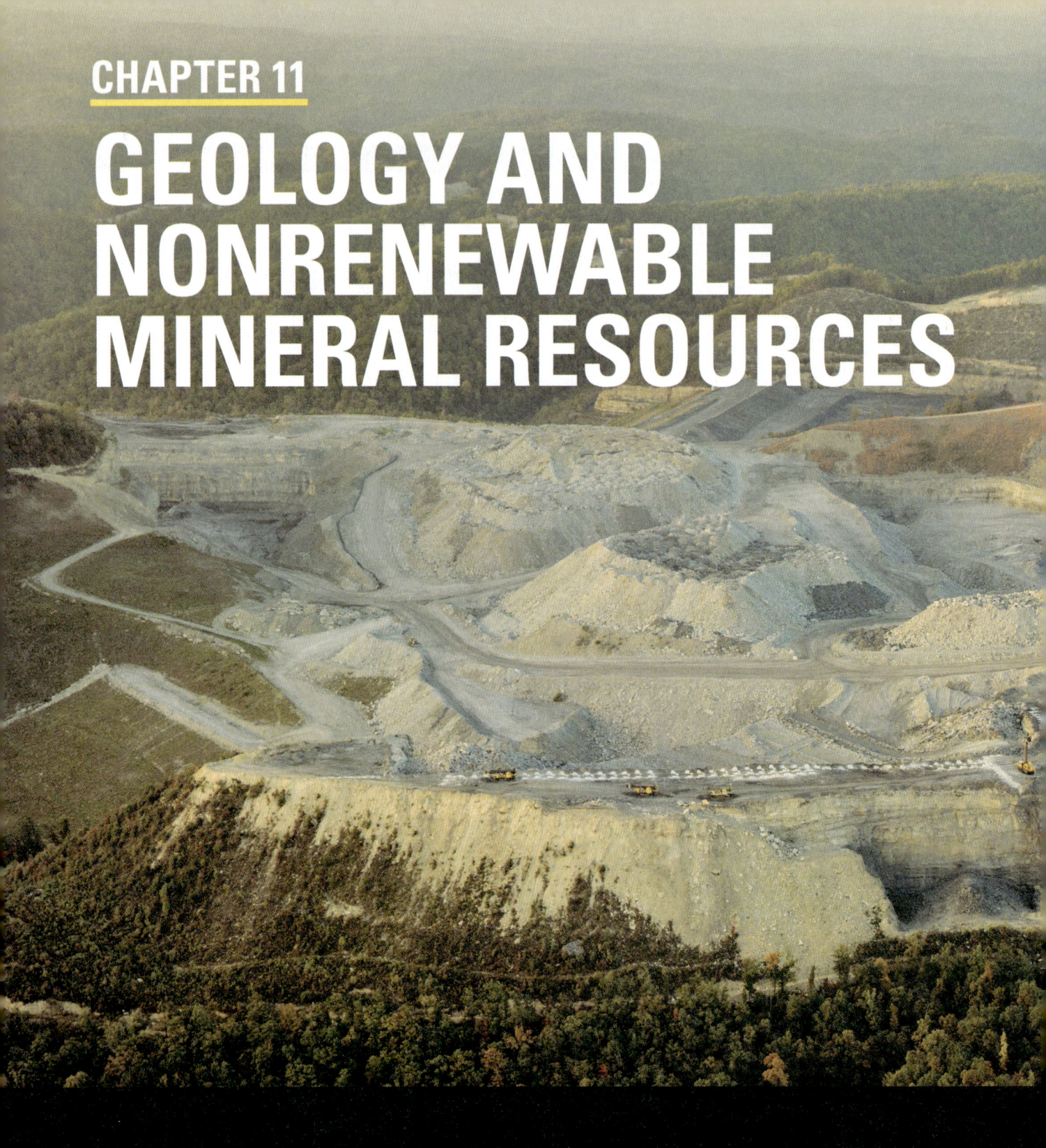

CHAPTER 11
GEOLOGY AND NONRENEWABLE MINERAL RESOURCES

MOUNTAINS are moved in pursuit of valuable resources found within Earth's crust. This mountaintop was removed to extract the coal beneath it. In mines around the world, land is blasted, scraped, and bulldozed away to obtain fossil fuels, gravel, sand, gold, silver, copper, and aluminum. Also sought is a dwindling supply of 17 rare earth metals that are crucial ingredients in everything from cell phones to jet engines.

KEY QUESTIONS

11.1 How do geological processes relate to society and the environment?

11.2 What are Earth's mineral resources and how long might reserves last?

11.3 What are the effects of using mineral resources?

11.4 How can society use mineral resources more sustainably?

A coal mining operation in West Virginia has removed the top of this mountain, with the exception of an island of trees that surrounds a cemetary.

Better Batteries for Electric Cars

with National Geographic Explorer Yu-Guo Guo

If you've ever run track or been on a swim team, you're probably very aware of how long a meter is. Now, take that meter and divide it into *billionths*. Scientists who deal in things that measure a billionth of a meter—called a nanometer by today's tech experts—believe there's a big future in the very small.

One of those scientists is National Geographic Explorer Yu-Guo Guo, a professor in the Institute of Chemistry at the Chinese Academy of Sciences in Beijing. Dr. Guo points out that "nanoscale isn't just small, it's a special kind of small." A nanometer is a unit of measure used to measure the tiniest parts of our world—molecules and atoms—that are only visible using very powerful microscopes. Guo sees real-world applications of technologies on this very small scale. He is working on energy conversion and storage systems that could revolutionize the way people live.

For Guo, nanotechnologies offer lots of exciting possibilities, starting with electric cars. While electric cars offer a sustainable alternative to fossil fuels, until recently the size and relative ineffectiveness of the lithium-ion batteries used in electric cars have made consumers hesitant to make the leap. Guo believes the key to improving performance and lowering costs of those batteries lies in nanoparticles. Nanoparticles have the potential to quickly absorb and store vast numbers of lithium ions. Using nanoparticles in lithium-ion batteries could improve their capacity and optimize their performance.

The nanostructures Guo has invented could make electric car batteries smaller, more powerful, and less expensive than ever before. They boost the performance of high-power batteries by enabling a more efficient flow of electric current. Called "3-D conducting nanonetworks," they are dense networks of high-conducting material that let electrons reach every lithium storage particle for a far more powerful result.

"This new high-power technology means batteries can be fully charged in just a few minutes, as quickly and easily as you can fill your car with gas," Guo says. "The advanced batteries recover more energy when cars stop, deliver more power when cars start, and enable vehicles to run longer."

Can technology based on thinking smaller and smaller rescue a human population growing bigger and bigger? With the majority of the planet's 7.9 billion people now living in urban areas, interest in expanding the clean green car economy has shifted into high gear. "I am glad to be part of work that can help reduce carbon emissions and use our planet's resources more sustainably," Guo states. It looks like some of the world's smallest technologies will drive some of its biggest changes.

Thinking Critically
Draw Conclusions What lessons learned from the use of nanotechnology in electric cars might be carried forward into other industries?

Yu-Guo Guo's work with nanonetworks inside lithium-ion batteries could revolutionize the electric car, making it more affordable and more efficient.

CASE STUDY
The Importance of Rare Earth Metals

Minerals extracted from Earth's crust are processed into an amazing variety of products, raising the standard of living and providing economic benefits and jobs. Some mineral resources are familiar, such as gold, copper, aluminum, sand, and gravel. Less well known, rare earth metals and oxides are crucial to the technologies that support modern-day lifestyles and economies.

The 17 rare earth metals, also known as rare earths, include scandium, yttrium, and 15 lanthanide chemical elements, such as lanthanum (Figure 11-1). Rare earths are a key ingredient used to make TVs, computers, tablets, and phones, as well as energy-efficient light-emitting diode (LED) light bulbs, solar cells, and fiber-optic cables.

They are part of batteries and motors for electric and hybrid cars (Figure 11-2), catalytic converters in car exhaust systems, jet engines, and the powerful magnets in wind turbine generators. Rare earths also go into missile guidance systems, smart bombs, aircraft electronics, and satellites.

Industrialized nations rely on affordable supplies of rare earths to develop cleaner energy technology, maintain national security, and create the high-tech products that are major sources of economic growth. However, like all natural resources, extracting rare earths can be potentially harmful to the environment.

This chapter explores the nature and supplies of mineral resources and the rock cycle that produces them. You will read about the environmental impacts of mineral extraction and processing and how people can use these resources more sustainably.

As You Read Think about the ways in which you use rare earth metals every day. What are the societal benefits and environmental costs of mining and using them?

FIGURE 11-1
Lanthanum The high reactivity of lanthanum (La) makes it useful in batteries and other technologies.

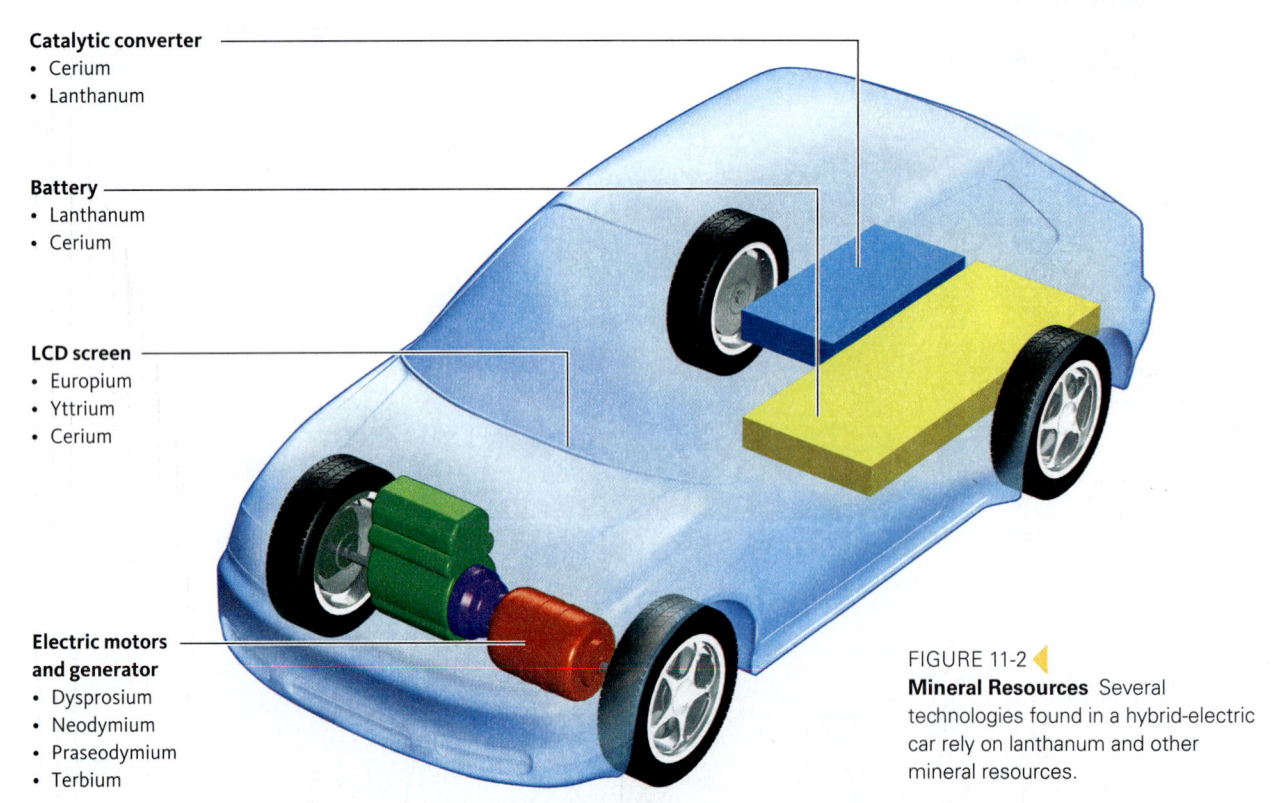

Catalytic converter
- Cerium
- Lanthanum

Battery
- Lanthanum
- Cerium

LCD screen
- Europium
- Yttrium
- Cerium

Electric motors and generator
- Dysprosium
- Neodymium
- Praseodymium
- Terbium

FIGURE 11-2
Mineral Resources Several technologies found in a hybrid-electric car rely on lanthanum and other mineral resources.

CHAPTER 11 GEOLOGY AND NONRENEWABLE MINERAL RESOURCES

11.1 How Do Geological Processes Relate to Society and the Environment?

CORE IDEAS AND SKILLS

- Identify the three major parts of Earth's geosphere.
- Describe the movement of tectonic plates.
- Explain the interaction between geologic events, human populations, and the environment.
- Understand that Earth's dynamic processes produce important benefits as well as potential threats.

KEY TERMS

core
mantle
asthenosphere
crust
lithosphere
tectonic plate
continental drift

Earth Is a Dynamic Planet

The geosphere is part of Earth's life-support system. Despite its visibly rigid surface, the geosphere is constantly moving. The arrangement of Earth's land masses has changed numerous times since the formation of the planet about 4.6 billion years ago. Geology is the scientific study of the dynamic processes that take place on Earth's surface and in its interior. Studying Earth's composition and processes helps scientists understand potential geological hazards, such as volcanoes and earthquakes. Geology also helps scientists identify potential natural resources. Minerals become available for society's use because of geological processes.

FIGURE 11-3 ▼

Our Dynamic Planet Earth's geosphere has a core, mantle, and crust. Within the core and mantle, dynamic forces affect what happens in the crust and on the surface.

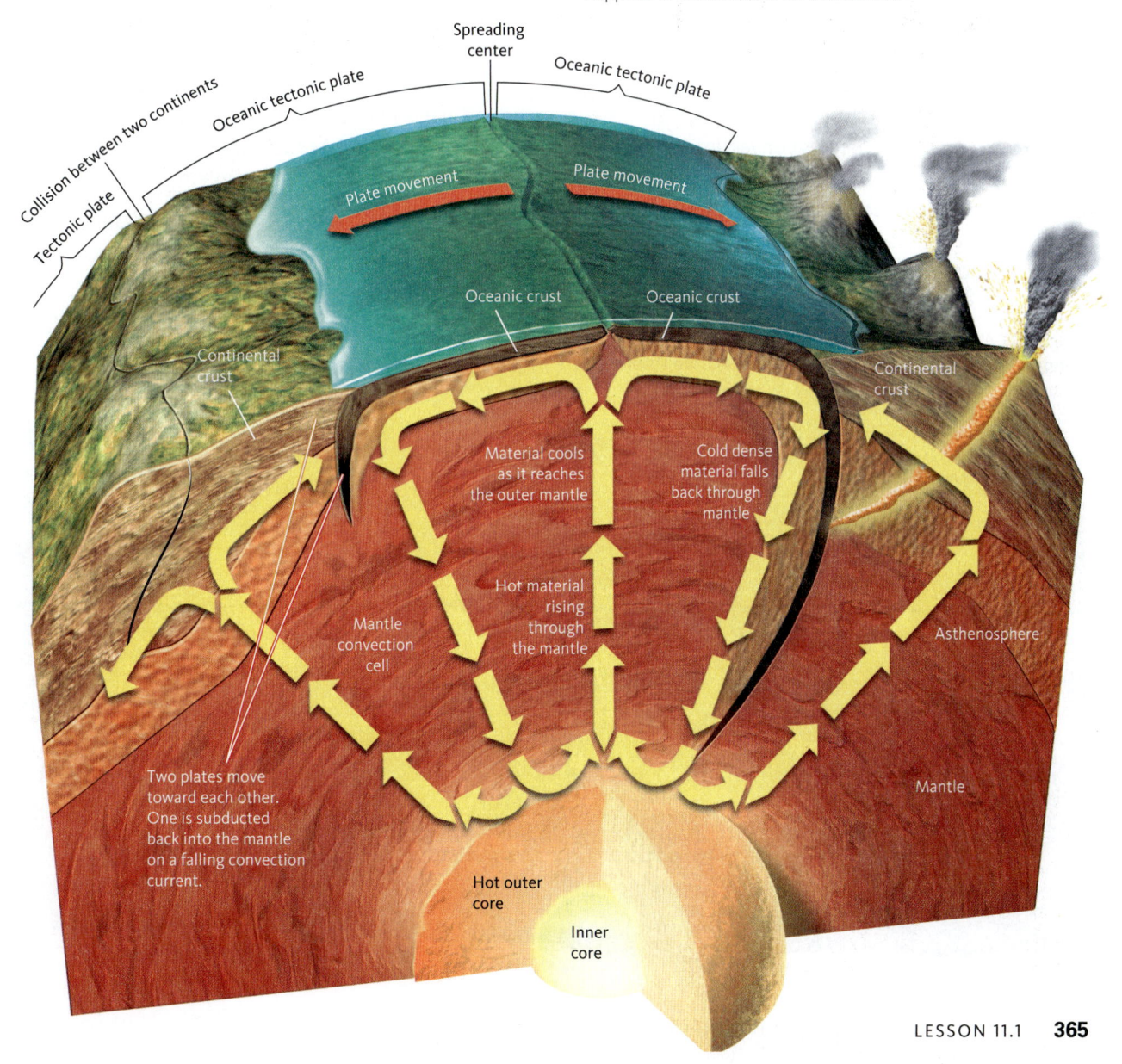

LESSON 11.1 **365**

As primitive Earth cooled over millions of years, its interior separated into three major layers: the core, the mantle, and the crust (Figure 11-3). The **core** is the geosphere's innermost layer. It is composed primarily of iron (Fe). The inner core is extremely hot and has a solid center surrounded by molten rock and semisolid material. The outer core is liquid iron and nickel (Ni).

Surrounding the core is a thick layer called the **mantle**. The outer part of the mantle is entirely solid rock. Beneath it is the **asthenosphere**—a volume of hot, partly melted, fluid rock.

The outermost and thinnest layer of solid materials is Earth's **crust**. It consists of the continental crust and the oceanic crust. The continental crust underlies the continents and includes the continental shelves extending into the ocean. The oceanic crust, which underlies the ocean basins, makes up most of Earth's crust. The combination of the crust and the rigid, outermost part of the mantle is called the **lithosphere**. The mineral resources on which society depends are found in the lithosphere.

Earth's liquid, semisolid, and solid components interact continuously. Tremendous heat within the core and mantle generates convection cells, or currents, that slowly move large volumes of rock and heat. The innermost material heats, rises, and begins to cool. As it cools, it becomes denser and sinks back toward the core where it is reheated, creating large, slow loops. Within the mantle, movement of some of the molten rock flows upward into the crust, where it is called magma. Magma that emerges onto Earth's surface is called lava. This cycling not only moves rock and minerals, it also helps transfer heat and energy throughout the planet.

The flows of energy and heated material within Earth's convection cells are so powerful they caused the lithosphere to break up into a dozen or so huge rigid pieces, called **tectonic plates** (Figure 11-4). The plates move across the asthenosphere at about the same speed that your fingernails grow. As tectonic plates have shifted, land masses have split apart and joined together, changing the shape of oceans and continents throughout Earth's history (Science Focus 4.2). The slow movement of the continents across Earth's surface is called **continental drift**.

The plates interact at their boundaries. At convergent plate boundaries, one plate is forced under the other toward Earth's core. Divergent plate boundaries are where plates are moving away from each other. At divergent boundaries, material from Earth's interior forms new land between the plates (Figure 11-6). As rocks and minerals cycle, they eventually make it to the lithosphere.

checkpoint What are the three major layers of the geosphere?

FIGURE 11-4 ▼

Tectonic Plates Earth's crust has been fractured into several major tectonic plates.

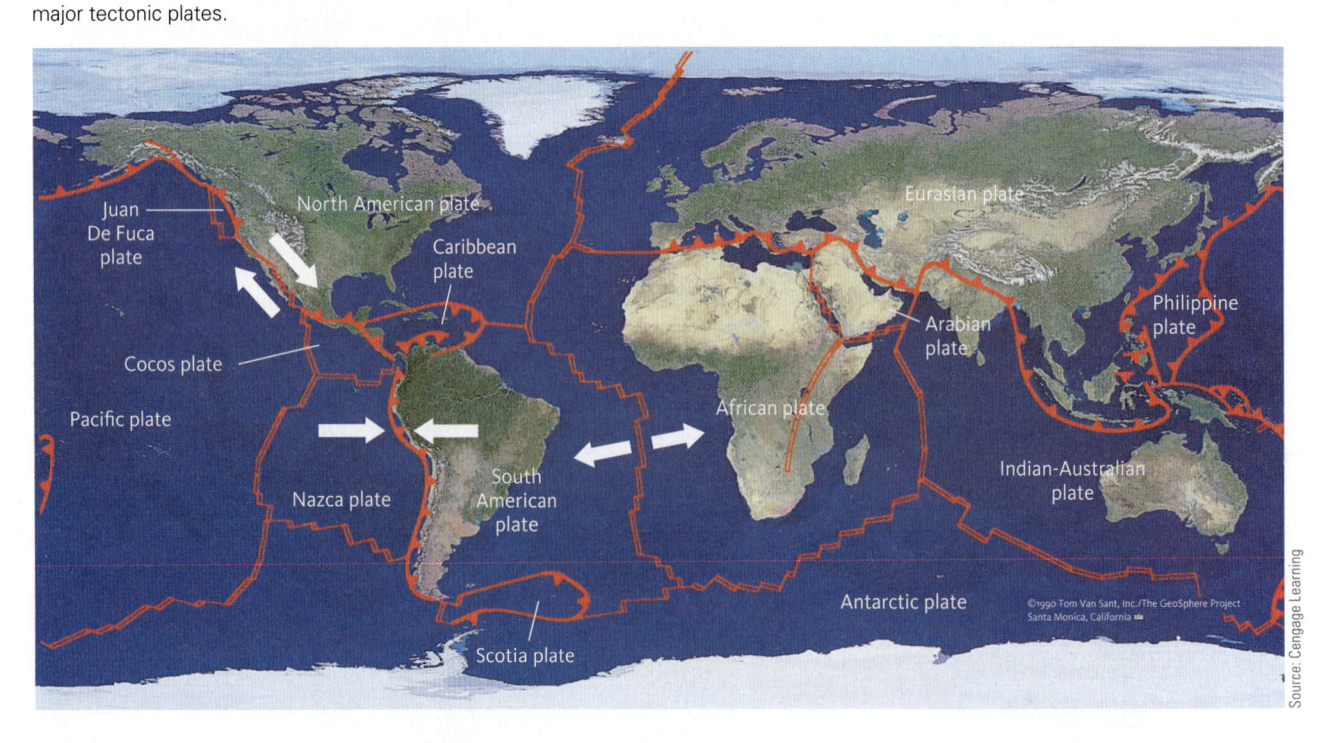

Geologic Processes Shape Earth's Surface

As tectonic plates separate, collide, or grind along against each other, the tremendous forces produced at tectonic plate boundaries reshape Earth's surface. Tectonic activity can trigger earthquakes and volcanic eruptions and can form or wear down mountains. An active volcano occurs where magma rises to Earth's surface through a central vent or long crack in the lithosphere (Figure 11-5).

Continental drift can alter ocean currents, affect local temperatures, change atmospheric circulation, and increase volcanic activity. These changes can influence and change climates, which affect Earth's biodiversity. Geologic processes can also lead to other geologic events. For example, earthquakes and volcanoes can trigger landslides. These events can have significant consequences for people and the environment.

Scientists analyze data from historical records, measurements that identify geologically active zones, and maps of high-risk areas to reduce the loss of life and property damage from earthquakes and volcanoes. Engineers and policy makers can establish building codes that regulate the placement and design of buildings in such areas and establish early warning and evacuation procedures. People can evaluate the risk of geologic events and factor it into their decisions about where to live and work.

Weathering and erosion are geologic processes that originate on and further impact Earth's surface. Over time, weathering and erosion break down mountains, form canyons, and build deltas. These processes pose problems for society when they wear away the ground around the built environment. Weathering and erosion can also damage structures, such as buildings and homes. Society contributes to and accelerates the rate of these processes, further affecting the environment.

In other instances, geological processes benefit humans and other organisms. New landforms provide habitats for living organisms. Geological processes also form deposits of natural resources, such as rocks and minerals, in Earth's crust where they become available for human use.

checkpoint Name three changes to Earth's surface that can result from shifting tectonic plates.

FIGURE 11-5
The internal pressure in a volcano can send lava, ash, and gases into the atmosphere or cause it to flow over land, posing a risk to local communities.

LESSON 11.1

CHAPTER 11   GEOLOGY AND NONRENEWABLE MINERAL RESOURCES

FIGURE 11-6
A diver swims in the rift valley between the Eurasian and American tectonic plates. This new land is forming in Nes Canyon in Northern Iceland.

11.1 Assessment

1. **Summarize** How are Earth's surface materials recycled?
2. **Synthesize** Look again at the yellow arrows in Figure 11-3. What is the term for this cyclic movement of heat? In which other illustration(s) in this book have you seen something similar?

SCIENCE AND ENGINEERING PRACTICES

3. **Using Models** Draw a model of Earth's interior. Include labels for mantle, crust, core, lithosphere, and asthenosphere.

CROSSCUTTING CONCEPTS

4. **Patterns** Use the internet to identify the earthquakes that have occurred in the last two hours worldwide. What do you notice about the location of these earthquakes?

11.2 What Are Earth's Mineral Resources and How Long Might Reserves Last?

CORE IDEAS AND SKILLS

- Define minerals and rocks and describe how they are formed during the rock cycle.
- Discuss the variety of mineral resources and some of their uses.
- Describe how mineral resources can become economically depleted.
- Explain how market prices affect mineral supplies.

KEY TERMS

mineral	igneous rock	rare earth metal
mineral resource	metamorphic rock	ore
rock	rock cycle	reserve
sedimentary rock		depletion time

The Rock Cycle Produces Rocks and Minerals

Earth's crust consists mostly of minerals and rocks. A **mineral** is a chemical element or inorganic compound that exists as a solid with a regularly repeating internal structure. A **mineral resource** is a concentration of minerals that is large enough to cover the cost of extracting and processing it into raw materials and useful products.

Rock is a solid combination of one or more minerals found in Earth's crust. Some kinds of rock, such as limestone (calcium carbonate, or $CaCO_3$) and quartz (silicon dioxide, or SiO_2), contain only one mineral (Figure 11-7). Most kinds of rock consist of two or more minerals. Granite, for example, is a mixture of mica, feldspar, and quartz crystals.

Types of Rock Based on the way they form, rocks are broadly classified as sedimentary, igneous, or metamorphic (Figure 11-8). **Sedimentary rock** is made of sediments—dead plant and animal remains and tiny particles of weathered and eroded rocks. Sediments are transported by water, wind, or gravity to other sites where they accumulate in layers. Eventually, the increasing weight and pressure on the underlying layers transform them into sedimentary rock.

Igneous rock forms below or on Earth's surface when magma wells up from the mantle and then cools and hardens. Igneous rock forms the bulk of Earth's crust, but it is usually buried beneath layers of sedimentary rock.

Metamorphic rock forms when an existing rock is subjected to high pressure, high temperatures that cause it to partially melt, or a combination of the two.

The interaction of physical and chemical processes that change Earth's rocks from one type to another is called the **rock cycle** (Figure 11-9). Rocks are recycled over millions of years by three processes—erosion, melting, and metamorphism—that produce sedimentary, igneous, and metamorphic rocks, respectively.

In these processes, rocks are broken down, melted, and fused together into new forms by heat and pressure. They are also cooled and sometimes recrystallized within Earth's interior. The rock cycle is the slowest of Earth's cyclic processes. It plays the major role in the formation of concentrated deposits of minerals that form mineral resources.

Types of Minerals Minerals exist as a single chemical element or as inorganic compounds formed by various combinations of elements. Examples of elements are gold and rare earth metals such as lanthanum (Figure 11-1). Examples of compounds include salt (sodium chloride, or NaCl), quartz (silicon dioxide, or SiO_2), and rare earth oxides (oxides that form when metals combine with oxygen). Most minerals are compounds rather than elements. Minerals can be further classified as metallic or nonmetallic. Single-element minerals, rare earths, and compounds that include rare earth oxides are all metallic minerals. Sand and limestone are two examples of nonmetallic minerals.

FIGURE 11-7 ▼

Minerals Calcium carbonate (left) and quartz (right) are two very common minerals. Calcium carbonate is commonly found in sedimentary rocks. Quartz is common in all rock types.

FIGURE 11-8 ▼

Types of Rocks		
Rock Type	**Formation Process**	**Examples**
Sedimentary	Weight and pressure compress layers of sediment	*Sandstone*—from sand *Shale*—from mud and clay *Dolomite* and *limestone*—from the remains of aquatic organisms *Lignite* and *bituminous coal*—from the remains of plants
Igneous	Magma cools and hardens	*Granite*—formed under ground *Basalt* and *pumice*—formed above ground
Metamorphic	High heat, pressure, and/or chemical interactions transform other rock	*Slate*—from superheated shale and mudstone *Marble*—from limestone exposed to heat and pressure

Minerals are used for many purposes, especially in technology industries. For example, about 60 of the 118 chemical elements in the periodic table are used for making computer chips. Figure 11-10 presents various products made with minerals. As discussed in the Case Study at the beginning of this chapter, **rare earth metals** have unique properties that lend themselves to use in technology products. Some rare earths have superior magnetic strength, the ability to purify water, and tolerance to extreme heat.

An **ore** is rock that contains a large enough concentration of a mineral—often a metal—to make it profitable for mining and processing. A high-grade ore contains a high concentration of the mineral. A low-grade ore contains a low concentration. Deposits of rare earth metals and oxides typically contain a mixture of the metals and their oxides that are difficult and costly to separate from one another and purify to an acceptable level.

checkpoint Explain the relationship between rocks, minerals, mineral resources, and ore.

Minerals Are Nonrenewable

The process by which minerals form through the rock cycle may take millions of years. Therefore, minerals are considered nonrenewable resources and their supplies can be depleted if they are removed faster than the rock cycle replenishes them. The portion of a mineral resource that is economically and technically feasible for mining is called its **reserve**. Whether or not existing mineral reserves can be expanded often depends on economic and technological factors, or on the discovery of new, accessible deposits.

Economic Depletion The future supply of any mineral resource depends on the actual or potential supply of the mineral and the rate at which it is used. Society has never run out of a mineral resource, but a mineral becomes economically depleted when it costs more than it is worth to find, extract, transport, and process. **Depletion time** is the time it takes to use up a certain proportion—usually 80%—of the reserves of a mineral at a given rate of use.

LESSON 11.2   **371**

FIGURE 11-9

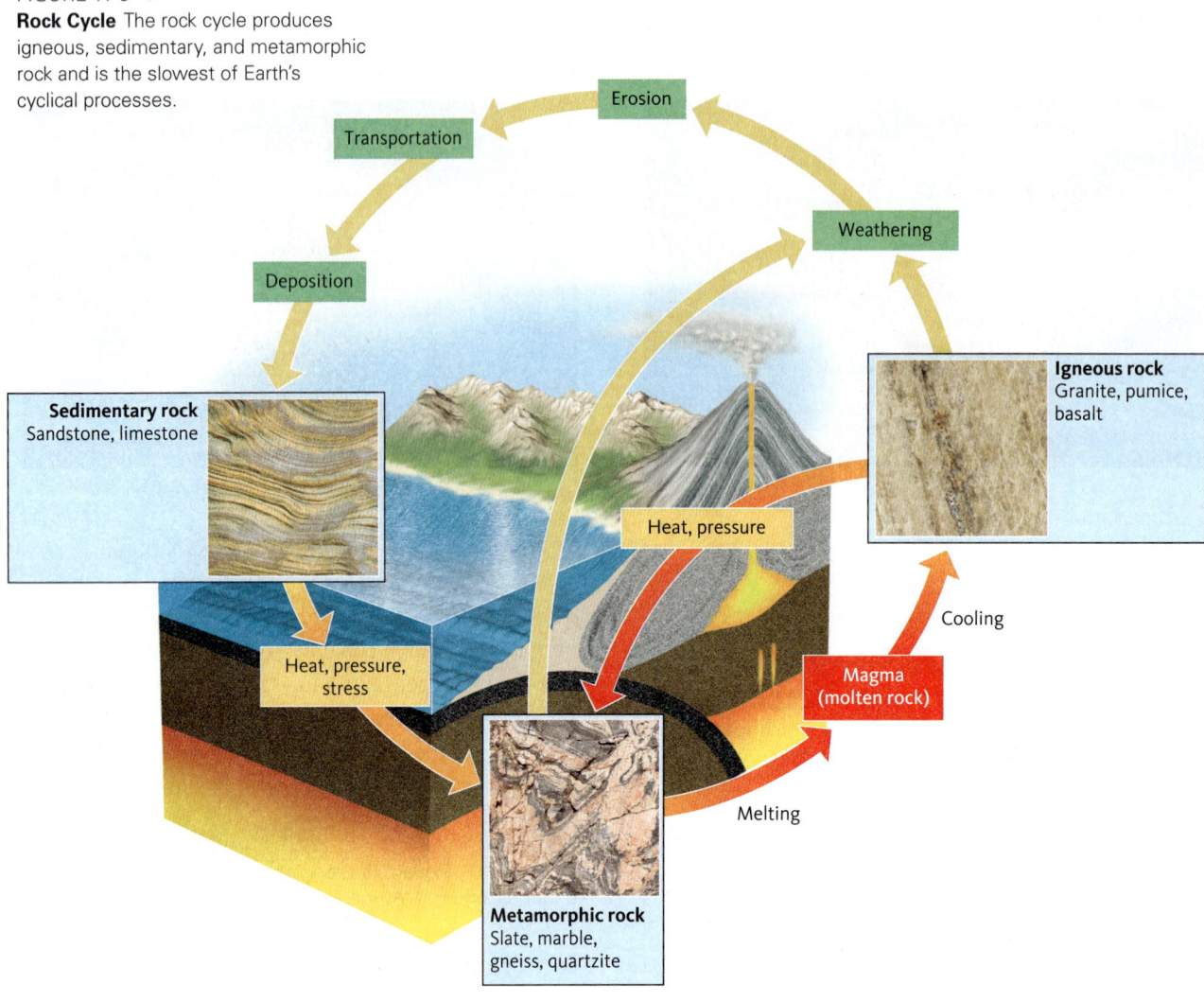

Rock Cycle The rock cycle produces igneous, sedimentary, and metamorphic rock and is the slowest of Earth's cyclical processes.

To extend the life of a mineral resource, people can recycle or reuse existing supplies, waste and use less, find a substitute, or do without.

When experts disagree about depletion times, it is often because they use different assumptions about supplies and rates of use. The shortest depletion-time estimate assumes no recycling or reuse and no increase in the reserve. (See Figure 11-11, curve A.) A longer depletion-time estimate assumes recycling will stretch the existing reserve (Figure 11-11, curve B). This scenario also assumes that better mining technology, higher prices, or new discoveries will increase the reserve. The longest estimate (Figure 11-11, curve C) makes the same assumptions as B, but also assumes that people will reuse and reduce consumption to expand the reserve further. Finding a substitute for a resource leads to a new set of depletion curves.

Earth's crust contains abundant deposits of minerals such as iron and aluminum. But concentrated deposits of important minerals such as manganese, chromium, cobalt, platinum, and some of the rare earths are relatively scarce. The United States, Canada, Russia, South Africa, and Australia are home to most of the mineral resources modern societies use.

In the United States, the total and per-person use of mineral resources and imports has risen sharply since 1950. According to the U.S. Geological Survey (USGS), each American uses an average of about 18 metric tons (about 20 tons) of minerals per year. As a result, the United States has economically depleted deposits of metals such as lead, aluminum, and iron. Currently, the United States is 100% import reliant on 19 minerals and more than 50% import reliant on 43 other minerals.

FIGURE 11-10 ▼

Mineral Resources and Their Uses	
Mineral	**Uses**
Aluminum (Al)	Beverage cans, motor vehicles, aircraft, buildings
Steel—An alloy of iron (Fe) and carbon (C). Manganese (Mn), cobalt (Co), and chromium (Cr) are added to make specialty steels.	Buildings, machinery, motor vehicles
Copper (Cu)	Electrical and communications wiring, plumbing pipes
Gold (Au)	Electrical equipment, tooth fillings, jewelry, coins, medical implants
Sand (SiO_2)	Glass, bricks, concrete for roads and buildings
Rare earth minerals	Technology equipment, computer chips
Gravel	Roadbeds, concrete
Limestone ($CaCO_3$)	Concrete, cement
Phosphate salts	Inorganic fertilizers, detergents

Most of the mineral imports to the United States come from reliable, politically stable countries. However, there are serious concerns about access to adequate supplies of four strategic metals—manganese, cobalt, chromium, and platinum (Figure 11-12). These metals are essential for the country's economic and military strength. The United States has little or no reserves of these metals.

Global and U.S. Supplies Not all rare earths are actually rare. However, they are hard to find in concentrations high enough to extract and process affordably. China has much of the world's known rare earth reserves and the United States has almost none.

In 2015, China produced 90% of the world's rare earth metals and oxides. One reason for this is that China does not strictly regulate the environmentally disruptive mining and processing of rare earths. This means Chinese companies have lower production costs than do companies in countries with stricter regulations.

The United States and Japan are heavily dependent on rare earths and their oxides. Japan has no rare earth reserves. The United States' only rare earth mine is in California. It used to be the world's largest supplier of rare earth metals. The mine closed for

FIGURE 11-11 ▼

Depletion Curves Each of these depletion curves for a mineral resource is based on a different set of assumptions. Dashed vertical lines represent 80% depletion.

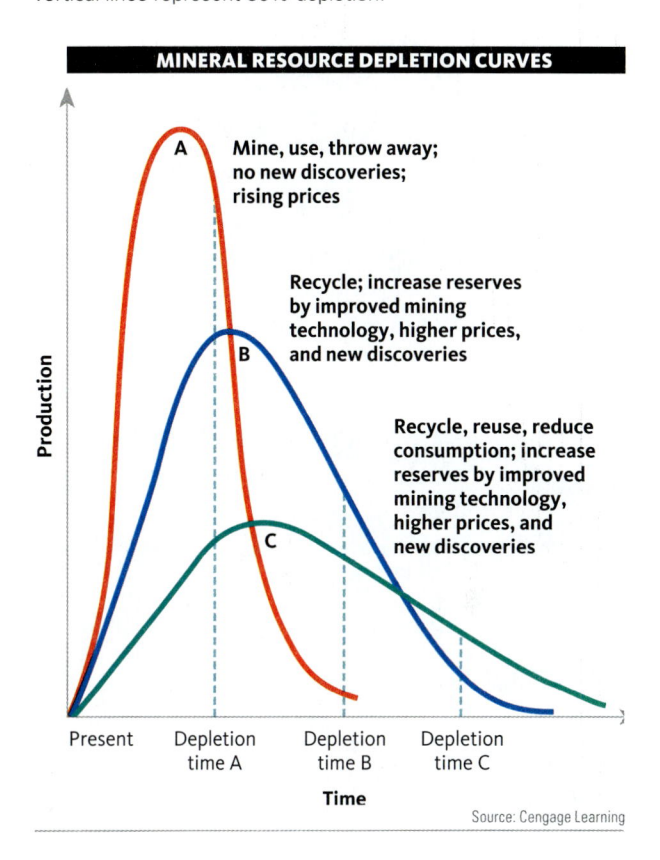

MINERAL RESOURCE DEPLETION CURVES

A: Mine, use, throw away; no new discoveries; rising prices

B: Recycle; increase reserves by improved mining technology, higher prices, and new discoveries

C: Recycle, reuse, reduce consumption; increase reserves by improved mining technology, higher prices, and new discoveries

Source: Cengage Learning

10 years in 2002 because of the expense of meeting pollution regulations and because China drove the prices of rare earths down to a point where the mine was too costly to operate. In 2014, the company that owned the mine declared bankruptcy.

China dominates the world in converting rare earth minerals to individual metals and oxides—a long, complex, and environmentally harmful chemical process. Since 2010, China has sharply raised the price of rare earth exports.

checkpoint Why are mineral resources nonrenewable?

Market Prices Affect Supplies

Geological processes determine the quantity and location of a mineral resource in Earth's crust. Economic factors determine what part of the known supply of a mineral resource is extracted and used. According to standard economic theory, when a resource becomes scarce in a competitive market system, its price rises. Higher market prices can encourage exploration for new deposits, stimulate development of better mining technology, and make it profitable to mine lower-grade ores.

FIGURE 11-12

Key Metals These four metals are critical to the U.S. economy. From top left, clockwise, they are manganese, cobalt, chromium, and platinum.

Higher market prices of mineral resources can also promote recycling, reduce resource waste, and encourage a search for substitutes.

However, some economists say the standard effect of supply and demand on price may no longer apply to the market for mineral resources in most more-developed countries. Governments use subsidies, tax breaks, and import tariffs to control the supply, demand, and prices of key mineral resources to such an extent that a truly free market does not exist. One example is a depletion allowance. Companies deduct the costs of developing and extracting mineral resources from their taxable income. These allowances amount to 5–22% of their gross income from selling the mineral resources.

Generally, mining companies maintain that they need taxpayer subsidies and tax breaks to keep the prices of minerals low for consumers. They argue that without these subsidies and tax breaks, they might move their operations to countries where they would not have to pay taxes or comply with strict mining and pollution regulations.

Expanding Reserves Some analysts believe people can increase supplies of some minerals by extracting them from lower-grade ores. They point to the development of new earth-moving equipment, improved techniques for removing impurities from ores, and technological advances in mineral extraction and processing that can make lower-grade ores accessible, sometimes at lower costs. In 1900, the copper ore mined in the United States was typically 5% copper by weight. Today, it is typically 0.5% copper, yet copper costs less (when prices are adjusted for inflation). Unfortunately, mining lower-grade ores requires mining and processing larger volumes of ore, which presents more environmental and economic challenges.

Minerals from the Oceans Most minerals in the oceans occur in such low concentrations that recovering them takes more energy and money than they are worth. Currently, only magnesium, bromine, and sodium chloride are abundant enough to be extracted profitably from seawater. On the other hand, sediments on the shallow continental shelf and shorelines contain significant deposits of minerals such as sand, gravel, phosphates, copper, iron, tungsten, silver, titanium, platinum, and diamonds.

One potential ocean source of minerals is hydrothermal ore deposits that form when superheated, mineral-rich water shoots out of vents in volcanic regions of the ocean floor. As the hot water comes into contact with cold seawater, particles of various metal sulfides precipitate out. The particles accumulate in chimney-like structures called hydrothermal vents (Figure 11-13). These deposits are especially rich in minerals such as copper, lead, zinc, silver, gold, and some of the rare earth metals. They also support exotic communities of marine life living in the darkness of the ocean floor (Explorers At Work).

Because of the rapidly rising prices of many of these metals, interest in deep-sea mining is growing. In 2011, China began using remote-controlled underwater equipment and a manned deep-sea craft to evaluate the mining of mineral deposits around geothermal vents. The UN International Seabed Authority, which was established to manage seafloor mining in international waters, began issuing mining permits in 2011.

According to some analysts, seafloor mining is less environmentally harmful than mining on land. Other scientists, however, are concerned that seafloor mining stirs up the sediment and can harm or kill organisms that feed by filtering seawater. Supporters of seafloor mining say that the number of potential mining sites is quite small and that many of these organisms can live elsewhere.

Another possible source of metals from the ocean is manganese nodules. These potato-sized nodules cover large areas of the Pacific Ocean floor and smaller areas of the Atlantic and Indian Ocean floors. These nodules also contain low concentrations of various rare earth minerals.

FIGURE 11-13
Hydrothermal vents, or "black smokers," are a potential source of important minerals but can also support a variety of marine life.

> **CONSIDER THIS**
>
> **Metal prices entice thieves.**
> Resource scarcity can promote theft. For example, copper prices have risen sharply in recent years because of increasing demand and limited supply. As a result, in many U.S. communities, thieves have been stealing copper to sell it. They strip abandoned houses of copper pipe and wiring and steal outdoor central air-conditioning units for their copper coils. Thieves also steal electrical wiring from beneath city streets and copper piping from farm irrigation systems. In 2015, copper wiring was stolen from New York City's subway system, temporarily shutting down two of the city's busiest lines.

To date, mining on the ocean floor has been hindered by the high costs involved, the potential threat to marine ecosystems, and arguments over rights to the minerals in deep-ocean areas that do not belong to a specific country.

checkpoint What affects whether a mineral is mined and how much of it is mined?

11.2 Assessment

1. **Compare and Contrast** Compare and contrast mineral resources and mineral reserves.
2. **Identify Main Ideas** How can a mineral resource become economically depleted?
3. **Restate** How does the high market price of a mineral affect how people use it?
4. **Apply** How has increased access to mineral resources influenced modern society?

SCIENCE AND ENGINEERING PRACTICES

5. **Engaging in Argument** Would you support mining at hydrothermal vents? Why or why not?

SCIENCE AND ENGINEERING PRACTICES

6. **Constructing Explanations** Recycling metals requires large inputs of water and energy. Do you think it is better to recycle or to mine for new sources to make metal products? Support your reasoning with evidence from the chapter and other sources.

11.3 What Are the Effects of Using Mineral Resources?

CORE IDEAS AND SKILLS
- Describe the major types of mining.
- Discuss the harmful environmental effects of mining.

KEY TERMS

surface mining	strip mining	subsurface
overburden	mountaintop	mining
spoils	removal	tailings
open-pit mining		smelting

Ore Type and Topography Dictate Mining Techniques

Every metal product has a life cycle that includes mining the mineral, processing it, manufacturing the product, and disposing of or recycling the product. The demand for these products drives mining companies and geologists to seek out deposits of minerals. When a geologist discovers a large deposit of minerals, a mining company determines the best method to remove the minerals, considering supply and profitability. Every method requires large amounts of water and energy. Additionally, every method produces waste and pollution at every step of the mineral's life cycle.

Surface Mining Several mining techniques are used to remove mineral deposits. Shallow deposits are removed by surface mining. In **surface mining**, vegetation, soil, and rock overlying a mineral deposit are cleared away. This waste material is called **overburden** and is usually deposited in piles called **spoils** (Figure 11-15). Surface mining is used to extract about 90% of the nonfuel mineral resources and 60% of the coal used in the United States.

Depending on the resource being sought and the local topography, mining companies can choose from different types of surface mining. In **open-pit mining**, machines dig very large holes and remove metal ores containing copper (Figure 11-16), gold or other metals, or sand, gravel, or stone.

Another type of surface mining is called strip mining. **Strip mining** is any form of mining involving the extraction of mineral deposits that lie in large horizontal beds close to the surface.

NATIONAL GEOGRAPHIC | EXPLORERS AT WORK
Robert Ballard Ocean Explorer

Plate tectonics—first introduced in the 1960s—revolutionized scientists' understanding of the way Earth's processes work and how they thought about them. Bob Ballard, an oceanographer and National Geographic Explorer, and his team began to question the cycle of rock, minerals, and ocean water on the sea floor.

If Earth's plates were spreading and moving under the ocean, then the emerging magma must be reacting with the cold temperatures and high pressures of the deep sea. In 1977, Bob Ballard and a team of scientists piloted a submersible to the sea floor near the Galapagos Islands (Figure 11-14). There they made a shocking discovery: Hot springs formed large chimney structures that were teeming with life (Figure 11-13).

The chimneys, or hydrothermal vents, formed from the chemicals that emerge from beneath the crust and mix with seawater. This answered their question about ocean circulation. When the hot magma cools, it contracts and cracks. Seawater penetrates the cracks, sinks down below Earth's surface, and interacts with the hot, sometimes molten, rock. The water pulls chemicals and minerals from the rock and brings them back to the surface when it emerges through the hydrothermal vents. The minerals that are pulled from the rock give the ocean its chemical makeup and make it salty.

Another great discovery at hydrothermal vents was the biological communities. The vents are far too deep for sunlight to reach. Previously, scientists thought all life depended on sunlight. The discovery of ecosystems based on chemicals from the vents as their energy source proved that certain producers can survive without light.

Ballard continued to study and research hydrothermal vent communities. He also discovered many famous shipwrecks, including the *Titanic*, *Lusitania*, and many others. He also founded JASON, an organization that brings live science to students all around the world.

FIGURE 11-14
Bob Ballard (left) and a colleague explore the depths of the Caribbean Sea in a submersible in the 1970s.

FIGURE 11-15
Spoils from a potash mine pile high near a town in Germany. Potash is a soluble form of potassium used in many fertilizers.

Area strip mining is used where the terrain is fairly flat. In this type of mining, a gigantic earthmover strips away the overburden. Then a power shovel—which can be as tall as a 20-story building—removes the mineral deposit or energy resource such as coal. The resulting trench is filled with overburden, and a new cut is made parallel to the previous one. This process is repeated over the entire site.

Contour strip mining is used mostly to mine coal and other mineral resources on hilly terrain. Huge power shovels and bulldozers cut a series of terraces into the side of a hill. Earthmovers remove the overburden, an excavator or power shovel extracts the coal, and the overburden from each new terrace is dumped onto the one below. Unless the land is restored, what is left is a series of spoils banks and a highly erodible hill of soil and rock called a highwall.

Another surface mining method is **mountaintop removal**. (See the photo at the start of the chapter.) In this method, the top of a mountain is removed to expose seams of coal that are then extracted. With this method, mining companies first clear large areas of tree cover, then dig holes for explosives used to blow the top off the mountain. Mountaintop removal destroys forests, buries mountain streams, and increases the risk of flooding. Wastewater and toxic sludge, produced when the coal is processed, are often stored behind dams in these valleys that can overflow or collapse, releasing toxic substances like arsenic and mercury.

More than 500 mountaintops in West Virginia and other Appalachian states have been removed to extract coal. According to the Environmental Protection Agency (EPA), the resulting spoils have buried more than 1,100 kilometers (700 miles) of streams—a total roughly equal to the distance between New York and Chicago.

Thousands of abandoned surface mines dot the U.S. landscape. The Surface Mining Control and Reclamation Act of 1977 requires surface mines to be restored. The government is required to take care of mines abandoned before 1977, while companies currently mining are required to restore the active mines. The program is greatly underfunded and many mines have not been reclaimed.

checkpoint What happens to the waste material in surface mining?

FIGURE 11-16
The Bingham Canyon Mine outside of Salt Lake City, Utah, is an open-pit copper mine that extends about 4 km (2.5 miles) wide and 1.2 km (0.75 miles) deep.

Subsurface Mining Deep deposits of minerals are removed by **subsurface mining**, in which minerals are removed through tunnels and shafts (Figure 11-17). This method is used to remove metal ores and coal too deep to be extracted by surface mining. Miners dig a deep, vertical shaft and blast open tunnels and chambers to reach the deposit. Then they use machinery to remove the resource and transport it to Earth's surface.

Subsurface mining disturbs less than one-tenth as much land as surface mining and usually produces less waste material. However, the environmental damage is significant and subsurface mines also lead to other hazards to humans such as cave-ins, explosions, and fires. Miners often get lung diseases caused by prolonged inhalation of mineral or coal dust in subsurface mines.

Underground mines can create subsidence, which is the collapse of the land above them. Subsidence can damage houses, crack sewer lines, break gas mains, and disrupt groundwater systems. Subsurface mining also requires large amounts of water and energy to process and transport the mined material.

Collectively, surface and subsurface mining operations produce three-fourths of all U.S. solid waste and cause major water and air pollution. Aerobic bacteria interacting with minerals in spoils piles produce sulfuric acid (H_2SO_4). When rainwater seeps through the spoils piles it produces acid mine drainage. This polluted runoff enters streams and groundwater.

Mining has polluted mountain streams in 40% of western watersheds in the United States. It also accounts for half of the country's emissions of toxic chemicals into the atmosphere.

Mining can be even more harmful to the environment where environmental regulations are lacking or not reliably enforced. In China, for instance, the mining and processing of rare earth metals and oxides has stripped much land of its vegetation and topsoil. It also has polluted the air, acidified streams, and left toxic and radioactive waste piles.

checkpoint What are some environmental hazards of subsurface mining?

LESSON 11.3

380 CHAPTER 11 GEOLOGY AND NONRENEWABLE MINERAL RESOURCES

FIGURE 11-17
ON ASSIGNMENT National Geographic photographer Robb Kendrick captured this image of work inside a coal mine in Meghalaya, India. Coal mining is big business here but there is little machinery or regulation. Miners work in dirty and dangerous conditions for scant pay.

CONSIDER THIS

The real cost of gold is pollution.

Some estimates suggest that mining enough gold to make a pair of wedding rings produces roughly enough mining waste to equal the total weight of more than three midsized cars. Many mining companies dig up massive amounts of rock containing only small concentrations of gold. About 90% of gold mines extract the mineral by spraying a solution of highly toxic cyanide salts onto piles of crushed rock. Until sunlight breaks down the cyanide, the settling ponds are extremely toxic to birds and mammals drawn to these ponds in search of water. The ponds can leak or overflow, posing threats to underground drinking water supplies and to organisms in lakes and streams. Some mining companies declare bankruptcy before cleaning up their operations, leaving behind ponds of poisoned water.

Processing Ore Has Harmful Effects

Ore extracted by mining typically has two components: the ore mineral that contains the desired metal and waste material. Removing the waste material from ores produces **tailings**—rock wastes that are left in piles or put into ponds where they settle out. Particles of toxic metals in tailings piles can be blown by the wind, washed out by rain, or leak from holding ponds, thus contaminating surface water and groundwater.

After the waste material is removed, heat or chemical solvents are used to extract the metals from the mineral ores. Heating ores to release metals is called **smelting**. Without effective pollution control equipment, a smelter emits large quantities of air pollutants. These pollutants include sulfur dioxide and suspended particles that damage vegetation and acidify soils in the surrounding area. Smelters also cause water pollution and produce liquid and solid hazardous wastes that require safe disposal. One study found that lead smelting is the world's second most toxic industry after the recycling of lead-acid batteries. Using chemicals to extract metals from their ores can also create numerous problems.

The environmental impacts of mining a metal ore are determined partly by the ore's percentage of metal content, or grade. The easily accessible higher-grade ores are usually mined first. Mining lower-grade ores takes more money, energy, water, and other resource. Several factors can limit the mining of lower-grade ores. First, it requires mining and processing larger volumes of ore, which takes more energy and costs more. A second factor is the dwindling supplies of fresh water needed for mining and processing some minerals, especially in arid and semiarid areas. Another factor is the growing environmental impacts of land disruption, along with waste material and pollution produced during mining and processing.

One way to improve mining technology and reduce its environmental impact is to use a biological approach, sometimes called biomining. In biomining, miners use natural or genetically engineered bacteria to remove desired metals from ores through wells bored into the deposits. This leaves the surrounding environment undisturbed and reduces the air and water pollution associated with removing the metal from ores by traditional means. However, biomining is slow. It can take decades to remove the same amount of material that current methods can remove in months or years. So far, biomining is economically feasible only for low-grade ores for which current techniques are too expensive.

checkpoint What are the harmful effects of mining on the environment?

11.3 Assessment

1. **Recall** Name and describe three types of mining.
2. **Summarize** What are the advantages and disadvantages of subsurface mining?

SCIENCE AND ENGINEERING PRACTICES

3. **Engaging in Argument** Did learning about gold mining change your perspective on gold? Support your reasoning with evidence from the lesson.

CROSSCUTTING CONCEPTS

4. **Patterns** Perform an internet search to find the location of copper mines in the United States according to the USGS. Using a map of the United States, roughly plot the mines on the map. What do you notice about their locations?

11.4 How Can Society Use Mineral Resources More Sustainably?

CORE IDEAS AND SKILLS

- Identify new materials that are replacing some metals for common uses.
- Explain how mineral resources can be used more sustainably.

KEY TERMS

materials revolution   nanotechnology

Substitutes for Some Minerals Exist

Because of the scarcity of some key minerals, scientists are seeking substitutes. In the current **materials revolution**, they are replacing some metals with silicon and other materials. They are also finding substitutes for scarce minerals through nanotechnology and other emerging technologies.

Nanotechnology uses science and engineering to create and manipulate materials out of atoms and molecules at the ultra-small scale of less than 100 nanometers. The diameter of the period at the end of this sentence measures about half a million nanometers. At the nanometer level, common materials have uncommon and unexpected properties. For example, scientists have learned to link carbon atoms together to form carbon nanotubes that are 60 times stronger than high-grade steel (Science Focus 11.1).

Other emerging technologies include fiber-optic glass cables that transmit pulses of light, which are replacing copper and aluminum wires in telephone cables. (Nanowires may eventually replace fiber-optic glass cables.) High-strength plastics and materials strengthened by lightweight carbon, hemp, and glass fibers are beginning to transform the automobile and aerospace industries. These new materials do not need painting (reducing pollution and costs), can be molded into any shape, and can increase fuel efficiency by greatly reducing the weight of motor vehicles and airplanes. Such new materials are also being used to build bridges.

Graphene (Figure 11-18) may be another breakthrough material. Created using high-purity graphite, graphene is one of the world's thinnest and strongest materials and is light, flexible, and stretchable. A single layer is 150,000 times thinner than a human hair and 200 times stronger than structural steel. Graphene is also a good conductor of electricity and heat—better than any other known material.

FIGURE 11-18
Graphene, which consists of a single layer of carbon atoms, is a revolutionary new material that could act as a substitute for some minerals in technology.

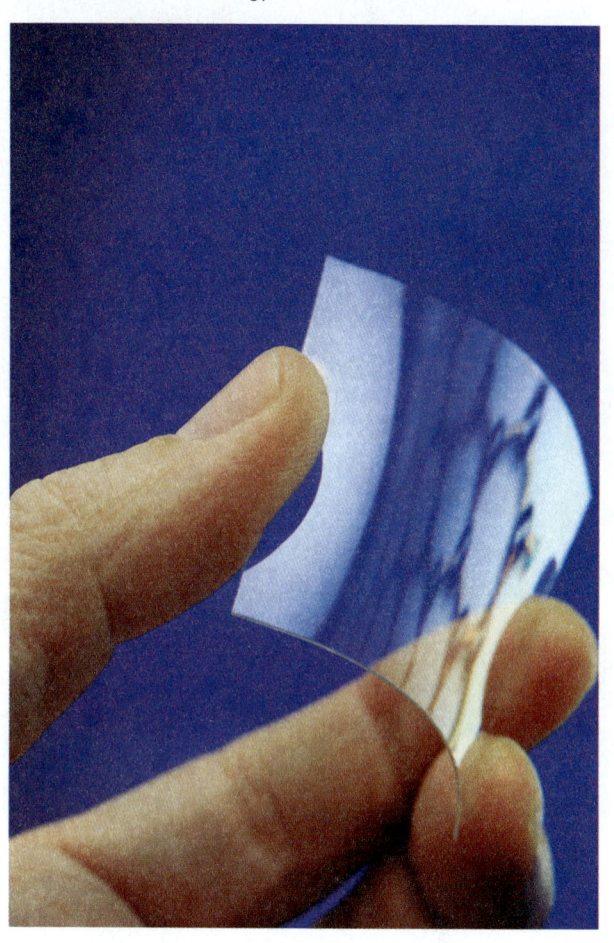

There are no known substitutes for some mineral resources, however. Platinum is widely used in industrial processes as a catalyst to speed up chemical reactions. Chromium is an essential ingredient of stainless steel. Finding acceptable and affordable substitutes for such scarce resources may not be possible.

checkpoint How can nanotechnology reduce the environmental impact of traditional metals?

Mineral Resources Can Be Used More Sustainably

There are ways to use mineral resources more sustainably. One strategy is to focus on recycling and reuse of nonrenewable mineral resources.

LESSON 11.4   383

FIGURE 11-19
A scientist at the Brookhaven National Laboratory in Upton, New York, uses advanced equipment to study materials on an atomic scale.

SCIENCE FOCUS 11.1

THE NANOTECHNOLOGY REVOLUTION

Materials produced with nanotechnology are used in more than 1,600 consumer products. These include batteries and electronics, clothes, foods, and more. Clothing could one day contain a new one-atom-thick nanotextile material that conducts electricity. A laptop could be charged by plugging into a T-shirt.

Nanotechnologists envision a wide range of technological innovations: a supercomputer smaller than a grain of rice, motors smaller than a human cell, biocomposite materials that would make bones and tendons super strong, nanovessels of medicines delivered to cells anywhere in the body, and nanomolecules designed to kill cancer cells. Nanotech could provide materials for a wide variety of products using atoms of abundant elements (such as hydrogen, oxygen, and carbon) as substitutes for scarcer elements (such as copper, cobalt, and nickel).

Nanotech has many potential environmental benefits. Designing, building, and altering products from the molecular level up would greatly reduce the need to mine many materials. It also would require less material and energy and reduce waste. Nanoparticles may be able to remove industrial pollutants in contaminated air, soil, and groundwater. Nanofilters might someday be used to purify seawater affordably.

Nanotechnology has some possible harmful environmental and health effects. Nanoparticles are potentially more toxic than conventional materials. The tiny particles can enter the environment during production, use, and after a product is discarded, potentially polluting water, soil, and air. Ongoing independent studies are focused on the impact of these particles on the environment and organisms but little is currently known about the long-term impact.

Many analysts say more information and regulation is needed before widely using nanoparticles.

Thinking Critically
Draw Conclusions Should the United States adopt the European Union's practice of requiring manufacturers to prove the safety of a product before allowing its use? Support your reasoning.

Redesigning manufacturing processes to use fewer resources would also enhance the sustainability of mineral resources. Additional strategies include finding substitutes for the minerals and reducing demand for the products. Also, mining subsidies could be shifted to increase subsidies for reuse and recycling programs.

Recycling is an application of the nutrient cycling factor of sustainability. It has a lower environmental impact than mining and processing metals from ores. Recycling aluminum beverage cans and scrap aluminum produces 95% less air pollution, 97% less water pollution, and uses 95% less energy than mining and processing aluminum ore. Cleaning up and reusing items instead of recycling has an even lower environmental impact.

Researchers are working hard to ensure adequate supplies of rare earths for the short term and to find alternatives to these materials for the long term (Case Study). One way to increase supplies is to extract and recycle rare earth metals from the massive amounts of electronic wastes that are being produced. Less than 1% of rare earths are recovered and recycled. The USGS and Department of Energy are evaluating mining wastes as a potential source of rare earths.

Companies that build batteries for electric cars are beginning to switch from making nickel-metal hydride batteries, which require the rare earth metal lanthanum, to manufacturing lighter-weight lithium-ion batteries (Explorers at Work).

Lithium (Li), the world's lightest metal, is a vital component of lithium-ion batteries used in cell phones, tablet and laptop computers, and a growing number of other products. Most countries do not have large supplies of lithium. More than 80% of global reserves are located within only two countries—Chile and China. The United States holds less than 1% of world reserves.

As a result, Japan, China, South Korea, and the United Arab Emirates are buying access to global lithium reserves to ensure their ability to sell lithium or lithium-ion batteries to the rest of the world. Within a few decades, the United States may be heavily dependent on expensive imports of lithium and lithium batteries. Researchers have been developing processes to extract lithium from the brine waste produced by geothermal power plants. One such plant is currently in operation in the United States. This process could reduce the United States' dependence on imported lithium. In 2021, the U.S. Department of Energy's Geothermal Technologies Office announced the American-Made Geothermal Lithium Extraction Prize. The $4 million prize competition is intended to fast-track improvements to lithium recovery profitability by inspiring entreprenuerial involvement and expansion of the extraction workforce. Winning teams are expected to be announced by early 2023.

Scientists are searching for substitutes for rare earth metals that could be used to make magnets and related devices. In Japan and the United States, researchers are developing a variety of devices that require no rare earths, are light and compact, and can deliver more power with greater efficiency at a lower cost. These include electric motors for hybrid vehicles and electromagnets for wind turbines.

To further reduce the demand for mineral reserves, people can choose to use less or do without products that require these minerals. Owning a product—such as a car, computer, or mobile device—until the end of its life rather than choosing to replace it with the newest model is one way to reduce demand for mineral resources. Choosing to own fewer electronics is another way.

checkpoint Describe three strategies for using rare earth minerals more sustainably.

11.4 Assessment

1. **Recall** What is the materials revolution?
2. **Identify Main Ideas** What is special about materials like graphene and fiber-optic glass cables?
3. **Apply** What is one way you could see yourself using mineral resources more sustainably?

CROSSCUTTING CONCEPTS

4. **Stability and Change** What do you think would happen if governments shifted subsidies and tax breaks away from mining toward recycling and research and development (R&D)?

TYING IT ALL TOGETHER
Rare Earth Metals and Sustainability

The Case Study that opened this chapter told about the importance of rare earth metals. These scarce metals are critical to the economies of the United States and other countries, as they are used in various modern technologies from cell phones to energy-efficient light bulbs.

However, the extraction, production, and use of minerals and their products have long-lasting and sometimes permanent effects on the environment. The nonrenewable nature of minerals also guarantees they will not always be available for use.

Technological developments could help society expand supplies of mineral resources and use them more sustainably. For example, a nanotechnology revolution could lead to new materials to replace rare earth minerals, greatly reducing the environmental impacts of mining and processing them.

Biomining can make use of microbes to extract mineral resources without disturbing the land or polluting air and water as much as conventional mining does. Another emerging technology uses graphene to replace conventional transistors and produce more efficient and affordable solar cells to generate electricity. This is an application of the solar energy factor of sustainability.

In addition, industries can mimic nature using a diversity of ways to reduce the harmful environmental impacts of mining and processing mineral resources, thus applying the biodiversity factor of sustainability.

Finally, people can use mineral resources more sustainably by reusing and recycling them, and by reducing unnecessary use and waste. This strategy applies the nutrient cycling factor of sustainability.

Work with a group to complete the following steps.

1. Use the internet to find the current average life expectancy, in years, of each of the following electronic products: TVs, laptops, tablets, phones. Tabulate your data.
2. Estimate the average life span of each device among your team members. Add a column with this information to your table. How do the numbers compare?
3. What are the implications of the life spans on the sustainable use of mineral resources?
4. Go to the EPA webpage on electronics donation and recycling. Where can these items be donated or recycled near you?

CHAPTER 11 SUMMARY

11.1 How do geological processes relate to society and the environment?

- Earth has a core, mantle, and crust.
- Geological processes break Earth's surface into slowly moving tectonic plates that split apart and move continents.
- Geological processes within Earth and on its surface produce the mineral resources we depend on.

11.2 What are Earth's mineral resources and how long might reserves last?

- Mineral resources are nonrenewable because it takes millions of years to produce them.
- The three types of rocks are sedimentary, igneous, and metamorphic. They are formed by the rock cycle.
- Metallic and nonmetallic minerals must be removed from ores that are mined.
- Mineral resources exist in finite amounts and can be economically depleted when it costs more than they are worth to find, extract, and process them.
- There are several ways to extend supplies of mineral resources, but each can be limited by economic and environmental factors.

11.3 What are the effects of using mineral resources?

- Extracting minerals from Earth's crust and converting them to useful products disturbs the land, erodes soils, produces large amounts of solid waste, and pollutes the air, water, and soil.
- Strip mining, open-pit mining, mountaintop removal, and subsurface mining all create environmental and safety problems.
- Processing ores has harmful effects. The resulting piles of rock wastes contain toxic metals that can contaminate surface water and groundwater.
- Smelting can pollute the air and water and produce liquid and solid hazardous wastes.

11.4 How can society use mineral resources more sustainably?

- People can try to find substitutes for scarce resources to reduce society's need for mineral resources.
- Reusing or recycling metal products and redesigning manufacturing processes to use fewer mineral resources will enhance sustainability. Reducing or ending mining subsidies and increasing subsidies for reuse, recycling, and finding substitutes also enhances sustainability.

MindTap If you have been provided with access to a MindTap course, additional resources are available at login.cengage.com.

CHAPTER 11 ASSESSMENT

Review Key Terms

Select the key term that best fits each definition. Not all terms will be used.

asthenosphere	mineral	rock cycle
continental drift	mineral resource	sedimentary rock
core	mountaintop removal	smelting
crust	nanotechnology	spoils
depletion time	open-pit mining	strip mining
igneous rock	ore	subsurface mining
lithosphere	overburden	surface mining
mantle	rare earth metal	tailings
metamorphic rock	reserve	tectonic plate
materials revolution	rock	

1. A solid of one mineral or a combination of minerals found in Earth's crust
2. The combination of Earth's crust and rigid outermost part of the mantle
3. The supply of a nonrenewable mineral resource that is economically feasible for extraction
4. Type of mining in which terraces are cut into the side of a slope
5. The soil and rock removed from a surface mine before minerals are extracted
6. A process in which ores are heated to release metals
7. One of a group of minerals with unique properties that lend themselves to use in technology products
8. The slow movement of continents across Earth's surface
9. The layer of Earth beneath the lithosphere in which convection currents flow
10. Rock that contains a large enough concentration of a particular mineral—often a metal—to make it profitable for mining and processing

Review Key Concepts

11. List five important mineral resources and their uses.
12. Describe the environmental effects of gold mining.
13. What are tectonic plates? What typically happens when they collide or move apart from each other?
14. What is depletion time and what factors affect it?
15. What are the advantages and disadvantages of biomining?
16. What problems could arise from the widespread use of nanotechnology?
17. Describe the conventional view of the relationship between the supply of a mineral resource and its market price.
18. Describe the potential of using graphene as a new resource.
19. What are the major safety hazards of working in a subsurface coal mine?

Think Critically

20. Consider the research efforts of Bob Ballard and Yu-Guo Guo. How does their work contribute to the environmental community? To the everyday needs of society?
21. Consider the processes you learned about in Lesson 11.1. Think about the ways they affect Earth's features. Create a model incorporating the key processes to show how they work together to shape Earth's surface.
22. How do plate tectonics, the rock cycle, and surface processes like weathering and erosion contribute to the availability of mineral resources for human extraction and use?
23. List three ways a nanotechnology revolution could benefit you and three ways it could harm you. Do the benefits outweigh the potential hazards or vice versa? Explain your reasoning.
24. Use the second law of thermodynamics to analyze the scientific and economic feasibility of each of the following processes:
    a. extracting certain minerals from seawater
    b. mining increasingly lower-grade mineral deposits
    c. continuing to mine, use, and recycle minerals at increasing rates

Chapter Activities

A. Investigate: Product Life Cycles STEM

Many products that require scarce or expensive resources can unnecessarily end up in landfills. These products are said to have a "cradle-to-grave" life cycle. Products that can be repurposed or recycled have a "cradle-to-cradle" life cycle. In this activity, you will research three products that could have a cradle-to-cradle life cycle and learn what it takes to make the product more sustainable.

Materials

| scissors | colored paper | old magazines |
| glue | markers | |

1. Use the internet to determine which mineral resources are used in the manufacture of each of the following items and how much of each mineral is required to make each item:
   a. cell phone
   b. smart TV
   c. large pickup truck

2. Pick three of the lesser-known minerals you have learned about in this exercise. Find out the geographic origin of these three minerals.

3. For each of the three minerals, what environmental effects result from mining and processing? Have there been any efforts to reduce those effects?

4. Consider the end of life for each of the three products you identified. Research what it would take to recycle or reuse the mineral that is used in the highest proportion. Based on a cost-benefit analysis, do you feel that the item can have a fate other than the landfill at the end of its life?

5. Use the materials to help you create a visual model of a cradle-to-cradle life cycle for one of the products listed above. The life cycle may not be linear. It may have multiple branches as components of the product are recycled or repurposed.

B. Take Action

National Geographic Learning Framework

Attitudes | Responsibility
Skills | Problem-Solving
Knowledge | Our Living Planet

Products with rare earths and other metals not only require a lot of resources and produce pollution during production, they can also be harmful when disposed of if they end up in a landfill. Many of these products can be recycled, but only at special facilities. What options exist in your community for recycling electronics like cell phones, TVs, and computers? Find out. Then start a program at your school to help students and their families recycle their electronics rather than throw them away.

Set up a collection site and plan to regularly deliver the items to a local e-waste facility. You may want to have a campaign to encourage others to bring in their old items. Consider campaign slogans, posters, and ads around your school. Keep a record of the number of items or weight of items you collected and how much you diverted from the landfill. If this program already exists locally or is quickly successful, expand the effort to include lithium-ion batteries, CF and halogen light bulbs, or other products.

CHAPTER 12
NONRENEWABLE ENERGY RESOURCES

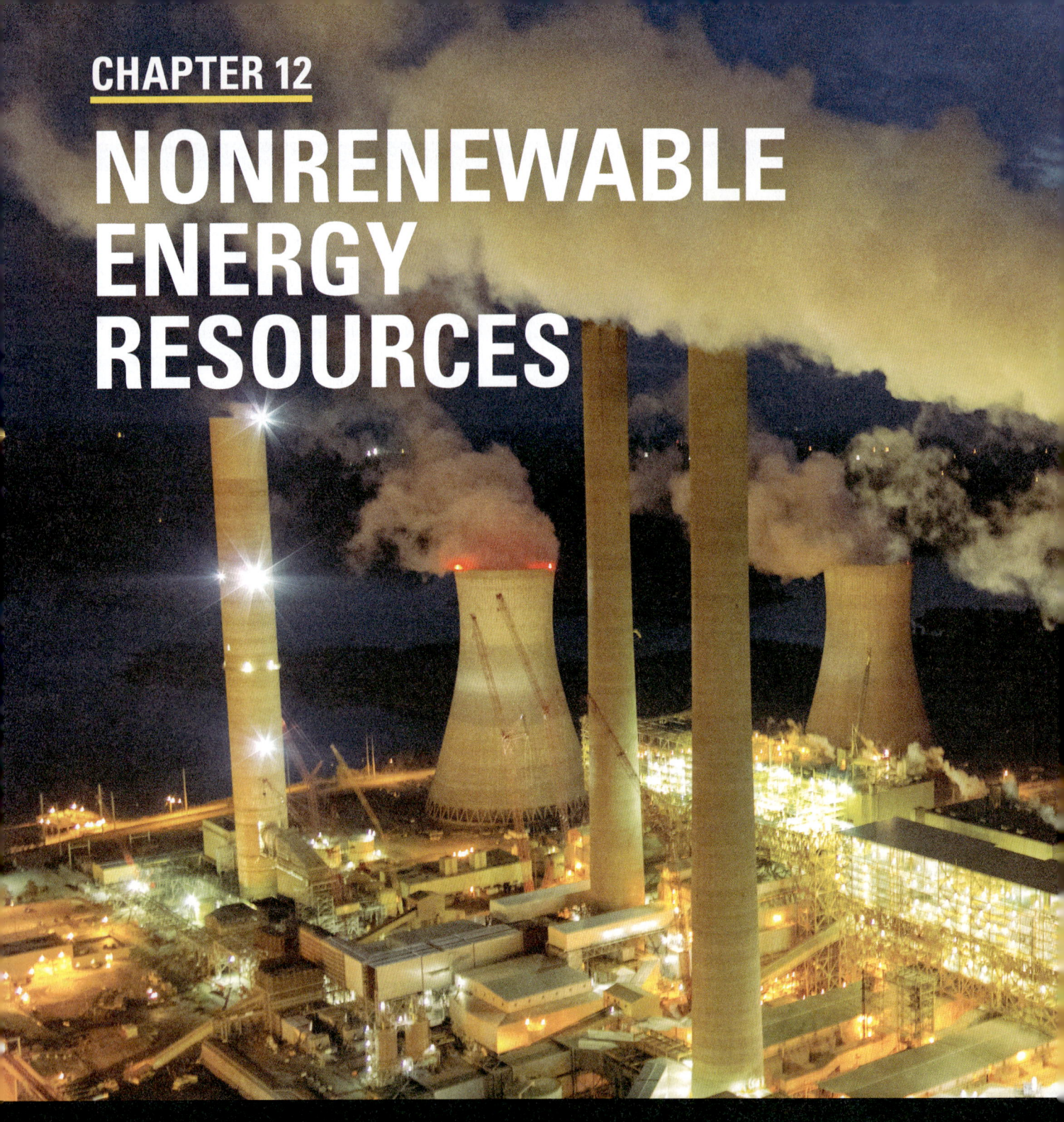

COMMERCIAL ENERGY helps society meet many needs, from industry, transportation, and technology to people's personal well being. Most commercial energy comes from nonrenewable resources including coal, oil, natural gas, and nuclear power. Nonrenewable resources are limited in supply. Using them to produce energy often harms the environment and human health. How will these issues impact energy use in the future?

KEY QUESTIONS

12.1 What is net energy and why is it important?

12.2 What are the advantages and disadvantages of using fossil fuels?

12.3 What are the advantages and disadvantages of using nuclear power?

Steam and smoke billow from a coal-fired power plant in the southeastern United States.

NATIONAL GEOGRAPHIC | EXPLORERS AT WORK

Hunting for Methane Leaks

with National Geographic Explorer Katey Walter Anthony

Trekking on thin ice and playing with fire is all in a day's work for Katey Walter Anthony. Dr. Walter Anthony hunts for methane gas bubbling up through thawing lake ice in the Arctic. When she suspects she's found a methane gas leak, Walter Anthony strikes a match. If flames erupt, she's hit the jackpot.

Walter Anthony's job is more risky—and less comfortable—than most. But she keeps logging hours in the frigid far north because her work there has global implications. Walter Anthony is exploring the connection between changes to natural gas deposits in the Arctic and changes to Earth's climate.

In the Arctic, methane has been trapped beneath a frozen layer of permafrost for tens of thousands of years. However, the atmosphere is warming and the permafrost is starting to thaw. As it does, methane from natural gas deposits is escaping. "All that carbon was locked up safely in the permafrost freezer," says Walther Anthony. "Now the freezer door is opening …and the lakes essentially burp out methane."

Walter Anthony sees this as both a problem and a solution. It is a problem because escaping methane, a greenhouse gas, contributes to climate change. In fact, it has 25 times the warming potential of carbon dioxide. Scientists, engineers, policy makers, and other decision makers need more information about emissions of methane and other greenhouse gases to prepare for the impacts of climate change.

Methane leaks also help fuel a vicious cycle. Greenhouse gas emissions contribute to climate change. Climate change thaws the frozen Arctic. Thawing leads to more methane leaks. In other words, methane leaks cause and are caused by climate change.

On the other hand, the seeping methane may also be part of a solution. Where Walter Anthony works it is freezing cold and isolated, but people still live there. Because of their isolation, people in the region often struggle to get the energy they need to heat and power their homes and businesses. Walter Anthony has begun to research how to use seeping methane as an energy source for these remote communities. Energy companies could harness the methane before it is released into the atmosphere and distribute it locally to people who need it.

Burning methane as a fuel source releases only small amounts of greenhouse gases compared to methane gas that seeps straight into the atmosphere. Furthermore, a local energy source would reduce local dependence on diesel fuel and make energy more affordable. For these reasons, Walter Anthony's research could make the future brighter for people locally *and* globally.

Thinking Critically

Analyze How is Katey Walter Anthony working to turn an environmental problem into an energy solution? What are some challenges to bringing about this solution?

A burst of flame tells Katey Walter Anthony (on right) that methane gas is leaking from this Arctic lake.

CASE STUDY
Fracking for Oil and Gas

Vast deposits of oil and natural gas are tightly trapped between layers of shale rock formations across the central United States. Until a few decades ago, these deposits remained untapped. Extracting the oil or natural gas from where it was tightly held between rock layers was too difficult and costly.

This situation changed in the late 1990s when oil and gas producers combined two existing extraction technologies (Figure 12-1). One is horizontal drilling, which involves drilling a vertical well deep into the ground, turning the flexible shaft of the drill, and then drilling horizontally. This technology provides a way to reach tightly held oil and natural gas deposits.

The second technology, called hydraulic fracturing or fracking, is used to free the trapped natural gas and oil. Fracking involves using high-pressure pumps to inject huge volumes of a mixture of water, sand, and chemicals into the horizontal well. The mixture then blasts through holes drilled in the underground well pipe.

The force of the blast fractures the surrounding shale rock and drives grains of sand into the cracks to hold them open. When the pressure is released, the oil or natural gas flows out of the cracks. Then it is pumped to the surface.

The mixture that is injected into the well contains naturally occurring salts, toxic heavy metals, and radioactive materials leached from the rock. Fracking companies also add potentially harmful drilling chemicals to the mix. After fracking is complete, about half the mixture remains below ground, while the other half becomes wastewater.

Processing the hazardous wastewater is a challenge. It may be stored in underground wells or in surface holding ponds, where it can leak. It may be sent to sewage treatment plants that are often ill-equipped to handle it properly. Or it can be cleaned up and used again for fracking—the best method but also the most rarely used because of its high cost.

Some say fracking marks a new era in oil and natural gas production. But the process also yields lower net energy compared to traditional production. (Net energy is discussed in this chapter.) Oil and natural gas prices must remain high to make it profitable to frack oil and natural gas.

In this chapter, you will learn about oil, natural gas, and other sources of nonrenewable energy. You will learn how using various types of nonrenewable energy impacts the environment and affects local and global economies.

As You Read Consider the impacts of making nonrenewable energy resources available to use. Think about how your lifestyle creates demand for various types of energy, as well as how such energy is produced.

FIGURE 12-1 ▼
This diagram illustrates how horizontal drilling and hydraulic fracturing are used to access oil and natural gas deposits trapped in layers of shale.

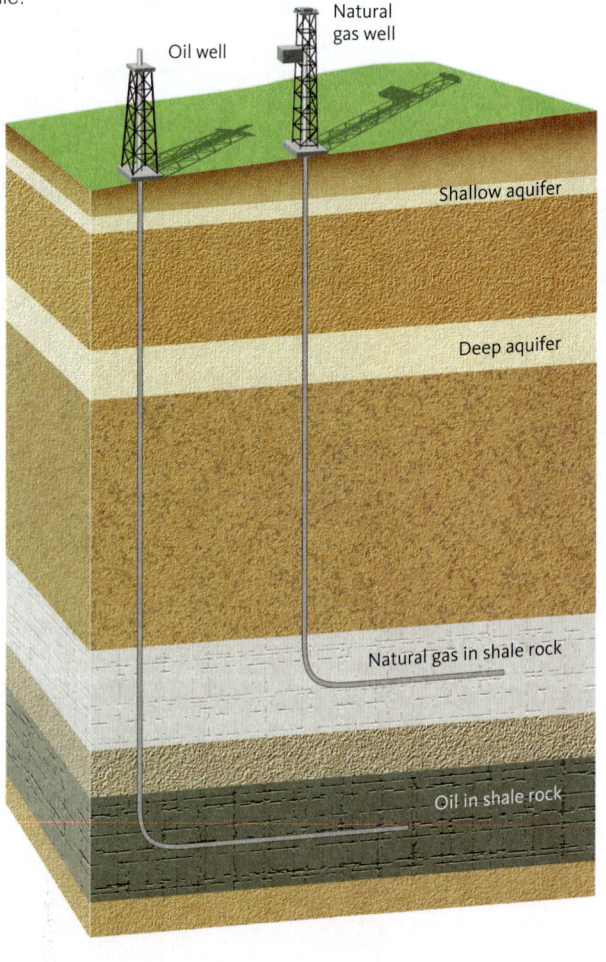

12.1 What Is Net Energy and Why Is It Important?

CORE IDEAS AND SKILLS

- Define commercial energy, and identify the types of nonrenewable energy resources.
- Understand that it takes energy to produce energy, and explain the significance of net energy.
- Explain why energy resources with a low net energy need government subsidies.

KEY TERMS

commercial energy   net energy

Energy Comes from Different Sources

The energy that heats Earth and makes life possible comes from the sun and is an application of the solar energy factor of sustainability. Without this free, inexhaustible input of solar energy, Earth's average temperature would be −240°C (−400°F). Life, if it existed, would not exist as it does now.

To supplement the sun's life-sustaining energy, society produces **commercial energy**—energy that is sold in the marketplace. Commercial energy is produced from either renewable or nonrenewable energy resources. Nonrenewable energy resources exist in fixed amounts. These resources took millions to billions of years to form and will be used more quickly than they can be replaced. They include fossil fuels (oil, coal, and natural gas) and the nuclei of certain atoms (nuclear energy).

Renewable energy resources (discussed in Chapter 13) are replenished by natural processes. They include wind, flowing water (hydropower), the sun's heat (solar energy), biomass (energy in plants), and heat in Earth's interior (geothermal energy).

Most of the world's commercial energy is produced using nonrenewable energy resources. In 2020, nearly 90% of the commercial energy used worldwide and in the United States came from nonrenewable energy resources. Just over 10% came from renewable energy resources (Figure 12-2).

checkpoint What are four nonrenewable energy resources?

FIGURE 12-2 ▼

Energy Use These two graphs show energy use by source throughout the world and within the United States in 2020.

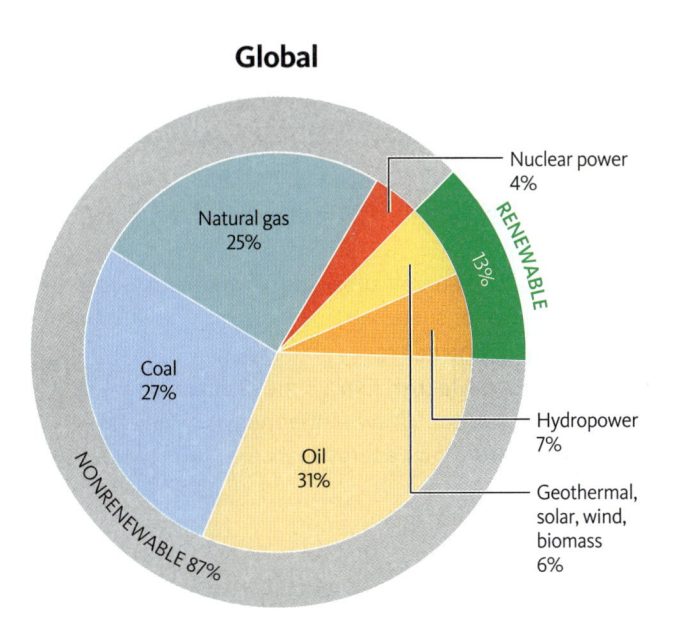

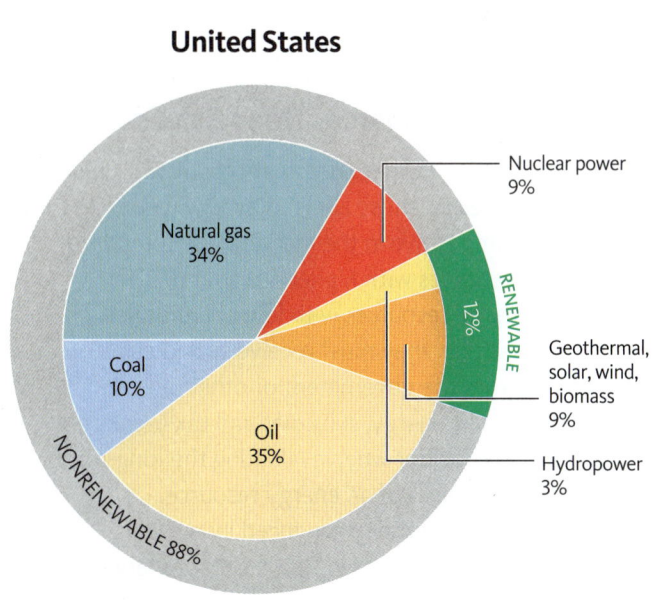

Global

- Natural gas 25%
- Coal 27%
- Oil 31%
- Nuclear power 4%
- Hydropower 7%
- Geothermal, solar, wind, biomass 6%
- RENEWABLE 13%
- NONRENEWABLE 87%

United States

- Natural gas 34%
- Coal 10%
- Oil 35%
- Nuclear power 9%
- Geothermal, solar, wind, biomass 9%
- Hydropower 3%
- RENEWABLE 12%
- NONRENEWABLE 88%

Sources: British Petroleum, U.S. Energy Information Administration, and the International Energy Agency

Only Net Energy Counts

It takes energy to produce energy. For example, before oil can be used, it must be located and pumped from beneath the ground or ocean floor. Then it must be transported to a refinery and converted to gasoline, diesel, and other fuels and chemicals. Finally, it is delivered to consumers. Each step uses high-quality energy obtained by burning fossil fuels, especially gasoline and diesel fuel. Due to the second law of thermodynamics (Chapter 2), some of the high-quality energy used in each step is always degraded to lower-quality energy. This energy is most often wasted as heat released into the environment.

Net energy is the amount of high-quality energy available from a given quantity of an energy resource minus the high-quality energy needed to make that energy resource available.

Net energy = energy output − energy input

Suppose that it takes about 9 units of energy to produce 10 units of energy from an energy resource. Then the net energy is 1 unit of energy. Net energy is like the net profit earned by a business after it deducts its expenses. If a business has $1 million in sales and $900,000 in expenses, its net profit is $100,000. Many analysts view net energy as the single most important measure that can be used to evaluate the long-term economic usefulness of energy resources.

checkpoint What is net energy?

Some Energy Resources Need Subsidies

Resources with a low net energy are costly to bring to market. As a result, these kinds of energy resources are not as competitive in the marketplace compared to energy resources with higher net energies. In order for energy resources with low net energies to be competitive, they must receive subsidies or tax breaks from the government or other outside sources. Like net energy, subsidies and tax breaks are important to consider when evaluating the long-term economic usefulness of an energy resource.

For example, electricity produced by nuclear power has a low net energy. This is because large amounts of high-quality energy are needed to carry out various steps involved in producing electricity from this energy resource. First, energy is required to extract and then process the raw materials used as fuel in a nuclear power plant. Energy is also needed to build and maintain the plant and, in time, to safely shut down the plant. Finally, energy is needed to store the radioactive waste that results from producing electricity using nuclear power. This series of steps is called the nuclear fuel cycle and is discussed later in this chapter.

The low net energy and thus the high cost for the whole nuclear fuel cycle is one reason why governments throughout the world heavily subsidize nuclear-generated electricity. Subsidies make this energy product available to consumers at an affordable price. However, such subsidies hide the true cost of using nuclear power to produce electricity. As a result, this practice is a violation of the full-cost pricing factor of sustainability.

Because of its low net energy, the current nuclear fuel cycle must be heavily subsidized. However, the government also invests in research and development to improve technology for producing energy from nuclear power and other energy resources. In time, government-supported or private research and development could be used to develop methods for producing nuclear power that would result in a higher net energy.

checkpoint How can an energy resource with a low net energy compete with those that have a higher net energy?

12.1 Assessment

1. **Define** What is commercial energy?
2. **Contrast** What is the difference between nonrenewable and renewable energy resources?
3. **Apply** How might improvements in technology for producing a given energy resource affect its net energy?
4. **Infer** How do energy subsidies violate the full-cost pricing factor of sustainability?

CROSSCUTTING CONCEPTS

5. **Energy and Matter** Explain the relationship between the second law of thermodynamics and the net energy of an energy resource.

12.2 What Are the Advantages and Disadvantages of Using Fossil Fuels?

CORE IDEAS AND SKILLS
- Discuss oil, natural gas, and coal as commercial energy sources.
- Understand the advantages and disadvantages of using fossil fuels for energy.

KEY TERMS

crude oil
refining
petrochemical
peak production
proven oil reserve
horizontal drilling
hydraulic fracturing
natural gas
coal

Crude Oil Is an Important Energy Source

Hydrocarbons are compounds of hydrogen and carbon atoms (Lesson 2.2). **Crude oil**, or petroleum, is a mixture of hydrocarbons and other compounds. It exists as a black, gooey liquid that can burn easily. Crude oil forms over millions of years from the decayed remains of ancient organisms, which become crushed beneath layers of rock and are subjected to intense pressure and heat. Consequently, oil is considered a type of fossil fuel.

Oil deposits form when extreme pressure caused by the weight of overlying rocks forces oil upward through cracks and pores in Earth's crust. Oil works its way toward the surface until it reaches a layer of impenetrable rock. The trapped oil collects there, often with natural gas. These underground deposits fill pores and cracks somewhat like water soaking into a sponge. They can form either on land or beneath the sea floor.

Geologists identify potential oil deposits by looking for certain geologic formations, such as layers of shale rock, where oil is likely to collect. To find these formations, they use large machines to pound the ground and send shock waves deep below the surface. They record how long it takes for the waves to be reflected back and enter this information into computers. It is converted to 3-D seismic maps that show locations and sizes of various rock formations.

When an oil deposit is located, oil companies drill holes and remove rock cores to determine if there is enough oil to be extracted profitably. If there is, one or more wells are drilled. The oil is drawn out of the rock pores by gravity. It flows to the bottom of the well and is pumped to the surface (Figure 12-3).

Crude oil must be processed at a refinery before it can be used for fuel. Oil gets transported to a refinery by pipeline, truck, rail, or ship. At the refinery, the crude oil is heated to separate it into various fuels and other components with different boiling points in a complex process called **refining** (Figure 12-4). Like all the steps in oil production and use, refining requires an enormous amount of energy. It decreases oil's net energy.

Refining crude oil results in usable by-products as well as undesirable pollutants. About 2% of the products of refining are called **petrochemicals**. Petrochemicals are used as raw materials to make other chemicals. They are also used as ingredients in products such as plastics, synthetic fibers, paints, medicines, and cosmetics.

A less desirable by-product of the refining process is petcoke, a black powder that accumulates in huge piles at refineries. Petcoke becomes an air pollutant if it is not properly stored and winds blow it into the atmosphere. Another pollutant of concern is carbon dioxide (CO_2), which is released into the atmosphere when oil or any carbon-containing fossil fuel is burned. Scientists have identified CO_2 as a greenhouse gas known to warm the atmosphere in what is called the natural greenhouse effect. As increasing quantities of oil and other fossil fuels are burned, the amount of CO_2 in the atmosphere is also increasing. The excess CO_2 magnifies the natural greenhouse effect, warms the atmosphere, and changes the climate. (Climate change is discussed in Chapter 16.)

After about a decade of pumping, the pressure in a well usually drops, and its rate of crude oil production starts to decline. This point in time is referred to as **peak production** for the well. The same decline in production can take place at an oil field, where many wells operate together.

checkpoint How do geologists locate potential oil deposits?

Society Relies Heavily on Oil

Today's society relies heavily on crude oil. In 2020, the world used roughly 91 million barrels of crude oil per day (one barrel of oil contains 159 liters or 42 gallons of oil). Global crude oil consumption is expected to continue to increase through 2035.

How much crude oil exists on Earth? The exact answer is unknown, although geologists have estimated amounts of crude oil in proven oil reserves.

CHAPTER 12 NONRENEWABLE ENERGY RESOURCES

FIGURE 12-3
ON ASSIGNMENT All the wells in this oil field, photographed by National Geographic photographer Sarah Leen, pump oil from a single proven oil reserve in southern California.

LESSON 12.2 399

Proven oil reserves are known deposits from which oil can be extracted profitably at current prices using current technology. The number of proven oil reserves and the amount of available oil are not fixed. A reserve is partially determined by pricing and available technology.

Higher prices and more efficient technology can turn previously unavailable oil deposits into proven oil reserves. Horizontal drilling and hydraulic fracturing are two such technologies (Case Study). Recall that **horizontal drilling** involves drilling a vertical well deep underground and then turning the shaft 90 degrees to drill horizontally into shale rock. **Hydraulic fracturing**, or fracking, uses high-pressure pumps to inject a mix of sand, water, and chemicals into the well to fracture shale rock so that tightly held gas and oil can be pumped out.

FIGURE 12-4
Refining Crude oil is boiled to form vapor that contains various components. The vapor rises through a distillation tower. Components condense back into liquids at different heights within the tower depending on their boiling points. Then they are collected. The tower may be as tall as a nine-story building.

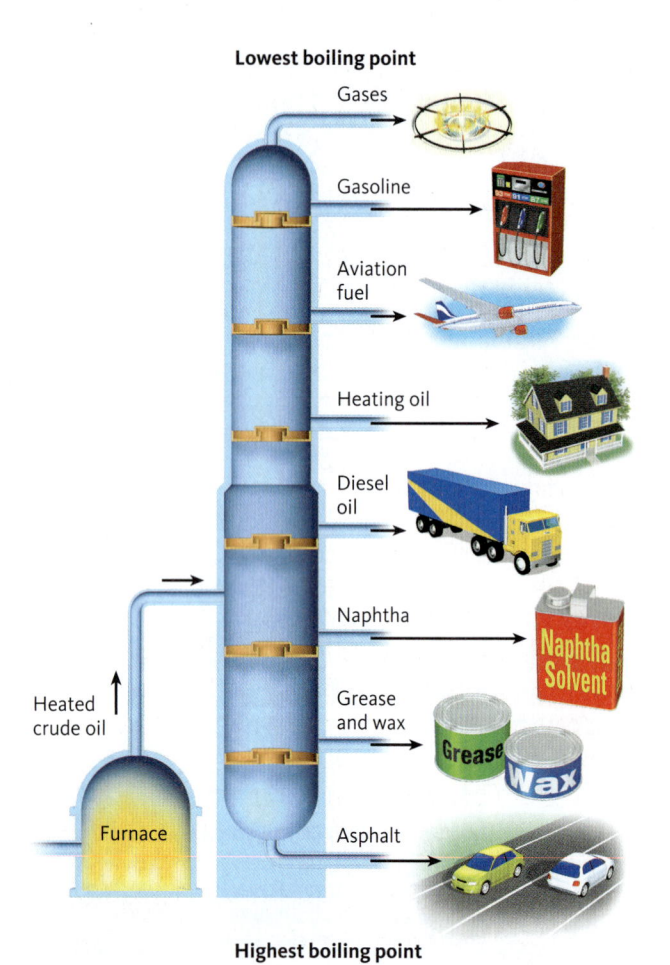

FIGURE 12-5

Crude Oil	
Advantages	**Disadvantages**
Ample supplies for many decades	Water pollution from oil spills and leaks
Medium net energy	Emits CO_2 and other air pollutants when produced and burned
Low land disruption	Environmental costs not included in market prices
Efficient distribution system	Vulnerable to international supply interruptions

There are various ways to produce more crude oil. Oil companies can drill for oil far offshore in deep ocean seabed deposits. They can also extract oil from remote areas such as the Arctic. There are also alternative types of oil, discussed later, that can be used as energy sources. But these remote or alternative oil sources are harder and more expensive to extract and process. Tapping into them often results in lower net energy, greater environmental impacts, and higher oil prices.

There is controversy over whether society has reached *global* peak production of oil. This would mean that half of the world's oil reserves have been used. Once global peak production is reached, it is expected that net energy from oil will decline, and oil prices will rise. Eventually the most accessible oil fields will become depleted. Companies will have to tap into the remaining half of the world's estimated oil reserves. Such reserves are smaller, less accessible, and less rich in oil.

Some analysts say there is not a global shortage of oil but rather a shortage of cheap oil, as the world's easy-to-reach, concentrated deposits are becoming depleted. The world still has ample oil supplies, but at some point they will run out. Figure 12-5 lists the advantages and disadvantages of using crude oil as an energy resource.

checkpoint How can oil companies increase proven oil reserves?

Oil Reserves, Production, and Usage Are Not in Sync

Oil reserves are not uniformly located throughout the world, and they are often not located where society most needs them (Figure 12-6). Thirteen countries hold about 81% of the world's proven

FIGURE 12-6

Crude Oil Economy This graph identifies the world's top three countries in terms of oil reserves, oil production, and oil consumption in 2020.

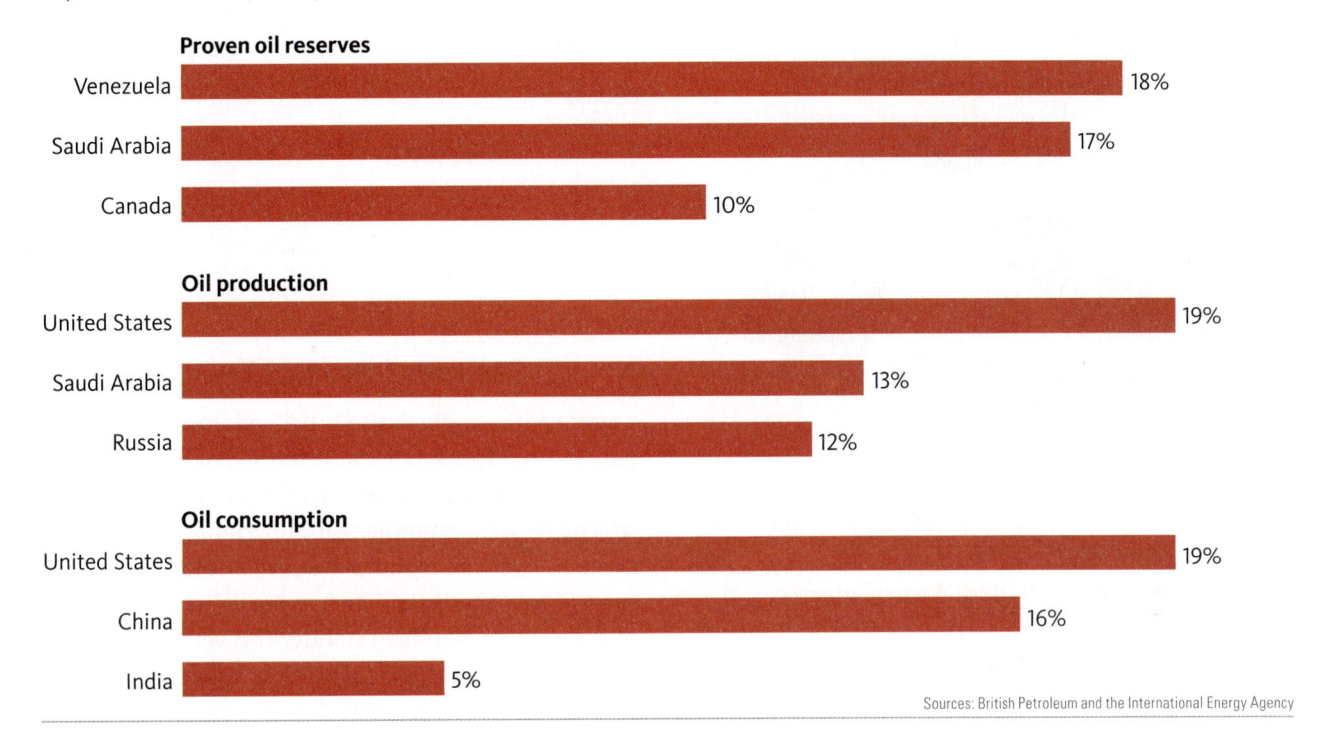

Proven oil reserves
- Venezuela: 18%
- Saudi Arabia: 17%
- Canada: 10%

Oil production
- United States: 19%
- Saudi Arabia: 13%
- Russia: 12%

Oil consumption
- United States: 19%
- China: 16%
- India: 5%

Sources: British Petroleum and the International Energy Agency

crude oil reserves. These 13 countries make up the Organization of Petroleum Exporting Countries (OPEC). OPEC is likely to control most of the world's crude oil supplies for many years to come. As of 2021, members of OPEC included Algeria, Angola, the Democratic Republic of the Congo, Equatorial Guinea, Gabon, Iran, Iraq, Kuwait, Libya, Nigeria, Saudi Arabia, United Arab Emirates, and Venezuela. Not included in OPEC are two of the three largest oil producers (Russia and the United States) and the three largest oil consumers (the United States, China, and India).

U.S. Production The United States gets about one-third of its commercial energy from crude oil (Figure 12-2, right). Since 1982, oil consumption in the United States has been greater than domestic production. The country has had to import some of the oil it uses as a result. However, the gap between U.S. oil production and consumption has narrowed.

Domestic oil production has grown enough to make the United States the largest producer of oil as of 2020. In fact, 2020 marked the first time the United States exported more petroleum than it imported. While this remained true for 2021, the country became a net petroleum importer again in 2022. The United States gets most of its imported oil from Canada, Mexico, Russia, Saudi Arabia, and Columbia. Overall, the United States uses about 19% of the world's oil, while it produces 19% and has 4% of the world's proven crude oil reserves.

Alternative Oil Sources As proven crude oil reserves begin to decline, the world can turn to heavy crude oil. This type of crude oil is thicker and more sticky than regular crude oil. Heavy crude oil comes from depleted crude oil wells and is extracted from two sources: oil shale rock and tar sands.

Heavy oil extracted from oil shale rock is called shale oil (Figure 12-7). Oil shale rock contains a mixture of hydrocarbons called kerogen, which can be used to produce shale oil. Kerogen is actually mixed in with the rock where it is found (as opposed to being trapped between layers of rock). As a result, producing shale oil requires mining, crushing, and heating the oil shale rock to extract the kerogen. This requires inputs of even more energy. About 72% of the world's estimated shale oil deposits are buried deep in rock formations located primarily in the U.S. states of Colorado, Wyoming, and Utah. Currently these deposits are too costly to develop.

Another expanding source of heavy oil is tar sands, which are also known as oil sands. Tar sands contain a mixture of clay, sand, water, and a thick, tar-like oil.

LESSON 12.2

FIGURE 12-7

Heavy Oil Shale oil extracted from oil shale rock is thick and does not flow easily.

Three-quarters of the world's tar sands lie under a vast area of remote boreal forest in Canada. Extracting and processing this heavy oil greatly reduces its net energy and causes significant environmental impacts (Figure 12-8). It is also more expensive to produce. New production of oil from tar sands has slowed in recent years because of the low price of conventional crude oil.

Those who favor using tar sands and other alternative oil sources say it would reduce U.S. dependence on foreign oil, create jobs, and keep oil prices low. Figure 12-9 lists the advantages and disadvantages of using heavy oil made from oil shale rock and tar sands.

checkpoint Why would the United States want to produce more of its own oil rather than relying on imports?

FIGURE 12-8

Environmental Impact The extent of environmental damage from open-pit mining for heavy oil is evident from aerial views like this one of Canada's Alberta Oil Sands site.

FIGURE 12-9

Heavy Oil from Oil Shale Rock or Tar Sands	
Advantages	Disadvantages
Large potential supplies	Low net energy
Efficient distribution system in place	High cost
Easily transported within and between countries	Releases CO_2 and other air pollutants when produced and burned
	Severe land disruption and high water use

Natural Gas Is an Important Energy Source

Commercial energy is also produced from natural gas, another fossil fuel that is formed over millions of years. **Natural gas** contains a mixture of gases, of which 50–90% is methane (CH_4). Natural gas also contains smaller amounts of heavier hydrocarbons such as propane and butane, as well as small amounts of highly toxic hydrogen sulfide. This versatile fuel is widely used for cooking, heating, and industrial processes. It is also used to fuel cars and trucks and to power turbines that produce electricity in power plants and emergency generators.

The versatility of natural gas helps explain why its use is increasing. Natural gas provided about one-quarter of the energy consumed worldwide and 34% of the energy consumed in the United States in 2020. When burned efficiently, natural gas is also cleaner than oil and much cleaner than coal.

Natural gas is often found above deposits of crude oil. It also exists in tightly held deposits between layers of shale rock and can be extracted through fracking (Case Study). When a natural gas deposit is tapped, propane and butane gases within the deposit can be removed first, by liquefying them under high pressure. This form of fuel is called liquefied petroleum gas (LPG). People who live in rural areas that are not served by natural gas pipelines can use LPG, which is stored in pressurized tanks.

The rest of the natural gas (mostly methane) is purified and pumped into pressurized pipelines. A network of pipelines carries the fuel over land to where it is needed. Natural gas can also be transported across oceans. First, it is converted to liquefied natural gas (LNG) at a high pressure and low temperature. This highly flammable liquid is carried in refrigerated tanker ships. At its destination port, LNG is heated and converted back to into gas for distribution by pipeline. Liquefying natural gas greatly reduces its net energy and increases its cost.

Natural gas reserves are not uniformly distributed. Figure 12-10 identifies countries with the largest reserves of natural gas as well as countries that are top natural gas producers and consumers. Currently, the United States does not have to rely on natural gas imports because domestic production has been increasing rapidly. This trend is mainly due to growing use of horizontal drilling and fracking to extract natural gas from shale rock. The United States now exports natural gas as LNG to other countries.

The demand for natural gas in the United States is projected to more than double between 2010 and 2050. If much of this demand is met by increased production of natural gas by fracking, the country could continue to meet its needs for natural gas with domestic resources. However, it is unclear how many U.S. natural gas reserves exist. Studies also show that natural gas production from shale rock tends to peak and drop off much faster than does production from conventional natural gas wells. Furthermore, using fracking to extract and produce natural gas reduces its net energy. Without effective regulation, it also does more harm to the environment.

checkpoint What are some uses for natural gas?

Fracking Has Benefits and Drawbacks

The U.S. Energy Information Administration (EIA) predicts that within one or two decades at least 100,000 more natural gas wells will be drilled and fracked in the United States. Companies that use fracking to produce natural gas argue that this is a beneficial trend. They point out that increased natural gas production from fracking has lowered U.S. natural gas prices and benefited those who use this fuel (about half of all U.S. consumers). It has also led to increased use of natural gas to generate electricity. The resulting drop in coal-generated electricity has cut emissions of CO_2 and other air pollutants released by burning coal.

Companies that use fracking also argue that fracking improves local economies. For example, the natural gas fracking boom in Pennsylvania created about 18,000 jobs. Producers paid millions of dollars to landowners for signing leases to the gas under their land, along with royalties on the gas removed.

Fracking may provide a number of economic benefits. However, some analysts warn that the greatly increased use of fracking to produce natural gas (and oil) leads to several harmful environmental effects. These include water contamination, air pollution, and increased earthquake activity.

FIGURE 12-10

Natural Gas Economy This graph identifies the world's top three countries in terms of natural gas reserves, natural gas production, and natural gas consumption in 2020.

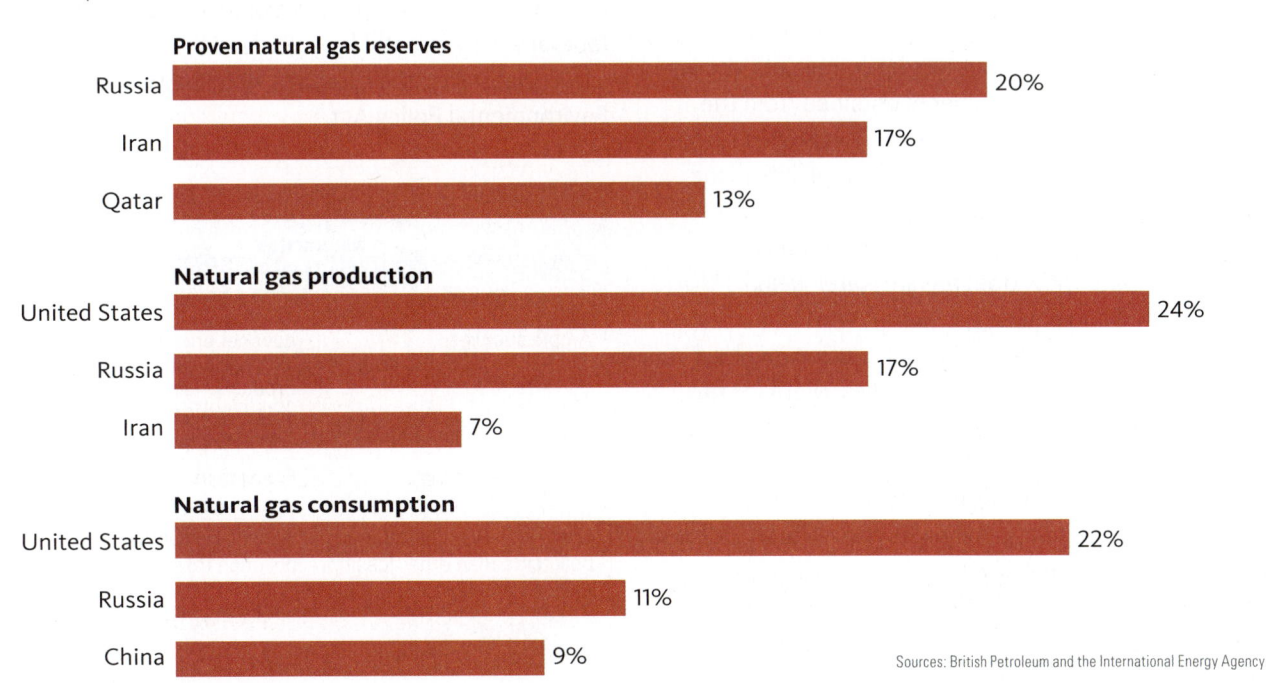

Proven natural gas reserves
- Russia: 20%
- Iran: 17%
- Qatar: 13%

Natural gas production
- United States: 24%
- Russia: 17%
- Iran: 7%

Natural gas consumption
- United States: 22%
- Russia: 11%
- China: 9%

Sources: British Petroleum and the International Energy Agency

FIGURE 12-11 ▼
Contaminated Water Fracking is suspected to cause high levels of explosive methane gas in some communities' water supplies. As proof, this homeowner literally lights her water on fire.

Water Issues The enormous volume of water used for fracking as well as the hazardous wastewater it generates can have serious environmental impacts (Case Study). Other aspects of natural gas production may also lead to water pollution. For example, faulty pipes used for drilling and transport can allow natural gas to leak and enter nearby water sources.

Fracking companies insist their operations don't contaminate water supplies. However, critics say fracking companies are suppressing complaints about water issues. In 2014, documents obtained from the state governments of Pennsylvania, Texas, Ohio, and West Virginia revealed hundreds of complaints made over the previous decade about water problems in these states. People reported contamination and diminished water flow from private water wells because of oil and natural gas production.

Most of the confirmed cases of pollution involved methane gas in drinking water supplies. Methane gas fizzing out of faucets in some Pennsylvania homes can be ignited just like a natural gas stove burner (Figure 12-11). People living in these homes must leave windows open partway at all times to prevent buildup of lethal, explosive methane gas. These problems began when an energy company drilled a fracking well, but the company has denied responsibility.

Air Pollution Because methane is a greenhouse gas, emission of this gas plays a role in warming the atmosphere and bringing about climate change. According to the U.S. Environmental Protection Agency (EPA), the natural gas and petroleum industry is the largest source of methane emissions released by human activities. Methane also leaks from improperly sealed wells abandoned by small producers who have gone bankrupt. As a result, methane can contaminate air and groundwater supplies. Exhaust fumes from heavy truck traffic and electrical generators on fracking sites have also led to increases in local levels of air pollution and greenhouse gas emissions.

Increase in Earthquakes Studies have suggested fracking activities may be the cause for hundreds of small earthquakes in areas that are not traditionally prone to seismic activity. The shifts were caused by the high-pressure injection of fracking wastewater into deep underground hazardous waste wells. In some areas, the fracking process itself is possibly causing small earthquakes. The largest U.S. earthquake attributed to hydraulic fracking occurred in 2018 in Texas. It measured 4.0 in magnitude.

There is currently little protection from the hazardous impacts of fracking on people or the environment. Natural gas companies are not required to identify or publicize the materials used in fracking. Pressure from natural gas suppliers has excluded natural gas fracking from EPA regulations under the Safe Water Drinking Act. Other loopholes have exempted natural gas production from parts of other federal environmental laws. Such laws include the Clean Water Act, the Clean Air Act, and the National Environmental Policy Act.

FIGURE 12-12 ▼

Natural Gas	
Advantages	**Disadvantages**
Ample supplies	Low net energy for LNG
Versatile fuel	Production and delivery may emit more CO_2 and CH_4 per unit of energy than coal.
Medium net energy	Fracking uses and pollutes large volumes of water.
Emits less CO_2 and other air pollutants than other fossil fuels when burned	Potential groundwater pollution from fracking

FIGURE 12-13 ▼
Major Types of Coal Over millions of years, several different types of coal have formed. Peat is a soft material made of moist, partially decomposed organic matter, similar to coal. It is not classified as coal, although it is used as a fuel. These major types of coal vary in the amounts of heat, carbon dioxide, and sulfur dioxide released per unit of mass when they are burned.

← Increasing moisture content | Increasing heat and carbon content →

Peat (not a coal) — Partially decayed plant matter in swamps and bogs; low heat content

→ Heat / Pressure →

Lignite (brown coal) — Low heat content; low sulfur content; limited supplies in most areas

→ Heat / Pressure →

Bituminous (soft coal) — Extensively used as a fuel because of its high heat content and large supplies; normally has a high sulfur content

→ Heat / Pressure →

Anthracite (hard coal) — Highly desirable fuel because of its high heat content and low sulfur content; supplies are limited in most areas

Concerns over fracking have led some states, cities, towns, counties, and groups to ban it. Before 100,000 more natural gas wells are drilled over the next 10–20 years, analysts call for the following protective measures:

- elimination of all exemptions from environmental laws for the natural gas industry
- stricter regulation and monitoring
- better data on impacts of fracking and gas leaks
- repair of existing leaks in production and distribution systems

Such efforts could limit air and water pollution risks and avoid public backlash against fracking.

Like any energy resource, natural gas has advantages and disadvantages. It is plentiful and is the cleanest burning fossil fuel. However, there are environmental impacts from using natural gas. Figure 12-12 lists the advantages and disadvantages of using natural gas as an energy resource.

checkpoint What water issues have been caused by fracking?

Coal Is Plentiful but Dirty

Coal is the world's most abundant fossil fuel. **Coal** is formed from the remains of land plants that are buried and exposed to intense heat and pressure for 300–400 million years. Figure 12-13 identifies the major types of coal that have formed by this process. Coal is burned to produce electricity in a typical power plant (Figure 12-14). It is also burned in industrial plants to make steel, cement, and other manufactured products.

Figure 12-15 identifies countries with the largest reserves of coal as well as countries that are top coal producers and consumers. In 2020, coal-burning power plants generated about 27% of the world's electricity and 10% of electricity used in the United States. Coal consumption peaked between 2005 and 2008 in this country. In the past decade, however, there has been a significant decline in the use of coal for electricity generation. This has been mainly due to the promotion of natural gas as a cheaper energy resource.

Coal is the dirtiest of all fossil fuels. Coal mining processes severely degrade land and pollute water and air. Burning coal also releases various pollutants into the atmosphere. For example, burning coal produces 46% of global emissions of CO_2, which contributes to atmospheric warming and climate change. Coal-burning power and industrial plants are the largest emitters of this greenhouse gas.

Burning coal also converts sulfur in coal to the air pollutant sulfur dioxide (SO_2). Sulfur dioxide contributes to acid rain and can cause serious human health problems (Lesson 16.1). Currently, China is the world's leading emitter of CO_2 and SO_2. The United States is the second-largest emitter of CO_2.

Finally, burning coal produces soot, which is a mixture of very fine particles of chemicals, acids, metals, soils, and dust. These pollutants are so small, they can collect deep in the lungs. They cause health issues ranging from breathing problems to premature death, heart attacks, and strokes.

Some coal-burning plants install air pollution control equipment called scrubbers. Scrubbers

LESSON 12.2   405

FIGURE 12-14

Electricity from Coal This diagram identifies parts of a coal-fired power plant. The plant burns crushed coal to boil water and produce steam. The steam spins a turbine to produce electricity. Which part of the diagram shows where pollution control technology is being used?

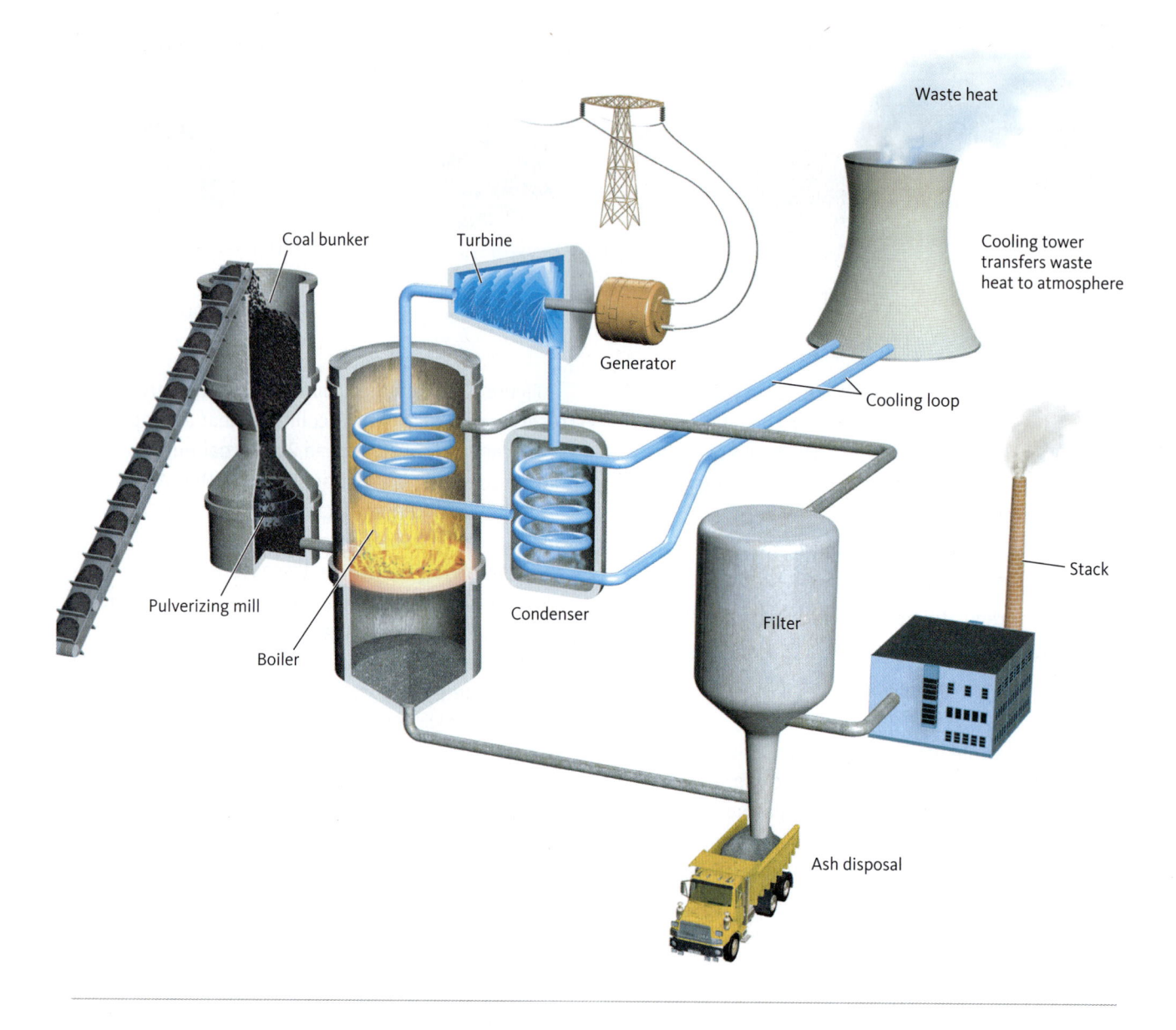

- Coal bunker
- Turbine
- Waste heat
- Cooling tower transfers waste heat to atmosphere
- Generator
- Cooling loop
- Pulverizing mill
- Condenser
- Filter
- Stack
- Boiler
- Ash disposal

remove certain pollutants and collect them as ash before they can exit through the smokestacks. Coal ash represents another pollution hazard. It may contain toxins such as arsenic, lead, mercury, cadmium, and radioactive radium. The ash must be stored safely, essentially forever (Chapter 15).

Proper control equipment can significantly reduce harmful emissions from burning coal. However, this type of pollution prevention reduces coal's net energy. And while many U.S. coal-burning power plants use scrubbers, most power plants in countries like China and India do not use such air pollution control equipment.

Coal faces increasing competition from cleaner-burning and cheaper natural gas and wind power (Chapter 13). Few new coal-burning power plants are being built in the United States, and some older ones are being closed. In addition, new coal plant projects worldwide have fallen off significantly, with 44 countries committing to stop constructing new ones.

Even so, coal will continue to be used in countries with large markets, such as China, Australia, and

FIGURE 12-15
Coal Economy This graph identifies the world's top three countries in terms of coal reserves, coal production, and coal consumption in 2020.

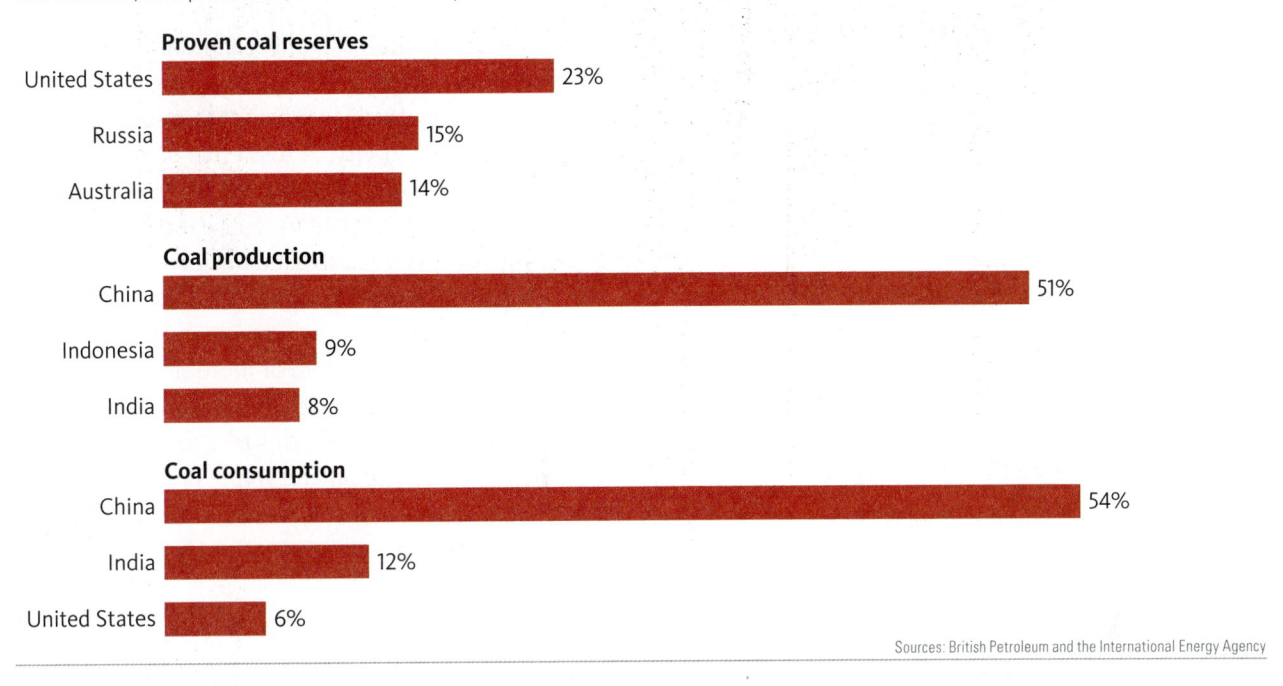

Proven coal reserves
- United States: 23%
- Russia: 15%
- Australia: 14%

Coal production
- China: 51%
- Indonesia: 9%
- India: 8%

Coal consumption
- China: 54%
- India: 12%
- United States: 6%

Sources: British Petroleum and the International Energy Agency

India, as long as its harmful environmental and health effects are not included in its market price. Figure 12-16 lists the advantages and disadvantages of using coal as an energy resource.

checkpoint What are the environmental problems associated with using coal as a source of energy?

FIGURE 12-16

Coal	
Advantages	**Disadvantages**
Ample supplies in many countries	Severe land disturbance and water pollution
Medium to high net energy	Fine particle and toxic mercury emissions that threaten human health
Low cost when environmental and health costs are not included	Emits large amounts of CO_2 and other air pollutants when produced and burned

12.2 Assessment

1. **Recall** How is liquefied natural gas (LNG) transported from the United States to other countries?

2. **Explain** Why is the use of coal declining in the United States?

3. **Calculate** Assuming an average 2020 price of about $42 per barrel of oil, approximately how much was spent on global oil consumption in 2020? (Hint: Refer to the beginning of this lesson.)

4. **Apply** How does the extraction and production process for heavy oil from tar sands affect its net energy?

SCIENCE AND ENGINEERING PRACTICES

5. **Engaging in Argument** You work for a power company that needs more electrical power for your customers. Comparing the advantages and disadvantages of using oil, coal, or natural gas, what recommendation would you give your supervisor regarding what type of fossil fuel to use? Include evidence from this section and other sources to support your recommendation.

LESSON 12.2

12.3 What Are the Advantages and Disadvantages of Using Nuclear Power?

CORE IDEAS AND SKILLS

- Understand how a nuclear fission reaction works, and describe the nuclear fuel cycle.
- Explain the advantages and disadvantages of using nuclear power.
- Discuss the future of nuclear power.

KEY TERMS

nuclear fission   nuclear fusion

Nuclear Fission Generates Power

Like fossil fuels, nuclear power falls under the category of nonrenewable energy resources. Nuclear power requires the mineral uranium-235, which is mined from limited ores in Earth's crust. The world's three leading producers of nuclear power are, in order, the United States, France, and China.

Inside both nuclear and fossil-fuel power plants, water is boiled to produce steam that spins a turbine and generates electricity. In fossil-fuel power plants, fossil fuels are burned to produce heat to boil water. Nuclear power production involves a more complex and costly process. A controlled nuclear fission reaction is carried out to provide the necessary heat.

Nuclear fission occurs when a neutron is used to split a large nucleus into two or more smaller nuclei. Each fission reaction releases neutrons, which results in a chain reaction that releases an enormous amount of energy in a short time. The whole process takes place inside the reactor of a nuclear power plant (Science Focus 12.1).

FIGURE 12-17 ▼

Nuclear Fuel Cycle Using nuclear power to produce electricity involves a sequence of steps and technologies that together are called the nuclear fuel cycle.

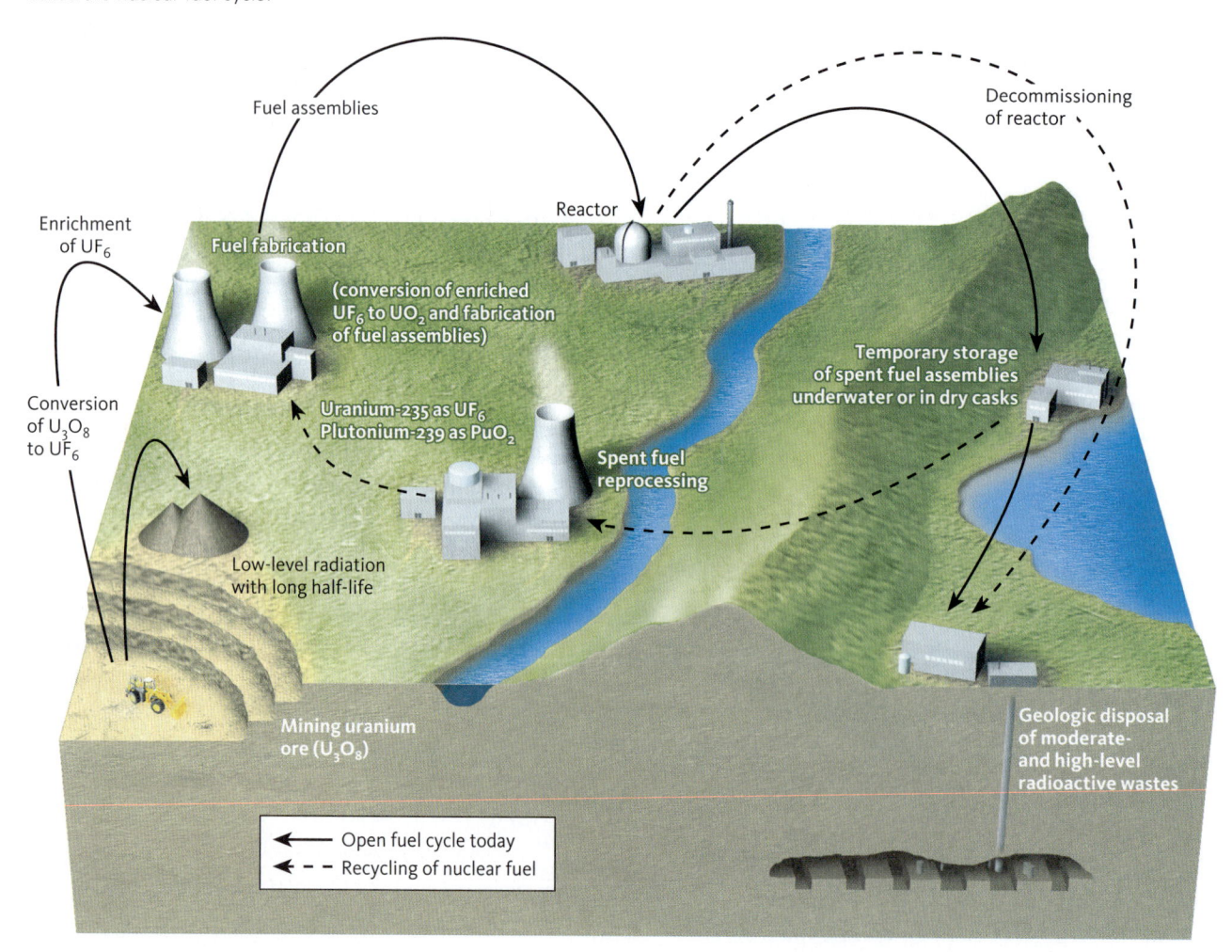

Labels in figure:
- Enrichment of UF_6
- Conversion of U_3O_8 to UF_6
- Fuel assemblies
- Fuel fabrication (conversion of enriched UF_6 to UO_2 and fabrication of fuel assemblies)
- Uranium-235 as UF_6
- Plutonium-239 as PuO_2
- Low-level radiation with long half-life
- Mining uranium ore (U_3O_8)
- Reactor
- Decommissioning of reactor
- Spent fuel reprocessing
- Temporary storage of spent fuel assemblies underwater or in dry casks
- Geologic disposal of moderate- and high-level radioactive wastes
- ← Open fuel cycle today
- ←- - Recycling of nuclear fuel

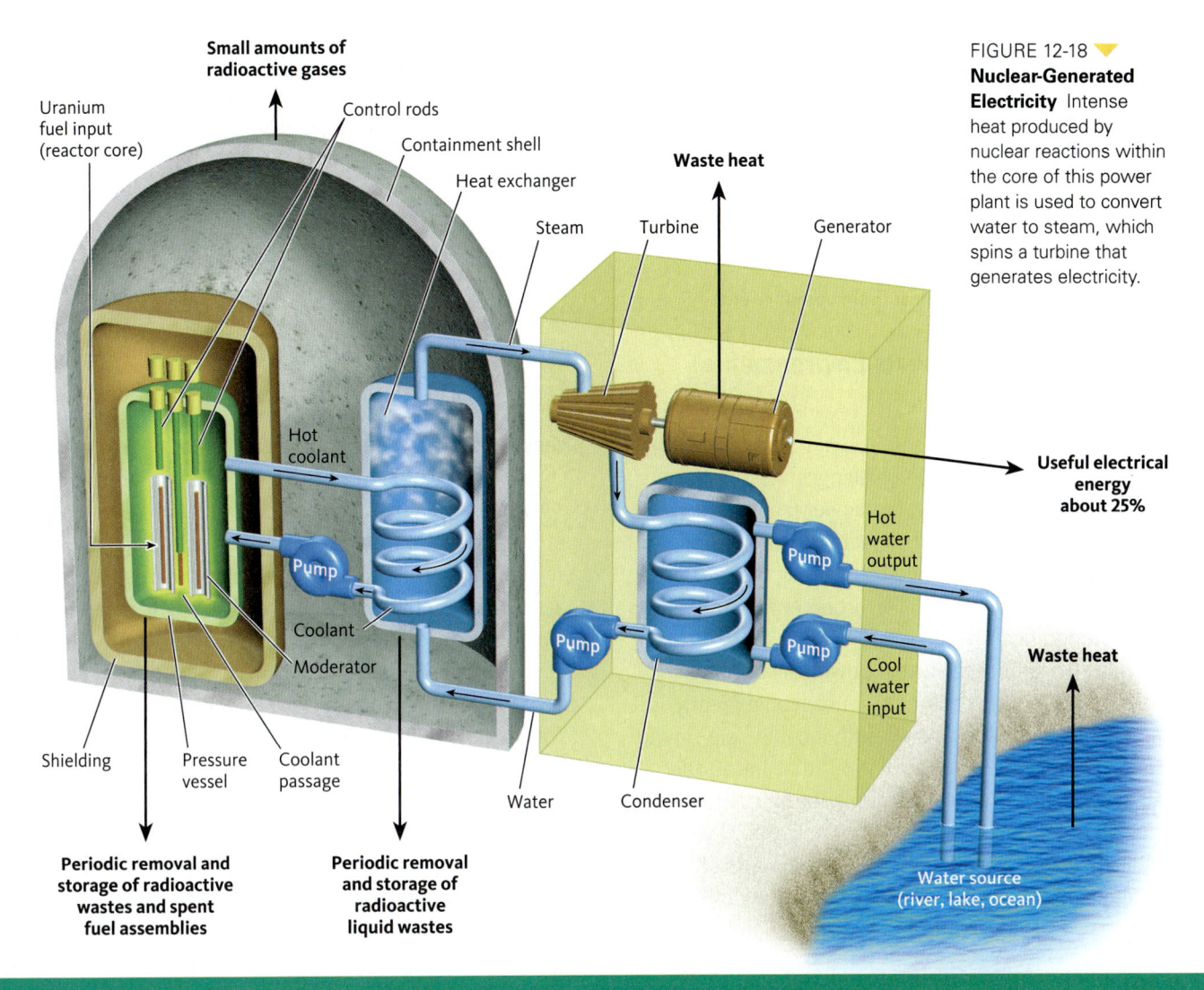

FIGURE 12-18

Nuclear-Generated Electricity Intense heat produced by nuclear reactions within the core of this power plant is used to convert water to steam, which spins a turbine that generates electricity.

SCIENCE FOCUS 12.1

NUCLEAR FISSION REACTORS

Most nuclear-generated electricity is produced by light-water nuclear fission reactors (Figure 12-18). The fuel for this type of reactor is made from uranium ore. After the uranium ore is mined, it is enriched to increase the concentration of its fissionable material (uranium-235) to about 5%.

Enriched uranium-235 is processed into small pellets of uranium dioxide. Each pellet, about the size of a pencil eraser, contains as much energy as a ton of coal. Large numbers of pellets are packed into closed pipes called fuel rods. The rods are bundled together in fuel assemblies and placed in the reactor core.

Plant operators use control rods to regulate how much power is produced. The rods are moved into and out of the reactor core to absorb more or fewer neutrons. This slows down or speeds up the fission reaction.

A coolant, usually fresh water, circulates through the core to prevent fuel rods and other components from melting and releasing massive amounts of radioactivity into the environment. An emergency core cooling system also protects against meltdowns.

A containment shell made of thick, steel-reinforced concrete surrounds the reactor core. It is designed to keep radioactive materials from escaping into the environment if an internal explosion or a core meltdown occurs. It also protects the core from external threats such as weather disasters. The need for all these safety features means building a nuclear power plant can cost $10 billion or more.

Thinking Critically

Make Judgments Would you feel comfortable living near a nuclear power plant? Why or why not?

LESSON 12.3

Building and running a nuclear power plant is only one part of the nuclear fuel cycle (Figure 12-17). This cycle begins with the mining of uranium. It continues with processing and enriching the uranium to make fuel for use in a reactor. It ends with safely storing the resulting highly radioactive wastes for thousands of years until their radioactivity falls to safe levels.

checkpoint How does nuclear fission produce energy?

Nuclear Power Presents Environmental Challenges

Using nuclear power has some advantages, but it also has disadvantages and challenges. As long as a reactor is operating safely, the power plant itself has a fairly low environmental impact and little risk of an accident. However, when considering the entire nuclear fuel cycle, potential environmental impacts increase significantly.

Those who support nuclear power claim that increased use of this energy resource could greatly reduce CO_2 emissions. As a result, it could help slow climate change. Scientists point out that this is only partially correct. While nuclear plants are operating, they do not emit CO_2. However, during the 10 years it typically takes to build a plant, large amounts of CO_2 are emitted, especially in the manufacturing of huge quantities of construction cement. Every other step in the nuclear fuel cycle also results in the release of CO_2. Such emissions may be much lower than emissions from coal-burning power plants. However, they still contribute to atmospheric warming and climate change.

Nuclear power plants do not emit air pollutants as long as the plant operates without problems. Modern plants perform with little risk, but many of the nuclear reactors in the United States are aging. The average age of nuclear plants in the United States is 39 years, and worldwide it is 30 years. Plants are licensed to operate for 40 years and can request a 20-year extension before the plant is shut down.

Another disadvantage of nuclear power is the potential threat of accidents. Major nuclear plant accidents are rare, but such accidents can release harmful radioactive materials into the environment. Human exposure to even low levels of radiation over a long period can cause cancer. High doses can cause immediate death or delayed radiation sickness. Symptoms of radiation sickness include weakness, burns, reduced organ function, nausea, and hair loss.

Safe handling of the nuclear waste produced by reactors also presents major challenges. The high-grade uranium fuel in a typical nuclear reactor lasts for three to four years before it becomes spent

FIGURE 12-19

Storing Nuclear Waste After three or four years in a reactor, spent fuel rods are removed and stored in a deep pool of water contained in a steel-lined concrete basin for cooling (left). After about five years of cooling, the fuel rods can be stored in dry casks (right). The casks are made of heat-resistant metal alloys and thick concrete.

(useless) and must be replaced. The spent fuel rods are so hot and radioactive that they cannot be thrown away. Researchers found that 10 years after being removed from a reactor, a single spent fuel rod assembly can still emit enough radiation to kill a person standing 1 meter (39 inches) away in less than three minutes.

After spent fuel rod assemblies are removed from reactors, they are stored in water-filled pools (Figure 12-19, left). After several years of cooling and decay of some of their radioactivity, the rods can be transferred to dry casks for storage (Figure 12-19, right). These casks are licensed for 20 years and could last 100 years or more. Even so, this is a fraction of the thousands of years required to safely store high-level radioactive waste.

Security at storage sites is also a concern. One study warns that the waste storage pools and dry casks at two-thirds of the United States' commercial nuclear reactors are especially vulnerable to sabotage or terrorist attack. This is due to the fact these facilities are located outside of the heavily protected reactor containment buildings.

Spent fuel rods can also be processed to remove their radioactive plutonium, which can then be used as nuclear plant fuel. This reprocessing reduces the storage time for the remaining wastes from 240,000 years to 10,000 years. (For comparison, modern humans appeared about 200,000 years ago.) However, reprocessing is costly. It also produces bomb-grade plutonium that nations or terrorists can use to make nuclear weapons. The U.S. government, after spending billions of dollars, abandoned this fuel recycling approach in 1977. A few countries still reprocess some of their nuclear fuel.

Most scientists and engineers agree in principle that the safest way to store high-level radioactive wastes from nuclear plants is to bury them deep underground for thousands of years. However, scientists are not able to verify that deep burial is the answer. After 60 years of research, there is no widely accepted or tested way to store such waste safely for thousands of years. As research continues, these deadly wastes are building up. In the United States, about 78% of nuclear waste is stored in pools and 22% is stored in dry casks. This practice takes care of waste for 100 years at most.

Another costly radioactive waste problem arises when a nuclear power plant reaches the end of its useful life after 40–60 years. At that point, the plant must be decommissioned, or closed and torn down.

A quarter of nuclear plants in operation in 2020 will have to be decommissioned by 2025. New nuclear reactors are not being built fast enough to replace the aging reactors that must be retired. This explains in part why production of electricity from nuclear power has not grown much since 2006 and is not expected to grow much between now and 2050.

Scientists have proposed three ways to decommission plants. One strategy is to remove and store the highly radioactive parts in a permanent, secure facility. A second approach is to install a physical barrier around the plant and set up full-time security for 30–100 years. A third option is to enclose the entire plant in a concrete and steel-reinforced tomb, called a containment structure.

Regardless of the method chosen, the high cost of retiring nuclear plants adds to the enormous cost of the nuclear fuel cycle. It also reduces its already low net energy. Even if all the nuclear power plants in the world were shut down tomorrow, their high-level radioactive wastes and components would need to be safely contained for thousands of years.

checkpoint What happens to uranium fuel rods after they become spent, or useless?

Nuclear Accidents Spread Uncertainty

The danger in operating nuclear reactors is that they may experience explosions or loss of coolant water. Without the cooling water, the core of the reactor can experience a meltdown. Explosions and meltdowns can release radioactivity into the environment. Between 1952 and 2015, 34 serious nuclear "incidents" or "accidents" occurred worldwide. Following are descriptions of three notable nuclear accidents.

Three Mile Island Three Mile Island (TMI) is a nuclear power plant located near Harrisburg, Pennsylvania. At 4:00 a.m. on March 28, 1979, the Unit 2 reactor was online and operating at 97% power when a small valve malfunctioned and failed to close. As a result, coolant water drained from the reactor.

The malfunction caused an immediate shutdown of the reactor. The control panel instruments, also malfunctioning, showed the reactor was full of coolant water when it wasn't. Operators chose not to use the emergency water cooling system as a result. The reactor overheated, which led to release of contaminated water and radioactive gas.

LESSON 12.3

NATIONAL GEOGRAPHIC | EXPLORERS AT WORK
Leslie Dewan Nuclear Engineer

Many people would place nuclear power at the bottom of the list of alternatives to fossil fuels. But not Leslie Dewan. She firmly believes that given the right process, nuclear power could serve as a safer, cleaner, less wasteful energy alternative.

Dr. Dewan is a nuclear engineer and the co-founder of a company called Transatomic Power. Dewan is researching ways to produce nuclear energy more safely and efficiently. Specifically, she is exploring an updated version of a molten salt reactor.

The molten salt reactor was developed in the 1960s. It was abandoned after being deemed bulky and expensive to build compared to the light-water reactor, which is commonly used today. That decision was made before the world witnessed the devastation of a nuclear accident.

The main advantage of molten salt reactors is that they burn liquid uranium fuel rather than solid fuel. This design reduces the chances of a meltdown. Light-water reactors require a continuous supply of electric power to constantly pump water over the core to keep it from overheating. In molten salt reactors, if electric power is lost, the fuel automatically drains into a separate tank and freezes solid in a few hours.

In addition to being safer, molten salt reactors are also more efficient. Dewan and her co-founder, Mark Massie, have designed a molten salt reactor that will burn up to 96% of its fuel. Light-water reactors use only 4% of their fuel.

In molten salt reactors, uranium salt serves as the fuel. When heated above 500°C (930°F), the fuel salt becomes liquid. "You basically simmer the reactor like a Crock-Pot™ for decades," Dewan says. "That's how we achieve a 96% burn rate. We're able to leave the uranium in and constantly remove the poisons that would otherwise shut it down." Even better, molten salt reactors can reuse radioactive nuclear waste as fuel. Over 80,000 tons of such waste exist in the United States alone.

Dewan and Transatomic Power still have to raise the funds needed to build a demonstration reactor to test their design. The process will take years. But she remains optimistic.

"At the most fundamental level I'm an environmentalist," she says. "I'm doing this because I think nuclear power is the best way of producing large amounts of carbon-free electricity. I think the world needs nuclear power, alongside solar, wind, hydro, and geothermal, if we want to have any hope of reducing fossil fuel emissions and preventing global climate change."

FIGURE 12-20
Leslie Dewan's offices for Transatomic Power are located on the campus of the Massachusetts Institute of Technology, where Dewan received her undergraduate and graduate degrees.

According to authorities, off-site readings for radioactivity during and after the accident were within safety limits. However, this was not clearly communicated to the public at the time of the accident. An evacuation was eventually ordered, and more than 140,000 people fled in a panic.

Cleanup at TMI took almost 12 years and cost $973 million to complete. The Unit 1 reactor is still producing electricity and operating well. The Unit 2 reactor was decommissioned. Better training and increased safety rules for operators have made the plant safer and more reliable. Construction of new plants used what was learned from the TMI accident. However, the public lost confidence in the use of nuclear power for the next two decades. This fear contributed to a slowdown in new plant construction.

Chernobyl On April 25, 1986, the number 4 reactor at the Chernobyl Nuclear Power Plant in Ukraine was undergoing a routine safety test. A similar test had been performed the previous year without any problems. As part of the test, operators disabled the automatic shutdown mechanisms.

Due to design flaws in the equipment, a power surge occurred inside the reactor during the test. Efforts to stabilize the reactor instead caused damage to several fuel assemblies and resulted in destruction of the reactor. Two explosions occurred, and fission products erupted into the atmosphere.

Two plant workers died in the explosions, and 29 more people died within days to months from injuries and radiation poisoning. Areas of Ukraine, Belarus, and Russia were heavily contaminated by radiation. More than 130,000 people living within a 30-kilometer (18-mile) radius around the plant were hastily evacuated and eventually relocated. This area came to be called the Exclusion Zone. The reactor was encased in a container of lead and concrete (which is being replaced because it is leaking).

Since the accident, more than 350,000 people have been evacuated and relocated. The death toll due to cancer from the accident is predicted to reach 4,000 for people exposed to high doses of radiation. Another 5,000 deaths are predicted among those exposed to lower levels. Today, the Exclusion Zone remains radioactive, although levels in some areas are higher than others. Tours are even conducted inside the Zone (Figure 12-21). A few hundred people also have returned to live there despite the risks.

Fukushima Daiichi A major accident occurred on March 11, 2011, at the Fukushima Daiichi Nuclear Power Plant on the northeast coast of Japan. The event began with a strong offshore earthquake, which caused a severe tsunami. A wall of seawater swept into shore, destroying coastal communities. The water also washed over the nuclear plant's protective seawalls.

FIGURE 12-21
Return to Chernobyl Tourists visit Pripyat, a town in the Chernobyl Exclusion Zone, 25 years after the 1986 accident. The man wearing the gas mask brought it along as a prop to have his picture taken in it. The area has been declared safe enough for day tours.

Seawater knocked out the circuits and backup generators of the emergency core cooling systems for three of the plant's reactors. That set off a chain of events that allowed hydrogen gas to build up. The hydrogen gas exploded, blowing the roofs off three reactor buildings. Radioactivity was released into the atmosphere and nearby coastal waters.

Evidence indicates that the cores of the three reactors suffered full meltdowns. They contaminated a large area with low to moderate levels of radioactivity. In 2013, radioactivity from contaminated groundwater and one of the plant's 3,785-liter (1,000-gallon) wastewater storage tanks leaked into the coastal waters near the plant.

Hundreds of thousands of people were evacuated due to property damage from the tsunami and in an effort to minimize their exposure to radiation. The tsunami killed almost 16,000 people, but no one has died from exposure to radiation from the nuclear plant accident. Despite these efforts, the World Health Organization (WHO) says that cases of cancer among people living in and around Fukushima are expected to increase due to radiation fallout from the explosions. The WHO also recommends that human health and the environment continue to be monitored for radiation.

Officials say the costly cleanup and decommissioning of the damaged reactors at Fukushima Daiichi will take several decades. The accident also caused Japanese citizens and people from all nuclear countries to seriously question the safety of nuclear power. In the aftermath, the Japanese government shut down all of its nuclear reactors and considered abandoning its use of nuclear power. The accident also prompted Germany, Switzerland, and Belgium to announce plans to phase out nuclear power.

Japan has come to rely more on imports of liquefied natural gas and coal to produce electricity. Resorting back to fossil-fuel electricity production has increased air pollution emissions in Japan. Fortunately, Japan has not had to replace all of its nuclear energy with energy from fossil fuels. Half the demand for energy produced from nuclear power has been eliminated by cutting energy waste and improving energy efficiency.

checkpoint How did the accidents at TMI, Chernobyl, and Fukushima Daiichi affect public perception of nuclear power?

Nuclear Power's Future Is Uncertain

The future of nuclear power is a subject of debate. Critics argue that its two biggest drawbacks are the high cost of the nuclear fuel cycle and the fact that its technology and its by-products can be used in making nuclear weapons.

Government Subsidies and Spending The U.S. government has provided large research and development subsidies, tax breaks, and loan guarantees to the nuclear industry for more than 50 years. It has assumed most of the financial burden of developing ways to store radioactive wastes. In addition, the government has provided accident insurance guarantees because insurance companies refuse to insure fully any nuclear reactor against the effects of a catastrophic accident.

Since 1948, the U.S. government has spent $95 billion on nuclear energy research and development. This is more than four times the amount spent on research and development for all forms of renewable energy combined. Many people question the need for continued taxpayer support for nuclear power, especially since its energy output is not increasing.

Safety Concerns Nuclear power plants in the United States and most other more-developed countries have multiple built-in safety features. As a result, the risk of exposure to radioactivity from these plants is very low. Even so, accidents have dampened public and investor confidence in nuclear power.

Another serious safety concern related to commercial nuclear power is the spread of nuclear weapons technology. In the international marketplace, the United States and eight other countries have been selling commercial and experimental nuclear reactors and uranium

FIGURE 12-22

The Nuclear Fuel Cycle	
Advantages	**Disadvantages**
Low environmental impact (without accidents)	Low net energy
Emits one-sixth as much CO_2 as coal	Higher overall cost
Low risk of accidents in modern plants	Produces long-lived, harmful radioactive wastes
	Promotes availability of nuclear weapons

fuel-enrichment and purification technology for decades. Much of this information and equipment can be used to produce bomb-grade uranium and plutonium for use in nuclear weapons.

The 60 countries with nuclear weapons or the knowledge to develop them have gained most of their information by using civilian nuclear power technology. Some critics view this threat as the single most important reason for not building more nuclear power plants.

Some scientists call for replacing today's uranium-based reactors with ones fueled by thorium. Such reactors would be less costly and safer because they cannot melt down. The nuclear waste they produce cannot be used for nuclear weapons. China plans to explore this option. Research also continues into other areas of nuclear energy (Explorers at Work).

Analysts suggest that any new nuclear technology must meet five criteria to be environmentally and economically acceptable.

1. Reactors must be built so that a runaway chain reaction is impossible.
2. Fuel used in reactors and methods of fuel enrichment and fuel reprocessing must not lend themselves to production of nuclear weapons.
3. Spent fuel and dismantled structures must be easy to dispose of without burdening future generations with harmful radioactive waste.
4. Taking its entire fuel cycle into account, nuclear power must generate a net energy high enough to eliminate the need for government subsidies, tax breaks, or loan guarantees to compete in the open marketplace.
5. The entire nuclear fuel cycle must generate fewer greenhouse gas emissions than other energy alternatives.

The advantages and disadvantages associated with the nuclear fuel cycle are summarized in Figure 12-22.

checkpoint What are two reasons for a lack of public support for subsidizing nuclear power?

Nuclear Fusion Could Be a Solution

Some proponents of nuclear power hope to develop nuclear fusion. In **nuclear fusion**, the nuclei of two isotopes of a light element, such as hydrogen, are forced together at extremely high temperatures. This causes them to fuse to form a heavier nucleus, releasing energy in the process. Some scientists hope controlled nuclear fusion can provide a limitless source of energy with fewer risks.

With nuclear fusion there would be no risk of meltdowns, no release of large amounts of radioactive materials, and little risk of the spread of nuclear weapons. Fusion power also might be used to destroy toxic wastes, to supply electricity for desalinating water, and to help produce hydrogen fuel as a clean-burning energy source.

In the United States, after more than 50 years of research and a $25 billion investment (mostly by the government), controlled nuclear fusion is still in its infancy. None of the tested approaches have been able to produce more energy than they used. In 2006, the United States, China, Russia, Japan, South Korea, India, and the European Union agreed to invest $12.8 billion in a joint effort to build a large-scale experimental nuclear fusion reactor by 2026. This will determine if fusion can produce net energy at an affordable cost. By late 2021, the project was 75% completed. Located in southern France, the reactor is being built from roughly 10 million specialized components manufactured and shipped in from countries around the world. It is on track to become operational in December 2025.

checkpoint What would be some benefits of using nuclear fusion to produce energy?

12.3 Assessment

1. **Recall** What fuel is used in a nuclear fission reactor?
2. **Compare and Contrast** How is the process of using fossil fuels to produce electricity similar to and different from using nuclear power to produce electricity?
3. **Summarize** How is the nuclear power industry related to the development of nuclear weapons?
4. **Explain** Nuclear-generated power plants are heavily subsidized. Do you think the market price of these plants should include the costs of the nuclear fuel cycle? Explain.

SCIENCE AND ENGINEERING PRACTICES

5. **Developing and Using Models** Use the internet to research basic fission and fusion reactions. Create a simple model of each reaction comparing their similarities and differences.

TYING IT ALL TOGETHER STEM
Trends in Energy Use

In this chapter, you learned about nonrenewable energy resources including oil, natural gas, coal, and nuclear energy. You also learned about variables that impact the long-term economic usefulness of an energy resource—including subsidies, market prices, feasibility of extracting the resource, the environmental impacts of producing and using the resource, and the resource's net energy. These factors were described in the Case Study about fracking for oil and gas in the United States.

The use of different energy sources varies over time. Researchers study trends in energy use to better understand demands that society will place on energy resources in the future. Figure 12-23 shows trends in energy use worldwide since 2010.

Use Figure 12-23 to answer the questions that follow.

1. Describe any general trends that you observe for each energy source. For example, between 2010 and 2020, has global consumption of natural gas increased, decreased, or remained relatively constant?
2. The use of which energy source grew the most between 2010 and 2020? By how much did it grow (in quadrillion Btus)?
3. Based on what you read in this chapter, what factors do you think explain the predicted energy usage in 2050?
4. Use the graph to estimate the total amount of energy used for all energy plotted on the graph. Use your calculations to add a line for "Total Energy Usage" to the existing graph.
5. Compare total energy usage in 2020 to the projected total energy usage in 2050. By what percentage is energy usage predicted to increase by 2050 compared to 2020?
6. Based on the bar chart data, what projection can you make with regard to the consumption of renewables through 2050? By approximately how much will the share of consumption change during that time? What projection can you make about the use of nonrenewable sources of energy? What can you conclude from these projected trends?

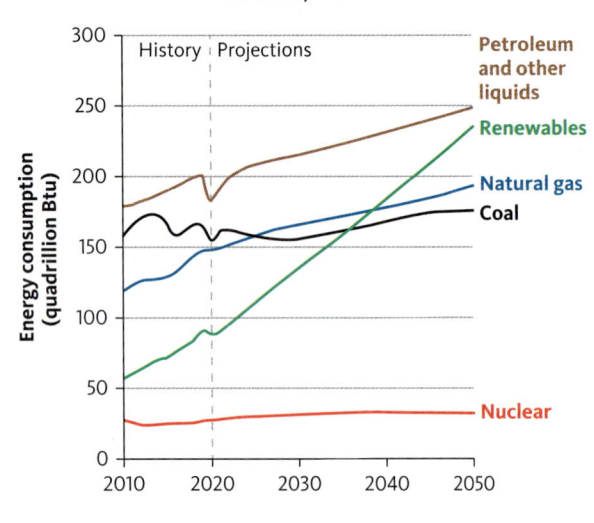

FIGURE 12-23 ▼
This graph shows energy consumption by fuel source in the United States from 2010–2020, with projections to 2050.

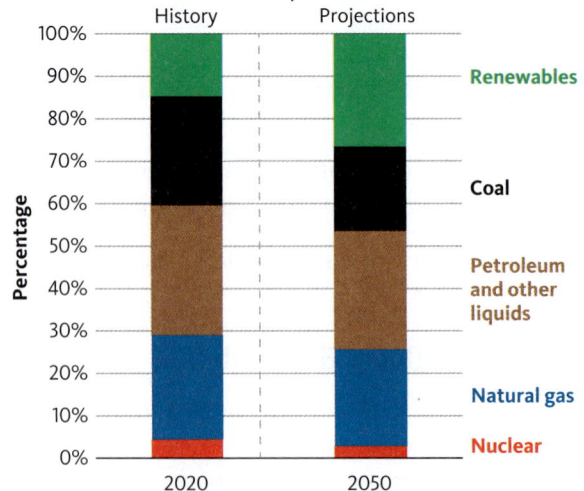

Source: U.S. Energy Information Administration *International Energy Outlook* 2021

416 CHAPTER 12 NONRENEWABLE ENERGY RESOURCES

CHAPTER 12 SUMMARY

12.1 What is net energy and why is it important?

- Roughly 90% of the world's commercial energy comes from nonrenewable resources—87% from carbon-containing fossil fuels and 4% from nuclear power.
- It takes energy to produce energy. Net energy is the amount of energy available from a resource minus the amount of energy needed to make it available.
- Energy resources with low net energies need subsidies or tax breaks from the government or other outside sources to compete in the marketplace with energy resources that have medium or high net energies.

12.2 What are the advantages and disadvantages of using fossil fuels?

- Oil is abundant and has a medium net energy. When burned, it produces CO_2 and other air pollutants that are costly to control. Extracting and transporting oil can pollute water from spills and leaks.
- Natural gas is abundant, has a medium net energy, and burns cleaner than oil. Fracking can be used to extract natural gas from rock deposits. Without careful regulation, fracking can result in air and groundwater contamination. It also produces a hazardous wastewater that must be stored safely.
- Coal is an abundant fuel with a medium to high net energy. Extracting coal disrupts land and can pollute groundwater. Burning coal at power and industrial plants causes more air pollution than burning oil or natural gas, unless the plants are equipped with expensive air-pollution-control equipment.

12.3 What are the advantages and disadvantages of using nuclear power?

- Nuclear fission of uranium-235 fuel in a power plant produces heat that can be used to produce electricity.
- The nuclear fuel cycle describes all aspects of producing energy from nuclear fission reactions.
- Nuclear power emits fewer air pollutants and much less CO_2 than burning coal.
- Because of its low net energy and high cost, the nuclear fuel cycle must be subsidized to compete in the marketplace.
- Nuclear power plant accidents are rare, but major accidents have released radioactive materials into the environment. Because of this risk and the high cost of the nuclear fuel cycle, some countries are phasing out their use of nuclear power.

MindTap If you have been provided with access to a MindTap course, additional resources are available at login.cengage.com.

CHAPTER 12 ASSESSMENT

Review Key Terms

Select the key term that best fits each definition. Not all terms will be used.

coal
commercial energy
crude oil
horizontal drilling
hydraulic fracturing
natural gas
net energy
nuclear fission
nuclear fusion
peak production
petrochemical
proven oil reserve
refining

1. The amount of usable energy left once the energy is made available
2. A black, gooey liquid containing a mixture of combustible hydrocarbons
3. Energy produced when the nuclei of two isotopes of a light element such as hydrogen are forced together at extremely high temperatures
4. The process by which oil is heated to separate it into various fuels and other components with different boiling points
5. Energy sold in the marketplace
6. A mixture of gases, of which 50–90% is methane
7. Method that uses pumps to blast huge volumes of a mixture of water, sand, and various chemicals into a well to fracture rock and release natural gas or oil
8. Product made from the refining of crude oil
9. A solid fossil fuel formed from the remains of land plants that are buried and exposed to intense heat and pressure for 300–400 million years

Review Key Concepts

10. Why are certain energy resources called fossil fuels?
11. What challenges and opportunities are associated with methane escaping from Arctic lakes?
12. What is net energy, and why is it important for evaluating energy resources?
13. Use the net energy concept to explain why some energy resources are subsidized, and give an example of such a resource.
14. How do horizontal drilling and hydraulic fracturing methods allow companies to access previously unavailable oil and gas deposits?
15. What are the major advantages and disadvantages of using crude oil as an energy resource?
16. What are the major advantages and disadvantages of using natural gas as an energy resource?
17. What are the major advantages and disadvantages of using coal as an energy resource?
18. What are the major advantages and disadvantages of using nuclear power as an energy resource?

Think Critically

19. What factors should governments consider when deciding to subsidize electric power plants? Which factor or factors should be weighed most carefully in decision making? Explain your thinking.
20. How would you respond to someone who claims that the only thing consumers should worry about when it comes to using an energy resource is how much it will cost them to use it? Refer to full-cost pricing and net energy in your response.
21. Some analysts argue that in order to continue using oil at the current rate, we must discover and add to global oil reserves the equivalent of two new Saudi Arabian reserves every seven years. Do you think this is possible? If not, what effects might the failure to find such supplies have on your life and on the lives of future generations?
22. Explain why you agree or disagree with the following proposals made by various analysts as ways to solve U.S. energy problems:
    a. Oil companies should find and develop more domestic supplies of crude oil to increase domestic oil production and reduce dependence on imported oil.
    b. The government should place a heavy tax on gasoline and imported oil to reduce the consumption of crude oil resources and encourage the use of alternative energy sources.
    c. There should be a switch to greater dependence on coal as an energy resource.
    d. There should be a switch to greater dependence on nuclear power as an energy resource.

Chapter Activities

A. Investigate: Fracking STEM

Fracking is a controversial practice. While it expands access to energy resources, it has many environmental consequences. Fracking requires large amounts of water, much of which is permanently removed from the water cycle as it is trapped underground. It can cause earthquakes and pollute the air and water. In this activity, you will investigate potential impacts of fracking on a community and decide if it should or should not be allowed there.

Materials
large map of your county (or another county in which gas drilling is possible)
small copies of the map for each group
colored pencils or markers

1. As a class, look at the map and brainstorm ideas about how the county would be impacted if fracking were to occur in the area. Consider social as well as environmental impacts.

2. Break into small groups. Each group chooses a feature from the map to research (homes, roads, fields, water bodies, etc.).

3. Create a slide show, poster, or other visual presentation to share with the rest of the class that explains how your feature could be impacted by fracking. Present your findings, and receive feedback from the rest of the class.

4. Consider the feedback provided from other groups and, back in small groups, indicate areas on the map that might be impacted. Create a key to communicate your new markings on the map.

5. As a whole class, discuss the changes each group made to their small maps. Discuss which changes to incorporate on the large map, and modify the large map to show the changes agreed upon by the class.

6. Individually, write a short recommendation to the county chairperson indicating whether fracking should be allowed in the county. Use evidence to support your recommendation.

B. Take Action

National Geographic Learning Framework
Attitudes | Responsibility
Skills | Communication
Knowledge | Our Living Planet

The U.S. Environmental Protection Agency (EPA) runs an interactive database on its website called MyEnvironment. Its purpose is to provide a cross-section of environmental data for any geographic location in the United States. To use MyEnvironment, a person simply types an address, zip code, or place name into the search field in order to access a variety of data for that location. Examples of data include energy use, air and water quality, and health hazards. Often, local data can also be compared to county, state, or national data.

Work with a group to access MyEnvironment. Use information from this chapter to develop a question you can investigate regarding nonrenewable energy use as it relates to your community. For example, which energy resource is most commonly used in your community or state? Do any energy facilities operate in your area? If so, what impacts, positive or negative, might they have on your community?

Create a presentation to share results from your investigation. As a class, assess your community's environmental health data. Identify agencies or companies that you could contact for further information about energy issues in your community.

CHAPTER 13
RENEWABLE ENERGY RESOURCES

THE SKY IS THE LIMIT for wind energy. Wind is a renewable energy resource—it is continually being replenished. Using wind and other sources of renewable energy can make energy more accessible, more reliable, and more sustainable. It can also protect human health and the environment. Transitioning to renewable energy requires governments, industries, and individuals to rethink their energy use.

KEY QUESTIONS

13.1 Why is energy efficiency an important energy resource?

13.2 What are sources of renewable energy?

13.3 How can society transition to a more sustainable energy future?

Spinning turbines produce electricity at an offshore wind farm near the coast of Yorkshire, UK.

NATIONAL GEOGRAPHIC | EXPLORERS AT WORK

The Boiling River
with National Geographic Explorer Andrés Ruzo

Can you imagine a river with water so hot that if you put your hand in it you would get third-degree burns in less than one second? Geologist Andrés Ruzo not only can imagine such a river, he almost fell into it.

Growing up in Peru, Andrés had heard his family talk about the "boiling river," which he assumed was just a legend. But in 2011 when he started working on his Ph.D. to help determine Peru's geothermal potential, he decided to investigate the legend further. His aunt actually offered to be his guide. She led him deep into the jungle until, Andrés says, "I heard a low surge that got louder and louder as we got closer. It sounded like ocean waves constantly crashing. I saw smoke and vapor coming up through the trees."

The boiling river was real. "The river flowed hot and fast," Ruzo recalls. "I grabbed my thermometer, and the average temperature was 86 degrees Celsius!" That's about 187°F.

Ruzo has returned every year since to study the river further. He has recorded water temperatures along the river ranging from 49–91°C (120–196°F). One time he got caught on a rock in the middle of the river in a rainstorm. The rain created so much steam it nearly blinded him. He dared not move until the storm ended for fear of falling into the boiling water.

So what causes the water to boil? Boiling rivers in other parts of the world are generally associated with volcanoes, but the closest volcano to this river is 644 kilometers (400 miles) away. Ruzo's research suggests it's the result of a large hydrothermal system that is unique in the world. Magma rising within Earth's crust interacts with the local water table to produce areas of heated groundwater. Geothermal energy released by the hydrothermal system is evidenced as boiling water at the surface.

Ruzo says he's always been interested in volcanoes. "As a boy I would spend my summers on the family farm in Nicaragua, which rests on top of a volcano called Casita Volcano. I was able to see firsthand the power of Earth's heat.

"Later, as an undergrad at Southern Methodist University, these childhood memories inspired me to take a volcanology class. The first time I opened my class textbook, there on the page was a photo of the Casita Volcano. This created a personal connection with the subject that awakened my passion for geology," Ruzo says.

Ruzo firmly believes that geothermal energy can solve some of the world's energy problems. This renewable resource is already being used to generate heat and electricity in many places around the world. Ruzo also believes geothermal scientists and engineers should partner with petroleum companies.

"The oil and gas industry has spent over a century mastering the production of resources from the subsurface," argues Ruzo. "Because the development of geothermal energy uses virtually the same tools, technology, skill sets, and personnel as oil and gas companies, geothermal represents an untapped market with minimal barriers of entry for the oil and gas industry."

Thinking Critically
Evaluate What are the benefits of geothermal and oil-and-gas firms joining forces? Do you think that they can work together effectively? Why or why not?

Andrés Ruzo travels the globe to hunt for geothermal energy, from dry deserts (top) to boiling rivers (bottom). Ruzo's work in renewable energy helps him fulfill one of his goals, which is to be a force of positive change in the world.

CASE STUDY
The Potential for Wind Power in the United States

Simply put, wind is air in motion. Earth's winds are caused by uneven heating of the planet's surface by the sun. Land near the Equator absorbs more solar energy than land near the Poles. The uneven heating of Earth's surface and its atmosphere, combined with Earth's rotation, causes winds to blow. (See Figure 6-4 in Lesson 6.1.)

The kinetic energy from blowing winds can be captured and converted to electrical energy by devices called wind turbines. As a turbine's blades spin, they turn a drive shaft that connects to the blades. The drive shaft then turns an electric generator, which produces electricity (Figure 13-1). Groups of wind turbines called wind farms transmit electrical energy to electrical grids. Wind farms can operate both on land and at sea.

Today's wind turbine towers can be as tall as an 80-story building and have blades as long as 70 meters (230 feet). This height allows them to tap into the strong, reliable winds found at high altitudes on land and at sea. A typical wind turbine can generate enough electricity to power more than 450 U.S. homes.

Texas, the nation's leading oil-producing state, is also the nation's leading producer of electricity from wind power. In 2020, wind energy produced 23% of the state's electricity.

Expanding the U.S. wind farm industry would reduce pollution, create thousands of new jobs, and boost the American economy. Because wind is an indirect form of solar energy, relying more on wind power is a means to implement the solar energy factor of sustainability.

In 2021, the Biden administration established a target of creating 30 gigawatts of offshore wind energy along every U.S. coastline by 2030. That would be enough electricity to power 10 million homes. In addition, the National Renewable Energy Laboratory (NREL) estimates winds off the Atlantic and Pacific coasts and the shores of the Great Lakes could generate four times the electricity currently used in the lower 48 states.

Some analysts claim that wind power has more benefits and fewer drawbacks than all other energy resources. Even so, it is not the most energy efficient. (Energy efficiency is discussed throughout this chapter.)

To maximize the potential for wind power in the United States, some experts call for more investment in building land-based and offshore wind farms. They also call for investing in an updated and expanded "smart" electrical grid. Such a grid would be able to more efficiently transmit electricity from wind farms to consumers. Over time, these investments would reduce U.S. dependence on coal and other fossil fuels and help offset air pollution. Pollutants released by burning fossil fuels cause an estimated 350,000 premature deaths annually in the United States.

This chapter will consider the importance of using energy more efficiently. It will explore the pros and cons of renewable energy resources such as wind power, solar energy, flowing water, and Earth's internal heat. And it will suggest ways to transition to a more sustainable energy future.

As You Read Think about your community's energy needs and its commitment to renewable energy use. What do you know about renewable energy options available in your community? How can you as a consumer make choices to promote energy efficiency and the use of renewable energy resources?

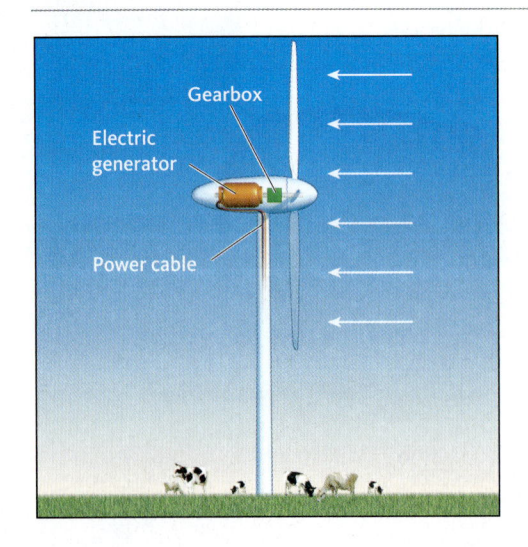

FIGURE 13-1
Wind Power Wind turbines convert the kinetic energy in wind to electricity. Wind power is an indirect form of solar energy.

13.1 Why Is Energy Efficiency an Important Energy Resource?

CORE IDEAS AND SKILLS

- Define energy efficiency and explain what makes a device energy efficient.
- Identify ways in which energy is used inefficiently.
- Describe ways to improve energy efficiency with regard to industry, transportation, and home building.

KEY TERMS

energy efficiency
cogeneration
hydrogen fuel cell

People Can Use Energy More Efficiently

People consume energy in many ways each day. Various sources of energy fuel their vehicles, power their lights and appliances, heat and cool their homes, and run equipment used to produce goods and services. Energy consumers may be surprised to know how much energy is wasted during these processes. They may be unaware of how much energy they could save by making some thoughtful choices. For example, people can save energy by using energy-efficient vehicles and appliances. They can design homes and buildings to be more energy efficient.

Energy efficiency is a measure of how much useful work we can get from each unit of energy we use. Improving energy efficiency means using less energy to provide the same amount of work. Consider a light-emitting diode (LED) light bulb. It uses much less energy to produce light compared to an incandescent light bulb, which wastes 90% of its energy as heat.

No energy-using device operates at 100% efficiency. Some energy is always lost as heat, as required by the second law of thermodynamics (Lesson 2.3). However, there are ways to improve energy efficiency so that less energy is wasted. Roughly 84% of all commercial energy used in the United States is wasted (Figure 13-2). About 41% of this energy unavoidably ends up as low-quality waste heat lost to the environment because of the second law of thermodynamics. The other 43% is mainly wasted due to the inefficiency of industrial motors, most motor vehicles, power plants, and numerous other energy-consuming devices.

FIGURE 13-2

U.S. Commercial Energy This diagram illustrates the flow of commercial energy through the U.S. economy. Only 38% of the country's high-quality energy ends up performing useful tasks.

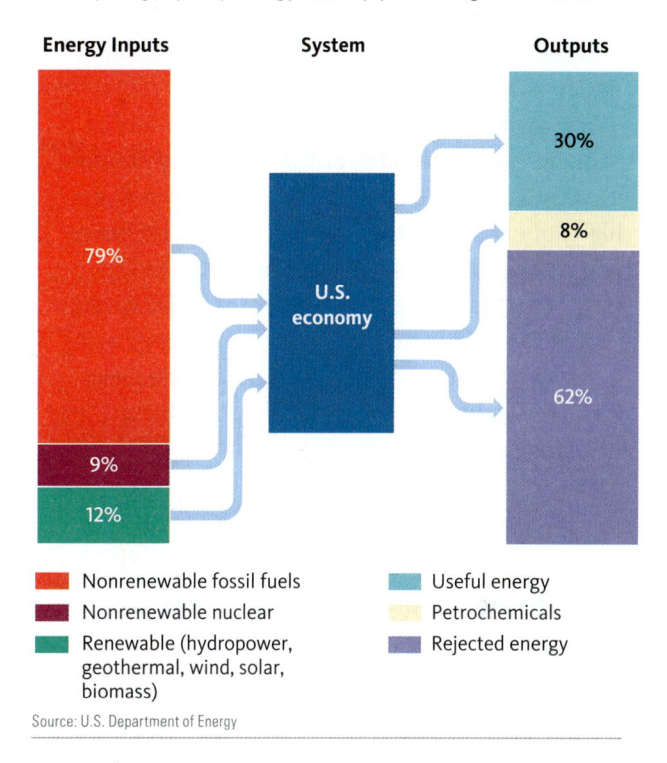

Energy Inputs: Nonrenewable fossil fuels 79%, Nonrenewable nuclear 9%, Renewable (hydropower, geothermal, wind, solar, biomass) 12%

System: U.S. economy

Outputs: Useful energy 30%, Petrochemicals 8%, Rejected energy 62%

Source: U.S. Department of Energy

Improving energy efficiency is especially important in the areas of industry and transportation, two top energy consumers in the U.S. economy. Data centers are one example. These facilities house computer servers that process all online information (such as data on social media sites) and provide cloud-based data storage for users. Most data centers are very energy inefficient. They typically use only 10% of the electrical energy they consume. The other 90% is lost as heat. Most also run 24 hours a day at their maximum capacities, regardless of demand. They require large amounts of energy for cooling to keep servers from overheating.

People also use energy inefficiently during their daily activities. They live and work in poorly designed buildings that require extra heating in cold weather and extra cooling in hot weather. Roughly three of every four Americans commute to work alone. Only 5% of commuters use more energy-efficient mass transit. Americans living in suburban areas mainly depend on cars to get around. Most car engines are inefficient, using only about 20% of the energy from burning gasoline to keep a car moving. The other 80% is lost as waste heat. In other words, only about 20% of the money that drivers spend on gasoline actually goes toward getting them somewhere.

LESSON 13.1

People can cut energy waste by using mass transit and driving more energy-efficient vehicles. They can choose to buy more energy-efficient devices. They can also make simple changes to their habits. For example, people can turn off lights and electronic devices when they are not using them.

Improving energy efficiency offers a variety of economic and environmental benefits. For example, it saves money and creates jobs. It results in a very high net energy, reduces oil imports, and improves energy security. It also reduces pollution and environmental degradation and helps slow climate change. Most energy analysts agree that improving energy efficiency provides the quickest, cleanest, and usually the least expensive way to achieve these economic and environmental benefits.

checkpoint Why is improving energy efficiency and reducing energy waste an important energy goal?

Industrial Processes Can Be More Energy Efficient

The industrial sector includes all facilities and equipment used to produce, process, or assemble goods. Industry accounts for 36% of the world's energy consumption and 33% of U.S. energy consumption. Industries that use the most energy are those that produce petroleum, chemicals, paper and wood products, steel, and aluminum.

One way utility companies and industries can save energy is to use **cogeneration** to produce two useful forms of energy from the same fuel source. For example, the steam used for generating electricity in a power or industrial plant is often released into the environment. It can be captured and used again to heat the plant or other nearby buildings instead of being released into the environment. The energy efficiency of these systems is 60–80%, compared to 25–35% for coal-powered and nuclear plants. The United States uses cogeneration to produce only 8% of its electricity.

Industries can also save money by using more energy-efficient, variable speed electric motors. They are designed to run at the minimum speed needed for each job. In contrast, standard motors run at full speed with their output throttled to match the task. This is somewhat like using one foot to push the gas pedal to the floorboard of your car while putting your other foot on the brake pedal to control its speed.

Recycling materials such as steel and other metals can also help industries save energy and money. Recycling also reduces negative environmental impacts. For example, producing steel from recycled scrap iron uses 75% less high-quality energy than producing steel from virgin iron ore and emits 40% less CO_2. Steel is the world's most recycled material.

Industries can also improve energy efficiency by making simple changes in the workplace. Replacing incandescent bulbs with LED lighting is one example. Businesses can adjust thermostat temperatures, limit air conditioner use, and install smart meters to monitor energy use. Workers can shut off computers, printers, and nonessential lights at the end of the workday.

A growing number of major corporations are saving money by improving energy efficiency. For example, Ford Motor Company saves $1 billion a year by turning off computers that are not in use. Its Go Green Initiative is helping nearly 2,000 U.S. Ford dealers upgrade to more energy-efficient technology. These upgrades can reduce each dealer's energy use by 27% and save them $33,000 annually on average.

checkpoint How does cogeneration improve energy efficiency?

Electrical Utilities Can Be More Energy Efficient

In the United States, electricity is delivered to consumers through an electrical grid. This network of transmission and distribution lines carries electricity from power plants, wind farms, and other electricity producers to homes, schools, offices, and other end users. Electricity used in your home may have been generated hundreds of miles away.

The U.S. electrical grid was designed more than 100 years ago when electricity demands were much lower. Today's huge demand for electricity is stretching this outdated grid to capacity. Little investment has been made in building more transmission lines. As former U.S. Secretary of Energy Bill Richardson observed, "We're a major superpower with a third-world electrical grid system."

Work is underway to reconfigure the current grid into a "smart" grid (Figure 13-3). This new grid will be a digitally controlled, ultra-high-voltage, high-capacity system with superefficient transmission lines. It will be less vulnerable to power outages. It will be able to quickly adjust for a major power loss in one area by automatically rerouting electricity from

FIGURE 13-3
Smart Grid This diagram shows how a smart grid would efficiently connect homes and businesses to energy sources.

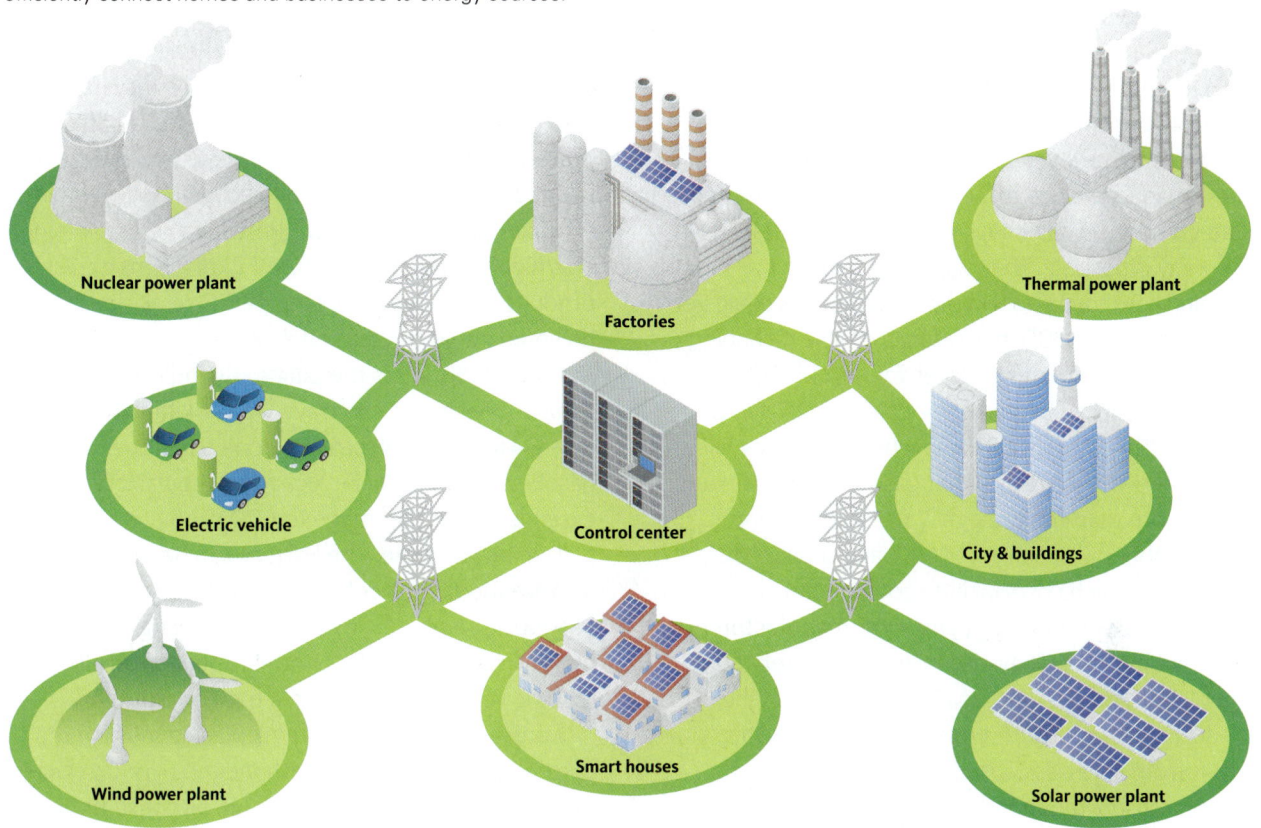

other parts of the country. A smart grid will also be better connected to sources of renewable energy throughout the country. As a result, it will allow power companies and consumers to more easily buy electricity produced from wind, solar, and other renewable forms of energy in areas where they are not directly available.

The smart grid will be more energy efficient, and it will help save consumers money. A smart grid will allow two-way communication between consumers and utility providers to increase energy efficiency. Smart meters will help people track how much electricity they use, when they use it, and what its cost is, essentially in real time. They can use this information to limit electricity use during times when rates are higher. According to the DOE, building such a grid will cost up to $800 billion over the next 20 years. However, it will save the U.S. economy $2 trillion during that period.

checkpoint How will a smart grid save consumers money?

Transportation Can Be More Energy Efficient

The EPA reported that in 2021, transportation (mostly by car) accounted for 29% of U.S. greenhouse gas emissions and more than 70% of U.S. oil use. More cars are on the road than ever before, and drivers are logging more miles too. Therefore, making vehicles more fuel efficient should be a priority in order to reduce the impacts of transportation on the environment and on human health.

CONSIDER THIS

Power outages are not just inconvenient, they're expensive.

The U.S. Department of Energy calculated the average cost for one hour of power interruption to various industries to be as follows:
- Cellular communications: $41,000
- Telephone ticket sales: $72,000
- Airline reservation system: $90,000
- Credit card operation: $2.58 million
- Stock trading: $6.48 million

Fuel Economy Between 1973 and 2020, the average fuel economy for new cars and light trucks in the United States increased from 5 kilometers per liter, or kpl (11.9 miles per gallon, or mpg), to 10.8 kpl (25.4 mpg). The government goal is for such vehicles to get 23.4 kpl (55 mpg) by 2026. Existing fuel economy standards for new vehicles in Europe, Japan, China, and Canada are much higher than those in the United States. Energy experts suggest even greater fuel economy is possible. They argue that by 2040, all new cars and light trucks sold in the United States could get more than 43 kpl (100 mpg) using available technology.

Even if a government pushes for more fuel-efficient vehicles, people do not always buy such vehicles. This is especially true when gas prices fall. Consumers do not realize that gasoline costs them much more than the price they pay at the pump.

A number of hidden costs get passed on to consumers. Such costs include government subsidies and tax breaks for oil companies, car manufacturers, and road builders. They also include costs related to pollution control and cleanup, as well as higher medical bills and health insurance premiums that result from illnesses caused by pollution related to the transportation industry. The Fuel Freedom Foundation estimated that in 2017, the hidden costs of gasoline for U.S. consumers amounted to $1.02 per liter ($3.88 per gallon). If people were more aware of these costs, they might be more motivated to purchase fuel-efficient vehicles.

One way to include more hidden costs in the market cost of gasoline is through higher gasoline taxes. This would implement the full-cost pricing factor of sustainability. However, higher gas taxes are not politically favorable in the United States. Some analysts call for increasing U.S. gasoline taxes while reducing payroll and income taxes to balance such increases. They say this would offset any additional financial burden to consumers. Another way for governments to encourage higher energy efficiency in transportation is to give consumers tax breaks or other economic incentives to encourage them to buy more fuel-efficient vehicles.

One federal tax that is meant to discourage people from buying fuel-inefficient vehicles is the Gas Guzzler Tax. This tax is applied to the domestic sale of certain new vehicles that don't get a combined city/highway mileage of at least 8 kpl (22.5 mpg). In contrast, federal tax credits are offered to people who buy qualifying all-electric or plug-in hybrid vehicles, which are very fuel efficient.

Alternative Fuel Technology There is growing interest in developing superefficient, ultra-light, and ultra-strong cars using existing materials and technology. One such vehicle is the energy-efficient, gasoline-electric hybrid car. Hybrid cars have a small, traditional gasoline-powered engine and a battery-powered electric motor that provides the energy needed for acceleration and hill climbing. The most efficient current models of these cars get a combined city/highway mileage of up to 27 kpl (58 mpg). They produce about 55% less tailpipe CO_2 emissions than do comparable conventional cars.

Another option is the plug-in hybrid electric vehicle. These cars can travel 48–97 kilometers (30–60 miles) on electricity alone. Then the small gasoline motor kicks in, recharges the battery, and extends the driving range to 600 kilometers (370 miles) or more. The battery can be plugged into a conventional 120-volt outlet and be fully charged in about three hours, or even faster using a 220-volt outlet (Figure 13-4). Yet another option is an all-electric vehicle that runs on a battery only.

According to a DOE study, replacing most of the current U.S. vehicle fleet with plug-in hybrid vehicles over three decades would cut U.S. oil consumption by 70–90%. In addition, such a move would eliminate the need for costly oil imports, save consumers money, and reduce CO_2 emissions by 27%. If batteries in the hybrid cars were recharged mostly by electricity produced using renewable energy (such as wind turbines, solar cells, or hydroelectric power), U.S. emissions of CO_2 would drop by 80–90%. This would reduce projected climate change. It would also save thousands of lives by reducing air pollution from motor vehicles and coal-burning power plants.

Unfortunately, hybrid, plug-in hybrid, and all-electric cars cost too much for most average consumers to buy. Their batteries are what make them so expensive, and they must be replaced over time. If a battery is not under warranty, replacing it costs an average of $5,500. It is critical to ramp up research and development of suitable batteries (Chapter 11, Explorers At Work). The ideal battery would be affordable, small, lightweight, easily rechargeable, and able to store enough energy for long-distance trips. Building a network of recharging stations within and between communities will also increase use of battery-powered vehicles.

FIGURE 13-4
This parking lot features charging stations for plug-in electric cars. What other actions are needed to make this alternative fuel technology practical?

The **hydrogen fuel cell** is another potential energy resource that could be used to power electric vehicles. This device uses hydrogen (H_2) as a fuel to produce electricity when it reacts with gas (O_2) in the atmosphere. The process emits harmless water vapor. Advantages and drawbacks of the hydrogen fuel cell are further discussed in Lesson 13.2.

Alternative Transportation Building or expanding mass transit systems within cities helps to make the transport of people less costly and more energy efficient. Such has been the case with commuter trains used on the East and West Coasts of the United States. Constructing high-speed rail lines between cities also saves money and energy. High-speed rail lines are common in Japan, much of Europe, and China. The United States has none. Building bike lanes along highways and city streets also encourages using alternative transportation. Moving freight by rail instead of in heavy trucks saves energy and money in commercial transportation.

checkpoint Why is the true price of gasoline higher than the price paid at the pump?

Buildings Can Be More Energy Efficient

Building construction and operation accounts for about one-third of the world's resource consumption and 25–40% of its energy use. It also contributes 30–40% of all CO_2 emissions and 30–40% of all solid waste. Making the building industry more sustainable over the next few decades will require using energy more efficiently. It will also require using more renewable energy resources, reusing and recycling materials, and reducing pollution and waste.

More Energy-Efficient Design Green architecture focuses on building design that is energy efficient, resource efficient, and cost efficient. In fact, it can even help make net zero energy buildings possible. A net zero energy building produces enough renewable energy onsite to meet its energy needs over the course of a year. Green architecture combines time-tested methods with modern materials and technology. For example, a building that is oriented to face the sun can get more of its heat from solar energy. That can cut heating costs up to 20%.

FIGURE 13-5
ON ASSIGNMENT National Geographic photographers Diane Cook and Len Jenshel shot this picture looking down on Chicago's City Hall building for a story about green roofs.

430 CHAPTER 13 RENEWABLE ENERGY RESOURCES

This is a simple application of the solar energy factor of sustainability that people have been using for centuries. Building orientation and window placement can also make use of natural light to reduce the need for electric lights during the day.

Some homes and urban buildings also have living roofs, or green roofs. A green roof is covered with specially formulated soil and selected vegetation (Figure 13-5). Green roofs can reduce the costs of cooling and heating a building in several ways. They absorb heat from the summer sun, insulate the structure, and retain heat in the winter.

Superinsulation is another important feature in energy-efficient building design. Heating costs can be cut by as much as 75% when a building is well insulated and does not experience any air leaks. Heat generated from direct sunlight, running appliances, and human bodies can sufficiently warm a superinsulated house. There is little or no need for a backup heating system, even in extremely cold climates. Efficient heat exchangers bring enough outdoor air into heavily insulated and airtight houses to prevent the buildup of indoor air pollution. One example of superinsulation is straw bale construction. The walls in a straw bale house are built with straw bales that are covered inside and out with mud-based adobe bricks. Such walls have insulating values of two to six times those of conventional walls.

Green architecture makes use of technology such as insulated windows and energy-efficient appliances and lighting. It also makes use of solar hot water heaters, electricity from solar cells, windows that darken automatically to deflect heat from the sun, and wastewater recycling.

Green building certification recognizes new construction that saves energy and protects the environment. Standards for this certification have been adopted in 70 countries, thanks to the efforts of the World Green Building Council. The U.S. Green Building Council's Leadership in Energy and Environmental Design (LEED) program was established in 1998. As of 2019, more than 69,000 U.S. buildings had been awarded silver, gold, and platinum LEED standard certificates for meeting various energy efficiency and environmental standards. That is largest number of any country. Certified green buildings typically reduce water use by 30–50%, CO_2 emissions by 35%, energy use by 30%, and lifetime operating costs by 50–90%.

Saving Energy in Existing Buildings Energy efficiency for existing buildings can be improved in a number of ways. Tips for improving energy efficiency in the home are summarized in Figure 13-7.

Prevent loss of heated or cooled air. About one-third of the heated air in typical U.S. homes and other buildings escapes through holes, cracks, and single-pane windows (Figure 13-6). During hot weather, these windows and cracks also let in heat, increasing the need for air conditioning. Single-pane windows can be replaced with energy-efficient windows.

FIGURE 13-6
Heat Loss The red and yellow colors in this thermogram, or infrared photo, indicate heat loss from a poorly insulated house.

FIGURE 13-7

Saving Energy You can save energy (and money) by taking steps to reduce heat and cooling loss in your home.

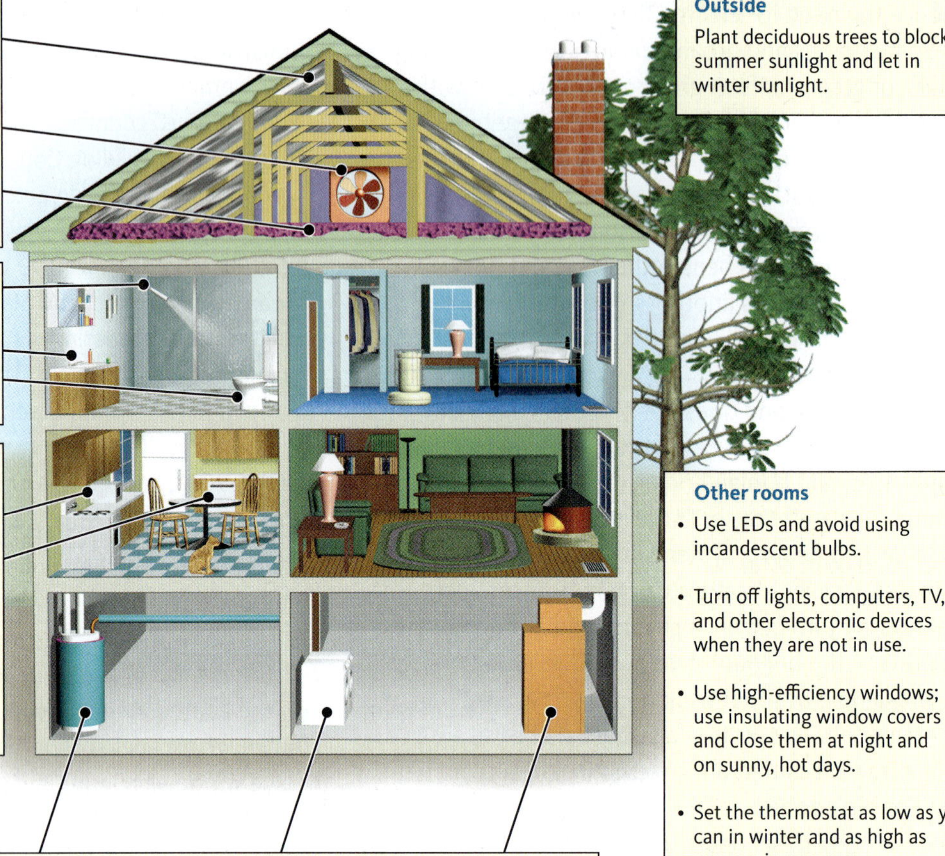

Attic
- Hang reflective foil near the roof to reflect heat.
- Use a house fan.
- Be sure attic insulation is at least 30 centimeters (12 inches) thick.

Bathroom
- Install water-saving toilets, faucets, and shower heads.
- Repair water leaks promptly.

Kitchen
- Use a microwave rather than a stove or oven as much as possible.
- Run only full loads in the dishwasher and use low- or no-heat drying.
- Clean refrigerator coils regularly.

Basement or utility room
- Use a front-loading clothes washer. If possible, run only full loads with warm or cold water.
- If possible, hang clothes on racks for drying.
- Run only full loads in the clothes dryer and use a lower heat setting.
- Set the water heater at 140°F if a dishwasher is used and 120°F or lower if no dishwasher is used.
- Use a water heater thermal blanket.
- Insulate exposed hot water pipes.
- Regularly clean or replace furnace filters.

Outside
Plant deciduous trees to block summer sunlight and let in winter sunlight.

Other rooms
- Use LEDs and avoid using incandescent bulbs.
- Turn off lights, computers, TV, and other electronic devices when they are not in use.
- Use high-efficiency windows; use insulating window covers and close them at night and on sunny, hot days.
- Set the thermostat as low as you can in winter and as high as you can in summer.
- Weather-strip and caulk doors, windows, light fixtures, and wall sockets.
- Keep heating and cooling vents free of obstructions.
- Keep the fireplace damper closed when not in use.
- Use fans instead of, or along with, air conditioning.

Energy-efficient windows cut heat loss by two-thirds and also cut summer cooling costs. Sealing cracks and leaky heating and cooling ducts also improves energy efficiency.

Heat interior spaces more efficiently. The most energy-efficient way to heat indoor spaces or conserve indoor heat is to use superinsulation and to plug any leaks. Other options, listed in order of energy efficiency, include a geothermal heat pump that draws heat stored underground into a home; passive solar heating; a high-efficiency conventional heat pump (in warm climates only); and a high-efficiency natural gas furnace.

Heat water more efficiently. Water can be heated efficiently using a roof-mounted solar hot water heater. Another option is a tankless instant water heater. These heaters use high-powered burners to quickly heat water as it runs through a heat exchanger. They deliver heated water as it is needed, rather than keeping water in a tank hot all the time.

Use energy-efficient appliances. A refrigerator with its freezer located in a bottom drawer uses about half as much energy as one with the freezer on top or on the side. Microwave ovens use 25–50% less electricity than electric stoves do for cooking and 20% less than convection ovens use. Clothes dryers with moisture sensors cut energy use by 15%. Front-loading clothes washers use 55% less energy and 30% less water than top-loading models and cut operating costs in half.

Use energy-efficient computers. The EPA's Energy Star program recognizes appliances and devices that use less energy than typical models. According to the EPA, if all computers sold in the United States met its Energy Star requirements, buyers would save $1.8 billion annually in energy costs. The resulting reduction in greenhouse gas emissions would be equal to taking 2 million vehicles off the road.

Use energy-efficient lighting. Energy-wasting incandescent light bulbs should be replaced with more energy-efficient and longer-lasting LED bulbs. Motion sensing lights, which turn on and off as needed, can also be installed to conserve energy.

Stop using the standby mode. According to the DOE, keeping TVs and other electronic devices on standby when they are not in use consumes 10% of electricity used by a typical household. Consumers can reduce their energy use and their monthly power bills by plugging standby electronic devices into a smart power strip. The strip cuts off power to a device when it detects that the device has been turned off.

checkpoint What is a net zero energy building?

13.1 Assessment

1. **Recall** What does it mean for a device to be energy efficient?
2. **Explain** Why don't car buyers always choose fuel-efficient vehicles, and how can government regulations encourage them to do so?
3. **Evaluate** Using Figure 13-7 as a guide, conduct an inventory of your home to identify energy losses. After identifying the energy losses, summarize the steps needed to reduce the loss of energy.

SCIENCE AND ENGINEERING PRACTICES

4. **Communicating Information** Use the text and your own research to create a diagram that shows the components of the current electrical grid and pathways it uses to deliver electricity.

13.2 What Are Sources of Renewable Energy?

CORE IDEAS AND SKILLS

- Explain why renewable energy resources have not been more widely adopted.
- Identify sources of renewable energy and their applications.
- Understand the advantages and disadvantages of each source of renewable energy.

KEY TERMS

passive solar heating system
active solar heating system
solar thermal system
photovoltaic cell
hydropower
biomass
biofuel
geothermal energy

Use Renewable Energy Resources to Produce Electricity

We can use renewable energy from the sun, wind, flowing water, Earth's interior (geothermal energy), and biomass to produce electricity. All of these sources of renewable energy are constantly replenished at no cost to us.

Studies show that adequate government funding for research and development, along with proper subsidies and tax breaks, would enable renewable energy to provide one-third of the world's electricity by 2025. That number could increase to 50% by 2050. As of 2020, nearly 50 countries supplied more than 50% of their electricity using renewable energy. Only 12% of U.S. electricity was supplied using renewable energy.

Why isn't renewable energy use in the United States expanding more rapidly? First, people's misperceptions play a part. People tend to think that solar and wind energy are too diffuse, too intermittent and unreliable, and too expensive to use on a large scale. Second, government support is lacking. Since 1950, government funding, tax breaks, and subsidies for research and development of renewable energy have been much lower than those for fossil fuels and nuclear power. Third, while government subsidies and tax breaks for renewables have been increasing, Congress must renew them every few years. That hinders investments in renewable energy. In contrast, subsidies and tax breaks for fossil fuels and nuclear power have essentially been guaranteed for decades, due in large part to political pressure from the fossil fuel industry.

Fourth, prices for nonrenewable energy resources don't include most of the harmful environmental and human health costs of producing and using them. As a result, they are shielded from free-market competition with renewable sources of energy. Finally, history shows it typically takes 50–60 years to make the transition from one dominant fuel to another, such as from wood to coal and coal to oil and natural gas. Renewable wind and solar energy are the world's fastest-growing sources of energy. Even so, it will likely take decades for them to supply 25% or more of the world's energy or electricity.

checkpoint What are two misconceptions people have about renewable energy?

Solar Energy

Solar energy has several applications in homes and businesses. Different types of solar heating systems can help heat interior spaces and water. Solar energy can also be used to produce electricity.

Solar Heating Systems
A building that has enough access to sunlight can get all or most of its heat through a **passive solar heating system** (Figure 13-8, left). Such a system absorbs and stores energy directly from the sun within a well-insulated, airtight structure. Water tanks and walls and floors of concrete, adobe, brick, or stone can store much of the collected solar energy as heat. The heat is then slowly released. A small backup heating system can be used to provide additional heat but it is usually not necessary.

An **active solar heating system** (Figure 13-8, right) captures energy from the sun by pumping a heat-absorbing fluid such as water or an antifreeze solution through special collectors. The collectors are usually mounted on a roof or on special racks that face the sun. Some of the collected heat can be used directly. The rest can be stored for later use in large insulated containers filled with gravel, water, clay, or a heat-absorbing chemical.

Rooftop active solar collectors are used to heat water in many homes and apartment buildings. One in ten houses and apartment buildings in China uses the sun to provide hot water with systems that cost the equivalent of $200. Once the initial cost is paid, the hot water is heated for free. In China and Israel, builders are required to install rooftop solar water heaters on all new buildings.

Direct solar energy is not useful for keeping a building cool. However, indirect solar energy (mainly wind) can be used to help cool buildings. People can open windows to take advantage of cooling breezes.

FIGURE 13-8 ▼

Passive and Active Solar Heating Systems Passive solar home heating systems (left) are used in buildings that have a good deal of access to sunlight. Active solar home heating systems (right) collect energy from the sun through racks on the roof.

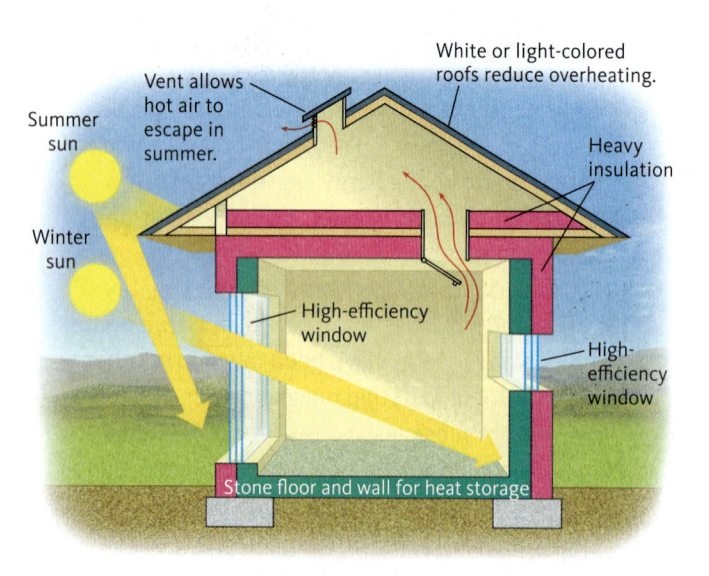

PASSIVE

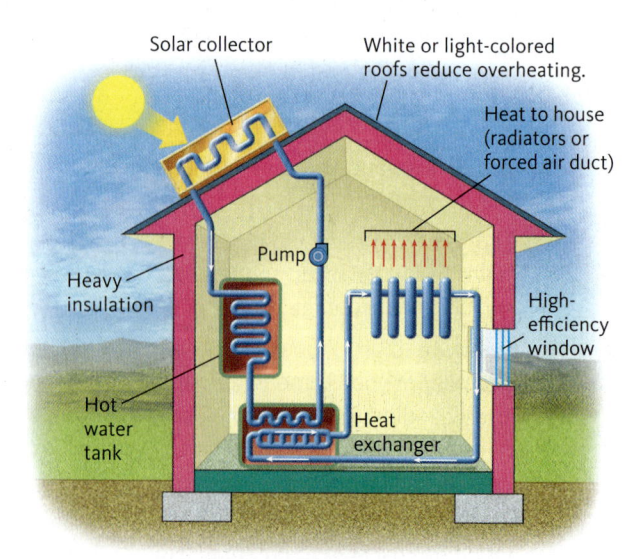

ACTIVE

FIGURE 13-9
The Ivanpah Solar Electric Generating System is located in California's Mojave Desert. More than 300,000 computer-controlled mirrors concentrate sunlight and reflect it toward three central power towers. The concentrated sunlight causes water to boil inside boiler pipes atop each tower. Steam from the boiling water is then used to produce electricity.

Turning on fans can keep the air moving. When there is no breeze, superinsulation and high-efficiency windows can help keep hot air outside.

Other ways to naturally keep buildings cool include blocking sunlight with shade trees, broad overhanging eaves, window awnings, or shades. A light-colored roof can reflect up to 90% of the sun's heat (compared to only 10–15% for a dark-colored roof). A green roof can absorb extra heat. Homes with geothermal heat pumps can use them to pump cool air from underground into buildings during summer. Figure 13-10 lists the advantages and disadvantages of solar heating systems.

FIGURE 13-10 ▼

| Passive and Active Solar Heating Systems ||
Advantages	Disadvantages
Medium net energy	Need access to sun 60% of time during daylight
Low emissions of CO_2 and other air pollutants	Blockage of sunlight by trees and other structures
Low land disturbance	High installation and maintenance costs for active systems
Moderate cost (passive)	Need backup system on cloudy days

Concentrating Solar Power **Solar thermal systems**, also known as concentrating solar power (CSP), use different methods to collect and concentrate solar energy in order to boil water and produce steam for generating electricity. These systems are used in deserts and other open areas with ample sunlight.

One such system uses rows of highly curved mirrors called parabolic troughs to collect and concentrate sunlight. Each trough focuses incoming sunlight on a pipe that runs through its center and is filled with synthetic oil. The oil heats to temperatures as high as 400°C (750°F). That heat is used to boil water and produce steam. The steam in turn powers a turbine that drives a generator to produce electricity.

Another solar thermal system uses an array of computer-controlled mirrors to track the sun and focus its energy on a central power tower (Figure 13-9). The concentrated heat is used to boil water and produce steam. The steam drives turbines to produce electricity. Heat produced by either of these systems also can be used to melt a certain kind of salt stored in a large insulated container.

The heat stored in this molten salt system can then be released as needed to produce electricity at night or on cloudy days.

The world's largest solar thermal plant, the Noor Ouarzazate Solar Power Complex, is located in Morocco. The $2.5 billion facility includes several power plants that can produce enough electricity to service more than one million homes.

People can use solar thermal systems on a smaller scale as well. In some sunny rural areas, people use inexpensive solar cookers to boil and sterilize water and to cook food. (See Engineering Project 4.) Solar cookers can replace wood and charcoal fires. They reduce indoor air pollution, a major killer of many people in less-developed nations. They also reduce deforestation by lowering the need for firewood and charcoal made from firewood.

Because solar thermal systems have a low net energy, they do require large government subsidies or tax breaks to be competitive in the marketplace. Figure 13-11 lists the advantages and disadvantages of solar thermal systems.

FIGURE 13-11 ▼

| Solar Thermal Systems ||
Advantages	Disadvantages
High potential for growth	Low net energy and high costs
No direct emissions of CO_2 and other pollutants	Need for backup or storage system for use on cloudy days
Lower costs with natural gas turbine backup	Substantial water use requirements
Source of new jobs	Can disrupt desert ecosystems

Solar Cells Solar energy can be converted directly into electrical energy using **photovoltaic cells**, commonly called solar cells. Solar cells are the world's fastest-growing technology for producing electricity. The largest solar-cell power plants are operating in Portugal, Spain, Germany, South Korea, the southwestern United States, and China. Between 2010 and 2019, the cost per watt of electricity produced by solar cells in the United States fell by roughly 90%. Costs are expected to keep falling.

Most solar cells are very thin transparent wafers of purified silicon (Si) or polycrystalline silicon with trace amounts of metals. They produce electricity (flowing electrons) when sunlight strikes them.

NATIONAL GEOGRAPHIC | EXPLORERS AT WORK
Xiaolin Zheng, Nanoscientist

What if "installing" solar energy technology were as simple as peeling a solar cell off a sheet and sticking it on a backpack, a mobile phone, or a windowpane? Thanks to nanoscientist and National Geographic Explorer Xiaolin Zheng, that scenario is on its way to becoming a reality.

Zheng led a team of students at Stanford University to develop a flexible, peel-and-stick solar cell. Zheng's solar stickers are much thinner than a sheet of plastic wrap. Yet they produce the same amount of electricity as standard solar cells. They are also much lighter and more flexible. That makes them less expensive to install and more versatile to use. Unlike rigid solar cells, solar stickers could be applied to a variety of surfaces to help power light bulbs, electronic devices, solar cars, and more.

Zheng's dream is to make solar stickers as easy to get as batteries. It is a dream inspired by an observation her father made years ago while gazing out the window of the family's apartment in Anshan, China. As Zheng explains, "In China, the rooftops of many buildings are packed with solar energy devices. One day my father mentioned how great it would be if a building's entire surface could be used for solar power, not just the roof, but also walls and windows."

To make the transition from laboratory to real world, Zheng is researching ways to make the solar stickers bigger and more efficient. She's also seeking to make the peel-off process more suitable for mass production.

Zheng believes her work is part of her duty as a scientist to make the world a better place. "For the future of our environment, we need to advance renewable energy rather than heavily relying on fossil fuels to meet growing demand," says Zheng. "Solar power has always been my favorite because sunshine is so clean, abundant, and has fewer limitations on where it can be used."

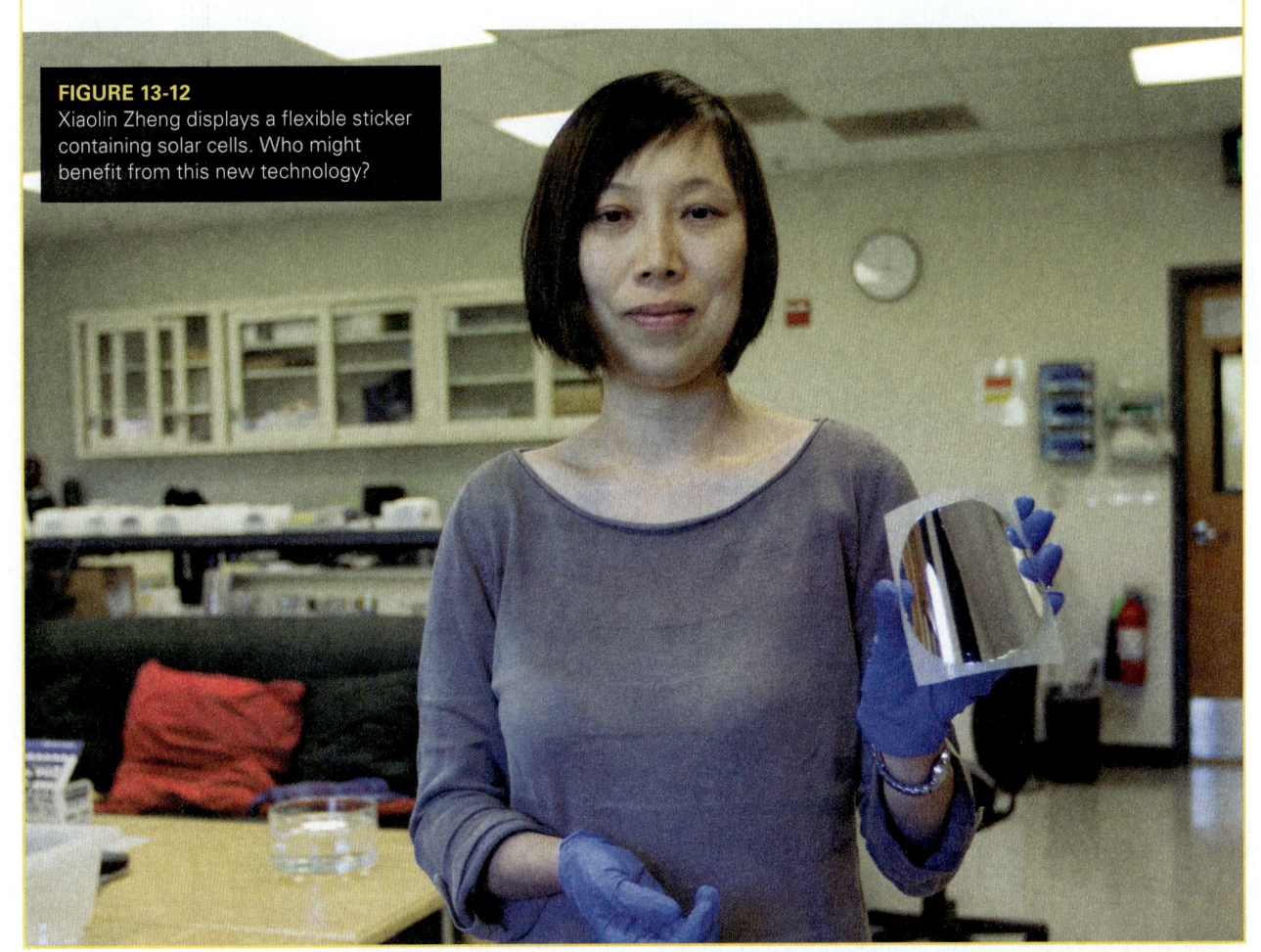

FIGURE 13-12
Xiaolin Zheng displays a flexible sticker containing solar cells. Who might benefit from this new technology?

438 CHAPTER 13 RENEWABLE ENERGY RESOURCES

FIGURE 13-13
ON ASSIGNMENT National Geographic photographer Annie Griffiths photographed these women displaying solar lanterns they have been trained to build and repair through Self Employed Women's Association. The women live in Rajat, India, and do not have access to electricity.

LESSON 13.2

Many cells wired together in a panel can produce large amounts of electrical power. Such systems can be connected to existing electrical grid systems. They can also be connected to batteries that store the electrical energy until it is needed.

People can mount arrays of solar cells on rooftops and incorporate them into almost any type of roofing material. Scientists are also developing ways to produce solar cells that can be attached to or embedded in other surfaces, such as outdoor walls, windows, drapes, and clothing (Explorers At Work). Solar power providers in Japan, Great Britain, India, Italy, and Australia are putting floating arrays of solar cell collectors on the surfaces of lakes, reservoirs, ponds, and canals. Engineers are also developing dirt- and water-repellent coatings to keep solar panels and collectors clean without having to use water.

Solar cells have great potential for providing electricity in less-developed countries. Nearly 940 million people in the world, nearly one of every eight, live in rural villages that are not connected to an electrical grid. A growing number of these people now use solar cells to generate electricity. They also use solar cells to power highly efficient LED lamps (Figure 13-13) that replace polluting kerosene lamps. As these small, off-grid systems reach more rural villages, they will help hundreds of millions of people to lift themselves out of poverty.

Solar cells have no moving parts, need no water for cooling, and operate safely and quietly without emitting pollutants or greenhouse gases. However, conventional solar cells do contain toxic materials that must be recovered when the cells wear out after 20–25 years of use, or when they are replaced by new systems. And solar cells are not a carbon-free option because fossil fuels are used to produce and transport solar panels. However, these emissions are low compared to emissions released during the process of generating electricity from nonrenewable energy sources.

Solar cells typically convert only 20% of the incoming solar energy into electricity, although their efficiency is rapidly improving. In 2020, the National Renewable Energy Laboratory developed a solar cell with an efficiency of 47%. Solar cells are becoming more cost effective and demand for them is rising. There are also incentives to switch to solar, including the U.S. solar investment tax credit. Costs to produce solar cells with thin-film nanotechnology, graphene (see Chapter 11), and other mineral materials are expected to fall to the point at which they will be competitive with fossil fuels. Figure 13-14 lists the advantages and disadvantages of solar cells.

FIGURE 13-14

Solar Cells	
Advantages	Disadvantages
Medium net energy	Need access to sun
Little or no direct emissions of CO_2 and other air pollutants	Need electricity storage system or backup
Ease of installation, transport, and expansion as needed	High costs for older systems (but dropping rapidly)
Competitive cost for newer cells	Solar power plants can disrupt desert ecosystems.

checkpoint How does a passive solar heating system work?

Hydropower

Technology that uses the kinetic energy of flowing or falling water to produce electricity is known as **hydropower**. This renewable energy resource is an indirect form of solar energy, produced as part of Earth's solar-powered water cycle. (See Figure 3-16, Lesson 3.4.) Heat from the sun causes surface water to evaporate into the atmosphere. Water returns to Earth in the form of rain or snow that may be deposited at higher elevations.

Hydropower is the most widely used renewable energy resource in the world. In 2020, hydropower produced about 17% of the world's electricity in 160 countries. That same year, about 7% of electricity in the United States was generated with hydropower, about half of it for use on the West Coast. In order, the world's top three producers of hydropower are China, Brazil, and the United States. The International Energy Association predicts that hydropower will be the dominant source of flexible electricity by 2050. Countries with the greatest potential include China, India, Brazil, Central Africa, and parts of the former Soviet Union.

The most common approach to harnessing hydropower is to build a high dam across a large river to create a reservoir. Some of the water stored in the reservoir is allowed to flow through large pipes at controlled rates. The flowing water causes blades on a turbine to turn, and the turbine spins a generator to produce electricity (Figure 13-15). Electric lines then carry the electricity to where it is needed. The

FIGURE 13-15
Flowing water causes turbines to spin inside each of these generators. The action of the spinning turbines turns a shaft that connects to a motor. The motor converts the mechanical energy to electricity that travels through power lines to where it is needed.

volume of water flowing through the system, along with how far the water drops, determines the amount of electricity that is generated.

Microhydropower generators are becoming an increasingly important way to produce low-cost electricity with minimal environmental impact. The generators are floating turbines, each about the size of an overnight suitcase. They can be placed in any stream or river without altering its course. According to the DOE, one of these systems can produce enough electricity to power a large home, a small resort, or a hobby farm. Furthermore, it can deliver electricity as far as 1.6 kilometers (1 mile) from where the generator is located.

Hydropower is the least expensive renewable energy resource. Once a dam is up and running, its source of energy—flowing water—is free and is annually renewed by snow and rainfall. The process by which hydropower plants generate electricity does not cause pollution. However, their reservoirs do release undesirable methane (CH_4) into the atmosphere as vegetation decomposes underwater. Warm climates enhance this decomposition process. Methane is a potent greenhouse gas.

Despite their potential, the use of large-scale hydropower plants is expected to fall slowly over the next several decades. Many existing reservoirs will fill with silt and become useless. New systems to replace them will not be built quickly enough. The electricity output of hydropower plants may also drop if atmospheric temperatures continue to rise. This will cause melting of mountain glaciers, a primary source of water for some facilities. Figure 13-16 lists the advantages and disadvantages of hydropower.

Energy in Oceans People can also produce electricity from flowing water by tapping into the energy from ocean tides and waves. In some coastal bays and estuaries, water levels can rise or fall by 6 meters (20 feet) or more between daily high and low tides. Dams can be built across the mouths of such bays and estuaries to capture tidal energy for hydropower, although currently such sites are rare.

LESSON 13.2   441

FIGURE 13-16

Hydropower	
Advantages	Disadvantages
High net energy	Large land disturbance and displacement of people
Large untapped potential	High CH_4 emissions from rapid biomass decay in shallow tropical reservoirs
Low-cost electricity	Disruption of downstream aquatic ecosystems
Low emissions of CO_2 and other air pollutants in temperate areas	

Three large tidal energy dams are currently operating worldwide. The largest one is in South Korea. Others are located in France and in Nova Scotia.

For decades, scientists and engineers have been trying to produce electricity by tapping into wave energy along seacoasts where there are almost continuous waves. Scientists estimate that harnessing the world's wave power at an affordable cost could provide more than twice the amount of electricity that the world uses. Production of electricity from tidal and wave systems is currently limited. Challenges include a lack of suitable sites, citizen opposition at some sites, high costs, and damage to equipment resulting from saltwater corrosion and storms.

checkpoint What are the pros and cons of hydropower?

Wind Power

Since 1990, wind power has been the world's second fastest-growing source of electricity after solar cells. In 2020, wind farms produced about 6% of the world's electricity—enough electricity to serve more than 530 million households. Experts predict that by 2050, this number could grow to 31%. Many energy analysts feel that wind power has more benefits and fewer serious drawbacks than any other energy resource. It also has huge promise. Studies suggest wind power could potentially produce 40 times the world's current use of electricity.

In 2020, the United States ranked second in the world in producing electricity from wind, behind China. That same year, electricity produced by wind turbines accounted for 8.4% of total U.S. electricity generation, and the United States set a record for wind turbine capacity additions.

The frontier for wind energy is offshore wind farms. Winds are generally much stronger and steadier over coastal waters than on land. Being able to capture their energy can reduce the cost of electricity production. It can offset the higher cost of building offshore compared to building on land. When located far enough offshore, wind farms are not visible from land. Building offshore also eliminates the need for negotiations among multiple landowners over the locations of turbines and electric transmission lines. More than 160 offshore wind farms currently produce electricity worldwide.

Experts predict that with expanded and sustained subsidies and a smart grid, wind farms off the Atlantic and Gulf coasts could generate enough electricity to more than replace all of the United States' coal-fired power plants. Many states in these regions plan to tap into this vast source of energy and boost their economies. Two commercial scale offshore wind farms are currently under construction, off the coasts of Massachusetts and Rhode Island.

Wind power has many advantages. Wind is abundant, widely distributed, and inexhaustible. Wind power is mostly carbon-free and pollution-free. A wind farm can be built within 9 to 12 months and expanded as needed. And although wind farms may cover large areas of land, the turbines themselves occupy only a small portion of the land.

Many U.S. landowners in favorable wind areas are investing in wind farms. Landowners typically receive $3,000–$10,000 a year in royalties for each wind turbine placed on their land. The land can still be used for activities such as growing crops or grazing cattle. Consider that an acre of land in northern Iowa planted in corn can produce about $1,000 worth of ethanol fuel. One wind turbine on the same site can produce $300,000 worth of electricity per year.

In addition, wind power has a medium-to-high net energy. The DOE and the Worldwatch Institute estimate that, if we were to apply the full-cost pricing factor of sustainability by including the harmful environmental and health costs of various energy resources in comparative cost estimates, wind energy would be the least costly way to produce electricity.

Wind power does have drawbacks. Land areas with the greatest wind power potential are often sparsely populated and remote. Roads are needed to deliver massive turbine parts to such areas (Figure 13-18). As a result, the proposed U.S. smart grid must connect with wind farms wherever they are located. Winds

FIGURE 13-17

Wind Power	
Advantages	**Disadvantages**
High net energy	Needs a widespread smart electrical grid
Wide availability	Needs backup or storage system when winds die down
Low electricity cost	Appearance (may be considered an eyesore)
Little or no direct emissions of CO_2 and other air pollutants	Low-level noise (may be considered bothersome)
Easy to build and expand	Potential to kill birds and bats if not properly designed and located

can be unpredictable, so a backup power source such as natural gas is required to generate electricity. A smart grid could use also combat this issue by using wind farms operating in other areas. Figure 13-17 lists the advantages and disadvantages of wind power.

checkpoint Why are offshore wind farms so promising?

Biomass and Biofuels

Energy can be produced by burning the solid **biomass**, or organic matter found in plants or plant-related material. Energy from biomass can also be converted to gaseous or liquid **biofuels**. Most biomass fuel is used for heating and cooking. It can also be used for industrial processes and generating electricity. Examples of biomass fuels include wood, wood wastes, charcoal made from wood, and agricultural wastes such as sugarcane stalks, rice husks, and corncobs.

Wood is a renewable fuel only if it is not harvested faster than it is replenished. The problem is that 2.7 billion people in nearly 80 less-developed countries face a fuelwood crisis. They often are forced to meet their fuel needs by harvesting trees faster than new ones can replace them.

One solution is to plant fast-growing trees, shrubs, and perennial grasses in biomass plantations. However, repeated cycles of growing and harvesting biomass plantations can deplete the soil of important nutrients. It can also allow for the spread of nonnative tree species that become invasive species.

FIGURE 13-18
A tractor-trailer hauls a set of wind turbine blades to a wind farm in China. China plans to triple its wind power capacity by 2030.

FIGURE 13-19

Biomass	
Advantages	**Disadvantages**
Wide availability in some areas	Potential increase in deforestation
Moderate costs	Damage to ecosystems from clear-cutting, soil erosion, water pollution, and loss of wildlife habitat
Medium net energy	Potential spread of invasive species
No net CO_2 increase if harvested, burned, and replanted sustainably	Increase in CO_2 emissions if harvested and burned unsustainably
Plantations can help restore degraded lands	

Clearing forests and grasslands to provide fuel also reduces biodiversity and the amount of vegetation that would otherwise capture climate-changing CO_2.

Burning wood and other biomass produces CO_2 and other pollutants such as fine particulates in smoke. Since 2015, the EPA has been phasing in stricter regulations to curb such pollution from new residential wood-burning stoves in the United States. Figure 13-19 lists the advantages and disadvantages of burning solid biomass as fuel.

Biomass can also be converted into liquid biofuels for use in motor vehicles. The two most common biofuels are ethanol (ethyl alcohol produced from plants and plant wastes) and biodiesel (produced from vegetable oils). The biggest biofuel producers are, in order, the United States (producing ethanol from corn), Brazil (producing ethanol from sugarcane residues), Indonesia (producing biodiesel primarily from palm oil), and Germany (producing biodiesel primarily from rapeseed oil).

Biofuels have three major advantages over gasoline and diesel fuel produced from oil. First, biofuel crops can be grown throughout much of the world. That in turn can help more countries reduce their dependence on imported oil. Second, if growing new biofuel crops keeps pace with harvesting them, there is no net increase in CO_2 emissions. However, that does not hold true if existing grasslands or forests are cleared to plant biofuel crops. Third, biofuels are easy to store and transport through existing fuel networks and can be used in motor vehicles at little additional cost.

In the United States, most ethanol is made from corn. The government heavily subsidizes this biofuel, in part because it has a low net energy. Fossil fuels are used to produce fertilizers, grow corn, and convert corn to ethanol.

Studies have shown that the corn-based ethanol program in the United States has taken more than 2 million hectares (5 million acres) of land out of the soil conservation reserve, an important topsoil preservation program. Other studies have concluded that burning corn-based ethanol adds at least 20% more greenhouse gases to the atmosphere per unit of energy than does burning gasoline. Growing corn also requires substantial amounts of water and land—resources that are in short supply in some areas.

Furthermore, scientists warn that large-scale biofuel crop farming could reduce biodiversity. It could also degrade soil quality and increase erosion. Small farmers could be pushed off their land and food prices could increase if it becomes more profitable to grow corn and other crops for biofuel rather than to feed livestock and people. Recent studies confirm most of these warnings. As a result, a number of scientists and energy economists call for withdrawing all government subsidies for corn-based ethanol production and sharply reducing the large amount of ethanol that is currently required in U.S. gasoline.

Efforts are underway to find alternatives to corn-based ethanol. One approach involves using algae to produce biofuels (Figure 13-20). As a crop, algae can grow rapidly year-round in various aquatic environments. The algae store their energy as natural oils within their cells. This oil can be extracted and refined to make a product very much like gasoline or biodiesel. A major benefit of this process is that algae use CO_2 as their fuel source to produce the oil. That makes the energy production carbon neutral. Currently, extracting and refining the oil from algae is too costly. More research is also needed on which types of algae are most suitable and which ways of growing them are the most successful and affordable.

Cellulosic ethanol is another possible alternative. This kind of ethanol is produced from the inedible cellulose that makes up most of the biomass of plants. Cellulose is found in plant leaves, stalks, and wood chips. One plant useful for cellulosic ethanol production is switchgrass, a prairie grass that grows faster than corn. Switchgrass is hardy, does not require nitrogen fertilizers, and produces ethanol with a medium net energy. However, growing

FIGURE 13-20
Bags of algae hang outside a power plant that uses the algae to make biodiesel. What are the advantages of using algae to produce this type of fuel?

switchgrass requires even more land than corn. The production of cellulosic ethanol has been slow to take off and still remains limited.

Brazil makes its ethanol from sugarcane residue, called bagasse. This sugarcane ethanol has a medium net energy that is about eight times higher than that of corn-based ethanol. Essentially all of Brazil's motor vehicles run on ethanol or ethanol-gasoline mixtures produced from sugarcane that is grown domestically. As a result, Brazil has greatly reduced its dependence on imported oil.

The ongoing challenge will be to grow biofuel crops using more sustainable agricultural methods. Such methods should require less irrigation. They should also limit land degradation, air and water pollution, greenhouse gas emissions, and threats to biodiversity. In addition, any system used to produce biofuel should have a moderate-to-high net energy. Only then can it compete in the energy marketplace without large government subsidies. Figure 13-21 lists the advantages and disadvantages of biofuels.

checkpoint Why does corn-based ethanol have a low net energy?

Geothermal Energy

Unlike the forms of renewable energy that come from the sun, **geothermal energy** is heat stored in soil, underground rocks, and fluids in Earth's mantle. Geothermal energy can be used to heat and cool

FIGURE 13-21

Biofuel	
Advantages	**Disadvantages**
Reduced CO_2 emissions for some crops	Threat to food supply when fuel crops compete with food crops for land and drive up food prices
Medium net energy for biodiesel from oil palms	Invasive spread of some fuel crops
Medium net energy for ethanol from sugarcane	Low net energy for corn-based ethanol
	Higher CO_2 emissions from corn-based ethanol

LESSON 13.2 445

FIGURE 13-22

Geothermal Energy Power plants can produce electricity from heat extracted from underground geothermal reservoirs.

2. Heat from underground spins a turbine to power a generator and produce electricity.

Generator
Steam turbine
Heat exchanger

3. Steam from the turbine condenses to water and is pumped back down to the geothermal reservoir.

Production well
Injection well

1. Hot water or steam is pumped under pressure to the surface from underground.

Geothermal reservoir

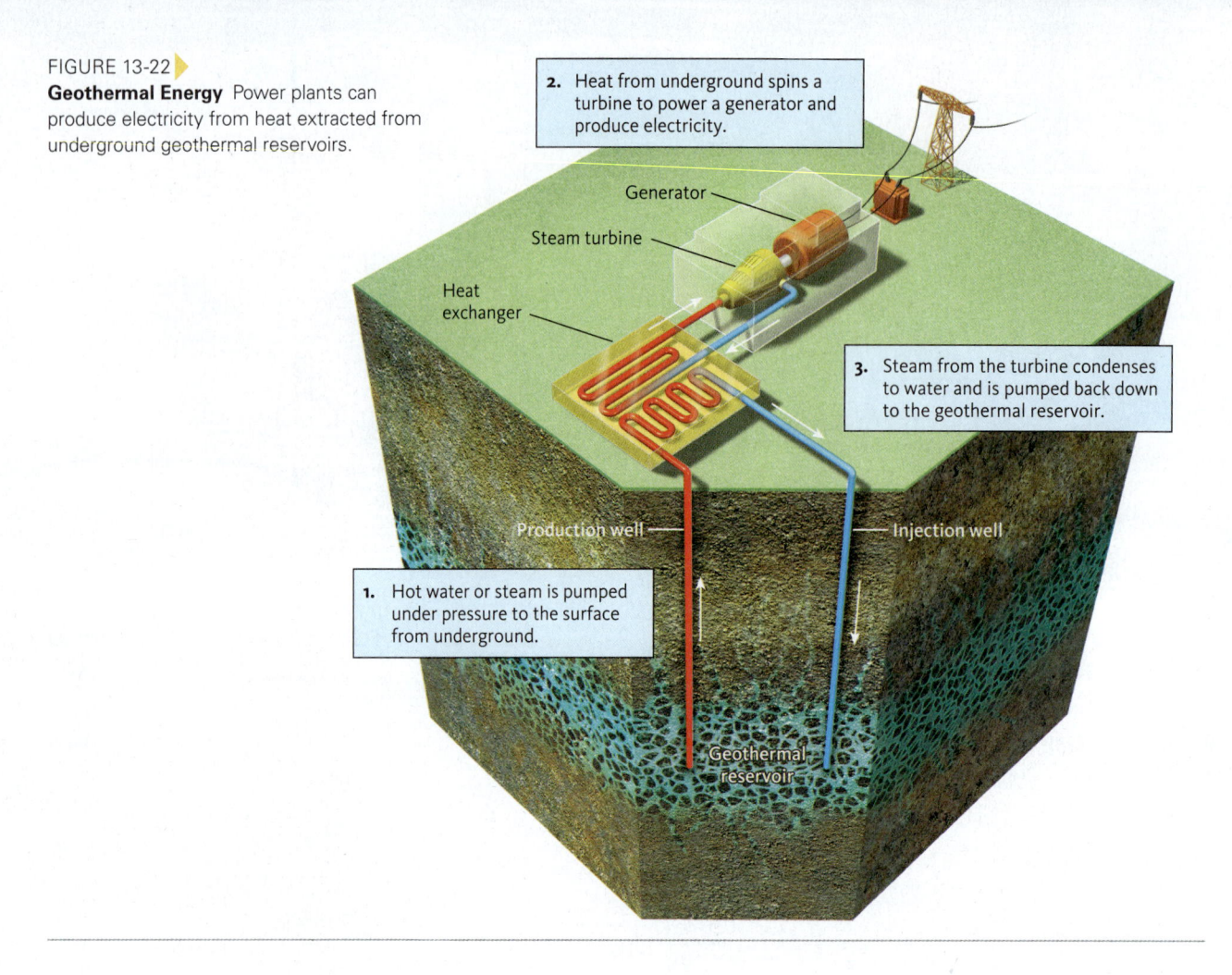

buildings and water and to produce electricity. Geothermal energy is available around the clock. However, it is only practical at sites with high enough concentrations of underground heat.

A geothermal heat pump system can heat and cool a house almost anywhere in the world. This system makes use of the temperature difference between Earth's surface and an underground depth of 3–6 meters (10–20 feet). At this depth, the temperature is typically 10–16°C (50–60°F) year-round. In winter, a closed loop of buried pipes circulates a fluid that extracts heat from the ground and carries it to a heat pump. The pump transfers the heat to a home's heat distribution system. In summer, the system works in reverse. Heat is removed from a home's interior and stored below ground.

The EPA estimates that a geothermal heat pump system can heat or cool a 190-square-meter (2,000-square-foot) house for as little as a dollar a day. The system usually pays for itself in about three to five years. And it offers the most energy-efficient, reliable, clean, and cost-effective way to heat or cool indoor spaces.

Engineers can also tap into deeper, more concentrated areas of geothermal energy called hydrothermal reservoirs. Wells are drilled down into the reservoirs to extract dry steam (with a low water content), wet steam (with a high water content), or hot water (Figure 13-22). The steam or water is then used for a variety of purposes. These include heating buildings, providing hot water, growing vegetables in greenhouses, raising fish in aquaculture ponds, and spinning turbines to produce electricity (Figure 13-23).

This source of energy is used in more than two dozen countries. The United States is the world's largest producer of electricity from hydrothermal reservoirs. Most of it is produced in California, Nevada, Utah, and Hawai'i. The Geysers, located north of San Francisco, is the world's largest hydrothermal reservoir. Thirteen power plants there generate enough electricity to service 725,000 homes.

Another source of geothermal energy is hot dry rock, which is found 5 kilometers (3 miles) or more underground almost everywhere on Earth. Water is injected through wells drilled into the hot dry rock.

FIGURE 13-23
The Nesjavellir power plant in Iceland uses geothermal energy to produce electricity and heat. The plant's wastewater also heats the waters of the nearby Blue Lagoon, a popular tourist destination.

FIGURE 13-24 ▼

Geothermal Energy	
Advantages	Disadvantages
Medium net energy and high efficiency at accessible sites	High cost except at concentrated and accessible sources
Lower CO_2 emissions than fossil fuels	Scarcity of suitable sites
Low cost at favorable sites	Noise and some CO_2 emissions

Some of the water absorbs the intense heat to become steam that is brought to the surface. The steam is used to spin turbines to generate electricity. According to the USGS, tapping just 2% of this resource in the United States could produce more than 2,000 times the amount of electricity currently used in the country. The limiting factor is cost, which could be brought down by more research and by improved technology. Some call for the use of geothermal fracking, based on the fracking technology used to produce oil and natural gas (Chapter 12). However, others worry earthquakes in the drilled areas could become more frequent. Figure 13-24 lists the advantages and disadvantages of geothermal energy.

checkpoint Why isn't geothermal energy more widely used?

Hydrogen Fuel

Some scientists say the fuel of the future is hydrogen gas (H_2). Most research has focused on developing hydrogen fuel cells. These devices combine H_2 and oxygen gas (O_2) to produce electricity and harmless water vapor, which is released into the atmosphere ($2 H_2 + O_2 \rightarrow 2 H_2O$ + energy).

Widespread use of hydrogen for powering motor vehicles, heating buildings, and producing electricity would eliminate most of the outdoor air pollution that comes from burning fossil fuels. Using hydrogen fuel would also greatly reduce the threats of climate change and ocean acidification. It does not lead to CO_2 emissions, as long as the H_2 is not produced with the help of fossil fuels or nuclear power. Hydrogen also provides more energy per gram than any other fuel, making it a lightweight fuel ideal for aircraft.

Turning hydrogen into a major fuel resource is challenging. First, there is hardly any hydrogen gas (H_2) in Earth's atmosphere. People can use several methods to produce H_2. They can heat water or pass electricity through water. They can strip it from the methane (CH_4) found in natural gas and gasoline molecules. Or they can carry out a chemical reaction involving coal, oxygen, and steam. However, hydrogen has a negative net energy. More high-quality energy is required to produce H_2 using these methods than is obtained by burning it.

Cost is another challenge. Fuel cells are the best way to use H_2 to produce electricity, but current versions of hydrogen fuel cells are expensive. Progress in the development of nanotechnology could bring down the price of hydrogen fuel cells.

Production methods for H_2 fuel can also be problematic. Electricity from coal-burning and nuclear power plants can be used to decompose water into H_2 and O_2. But using coal and the nuclear fuel cycle is harmful to the environment. Making H_2 from coal or stripping it from methane or gasoline also adds much more CO_2 to the atmosphere per unit of heat generated than does burning the coal or methane directly. Finally, hydrogen gas is flammable so safety is a concern.

For now, hydrogen's negative net energy is a serious limitation. Hydrogen fuel will have to be subsidized for it to compete in the open marketplace. There is promising new technology, however. The "artificial leaf" is a credit card-sized silicon wafer that produces H_2 and O_2 when placed in a glass of tap water and exposed to sunlight. An electrochemical cell is also being developed that mimics photosynthesis. The cell uses the sun's energy to split water molecules and to produce H_2 fuel.

Scaling up either of these processes over the next several decades could produce large amounts of H_2 at an affordable price with an acceptable net energy. That could help implement the solar energy factor of sustainability on a global scale.

FIGURE 13-25 ▼

Hydrogen Fuel	
Advantages	Disadvantages
Can be produced from plentiful water at some sites	Negative net energy
No CO_2 emissions if produced with use of renewables	CO_2 emissions if produced from carbon-containing compounds
Good substitute for oil	Need for subsidies due to negative net energy
High efficiency in fuel cells	Need for H_2 storage and distribution system

Figure 13-25 lists the advantages and disadvantages of hydrogen fuel.

checkpoint How is hydrogen obtained for fuel cells?

13.2 Assessment

1. **Compare and Contrast** How are passive and active solar systems alike and different?
2. **Explain** How can landowners benefit from allowing wind turbines to be placed on their property?
3. **Analyze** In what ways are subsidies for renewable energy resources helpful and harmful?
4. **Evaluate** Of all the types of renewable energy described in this lesson, which do you think holds the most promise? Consider social, economic, and environmental factors in your decision.

SCIENCE AND ENGINEERING PRACTICES

5. **Engaging in Argument** Write a letter to your congressional representative urging the government to step up its support for renewable energy resources. Argue for the benefits of renewable energy and suggest measures the government can take to promote them. Use details from the text and your own research to support your argument.

13.3 How Can Society Transition to a More Sustainable Energy Future?

CORE IDEAS AND SKILLS
- Identify the challenges associated with transitioning to a more sustainable energy future.
- Describe paths society can take to transition to more sustainable energy use.

KEY TERMS
decarbonization   district heating

A More Sustainable Energy Future Is Possible

Society faces a number of challenges when it comes to transitioning to a more sustainable energy future. One such challenge is the fact that fossil fuels and nuclear power are artificially cheap, mainly as a result of government subsidies and tax breaks. In addition, the true cost of these energy resources to the environment and human health is not calculated into their market prices. In 2020, governments around the world spent approximately $444 billion on subsidies for the fossil fuels industry compared to $165 billion on subsidies for renewable energy. This distortion of the energy marketplace violates the full-cost pricing factor of sustainability.

Another challenge is the lack of long-lasting government tax breaks, rebates, low-interest loans, and other economic incentives for investing in energy-efficient fuels and technology. There is also a lack of commitment by some governments to educate the public about the environmental and money saving advantages of improving energy efficiency and switching to renewable resources.

Choosing Energy Paths In considering the transition to renewable energy, scientists have reached several general conclusions. Experts predict that during this century there will likely be a gradual shift in dependence from nonrenewable fossil fuels to a mix of renewable energy from the sun, wind, flowing water (hydropower), and Earth's heat (geothermal energy). The use of renewable resources will enable many people to depend less on large centralized power systems and more on decentralized systems. These systems include rooftop water heaters, single wind turbines, and solar cell panels.

> **CONSIDER THIS**
>
> **You can take steps toward energy efficiency.**
>
> Try these tips for saving energy. Get your family and friends involved, if possible.
> - Walk, bike, or use mass transit or a carpool to get to school or other activities.
> - Conduct a home energy audit and discuss ways to make your home energy efficient.
> - Open windows and use fans to keep rooms cool instead of using air conditioning.
> - Use a programmable thermostat and energy-efficient heating and cooling systems, lights, and appliances.
> - Check to see that your home's water heater thermostat is set to 43–49°C (110–120°F).
> - Turn off lights, TVs, computers, and other electronics when they are not in use.
> - Wash laundry in cold water and let it air-dry.

FIGURE 13-26
Reykjavik is Iceland's largest city. Even so, only about 130,000 people live there. Why has the city had such success with renewable energy resources?

ENGINEERING FOCUS 13.1

GREEN CITY

Iceland's capital city Reykjavik has the distinction of being the world's northernmost national capital (Figure 13-26). Reykjavik also has the distinction of being one of the world's cleanest cities, thanks to virtual independence from fossil fuels for energy. The city—indeed the whole nation of Iceland—is a leader in renewable energy technology.

Iceland's geography and geology have been key to its success with renewable energy. Glaciers cover about 10% of the island. As the glaciers slowly melt, they feed rivers across the nation. Hydroelectric dams use the energy of flowing river water to supply three-quarters of Iceland's electricity. Iceland is also home to 35 active volcanoes, which provide an abundant source of geothermal energy. Geothermal power plants extract this energy and use it to produce electricity and heat (Figure 13-23). About one-quarter of Iceland's electricity is supplied by geothermal energy. Almost 90% of the heating and hot water in Iceland is provided via this renewable energy resource as well.

Geothermal energy is used to heat all of Reykjavik. The city uses **district heating**, a system designed to distribute heat (and sometimes power) to multiple buildings from a central plant. Residents served by district heating don't have to maintain their own heating systems, like boilers and furnaces.

In a geothermal district heating system, hot water is distributed from the central plant through a set of main pipes. Connecting pipes that branch off the main carry the water to individual buildings. Return water is still warm enough to be useful. It can be directed through other pipes installed beneath sidewalks, parking spaces, and roads to melt snow. Cogeneration is being used at two geothermal energy plants to generate electricity for the city as well.

Renewable energy use has dramatically reduced CO_2 emissions and improved standards of living in Iceland. Still, the nation remains dependent on fossil fuels for transportation and its fishing fleet. Those two categories represent almost 90% of the total fuel use and about one-third of the country's total CO_2 emissions.

Iceland is hard at work investing in development of alternative fuels. It operates a biodiesel production facility and has been testing the use of electric cars as well as hydrogen-powered vehicles. Iceland has in fact made a commitment to be free of fossil fuel use by 2050, which makes it a model for sustainable energy use.

Thinking Critically
Infer As an island nation, why might Iceland be motivated to end its dependence on fossil fuel energy sources?

Experts also project there will be a shift from gasoline-powered motor vehicles to hybrid and plug-in electric cars and perhaps to all-electric cars, with major improvements in battery technology.

A combination of improved energy efficiency and carefully regulated use of natural gas may be the best way to make the transition to using renewable energy resources during this century. To be successful, regulation will need to include tighter controls on emissions of methane and other greenhouse gases throughout the entire natural gas production and distribution system.

Because fossil fuels are still abundant and artificially cheap, society will continue to use them in large quantities. This presents two major challenges. The first is to successfully reduce the harmful environmental and health impacts of widespread fossil fuel use, in particular outdoor air pollution and greenhouse gas emissions. The second is to successfully include more of the harmful environmental and health costs of using fossil fuels in their market prices and implement the full-cost pricing factor of sustainability.

One approach calls for **decarbonization**, or a global transition away from fossil fuels. The goal of decarbonization is to completely phase out fossil fuel use by the end of the century. In 2015, the world's top seven industrialized countries—the United States, Japan, Germany, United Kingdom, France, Italy, and Canada—met to discuss ways to reduce their dependence on fossil fuels. Combined, these countries' carbon dioxide emissions total 25% of the world's output. These nations' ongoing efforts toward decarbonization require investing in new low-carbon technology without disrupting the economy or the ability to meet global energy needs.

Figure 13-27 summarizes several strategies for making the transition to a more sustainable energy future over the next 50 years. This shift is gradually taking place.

checkpoint How do subsidies for fossil-fuel-based energy compare to subsidies for renewable energy?

FIGURE 13-27 ▼

Making the Transition to a More Sustainable Energy Future
Improve Energy Efficiency
Increase fuel-efficiency standards for vehicles, buildings, and appliances.
Provide large tax credits for buying energy-efficient cars, houses, and appliances.
Reward utilities for reducing demand for electricity.
Greatly increase research on energy efficiency and battery technology.
Expand Renewable Energy
Greatly increase the use of renewable energy.
Provide large subsidies and tax credits for the use of renewable energy.
Greatly increase renewable energy research and development.
Reduce Pollution and Health Risk
Phase out coal subsidies and tax breaks.
Levy taxes on coal and oil use.
Greatly increase research and development on reducing emissions of carbon dioxide and other pollutants.

13.3 Assessment

1. **Recall** What are the challenges of transitioning to a more sustainable energy future?
2. **Explain** What advances in technology will aid the transition to a more sustainable energy future?

CROSSCUTTING CONCEPTS

3. **Patterns** Use the internet to locate data about world energy consumption. The data should cover at least 25 years and should break down consumption by energy sources, such as coal, oil, gas, wind power, and solar energy. What patterns do you notice about the use of each energy resource? What does this suggest about challenges and opportunities for transitioning to a more sustainable energy future?

TYING IT ALL TOGETHER STEM
Wind Power and Sustainability

In this chapter's Case Study, you learned about the immense potential of wind energy—an indirect form of solar energy. You learned that wind power is an important renewable energy resource, particularly in the United States. This country is the world's largest producer of wind power. Many analysts argue that wind power offers more benefits and fewer drawbacks than any other renewable energy resource. They call for greater investment in building land-based and offshore wind farms.

If we make the shift toward wind power and other direct and indirect forms of renewable solar energy, we could implement the three scientific factors of sustainability during this century. We could limit our harmful environmental impact. Making this shift would also follow the three social science factors of sustainability, including acting responsibly for future generations.

Use Figure 13-28 to answer the questions that follow.

1. Find the state where you live on the map. What is the general wind energy potential where you live?
2. Consider the areas where wind energy potential is excellent, outstanding, or superb. What do you notice about the geography of these areas and how it might relate to wind energy potential?
3. Next, consider these same areas in terms of access: Would wind farms located in these areas be located near populated areas? What might be some challenges of delivering the electricity produced in these areas?
4. Texas is the leading wind power producer in the United States. What is this state's wind power potential? What can you conclude about why Texas has had such success with producing wind power?
5. Use the internet to locate a map that shows existing wind farms in the United States. Use the map to evaluate how well the United States is making use of its wind energy potential.

FIGURE 13-28 ▼
This map shows the potential supply of land- and ocean-based wind energy in the United States.

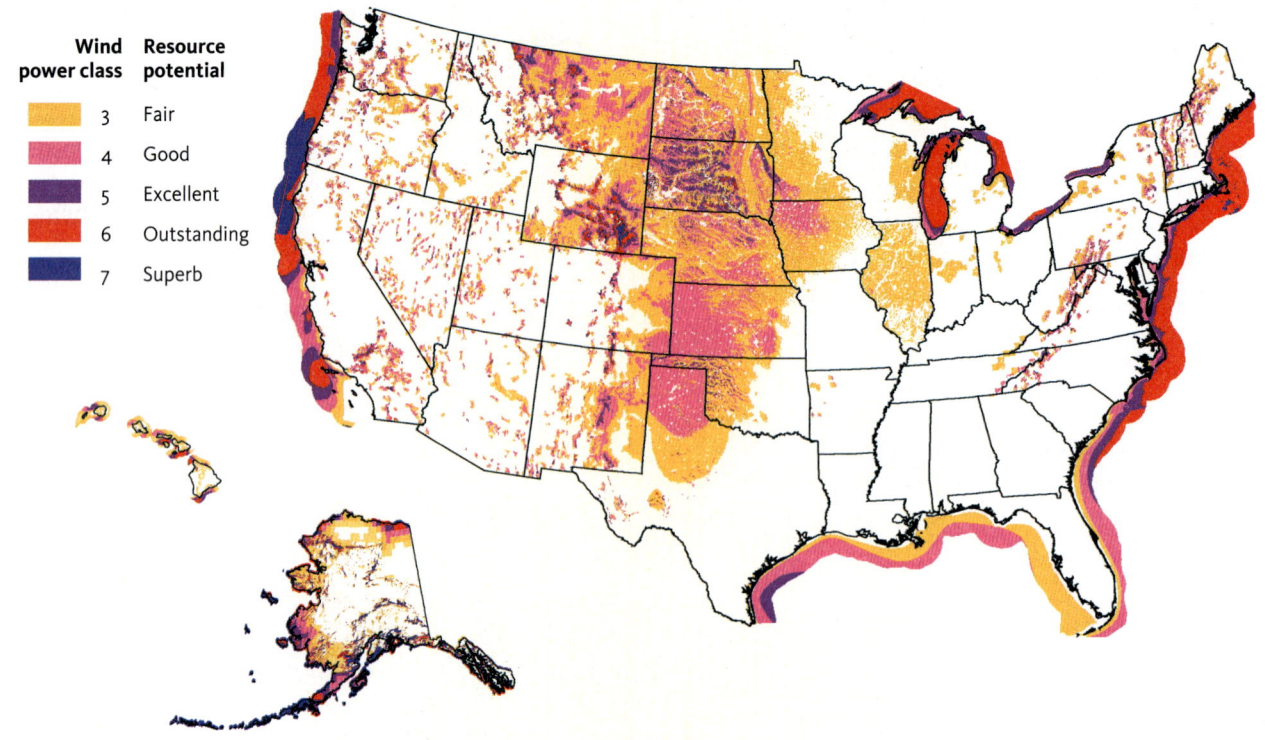

Wind power class	Resource potential
3	Fair
4	Good
5	Excellent
6	Outstanding
7	Superb

Sources: U.S. Geological Survey and U.S. Department of Energy

CHAPTER 13 SUMMARY

13.1 Why is energy efficiency an important energy resource?

- Energy efficiency is a measure of how much useful work can be obtained from each unit of energy used.
- One way to save energy is to use cogeneration, which involves producing two forms of useful energy from the same fuel source.
- Improving the transmission of energy through an updated electrical (smart) grid will save money, make delivery more efficient and reliable, better incorporate renewable energy resources, and provide more control to consumers.
- The use of energy-efficient vehicles and alternative modes of transportation will improve energy efficiency in the transportation sector.
- Green architecture incorporates a number of methods that improve energy efficiency, including using solar energy, adding green roofs, and using superinsulation techniques.
- More efficient maintenance of homes and businesses involves using energy-efficient windows and lighting, preventing air leaks, replacing heating and cooling equipment with energy-efficient appliances, using instant water heaters, and stopping the use of standby mode for electronic devices.

13.2 What are sources of renewable energy?

- Subsidies for nonrenewable energy such as fossil fuels give buyers a false impression of the true cost of energy.
- Renewable energy resources—solar, wind, hydropower, geothermal heat, biomass, biofuel, and hydrogen fuel—reduce dependence on fossil fuels and reduce harmful environmental impacts.
- Solar energy comes directly from the sun. Passive and active use of solar energy can heat buildings and water. Solar thermal systems can concentrate sunlight to produce high-temperature heat and electricity. Photovoltaic cells convert sunlight into electricity and are the world's fastest-growing technology for producing electricity.
- Hydropower is any technology that uses kinetic energy from flowing or falling water to produce electricity.
- Wind turbines capture energy from blowing winds over land or sea and convert it to electricity. Wind power is the second fastest-growing technology for producing electricity.
- Biomass can be burned for fuel. Biomass can also be converted to liquid biofuels such as ethanol and biodiesel.
- Geothermal energy is heat stored in soil, underground rocks, and fluids in Earth's mantle.
- Using hydrogen gas as a fuel is not yet a good alternative to fossil fuels because it has a negative net energy.

13.3 How can society transition to a more sustainable energy future?

- Making the transition to a more sustainable energy future will require greatly improving energy efficiency, using a mix of renewable energy resources, and including the environmental and health costs of energy in market prices.
- Decarbonization, or a shift away from fossil fuel use, is key to stabilizing Earth's climate.

CHAPTER 13 ASSESSMENT

Review Key Terms

Select the key term that best fits each definition. Not all terms will be used.

active solar heating system	district heating	hydropower
biofuel	energy efficiency	passive solar heating system
biomass	geothermal energy	photovoltaic cell
cogeneration	hydrogen fuel cell	solar thermal system
decarbonization		

1. System that absorbs and stores heat from the sun directly within a well-insulated structure
2. Thin wafer of purified silicon or polycrystalline silicon with trace amounts of metals that allow it to produce electricity
3. Device that uses hydrogen gas as a fuel to produce electricity when it reacts with oxygen gas in the atmosphere and emits harmless water vapor
4. Plant materials that can be burned as a solid fuel or converted into gaseous or liquid biofuels
5. Heat stored in soil, underground rocks, and fluids in Earth's mantle
6. Any technology that uses the kinetic energy of flowing or falling water to produce electricity
7. A measure of how much useful work each unit of input energy provides
8. Uses different methods to collect and concentrate solar energy in order to boil water and produce steam for generating electricity
9. Captures energy from the sun by pumping a heat-absorbing fluid such as water or an antifreeze solution through special collectors, usually mounted on a roof or on special racks that face the sun
10. System used to produce two useful forms of energy from the same fuel source
11. Involves a global transition away from dependence on fossil fuels

Review Key Concepts

12. What makes one appliance more efficient than another model of the same appliance?
13. Give an example of cogeneration and explain how it reduces fossil fuel use.
14. What is a smart grid and how is it more efficient than existing grids?
15. What are the reasons why energy is unnecessarily wasted in industrial processes?
16. List three ways to save energy and money in transportation.
17. List four ways to save energy and money in new buildings and in existing buildings.
18. List four reasons why renewable energy is not more widely used.
19. What are the major advantages and disadvantages of using hydropower to produce electricity?
20. What are the major advantages and disadvantages of using geothermal energy as a source of heat and to produce electricity?

Think Critically

21. Congratulations! You are in charge of the world. List the five most important features of your energy policy and explain why each of them is important and how they relate to each other. Use references from the chapter to justify each feature of your plan.
22. List five ways in which you waste energy during a typical day. Explain how these actions violate each of the three scientific factors of sustainability.
23. How can renewable energy resources better serve people in less-developed countries?
24. Suppose a developer has proposed building a wind farm near where you live. Write a letter to your local newspaper or an online blog explaining your position for or against the project. Include the concept of net energy in your argument. Research how the electricity you use now is generated and where the power plant is located. Include this information in your argument.

Chapter Activities

A. Develop Models: Wind Turbines STEM

In this chapter, you learned about using wind turbines to produce electricity. The basic components of a wind turbine are the blades, the gearbox, the electric generator, and the power cable (Figure 13-1). You will work with a team to build a model wind turbine. You will test how much electricity the model wind turbine produces as measured by a voltmeter.

Materials

1.5-V DC motor	wire stripper	craft stick
ruler	scissors	voltmeter
rubber band	tape	safety glasses
thin electric wire	alligator clips	small fan or hair dryer
	straight pins	
	paper plate	

1. Use a rubber band to attach the small motor to one end of the ruler, so its spindle side faces out.
2. Cut 2 pieces of electric wire 30 cm (12 in.) long. Strip the insulation off the ends of both wires using scissors or a wire stripper.
3. Connect the wires to the motor by twisting one end of each wire around each of the motor's terminals.
4. Attach an alligator clip to the free end of each wire. You will use the clips to connect to the voltmeter.
5. Cut a paper plate to make 6 blades for your model turbine. Each blade should measure about 1 cm x 6 cm (about ¼ in. x 2 in.). You can trace the shape of a craft stick and cut it in half to make a set of blades that are the right size.
6. Push 6 straight pins into one end of the cork, placing them at evenly spaced intervals around its circumference. Push the other end of the cork onto the motor's spindle. Be sure the cork is centered and spins freely. The spindle should also turn.
7. Attach the blades. Tape the head of each pin to the bottom edge of each blade, leaving the same amount of space between each blade and the cork. Tilt each blade so all the blades are oriented the same way.
8. Test your turbine. Your teacher will show you how to attach and use the voltmeter. Turn on your air source. Adjust the blade angles if needed to get the turbine to spin. Record your results.
9. Choose a variable to test, such as blade size, shape, or number. Repeat the test using your new design. Record your results.
10. Which design produced better results? What might be the reason for this? How did the parts of your model correspond to the components of an actual wind turbine? Draw a diagram of your model and label these corresponding parts.

B. Take Action

National Geographic Learning Framework

Attitudes | Responsibility
Skills | Problem Solving
Knowledge | Our Living Planet

Find ways to improve energy efficiency and save energy at your school. Consider the following areas:

- Lighting: Are lights turned off in rooms when they are not in use? How many lights are left on overnight? Can natural light be used instead of artificial lights? What percentage of lighting is provided by energy-efficient bulbs?
- Heating and cooling: What are thermostats set at in the daytime and overnight? Can the heating or cooling in hallways be turned down? Are doors in unused classrooms kept closed? Are there any drafty areas? Are any vents blocked by furniture?
- Computers: Are computers set to go to sleep when not in use? Is computer equipment turned off overnight and on weekends? Is computer equipment Energy Star rated?

Write a proposal for improving your school's energy efficiency and submit it to school officials.

STEM ENGINEERING PROJECT 4
DESIGN A WIND-POWERED GENERATOR

Many people worldwide want to cut down on their need for "grid" electricity. You may be surprised to learn that there's still plenty of room for improvement in today's machines and electrical devices. Here's why: Untapped kinetic energy is all around you. Whether in the form of a person walking, a machine vibrating, water flowing, or wind blowing, kinetic energy can be "harvested" and converted into electricity. Machines can be redesigned or modified to capture energy from these "micro-sources" and transfer that energy back into their electrical supply.

For example, suppose every washing machine and dryer had small kinetic-energy generators that converted energy from the appliance's vibrations into electrical energy. These appliances could require less electricity than even the most energy-efficient models currently on the market. Or suppose every coffee maker and rice cooker had batteries that recharged by harvesting energy from rising steam. Once you start looking for opportunities, you will begin seeing the possibilities of untapped energy all around you.

In this challenge, your team will use the practices of engineering to design a solution to a real problem. You will design, build, and test a small generator that captures energy from wind, steam, or flowing water and converts it into electricity.

Engineering DESIGN CYCLE
- Defining problems
- Developing and using models
- Planning and carrying out investigations
- Analyzing data and using math
- Designing solutions
- Forming arguments from evidence
- Obtaining, evaluating, and communicating information

Defining Problems

1. Define the problem.
   - What are the problems with relying on large-scale sources of electricity?
2. Identify a need that underlies the problem.
   - What need can small-scale energy generation help to fill?
3. A successful small-scale generator must meet the following criteria. What else can you add to this list?
   - The generator must be powered by kinetic energy.
   - The generator must produce at least 4 volts and at least 1 ampere of current for 5 minutes. This is enough electricity to give a boost charge to a standard lithium-ion cell phone battery.
   - The generator must be safe and easy to use.
4. What constraints will your team be working within? Constraints can include materials, time, and budget.

Developing and Using Models

5. Locate a reliable source of wind, steam, or running water to base your design on. Be careful to avoid placing your hand near steam, as it can cause burns. You can use a pinwheel to help you find a steady wind source.
6. Review the materials presented by your teacher. As a team, select a motor and blades.
7. Connect your motor to a battery and make note of the direction and speed in which the blades rotate. Your generator will need to move the blades in the opposite direction.
8. Brainstorm ways to modify the blades. Weigh the pros and cons of each material. Apply what you discovered about blade design in the Chapter 13 activity, Develop Models: Wind Turbines.
9. Draw a labeled diagram of your team's design. Keep in mind that it will need to meet the criteria listed in step 3.

Planning and Carrying Out Investigations

10. Plan a test of your design with your team. Write out your plan in steps.
11. Gather your materials and build your wind-powered generator.
12. Conduct your test. Use the multimeter to measure both voltage and current. Record your data.
13. Modify your design and test it again. Record your modifications and your new data.

Analyzing Data and Using Math

14. Analyze your data. Did your modification improve your design? Explain your findings.
15. What limitations exist in your data?
16. How well does your solution meet each of the criteria you listed in step 3?

Designing Solutions

17. What were the strengths and weaknesses of your design?
18. Make further improvements to optimize your design. Revise your description and diagram accordingly.

Forming Arguments from Evidence

19. Explain how you improved your wind-powered generator.
20. What evidence justified the change you made?
21. Make a claim about the effectiveness of your generator. Use evidence to support your claim.

Obtaining, Evaluating, and Communicating Information

22. Present your team's wind-powered generator to the class. Describe how your design works and how it was optimized to meet the criteria.
23. Provide respectful and specific feedback to other teams.
24. Conduct follow-up research. Apply your design to solve a real problem, such as providing a boost charge to a cell phone or powering a camp light. Have your teacher approve your plan before you carry it out. Do not open any electrical devices without the help of an adult.
25. Write a final report. Include an explanation of how kinetic energy is transformed into electrical energy in your design.

NATIONAL GEOGRAPHIC

Citizen Science

Of the People, By the People,
For the People

"Even [Charles] Darwin was a citizen scientist…"

—Mary Ellen Hannibal, author and environmental journalist

What do bird watching, online gaming, stargazing, and monitoring a rain gauge have in common? They are all examples of ways that ordinary people can participate in scientific research and advance scientific knowledge. In September 2015, the White House sponsored a live webcast called "Open Science and Innovation: Of the People, By the People, For the People." The event brought together researchers, government officials, and members of the public to acknowledge and celebrate citizen science and crowdsourcing efforts.

Citizen science is the collection and analysis of data by members of the general public in collaboration with professional scientists and scientific institutions. When scientists and non-scientists work together, science advances more quickly. Citizen science in the digital age uses "crowdsourcing," a method of gathering information and ideas by enlisting the participation of a large number of people online.

The term "citizen science" is relatively new, but the basic concept is not. One of the oldest citizen science projects is the Christmas Bird Count (CBC) sponsored by the National Audubon Society. Inaugurated in 1900, the count runs annually from December 14 through January 5. Birding circles across the United States and Canada, led by experienced members, collect information about local bird populations. More than 2,000 birding circles participate. The data they collect help biologists and other scientists monitor bird counts and inform conservation efforts for bird populations that may be in trouble.

Anyone can be a citizen scientist. Formal training in the sciences to become one isn't necessary. The advent of smartphones with built-in GPS and cameras has given people more accurate tools for collecting field data. But all you really need is an interest in a project and the desire to follow through with data collection. If the project happens to be ecological, then your participation is another way to learn about, protect, and preserve the environment.

The CBC is the world's longest-running bird-oriented citizen science project. The data collected in its more than 115-year history are invaluable.

Luckily for aspiring citizen scientists, there is no shortage of citizen science projects in which to participate, and new ones are in development all the time. Some projects seek to collect data on a specific species or habitat. Others gather information about the human footprint on the environment, or about human behavior itself. The range of topics for citizen science projects is vast and inclusive.

A NEW COCKROACH?

The National Cockroach Project (NCP) enlists high schoolers and other citizen scientists across the United States to locate and collect dead cockroaches and ship them to the project, where they are assigned a DNA barcode and cataloged. DNA barcoding is a process of identifying species by isolating a single gene of an organism. By participating in the project, citizen scientists are helping to populate DNA databases, such as Barcode of Life Database and GenBank, that are made available to researchers interested in comparing and identifying species. In addition, researchers collecting the cockroach data sent to the NCP are asking three questions.

- Do American cockroaches differ genetically between cities?
- Do U.S. genetic types match those in other parts of the world?
- Are there genetic types that represent undiscovered look-alike species?

DNA barcoding and analysis of cockroaches submitted by citizen scientists to the NCP demonstrated evidence of 95 species of American cockroaches. It also potentially uncovered a genetically distinct cockroach that may represent a new subspecies of everyone's favorite insect.

Despite what its name suggests, the American cockroach is not native to North America. It was most likely brought to the New World on ships from Africa in the early 17th century.

ANALYZING CYCLONE DATA

Whether they are called cyclones, typhoons, or hurricanes, powerful tropical storms can have potentially devastating consequences. Moreover, the number of hurricanes in the Atlantic and northern Pacific has increased dramatically since the early 1970s. Because of the havoc hurricanes can wreak, meteorologists and other scientists are trying to better gauge the path and strength of these storms. One way to do that is to analyze historical satellite images, which is just what the Cyclone Center project is doing.

Tropical cyclones take shape over open ocean, and most of the historic imagery is satellite data, as opposed to aircraft reconnaissance data, which are more reliable. So meteorologists need help in analyzing long-term trends in cyclone behavior. That is where citizen scientists come in. The Cyclone Center has undertaken the analysis of 32 years' worth of tropical cyclone imagery from 1978 to 2009 in the North Atlantic Ocean. That is about 300,000 images that would take one person 12 years to analyze—and that's without taking a vacation! Not only are there hundreds of thousands of images to analyze, but the scientists at Cyclone Center know that the human eye is much better at finding information in the images than the best computers. For that reason, Cyclone Center sought the help of members of the public.

Cyclone Center is a partnership of scientists from the National Oceanic and Atmospheric Administration's (NOAA's) National Centers for Environmental Information and other groups, including the Citizen's Science Alliance. Cyclone Center is hosted on the Zooniverse website. Once participants enter the Cyclone Center site, they are presented with questions about weather-related images. Participants are asked to match cloud patterns with cyclone types, which images look like stronger storms, where the center of a storm lies, and what colors are involved in the storm centers.

As of July 2015, nearly 20,000 images had been analyzed by 9,800 participants. The scientists in charge of the project have been pleased—and impressed—with the accuracy of the citizen scientists' analyses. Ultimately, the goal of this project is to collect information that will help meteorologists and other weather scientists better predict and track violent storms and reduce devastation to coastal communities.

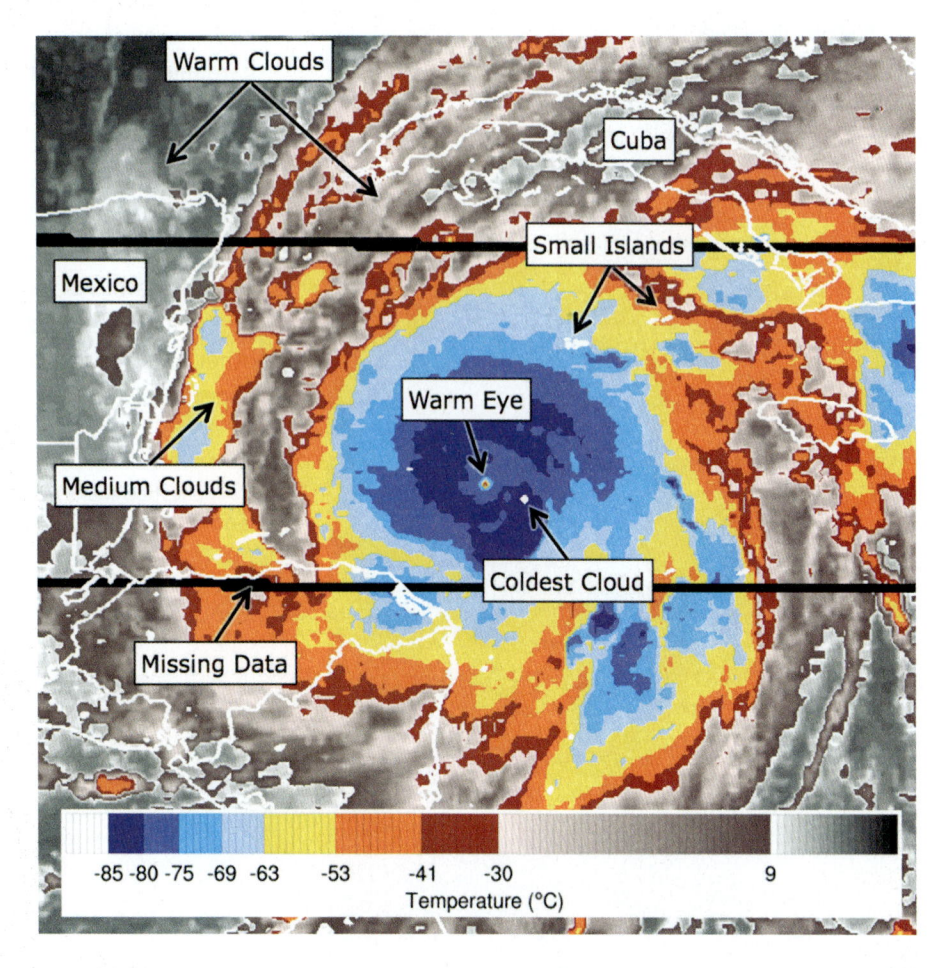

> Are you good at picking out visual details and seeing patterns? Consider helping weather scientists by participating in Cyclone Center's massive storm-cataloging project.

WHAT ALL THE BUZZ IS ABOUT

If you haven't heard, then you aren't paying attention: Honeybees are in trouble. A puzzling phenomenon called Colony Collapse Disorder, or CCD, has led to dramatic declines in honeybee populations across the United States and worldwide. (See the Case Study in Chapter 7.) The main characteristic of CCD is complete abandonment of the hive. The worker bees simply fly away and do not return, leaving behind the queen and young bees. Scientists are unsure what leads to CCD but think it may result from multiple causes, including pathogens, exposure to pesticides, and habitat loss. One thing is for certain. Without honeybees, healthy foods such as cashews, almonds, melons, pumpkins, pears, peaches, avocados, raspberries, apples, and many other kinds of fruits and nuts could not be farmed commercially. They would likely disappear from grocery store shelves forever.

Citizen scientists can help in efforts to understand and stop CCD by participating in programs like the Great Sunflower Project. The Great Sunflower Project offers three ways to gather data and evaluate and improve honeybee habitats. For those who like gardening, the Safe Gardens for Pollinators program encourages participants to plant Lemon Queen sunflowers—a confirmed honeybee favorite—to monitor the effects of pesticides on pollinators. The Pollinator Friendly Plants program helps identify specific plants that support pollinator communities. The Pollinator Habitat Challenge gives citizen scientists the chance to improve gardens, parks, and other green spaces in their communities in ways that benefit pollinators.

A honeybee seeks nectar from a flower on an almond tree.

These neurons were mapped by gamers playing EyeWire.

MAPPING THE MOUSE BRAIN

Citizen scientists collect data on living organisms, weather phenomena, and even the brain. Although people have studied it for hundreds of years, the human brain largely remains a mystery. It holds 80 billion neurons connected by 100 trillion synapses—all of which are invisible to the naked eye. How do scientists even begin to study it?

They start small—by mapping the mouse brain. The innovators at Princeton University created an online social game called EyeWire that helps neuroscientists interpret brain scans to better understand how neurons and neural circuits behave. The game is free for the general public and includes video tutorials to coach players.

The objective of EyeWire is to map out how neurons connect by clicking through a razor-thin cross-section of a mouse retina only 20 nanometers thick. Once in the game, players can manipulate the 3-D, puzzlelike image to guess which way the neurons move and connect. Gamers, students, teachers, and scientists work their way through 256 cross-sections.

By clicking in strategic patterns, players determine the continuous path and branches of the cells, called dendrites, moving through the neuron. Once one cross-section is finished, the collaborative results from a number of players help determine which result is the most accurate.

The citizen scientists who have participated in EyeWire have collectively helped build a valuable database for understanding neuron behavior that will aid in mapping the human brain too. More than 250,000 people from around the world have played the game since it launched in 2012. Citizen scientists are able to interpret images in ways that computers cannot. Simply put, your human brain can identify patterns and cell types much better than any software program. The developers of EyeWire knew of other crowdsourced projects and the value that ordinary people can contribute to scientific inquiry and so purposely created a game that would engage players. Little did they know how popular EyeWire would become.

CITIZEN SCIENCE: OF THE PEOPLE, BY THE PEOPLE, FOR THE PEOPLE

CITIZEN SCIENTISTS IN THE NATIONAL PARKS

For citizen scientists who prefer getting out into nature, the National Parks Service Research Learning Centers (RLCs) are a perfect fit. What better place to study biodiversity, collect data, and learn about the world around us than in the many national parks across the United States?

The RLCs in the national parks offer citizen scientists many opportunities to participate in data collection for specific studies. National Park RLCs partner with schools and other organizations to offer field trips, camps, and workshops. In return, citizen scientists provide valuable information for various studies that helps park rangers plan and manage resources and devise solutions to problems facing the parks and the environment in general.

RLCs in national parks across the country sometimes coordinate initiatives. Starting in 2011, RLCs such as the Continental Divide Research Learning Center in Colorado's Rocky Mountain National Park engaged citizens for help with sampling mercury levels in the lakes, streams, and wetlands. Mercury deposited by polluted air from coal-burning power plants is toxic to humans and wildlife, and mercury contamination can cause reproductive and neurological problems.

Dragonflies are a sentinel species, which means they are a good indicator of mercury levels. Their larvae live up to 5 years underwater, and once they mature into adults, dragonflies occupy a high-middle part of the food chain. They eat many kinds of smaller insects, and are in turn eaten by fish that are eaten by other animals. Gauging mercury levels in dragonflies helps scientists determine the health of food webs and ecosystems. In the more than 4 years the Dragonfly Mercury Project has been underway, more than 800 citizen scientists in dozens of national parks have collected and identified dragonfly larvae and sent them to research labs for analysis.

BIOBLITZES

The first BioBlitz was held by the U.S. National Park Service in 1996 in Washington, D.C. It was such a huge success that other organizations around the world quickly adopted the concept. A BioBlitz is a whirlwind inventory of the organisms living within a predefined area. BioBlitzes usually take place in a 24- or 48-hour period. People of all ages and backgrounds descend to observe, count, and collect as much data as they can about the plants and animals living there. Scientists use the data to create a snapshot of the biodiversity in the host national park or other natural area. BioBlitzes offer scientists and the public an opportunity to do fieldwork side-by-side. The data collected by participants often add to the park's official species list. BioBlitz events highlight the importance of protecting the biodiversity of extraordinary places.

Since 2006, the National Park Service and the National Geographic Society have worked together to host BioBlitzes. In celebration of the 100-year anniversary of service and conservation in the national parks, the 2016 BioBlitz, co-hosted by the National Park Service and the National Geographic Society, took place at national park sites in and around Washington, D.C. The capital celebration was the cornerstone of concurrent BioBlitzes and biodiversity events at more than 400 U.S. National Parks. The 2016 event also featured a Biodiversity Festival on the National Mall at Constitution Gardens.

> Left and below: Students, park rangers, teachers, and naturalists participate in a BioBlitz at Rocky Mountain National Park in Colorado. The word "BioBlitz" (also spelled "bioblitz") was coined by U.S. National Park Service naturalist Susan Rudy in 1996. Since then, BioBlitzes have sprung up around the world, from Taiwan to Portugal.

CONDUCT YOUR OWN BIOBLITZ

Have you ever wondered what lives in the environment where you live? Perhaps there is a park or open space near your school, or a lake, woodlands, or riverbank near your home that you'd like to explore. You don't have to wait for a nationally sponsored event to find out. Capitalize on your curiosity. Use the following steps to guide you in your own BioBlitz.

Students in Wales conduct an inventory of plant and animal species living within a quadrat, or one square meter of habitat.

1 Review the BioBlitz concept.

You may want to watch a video of a National Geographic BioBlitz from years past. Remember that a BioBlitz is an event where teams of citizen scientists help identify as many species as possible in a natural area. As you watch the video and review concepts of biodiversity and BioBlitz, consider the following question: In what ways would taking an inventory of all of the species in a natural area be useful?

2 Determine areas with potential for biodiversity.

Consider places in your community or near your school where you might look for biodiversity. Where have you seen a variety of plants or animals? What habitats and what conditions enable animals and plants to survive?

3 Use an interactive map tool online to explore the study area.

Find and create a map of the selected area where you will conduct your BioBlitz. Employ different layers in the map tool to determine important features of your area. Print the map and label it. While making your map, consider the following questions: What physical features can you identify? In what areas do you expect to find a variety of species? What developed areas might affect the biodiversity you will inventory as part of the BioBlitz?

4 Plan your fieldwork before you start.

Make plans about how to work efficiently with the time you'll have to conduct your BioBlitz. Work in groups of two or three. Mark maps with where you will likely be. Bring a notebook and pencil. Each group should have a length of rope or tape to mark your study area. Take a hand lens and a digital camera.

5 Conduct your BioBlitz.

Begin by spending 5 minutes sitting silently and observing your surroundings. In your notebook, draw or describe in words any living things you see, hear, or smell. If you notice any animals, record notes and take a photograph if possible. Before or after your silent observation, choose an area to study and place your rope or tape to mark it. Do not take any species from the study site, and be sure to leave the site as you found it.

As you conduct your BioBlitz, mark your findings on a map of the study area and also record as much information as possible.

6 Identify species.

When you are finished with your inventory, return to your classroom. Consult expert resources, such as field guides, to identify organisms you observed and add more information to your notes. Your goal is to create an inventory representing the diversity of the area you studied.

7 Compile the results on a map and share your data.

Create a large map showing the distribution of species within your study area.

8 Present your findings.

Discuss biodiversity within and among the areas you and your classmates inventoried. Consider the following questions as you present and discuss your findings:

How many species were found?

What species were found?

In what types of habitats were species found?

What species were found near one another?

What abiotic factors may have had an effect on species found?

How could the group's research methods have impacted the species found?

What would you do differently if you were to conduct another BioBlitz?

Discuss any challenges you encountered—such as difficulty sampling very small or flying organisms, or poor weather conditions—and the possible effects they may have had on your results.

An owl butterfly
(Caligo memnon)

UNIT 5
Environmental Concerns

People wait in line for COVID-19 vaccines at the Taipei Main Rail Station in Taiwan.

As society becomes increasingly globalized and the human population continues to grow, our dependence on one another becomes ever more apparent. The actions of people on one side of the planet can have devastating consequences to populations on the other side of the world. The more our lives become entangled on a global scale, hazards that may have once affected a relatively small percentage of humans are more likely to spread to a far greater portion of the population.

In 2020, we observed this phenomenon as the COVID-19 virus became a global pandemic. Solutions to the coronavirus required the consideration of multinational efforts, such as the distribution of vaccines and restrictions to cross-border travel. Attempts to contain the virus's spread, such as masking mandates and school and business closings, became political flashpoints across the globe.

In this unit, you will consider various risks associated with human activity. You will evaluate different points of view for how we approach the world, our place in it, and our responsibilities to one another as global citizens.

CHAPTER 14
HUMAN POPULATION AND URBANIZATION

CHAPTER 15
ENVIRONMENTAL HAZARDS AND HUMAN HEALTH

CHAPTER 16
AIR POLLUTION, CLIMATE CHANGE, AND OZONE DEPLETION

CHAPTER 17
SOLID AND HAZARDOUS WASTE

CHAPTER 18
ENVIRONMENTAL ECONOMICS, POLITICS, AND WORLDVIEWS

ENGINEERING PROJECT 5: Design a Carbon-Capturing Device

CHAPTER 14
HUMAN POPULATION AND URBANIZATION

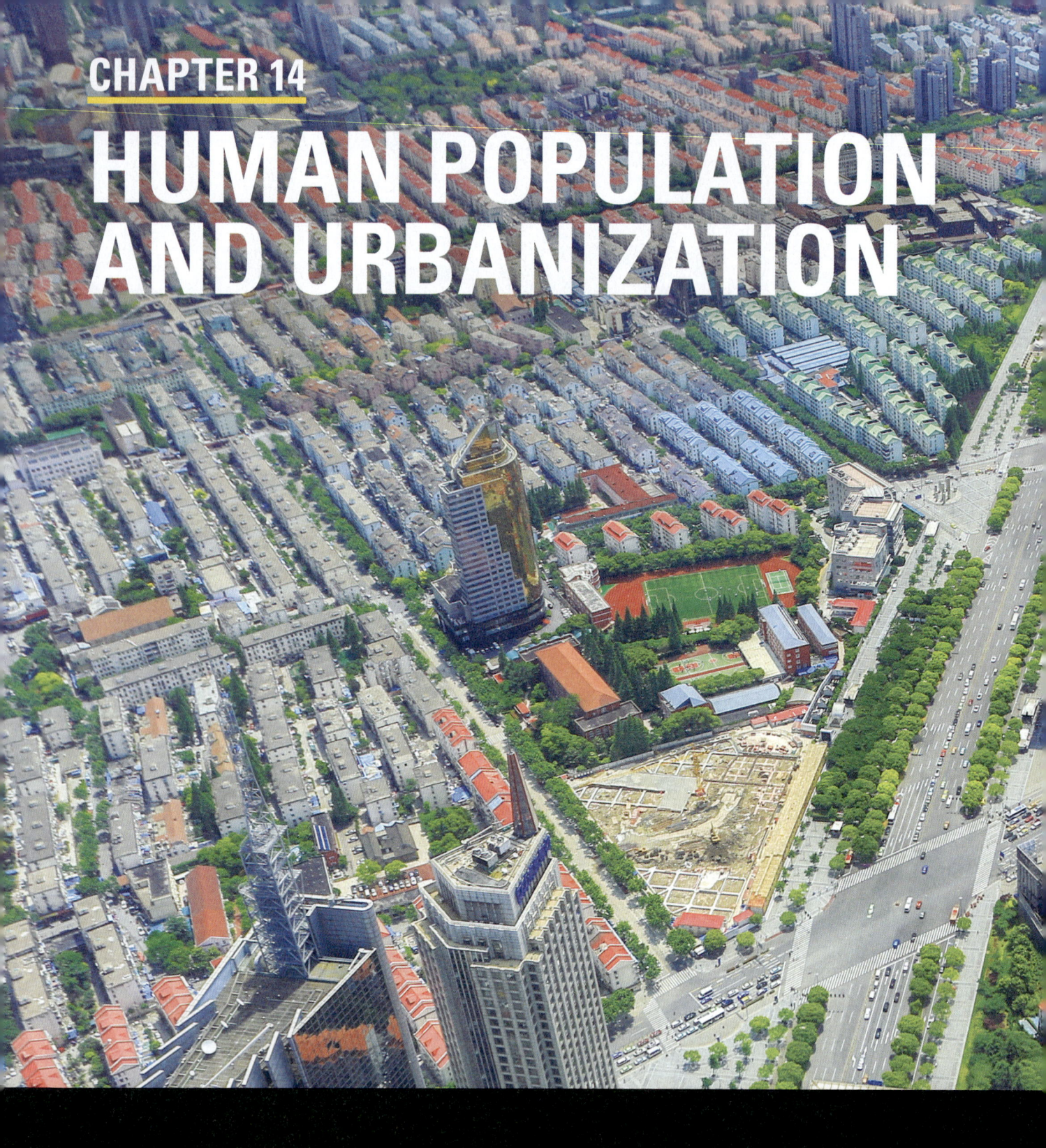

WITH A POPULATION of over 26 million people, Shanghai is the largest city in China. Like many other cities around the world, Shanghai has undergone rapid urban growth in recent decades. By 2030, two-thirds of all people are expected to live in urban areas. As cities swell, humanity is faced with the question: How can cities grow sustainably?

KEY QUESTIONS

14.1 How many people can Earth support?

14.2 What factors influence the size of the human population?

14.3 What are the effects of urbanization on the environment?

14.4 How can cities become more sustainable?

NATIONAL GEOGRAPHIC | EXPLORERS AT WORK

Healthy Planet, Healthy People

with National Geographic Explorer Christopher Golden

The course of Christopher Golden's life was set when he was just a third-grade student writing a report for school. After flipping through an encyclopedia, he chose the ring-tailed lemur. Learning about the lemur led to learning about Madagascar, the animal's home. The young Christopher was hooked. Long after he turned in his assignment, he continued to read all he could about the lemur and its African island home. Then, at the age of 16, he spent a summer working on a research project in Madagascar with National Geographic Explorer Luke Dollar. That experience sealed the deal. Today, Dr. Golden spends half of every year researching and exploring in Madagascar.

Golden's time in the island's remote forested areas has enabled him to learn about the human population there. Golden discovered that hunters were killing off the wildlife at a rapid rate. Furthermore, the animals carry diseases that could infect the people who ate them. The answer seemed obvious: Protect the wildlife through conservation.

But the problem—and its solution—wasn't so simple. The hunters depended on the meat for nutrients that their diet didn't otherwise provide. Wildlife in Madagascar is limited. Once people began to arrive on the island's shores, they eradicated many of its native species. As the human population grew, the problem worsened. So Golden came up with the idea of replacing wildlife hunting with raising chickens. Golden also worked with other scientists to develop a vaccine to make the chickens healthier and more productive.

It is this approach that distinguishes Golden. Golden combines the methods of an epidemiologist with the environmental concern of an ecologist. Epidemiologists study empirical (firsthand) data to find the causes and effects of disease outbreaks in populations. Golden uses empirical data and models to link environmental problems, such as wildlife depletion and climate change, to human health. This leads to action and policy change.

But Golden knows he can't do it alone. As a result, he created a research organization called MAHERY, or Madagascar Health and Environmental Research, in 2004. (*Mahery* means "powerful" in Malagasy.) The organization trains local university students in fields including animal biology and economics.

Recently, Golden has expanded his efforts beyond Madagascar and become director of the HEAL (Health and Ecosystems: Analysis of Linkages) program. This global program conducts research on the connection between environmental changes and the health of human populations around the world.

Golden embodies the attitudes of responsibility, problem-solving, and creativity. He says, "I think that connecting with natural environments is so important. I do my best thinking, innovation, and creation while lost in the woods." What about you?

Thinking Critically
Draw Conclusions What makes Golden's approach to protecting the environment unique?

Christopher Golden teaches a community of people in Madagascar how to prepare chickens for cooking. Raising chickens offers an alternative to hunting native species.

CASE STUDY
Population 7.9 Billion

It took roughly 200,000 years for the number of people on Earth to reach 2 billion. That happened in the 1920s. It took less than 50 years to add the second 2 billion and only 25 years to add the third 2 billion. In 2022, more than 7.9 billion people inhabited Earth. Experts' projections of the future population vary from 8 to 16 billion by 2100. (See the growth curve of the human population, Figure 1-13.)

Why does it matter that there are now almost three times as many people than there were in 1950? Some say it doesn't. They contend that humans will continue to develop new technologies, effectively pushing Earth's carrying capacity for humans higher and higher. However, many scientists argue that an ever-growing human population is unsustainable. They point out that people's ecological footprints are also expanding, degrading the natural capital that keeps them alive and supports their lifestyles.

A team of internationally renowned scientists has estimated that humans have likely exceeded the global boundary limits of four of Earth's major life-sustaining systems and are close to the boundary limits for five other systems (Science Focus 3.4).

Three major factors account for the rapid rise in the human population. First, the emergence of early and modern agriculture about 10,000 years ago gave people more food with each passing century. Second, people have developed technologies that have enabled them to expand into almost every climate zone and habitat on the planet. Third, death rates overall have dropped sharply because of improved sanitation and health care and the development of antibiotics and vaccines to help control infectious diseases.

The key questions are: How long can the human population keep growing? If the population stops growing, how long can the planet continue to support over 7 billion people? What can people do to prevent a potential population crash and improve the quality of life for the existing population?

In this chapter, you will learn more about patterns of human population change and the environmental impacts of a growing population. You will learn about the factors that influence population growth and consider whether and how people should try to slow it. You will also examine ways to increase the sustainability of cities, where most people live.

As You Read Think about the city, town, or suburb where you live. How could your city, town, or suburb become more sustainable?

FIGURE 14-1 A crowd of people waits at a bus station in China. Earth has never supported such a large number of people before.

14.1 How Many People Can Earth Support?

CORE IDEAS AND SKILLS
- Identify trends in human population growth.
- Calculate population change.

KEY TERMS

fertility
mortality
population change

Recent Trends in the Human Population

Each year, the world's human population changes. It grows when there are more births than deaths and declines when there are more deaths than births. The human population has grown very rapidly in recent history (Figure 1-13). Population analysts recognize three important trends in the human population.

One overall trend is the slowing of the growth rate since 1960 (Figure 14-2). Although the human population is still growing, it is not growing as rapidly as it was before. The current population growth rate is about 1.1%. This may not seem like much, but consider the numbers involved. It added about 68 million people in 2021, which is an average of 187,000 people every day.

Another trend in human population growth is its uneven distribution. The populations of less-developed countries are growing 14 times faster than the populations of more-developed countries. Approximately 98% of human population growth occurs in the world's less-developed countries and only 2% occurs in more-developed countries. Between 2022 and 2050, billions of people are likely to be born, most of them in less-developed countries. Most of these countries are not equipped to handle the pressures of such rapid growth.

Finally, analysts note that people are moving from rural areas to urban areas in large numbers. Over half of the world's people now live in urban areas, and this trend is increasing. More people live in or around Tokyo, Japan—the world's most populous urban area—than in the entire country of Canada.

checkpoint Summarize three trends in human population growth.

Calculating Population Change

The basics of global population change are quite simple. If there are more births than deaths during a given period, the human population increases. If there are more deaths than births, the population decreases.

FIGURE 14-2 ▼
Human Population Growth Rate The global human population's annual growth rate is slowing, as shown in this graph. The projected growth rate is also shown. How is the human population expected to change by 2050? (Hint: This graph shows rate of change.)

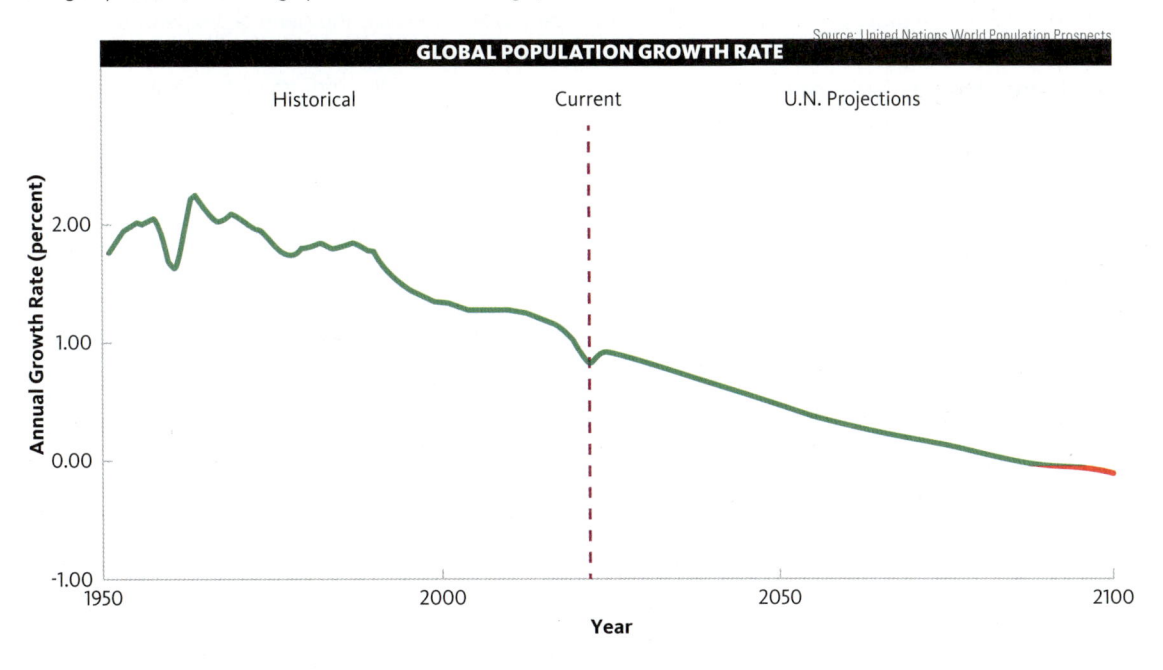

LESSON 14.1   475

SCIENCE FOCUS 14.1

THE POPULATION DEBATE

Are there physical limits to human population growth and economic growth on a finite planet? Some say *yes*. Others say *no*.

The debate has gone on for hundreds of years. Meanwhile, humans are degrading much of Earth's irreplaceable natural capital as people's ecological footprints grow and planetary boundaries are exceeded. Exceeding planetary boundaries can lead to long-term changes, social disruption, and the possibility of a sharp decline in the human population from increasing death rates.

To some analysts, the key problem is the large and rapidly growing number of people in less-developed countries. To others, the key factor is overconsumption in more-developed countries because of their high rates of resource use per person. To others, it is both.

Another view of population growth is that technology allows people to overcome the environmental limits that populations of other species face. This has increased Earth's carrying capacity for the human species. Some analysts believe people can continue increasing economic growth and avoid serious damage to our planet. People can make technological advances in areas such as food production and medicine, and depleted resources can be replaced. As a result, they see no need to slow population growth or resource consumption.

Proponents of slowing or stopping population growth point out that in addition to degrading Earth's life-support systems, hundreds of millions of people struggle to live. One of every eight people on the planet struggles to survive on less than $1.90 a day. This raises the question: How will humanity meet the basic needs of the additional 2 billion or more people that may be added to the population between 2022 and 2050?

No one knows for certain how soon limiting factors may stabilize, reduce, or sharply reduce the human population. Analysts call for people to confront this vital scientific, political, economic, and ethical issue.

Thinking Critically
Predict Do you think there are environmental limits to human population growth? If so, what do you think will cause the limits and how close do you think we are?

The population size of a given area includes **fertility** (births), **mortality** (deaths), and migration. *Migration* refers to people moving into an area (immigration) and out of it (emigration). You can calculate the **population change** of an area by subtracting the total number of people leaving from the number of people entering the area during a specific time, such as one year.

> population change = (births + immigration) − (deaths + emigration)

When births plus immigration exceed deaths plus emigration, the population of a particular area grows. When the reverse is true, the population declines.

checkpoint What factors determine whether the population of the country where you live will increase or decrease this year?

14.1 Assessment

1. **Describe** Describe the general shape of human population growth as graphed in Figure 1-13. During which years was the growth exponential?
2. **Summarize** What three trends have demographers identified in human population growth?
3. **Recall** What factors do you need to know in order to calculate a region's population growth?

CROSSCUTTING CONCEPTS

4. **Stability and Change** Look at the graph of the human population growth rate in Figure 14-2. After seeing this graph, suppose a friend concludes that the human population is currently declining. Explain the error in your friend's reasoning.

14.2 What Factors Influence the Size of the Human Population?

CORE IDEAS AND SKILLS
- Identify total fertility rate as a key factor affecting human population growth or decline.
- Describe the effect of age structure on a population's growth rate.
- Discuss ways to slow human population growth.

KEY TERMS
total fertility rate (TFR) infant mortality rate
life expectancy

Several Factors Affect Human Population Growth

Women are having fewer babies, but the world's population is still growing. A key factor affecting a population's growth and size is its **total fertility rate (TFR)**. The TFR is the average number of children born to women of childbearing age. Between 1955 and 2022, the global TFR dropped from 5 to 2.4. Those who support slowing population growth see this as good news. However, to stop population growth, the global TFR must drop to 2.1—the rate necessary for replacing both parents after taking infant mortality into account—and it must stay at that rate.

Factors That Affect Birth and Fertility Rates

Many factors affect a country's average birth rate and TFR.

- One factor is the importance of children as a part of the labor force, especially in less-developed countries. Many poor families struggle to survive on less than $1.90 a day. Some of these families include a large number of children to help them haul drinking water, gather wood for heating and cooking, and grow or find food. Many children also work for wages to help their families survive (Figure 14-3). In more-developed countries, children usually don't enter the workforce until they are in their late teens or early twenties.

- Another economic factor related to TFR is the cost of raising and educating children. In the United States, the cost of raising a child born in 2013 to age 18 will range from an estimated $169,000 to $390,000, depending on household income.

- The availability of pension systems can influence the number of children couples have, especially in less-developed countries. If people know they will have a pension for financial support after retirement, they may have fewer children because they will not need to depend on the children as much for support in old age.

- People living in urban areas usually have better access to family planning services. They tend to have fewer children than do those living in the rural areas of poorer countries.

- Total fertility rates tend to be low when women have access to education and paid employment outside the home. In less-developed countries, a woman with no education typically has two more children than a woman with a high school education.

- The average age at marriage also plays a role. Women normally have fewer children when they marry at age 25 or older.

- The availability of reliable birth control is another factor that influences a country's TFR (Figure 14-4). Each year more than 208 million women become pregnant. It is estimated that 40 million (20%) are unwanted pregnancies. In Uganda, where birth control is not widely available, women have an average of two more children than they want to have. Uganda is also one of the world's poorest countries. The availability of reliable birth control allows women to control the number and spacing of their children.

FIGURE 14-3
A young boy works in a shipyard in Dhaka, Bangladesh.

LESSON 14.2 477

FIGURE 14-4
An Ugandan woman and her children sit outside their village home. The TFR in Uganda is six, and women have little access to family planning resources.

- Religious beliefs, traditions, and cultural norms also play a role. In some countries, these factors favor large families. Many people oppose abortion and some forms of birth control.

Factors That Affect Death Rates The rapid growth of the world's population over the past 100 years is not primarily the result of a rising birth rate. Instead, it is largely the result of declining death rates, especially in less-developed countries. More people live longer and fewer infants die because of larger food supplies, improvements in food distribution, better nutrition, improved sanitation, safer water supplies, and medical advances such as immunizations and antibiotics.

- Life expectancy is a useful indicator of the overall health of people in a country or region. **Life expectancy** is the average age a person born in a certain year can be expected to live. Between 1955 and 2022, global life expectancy increased from 48 years to 73 years. Between 1900 and 2022, U.S. life expectancy rose from 47 years to 79 years. In the world's poorest countries, life expectancy in 2022 was 62 years or less. Research indicates that poverty, which reduces the average life span by seven to ten years, is the single most important factor affecting life expectancy.

- Another important indicator of overall health in a population is its infant mortality rate. **Infant mortality rate** reflects the number of babies out of every 1,000 born who die before their first birthday. Infant mortality rate can be considered a measure of a society's quality of life because it indicates the general level of nutrition and health care. A high infant mortality rate usually indicates insufficient food, poor nutrition, and a high incidence of infectious disease. Infant mortality also affects the TFR. In areas with low infant mortality rates, women tend to have fewer children because not as many children die at an early age.

Infant mortality rates in most countries have declined dramatically since the 1960s. Still, every year millions of infants die of preventable causes during their first year of life. Most of these deaths occur in less-developed countries. In 2022, the infant mortality rate was 5.5 in the United States, 40 in less-developed countries, and as high as 77 in the least-developed countries.

Factors that Affect Migration People migrate to another area within their country or to another country to seek jobs and economic improvement. Other migrants are driven by religious persecution, ethnic conflicts, political oppression, or war. There are also environmental refugees—people who leave their homes because of water or food shortages, soil erosion, or some other environmental problem.

checkpoint What is TFR and how does it relate to population growth?

Age Structure Affects Growth Rate and Decline

A population's age structure is an important factor that affects how it changes. Age-structure diagrams show the percentages or numbers of males and females in the total population by age. These diagrams can be used to compare age structures among different populations. For example, Figure 14-5 shows age-structure diagrams for more- and less-developed countries.

A population's age structure can also be used to make projections. A country with a large percentage of people younger than age 15 is likely to experience rapid population growth. The number of births in such a country is likely to rise because of the large number of girls entering reproductive ages. Most future population growth will take place in less-developed countries because of their typically young age structure and rapid population growth rates.

> **CONSIDER THIS**
>
> **The United States has the third-largest population.**
>
> The population of the United States grew from 76 million in 1900 to 332 million in 2022. The United States has the world's third-largest population after China with 1.5 billion people and India with 1.4 billion people. During the period of high birth rates between 1946 and 1964, known as the baby boom, 79 million people were added to the U.S. population. At the peak of the baby boom in 1957, the average TFR was 3.7 children per woman. In most years since 1972, it has been at or below 2.1 children per woman.
>
> The drop in the TFR has slowed the rate of population growth in the United States, but the country's population is still growing, mainly because of immigration. The U.S. Census Bureau projects that between 2022 and 2050, the U.S. population will grow from 332 million to 398 million—an increase of 66 million people. Because of a high per-person rate of resource use, each addition to the U.S. population has an enormous environmental impact.

Changes in the distribution of a country's age groups have long-lasting economic and social impacts. The American baby boom added 79 million people to the United States between 1946 and 1964.

FIGURE 14-5 ▼

Age Structure Diagrams Population structure by age and gender differs greatly between more- and less-developed countries.

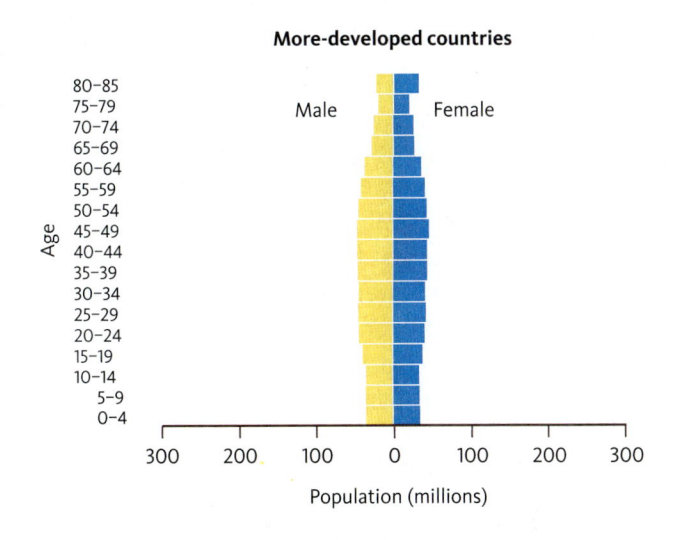

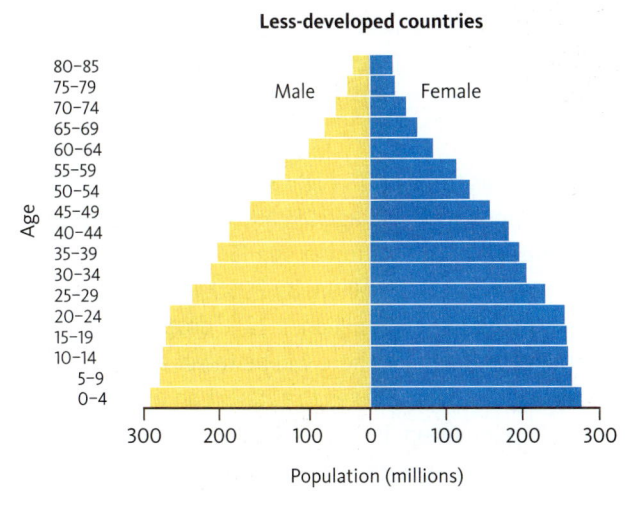

Sources: UN Population Division and Population Reference Bureau

For decades, the baby boom generation has strongly influenced the U.S. economy because they make up a signifcant percentage of all adult Americans. The baby boom generation plays an important role in the outcome of public elections and the passage or weakening of laws.

Aging Populations Can Decline Rapidly The global population of people 65 and older is projected to triple between 2022 and 2050. By then, one of every six people will be a senior citizen. This segment of the population has grown because of declining birth rates and medical advances that have extended life spans. As the percentage of people age 65 or older increases, more countries will begin experiencing population declines. If population decline is gradual, its harmful effects usually can be managed. However, countries experiencing rapid declines feel the effects more severely.

Figure 14-6 lists some of the problems associated with rapid population decline. Countries with rapidly declining populations include Japan, Germany, Italy, Bulgaria, Hungary, Serbia, Greece, and Portugal. Population declines are difficult to reverse because of the low numbers of people under the age of 15 and people in their reproductive years.

FIGURE 14-6 ▼

Problems with Rapid Population Decline
Can threaten economic growth
Labor shortages
Less government revenues with fewer workers
Less entrepreneurship and new business formation
Less likelihood for new technology development
Increasing public deficits to fund higher pension and health-care costs
Pensions may be cut and retirement age increased

Japan has the world's highest percentage of elderly people (above age 65) and the world's lowest percentage of people under age 15. In 2022, Japan's population was 124 million. By 2050, its population is projected to be 97 million, a 22% drop. As its population declines, there will be fewer adults working and paying taxes to support an increasing elderly population. Japan discourages immigration, and this could threaten its economic future.

checkpoint What do age structure diagrams show?

CONSIDER THIS

China curbed its population growth.
In the 1960s, China's population was growing so rapidly that there was a serious threat of mass starvation. To avoid this, the government enacted a strict population control program. China's goal was to sharply reduce population growth by promoting one-child families. Couples who broke the one-child pledge lost benefits including improved housing, salaries, and healthcare. Some criticized China's population program for violating human rights. However, China cut its birth rate in half between 1972 and 2012. In 2015, married couples were allowed to have two children.

Ways to Slow Population Growth

Controversy exists over whether people should try to slow population growth (Science Focus 14.1). Some analysts argue that people should try to slow population growth to avoid degrading Earth's life-support systems. Experience indicates that people can slow population growth by reducing poverty through economic development, elevating the status of women, and encouraging family planning.

Promoting Economic Development
One way to reduce population growth is to reduce poverty. Experts examined the birth and death rates of western European countries that became industrialized during the 19th century. They developed a hypothesis on population change known as demographic transition. According to this hypothesis, as countries become industrialized and economically developed, their per capita incomes rise and poverty declines. Their populations then grow more slowly and may eventually stop growing or decline. By 2015, 31 more-developed countries had stable or declining population sizes.

Some analysts believe most of the world's less-developed countries will make a demographic transition over the next few decades. The transition will occur because newer technologies will help them develop economically and reduce poverty. Other analysts fear rapid population growth, extreme poverty, increasing environmental degradation, and resource depletion could leave some less-developed countries stuck in the rapid-growth stage. This highlights economic development as a key to improving human health and stabilizing populations.

FIGURE 14-7
Empowerment of women slows rapid population growth. Here, a self-employed Ugandan woman cooks in her restaurant.

LESSON 14.2   481

Empowerment of Women A number of studies show women tend to have fewer children if they are educated, control their own fertility, earn an income of their own, and live in societies that do not suppress their rights. In most societies, women have fewer rights and fewer educational and economic opportunities than men have. Women do almost all of the world's domestic work and child care for little or no pay. They provide more unpaid health care (within their families) than all of the world's organized health-care services combined. In rural areas of Africa, Latin America, and Asia, women do 60–80% of the work associated with growing food, hauling water, and gathering and hauling wood and animal dung for use as fuel.

While women account for 66% of all hours worked, they receive only 10% of the world's income and own just 2% of the land. They also make up over half of the world's poor and most of its 800 million illiterate adults. Poor women who cannot read have an average of five to seven children, compared with one or two children in societies where almost all women can read. This identifies the need for all children to get at least an elementary school education. Programs that assist impoverished families (such as free school lunch programs) enable families to educate their children and help raise them out of extreme poverty.

A growing number of women in less-developed countries are taking charge of their lives (Figure 14-7) and reproduction. As this number expands, such change driven by individual women will play an important role in stabilizing populations. This change will also improve human health, reduce poverty and environmental degradation, and allow more people access to basic human rights.

Family Planning Programs that offer education and clinical services to help couples choose how many children to have and when to have them are known as family planning. Such programs vary from culture to culture, but most of them provide information on birth spacing, birth control, and health care for pregnant women and infants.

Family planning enables women to limit the size of their families, if they wish to do so, and plan their pregnancies. According to studies by the UN Population Division and other population agencies, family planning has been a major factor in reducing the number of unintended pregnancies and births and the number of abortions worldwide.

In addition, family planning has reduced the number of mothers and fetuses who die during pregnancy, rates of infant mortality, and rates of HIV and AIDS. Family planning also has financial benefits. Studies show that each dollar spent on family planning in countries such as Thailand, Egypt, and Bangladesh saves $10–$16 in health, education, and social service costs by preventing unwanted births.

checkpoint What is the demographic transition hypothesis?

14.2 Assessment

1. **Identify** What is the key factor that determines the growth rate of a human population?
2. **Explain** How does a population's age structure help analysts make predictions about how its size might change?
3. **Contrast** In general, how do age structures differ between more- and less-developed countries?
4. **Describe** Describe three ways to slow population growth.

SCIENCE AND ENGINEERING PRACTICES
5. **Using Math** How can a country with a high infant mortality rate experience rapid population growth? Use the equation for population change.

14.3 What Are the Effects of Urbanization on the Environment?

CORE IDEAS AND SKILLS
- Describe three trends in urbanization and the effects of urban sprawl.
- Explain the advantages and disadvantages of urbanization.
- Recognize the plight of poor people in urban areas.

KEY TERMS

urbanization · urban sprawl

Population Experts Identify Three Urban Trends

About 56% of the world's people and 83% of Americans live in urban areas. Urban growth is occurring worldwide. It is happening the fastest in less-developed countries (Figure 14-8). This trend

FIGURE 14-8
Traffic congestion is a major problem in Dhaka, Bangladesh, and other cities that have seen rapid urbanization.

toward living in urban areas is called **urbanization**. Urban areas grow in two ways—by natural increase (more births than deaths) and by immigration, mostly from rural areas. People move from rural to urban areas in search of jobs, food, housing, educational opportunities, better health care, and entertainment. Others relocate to urban areas because of factors such as famine, loss of land for growing food, deteriorating environmental conditions, war, and religious, racial, or political conflicts.

Population experts identify three major trends related to urban populations. First, the percentage of people living in urban areas has grown sharply, and this trend is projected to continue. Between 1850 and 2022, the percentage of the world's people living in urban areas increased from 2% to 56% and is likely to reach 67% by 2050, with most new urban dwellers living in less-developed countries.

Second, the numbers and sizes of urban areas are mushrooming. In 2020 there were 34 megacities—cities with 10 million or more people—a majority of them in less-developed countries. Nine of these urban areas are hypercities with more than 20 million people. The largest hypercity is Tokyo, Japan, with 37.4 million people. By 2025, the number of megacities is expected to reach 37, with 21 of them in Asia. Some megacities and hypercities are merging into vast urban megaregions, each with more than 100 million people. The largest megaregion is the Hong Kong–Shenzhen–Guangzhou region in China, with about 120 million people. That is six times more people than live in or near New York City.

Finally, poverty is becoming increasingly urbanized, mostly in less-developed countries. The UN estimates that at least 1 billion people live in the slums of major cities in less-developed countries.

LESSON 14.3   483

Urban Sprawl In the United States and several other countries, **urban sprawl** is eliminating agricultural and wild lands. Urban sprawl is the growth of low-density development on the edges of cities and towns. The result is a dispersed jumble of housing developments, shopping malls, parking lots, and office complexes loosely connected by multilane highways.

Urban sprawl is the product of ample affordable land, cars, government funding of highways, and lack of urban planning. Many people prefer living in suburbs. Compared with central cities, these areas provide lower-density living and access to single-family homes on larger lots. Often these areas also have newer schools and lower crime rates. On the other hand, urban sprawl has caused or contributed to a number of environmental problems. These problems include loss of cropland and forests, habitat fragmentation, increased water pollution, increased runoff and flooding, increased energy use, and increased air pollution.

checkpoint Give three factors that contribute to urbanization.

Urbanization Has Advantages

Urbanization has many benefits. Cities are centers of economic development, innovation, education, social and cultural diversity, and employment. Urban residents in many parts of the world tend to live longer than rural residents and have lower infant mortality and fertility rates. They usually have better access to medical care, family planning, education, and social services than do rural residents.

Urban areas also have environmental advantages. Recycling is more economically feasible because of the high concentrations of recyclable materials in urban areas. Concentrating people in cities preserves biodiversity by reducing stress on wildlife habitats. Heating and cooling multistory buildings in central cities takes less energy per person than heating and cooling single-family homes and smaller office buildings. Central-city dwellers tend to drive less and rely more on mass transportation, carpooling, walking, and bicycling.

checkpoint What are the advantages of urbanization?

Urbanization Has Disadvantages

Most urban areas are unsustainable systems. Urban populations occupy only 3% of Earth's land area, but they consume 75% of the world's resources and produce 75% of its pollution and wastes. Because of this high input of food, water, and other resources, and the resulting high-waste output, most of the world's cities have huge ecological footprints that extend beyond their boundaries.

Lack of Vegetation In urban areas, most trees, shrubs, grasses, and other plants are cleared to make way for buildings, roads, parking lots, and housing

FIGURE 14-9
The High Line is a public park built on a reclaimed section of an old freight rail line in New York, New York.

> **CONSIDER THIS**
>
> **Over 65 million people in India live in slums.**
>
> With a population of 1.4 billion, India has the second largest population on Earth. A growing middle class is thriving in India. However, the country also faces serious poverty, malnutrition, and environmental problems. About one-fourth of all people in India's cities live in slums, and progress and prosperity have not reached many of the nearly 650,000 rural villages where much of India's population lives. Nearly half of the country's labor force is unemployed or underemployed, and 42% of its population lives in extreme poverty.

developments. Most cities do not benefit from the free ecosystem services provided by green spaces. These services include air purification, generation of oxygen, removal of atmospheric CO_2, control of soil erosion, and wildlife habitat. Some cities are beginning to address this problem by reclaiming green spaces (Figure 14-9).

Water Problems As urban areas grow and their water demands increase, expensive reservoirs and canals are often built, and deeper wells must be drilled. This can deprive rural and wild areas of surface water and can deplete groundwater supplies. Flooding also tends to be worse in cities built on floodplains near rivers or coastlines. In most cities, buildings and paved surfaces cause precipitation to run off quickly and overload storm drains. Urban development has destroyed or degraded large areas of wetlands that served as natural sponges to absorb excess storm water. Many coastal cities will very likely face increased flooding in this century as sea levels rise because of climate change.

Pollution and Health Problems Cities produce most of the world's air pollution, water pollution, and solid and hazardous wastes. Pollutant levels are generally higher in cities because the pollution is produced in a confined area and cannot be dispersed as readily as pollution produced in rural areas. High population densities in urban areas promote the spread of infectious diseases, especially without adequate drinking water and sewage systems.

Excessive Noise Most urban dwellers are subjected to noise pollution. Noise pollution is defined as any unwanted, disturbing, or harmful sound. Sound becomes unwanted when it interferes with normal activities such as sleeping or conversation, or disrupts or diminishes your quality of life. Noise levels are measured in decibel-A (dBA) sound pressure units that vary with different human activities. Prolonged exposure to sound levels above 85 dBA causes permanent hearing damage. Just 1.5 minutes of exposure to 110 decibels or more can cause hearing damage. Noise pollution can be reduced by shielding noisy activities or processes, moving noisy operations or machines away, and using anti-noise technologies.

Effect on Local Climates Cities tend to be warmer, rainier, foggier, and cloudier than suburbs and nearby rural areas. In cities, the enormous amount of heat generated by cars, factories, furnaces, lights, air conditioners, and heat-absorbing dark roofs and streets creates an urban "heat island" surrounded by cooler suburban and rural areas. As cities grow and merge, these heat islands merge and can reduce the natural cleansing of polluted air. The urban heat island effect can also increase dependence on air conditioning. This in turn leads to higher energy consumption, greenhouse gas emissions, and other forms of air pollution.

checkpoint Describe five problems with cities.

Life in Slums Is a Desperate Struggle

Poverty is a way of life for many urban dwellers in less-developed countries. Approximately 1.6 billion urban residents, or one-quarter of the world's urban population, live in poverty.

Some poor people live in slums—overcrowded neighborhoods dominated by dilapidated housing where several people might live in a single room (Figure 14-10). Other poor people live in squatter settlements on the outskirts of cities. Some build shacks from corrugated metal, plastic sheets, scrap wood, and other scavenged building materials. Others live in rusted shipping containers and junked cars.

Poor people living in slums, squatter settlements, or on the streets usually lack clean water, sewers, electricity, and roads. They are subject to severe air and water pollution and hazardous wastes from nearby factories. Many of these settlements are in locations prone to landslides, flooding, earthquakes, or fire.

FIGURE 14-10
ON ASSIGNMENT National Geographic photographer Robin Hammond shot this image of the sawmill district in Lagos, Nigeria. Thousands of people live and work in shanties and workshops stretched across the outskirts of this vast city.

LESSON 14.3

Some governments address these problems by legally recognizing slums and granting legal titles to the land. They base this on evidence that poor people usually improve their living conditions once they know they have a permanent place to live.

Large cities in less-developed countries struggle to deal with poverty, population, and pollution. For example, Mexico City is one of the world's megacities. More than one-third of its residents live in slums or in squatter settlements without running water and electricity. At least 3 million people have no sewage facilities. Human waste is deposited in gutters, vacant lots, and open ditches, attracting rats and flies.

Mexico City has serious air pollution problems because of too many cars, polluting factories, and a warm, sunny climate that increases smog. In 1992, the UN named Mexico City the most polluted city on the planet. Since then, Mexico City has made progress in reducing the severity of some of its air pollution problems. The city moved refineries and factories out of the city, banned cars in its central zone, and required air pollution controls on all cars made after 1991. It also phased out the use of leaded gasoline and replaced some old buses, taxis, and delivery trucks with vehicles that produce fewer emissions. As a result, air pollution has decreased.

In 2013, Mexico City won a Sustainable Transportation Award for expanding its rapid transit system and expanding the bike-sharing program and bike lanes. Mexico City still has a long way to go. However, this example shows what a city can do to improve environmental quality once a community decides to act (Figure 14-11).

checkpoint What are conditions like for people living in squatter settlements?

14.3 Assessment

1. **Recall** What are the three major trends related to urban populations?
2. **Explain** What are the advantages and disadvantages of urbanization?
3. **Make Judgments** Should it be the responsibility of city governments to provide clean water, sewers, electricity, and roads to squatter settlements, or should local governments have the right to bulldoze these areas? As mayor of a large city, how would you handle this problem?

SCIENCE AND ENGINEERING PRACTICES
4. **Engaging in Argument** Do you think the advantages of urban sprawl outweigh the disadvantages? Why or why not?

FIGURE 14-11
Mexico City has taken steps to become more sustainable. The city rebuilt its transit system and parks and reduced its air pollution.

14.4 How Can Cities Become More Sustainable?

CORE IDEAS AND SKILLS
- Analyze the impact of cars in cities and ways to reduce that impact.
- Describe the eco-city concept.

KEY TERMS

eco-city smart growth

Transportation and Urban Sprawl

The modes of transportation used in cities depend on population density. The denser the population, the fewer motor vehicles are used. Residents of widely spread urban areas depend on cars for transportation. This greatly expands their ecological footprints.

If a city cannot spread outward, it must grow vertically—upward and below ground—so it occupies a small land area with a high population density. Most people living in compact cities such as Hong Kong and Tokyo get around by walking, biking, or using mass transit like rail and bus systems.

In other parts of the world, a combination of plentiful land and networks of highways has produced dispersed cities whose residents depend on cars for most travel. Such cities are found in the United States, Canada, and Australia. The resulting urban sprawl has a number of undesirable effects.

The United States is an example of a car-centered nation. The United States has 4.25% of the world's people and 19% of the world's cars. In its dispersed urban areas, cars are used almost exclusively, and 86% of residents drive alone to work every day.

Advantages and Disadvantages of Cars Cars offer a convenient and comfortable way to move people around. Much of the world's economy is built on producing cars and supplying fuel, roads, services, and repairs for them. Despite their benefits, cars have harmful effects on people and the environment. Globally, car accidents kill 1.2 million people a year and injure 50 million more. They also kill about 50 million wild animals and family pets every year.

Cars are the world's largest source of outdoor air pollution. This pollution kills about 100,000 people each year in the United States. Cars are also the fastest-growing source of climate-changing CO_2 emissions. A third of the world's urban land and half of U.S. urban land is devoted to roads, parking lots, gas stations, and other car-related uses.

Another problem with cars is traffic congestion. If current trends continue, U.S. motorists will spend an average of two years of their lives in traffic jams. Traffic congestion in some cities in less-developed countries is much worse (Figure 14-8). Building more roads is not the answer because more roads encourage more people to drive.

Reducing Use of Cars Various methods to reduce the use of cars have been proposed. Some environmental scientists and economists suggest that people can reduce the harmful effects of cars by following a "user-pays approach." This means making drivers pay directly for most of the environmental and health costs they cause. This is an example of the full-cost pricing factor of sustainability.

One way to phase in a user-pays approach would be to charge a tax or fee on gasoline to cover the estimated harmful costs of driving. According to a recent study, such a tax would amount to about $3.18 per liter ($12 per gallon) of gasoline in the United States. These costs include higher medical and health insurance bills, higher taxes to support regulating and reducing air pollution, and military defense of the Persian Gulf oil supplies.

Gradually phasing in such a tax, as has been done in many European nations, could spur the use of more energy-efficient motor vehicles and mass transit. It would also reduce pollution and slow projected climate change and ocean acidification. Proponents of higher gasoline taxes urge governments to do two major things. One is to fund programs to educate people about the hidden costs they are paying for their gasoline. Another is to use gasoline tax revenues to finance bus and rail systems, bike lanes, and sidewalks. Taxes on income, wages, and wealth could be reduced to offset the increased taxes on gasoline. Such a tax shift would make higher gasoline taxes more acceptable.

Taxing gasoline heavily would be difficult in the United States for three reasons. First, it faces strong opposition from people who feel they are already overtaxed. Another opposition group includes powerful industries such as carmakers, oil and tire companies, road builders, and many real estate developers. Second, the dispersed nature of most U.S. urban areas makes people dependent on cars, and higher taxes would be an economic burden. Third, mass-transit options, bike lanes, and sidewalks

LESSON 14.4 489

FIGURE 14-12
The U.S. city of Portland, Oregon, offers its residents viable alternatives to driving with its extensive network of bike lanes and well-planned light-rail and bus lines.

are not widely available in the United States. Most of the money from gasoline taxes is used for building and improving highways for cars.

Another user-pays approach to reduce car use and urban congestion is to raise parking fees and charge tolls on roads, tunnels, and bridges leading into cities. Densely populated Singapore is rarely congested because it auctions the rights to buy a car, and drivers are charged a fee every time they enter the city. Several European cities have also imposed stiff fees for driving in their central districts, while others have banned parking on city streets and established networks of bike lanes.

More than 300 European cities have car-sharing networks that provide short-term rental of cars. Portland, Oregon, was the first U.S. city to develop a car-sharing system. Network members reserve a car in advance or contact the network and are directed to the closest car. In Berlin, Germany, car sharing has cut car ownership by 75%. Car sharing in Europe has reduced the average driver's CO_2 emissions by 40–50%. Car-sharing networks have sprouted in several U.S. cities, and some large car-rental companies now rent cars by the hour.

Bicycling accounts for a third of all urban trips in the Netherlands and in Copenhagen, Denmark, compared with less than 1% of trips in the United States. Large, bike-friendly cities in the United States include New York, Chicago, Baltimore, Minneapolis, Philadelphia, and Portland (Figure 14-12). More than 712 cities in 50 countries, including 78 U.S. cities, have bike-sharing systems. These systems allow individuals to rent bikes as needed from widely distributed stations.

checkpoint Give an example of a "user-pays" approach to reducing car use.

Developing More Sustainable Cities

Many environmental scientists and urban planners call for new and existing urban areas to be made more sustainable and livable through ecological design. They say doing so is an important way to increase people's beneficial environmental impact.

NATIONAL GEOGRAPHIC | EXPLORERS AT WORK
Caleb Harper — Urban Agriculturalist

Caleb Harper tweets with plants. He checks in regularly with his 2,000 or so food crops just to see how they are doing, and they tell him. They tell him via data gathered by sensors and reported through an online interface. If that sounds futuristic, it is.

Harper is a research scientist at the Massachusetts Institute of Technology (MIT), a National Geographic Explorer, and a creative visionary. Harper imagines a future in which more farmers grow more food with less water and without the need for sunlight, pesticides, or fertilizer. He sees all this occurring in the places where most people live and eat—in cities.

Growing food in cities would cut down on the environmental costs of transport and reduce food waste. According to Harper, an old urban warehouse or even the side of a skyscraper could become a super-efficient farm.

Harper founded the CityFARM project at MIT to work toward making his dream a reality. Technically, Harper's vision is possible. The only factors plants require to survive and grow are certain nutrients including water and certain wavelengths of light. All of these needs can be fulfilled without soil or sunlight.

For example, aeroponics is a system for growing plants in moist air that was designed by NASA for use on space stations. Aeroponic systems use up to 90% less water than traditional agriculture. Additionally, food plants grown with systems such as aeroponics grow several times faster and contain 2–3 times more nutrition than their traditionally grown counterparts. These plants get exactly what they need in exactly the amounts needed.

However, determining "exactly what plants need" is a significant challenge. Variables—such as oxygen, minerals, acidity, and moisture—can be combined in an endless number of ways. Furthermore, different crops and their many varieties each thrive at slightly different combinations.

Harper contends that to optimize the growth, nutrition, and flavor of food plants, much more data are needed. Harper started an open-source library of "recipes" for growing food plants to address this need. Through computer networks, many people can experiment, tweak, and share their findings, moving the technology forward in a giant collaboration.

Harper got his love of growing plants from the farm where he grew up. Harper found a way to combine his passion for growing food with a love of computers, architecture, and engineering. The result could be a new way to feed cities.

FIGURE 14-13
Caleb Harper shows off some healthy greens grown in his lab at MIT.

FIGURE 14-14

Sustainable Transit Curitiba's bus rapid-transit system has greatly reduced car use in this Brazilian city.

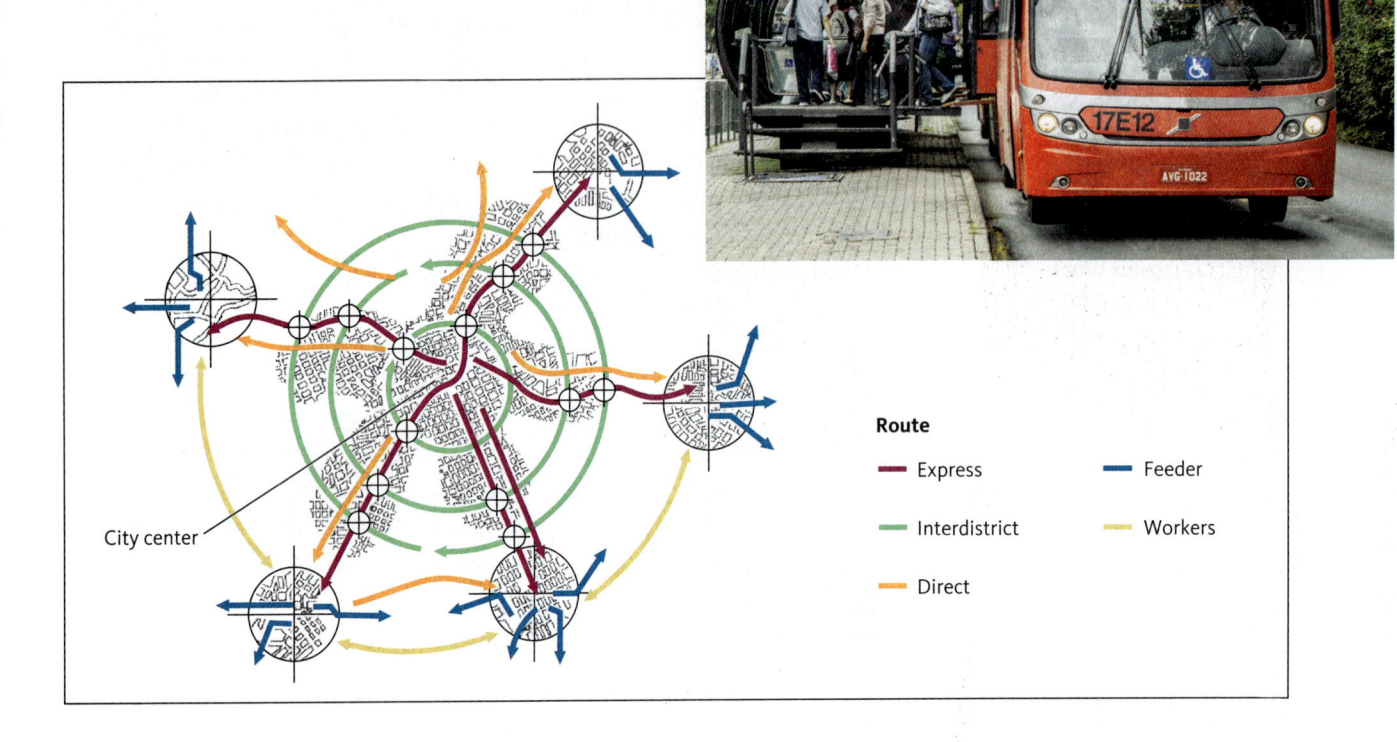

Route
- Express
- Interdistrict
- Direct
- Feeder
- Workers

City center

ENGINEERING FOCUS 14.2

CURITIBA'S TRANSPORTATION SYSTEM

An example of an eco-city is Curitiba (koor-ee-CHEE-ba), a city of over 3.2 million people known as the "ecological capital" of Brazil. In 1969, planners in this city decided to focus on an inexpensive, efficient mass-transit system rather than on the car.

Curitiba's superb bus rapid-transit (BRT) system moves large numbers of passengers efficiently, including 72% of commuters (Figure 14-14). Each of the system's five major "spokes," connecting the city center to outlying districts, has two express lanes for buses only. Double- and triple-length bus sections, coupled together as needed, carry up to 300 passengers. Extra-wide bus doors and covered boarding platforms where passengers can pay before getting on the bus speed up the boarding process.

In Curitiba, only high-rise apartment buildings are allowed near major bus routes, and the bottom two floors of each building must be devoted to stores. This reduces the need for residents to travel. Cars are banned from 49 blocks downtown. The downtown has a network of pedestrian walkways connected to bus stations, parks, and bicycle paths running throughout most of the city. The city uses old buses as roving classrooms to train its poor in basic job skills. Other retired buses have become health clinics, soup kitchens, and day-care centers that are free for low-income parents.

Curitiba uses less energy per person than most comparably sized cities. It also emits lower amounts of greenhouse gases and other air pollutants and has less traffic congestion than most comparably sized cities.

Thinking Critically
Evaluate Look at the shape of Curitiba's rapid-transit scheme. How does its structure suit its function?

492 CHAPTER 14 HUMAN POPULATION AND URBANIZATION

The Eco-City Concept An **eco-city** is a people-oriented city rather than a car-oriented city. Its residents are able to walk, bike, or use low-polluting mass transit for most of their travel. Its buildings, vehicles, and appliances meet high energy-efficiency standards. Much of its energy comes from solar cells on the rooftops and walls of buildings, solar hot-water heaters on rooftops, and geothermal heating and cooling systems. Buildings, vehicles, and appliances meet high energy-efficiency standards. People recycle, reuse, or compost most of their wastes, grow much of their food, and protect biodiversity by preserving surrounding land. Trees and plants adapted to the local climate and soils are planted throughout the city to provide shade, beauty, and wildlife habitats, and to reduce air pollution, noise, and soil erosion.

An eco-city cleans up and uses abandoned lots and industrial sites and preserves nearby forests, grasslands, wetlands, and farms. Much of the food people eat comes from nearby organic farms, solar greenhouses, community gardens, and individual gardens on rooftops, in yards, and in window boxes. Everyone has easy access to parks. Eco-cities serve residents without regard to race, creed, income, or any other factor.

The eco-city model is not a futuristic dream but a growing reality. Many U.S. cities are striving to become more environmentally sustainable and livable, including Portland, Oregon; Davis, California; Olympia, Washington; and Chattanooga, Tennessee. Examples outside the United States include Curitiba, Brazil; Bogotá, Colombia; Waitakere City, New Zealand; Stockholm, Sweden; Helsinki, Finland; Copenhagen, Denmark; Melbourne, Australia; Vancouver, Canada; Leicester, England; and Amsterdam, Netherlands.

Eco-cities also plan for smart growth. **Smart growth** entails a set of policies and tools that encourage more environmentally sustainable urban development with less dependence on cars. It uses zoning laws and other tools to channel growth in order to reduce the ecological footprint.

Some critics contend that by limiting urban expansion, smart growth can lead to higher land and housing prices. Smart growth supporters counter that it controls and directs sprawl, protects ecologically sensitive and important lands and waterways, and results in neighborhoods that are more enjoyable places to live. Figure 14-15 lists some widely used smart growth tools.

FIGURE 14-15

Smart Growth Smart growth tools can be used to prevent or control urban growth and sprawl. Which five of these tools do you think would be best for preventing or controlling urban sprawl?

Smart Growth Tools	
Limits and regulations	Limit building permits.
	Draw urban growth boundaries.
	Create greenbelts around cities.
Zoning	Promote mixed use of housing and small businesses.
	Concentrate development along mass-transit routes.
Planning	Use ecological land-use planning.
	Use environmental impact analysis.
	Use integrated regional planning.
Protection	Preserve open space.
	Buy new open space.
	Prohibit certain types of development.
Taxes	Tax land, not buildings.
	Tax land on value of actual use instead of on highest value as developed land.
Tax breaks	Give breaks to owners agreeing not to allow certain types of development.
	Give breaks for cleaning up and developing abandoned urban sites.
Urban renewal and growth	Update existing towns and cities.
	Build well-planned new towns and villages within cities.

The three scientific factors of sustainability—reliance on solar energy, nutrient cycling, and biodiversity—can guide people in dealing with the problems brought on by population and urban growth. Solar, wind, and other renewable energy technologies can help cut pollution that is increasing as the population, resource use per person, and urban areas grow. Reusing, recycling, and composting more materials can cut solid wastes and reduce a population's ecological footprint. Preserving biodiversity can help sustain the life-support systems on which people and all other species depend.

Making this transition to sustainability is also in keeping with the social science factors of sustainability. Full-cost pricing requires that the harmful environmental costs of urbanization be included in the market prices of goods and services. To achieve this, people must work together to find win-win solutions to population and urban problems.

checkpoint What are the main features of an eco-city?

LESSON 14.4   493

14.4 Assessment

1. **Restate** Describe four ways a city can reduce car use.
2. **Explain** Which method of reducing car use is the most plausible in the United States? Explain your reasoning.
3. **Elaborate** Can a dispersed city be an eco-city? Explain.

SCIENCE AND ENGINEERING PRACTICES

4. **Designing Solutions** Evaluate each method of reducing car use for its potential to improve your own city, town, or suburb. Discuss the trade-offs of each.

TYING IT ALL TOGETHER STEM
Analyzing Population Growth

In this chapter's Case Study, you read about the population explosion that resulted in today's 7.9 billion people worldwide. Many environmental scientists believe that continued growth will be unsustainable in the long run.

When making projections, it helps to study the past. Analysts studying a vast body of population data have identified several global trends. One striking trend is the uneven distribution of population growth between more-developed and less-developed countries. (More- and less-developed countries are defined in Lesson 1.2.) A few of these data are given here for your own consideration and analysis.

Use the table to help you answer the questions that follow.

1. What was the average rate of population growth per decade in more-developed countries between 1950 and 2010?
2. What was the average rate of population growth per decade in less-developed countries between 1950 and 2010?
3. Copy these data into your own table and add 4 rows at the bottom. Make projections for 2020, 2030, 2040, and 2050 using the growth rates you found in items 1–2.
4. Make a two-line graph to display the data. How do the lines compare?
5. Make a claim about the difference between population growth in more-developed and less-developed countries. Support your claim with quantitative evidence.
6. What are the implications of your projections on the environment?
7. Which of the factors that you read about in Lesson 14.2 do you think could make the biggest impact on future population growth? Explain.

FIGURE 14-16

Year	World Population Growth Population in More-Developed Countries (billions)	Population in Less-Developed Countries (billions)
1950	0.8	1.7
1960	0.9	2.1
1970	1.0	2.7
1980	1.0	3.3
1990	1.0	4.2
2000	1.0	5.0
2010	1.0	5.7

Sources: UN Population Division and Population Reference Bureau

CHAPTER 14 SUMMARY

14.1 How many people can Earth support?

- Three trends related to human population growth are: the growth rate has slowed since 1960; the growth is distributed unevenly; and large numbers of people have moved from rural to urban areas.
- Population change is the total number of people entering an area during a specific time subtracted by the number of people leaving it during that time. People enter a population through fertility and immigration. People leave a population through mortality and emigration.

14.2 What factors influence the size of the human population?

- Total fertility rate (TFR) is the key factor that determines the size of a human population. It is the average number of children born to the women in that population. Many factors affect a country's average birth rate and TFR.
- Life expectancy and infant mortality are factors that affect death rates.
- A population's age structure is based on the percentages or numbers of males and females in the total population by age. Age structure affects how fast a population grows or declines. Age structure differs between more- and less-developed countries.
- Population growth can be slowed by promoting economic development (reducing poverty), empowering women, and promoting family planning.

14.3 What are the effects of urbanization on the environment?

- Trends related to urban populations include: the percentage of the global population living in urban areas has grown sharply and growth is expected to continue; the number and sizes of urban areas are growing; and poverty is becoming increasingly urbanized.
- While urbanization has economic and environmental advantages, it also has disadvantages. Most cities are unsustainable because of high levels of resource use, waste, pollution, and poverty.
- Urban sprawl is the growth of low-density development on the edges of towns and cities. It has caused or contributed to many environmental problems. Rapid population declines can also cause economic and other problems.
- Poor people in urban areas live in crowded, unsanitary conditions and are subject to air and water pollution and hazardous wastes from nearby industries.

14.4 How can cities become more sustainable?

- In some countries, many people live in widely dispersed urban areas and depend mostly on cars for their transportation, which greatly expands their ecological footprints.
- Following a user-pays approach may reduce the harmful effects of car use. This approach involves users paying gasoline taxes, parking fees, and tolls. Other ways to reduce car use are car-sharing networks and bicycling.
- Eco-friendly cities (eco-cities) enable people to choose walking, biking, or mass transit for most transportation needs. Eco-cities recycle or reuse most of their wastes, grow most of their food, and protect biodiversity by preserving surrounding land.

CHAPTER 14 ASSESSMENT

Review Key Terms

Select the key term that best fits each definition.

- eco-city
- fertility
- infant mortality rate
- life expectancy
- mortality
- population change
- smart growth
- total fertility rate
- urban sprawl
- urbanization

1. The average number of children born to women in a population
2. The number of births in a population during a specified time
3. The growth of low-density developments on the edges of cities and towns, eliminating agricultural and wild lands
4. The number of people entering a population minus the number of people leaving a population
5. The number of babies out of every 1,000 born who die before the age of one
6. Using a set of policies and tools that encourage environmentally sustainable development and reduce or eliminate dependence on cars
7. The number of deaths in a population during a specified time
8. The average number of years a person born in a specified year is expected to live
9. A population shift from rural to urban areas
10. A people-oriented city planned using smart growth

Review Key Concepts

11. List three factors that account for the rapid increase in the world's human population over the past 200 years.
12. About how many people are added to the world's population each year?
13. Summarize the debate over how long the human population can keep growing.
14. List three variables that affect the growth and decline of human populations.
15. How has the global TFR changed since the 1950s?
16. List eight factors that affect fertility rates.
17. How do life expectancy and infant mortality affect the population size of a country?
18. What are some problems related to rapid population decline due to an aging population?
19. Describe five things cities can do to become more sustainable.

Think Critically

20. Identify a major environmental problem, and describe the role that population growth plays in this problem.
21. Some people think our most important environmental goal should be to curb population growth in less-developed countries, where much growth is expected to take place between now and 2050. Others argue that the most serious environmental problems stem from high levels of resource consumption per person in more-developed countries, which have much larger ecological footprints per person than do less-developed countries. What is your view? Explain.
22. If you could reverse one of the three urban trends discussed in this chapter, which one would it be? Explain.
23. What, if anything, could you change about the way you get to school to reduce your environmental impact?
24. Consider the characteristics of an eco-city. How close to this eco-city model is your city or the city nearest to where you live? List three ways to make your city or the city nearest you more sustainable and livable.

Chapter Activities

A. Investigate: Transportation Trade-Offs STEM

People need efficient ways of getting around in urban areas. Without urban planning, cities can grow dispersed and car-heavy, especially where the land allows for it. Planning efficient and reliable transportation systems—including bike lanes, buses, and trains—can greatly reduce a city's ecological footprint.

Materials

4 index cards   easel paper   colored markers

1. As a class, list the trade-offs of each of the following modes of urban transportation: buses, bicycles, commuter trains, and high-speed trains.

2. Label each card with one of the transportation types. Write the pros on side and the cons on the other.

3. Your group is designing the transportation system for a mid-sized city from scratch. Weigh the advantages and disadvantages on each card. Then place the cards in order from highest to lowest priority.

4. Design the transportation system for the city. Decide how to arrange bike lanes, train tracks, and bus routes. Consider how the different pieces of the system will connect together for people who need to use more than one mode. Sketch out ideas in pencil.

5. Use the colored markers and large paper to make a diagram of your plan.

6. Present your transportation design to the class, explaining its trade-offs.

Questions

1. How did your group prioritize the modes of transportation?

2. Which part of your plan do you think would cost the most to implement?

3. Which part of your plan do you think would make the greatest positive impact on the environment in comparison to driving cars?

4. Which of the transportation modes could be added to an existing city that is currently designed for cars only? How might this be accomplished?

5. Describe a good idea presented by another group.

B. Take Action

National Geographic Learning Framework

Attitudes | Responsibility
Skills | Communication
Knowledge | Our Human Story

Select one of the options below. Gather information about the population in your area and how it has changed over time.

- Interview someone who has lived in your area for three decades or more.
- Visit a local historical society and ask to see historic photos of your area.
- Obtain satellite images of your area that show urban sprawl using an online mapping tool.

1. In what ways has the population changed in your area or nearby? Specify the amount of time over which the changes have occurred.

2. Which changes have harmed the environment? Which changes have been beneficial?

3. Predict future changes to your population based on your findings.

4. Refer to the table of smart growth (Figure 14-15). Which tool(s) could be implemented in your area or nearby?

5. Compile a list of recommendations and send them to a local official.

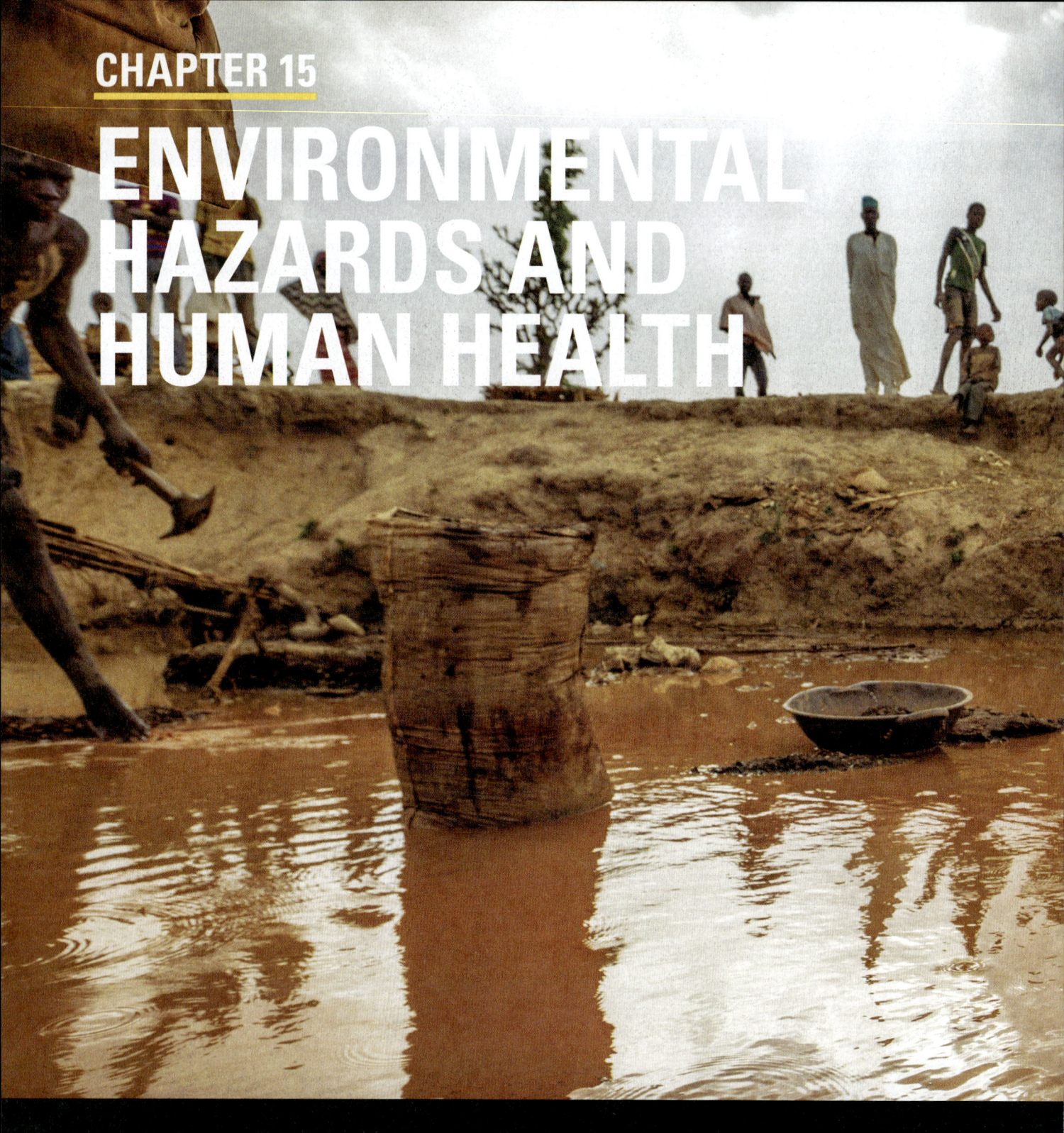

CHAPTER 15
ENVIRONMENTAL HAZARDS AND HUMAN HEALTH

PEOPLE EVERYWHERE ARE EXPOSED to environmental hazards. Some are biological, such as disease. Others are chemical, such as air and water pollution caused by the manufacturing, energy, and mining industries. Some hazards are hidden in the products we use. Others result from lifestyle choices or from poverty and poor living conditions. Yet many of these health threats can be reduced when people take appropriate action.

KEY QUESTIONS

15.1 What are the major types of health hazards?

15.2 How do biological hazards threaten human health?

15.3 How do chemical hazards threaten human health?

15.4 How can people evaluate risks from chemical hazards?

15.5 How do people perceive and avoid risks?

NATIONAL GEOGRAPHIC | EXPLORERS AT WORK

Health Testing Made Simple

with National Geographic Explorer Hayat Sindi

A piece of paper about the size of a postage stamp is now helping medical teams to test people's health. Dr. Hayat Sindi developed this simple, low-tech diagnostic tool. Dr. Sindi hopes it will save millions of lives in areas that have little access to high-tech medical support.

Sindi and her co-founder, Dr. George Whitesides, created the nonprofit organization Diagnostics for All. Its purpose is to develop and distribute low-cost diagnostic tools. "My mission is to find simple, inexpensive ways to monitor health that are specifically designed for remote places and harsh conditions," says Sindi.

How do the paper-based tests work? Each paper is etched with microchannels and wells, which are filled with various diagnostic chemicals. A drop of saliva, urine, or blood is placed on the paper. The fluid travels through the channels, and a chemical reaction occurs that makes the chemicals change color. Results show up in less than a minute. A diagnosis can be made easily based on the colors observed.

The first application of Sindi's device is to test liver function. Some drugs used to combat diseases such as HIV/AIDS, tuberculosis, and hepatitis can cause liver damage or failure. Monitoring liver function in patients taking these drugs is important. In more-developed countries, doctors monitor liver function frequently. In isolated areas or in less-developed countries, such monitoring rarely happens.

That is changing thanks to the paper-based technology Sindi has pioneered. "Paper is very inexpensive, universally available, lightweight, and easy to carry," Sindi explains. "It's a tool that allows even the poorest people in the most medically challenged places to get the tests they need."

Sindi has long been a pioneer. She was the first Saudi Arabian woman accepted to study biotechnology at Cambridge University, despite the fact that she spoke little English. She became an innovator and inventor, developing diagnostic tools including one used for early detection of breast cancer.

In 2001, Sindi became the first woman to earn her Ph.D. in Biotechnology at Cambridge. Later, she worked as a visiting scholar at Harvard University. She was named a UNESCO Goodwill Ambassador for science education and helps inspire women and girls, particularly in the Middle East, to pursue science.

Sindi gives lectures and has appeared on talk shows to discuss her work. She even bicycled across the Middle East with hundreds of professional women to highlight the plight of women and children. As she explains, "I want all women to believe in themselves and know they can transform society."

Thinking Critically
Evaluate What challenges and obstacles do you think Hayat Sindi has experienced in her career as a scientist? How do you think she overcame them?

Hayat Sindi invented a paper-based health test technology that does not require access to electricity, special equipment, or a lab. With minimal training, health care workers can travel house-to-house to perform diagnostic tests that provide results right away.

CASE STUDY
Mercury's Toxic Effects

The metal mercury (Hg) and its compounds are toxic to humans. Long-term exposure to high levels of mercury can be harmful. It can permanently damage the nervous system, brain, kidneys, and lungs. Fairly low levels of mercury can also harm fetuses and cause birth defects. Pregnant women, nursing mothers and their babies, women of childbearing age, and young children are especially vulnerable to mercury's harmful effects.

Mercury is naturally released into the air from rocks, soil, and volcanoes, and through vaporization from oceans. Such natural sources account for about a third of the mercury that enters the atmosphere each year. The remaining two-thirds comes from human activities.

Small-scale mines are the largest source of airborne mercury pollution. Thousands of these mines operate in Asia, Latin America, and Africa. (See the photo that opens this chapter.) Miners use mercury to separate gold from its ore. Then they heat the mixture of gold and mercury to release the gold. The process releases mercury vapors.

Mercury can also contaminate water and soil. Coal-burning power and industrial plants, cement kilns, smelters, and solid-waste incinerators also emit mercury.

Elemental mercury cannot be broken down or degraded. As a result, it builds up in soil, water, and tissues of humans and other organisms. Under certain circumstances, mercury can be converted to methylmercury, which is even more toxic.

Like DDT (Figure 7-14), methylmercury can be biologically magnified in food chains and food webs. High levels of methylmercury are often found in the tissues of large fishes that feed at high trophic levels. Tuna, swordfish, and sharks are a few such fishes.

People are exposed to mercury in two major ways. They may eat fish and shellfish contaminated by methylmercury (Figure 15-1). They may also inhale mercury vapors or particles present in the air.

The greatest risk from exposure to low levels of methylmercury is brain damage in fetuses and young children. Studies estimate that 100,000–200,000 children born each year in the United States are exposed in utero to levels that exceed EPA safety standards. They are at risk for reduced IQs and possible nervous system damage. Other health risks include poor balance and coordination, muscle weakness, tremors, memory loss, insomnia, hearing loss, hair loss, and peripheral vision loss.

This problem raises two important questions: How do scientists determine the potential harm from exposure to mercury and other chemicals? And how serious is the risk of harm from a particular chemical compared to other risks?

In this chapter, you will look at how scientists try to answer these and other questions about human exposure to environmental and health risks. You will also learn about ways to evaluate and avoid some risks.

As You Read Think about health hazards and risks you are exposed to every day. Which ones are unavoidable and which can you control to reduce them?

◀ FIGURE 15-1
Mercury has contaminated fish in many bodies of water such as this creek in the U.S. state of Oregon. Warning signs help educate potential consumers.

15.1 What Are the Major Types of Health Hazards?

CORE IDEAS AND SKILLS
- Understand the types of risks people encounter.
- Compare the processes of risk assessment and risk management.

KEY TERMS
risk
risk assessment
risk management
pathogen

Society Faces Many Types of Hazards

Hazards to human health may result from a variety of environmental factors. They may also result from the choices people make. A **risk** is the probability of suffering harm from a hazard that can cause injury, disease, death, economic loss, or damage.

Scientists will often describe a risk in terms of its probability to cause harm. For example: "The lifetime probability of developing lung cancer from smoking one pack of cigarettes per day is 1 in 250." This means that 1 of every 250 people who smoke a pack of cigarettes every day will likely develop lung cancer in his or her lifetime. (A typical lifetime is usually considered to be 70 years.) Probability can also be expressed as a percentage, as in a 30% chance of developing a certain kind of cancer. The greater the probability of harm, the greater the risk.

In **risk assessment**, statistical methods are used to estimate how much harm a particular hazard may cause to human health or the environment. In **risk management**, decisions are made about whether and how to reduce a particular risk to a certain level and at what cost. Figure 15-2 summarizes how risks are assessed and managed.

People take unnecessary risks every day. For example, people may choose to drive or ride in a car without a seatbelt. They may also text while driving. People may choose to eat foods that are high in cholesterol or have too much sugar. They may drink too much alcohol or smoke cigarettes. Even choosing to live in an area prone to tornadoes, hurricanes, or floods exposes people to greater risk.

No one can live a risk-free life, but people can reduce their exposure to risks. When assessing risks, it is important to understand how serious the risks are and whether the benefits of certain activities outweigh the risks. Five major types of hazards pose risks to human health.

- Biological hazards stem from more than 1,400 pathogens. **Pathogens** are microorganisms, viruses, or other agents that cause disease.
- Chemical hazards arise from certain harmful chemicals found in air, water, soil, food, and human-made products.
- Natural hazards include fire, earthquakes, volcanic eruptions, floods, and severe storms.
- Cultural hazards exist in daily life, such as unsafe working conditions, risks of criminal assault, and living in poverty.
- Lifestyle choices such as smoking, making poor dietary choices, and having unsafe sex can also be hazardous.

FIGURE 15-2
Evaluating Risks Risk assessment and risk management are used to estimate the seriousness of various risks and to help reduce such risks.

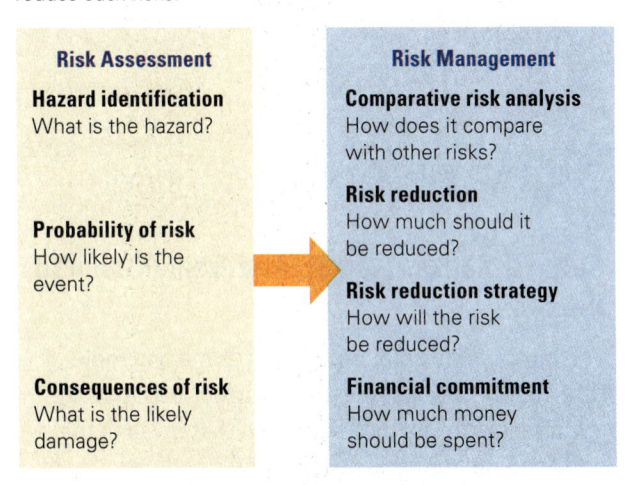

Risk Assessment

Hazard identification
What is the hazard?

Probability of risk
How likely is the event?

Consequences of risk
What is the likely damage?

Risk Management

Comparative risk analysis
How does it compare with other risks?

Risk reduction
How much should it be reduced?

Risk reduction strategy
How will the risk be reduced?

Financial commitment
How much money should be spent?

checkpoint What do scientists calculate in order to make a statement about a given risk?

LESSON 15.1   503

15.1 Assessment

1. **Restate** What are the five major types of hazards?
2. **Recall** What are three risky lifestyle choices that people may make?
3. **Explain** How do scientists use probability when describing risks?

CROSSCUTTING CONCEPTS

4. **Scale, Proportion, and Quantity** Review Figure 15-2. Explain how you apply risk assessment and risk management in your daily living.

15.2 How Do Biological Hazards Threaten Human Health?

CORE IDEAS AND SKILLS

- Discuss infectious diseases caused by pathogens, and explain the difference between transmissible and nontransmissible disease.
- Explain the difference between bacteria and viruses, how they are spread and treated, and what role parasites play in disease.
- Understand how disease can be reduced or prevented.

KEY TERMS

infectious disease
bacterium
virus
parasite
transmissible disease
nontransmissible disease

Some Diseases Can Spread from Person to Person

The most serious biological hazards many people face are infectious diseases. An **infectious disease** is a disease caused by a pathogen that invades the body and multiplies in cells and tissues. Examples of pathogens include bacteria, viruses, and parasites.

Bacteria are single-cell organisms that are found everywhere and that can multiply very rapidly on their own. Most bacteria are harmless and some are even beneficial. However, some bacteria cause diseases such as strep throat and tuberculosis.

Viruses are smaller than bacteria and work by invading a cell. A virus takes over a cell's genetic machinery to copy itself. In that way, it is able to spread throughout the body. Viruses may cause diseases such as flu and AIDS.

Parasites are organisms that live on or inside other organisms and feed on them. The organism a parasite feeds on is called its host. Parasites can cause infectious diseases such as diarrhea and malaria.

A **transmissible disease** is an infectious disease that can be transmitted from one person to another. Transmissible diseases may be bacterial diseases, such as tuberculosis, many ear infections, and gonorrhea. They may also be viral diseases, such as the common cold, flu, and AIDS. Transmissible diseases can be spread through air, water, and food. They can also be transmitted through body fluids such as feces, urine, blood, semen, and droplets sprayed by sneezing and coughing.

A **nontransmissible disease** is caused by something other than a living organism. It does not spread from person to person. Nontransmissible diseases include cardiovascular diseases, most cancers, asthma, and diabetes.

In 1900, infectious disease was the leading cause of death in the world. Since then, and especially since 1950, infectious disease rates have dropped significantly. So have deaths from infectious disease. A combination of advances made this possible. They include improved sanitation, better health care, the use of antibiotics to treat bacterial diseases, and the development of vaccines to prevent the spread of some viral diseases.

Infectious diseases are less threatening to society today. However, they do remain a serious health threat, especially in less-developed countries. One reason for this is that many disease-carrying bacteria have developed genetic immunity to widely used antibiotics (Science Focus 15.1). Also, many disease-transmitting species of insects have become resistant to widely used pesticides like DDT that once helped control their populations. Another factor that will likely add to the future impact of infectious diseases is projected climate change due to atmospheric warming. Many scientists warn that warmer temperatures will likely allow some diseases to spread more easily from tropical to temperate areas.

A large-scale outbreak of an infectious disease within an area or a country is called an epidemic. A global epidemic, like COVID-19 or AIDS, is called a pandemic. Figure 15-3 shows annual death tolls caused by major infectious diseases worldwide.

The TB Threat Tuberculosis (TB) is a highly contagious bacterial infection that destroys lung tissue and can ultimately lead to death (Figure 15-4).

SCIENCE FOCUS 15.1

ANTIBIOTIC RESISTANCE

It is becoming increasingly harder to stop the spread of infectious bacterial diseases. The problem is bacteria are becoming genetically resistant to the antibiotics that have long been used to kill them. The reason for this lies in the astounding reproductive rate of bacteria. Some types of bacteria can grow from a population of 1 to well over 16 million in 24 hours. As a result, they can quickly become genetically resistant to an increasing number of antibiotics through natural selection. (See Figure 4-13.)

Other factors can also promote genetic resistance to antibiotics. One is the spread of bacteria worldwide by human travel and international trade. Another is the overuse of pesticides. Yet another factor is the overuse of antibiotics.

Antibiotics are often prescribed for colds, flu, and sore throats. Most of these illnesses are caused by viruses and so do not respond to treatment with antibiotics. In many countries, antibiotics are available without a prescription, which also promotes excessive and unnecessary use. The growing use of antibacterial hand soaps and other antibacterial cleansers could also be promoting antibiotic resistance in bacteria.

Antibiotics are also widely used to control disease and promote growth in livestock that are raised in crowded conditions. This practice has led to bacterial resistance to these drugs. According to the Centers for Disease Control and Prevention (CDC), 22% of antibiotic-resistant illness in humans is linked to food, especially meat from livestock treated with antibiotics.

Every major type of disease-causing bacterium has developed strains that are resistant to at least one of the roughly 200 antibiotics in use. Furthermore, "superbugs" are emerging that resist all such antibiotics. Researchers are investigating ways to make antibiotics more effective and to directly manipulate germs so they are less able to attack.

Thinking Critically

Apply What are three steps society could take to slow the rate at which disease-causing bacteria are developing resistance to antibiotics?

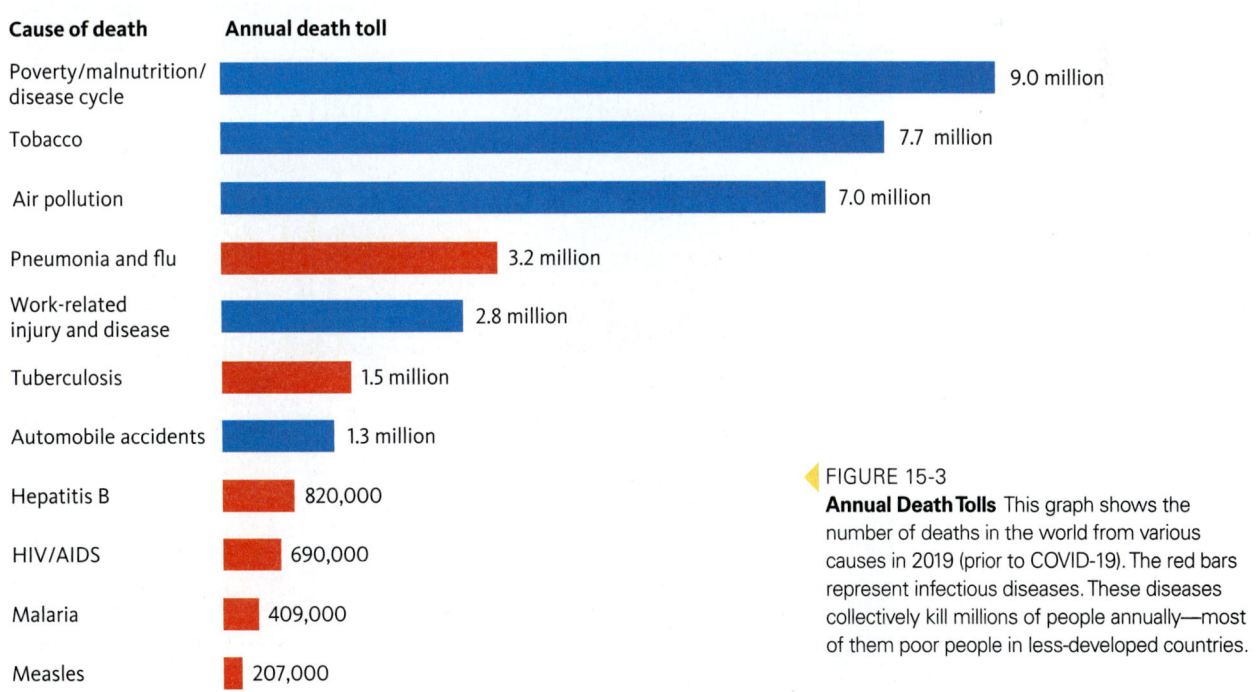

Cause of death	Annual death toll
Poverty/malnutrition/disease cycle	9.0 million
Tobacco	7.7 million
Air pollution	7.0 million
Pneumonia and flu	3.2 million
Work-related injury and disease	2.8 million
Tuberculosis	1.5 million
Automobile accidents	1.3 million
Hepatitis B	820,000
HIV/AIDS	690,000
Malaria	409,000
Measles	207,000

Sources: World Health Organization, Institute for Health Metrics and Evaluation, UN International Labor Organization, and U.S. Centers for Disease Control and Prevention

FIGURE 15-3
Annual Death Tolls This graph shows the number of deaths in the world from various causes in 2019 (prior to COVID-19). The red bars represent infectious diseases. These diseases collectively kill millions of people annually—most of them poor people in less-developed countries.

Tuberculosis strikes about 9 million people per year and kills 1.5 million. More than 95% of these deaths occur in less-developed countries. Because TB is highly contagious and symptoms include a chronic cough, it is spread easily.

Several factors account for the spread of TB since 1990. One is a lack of TB screening and control programs, especially where the disease is most widespread. Population growth, urbanization, and air travel have also aided the spread of TB. They have greatly increased person-to-person contact, especially in poor, crowded areas. A person with active TB might infect several people during a single bus or plane ride. Another problem is that most strains of the TB bacterium have developed genetic resistance to many antibiotics.

Slowing the spread of the disease requires early identification and treatment of people with active TB. This can be challenging. Many people don't show symptoms, so they don't know they are infected. The treatment itself is inexpensive but does take time. Symptoms often disappear after a few weeks, so many patients think they are cured and stop taking the drugs. This can allow TB to recur, possibly in drug-resistant form, and spread to others.

There is little financial incentive for large drug companies to invest a great deal of money on developing drugs to treat TB. Most people who have active TB live in the world's poorest countries. They cannot afford treatment. However, efforts to develop more effective antibiotics and vaccines are being undertaken by governments and private groups. Researchers are also developing new and easier ways to detect TB and to monitor patients for loss of liver function as a possible side effect of taking anti-TB drugs (Explorers At Work).

checkpoint What kind of pathogen is tuberculosis?

Viruses and Parasites Are Killers

Viruses are not affected by antibiotics and can be deadly (Figure 15-5). The biggest viral killer is influenza, as it often leads to fatal pneumonia in less-developed countries. Also known as flu, the influenza virus is transmitted by body fluids. It is also spread through airborne droplets released when an infected person coughs or sneezes. Flu viruses are so easily transmitted that an especially potent flu virus could spread around the world in only a few months. It could cause a pandemic and kill millions of people.

FIGURE 15-4
Effects of TB The red-colored areas in this x-ray show where TB bacteria have destroyed lung tissue.

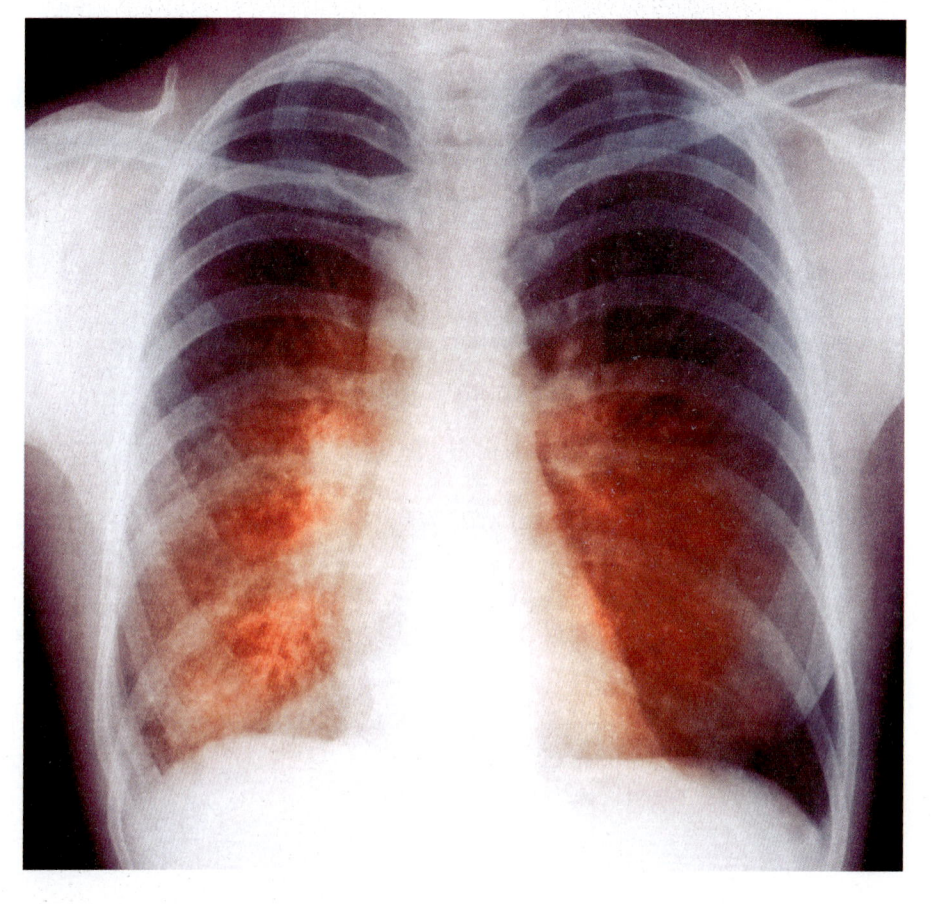

FIGURE 15-5
ON ASSIGNMENT Vicki Jensen is a virologist who researches some of the world's deadliest diseases. Dr. Jensen posed in her protective suit for National Geographic photographer Marco Grob when he visited her lab at the National Interagency Biodefense Campus in Fort Detrick, Maryland.

COVID-19 Coronavirus disease (COVID-19) is a highly infectious disease caused by the SARS-CoV-2 virus. COVID-19 is transmitted via close contact, through respiratory droplets and smaller aerosols. Its symptoms range from fever, cough, and body aches to difficulty breathing, loss of taste and smell, headache, and fatigue. COVID-19 causes mild to moderate illness in most people. However, it may cause severe illness in those with underlying diseases or advanced age (or who are unvaccinated).

This virus was discovered in December 2019 in Wuhan, China, but quickly spread to other countries. On March 11, 2020, the World Health Organization declared COVID-19 a global pandemic. To fight the pandemic, countries around the world encouraged health measures such as social distancing, handwashing, and masking. Travel restrictions and stay-at-home orders soon followed. An international effort also began to develop a vaccine. On December 31, 2020, the WHO issued its first emergency use authorization for a COVID-19 vaccine.

Vaccines have proven to be a very effective tool in battling the spread of COVID-19, including several variants that have subsequently emerged. Even so, by the end of 2021, COVID-19 had claimed 5.3 million lives worldwide, including more than 800,000 in the United States. Health experts were hopeful the virus would become endemic in 2022.

HIV/AIDS The human immunodeficiency virus (HIV) is the virus that causes AIDS. First identified in 1981, HIV and AIDS are major threats to human health worldwide. HIV is transmitted by unsafe sex and exposure to infected blood. Such exposure often happens when infected drug users share needles or when infected mothers pass the virus to their babies during pregnancy or birth. HIV impairs the immune system and leaves the body vulnerable to infections. A person with HIV can live a normal life, especially with proper treatment. In time, HIV can develop into AIDS, which is much more serious.

In 2019, 37.7 million people worldwide (including about 1.2 million in the United States) were living with HIV (Figure 15-6). That same year, 1.7 million new cases of HIV were diagnosed, including roughly 34,000 in the United States. One-third of new cases involved people aged 25–34. AIDS-related deaths in 2019 totalled 690,000.

The treatment for those who become infected with HIV includes a combination of antiviral drugs that can slow the progress of the virus. Better access to these theraputic drugs is thought to have prevented an estimated 12.1 million AIDS-related deaths since 2010. But antiviral drugs have drawbacks, such as possible loss of liver function. In addition, people don't always know to seek treatment. It is estimated that about one of every five people infected with HIV is not aware of it.

AIDS has had a serious impact in sub-Saharan Africa. At its peak, it reduced the life expectancy of the 750 million people living there from 62 years to 47 years on average. Thanks to improved treatments, life expectancy has risen to 61 years today. However, several sub-Saharan countries continue to struggle. For example roughly 21% of Botswana's adult population (aged 15–49) is infected with HIV.

Malaria Almost half of the world's people—most of them living in poor African countries—are at risk of getting malaria (Figure 15-7). People traveling to malaria-prone areas are also at risk because there is no vaccine that can prevent this disease.

Malaria is caused by a parasite that is spread by certain species of mosquitoes. A mosquito first bites an infected person. It picks up the parasite in the person's blood and passes it to the next person it bites. The parasite enters that person's bloodstream and multiplies in the person's liver. In doing so, it destroys its victim's red blood cells.

Symptoms of malaria include intense fever, chills, drenching sweats, severe abdominal pain, vomiting, headaches, and greater susceptibility to other diseases. More than 90% of all malaria cases occur in sub-Saharan Africa. Most involve children younger than five years old. Many children who survive malaria suffer brain damage or impaired learning ability. In 2019, malaria killed about 558,000 people.

Over the course of human history, malarial parasites probably have killed more people than all of the wars ever fought. The spread of malaria did slow during the 1950s and 1960s, a time when widespread draining of swamps and marshes eliminated mosquito breeding areas. These areas were also sprayed with insecticides. Drugs were developed to kill the parasites in people's bloodstreams as well. Since 1970, however, malaria has come roaring back.

There are several reasons for this. Most mosquito species that transmit malaria have become genetically resistant to most insecticides. The parasites themselves have become genetically resistant to common antimalarial drugs. Climate change is also expected to aid the spread of malaria

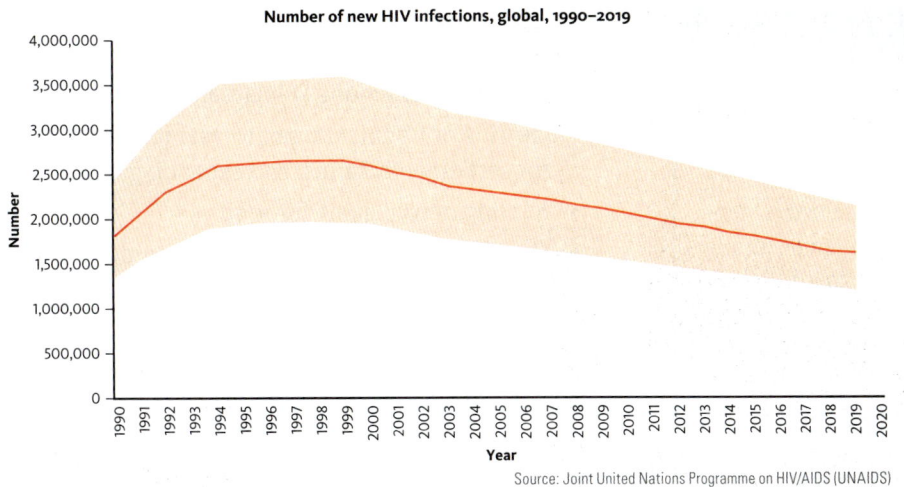

FIGURE 15-6 ▼

Global Prevalence of HIV This figure shows the estimated global total number of new HIV infections each year from 1990 through 2019. The rate of HIV infection has declined 23% in the last decade. Even so, this disease remains classified as a global epidemic.

Number of new HIV infections, global, 1990–2019

Source: Joint United Nations Programme on HIV/AIDS (UNAIDS)

by allowing malaria-carrying mosquitoes to move into warmer temperate areas.

Researchers are working on more effective ways to test, treat, and prevent the spread of malaria:

- As of 2021, the first-ever WHO-recommended vaccine is being used to protect children from 5 months of age.
- Poor people in malarial regions have received free or inexpensive insecticide-treated bed nets (Figure 15-8) and window screens.
- Surfaces inside homes have been sprayed with low concentrations of certain pesticides to reduce mosquitoes.
- Researchers have created a microneedle skin patch as a rapid test for malaria that would eliminate the need to draw blood. This limits risk of infection. They have also developed effective combinations of antimalarial drugs. Genetic engineering is also being researched for its potential to reduce mosquito populations or make them more resistant to malaria parasites.

FIGURE 15-7 ▼

Areas with Malaria About 47% of the world's population lives in areas in which malaria is common.

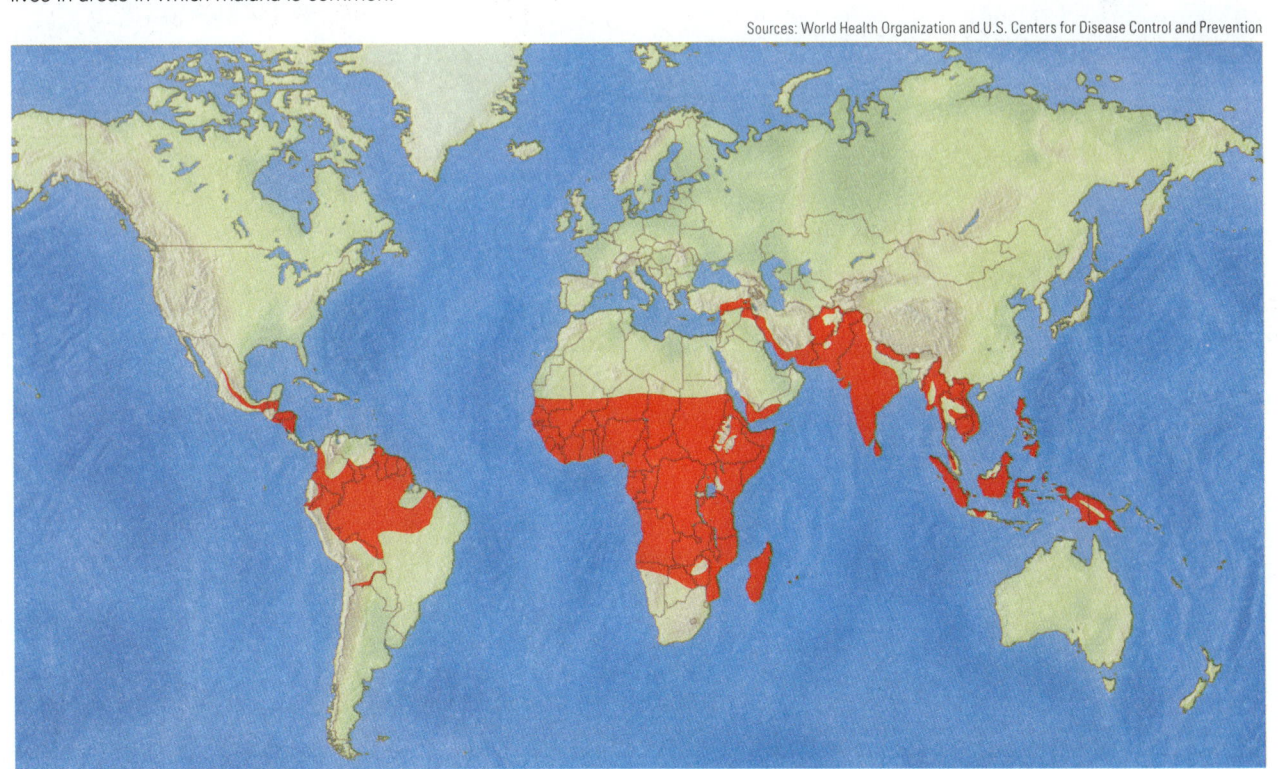

Sources: World Health Organization and U.S. Centers for Disease Control and Prevention

LESSON 15.2   **509**

CONSIDER THIS

Deforestation can help spread malaria.

The clearing and development of tropical forests has led to the spread of malaria among workers and the settlers who follow them. One study found that a 5% loss of tree cover in one part of Brazil's Amazon forest led to a 50% increase in malaria in that study area. The researchers hypothesized that deforestation creates partially sunlit pools of water. The pools make ideal breeding sites for mosquitoes that spread malaria.

Scientists estimate that throughout history more than half of all infectious diseases were originally transmitted to humans from wild or domesticated animals. The West Nile virus (spread by mosquitoes), HIV (originating from chimpanzees), and the avian flu (a flu strain from birds) all fall into this category.

Research into the spread of disease from animals to humans has led to a relatively new field called ecological medicine. Scientists in this field have been able to identify several human practices that aid the spread of diseases among animals and people:

- clearing or fragmenting of forests to make way for settlements, farms, and expanding cities
- hunting of wild game for food
- illegal trade in wild species across borders
- contamination of meat as part of industrialized production processes

Researchers are now monitoring viral "hot spots," areas where people are highly exposed to animals and most at risk, to help control disease outbreaks.

checkpoint What virus kills the most people globally?

Society Can Reduce the Incidence of Infectious Diseases

The percentage of all deaths worldwide from infectious diseases dropped from 35% to 14% between 1970 and 2019. The drop occurred mainly because a growing number of children were immunized against the major infectious diseases.

FIGURE 15-8
Malaria Prevention This baby in Senegal, Africa, is sleeping under an insecticide-treated mosquito net. The net provides a simple way to protect against malaria.

> **CONSIDER THIS**
>
> **Improved sanitation reduces infection.**
> More than one-third of the world's people—2.6 billion—do not have sanitary bathroom facilities. Nearly 1 billion people get their water for drinking, washing, and cooking from sources polluted by animal or human feces. A key to reducing sickness and premature death due to infectious disease is to focus on providing clean latrines and access to safe drinking water.

Between 1990 and 2017, the estimated annual number of children younger than age five who died from infectious diseases fell from 6.5 million to 2.4 million.

Health scientists and public health officials promote the following measures to help prevent or reduce the occurrence of infectious diseases—especially in less-developed countries:

- Increase research on tropical diseases and vaccines.
- Reduce poverty and malnutrition.
- Improve drinking water quality.
- Reduce the unnecessary use of antibiotics.
- Sharply reduce the use of antibiotics on livestock.
- Immunize children against major viral diseases.
- Provide oral rehydration for diarrhea patients.
- Conduct a global campaign to reduce HIV/AIDS.

The WHO estimates that these measures could save the lives of as many as 4 million children under age five each year. Some of these measures are proven, simple, and inexpensive. Improving people's health in less-developed countries would help stabilize their societies. It would also help them move toward more sustainable economies.

Good hygiene can greatly reduce the spread of infectious diseases. For example, people should wash their hands with plain soap frequently and thoroughly (for at least 20 seconds). They should not share personal items such as razors or towels. They should cover cuts and scrapes until they heal. And they should avoid contact with people when they are sick with flu or other transmissible diseases.

checkpoint What is the main reason for the drop in deaths due to infectious diseases since 1970?

15.2 Assessment

1. **Recall** What types of biological hazards does society face?
2. **Contrast** What is the difference between a pandemic and an epidemic?
3. **Explain** Why is tuberculosis difficult to treat and to keep from spreading?
4. **Summarize** What is the process by which a mosquito infects a person with malaria?
5. **Apply** Why is hand washing with plain soap more effective than using antibacterial or antimicrobial soaps and sanitizers?

CROSSCUTTING CONCEPTS

6. **Systems and System Models** How does the preventive measure shown in Figure 15-8 interrupt the system of transmission of malaria?

15.3 How Do Chemical Hazards Threaten Human Health?

CORE IDEAS AND SKILLS
- Understand how chemicals in the environment can harm the human body.
- Explain how effects of toxic chemicals can be reduced or avoided.

KEY TERMS
toxic chemical   mutagen
carcinogen   teratogen

Some Chemicals May Cause Cancers, Mutations, and Birth Defects

There is growing concern about the effects of toxic chemicals on human health. A **toxic chemical** is an element or compound that can cause temporary or permanent harm or death to humans. Toxic chemicals may cause cancers and birth defects and disrupt the human immune, nervous, and endocrine systems.

The U.S. Environmental Protection Agency (EPA) has identified the five toxic chemicals that are the most harmful to human health. They include arsenic, lead, mercury (Case Study), vinyl chloride (used to make PVC plastics), and polychlorinated biphenyls (PCBs). All toxic substances are classified into three major categories based on their effects on humans.

LESSON 15.3   511

Carcinogens are chemicals, some types of radiation, and certain viruses that can cause or promote cancer. Cancer is a disease in which malignant cells multiply uncontrollably and create tumors, or masses of abnormal cells. Tumors can damage the body and often lead to premature death. Examples of carcinogens are arsenic, benzene, formaldehyde, gamma radiation, PCBs, radon, ultraviolet (UV) radiation, vinyl chloride, and certain chemicals in tobacco smoke.

Typically, 10 to 40 years may pass between the initial exposure to a carcinogen and the appearance of detectable cancer symptoms. This time lag helps explain why many healthy teenagers and young adults have trouble believing that their own unhealthy habits, such as smoking, drinking, and poor diet, could lead to some form of cancer before they reach age 50.

Mutagens are the second major type of toxic substance. **Mutagens** include chemicals or forms of radiation that cause or increase the frequency of mutations, or changes, in the DNA molecules found in cells. Most mutations cause no harm, but some can lead to cancers and other disorders. For example, nitrite (NO_2^-) preservatives are used in processed foods such as cured meats. When digested, they form nitrous acid (HNO_2). Nitrous acid can cause mutations linked to increases in stomach cancer in people who consume large amounts of processed foods. Harmful mutations occurring in reproductive cells can be passed on to offspring and to future generations.

Teratogens are a third type of toxic substance. **Teratogens** are chemicals that harm a fetus or embryo or cause birth defects. Ethyl alcohol is a teratogen. It is an ingredient in alcoholic beverages. Women who drink alcoholic beverages during pregnancy increase their risk of having babies with low birth weight. Their babies are also more likely to have a number of physical, developmental, behavioral, and mental complications. Other teratogens include phencyclidine (a drug also known as PCP or angel dust), benzene, formaldehyde, lead, mercury, PCBs, thalidomide, vinyl chloride, and phthalates (THALL-eights).

checkpoint What are the harmful effects of carcinogens?

Some Chemicals May Affect Body Systems

Research involving wildlife and laboratory animals, as well as some studies of humans, has suggested that long-term exposure to some chemicals in the environment can disrupt important body systems.

The Immune System Toxic chemicals may affect the immune system. The immune system consists of specialized cells and tissues that protect the body against disease and harmful substances. It does so by forming antibodies, or specialized proteins that detect and destroy invading agents. Some chemicals such as arsenic and methylmercury can weaken the human immune system. The body then becomes vulnerable to attacks by allergens as well as infectious diseases.

The Nervous System Neurotoxins are natural and synthetic chemicals known to harm the nervous system. The nervous system includes the brain, spinal cord, and peripheral nerves. Neurotoxins can cause behavioral changes, learning disabilities, attention-deficit disorder, paralysis, and death. Examples of neurotoxins include PCBs, arsenic, lead, and certain types of pesticides.

The form of mercury known as methylmercury is an especially dangerous neurotoxin (Figure 15-9). People

FIGURE 15-9 ▼
Deadly Mercury The top two images of smaller brains show tissue damage from mercury poisoning that ultimately killed two young girls. The image on the bottom shows a healthy brain.

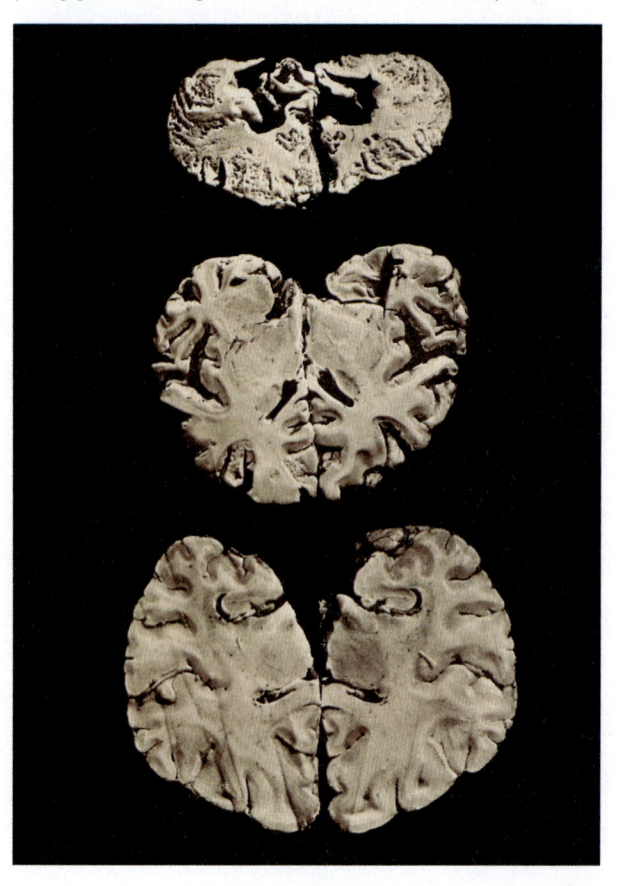

do not have much control over natural sources of mercury. However, they can take steps to prevent mercury emissions from human activities. Examples include phasing out waste incineration, removing mercury from coal before it is burned, and switching from coal to natural gas and renewable energy resources. Mercury can also be kept out of the environment by labeling products that contain mercury, such as batteries. These products can be recycled or repurposed so they do not end up in landfills.

The Endocrine System The body's endocrine system uses chemicals called hormones to regulate sexual reproduction, growth, development, learning ability, and behavior. Each hormone has a unique molecular structure that allows it to attach to certain parts of cells called receptors. Once a hormone attaches to a receptor, it sends a chemical signal. The signal causes the cell to respond in a certain way.

Some synthetic chemicals have molecules with structures similar to those of natural hormones. They can mimic or block natural hormones if they enter the body. They can interfere with the production and function of hormones as a result.

These molecules are called hormonally active agents (HAAs). They are also known as endocrine disruptors. Examples of HAAs include some herbicides and pesticides, lead, mercury, phthalates, various fire retardants, and BPA (Science Focus 15.2). They are ingredients in toiletries, cosmetics, and many hard and soft plastics. For example, the Food and Drug Administration (FDA) identified two chemicals widely used in antibacterial soaps and deodorants as possible endocrine disruptors.

Scientists are especially concerned about HAAs called phthalates. These chemicals are used to make plastics more flexible and to make cosmetics easier to apply to the skin. They are found in a variety of products, including many detergents, toiletries, cosmetics, baby powders, sunscreens, and the coatings on many timed-release drugs. They are also found in polyvinyl chloride (PVC) plastic products such as soft vinyl toys, teething rings, certain medical supplies, shower curtains, and swimming pools (Figure 15-10). Some food and drink containers also contain phthalates, which can leach out when heated.

Exposure of laboratory animals to high doses of various phthalates has caused birth defects, cancer, kidney and liver damage. It has also caused immune system suppression, and abnormal sexual development in these animals. Studies have linked exposure of human babies to phthalates with early puberty in girls and infertility problems in men.

More than a dozen countries—including the United States—have banned certain phthalates from use in children's toys and products. In the United States, use of phthalates in food contact materials is also restricted but it remains unregulated in other products, such as personal care products.

Concern is growing that increasing levels of hormone disrupters in our bodies may be causing more frequent health problems. Such problems include decreased sperm counts and mobility; rising rates of testicular cancer and genital birth defects in men; and rising rates of breast cancer in women. Even so, some scientists argue more research is needed to confirm a link between these medical problems and HAA levels in humans.

Concerns about BPA, phthalates, and other HAAs show how difficult it can be to assess the potential risk from exposure to very low levels of various chemicals.

FIGURE 15-10
Synthetic Chemicals The surfaces of many soft plastic products like this swimming pool often contain phthalates.

SCIENCE FOCUS 15.2

THE BPA CONTROVERSY

Bisphenol A (BPA) serves as a hardening agent in certain plastics that are used in a variety of products. People use many of these products daily, including certain types of reusable water bottles, microwave dishes, and food storage containers, as well as certain types of pacifiers, sipping cups, and baby bottles.

BPA is also used to make the plastic resin coatings that line nearly all food and soft drink cans—including those for baby formula and baby food. The coatings help preserve food and protect cans from high temperatures and rust. However, BPA in the coatings can leach into food. People can also be exposed to BPA by touching thermal paper used to produce some cash register receipts.

A CDC study indicated that nearly all Americans age six and older had trace levels of BPA in their urine. Children and adolescents generally had higher levels than adults. These levels were well below the acceptable level set by the EPA. However, that level was established in the late 1980s, when little was known about the potential effects of BPA on human health.

Research indicates that BPA in plastics can leach into water or food when the plastic is heated to high temperatures, microwaved, or exposed to acidic liquids. One study done by Harvard University Medical School had participants drink regularly from water bottles made with BPA for one week. In that short time, participants' levels of urinary BPA increased by 66%.

Studies by independent laboratories have reported significant adverse effects on test animals from exposure to very low levels of BPA. These effects included brain damage, early puberty, decreased sperm quality, certain cancers, heart disease, obesity, liver damage, impaired immune function, type 2 diabetes, hyperactivity, and impaired learning. On the other hand, there have been studies funded by the chemical industry that found no evidence or only weak evidence of adverse effects in test animals exposed to very low levels of BPA.

These conflicting findings may be confusing for consumers trying to decide which products to buy. Fortunately, most manufacturers offer BPA-free alternatives for the above-mentioned products. Also, many consumers can avoid plastic containers with a #7 recycling code (which indicates BPA may be present). People can also buy powdered infant formula instead of liquid formula from metal cans. They can choose glass bottles and food containers. And they can use glass, ceramic, or stainless steel mugs instead of plastic cups.

Many manufacturers have replaced BPA with bisphenol S (BPS). However, studies indicate that BPS can have effects similar to those of BPA, and BPS is now showing up in human urine at levels similar to those of BPA. There are substitutes for the plastic resins containing BPA or BPS but these replacements are more expensive. Potential health effects of some chemicals they may contain also need to be further studied.

The European Union, Canada, the United States, and a number of other countries ban the sale of plastic baby bottles and sipping cups containing BPA. Some countries, as well as more than a dozen U.S. states, also ban or restrict BPA in food contact materials intended for children. Recently, there have been calls for the FDA to strictly limit BPA in all plastics that contact food.

FIGURE 15-11
Many manufacturers have voluntarily chosen to stop using BPA. They label their products as BPA Free.

Thinking Critically

Form an Opinion Should plastics that contain BPA or BPS be banned from use in all children's products and in food containers? Explain your reasoning.

Resolving these uncertainties will take decades of research. Some scientists argue that until that happens, people should take the precaution of sharply reducing their use of products that contain potentially harmful endocrine disrupters. They recommend pregnant women, infants, young children, and teenagers be especially cautious.

People can reduce their exposure to HAAs a number of ways. First, people can choose to eat organic foods rather than processed, pre-packaged, or canned foods. They can also avoid using plastic cookware and plastic food and drink containers. Consumers can opt for natural cleaning and personal care products and avoid artificial air fresheners, fabric softeners, and dryer sheets. They can buy shower curtains made with natural fabric instead of vinyl. Finally, people can buy BPA- and phthalate-free sipping cups, pacifiers, and toys for babies and children.

checkpoint What problems arise from hormonally active agents (HAAs)?

15.3 Assessment

1. **Compare and Contrast** What similarities and differences are there between carcinogens, mutagens, and teratogens?
2. **Explain** How can toxic chemicals affect human immune, nervous, and endocrine systems?
3. **Evaluate** What steps could you take personally to reduce exposure to hormone-disrupting chemicals such as BPA?

SCIENCE AND ENGINEERING PRACTICES

4. **Engaging in Argument** If you were to perform a cost-benefit analysis of using HAAs in food containers, children's toys, and other plastic products, what information would you need to consider? Would this be enough information to determine whether HAAs should be used in these products? Why or why not?

15.4 How Can People Evaluate Risks from Chemical Hazards?

CORE IDEAS AND SKILLS
- Understand how scientists determine and measure toxicity.
- Understand the potential threat of trace chemicals and the controversy surrounding their regulation.
- Explain the precautionary principle and how it can be applied.

KEY TERMS

| toxicology | dose | precautionary |
| toxicity | response | principle |

Many Factors Determine the Toxicity of Chemicals

Toxicology is the study of the harmful effects of chemicals on humans and other organisms. **Toxicity** is a measure of the ability of a substance to cause injury, illness, or death to a living organism. A basic principle of toxicology is that any synthetic or natural chemical can be harmful if ingested or inhaled in a large enough quantity. But the critical question is: "What level of exposure to a particular toxic chemical will cause harm?"

This is a difficult question to answer because many variables must be considered when estimating the effects of human exposure to chemicals. One key variable is the dose. The **dose** is the amount of a harmful chemical that a person has ingested, inhaled, or absorbed through the skin at any one time.

Age is another variable that impacts how a person is affected by chemical exposure. Toxic chemicals usually have a greater effect on elderly adults. Fetuses, infants, and children are also more vulnerable to chemical exposure. They are exposed to proportionately more toxins per unit of body weight.

Infants and children are exposed to toxins in dust and soil when they put their fingers and other objects in their mouths. They have less-developed immune systems to protect against these toxins. Current research suggests exposure to chemical pollutants in the womb may be related to increasing rates of autism, childhood asthma, and learning disorders.

The EPA proposes that in determining any risk, regulators should assume that the risk factor for children is 10 times higher than the risk for adults.

FIGURE 15-12
Nonhuman primates like these common marmosets are used for biomedical research because they are genetically similar to humans. The use of primates for such research has sharply declined as other methods such as computer modeling and the use of tissue cultures have become more reliable.

Some health scientists say that to be on the safe side, regulators should assume that the risk for children is 100 times the risk for adults.

A person's genetic makeup is another factor that determines his or her sensitivity to a particular toxin. So does the condition of the body's detoxification systems, including the liver, lungs, and kidneys. Individuals suffering from immune deficiencies or other illnesses may be more susceptible to toxins.

A number of characteristics of the chemical itself also play a role in determining its toxicity:

Solubility: Water-soluble toxins can get into the aqueous solutions that surround cells inside the body. Oil- or fat-soluble toxins can actually penetrate cells' membranes. While the body may pass water-soluble toxins as waste, oil- or fat-soluble toxins can accumulate in body tissues and cells.

Persistence: Some chemicals are resistant to breaking down. Persistent chemicals are more likely to remain in the body and have long-lasting harmful health effects.

Bioaccumulation and biomagnification: As you learned in Chapter 7, animals that eat higher on the food chain are more susceptible to the effects of fat-soluble toxic chemicals. This is due to the magnified concentrations of the toxins in their bodies. Examples of chemicals that can be biomagnified include DDT, PCBs, and methylmercury (Case Study).

The health damage resulting from exposure to a chemical is called the **response**. An *acute effect* is an immediate or rapid response ranging from dizziness to death. A *chronic effect* is a permanent or long-lasting response and results from exposure to a single dose or to repeated lower doses of a harmful substance. Kidney and liver damage are examples of chronic effects.

Natural and synthetic chemicals are both potentially safe or toxic. In fact, many synthetic chemicals, including many of the medicines we take, are quite safe if used as intended. Conversely, many natural chemicals such as lead and mercury are deadly.

checkpoint What factors determine the toxicity of a chemical?

Scientists Use Various Tests to Estimate Toxicity

The most widely used method for determining toxicity involves tests with live laboratory animals (Figure 15-12). Scientists expose a population of such animals to measured doses of a specific substance under controlled conditions. Laboratory-bred mice and rats are widely used because, as mammals, their systems function similarly to human systems. Also, they are small and can reproduce rapidly under controlled laboratory conditions.

Scientists estimate the toxicity of a chemical by testing the effects of various doses of the chemical on test organisms. They plot the results in a dose-response curve (Figure 15-13). Part of the process involves determining the lethal dose—the dose that will kill an animal. A chemical's median lethal dose (LD50) is the dose that can kill 50% of the animals in a test population within a given time period. Scientists use mathematical models to extrapolate, or estimate, the effects of a chemical on humans based on results from testing that chemical on lab animals.

Chemicals vary widely in toxicity (Figure 15-14). Some cause serious harm or death after a single, very low dose. Others require such huge dosages to be harmful it is nearly impossible to get the required dose into the body. Most chemicals fall between these two extremes.

Animal tests have drawbacks. They typically take two to five years to complete and involve hundreds to thousands of test animals. They can cost as much as $2 million per substance tested. Some tests are painful for test animals and can harm or kill them. Animal welfare groups want to limit or ban the use of test animals and, at the very least, make sure they are treated humanely. Some scientists also challenge the validity of extrapolating data from laboratory animals.

FIGURE 15-13 ▼
Lethal Dose This hypothetical dose-response curve illustrates how scientists can estimate the LD50 for a chemical.

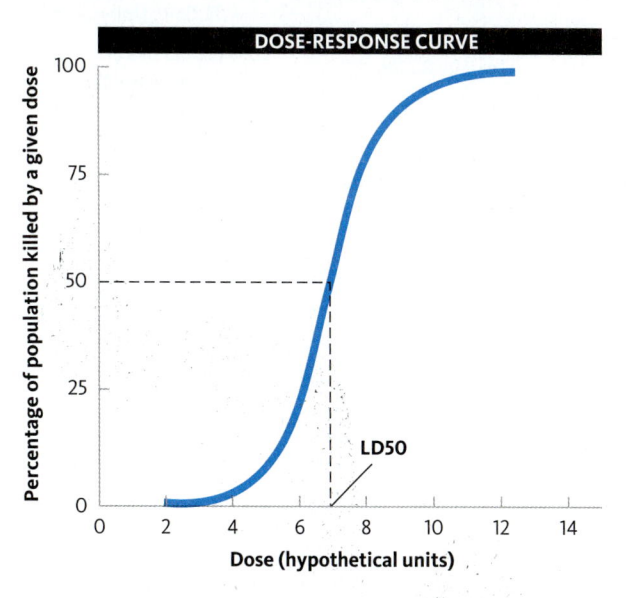

Source: Cengage Learning

LESSON 15.4   517

They argue that important differences exist between human and animal body systems.

Other scientists say that such tests and models can work fairly well when the correct experimental animal is chosen or when a chemical is toxic to several different test-animal species. This is especially true for testing that reveals cancer risks. Even so, other more humane methods for toxicity testing are increasingly being used in place of live animals. They include running computer simulations and using individual animal cells instead of whole, live animals. High-speed robot testing devices can now screen the biological activity of more than 1 million compounds a day to help determine their possible toxic effects.

Estimating toxicities via laboratory experiments is complicated. In real life, each of us is exposed to a variety of chemicals, some of which can interact in ways that decrease or enhance their individual effects. Toxicologists already have great difficulty in estimating the toxicity of a single substance. Evaluating mixtures of potentially toxic substances, isolating the culprits, and determining how they can interact with one another is overwhelming. For example, just studying the interactions among three of the 500 most widely used industrial chemicals would take 20.7 million experiments—a physical and financial impossibility.

Scientists use several other methods to determine toxicity. For example, case reports, usually made by physicians, provide information about people who have become sick or died after exposure to a chemical. Such information often involves accidental or deliberate poisonings, drug overdoses, homicides, or suicide attempts. Case reports can provide clues about environmental hazards and suggest the need for laboratory investigations. However, they are not entirely reliable because the actual dosage and the exposed person's health status may be unknown.

Epidemiological studies can also be useful. These studies compare the health of people exposed to a particular chemical (the experimental group) with the health of a similar group not exposed to the chemical (the control group). The goal is to determine whether the statistical association between exposure to a toxic chemical and an observed health problem is strong, moderate, weak, or undetectable.

The results of epidemiological studies can be limited. In many cases, too few people have been exposed to high enough levels of a toxic agent to detect statistically significant differences. The studies usually take a long time. It is also a challenge to isolate one chemical culprit. Linking an observed effect with exposure to a particular chemical is difficult because people are exposed to many different toxic agents throughout their lives. They can also vary in their sensitivity to such chemicals. Finally, epidemiological studies cannot help to evaluate hazards from new technologies or chemicals to which people have not yet been exposed.

checkpoint What is a dose-response curve?

FIGURE 15-14

Toxicity Ratings and Average Lethal Doses for Humans			
Toxicity Rating	**LD50 (milligrams per kilogram of body weight)***	**Average Lethal Dose†**	**Examples**
Supertoxic	Less than 5	Less than 7 drops	Nerve gases, botulism toxin, mushroom toxin, dioxin (TCDD)
Extremely toxic	5–50	7 drops to 1 teaspoon	Potassium cyanide, heroin, atropine, parathion, nicotine
Very toxic	50–500	1 teaspoon to 1 ounce	Mercury salts, morphine, codeine
Moderately toxic	500–5,000	1 ounce to 1 pint	Lead salts, DDT, sodium hydroxide, sodium fluoride, sulfuric acid, caffeine, carbon tetrachloride
Slightly toxic	5,000–15,000	1 pint to 1 quart	Ethyl alcohol, household cleansers, soaps
Essentially nontoxic	15,000 or greater	More than 1 quart	Water, glycerin, table sugar

*Dosage that kills 50% of individuals exposed

†Amounts of substances in liquid form at room temperature that are lethal when given to a 70-kg (150-lb) human

Source: Cengage Learning

Are Trace Levels of Toxic Chemicals Harmful?

Almost everyone who lives in a more-developed country is now exposed to potentially harmful chemicals in their environment (Figure 15-15). Many of these chemicals build up to trace levels in their blood and in other parts of their bodies. CDC studies have found that the average American's blood contains traces of 212 different chemicals. Some are potentially harmful, such as arsenic and BPA. It is unclear if trace amounts of various chemicals in air, water, food, and the human body are cause for concern. In most cases there are too few data to determine the effects of exposures to low levels of these chemicals.

Some scientists view exposures to trace amounts of synthetic chemicals with alarm, especially because of their potential long-term effects on human body systems. Other scientists view the threats from such exposures as minor. These scientists contend that the concentrations of such chemicals are so low that they are harmless. They also point out that average life expectancy has been increasing in most countries for decades, especially in more-developed countries.

The U.S. National Academy of Sciences estimates that only 10% of the more than 85,000 registered synthetic chemicals in commercial use have been thoroughly screened for toxicity. Only 2% of these chemicals have been adequately tested to determine whether they are carcinogens, mutagens, or teratogens. Hardly any of the chemicals in commercial use have been screened for possible damage to the human nervous, endocrine, and immune systems.

Lack of data and high costs make regulation difficult. In fact, federal and state governments do not supervise the use of nearly 99.5% of the commercially available chemicals in the United States. The problem is significantly worse in less-developed countries.

Most scientists call for more research on the health effects of trace levels of synthetic chemicals.

FIGURE 15-15

Potential for Exposure A number of potentially harmful chemicals are found in many homes.

Shampoo Perfluorochemicals to add shine

Teddy bear Some stuffed animals contain flame retardants and/or pesticides

Clothing Can contain perfluorochemicals

Nail polish Perfluorochemicals and phthalates

Baby bottle Can contain bisphenol A

Perfume Phthalates

Mattress Flame retardants in stuffing

Hairspray Phthalates

Carpet Padding and carpet fibers can contain flame retardants, perfluorochemicals, and pesticides

Food Some food can contain bisphenol A

Milk Fat can contain flame retardants

TV Wiring and plastic casing contain flame retardants

Frying pan Nonstick coating can contain perfluorochemicals

Sofa Foam padding can contain flame retardants and perfluorochemicals

Tile floor Can contain perfluorochemicals, phthalates, and pesticides

Fruit Imported fruit may contain pesticides banned in the United States

Water bottle Can contain bisphenol A

Computer Flame retardant coatings of plastic casing and wiring

Toys Vinyl toys contain phthalates

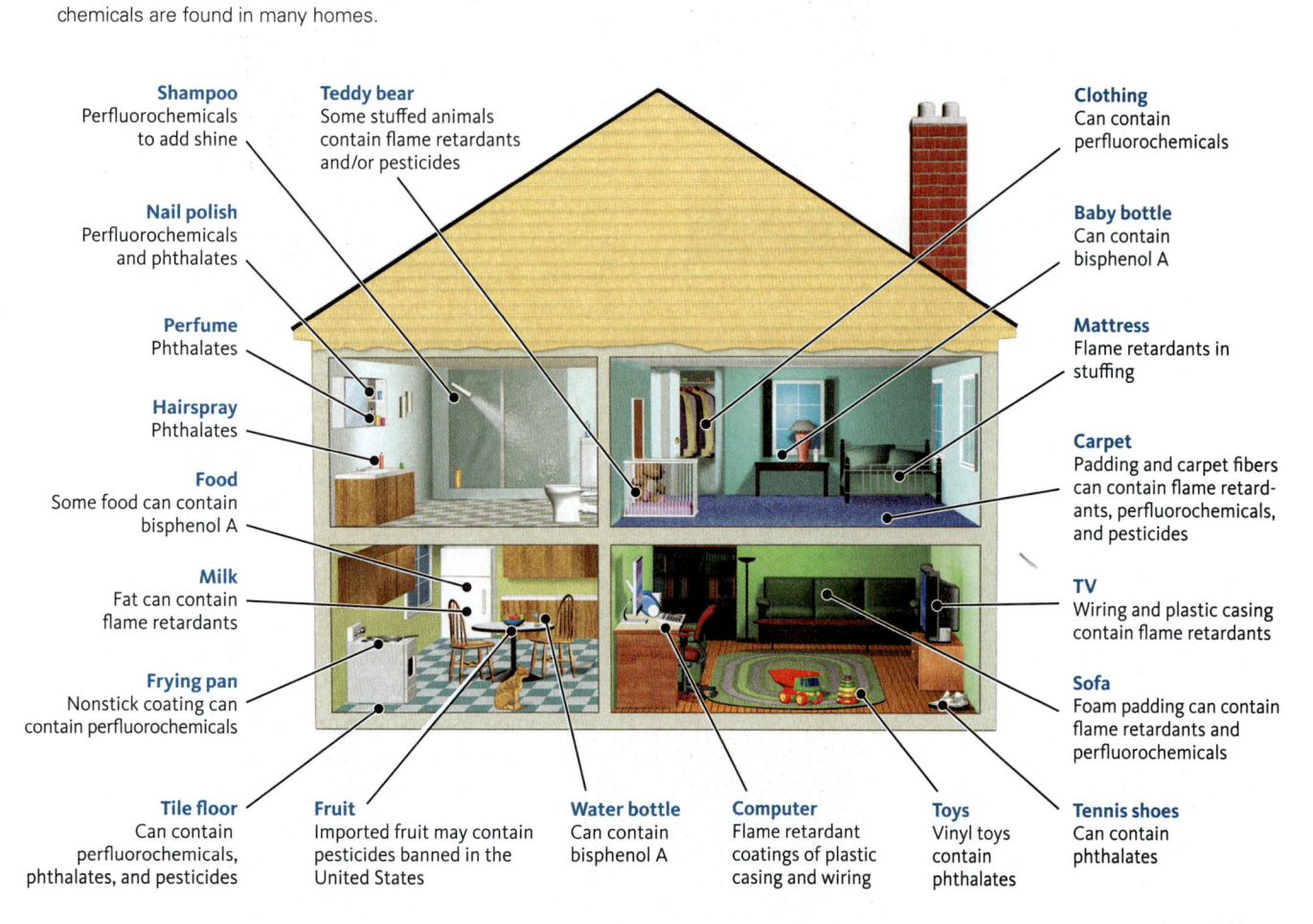

Tennis shoes Can contain phthalates

Scientists and regulators work to minimize harm and to account for some of the uncertainty regarding health effects. They usually set allowed levels of exposure to toxic substances at 1/100th or even 1/1,000th of estimated harmful levels.

checkpoint Why don't scientists have more concrete information about the toxicity of many chemicals?

How Widely Should Society Apply the Precautionary Principle?

Some scientists and health officials are pushing for much greater emphasis on pollution prevention to protect human health. They say chemicals that are known or suspected to cause significant harm should not be released into the environment as pollutants. Pollution prevention requires finding harmless or less harmful substitutes for toxic and hazardous chemicals. It also requires recycling toxic chemicals within production processes to keep them from reaching the environment.

Pollution prevention is a strategy for implementing the precautionary principle. According to the **precautionary principle**, when there is substantial preliminary evidence that an activity, technology, or chemical can harm living things or the environment, decision makers should take measures to prevent or reduce such harm, rather than waiting for more conclusive evidence. The question of how far to apply the cautionary principle is a controversial one.

> ### CONSIDER THIS
>
> **Pollution prevention pays.**
>
> The U.S.-based 3M Company makes 60,000 different products in 151 sites around the world. In 1975, 3M began a Pollution Prevention Pays (3P) program. The company rewards employees who come up with plans to reduce pollution and energy and material use. 3M reported that as of 2020, its 3P projects had prevented more than 2.4 million metric tons (2.7 million tons) of pollutants from reaching the environment. Their efforts had saved the company an estimated $2.3 billion.

Those in Favor People who favor using the precautionary principle argue that anyone proposing to introduce a new chemical or technology should bear the burden of establishing its safety. They also point out that using the precautionary principle is proactive. It focuses efforts on finding solutions to pollution problems that are based on prevention rather than on cleanup.

Applying the precautionary principle can also be good for business. It reduces health risks for employees. It frees businesses from having to deal with pollution regulations and reduces the threat of lawsuits from injured parties. Companies may increase their profits from sales of safer products and innovative technologies. They may improve their public image by operating in this manner.

Finally, proponents also argue that society has an ethical responsibility to reduce serious risks to human health, to the environment, and to future generations. This is in keeping with the ethics factor of sustainability.

Those Opposed Opponents of the precautionary principle argue that chemicals and technology should be used freely if they can further human health, comfort, or the economy. Manufacturers and businesses contend that widespread application of the much more conservative precautionary approach would make it too expensive and almost impossible to introduce any new chemical or technology. Some chemical producers argue that it does not take into account that there is inherent uncertainty in any scientific assessment of risk.

Some progress is being made toward applying the precautionary principle. One example is a 2011 rule issued by the U.S. EPA to control emissions of certain pollutants from older coal-burning plants in 28 states. The EPA recognized that these emissions have far-reaching effects. Prevailing winds carry emissions of mercury and harmful fine-particle pollutants from midwestern states, where the plants are located, to eastern states, where they ultimately get deposited (Figure 15-16). Thanks to this EPA rule, mercury emissions from power plants fell by 86% between 2011 and 2017. Roughly 11,000 premature deaths have been averted annually. In 2020, the Trump adminstration weakened the rule's legal standing, declaring the cost to industry for compliance was "not appropriate." A proposal to revoke this finding was introduced by the EPA in early 2022.

Movement is also happening on an international scale. A United Nations treaty known as the Minamata Convention seeks to curb most human-related inputs of mercury into the environment. The Convention entered into force in 2017. By the end of

FIGURE 15-16 ▼

Mercury Emissions This map shows levels of atmospheric wet mercury (Hg) in the lower 48 states in 2020. Why do the highest levels occur mainly in the eastern half of the United States?

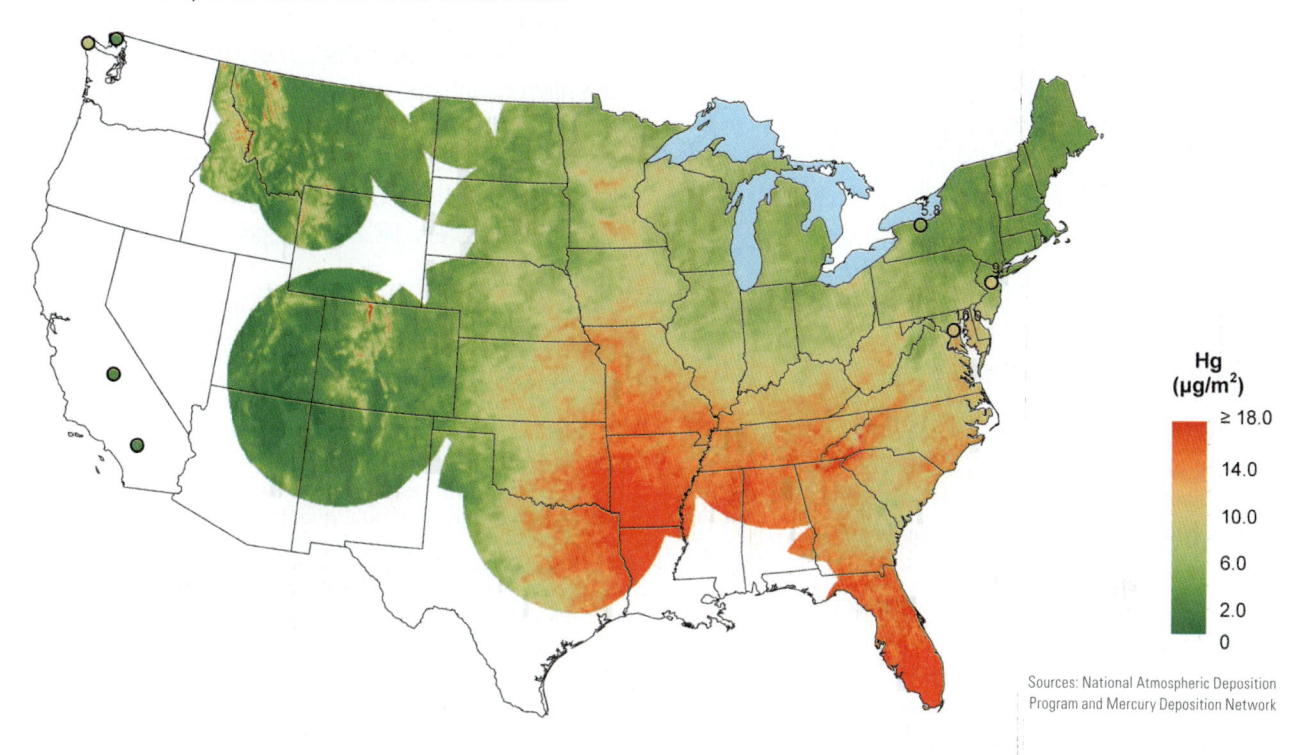

Hg (µg/m²)
≥ 18.0
14.0
10.0
6.0
2.0
0

Sources: National Atmospheric Deposition Program and Mercury Deposition Network

2021, 128 countries had signed and more than 100 countries had ratified the treaty, including the United States. Participating nations must require use of the best available mercury emission-control technologies within five years. The treaty also restricts the use of mercury in common household products and measuring devices.

There are major challenges to applying the precautionary principle more widely in the United States. The current regulatory system makes it virtually impossible for the government to limit or ban the use of toxic chemicals. However, there has been progress here as well. In 2016, the EPA was given new powers to review and regulate chemicals. In 2019, the EPA designated 20 chemicals commonly found in consumer products as "high priority" for risk evaluation and possible regulation and began evaluating these chemicals.

checkpoint What is the precautionary principle?

15.4 Assessment

1. **Define** What is LD50 and why is it important to researchers?
2. **Recall** Describe ways that researchers can test chemicals for their toxicity.
3. **Explain** Why are infants and children often more susceptible to the effects of toxic substances than adults?
4. **Apply** How is the EPA's effort to control emissions from midwestern coal-burning power plants a way to apply the precautionary principle?

SCIENCE AND ENGINEERING PRACTICES

5. **Engaging in Argument** Some say the precautionary principle is analogous to the judicial concept "guilty until proven innocent." In the United States, one is "innocent until proven guilty" in the court system. Do you think regulators in the United States should apply the precautionary principle to chemicals for which toxicity levels have not been determined? Explain your reasoning.

15.5 How Do People Perceive and Avoid Risks?

CORE IDEAS AND SKILLS
- Understand how to analyze and evaluate risks.
- Identify ways to more rationally face risks.
- Recognize that while some hazards are unavoidable, others can be reduced through lifestyles and choices.

KEY TERM
risk analysis

Sources of Health Risks

People can reduce the major risks they face by becoming informed, thinking critically about risks, and making careful choices. **Risk analysis** involves identifying hazards and evaluating their risks (risk assessment). It involves ranking risks and deciding on ways to manage them (risk management). Finally, it involves communicating information about risks with decision makers and the public.

In terms of the number of deaths caused each year, the greatest risk by far is poverty. Deaths due to poverty are the result of malnutrition and increased susceptibility to normally nonfatal infectious diseases and often-fatal infectious diseases transmitted by unsafe drinking water (Figure 15-17).

Studies show the four greatest risks that shorten people's lives are living in poverty, being born male, smoking, and being obese. Some risks that impact human life expectancy are unavoidable, such as geographic location or gender. But some of the risks that are most likely to cause premature death stem from choices people make and the lifestyles they follow (Figure 15-18).

Cigarettes and E-Cigarettes Cigarette smoking is the world's leading cause of suffering and premature death among adults. It is also the most preventable. Annually, tobacco contributes to the premature deaths of about 7 million people, resulting from 25 illnesses. These illnesses include heart disease, stroke, type 2 diabetes, lung and other cancers, memory impairment, bronchitis, and emphysema (Figure 15-19). By 2030, the annual death toll from smoking-related diseases is projected to reach more than 8 million. About 80% of these deaths are expected to occur in less-developed countries.

The World Health Organization estimates that tobacco use contributed to the deaths of 100 million people during the 20th century. Estimates suggest tobacco use could kill 1 billion people during this century unless individuals and governments act now to dramatically reduce smoking.

Overwhelming scientific consensus states that the nicotine inhaled in tobacco smoke is highly addictive. A British government study showed that adolescents who smoke more than one cigarette have an 85% chance of becoming long-term smokers. Cigarette smokers die, on average, 10 years earlier

FIGURE 15-17
Poverty Risks This graph shows the number of people, often living in poverty, who lack access to various basic amenities.

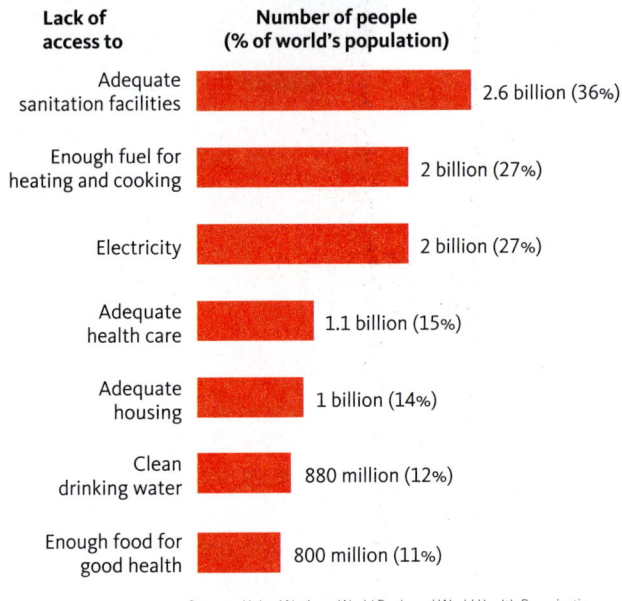

Lack of access to	Number of people (% of world's population)
Adequate sanitation facilities	2.6 billion (36%)
Enough fuel for heating and cooking	2 billion (27%)
Electricity	2 billion (27%)
Adequate health care	1.1 billion (15%)
Adequate housing	1 billion (14%)
Clean drinking water	880 million (12%)
Enough food for good health	800 million (11%)

Sources: United Nations, World Bank, and World Health Organization

FIGURE 15-18
Lifestyle Risks Some of the leading causes of death in the United States are preventable to the extent that they result from lifestyle choices (data collected prior to COVID-19).

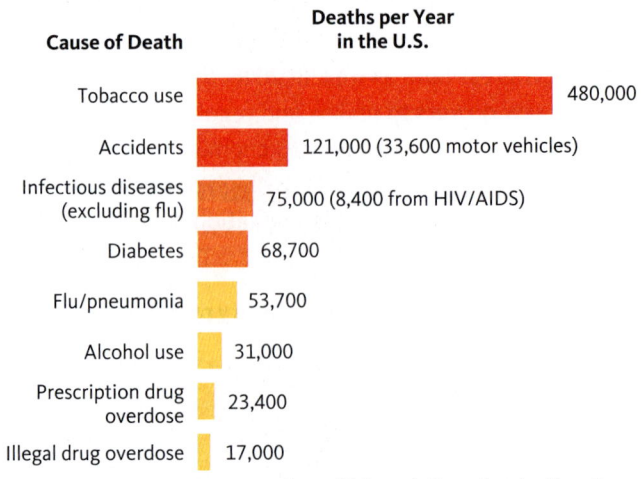

Cause of Death	Deaths per Year in the U.S.
Tobacco use	480,000
Accidents	121,000 (33,600 motor vehicles)
Infectious diseases (excluding flu)	75,000 (8,400 from HIV/AIDS)
Diabetes	68,700
Flu/pneumonia	53,700
Alcohol use	31,000
Prescription drug overdose	23,400
Illegal drug overdose	17,000

Source: U.S. Centers for Disease Control and Prevention

FIGURE 15-19 ▼
Signs of Lung Disease There is a startling difference between healthy human lungs (left) and the lungs of a person who died of emphysema (right). The major causes of emphysema are prolonged smoking and exposure to air pollutants.

than nonsmokers. But kicking the habit—even at 50 years of age—can cut such a risk in half. If people quit smoking by the age of 30, they can avoid nearly all the risk of dying prematurely.

Breathing secondhand smoke also poses health hazards. Estimates found that secondhand smoke contributes to 7,300 lung cancer deaths and 34,000 deaths from heart disease in the United States each year. Worldwide, secondhand smoke contributes to about 1 million deaths each year.

New studies identify another risk associated with smoking. Called thirdhand smoke, it is the residue and chemical pollutants left behind after a cigarette is extinguished. The chemicals can cling to clothes and furniture or combine with other molecules in the air to create other carcinogenic chemicals. Thirdhand smoke contains more than 250 chemicals and is particularly harmful to children, babies, and pets.

Smoking is actually on the decline in the United States. Several factors are credited for this decline.

The media has highlighted the harmful health effects of smoking. Mandatory health warnings also appear on cigarette packs. Many states have sharply increased cigarette taxes, and sales of cigarettes to minors are banned. Smoking has also been banned in workplaces, bars, restaurants, and public buildings.

More people are switching from traditional cigarettes to various forms of electronic nicotine delivery systems, or e-cigarettes. These devices contain pure nicotine dissolved in a syrupy solvent that contains flavoring to enhance taste and smell. A battery in the device heats the nicotine solution and converts it to a vapor as the user inhales. Smoking e-cigarettes is called vaping.

E-cigarettes do reduce or eliminate inhalation of some harmful chemicals found in regular cigarettes. However, they still expose users to nicotine, heavy metals, carcinogens, and other toxic particles. In addition, their vapor contains toxic nanoparticles that can travel deep into the lungs.

LESSON 15.5

NATIONAL GEOGRAPHIC | EXPLORERS AT WORK
Dan Buettner, Author

For the first time, Americans will live shorter lives than the generation before them. National Geographic Fellow Dan Buettner blames this backward step on the lifestyles Americans follow. Today, such lifestyle choices lead to increased rates of childhood obesity, which lead to illnesses such as heart disease and diabetes, which in turn lead to premature death.

Buettner is on the hunt for the secrets to longevity. He has traveled around the world to places where people regularly live past the age of 100. Buettner's travels have shed light on the secrets to long life. He shares what he's learned in two best-selling books.

Buettner calls regions with long-lived populations "Blue Zones." People in the various Blue Zones have different secrets to living longer. Some drink tea, others go to church, and yet others try to be "likable." But every group has five lifestyle choices in common: don't smoke, stay physically active, keep socially engaged, cherish family, and eat a plant-based diet.

Some people claim they don't want to grow old because for them it means a life full of loneliness, discomfort, disability, and other difficulties. But the lessons that Buettner shares from the Blue Zones may help many people instead live longer, happier, and healthier lives.

Recent studies also found that many flavorings in e-cigarettes contain several chemicals that may cause severe respiratory disease. Unfortunately, adolescents find these flavors appealing, and vaping by teens has become "an epidemic," according to the U.S. Surgeon General. The 2021 National Youth Tobacco Survey found that more than 2 million middle and high school students were currently using e-cigarettes. Any level of nicotine use by adolescents (a developmental phase now estimated to last until about age 26) can impact learning and attention and can lead to addiction.

The European Union (EU) has banned advertising and selling e-cigarettes and tobacco products to minors. In the United States, sales to people under 21 years of age are banned. Certain types of vaping devices containing kid-friendly flavors are also being banned if they cannot become authorized by the FDA. While the FDA has banned the sales of more than 50,000 flavored e-cigarette products, many still are under review and remain on the market, including those made by companies with the largest market share. Young people are also turning to other flavored products, such as disposable e-cigarettes, which are not yet subject to the same regulations.

checkpoint Why is poverty the greatest health risk?

People Evaluate Risks Irrationally

Most people improperly assess the relative risks from hazards that they face. Many people deny or shrug off the high-risk chances of death or injury from the activities they enjoy. Examples include motorcycling (death by motorcycle crash 30 times more likely than death by car crash), smoking cigarettes (1 in 250 by age 70 for a pack-a-day smoker), driving (1 in 3,300 without a seatbelt and 1 in 6,070 with a seatbelt), and hang gliding (1 in 1,250, see Figure 15-20). Yet some of these same people may truly fear their chances of being killed by influenza (1 in 130,000), lightning (1 in 3 million), a commercial airplane crash (1 in 9 million), or a shark attack (1 in 281 million).

Five factors can cause people to see an activity, technology, or product as being more or less risky than experts judge it to be. The first factor is fear. Research shows that fear causes people to overestimate risks and to worry more about unusual risks than they do about common, everyday risks. People tend to overestimate numbers of deaths caused by tornadoes, floods, fires, homicides, cancer, and terrorist attacks. They underestimate deaths from flu, diabetes, asthma, heart attack, stroke, and automobile accidents.

FIGURE 15-20

Risky Activities Certain activities, such as hang gliding, expose participants to greater risks of injury or death.

The second factor clouding risk evaluation is the degree of control individuals have over a given situation. Many people have a greater fear of things over which they do not have personal control. The third factor influencing risk evaluation is whether a risk is catastrophic or chronic. People usually are more frightened by news of a catastrophic accident such as a plane crash than news about a cause of death such as smoking, which has a much higher death toll spread out over time.

Fourth, some people have an optimism bias, or the feeling that they are invincible. These people tend to believe risks apply to other people but not to them. They may get upset when they see others driving erratically while talking on a cell phone or texting, for example, but they believe they can do so without impairing their own driving ability.

A fifth factor affecting risk analysis is that many of the risky things people do are highly pleasurable and give instant gratification. The potential harm from such activities comes later. Examples are smoking cigarettes, eating unhealthy foods, and tanning.

People can take steps to better evaluate and reduce risks. By following these guidelines, they can think about risks more rationally and make better lifestyle choices.

- *Compare risks*. In evaluating a risk, the key question is not "Is it safe?" but rather "How risky is it compared to other risks?"

- *Determine how much risk you are willing to accept*. For most people, a 1 in 100,000 chance of dying or suffering serious harm from exposure to an environmental hazard is a threshold for changing behavior. However, in establishing standards and reducing risk, the EPA generally assumes that a 1 in 1 million chance of dying from an environmental hazard is acceptable.

- *Evaluate the actual risk involved*. The news media usually exaggerate the daily risks society faces in order to capture interest and attract more readers, listeners, or television viewers. As a result, most people who are exposed to a daily diet of such exaggerated reports believe that the world is much more dangerous and risk filled than it really is.

- *Concentrate on evaluating and carefully making important lifestyle choices*. When evaluating a risk, it is important to ask, "Do I have any control over this?" There is no point worrying about risks over which we have little or no control. Individuals do have some or complete control over a variety of factors. They can act to reduce risks from heart attack, stroke, and certain cancers by deciding whether to smoke, what to eat, how much alcohol to drink, how much to exercise, and how safely to drive.

checkpoint What is optimism bias?

15.5 Assessment

1. **Recall** What five factors cause people to evaluate risk irrationally?
2. **Analyze** Are the characteristics of people in Blue Zones related to lifestyle, environment, or both?
3. **Evaluate** Are e-cigarettes a safe alternative to traditional cigarettes? Consider the severity and probability of the consequences of using them.
4. **Explain** What causes people to overestimate certain risks?

SCIENCE AND ENGINEERING PRACTICES

5. **Engaging in Argument** Do you think people should be able to make their own lifestyle choices regardless of risk? Why or why not?

TYING IT ALL TOGETHER STEM
What Makes a Healthy Community?

This chapter began with a Case Study describing how exposure to mercury and its compounds can permanently damage the human body. Mercury is an example of a chemical hazard and is just one type of hazard that threatens human health. This chapter also discussed biological, physical, cultural, and lifestyle hazards.

You also learned that some threats are unavoidable or difficult to avoid. Other threats can be reduced by properly assessing risks, following a healthy lifestyle, and making careful personal choices.

Dan Buettner has identified communities of people who live longer lives by reducing their environmental and health risks. Buettner calls these communities Blue Zones. Buettner's website describes the characteristics of Blue Zones and lessons that can be learned from them.

Work with a group to complete the following steps.

1. Visit Dan Buettner's website about Blue Zones. Read about the five Blue Zone communities on the Explorations page, including background and lessons learned.
2. Brainstorm five lifestyle-related questions to ask your classmates based on what you have learned about Blue Zones.
3. Gather as a class to review all the groups' questions. Choose the top 15 questions to create a class lifestyle quiz. Decide how you will rate students' answers, perhaps by providing a scale for them to indicate how best their answer matches the question.
4. Have each member of the class take the quiz. As a class, graph the results.
5. What trends and patterns can you see in the results? How "healthy" is your class? How can you apply these results to measure the health of your community?
6. As a class, develop a "Blue Zone" campaign to encourage members of your school or community to adopt healthier lifestyle habits.

CHAPTER 15 SUMMARY

15.1 What are the major types of health hazards?
- Risk is the probability of suffering from harm. Risk assessment uses statistical methods to estimate how much harm a hazard can cause. Risk management considers how to reduce the risk and the subsequent harm.
- The types of risk are biological, chemical, natural, cultural, and lifestyle choices.

15.2 How do biological hazards threaten human health?
- Infectious diseases are caused by pathogens such as bacteria, viruses, and parasites.
- Genetic resistance to antibiotics is increasing because bacteria reproduce rapidly, antibiotics are used unnecessarily, bacteria are spread by human travel, and antibiotics are overused on livestock.
- Viruses are not affected by antibiotics and can be easily transmitted.
- Parasites can be transmitted by insects and are a major problem in less-developed countries.

15.3 How do chemical hazards threaten human health?
- Toxic chemicals found in the environment can be carcinogenic, mutagenic, and teratogenic.
- Toxins can affect human body systems.
- The impact of toxins cannot be fully known through science and can be controversial.

15.4 How can people evaluate risks from chemical hazards?
- Numerous factors must be taken into account in determining the toxicity of chemicals.
- Pollution prevention is the safest way to introduce the use of chemicals and prevent them from entering the environment.
- The precautionary principle states that until society understands how harmful a toxin is, it should be avoided, used sparingly, or heavily regulated.

15.5 How do people perceive and avoid risks?
- The greatest risks to human health are connected to poverty, gender, and lifestyle choices.
- Most people do not evaluate risks very accurately. Guidelines can help individuals do so more rationally.

CHAPTER 15 ASSESSMENT

Review Key Terms

Select the key term that best fits each definition. Not all terms will be used.

bacterium	parasite	risk management
carcinogen	pathogen	teratogen
dose	precautionary	toxic chemical
infectious disease	principle	toxicity
mutagen	response	toxicology
nontransmissible disease	risk	transmissible disease
	risk analysis	virus
	risk assessment	

1. Type of chemical that harms or causes birth defects in a fetus or embryo
2. A disease caused by a pathogenic agent
3. An infectious disease that can be passed from one person to another
4. The view that when there is substantial preliminary evidence that an activity, technology, or chemical substance can harm humans, other organisms, or the environment, measures should be taken to prevent or reduce such harm rather than wait for conclusive scientific evidence
5. An element or compound that can cause temporary or permanent harm or death to humans
6. Type of chemical, radiation, or virus that causes or promotes cancer
7. An agent that can cause disease in other organisms
8. The study of the harmful effects of chemicals on humans and other organisms
9. The amount of a harmful chemical a person has ingested, inhaled, or absorbed through the skin at any one time
10. Pathogen that replicates by invading a cell, taking over its genetic machinery, and then spreading itself throughout the body

Review Key Concepts

11. Define and distinguish among risk, risk analysis, risk assessment, and risk management.
12. Give an example of risk from each of the following: biological hazard, chemical hazard, natural hazard, cultural hazard, and lifestyle choice.
13. Compare and contrast bacteria, viruses, and parasites, and give one example of a disease that each can cause.
14. List four steps people can take to better evaluate and reduce risks.
15. What is a neurotoxin, and why is methylmercury an especially dangerous neurotoxin?
16. How do hormonally active agents (HAAs) affect the body? What is one way to reduce exposure to them?
17. Describe how the toxicity of a substance can be estimated by testing laboratory animals, and explain the limitations of this approach.

Think Critically

18. Explain why you agree or disagree with each of the following statements:
    a. We should not worry much about exposure to toxic chemicals because almost any chemical, at a large enough dosage, can cause some harm.
    b. We should not worry much about exposure to toxic chemicals because, through genetic adaptation, we can develop immunity to such chemicals.
    c. We should not worry about exposure to a chemical such as BPA because there is not absolute scientific proof that this synthetic chemical has killed anyone.
19. Some workers in specific industries, such as shipbuilding or automobile detailing, are exposed to high levels of toxic substances. Should regulations reduce the allowable levels of these chemicals in the workplace? What economic effects might this have?
20. In Chapter 9 you answered the question, "If mosquitoes in your area were proven to carry malaria or a dangerous virus, how would you combat them?" After reading this chapter, would your answer to this question change? Why or why not?

Chapter Activities

A. Investigate: Disease Transmission STEM

In this chapter, you learned that an infectious disease is caused by pathogens such as bacteria, viruses, or parasites. Pathogens infect the body's cells and tissues. Transmissible infectious diseases can spread by air, water, food, and exchange of body fluids. In this activity, you will use the materials listed to model how contact with others can spread infection.

> **Materials**
> 1 cup of solution marked with a number
> empty plastic cup labeled "initial"
> empty plastic cup labeled "after 2"
> index card

1. Record your name and cup number on an index card. Pour one-fourth of the solution from your numbered cup into the "initial" cup. Set the "initial" cup aside.

2. Find a partner and record that person's cup number on your card. Then "exchange" body fluids by pouring your solution into theirs and taking half of it back. Repeat this procedure with one more partner, making sure to record their information.

3. Return to your "after 2" cup and fill it one-quarter with solution from your numbered cup. Using the solution remaining in the numbered cup, follow the same procedure to exchange body fluids with two more partners. How many people's "body fluids" are now mixed in your numbered cup?

4. Your teacher will add phenolphthalein solution to your numbered and "after 2" cups. Record the color of each solution after this step. Solutions that change colors show those who contracted a "virus" through contact with others. Compile your data as a class.

5. Looking at the data from the entire class, what changed between the cups labeled "after 2" and the final mixtures in the numbered cups? Why do you think there was a difference?

6. As a class, discuss how this activity modeled the spread of disease. Devise a way to determine who might be the original virus carriers. (Hint: Consider how to test the "initial" cups.)

B. Citizen Science

National Geographic Learning Framework
Attitudes | Curiosity
Skills | Collaboration
Knowledge | Critical Species

As you read in this chapter, influenza (flu) is one of the world's deadliest viruses. The ability to track the incidence of flu and predict flu outbreaks is critical to managing this illness. Flu Near You is a citizen science project developed by epidemiologists at Harvard, Boston Children's Hospital, and The Skoll Global Threats Fund. Volunteers from around the United States briefly report to Flu Near You each week to indicated if they or their family members have felt healthy or sick. Data are used to create a real-time map that shows local and national trends in flu-like illness.

Working with a small group, visit Flu Near You on the internet. View the most recent statistics listed on the map for flu activity in the United States, and look up information for your state as well. Discuss any trends you notice. Also compare the information to your own recent personal observations of illness in your school or community. If possible, join the project to submit anonymous reports each week to indicate whether you feel ill or healthy. Discuss with your group the benefits of using citizen science in this way to gather health data.

CHAPTER 16
AIR POLLUTION, CLIMATE CHANGE, AND OZONE DEPLETION

HUNDREDS OF MILLIONS of people live in coastal cities. During the last few decades, Earth's atmosphere has been warming and causing sea levels to rise. If this trend continues—and mounting evidence suggests it is likely—parts of many of these coastal cities could be under water by the end of this century.

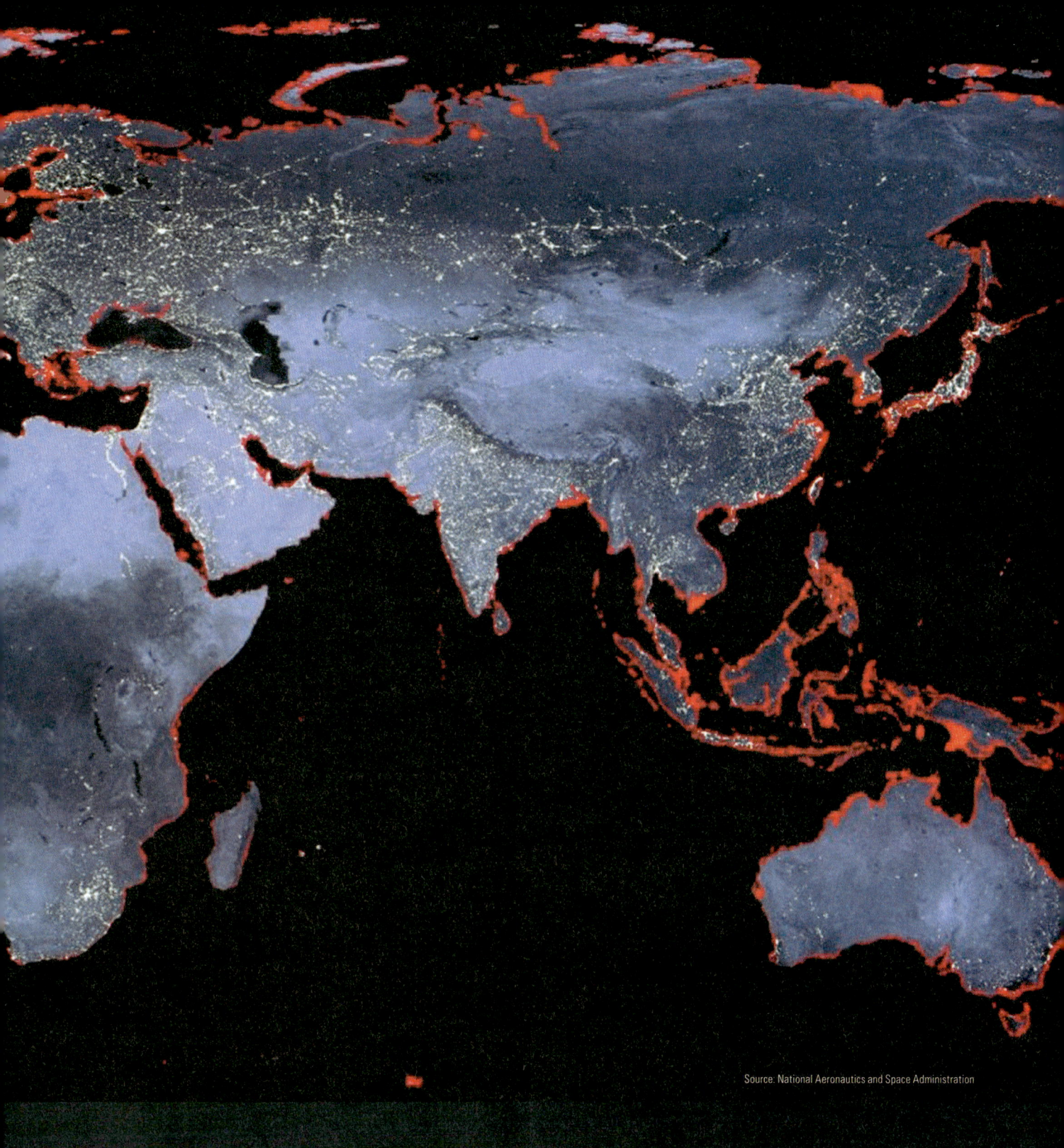

Source: National Aeronautics and Space Administration

KEY QUESTIONS

16.1 What are the major air pollution problems?

16.2 What are the effects of climate change?

16.3 How can people slow climate change?

16.4 How can people reverse ozone depletion?

The areas shown in red could be flooded by the end of this century due to a projected 1-meter (3-foot) rise in sea levels.

NATIONAL GEOGRAPHIC | EXPLORERS AT WORK

Conversing with Glaciers

with National Geographic Explorer Erin Pettit

Some people talk to their pets or plants. Erin Pettit carries on conversations with glaciers. Mainly, though, she listens.

Pettit is a glaciologist and National Geographic Explorer who loves to examine the boundary between glaciers and the oceans. The dynamic nature of this boundary may reveal what triggers glacial change—warmer water, the atmosphere above, or as yet unknown forces.

But that boundary is a very dangerous place. Scientists have long looked for ways to safely make measurements amid disintegrating ice shelves and iceberg-jammed waters. Pettit believes she has found a way. She uses underwater microphones to record the noisy boundary. "Acoustic research has already contributed much of what we know about the ocean," Pettit observes. "But no one had applied the technique to studying glaciers."

Glaciers offer a lot to hear. Glaciers meet oceans in an explosion of sound. When ice calves, or breaks off a glacier, it slams into waves. Then water gushes out from below, and air bubbles pop. Pettit listens to the sounds and figures out what they have to say about sea-level rise, climate change, and how critical processes like ocean circulation are being affected.

Ice speaks to Pettit in other ways as well. Her fieldwork has taken her to ice sheets in Antarctica where she has camped in a tent in −29°C (−20°F) weather and drilled and removed ice core samples. Examining these samples tells her how the ice sheets have grown and shrunk over hundreds of thousands of years. Ice core samples are like frozen history books of Earth's year-by-year environment.

Each layer of an ice core is like a tree ring, yet provides even richer detail. As annual snowfall compresses and turns to ice, snow captures impurities blown in by the wind. As a result, the sample shows yearly color differences and tells how much snow fell. Ice cores also reveal dark layers caused by volcanic activity or forest fires and specific carbon dioxide and nitrogen levels in the atmosphere.

Who will follow in Pettit's snowy boot prints? Perhaps one of the girls who participates in the free wilderness science program she created—Girls on Ice. Each summer, high school girls enter glacial landscapes, where they learn research techniques, perform experiments, and develop wilderness savvy. Pettit challenges herself to think creatively and inspires her students to do the same. "That's how you discover the most interesting things about yourself, other people, and the world around you," she says. "Curiosity and risk-taking drive many of the best scientific breakthroughs. You may feel scared, be questioned, or get cold before you learn how to keep warm, but the more you push yourself, the more discoveries you will make throughout life."

Thinking Critically
Infer What qualities do you think a glaciologist like Erin Pettit needs to possess?

Erin Pettit uses radar equipment to measure the thickness of a glacier in Alaska.

CASE STUDY
Melting Ice in Greenland

Greenland is the world's largest island with a population of 57,000 people. The ice that covers most of this mountainous island lies in glaciers that are up to 3.2 kilometers (2 miles) thick.

Areas of the island's ice have been melting at an accelerated rate during Greenland's summers (Figure 16-1). Some of this ice is replaced by snow during winter months. However, the annual net loss of Greenland's ice has increased during recent years.

Why does it matter that ice in Greenland is melting? It matters because considerable scientific evidence indicates that atmospheric warming is a key factor behind this melting. Atmospheric warming is the gradual rise in the average temperature of the atmosphere near Earth's land and water surfaces that has been occurring for at least 30 years.

Climate models predict that atmospheric warming will continue and lead to dramatic climate change during this century. The effects of climate change are far-reaching and leave no part of the biosphere untouched. If no action is taken, climate models project that Earth's systems could reach climate change tipping points, which would result in damage to most of Earth's ecosystems, people, and economies.

Greenland's glaciers contain enough water to raise the global sea level by as much as 7 meters (23 feet) if all of it melts and drains into the sea. It is highly unlikely that this will happen. But even a moderate loss of this ice over one or more centuries would raise sea levels considerably. Already, Greenland's ice loss has been responsible for nearly one-sixth of the global sea-level rise over the past 20 years. According to some scientists, Greenland's melting ice is an early warning that human activities are likely to disrupt Earth's climate in ways that could threaten life as we know it, especially during the latter half of this century.

This chapter will examine the nature of the atmosphere, air pollution, the likely causes and effects of projected climate change, and the depletion of ozone in the stratosphere. It will also look at some possible solutions to these serious environmental, economic, and political challenges.

As You Read Identify the types of outdoor and indoor air pollution that most affect your region, home, and school. What actions can you take to reduce these problems? What actions can you take to address the global issue of atmospheric warming?

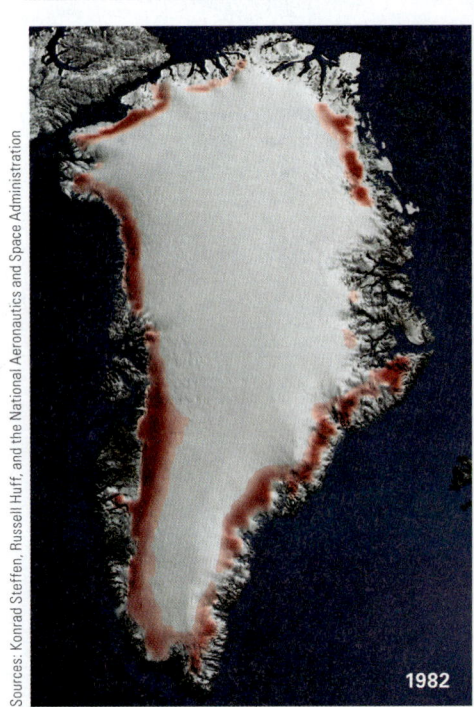

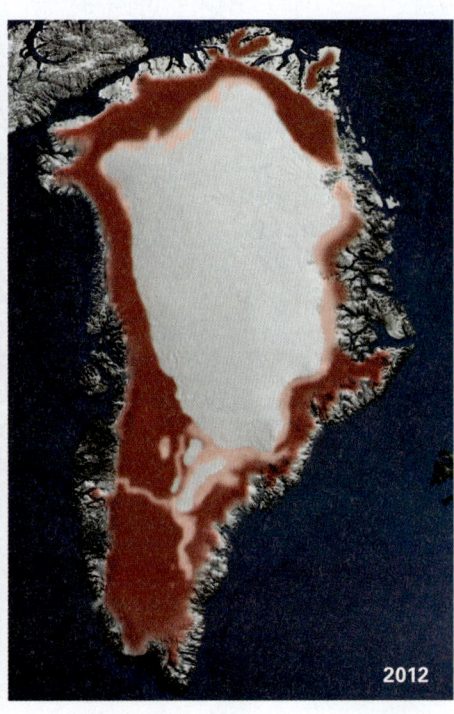

◀ FIGURE 16-1
Ice Loss in Greenland The total area of Greenland's glacial ice that melted by July 8, 2012 (red area in right image), was much greater than the amount that melted during the summer of 1982 (left).

16.1 What Are the Major Air Pollution Problems?

CORE IDEAS AND SKILLS
- Identify the layers of the atmosphere.
- Explain the causes and effects of outdoor and indoor air pollution.
- Describe factors that increase and decrease air pollution.
- Identify actions that people and governments can take to reduce air pollution.

KEY TERMS
ozone layer temperature inversion smog
air pollution acid deposition

The Atmosphere Consists of Layers

Life exists under a thin blanket of gases surrounding Earth, called the atmosphere. As you read in Lesson 3.1, the atmosphere consists of several layers. About 75–80% of Earth's air mass is found in the troposphere, the layer closest to Earth's surface (Figure 16-2). The troposphere extends about 17 kilometers (11 miles) above sea level at the Equator and about 6 kilometers (4 miles) above sea level over the Poles. If Earth were the size of an apple, this lower layer containing the air you breathe would be no thicker than the apple's skin.

About 99% of the volume of air consists of nitrogen (78%) and oxygen (21%). The remainder is argon (Ar), carbon dioxide (CO_2), and small amounts of water vapor (H_2O), dust particles, and other gases, including methane (CH_4), ozone (O_3), and nitrous oxide (N_2O). Several of these gases, including H_2O, CO_2, CH_4, and N_2O, are called greenhouse gases because they absorb and release energy that warms the troposphere. These gases play an important role in climate.

The atmosphere's second layer is the stratosphere. The stratosphere extends from 17 to 48 kilometers (11 to 30 miles) above Earth's surface. Although the stratosphere contains less matter than the troposphere, its composition is similar, with two notable exceptions. The stratosphere has a much lower volume of water vapor and a much higher concentration of ozone (O_3).

Much of the atmosphere's O_3 is concentrated in a portion of the stratosphere called the **ozone layer**.

FIGURE 16-2 ▼
Layers of the Atmosphere Earth's atmosphere is a dynamic system that includes four layers. The average temperature of the atmosphere varies with altitude (red line) and with differences in the absorption of incoming solar energy.

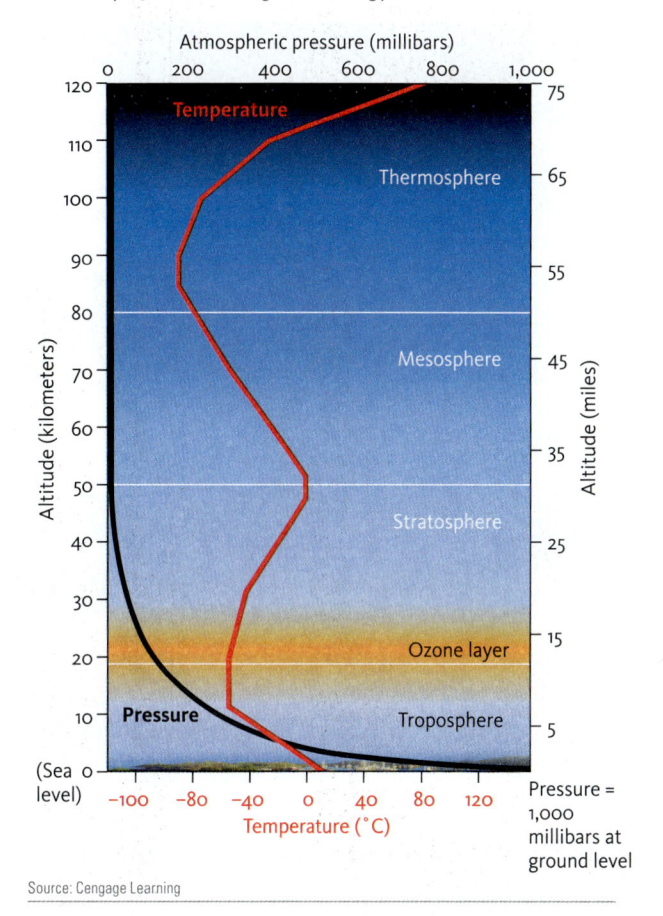

Source: Cengage Learning

Most O_3 is produced in this layer when oxygen molecules interact with ultraviolet (UV) radiation from the sun.

$$3 O_2 + UV \longleftrightarrow 2 O_3$$

This UV filtering effect of O_3 in the ozone layer acts as a "global sunscreen" that keeps 95% of the sun's harmful UV radiation from reaching Earth's surface. The stratospheric ozone layer allows life to exist on land and helps protect you from sunburn, skin and eye cancers, cataracts, and damage to your immune system. It also prevents much of the oxygen in the troposphere from being converted to photochemical O_3, a harmful air pollutant when found near the ground. The ozone layer is a vital part of Earth's natural capital that keeps humans and other species alive.

checkpoint Why is the ozone layer important to life?

LESSON 16.1 535

FIGURE 16-3 ▼
Primary and Secondary Air Pollutants Air pollution comes from mobile and stationary sources. Some primary air pollutants react with other chemicals in the air to form secondary air pollutants.

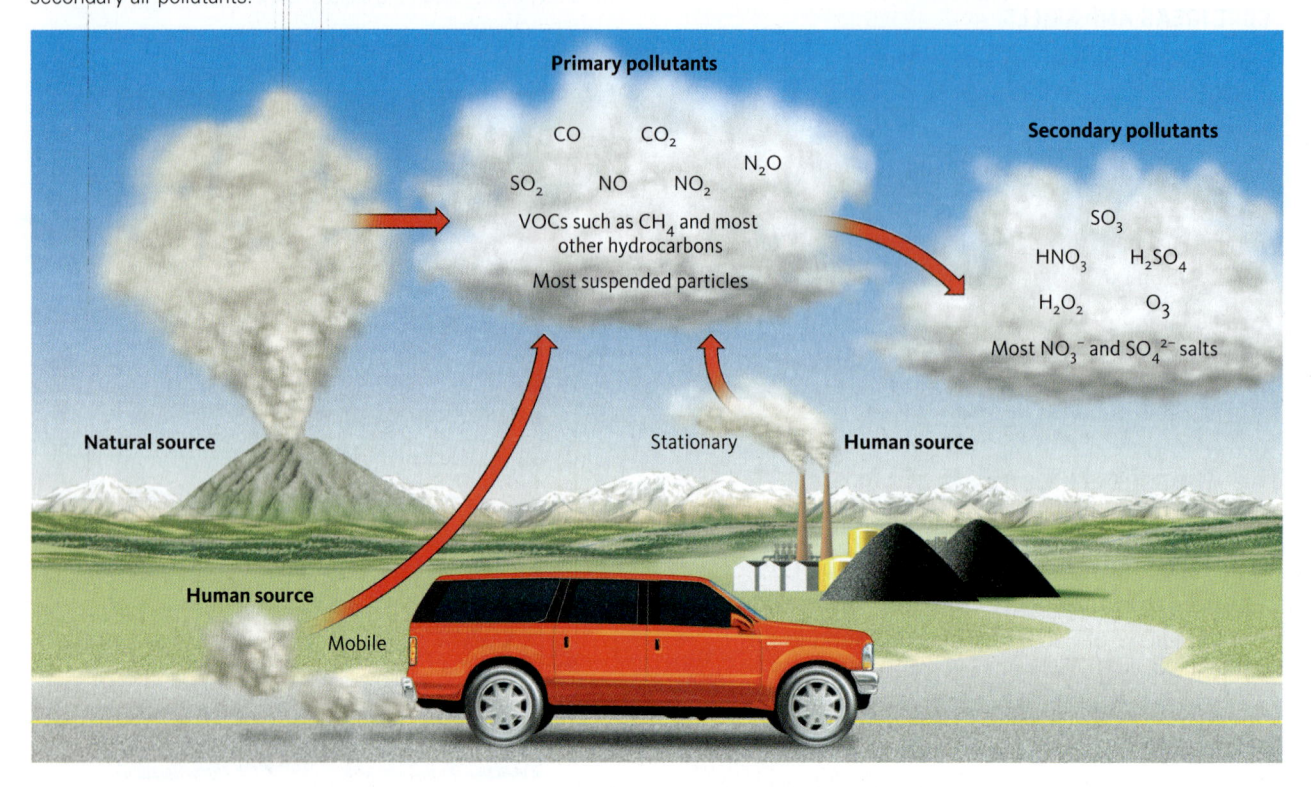

Primary pollutants

CO CO_2
SO_2 NO NO_2 N_2O
VOCs such as CH_4 and most other hydrocarbons
Most suspended particles

Secondary pollutants

SO_3
HNO_3 H_2SO_4
H_2O_2 O_3
Most NO_3^- and SO_4^{2-} salts

Natural source

Stationary   Human source

Human source
Mobile

Air Pollution Comes from Natural and Human Sources

Air pollution is any gaseous or solid material in the atmosphere that occurs in concentrations high enough to harm organisms, ecosystems, or human-made materials, or to alter climate. The effects of air pollution range from annoying to lethal.

Air pollutants come from natural and human sources. Natural sources include wind-blown dust, solid and gaseous pollutants from wildfires and volcanic eruptions, and volatile organic compounds released by some plants. Most natural air pollutants spread out over the globe and become diluted or are removed by chemical cycles, precipitation, and gravity. Pollutants emitted by volcanic eruptions or forest fires can temporarily reach harmful levels.

Most human sources of outdoor air pollution occur in industrialized and urban areas where people, cars, and factories are concentrated. The burning of fossil fuels in power plants, industrial facilities, and motor vehicles is the main human source of air pollution. Farming practices also add contaminants to the surrounding air.

Scientists classify outdoor air pollutants as primary or secondary (Figure 16-3). Primary pollutants are chemicals or substances emitted directly into the air from natural processes and human activities at concentrations high enough to cause harm. While in the atmosphere, some primary pollutants react with one another and with other natural components of air to form secondary pollutants.

Over the past 40 years, the quality of outdoor air in most of the more-developed nations has significantly improved. This occurred thanks mostly to grassroots pressure from citizens that led to air-pollution-control laws. However, according to the World Health Organization (WHO), one of every seven people in the world lives in urban areas where outdoor air is unhealthy to breathe. Most of them live in densely populated cities in less-developed countries where air-pollution-control laws do not exist or are poorly enforced.

Prolonged or high exposure to air pollution overloads the body's natural defense mechanisms. Very fine particles can get lodged deep in the lungs and contribute to cancer, asthma, heart attack, and stroke. In 2014, the WHO estimated that outdoor and indoor air pollution kills about 7 million people each year.

Carbon Oxides Carbon monoxide (CO) is a colorless, odorless, and highly toxic gas that forms during the incomplete combustion of carbon-containing materials. Major sources of CO are motor vehicle exhaust, burning forests and grasslands, smokestacks of fossil fuel-burning power plants and industries, tobacco smoke, open fires, and inefficient stoves used for cooking or heating. It is important to have CO detectors in your home. In the body, CO can reduce the ability of blood to transport oxygen to cells. Long-term exposure can trigger heart attacks and aggravate lung diseases such as asthma and emphysema. At high levels, CO can cause headaches, nausea, drowsiness, confusion, collapse, coma, and death.

Carbon dioxide (CO_2) is another colorless, odorless gas. About 93% of the CO_2 in the atmosphere is the result of the natural carbon cycle, and the rest comes from human activities. There is considerable scientific evidence that the rapid rise in atmospheric CO_2 levels since 1950 is largely due to human activities, especially the burning of fossil fuels and the removal of forests and grasslands, which absorb CO_2. This increase in CO_2 levels is a major cause of atmospheric warming and will likely lead to more climate change unless CO_2 emissions are sharply reduced.

Nitrogen Oxides and Nitric Acid Nitric oxide (NO) is a colorless gas that forms when nitrogen and oxygen gases react under high temperatures in car engines and coal-burning power and industrial plants. Lightning and certain bacteria in soil and water also produce NO as part of the nitrogen cycle

FIGURE 16-4
Patterns of lichen growth (and its absence) paint a picture of air quality in a region. This lichen is called old-man's beard. Wherever you see this lichen growing, the air is likely to be clean and have low sulfur dioxide levels.

(Figure 3-19). In the air, NO reacts with oxygen to form nitrogen dioxide (NO_2), a reddish-brown gas. Collectively, NO and NO_2 are called nitrogen oxides (NO_x). Some of the NO_2 reacts with water vapor in the air to form nitric acid (HNO_3) and nitrate salts (NO_3^-), components of harmful acid deposition. Nitrous oxide (N_2O) is a greenhouse gas that is emitted from fertilizers and animal wastes. Nitrous oxide is also produced by burning fossil fuels.

At high enough levels, nitrogen oxides can irritate the eyes, nose, and throat and aggravate lung ailments such as asthma and bronchitis. Nitrogen oxides can also suppress plant growth and reduce visibility when they are converted to nitric acid and nitrate salts.

Sulfur Dioxide and Sulfuric Acid Sulfur dioxide (SO_2) is a colorless gas with an irritating odor. About one-third of the SO_2 in the atmosphere comes from natural sources such as volcanoes. The other two-thirds of the SO_2 (and as much as 90% in some urban areas) comes from human sources. Human sources of SO_2 include the combustion of sulfur-containing coal in power and industrial plants, oil refining, and the smelting of sulfide ores.

In the atmosphere, SO_2 can be converted to aerosols and return to Earth as a component of acid deposition. Sulfur dioxide and the sulfuric acid droplets and sulfate particles that form reduce visibility and aggravate breathing problems. These substances can damage crops, trees, soils, and aquatic life in lakes. Sulfur substances also corrode metals and damage paint, paper, leather, and the stone used to build walls, statues, and monuments.

> **CONSIDER THIS**
>
> **Lichens can act as air pollution indicators.**
>
> A lichen consists of a fungus and an alga living together, usually in a mutualistic relationship. These hardy pioneers are good biological indicators of air pollution because they continually absorb air as a source of nourishment. Some lichens are sensitive to specific air-polluting chemicals. For example, old-man's beard (*Usnea trichodea*) sicken or die in the presence of excessive sulfur dioxide, even when the pollutant originates far away (Figure 16-4).

LESSON 16.1

Particulates Suspended particulate matter (SPM) consists of a variety of solid particles and liquid droplets that are small and light enough to remain suspended in the air for long periods. Most of the SPM in outdoor air comes from natural sources such as dust, wildfires, and sea salt. The remaining SPM comes from human sources such as burning coal, motor vehicles, wind erosion of topsoil, and road construction.

Particulate matter can irritate the nose and throat, damage the lungs, aggravate asthma and bronchitis, and shorten life spans. Toxic particulates such as lead and cadmium can cause genetic mutations, reproductive problems, and cancer. Particulates also reduce visibility, corrode metals, and discolor clothing and paint.

Ozone One of the ingredients of smog is ozone (O_3), a colorless and highly reactive gas. Ozone can cause coughing and breathing problems, aggravate lung and heart diseases, reduce resistance to colds and pneumonia, and irritate the eyes, nose, and throat. Ozone also damages plants, rubber in tires, fabrics, and paints. Scientific measurements indicate that human activities have decreased the amount of beneficial O_3 in the stratosphere and have increased the amount of harmful O_3 near ground level—especially in some urban areas.

Volatile Organic Compounds (VOCs) Organic compounds that exist as gases in the atmosphere are called volatile organic compounds (VOCs). Examples of VOCs are methane (CH_4) and hydrocarbons, which are emitted by the leaves of many plants. As a greenhouse gas, CH_4 is about 25 times stronger than CO_2 at warming the atmosphere. About a third of global CH_4 emissions comes from natural sources, such as plants, wetlands, and termites. The rest comes from human sources such as rice paddies, landfills, natural gas wells and pipelines, and cows (mostly from their belching). Other VOCs are liquids that evaporate quickly. Examples of liquid VOCs are benzene and other liquids used as industrial solvents, dry-cleaning fluids, and various components of gasoline, plastics, and other products.

checkpoint What are the causes and effects of particulate pollution?

Several Factors Affect Air Pollution

Factors that Decrease Air Pollution Five natural factors help reduce outdoor air pollution. First, particles heavier than air settle out as a result of gravitational attraction to Earth. Second, rain and snow partially cleanse the air of pollutants. Third, salty sea spray from the oceans washes out many pollutants from air that flows from land over the oceans. Fourth, winds sweep pollutants away and dilute them by mixing them with cleaner air. And fifth, some pollutants are removed by chemical reactions. For example, SO_2 can react with O_2 in the atmosphere to form SO_3. In turn, SO_3 reacts with water vapor to form droplets of H_2SO_4 that fall out of the atmosphere as acidic precipitation.

Factors that Increase Air Pollution Six factors can increase outdoor air pollution. First, urban buildings slow wind speed and reduce the dilution and removal of pollutants. Second, hills and mountains reduce the flow of air in valleys below them and allow pollutant levels to build up at ground level. Third, high temperatures promote the chemical reactions leading to the formation of photochemical smog. Fourth, emissions of VOCs from certain trees and plants in urban areas can promote the formation of photochemical smog.

The fifth factor is known as the grasshopper effect. This effect occurs when air pollutants are transported at high altitudes by evaporation and winds. Air pollutants travel from tropical and temperate areas to Earth's polar areas as part of Earth's global air circulation system. This happens mostly during winter. The grasshopper effect explains why pilots have reported seeing a reddish-brown haze over the Arctic for decades. It also helps explain why polar bears, sharks, and native peoples in remote Arctic areas have high levels of various harmful pollutants in their bodies.

The sixth factor that increases air pollution has to do with the vertical movement of air. During the day, the sun warms the air near Earth's surface. Normally, this warm air rises to mix with the cooler air above it. As that happens, most of the pollutants in the air disperse. However, under certain atmospheric conditions, a layer of warm air can temporarily lie atop a layer of cooler air nearer the ground. This is called a **temperature inversion** (Figure 16-5). Because the cooler air near the surface is denser than the warmer air above, it does not rise and mix with it. If this condition persists, pollutants can build up to harmful and even lethal concentrations in the stagnant layer of cool air near the ground.

checkpoint Explain how a temperature inversion can worsen the air pollution of a region.

FIGURE 16-5
Temperature Inversion Surrounded by mountains on almost all sides, the topography of Salt Lake valley makes it especially prone to prolonged temperature inversions.

Warm air layer

Inversion layer

Smog

Smog Is a Major Health Hazard

Seventy years ago, cities such as London, England, and the U.S. cities of Chicago, Illinois, and Pittsburgh, Pennsylvania, burned large amounts of coal in power plants and factories. They also burned coal for heating homes and often for cooking food. People in those cities were exposed to unhealthy industrial smog, especially during winter. Industrial **smog** consists mostly of a mix of sulfur dioxide, suspended droplets of sulfuric acid, and a variety of suspended solid particles in outside air. Those who burned coal inside their homes were often exposed to dangerous levels of indoor air pollutants.

Today, urban industrial smog is rarely a problem in most of the more-developed countries, where coal is burned only in large power and industrial plants with reasonably good air pollution control. However, industrial smog remains a problem in industrialized urban areas of China (Figure 16-6), India, Ukraine, Czechoslovakia, Bulgaria, and Poland. In these areas, large quantities of coal are still burned in houses, power plants, and factories with inadequate pollution controls.

Because of its heavy reliance on coal, China has some of the world's highest levels of industrial smog and 16 of the world's 20 most polluted cities. According to a 2014 Chinese government report, 92% of all Chinese cities did not meet the government's outdoor air quality standards in 2013.

Another type of smog is photochemical smog. A photochemical reaction is any chemical reaction activated by light. Photochemical smog is a mixture of primary and secondary pollutants formed under the influence of ultraviolet radiation from the sun. In greatly simplified terms:

$$\text{VOCs} + \text{NO}_x + \text{heat} + \text{sunlight} \longrightarrow \text{ground-level ozone } (O_3) \\ + \text{ other photochemical oxidants} \\ + \text{ aldehydes} \\ + \text{ other secondary pollutants}$$

The formation of photochemical smog begins when exhaust from morning commuter traffic releases large amounts of NO and VOCs into the air over a city. This smog is also referred to as brown-air smog as the NO converts to a reddish-brown NO_2.

LESSON 16.1

FIGURE 16-6
Commuters in many parts of China wear masks to protect themselves from heavy smog.

When exposed to UV radiation from the sun, some of the NO_2 reacts in complex ways with VOCs released by certain trees (such as some oak species, sweet gums, and poplars), motor vehicles, and businesses (especially bakeries and dry cleaners). The resulting mixture of pollutants, dominated by ground-level O_3, usually builds up to peak levels by late morning, irritating people's eyes and respiratory tracts. Some of these pollutants, known as photochemical oxidants, can damage people's lung tissue.

All modern cities have some photochemical smog, but it is much more common in cities with sunny climates and a great number of motor vehicles. Examples of cities with photochemical smog problems are Los Angeles, California, and Salt Lake City, Utah, in the United States; Sydney, Australia; São Paulo, Brazil; Bangkok, Thailand; and Mexico City, Mexico.

checkpoint What is the difference between industrial smog and photochemical smog?

Acid Deposition Is a Serious Regional Problem

Most coal-burning power plants, metal ore smelters, oil refineries, and other industrial facilities emit sulfur dioxide (SO_2), particulates, and nitrogen oxides. In more-developed countries, these facilities usually use tall smokestacks to vent their exhausts high into the atmosphere where wind can dilute and disperse these pollutants. This reduces local air pollution, but it can increase regional air pollution in downwind areas. Prevailing winds can transport the pollutants as far as 1,000 kilometers (600 miles). In the air, these compounds form secondary pollutants such as droplets of sulfuric acid (H_2SO_4), nitric acid vapor (HNO_3), and particles of acid-forming sulfate (SO_4^{2-}) and nitrate (NO_3^-) salts (Figure 16.3).

These acidic substances remain in the atmosphere for 2–14 days. During this period, they descend to Earth's surface in two forms: wet and dry deposition. Wet deposition consists of acidic rain, snow, fog, and cloud vapor. Dry deposition consists of acidic particles. The resulting mixture is **acid deposition**—

FIGURE 16-7

Acid Deposition Commonly called acid rain, acid deposition consists of rain, snow, dust, and other particles with a pH lower than 5.6.

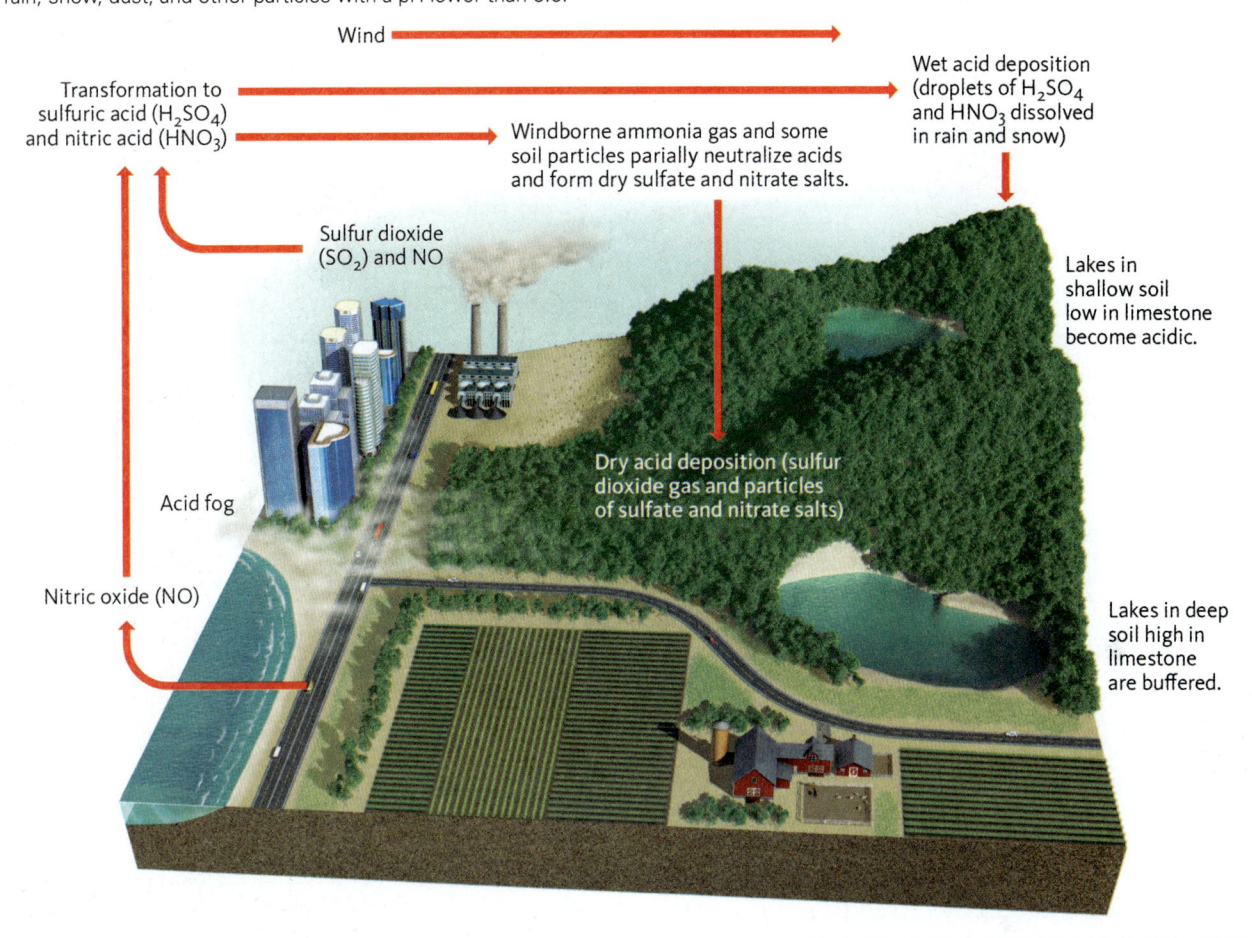

Wind

Transformation to sulfuric acid (H_2SO_4) and nitric acid (HNO_3)

Windborne ammonia gas and some soil particles parially neutralize acids and form dry sulfate and nitrate salts.

Wet acid deposition (droplets of H_2SO_4 and HNO_3 dissolved in rain and snow)

Sulfur dioxide (SO_2) and NO

Lakes in shallow soil low in limestone become acidic.

Acid fog

Dry acid deposition (sulfur dioxide gas and particles of sulfate and nitrate salts)

Nitric oxide (NO)

Lakes in deep soil high in limestone are buffered.

often called acid rain (Figure 16-7). Most dry deposition occurs within 2–3 days of emission, relatively close to the industrial sources. On the other hand, most wet deposition takes place within 4–14 days in more distant downwind areas.

Acid deposition is a problem in areas that lie downwind from coal-burning facilities and from urban areas with large numbers of cars. In some areas, soils contain compounds that can react with and help neutralize, or buffer, some inputs of acids. The areas most sensitive to acid deposition are those with thin, acidic soils that provide no natural buffering and those where the buffering capacity of soils has been depleted by decades of acid deposition. The map in Figure 16-8 shows areas of the world where acid deposition is or could become a problem.

Acid deposition (often along with other air pollutants such as O_3) can harm crops and reduce plant productivity. A combination of acid deposition and other air pollutants can affect forests by draining essential plant nutrients such as calcium and magnesium from forest soils. They also cause the soils to release ions of aluminum, lead, cadmium, and mercury, which are toxic to trees. These two effects rarely kill trees directly, but can weaken trees and leave them vulnerable to stresses such as severe cold, diseases, insect attacks, and drought.

Metals released from soils and rocks pose a threat to animals too. Acid deposition causes lead and mercury to leach into lakes used as sources of drinking water. These toxic metals can accumulate in the tissues of fish and of the animals that eat them, including people. This presents a serious health threat, especially for pregnant women. Because of excess acidity due to acid deposition, several thousand lakes in Norway and Sweden, and 1,200 lakes in Ontario, Canada, contain few if any fish. In the United States, several hundred lakes (most of them in the Northeast) are similarly threatened. Acid deposition also contributes to human respiratory diseases and causes damage to buildings and statues.

LESSON 16.1

FIGURE 16-8 ▼
Acid Deposition Problem Areas This map shows regions where acid deposition is or may become a problem. These regions have large inputs of air pollution or are sensitive areas with naturally acidic soils and bedrock. Do you live in an area that is affected by acid deposition or that is likely to be affected by acid deposition?

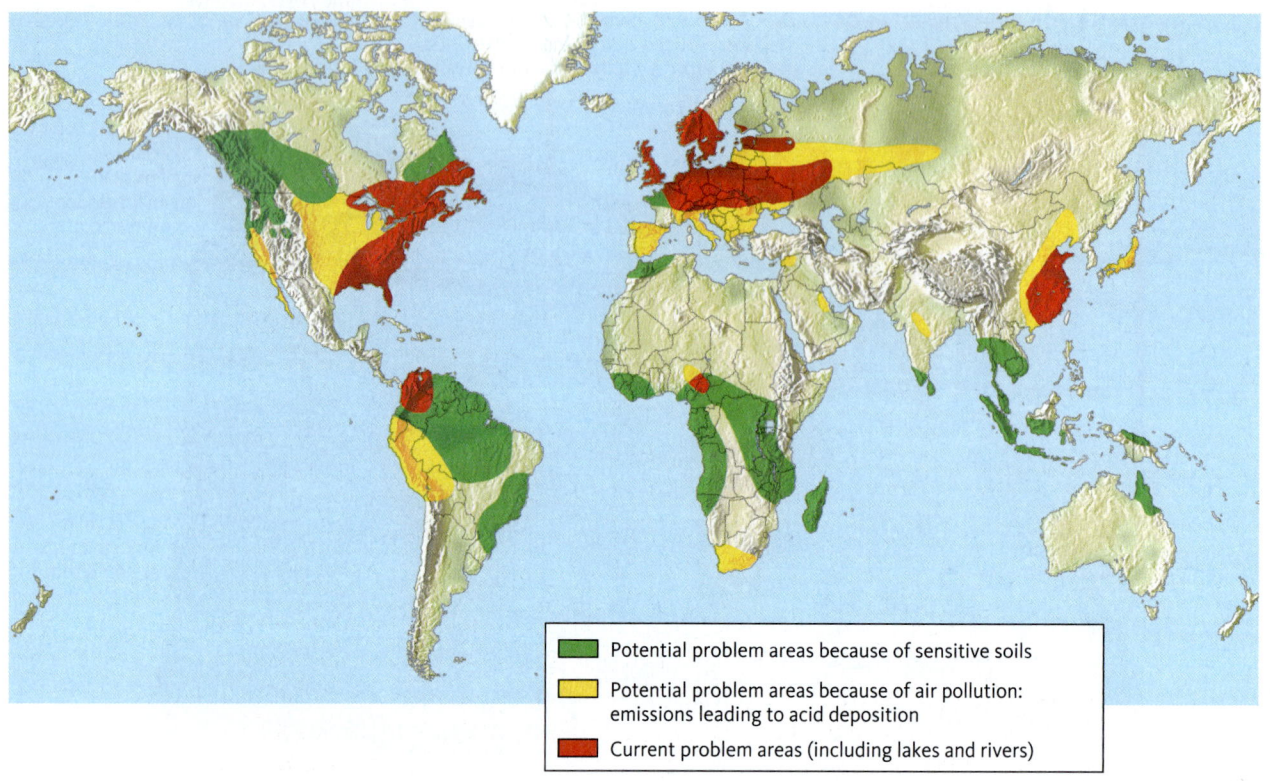

- 🟩 Potential problem areas because of sensitive soils
- 🟨 Potential problem areas because of air pollution: emissions leading to acid deposition
- 🟥 Current problem areas (including lakes and rivers)

Sources: World Resources Institute and U.S. Environmental Protection Agency

In the United States, older coal-burning power and industrial plants without adequate pollution controls, especially in the Midwest, emit the largest quantities of SO_2, particulates, and other pollutants that cause acid deposition. Prevailing winds in the United States carry these pollutants eastward, as you can see in Figure 16-8. As a result, typical precipitation in parts of the eastern United States can be 10 times more acidic than precipitation in the western United States. Some mountaintop forests in the eastern United States are bathed in fog and dews that are about 1,000 times as acidic as normal precipitation.

Acid deposition has also become an international problem wherever acid-producing emissions from one country are transported to other countries by prevailing winds. The worst acid deposition occurs in Asia, especially in China, which in 2020 got approximately 57% of its energy from burning coal. Some of eastern Asia's emissions travel on strong winds all the way across the Pacific Ocean to the west coast of North America.

Reducing Acid Deposition According to most scientists who study the problem of acid deposition, the best solutions are preventive approaches that reduce or eliminate emissions of SO_2, nitrogen oxides, and particulates. Since 1994, acid deposition has decreased sharply in the United States and especially in the eastern half of the country.

FIGURE 16-9 ▼

| Acid Deposition Solutions ||
Prevention	Cleanup
Reduce coal use and burn only low-sulfur coal.	Add lime to neutralize acidified lakes.
Use natural gas and renewable energy resources in place of coal.	Add phosphate fertilizer to neutralize acidified lakes.
Remove SO_2 and NO_x from smokestack gases and remove NO_x from motor vehicular exhaust.	Add lime to neutralize acidified soils.
Tax SO_2 emissions.	

This decrease is partly owing to significant reductions in SO_2 and nitrogen oxide emissions from coal-fired power and industrial plants under the 1990 amendments to the U.S. Clean Air Act. Figure 16-9 lists several ways to reduce acid deposition.

Implementing acid deposition prevention solutions is politically difficult. One problem is that the people and ecosystems affected by acid deposition often are quite far downwind from the sources of the problem. Also, countries with large supplies of coal (such as China, India, Russia, Australia, and the United States) have a strong incentive to use it. However, in the United States, the increasing use of affordable wind and solar power and cleaner-burning natural gas for generating electricity has reduced the use of coal to some extent.

checkpoint How does acid deposition affect humans and the environment?

Laws and Regulations Can Reduce Air Pollution

The United States provides an excellent example of how governments can reduce air pollution. The U.S. Congress passed the Clean Air Acts in 1970, 1977, and 1990. With these laws, the federal government established air pollution regulations for key outdoor air pollutants that are enforced by states and major cities.

Congress directed the EPA to establish air quality standards for six major outdoor pollutants—CO, NO_2, SO_2, SPM, O_3, and lead (Pb). Each standard specifies the maximum allowable level for a pollutant, averaged over a specific period. The EPA has also established national emission standards for more than 188 hazardous air pollutants (HAPs) that are thought to contribute to serious health and ecological problems.

According to the EPA, the combined emissions of the six major outdoor air pollutants decreased by about 77% between 1970 and 2020. The reduction of outdoor air pollution in the United States since 1970 has been a remarkable success story, mostly because of two factors. First, during the 1970s, U.S. citizens insisted that laws be passed and enforced to improve air quality. Prior to 1970, when Congress passed the Clean Air Act, air-pollution-control equipment did not exist. Second, the country was affluent enough to afford such controls and improvements. For example, as a result of these factors, a new car today in the United States emits 75% less pollution than did a pre-1970 car.

Environmental scientists applaud this success, but they call for strengthening U.S. air pollution laws by:

- Putting much greater emphasis on air pollution prevention. The power of prevention was made clear by the 99% drop in atmospheric lead emissions after lead in gasoline was banned in 1976.
- Sharply reducing emissions from approximately 20,000 older coal-burning power and industrial plants, cement plants, oil refineries, and waste incinerators that have not been required to meet the air pollution standards for new facilities under the Clean Air Acts.
- Ramping up controls on atmospheric emissions of toxic pollutants such as mercury.
- Emphasizing reduction of emissions of air pollutants that blow across state boundaries.
- Continuing to improve fuel-efficiency standards for motor vehicles, thereby also saving consumers money.
- Regulating more strictly the emissions from motorcycles and two-cycle gasoline engines used in devices such as chainsaws, lawnmowers, generators, scooters, and snowmobiles. The EPA estimates that running a typical gas-powered riding lawn mower for an hour creates as much air pollution as driving 34 cars for an hour.
- Setting much stricter air pollution regulations for airports and oceangoing ships.
- Sharply reducing indoor air pollution.

Emissions Trading One approach to reducing pollutant emissions has been to allow producers of air pollutants to buy and sell government air pollution allotments in the marketplace. For example, with the goal of reducing SO_2 emissions, the Clean Air Act of 1990 authorized an emissions trading, or "cap-and-trade" program. The cap-and-trade program enables the 110 most polluting coal-fired power plants in 21 states to buy and sell SO_2 air pollution rights.

Under this system, each plant is given a number of pollution credits annually, which allow it to emit a certain amount of SO_2. A utility that emits less than its allotted amount has leftover pollution credits. That utility can use its credits to offset SO_2 emissions at its other plants, keep them for future plant expansions, or sell them to other utilities or private citizens or groups.

FIGURE 16-10
Indoor Air Pollutants Indoor pollution is more common than many people realize.

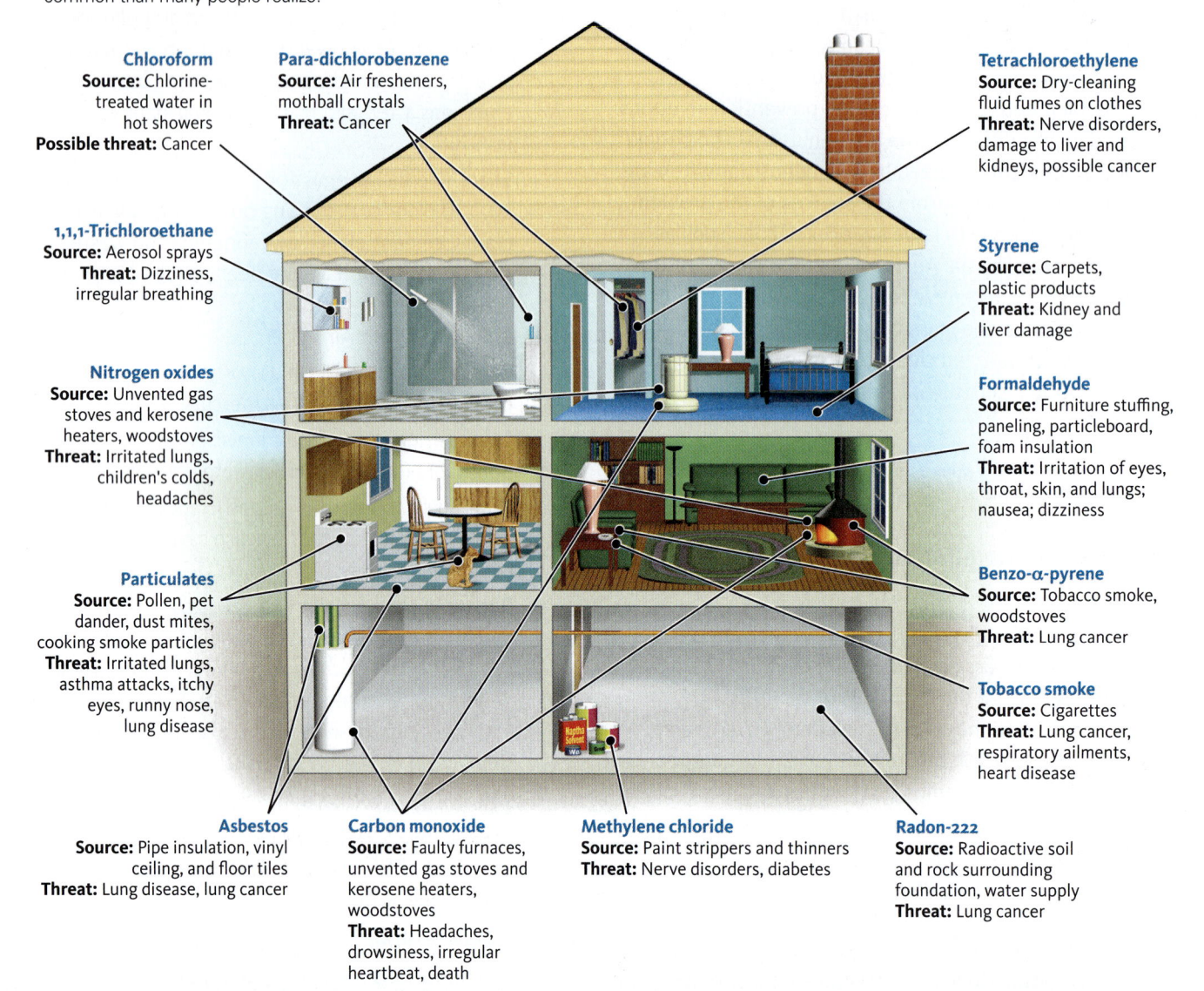

Chloroform
Source: Chlorine-treated water in hot showers
Possible threat: Cancer

Para-dichlorobenzene
Source: Air fresheners, mothball crystals
Threat: Cancer

Tetrachloroethylene
Source: Dry-cleaning fluid fumes on clothes
Threat: Nerve disorders, damage to liver and kidneys, possible cancer

1,1,1-Trichloroethane
Source: Aerosol sprays
Threat: Dizziness, irregular breathing

Styrene
Source: Carpets, plastic products
Threat: Kidney and liver damage

Nitrogen oxides
Source: Unvented gas stoves and kerosene heaters, woodstoves
Threat: Irritated lungs, children's colds, headaches

Formaldehyde
Source: Furniture stuffing, paneling, particleboard, foam insulation
Threat: Irritation of eyes, throat, skin, and lungs; nausea; dizziness

Particulates
Source: Pollen, pet dander, dust mites, cooking smoke particles
Threat: Irritated lungs, asthma attacks, itchy eyes, runny nose, lung disease

Benzo-α-pyrene
Source: Tobacco smoke, woodstoves
Threat: Lung cancer

Tobacco smoke
Source: Cigarettes
Threat: Lung cancer, respiratory ailments, heart disease

Asbestos
Source: Pipe insulation, vinyl ceiling, and floor tiles
Threat: Lung disease, lung cancer

Carbon monoxide
Source: Faulty furnaces, unvented gas stoves and kerosene heaters, woodstoves
Threat: Headaches, drowsiness, irregular heartbeat, death

Methylene chloride
Source: Paint strippers and thinners
Threat: Nerve disorders, diabetes

Radon-222
Source: Radioactive soil and rock surrounding foundation, water supply
Threat: Lung cancer

SCIENCE FOCUS 16.1

INDOOR AIR POLLUTION

Problems with air quality are not restricted to the outdoors. In fact, common air pollutants have been found at levels 2–100 times higher inside homes, buildings, and cars than outside. The health risks from exposure to such chemicals are growing because most people in more-developed urban areas spend 70–98% of their time indoors or inside vehicles.

Smokers, young children, elderly people, pregnant women, people with respiratory or heart problems, and factory workers are especially at risk from indoor air pollution.

A main culprit of indoor air pollution is the chemicals used to make building materials and products such as furniture and paneling. Figure 16-10 shows some typical sources of indoor air pollution in a modern home.

Some of the most dangerous indoor air pollutants are tobacco smoke, formaldehyde emitted from many building materials and various household products, and radioactive radon-222 gas, which can seep into houses from underground rock deposits.

Thinking Critically
Apply Describe three ways to reduce your exposure to indoor air pollution. Use the illustration for help.

Between 1990 and 2012, this emissions trading program helped to reduce SO_2 emissions from power plants in the United States by 76%. The program cost less than one-tenth of the cost projected by the utility industry, according to the EPA. Proponents of this market-based approach say it is cheaper and more efficient than government regulation of air pollution. Critics of this approach contend that it allows utilities with older, dirtier power plants to buy their way out of their environmental responsibilities and to continue to pollute.

The ultimate success of any emissions trading approach depends on two factors: how low the initial "cap" is set and how often it is lowered. The cap must be lowered to promote continuing innovation in air pollution prevention and control. Without these considerations, emissions trading programs can shift pollution problems from one area to another without achieving an overall improvement in air quality.

Emissions Reduction Strategies Figure 16-11 summarizes several ways to reduce emissions from stationary sources such as coal-burning power plants and industrial facilities—the primary contributors to industrial smog. Also listed are several ways to prevent and reduce emissions from motor vehicles, the primary contributors to photochemical smog.

FIGURE 16-11 ▼

Stationary Source Air Pollution	
Prevention	**Reduction or Dispersal**
Burn low-sulfur coal or remove sulfur from coal.	Disperse emissions using tall smokestacks (increases downwind pollution).
Convert coal to a liquid or gaseous fuel.	Remove pollutants from smokestack gases.
Switch from coal to natural gas and renewables.	Tax each unit of pollution produced.

Motor Vehicle Air Pollution	
Prevention	**Reduction**
Walk, bike, or use mass transit.	Require emission control devices.
Improve fuel efficiency.	Inspect car exhaust systems twice a year.
Get older, polluting cars off the road.	Set strict emission standards.

In more-developed countries, many of these solutions have been successful. However, the already poor air quality in urban areas of many less-developed countries is worsening as the numbers of motor vehicles in these nations rise. Over the next 10–20 years, technology could help all countries to clean up the air through improved engine and emission systems and hybrid-electric, plug-in hybrid, and all-electric vehicles.

checkpoint Explain how regulations have reduced air pollution in the United States.

16.1 Assessment

1. **Contrast** What is the difference between ozone in the troposphere and ozone in the stratosphere?
2. **Restate** Identify six air pollutants and explain the health effects of each.
3. **Classify** Would you classify acid deposition as a primary or secondary air pollution? Explain your reasoning.
4. **Summarize** What are the pros and cons of a cap-and-trade approach to reducing air pollution?
5. **Draw Conclusions** Describe one thing you can do to reduce stationary air pollution and one thing you can do to reduce mobile air pollution.

SCIENCE AND ENGINEERING PRACTICES

6. **Engaging in Argument** Evaluate the reasoning and evidence behind the following claim: Government regulations can reduce air pollution.

CROSSCUTTING CONCEPTS

7. **Patterns** What patterns of air pollution might be observable on a very large scale that are not observable on a smaller scale? What problem does this pose in terms of potential government action?

16.2 What Are the Effects of Climate Change?

CORE IDEAS AND SKILLS
- Define climate change.
- Describe evidence that indicates Earth's climate is undergoing rapid change.
- Explain how models are used to estimate future climate change.
- Describe the effects of present and projected future climate change.

KEY TERMS

climate change   carbon footprint   drought

Climate Change Has Accelerated

You may have heard the term *global warming* used in place of *atmospheric warming* or *climate change*. But global warming is a misleading term. Although the average global temperature has been getting warmer, some areas of Earth have warmed and others have cooled. **Climate change** is a change in the weather conditions of Earth or a particular area over a period of at least three decades. Climate change is neither new nor unusual. Over the past 3.5 billion years, many factors have altered the planet's climate. These factors include large-scale volcanic eruptions, changes in solar input, continents moving slowly atop shifting tectonic plates, impacts by large meteors, and slight changes in the planet's wobbly orbit around the sun. Earth's climate also is affected by global air circulation patterns, changes in the size of large areas of ice that reflect incoming solar energy and cool the atmosphere, varying concentrations of the greenhouse gases found in the atmosphere, and occasional changes in ocean currents.

Earth's climate has fluctuated over the past 900,000 years, slowly swinging back and forth between long periods of atmospheric warming and long periods of atmospheric cooling. The cold periods led to the ice ages (Figure 16-12). These periods of freezing and thawing are also known as glacial and interglacial periods.

For roughly 10,000 years, Earth has experienced an interglacial period characterized by a generally steady climate based on a relatively consistent global average surface temperature. This allowed the human population to grow as agriculture developed and later as cities grew. And for the past 1,000 years,

FIGURE 16-12
Ice Ages Over the past 900,000 years, the average global atmospheric temperature near Earth's surface has fluctuated widely. This graph is based on a body of scientific evidence that contains gaps but reveals trends. What are two conclusions you can draw from this graph?

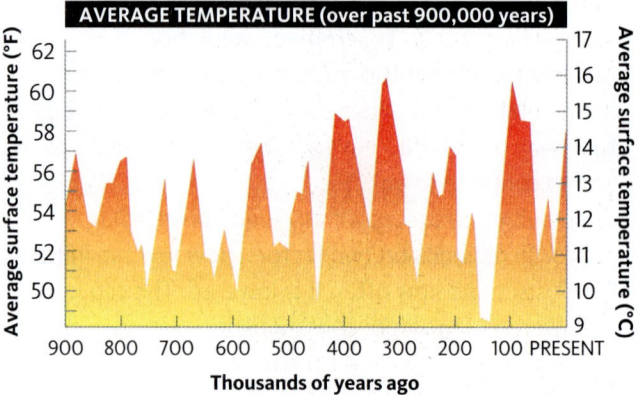

Sources: Goddard Institute for Space Studies, Intergovernmental Panel on Climate Change, National Academy of Sciences, National Aeronautics and Space Administration, National Center for Atmospheric Research, and National Oceanic and Atmospheric Administration

FIGURE 16-13
Average Temperature This graph shows the change in average global atmospheric temperature near Earth's surface between 1880 and 2015.

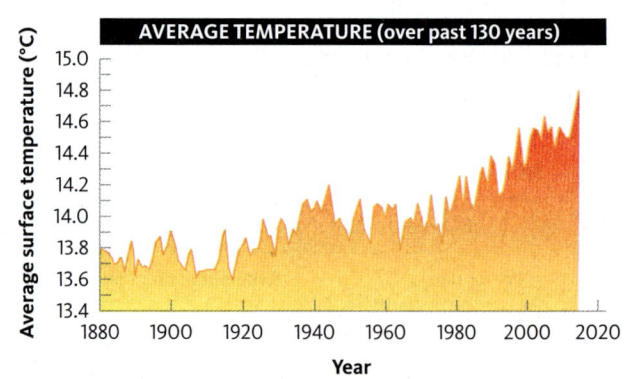

Sources: Goddard Institute for Space Studies, Intergovernmental Panel on Climate Change, National Academy of Sciences, National Aeronautics and Space Administration, National Center for Atmospheric Research, and National Oceanic and Atmospheric Administration

the average temperature of the atmosphere near Earth's surface has remained fairly stable. During the 19th and 20th centuries, temperatures began to rise. Human activities such as clearing forests and grasslands and burning fossil fuels increased atmospheric levels of greenhouse gases that play a key role in Earth's atmospheric temperature.

Most of the recent, sharp rise in the global average atmospheric temperature on land has taken place since 1975 (Figure 16-13). According to a study on climate change by the American Association for the Advancement of Science (AAAS), evidence from numerous scientific studies indicate that the

546   CHAPTER 16   AIR POLLUTION, CLIMATE CHANGE, AND OZONE DEPLETION

FIGURE 16-14
Ice cores are extracted from deep holes drilled into ancient glaciers, such as this one near the South Pole in Antarctica. Ice core analysis yields information about the past composition and temperature of the lower atmosphere, as well as solar activity, snowfall, and forest fire frequency.

rising inputs of greenhouse gases from human activities are overwhelming the natural factors that regulated gradual climate change in the past. Past dramatic changes in climate took place over periods of thousands to hundreds of thousands of years. Present climate change of the same magnitude is happening within several decades.

Tens of thousands of research studies by Earth's climate scientists indicate that this rapid climate change is caused mostly by human activities. You may have heard about the *climate change debate* in the media. This term is somewhat misleading, as no significant debate exists among 97% of the world's climate scientists about climate change or what is causing it. Debates about climate change are occurring in the political arena among citizens and politicians and center around which, if any, political actions should be taken to deal with climate change.

How do scientists know about how Earth's climate changed in the past? Scientists estimate past temperature and climate changes by analyzing evidence. Findings are strengthened when multiple, different lines of evidence point to the same conclusions. Sources of evidence include isotopes in rocks and fossils; plankton and isotopes in ocean sediments; tiny bubbles, layers of soot, and other materials trapped in layers of ancient air found in ice cores from glaciers (Figure 16-14); pollen from the bottoms of lakes and bogs; tree rings; and atmospheric temperature measurements taken regularly since 1861. Temperature measurements now include data from more than 40,000 measuring stations around the world, as well as satellites.

Using this evidence, the world's leading scientific organizations—including the Intergovernmental Panel on Climate Change (IPCC), U.S. National Academy of Sciences (NAS), British Royal Society, NOAA, NASA, and AAAS have reached these conclusions:

1. Climate change is happening now. It is caused mostly by human activities (especially deforestation and the burning of carbon-containing fossil fuels). Climate change will very likely get worse—how much worse depends on whether people act now to slow it.

2. Immediate and sustained action to curb climate change is possible and affordable and would bring major benefits for human health and economies as well as for the environment.

3. The sooner people act to slow climate change, the lower the risks and costs of significant climate disruption.

CHAPTER 16 AIR POLLUTION, CLIMATE CHANGE, AND OZONE DEPLETION

FIGURE 16-15
ON ASSIGNMENT National Geographic photographer Frans Lanting captured this shot of a melting glacier in Wrangell-St. Elias National Park, Alaska. A longer and warmer melting season is rapidly shrinking this natural resource and others like it in the Northern Hemisphere.

Data from tens of thousands of peer-reviewed scientific studies conducted over decades have led to the conclusion that climate change is happening now. Below are a few pieces of evidence.

- Between 1900 and 2020, Earth's average global surface temperature rose by 1°C (1.8°F). Much of this increase took place since the mid 1970s.
- The seven warmest years on record since 1861 have taken place since 2015. 2016 and 2020 are tied as the top two warmest years on record.
- In some parts of the world, glaciers that have existed for thousands of years are melting.
- The melting of Greenland's ice sheets has accelerated (Case Study).
- In Alaska, glaciers (Figure 16-15) and frozen ground (permafrost) are melting, loss of sea ice and rising sea levels are eating away at coastlines, and communities are being relocated inland.
- In the Arctic, floating summer sea ice has been shrinking in most years since 1979.
- The world's average sea level has been rising at an accelerated rate, especially since 1975. This rise is mostly due to the expansion of ocean water as its temperature increases and to increasing runoff from melting land-based ice.
- Atmospheric levels of CO_2 and other greenhouse gases that warm the troposphere have been rising sharply.
- As temperatures rise, many terrestrial, marine, and freshwater species have migrated toward the Poles and to higher elevations. Species that cannot migrate face extinction.

The Greenhouse Effect The greenhouse effect is a natural process that plays a major role in determining Earth's average atmospheric temperature and, thus, its climate. It occurs when some of the solar energy absorbed by Earth radiates into the atmosphere as infrared radiation (heat). As this radiation interacts with molecules of greenhouse gases in the air, it increases their kinetic energy and warms the lower atmosphere and Earth's surface. Four major greenhouse gases are H_2O, CO_2, CH_4, and N_2O. Life on Earth and human economic systems are dependent on the greenhouse effect because it keeps the planet at an average temperature of around 15°C (58°F).

Since the Industrial Revolution in the mid 1700s, humans have greatly increased agriculture, deforestation, and the burning of fossil fuels. These activities correlate with increases in the concentrations of greenhouse gases in the lower atmosphere. The average atmospheric concentration of CO_2 has risen by at least 40% since 1880. More than half of the increase has taken place since 1970 (Figure 16-16). This is a long-lasting increase because CO_2 typically remains in the atmosphere for 100 years or more. After staying between 180 and 280 parts per million (ppm) for 400,000 years, the concentration of CO_2 in the atmosphere averaged 415 ppm in 2021.

Atmospheric methane levels have also increased greatly since the mid 1700s. Ice core analysis reveals that about 70% of the global emissions of methane during the last 275 years were likely caused by human activities, and 30% came from natural sources.

In 2020, the three largest emitters of energy-related CO_2 were China, the United States, and India. In comparing CO_2 emissions sources, scientists use the concept of a carbon footprint. A **carbon footprint** is the amount of CO_2 generated by an individual, organization, country, or any other entity over a given period of time. China has the largest national carbon footprint. Americans have the largest per-person carbon footprint.

FIGURE 16-16

Atmospheric Carbon Dioxide This graph shows the concentration of atmospheric CO_2 measured at an atmospheric research center in Mauna Loa, Hawai'i, between 1960 and 2022.

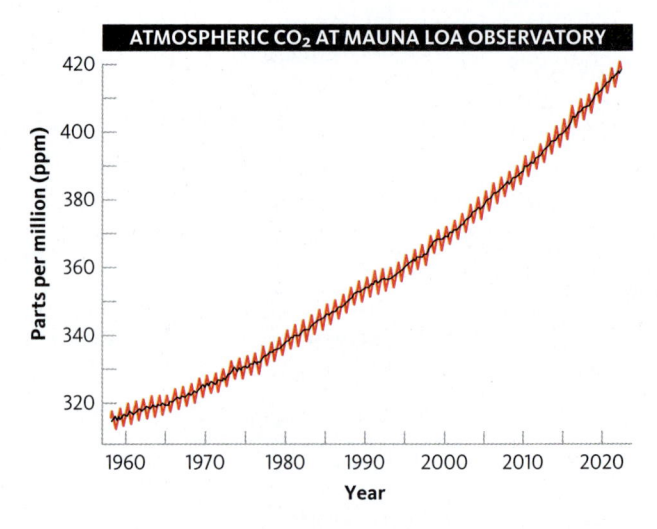

Sources: NOAA, Scripps Institute of Oceanography, UC San Diego

Climate models project that rising levels of CO_2, water vapor, and atmospheric temperatures will cause major ecological and economic disruption during this century unless deforestation and the burning of fossil fuels are sharply reduced (Science Focus 16.2).

The Role of Oceans The world's oceans absorb CO_2 from the atmosphere as part of the carbon cycle and help moderate Earth's climate. The oceans remove an estimated 25% of the CO_2 from the lower atmosphere. About 93% of Earth's CO_2 has been stored as carbon compounds in marine algae, vegetation, coral reefs, and bottom sediments for millions of years.

The oceans also absorb heat from the lower atmosphere. An estimated 80–90% of the heat held in the lower atmosphere by greenhouse gases since the 1970s has ended up in the oceans. Partly driven by this heat, ocean currents slowly transfer some of the absorbed CO_2 to the deep ocean where it is buried in carbon compounds in bottom sediments. The average temperature of the oceans has risen since 1970 because of these exchanges, with half of this warming occurring since 1997. But the oceans have warmed to a lesser degree and more slowly than the atmosphere owing to their huge mass and volume. The uptake of CO_2 and heat by the world's oceans has actually slowed the rate of climate change. However, this has also resulted in the growing and serious problem of ocean acidification.

Cloud Cover Warmer temperatures increase evaporation of surface water, which raises the humidity of the atmosphere in parts of the world. This results in more clouds. Different types of clouds have different effects. An increase in thick and continuous cumulus clouds at low altitudes can have a cooling effect by reflecting more sunlight back into space. An increase in thin, wispy cirrus clouds at high altitudes leads to a warming effect by preventing some heat from escaping into space.

Climate scientists are working with climate models to understand more about the role of clouds in climate and the causes of cloud formation. The latest scientific research indicates that the net global effect of cloud cover changes is likely to increase atmospheric warming. More research is needed in order to evaluate this effect.

Aerosols Air pollutants in the form of aerosols can affect the atmosphere in different ways. Aerosols are microscopic particles and droplets suspended in air. They are released or formed in the troposphere by volcanic eruptions and human activities. Most aerosols, such as the light-colored sulfate particles produced by fossil fuel combustion, tend to reflect sunlight and cool the lower atmosphere. Black carbon particles (soot) warm the atmosphere. Soot comes from coal-burning power and industrial plants, diesel exhaust, open cooking fires, and burning forests. Climate scientists do not expect aerosols to affect projected climate change very much in the next 50 years. These particles and droplets fall back to the ground or are washed out of the lower atmosphere within weeks, whereas CO_2 remains in the lower atmosphere for 100 years or longer. Also, people are reducing their inputs of aerosol and soot because of their harmful impacts on plants and human health.

checkpoint What role do the oceans play in regulating climate?

Rapid Atmospheric Warming Could Have Serious Effects

Most of the historical changes in the lower atmospheric temperature took place over thousands of years. What makes the current problem urgent is that we face a rapid projected increase in the average temperature of the lower atmosphere during this century. This will likely change the fairly mild climate we have had for the past 10,000 years. According to AAAS, "The rate of climate change now may be as fast as any extended warming period over the past 65 million years, and is projected to increase in coming years."

Worse-case climate model projections include the following effects within this century:

- floods in low-lying coastal cities from a rise in sea levels
- forests being consumed in vast wildfires
- grasslands turning into dust bowls
- rivers drying up
- ecosystems collapsing
- the extinction of up to half the world's species
- more intense and longer-lasting heat waves
- more destructive storms and flooding
- the rapid spread of infectious tropical diseases

SCIENCE FOCUS 16.2

USING MODELS TO PREDICT CHANGE

Climate scientists have developed complex computer models to help predict the effects of increasing levels of greenhouse gases. Climate models simulate interactions among incoming sunlight, clouds, landmasses, ocean currents, concentrations of greenhouse gases and other air pollutants, and other factors within Earth's complex climate system.

Scientists run continually improving models on supercomputers and compare the results to known past climate changes. They use this data to project future changes in Earth's atmosphere. Figure 16-17 gives a greatly simplified summary of some key interactions in the global climate system.

Climate models provide projections of changes in the average temperature of the lower atmosphere. Inputs include available data and different assumptions about future changes such as CO_2 and CH_4 levels in the atmosphere. How well the projections match what actually happens depends on the validity of the assumptions, the variables that are built into the models, and the accuracy of the data used.

Scientific research does not provide absolute proof or certainty. Science does, however, provide varying levels of certainty. When most experts in a particular scientific field generally agree on a level of 90% certainty about a set of measurements or model results, they say that their projections are very likely to be correct. When the level of certainty reaches 95% (a rarity in science), they contend that their projections are extremely likely to be correct.

In 1990, 1995, 2001, 2007, and 2014, the IPCC published reports on Earth's climate. A sixth report is planned for release in 2022. The reports summarize how global temperatures have changed in the past, how temperatures are projected to change during this century, and how these changes are likely to affect Earth's climate.

The 2014 IPCC report was based on analysis of past climate data and the use of more than two dozen climate models. This report concluded that it is extremely likely (95% certainty) that human activities have played the dominant role in the observed atmospheric warming since 1975. After running the models thousands of times, they found that the only way to get the model results to match actual measurements is by including human activities (Figure 16-18).

The current results of more than two dozen climate models now in use suggest it is very likely (with at least 90% certainty) that Earth's mean surface temperature will increase by 1.5–4.5°C (2.7–8.1°F) between 2013 and 2100. That is, unless people can sharply reduce deforestation along with emissions of CO_2 and other greenhouse gases.

While there is an extremely high degree of certainty about present atmospheric warming, there is a high degree of uncertainty about projected future changes in the average atmospheric temperature. This is indicated by the wide range of projections shown in Figure 16-19 (tan, orange, and red). Climate experts are working to reduce the uncertainty by learning more about Earth's climate system and improving climate data and models. Despite their limitations, these models are the best and only tools available for projecting likely changes.

Thinking Critically

Predict If the highest possible projected temperature increase shown in Figure 16-19 takes place, what are three major ways in which this will likely affect your lifestyle? How might it affect future generations?

FIGURE 16-17

Climate Interactions
Warming Processes
Heat and CO_2 emissions from oceans
CO_2 emissions from land-clearing, fires, and decay
Greenhouse gas emissions from natural and human sources
Cooling Processes
Heat and CO_2 removal and long-term storage in deep ocean
CO_2 removal by plants and organisms in soil
Aerosol emissions from natural and human sources

FIGURE 16-18 ▼
The Contribution of Human Factors These graphs show how the actual climate data compares with modeled projections between the period of 1860 and 2000. Scientists found that the actual data match projections far more closely when human factors are included in the models (right) than when only natural factors are included (left).

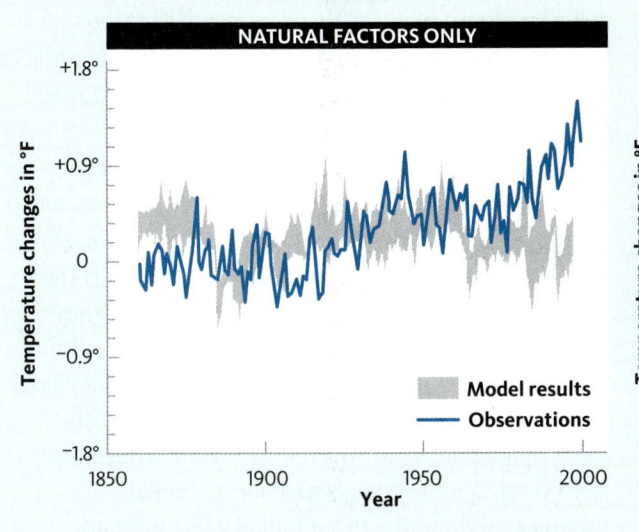

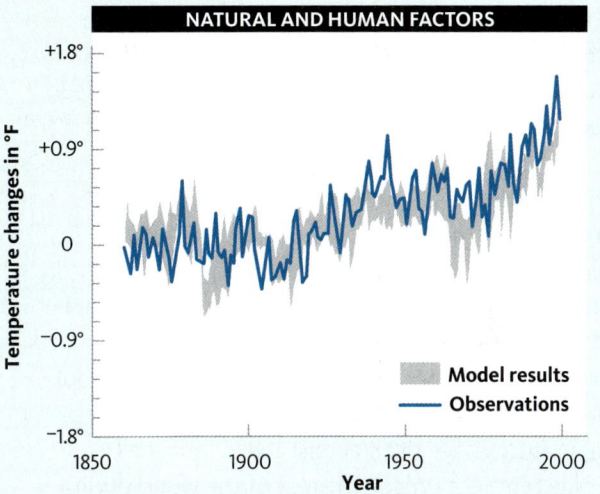

Sources: Intergovernmental Panel on Climate Change, 2001 and 2007

FIGURE 16-19 ▼
Projected Future Change This graph shows estimated changes in the average temperature of the atmosphere near Earth's surface (yellow area) and measured changes (black line). The projected change in temperature is given as a range of possibilities.

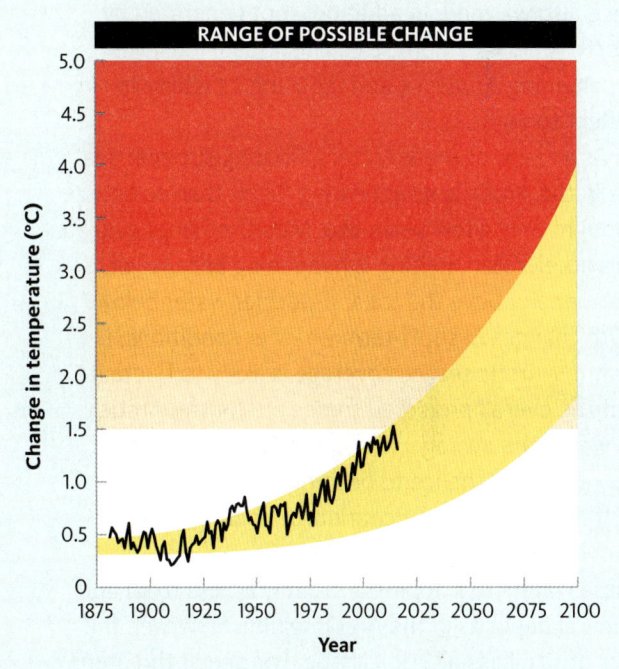

Sources: National Academy of Sciences, National Center for Atmospheric Research, Intergovernmental Panel on Climate Change, and Hadley Center for Climate Prediction and Research

LESSON 16.2

Climate models show that atmospheric warming and its harmful effects will be unevenly distributed. Temperatures in the tropics could reach extremes that have not occurred in over 10,000 years. The sea-level rise would be 15–20% higher in tropical areas than the global average. Droughts could also be much worse in the tropics. In a global economy, such effects would ripple to other parts of the world.

Snow and Ice Melt Models project that climate change will be severe in the world's polar regions. Light-colored ice and snow in these regions cool Earth by reflecting incoming solar energy back into space. This positive feedback loop is called the albedo effect. The melting of such ice and snow exposes much darker land and sea areas, which reflect less sunlight and absorb more solar energy. This has warmed the atmosphere above the Poles more rapidly than the atmosphere is warming at lower latitudes. The result is likely to be more melting snow and ice that cause further atmospheric warming above the Poles in a runaway positive feedback loop.

According to the 2014 IPCC report, Arctic air temperatures have risen almost twice as fast as average temperatures in the rest of the world during the last 50 years. They are now warmer than they have been in 44,000 years. Arctic ocean waters have also warmed. In addition, soot generated by North American, European, and Asian industries is darkening Arctic ice and reducing its ability to reflect sunlight.

As a result of these factors, floating summer sea ice in the Arctic is disappearing faster than scientists thought only a few years ago. Measurements indicate this accelerated melting is happening because of warmer air above the ice and warmer water below. With changes in short-term weather conditions, summer Arctic sea ice coverage is likely to fluctuate. But the overall projected trends are for the Arctic to warm, the average summer sea ice coverage to decrease, and the ice to become thinner.

If the current trend continues, floating summer Arctic ice may be gone by 2050. This would open these waters to shipping and allow access to oil and mineral deposits in the Arctic region. However, the loss of sea ice is a serious issue. For one, it threatens the survival of species that depend on the ice for migration and to find food, such as some seals, walruses, seabirds, and polar bears (Figure 16-20). But the problem goes beyond a concern for biodiversity. The loss of summer Arctic ice could also affect global air currents. A change in global air patterns would lead to dramatic and long-lasting changes in weather and climate across the planet.

Another effect of Arctic warming is faster melting of polar land-based ice, including that in Greenland (Case Study). This melting adds fresh water to the northern seas and will likely contribute to the projected rise in sea levels during this century. Earth's mountain glaciers are another great storehouse of ice. During the past 25 years, many of these glaciers have been shrinking wherever summer melting exceeds the winter snowpack. Glacier National Park in Montana once had 150 glaciers, but by 2022 only 25 remained.

Mountain glaciers play a vital role in the water cycle by storing water as ice during cold seasons and releasing it slowly to streams during warmer seasons. A prime example of high-elevation reservoirs is the glaciers of the Himalayan Mountains in Asia. They are a major source of water for large rivers like the Ganges that provide water for more than 400 million people in India and Bangladesh. These waters also feed China's Yangtze and Yellow Rivers, whose basins are home to more than 500 million people.

About 80% of the mountain glaciers in South America's Andes range are slowly shrinking. If this continues, 53 million people in Bolivia, Peru, and Ecuador who rely on meltwater from the glaciers for irrigation and hydropower could face severe water, power, and food shortages. In the United States, people living in the Columbia, Sacramento, and Colorado River basins face similar threats as the winter snowpack that feeds these rivers is projected to shrink by as much as 70% by 2050.

Permafrost Melt Permafrost exists in soils beneath 25% of the exposed land in Alaska, Canada, and Siberia in the Northern Hemisphere. Huge amounts of carbon are locked in permafrost soils. Climate change is projected to thaw significant amounts of permafrost. This thaw is already happening in parts of Alaska (Figure 16-21) and Siberia. If this trend continues, a great deal of organic material found below the permafrost will rot and release huge amounts of CH_4 and CO_2 into the atmosphere. This would accelerate projected atmospheric warming, melt more permafrost, and lead to atmospheric warming in a worsening spiral of change as part of a positive feedback loop.

Sea-Level Rise The IPCC estimate that the average global sea level is likely to rise by 0.4–0.6 meters

FIGURE 16-20 As floating summer sea ice dwindles in the Arctic, species such as polar bears lose important habitat. The bears use the ice as platforms for hunting prey. Without it, they starve. Polar bears are now listed as a threatened species.

(1.3–2.0 feet) by the end of this century. That would be about 10 times the rise that occurred in the last century. Much of this rise will likely come from Greenland's melting ice (Case Study). Accelerated melting could lead to seas rising by as much as 0.9–2.0 meters (3.0–7.0 feet), depending on the quantity of land-based ice in Greenland and West Antarctica. In 2016, researchers from the University of Massachusetts predicted a sea-level rise this century within the higher estimate (more than a meter, or between 5 and 6 feet) when they accounted for the complex physical processes of melting ice sheets. The sea-level rise will likely not be uniform around the world, based on factors such as ocean currents and winds. For example, Bangladesh's sea level could rise by as much as 4 meters (13 feet) by 2100.

According to the IPCC and NAS, a 1-meter (3-foot) rise in sea level during this century would cause:

- Degradation of at least one-third of the world's coastal estuaries, wetlands, coral reefs, and deltas where much of the world's rice is grown.
- Disruption of the world's coastal fisheries.
- Saltwater contamination of freshwater coastal aquifers resulting in degraded supplies of groundwater used for drinking and irrigation.
- Flooding of some of the world's largest coastal cities such as Venice, London, and New Orleans and displacement of at least 150 million people—an amount equal to nearly half of the current U.S. population (chapter opening photo).
- Flooding and erosion of low-lying barrier islands and gently sloping coastlines, especially in U.S. coastal states like Texas, Louisiana, New Jersey, South Carolina, North Carolina, and Florida (Figure 16-22).
- Flooding in large areas of low-lying countries such as Bangladesh (Figure 16-24).
- Submersion of low-lying island nations such as the Maldives and Fiji.

LESSON 16.2 555

FIGURE 16-21
In Alaska, huge chunks of coastal land are eroding as more of the permafrost layer melts.

More Extreme Weather Atmospheric warming affects different parts of Earth in different ways. A study by the National Center for Atmospheric Research has found that severe and prolonged drought was affecting at least 30% of Earth's land (excluding Antarctica). A **drought** is a prolonged period of dry weather. According to a study by climate researchers at NASA, up to 45% of the world's land area could experience extreme drought by 2059. It is difficult to tie a specific drought to atmospheric warming because natural cyclical processes also cause extreme droughts. However, extra heat in the atmosphere evaporates more water from soils. This depletion of soil moisture prolongs droughts and makes them more severe, regardless of their causes.

Atmospheric warming also results in longer, more frequent, and more intense heat waves. More heat waves could lead to a greater number of heat-related deaths, reduce crop production, and expand deserts. Since 1950, heat waves have been longer and more frequent (Figure 16-23).

FIGURE 16-22
Sea-Level Rise If the average sea level rises by 1 meter (3 feet), the areas shown here in red will be flooded. Even if the actual sea-level rise is only a fraction of this projection, storm surge damage will likely be significant in these areas.

Sources: Jonathan Overpeck, Jeremy Weiss, and U.S. Geological Survey

556 CHAPTER 16 AIR POLLUTION, CLIMATE CHANGE, AND OZONE DEPLETION

FIGURE 16-23 ▼
Extreme Weather As atmospheric temperature rises, so does the chance of heat waves, heavy rains, and other extreme weather.

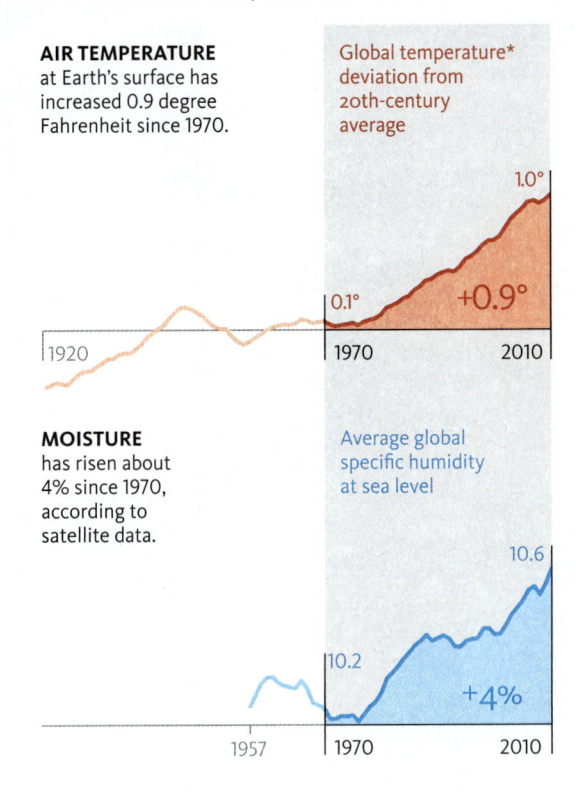

AIR TEMPERATURE at Earth's surface has increased 0.9 degree Fahrenheit since 1970.

Global temperature* deviation from 20th-century average: 0.1° (1970) → 1.0° (2010), +0.9°

MOISTURE has risen about 4% since 1970, according to satellite data.

Average global specific humidity at sea level: 10.2 (1970) → 10.6 (2010), +4%

HEAT WAVES—of which nighttime lows are one indicator—are striking a growing portion of the United States.

Percentage of United States experiencing summer minimum temperatures much above normal: 4% (1970) → 35% (2010), +31%

EXTREME RAINFALLS are now affecting larger areas of the United States as well.

Percentage of United States getting an elevated portion of precipitation from 16% extreme events: 9% (1970) → +7% (2010)

GRAPHS ARE SMOOTHED USING A TEN-YEAR MOVING AVERAGE.
*AVERAGE TEMPERATURE OVER LAND AND OCEAN

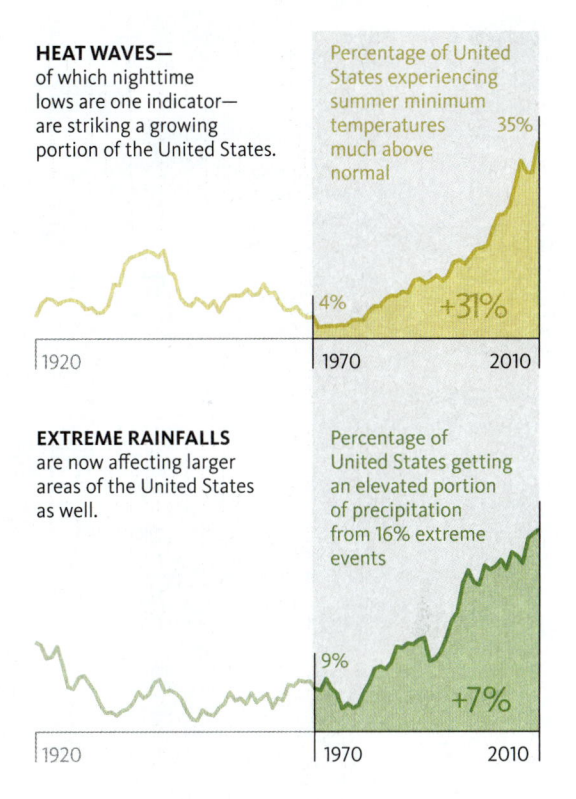

Sources: Jeff Masters, Weather Underground, National Climatic Data Center, and National Oceanic and Atmospheric Administration

At the same time, some areas, such as the eastern half of the United States, will likely experience increased flooding from heavy and prolonged precipitation because a warmer atmosphere can hold more moisture.

In other areas, atmospheric warming will likely lead to colder winter weather because of changes in global air circulation patterns. In 2010, a World Meteorological Organization panel of experts concluded that atmospheric warming is likely to lead to fewer but stronger hurricanes and typhoons. These systems could cause more damage in coastal areas where urban populations have grown rapidly. Although there is not sufficient evidence to link a specific extreme weather event to climate change, climate change is likely to increase the overall chances of extreme weather events over the next 30 years or more.

Biodiversity Loss According to the 2007 and 2014 IPCC reports, projected climate change is likely to alter ecosystems and take a toll on biodiversity on every continent. Up to 85% of the Amazon rain forest—one of the world's major centers of biodiversity—could be lost. If the global atmospheric temperature rises by the highest projected amount, this land would be converted to tropical savanna. As the atmosphere warms, 35–70% of the world's species could face extinction by 2100. The hardest hit species will be:

- Cold-climate plants and animals, including the polar bear in the Arctic, walruses in Alaska, and penguins in Antarctica.
- Species that live at higher elevations.
- Species with limited tolerance for temperature change, such as corals.
- Species with limited ranges.

The primary cause of these extinctions would be loss of habitat. On the other hand, populations of plant and animal species that thrive in warmer climates could increase. The most vulnerable ecosystems are coral reefs, polar seas, coastal wetlands, high-elevation forests, and tundra.

LESSON 16.2

FIGURE 16-24
ON ASSIGNMENT National Geographic photographer Michael Coyne snapped this photo of a rickshaw passing a general store on a submerged street in Bangladesh. People in low-lying countries like Bangladesh have already had to adapt to increased flooding.

559

FIGURE 16-25
With warmer winters, populations of mountain pine beetles have expanded and killed large numbers of trees in this lodgepole pine forest in the Canadian province of British Columbia.

Primarily because of drier conditions, forest fires have already become more frequent and intense in the southeastern and western United States.

A warmer climate could greatly increase populations of insects and fungi that damage trees, especially in areas where winters are no longer cold enough to control their populations. This explains the recent damage to pine forests in Canada (Figure 16-25) and the American West. Climate change also threatens many existing state and national parks, wildlife reserves, wetlands, and wilderness areas —and the biodiversity they contain.

Decline in Food Production Farmers face dramatic changes from shifting climates and an intensified hydrologic cycle. Crop productivity is projected to increase slightly with moderate warming at middle to high latitudes in areas like midwestern Canada, Russia, and Ukraine. The projected rise in crop productivity might be limited because the soils in these northern regions generally lack sufficient plant nutrients. Crop production will decrease if warming goes too far.

Climate change models project a decline in agricultural productivity and food security in tropical and subtropical regions, especially in Southeast Asia and Central America, largely because of excessive heat. Flooding of river deltas by rising sea levels could reduce crop production, partly because aquifers that supply irrigation water will be infiltrated by salt water. This flooding would also affect fish production in coastal aquaculture ponds. Food production in farm regions dependent on rivers fed by melting glaciers will drop. In arid and semiarid areas, food production will be lowered by more intense and longer droughts.

According to the IPCC, food is likely to be plentiful for a while in a warmer world. But during the latter half of this century, several hundred million of the world's poorest people could face starvation and malnutrition as food production drops.

NATIONAL GEOGRAPHIC | EXPLORERS AT WORK
Sylvia Earle Oceanographer

No one knows the importance of oceans more than "Her Deepness," Dr. Sylvia Earle. Sylvia Earle is one of the world's most respected oceanographers and is a National Geographic Society Explorer-in-Residence. *Time* magazine named her the first Hero for the Planet and the U.S. Library of Congress calls her "a living legend."

Earle's accolades are well earned. She has led more than 100 ocean expeditions and spent more than 7,000 hours under water, either diving or descending in research submarines to study ocean life (Figure 16-26). Her research has focused on the ecology and conservation of marine ecosystems, with an emphasis on developing deep-sea exploration technology.

Earle has authored more than 175 publications and has participated in numerous radio and television productions. During her long career, Earle has also been the Chief Scientist of the U.S. National Oceanic and Atmospheric Administration (NOAA), and she has founded three companies devoted to developing submarines and other devices for deep-sea exploration and research. Earle has also received more than 100 major international and national honors, including a place in the National Women's Hall of Fame.

These days, Earle is leading a campaign called Mission Blue to finance research and ignite public support for a global network of marine protected areas.

In regard to climate change, Earle wisely remarks that people should look not only at the effects of climate on the oceans, but also at the effects of the oceans on climate. Earth's climate patterns are driven largely by what is happening in the oceans. Reviving and maintaining healthy marine ecosystems, including coral reef systems, is key to ensuring the stability and resiliency of Earth's other systems.

According to Earle, a shift must occur in the general public's view of the oceans if they are to be made a priority. The oceans must be seen as an important life-support system for Earth, not just as a source of fishing and recreation. To preserve oceans, Earle advocates for creating marine protected areas similar to national parks and other protected areas on land. She aptly refers to these areas as "hope spots."

FIGURE 16-26
Sylvia Earle has been studying marine ecosystems since the 1960s. In 1970, she led the first all-female team to study the ecology of coral reefs in a self-contained laboratory under the sea.

Threats to Human Health, National Security, and Economies According to the IPCC and other reports, more frequent and prolonged heat waves will increase deaths and illnesses, especially among older people, people in poor health, and the urban poor who cannot afford air conditioning.

A warmer and more carbon dioxide-rich atmosphere will likely favor rapidly multiplying insects, including mosquitoes and ticks. These insects transmit diseases such as West Nile virus, Lyme disease, and dengue fever. Warming also will favor microbes, toxic molds, and fungi, as well as pollen-producing plants that cause allergies and asthma attacks. Insect pests and weeds are likely to multiply, spread, and reduce crop yields. Higher atmospheric temperatures and levels of water vapor in urban areas will also contribute to increased concentrations of photochemical smog. This may cause more pollution-related deaths and illnesses from heart ailments and respiratory problems.

Recent studies by the U.S. Department of Defense and the National Academy of Sciences warn that climate change could affect U.S. national security. These effects include food and water scarcity, poverty, environmental degradation, increased unemployment, social unrest, mass migration of environmental refugees, political instability, and the weakening of fragile governments.

Economies are also affected. More drought, flooding, and extreme weather events lower economic productivity, disrupt supply chains, and increase the costs for food and other commodities. According to a 2015 survey of 750 risk experts conducted by the World Economic Forum, failure to reduce and adapt to climate change tops the list of threats to the global economy, and the risks are increasing. Some researchers estimate the cost of projected climate change, if current trends continue, would be 5–20% per year of the world's economic activity as measured by gross domestic output.

Other researchers say the cost could be higher because most models used to calculate such costs have not accounted for everything. Some factors not fully accounted for include economic losses from projected ocean acidification, rising sea levels, and biodiversity losses.

checkpoint Describe three ways in which economies are likely to be affected by climate change.

16.2 Assessment

1. **Define** Explain the difference between average atmospheric temperature and climate.
2. **Contrast** How does current climate change differ from past climate change as it has been recorded within glaciers?
3. **Restate** Provide three pieces of evidence that indicate Earth's climate is undergoing rapid change.
4. **Generalize** What are the benefits and limitations of climate change models?
5. **Summarize** Describe six effects of present and projected climate change.

CROSSCUTTING CONCEPTS

6. **Cause and Effect** Is it more accurate to say that a specific drought, such as the North American drought of 2020–2022, was caused by climate change or correlated with climate change? Explain your choice. Then write a causal statement about climate change and drought.

SCIENCE AND ENGINEERING PRACTICES

7. **Constructing Explanations** Study the graphs shown in Figure 16-18. To what extent do the data support the conclusions given in Science Focus 16.2?

16.3 How Can People Slow Climate Change?

CORE IDEAS AND SKILLS
- Explain the concept of a climate change tipping point.
- Describe ways people and governments can slow atmospheric warming.
- List the pros and cons of geoengineering strategies to counteract climate change.

KEY TERMS

mitigation
climate change tipping point
carbon capture and storage (CCS)
geoengineering

Dealing with Projected Climate Change Is Difficult

Many climate scientists and other analysts think that reducing the threat of projected climate change is one of the most urgent scientific, political, economic, and ethical issues that humanity faces. However, the following characteristics of this complex problem make it difficult to tackle:

- The problem is global. Dealing with this threat will require unprecedented and prolonged international cooperation.
- The problem is a long-term political issue. Climate change is happening now and is already having harmful impacts, but it is not viewed as an urgent problem by most voters and elected officials. In addition, most of the people who will suffer the most serious harm from projected climate change during the latter half of this century have not been born yet.
- The current and projected harmful and beneficial impacts of climate change are not spread evenly. Higher-latitude nations, such as Canada, Russia, and New Zealand, may temporarily have higher crop yields, fewer deaths in winter, and lower heating bills. But other, mostly poor nations could see more flooding and higher death tolls.
- Proposed solutions, such as phasing out the use of fossil fuels, are controversial. They could disrupt economies and lifestyles and threaten the profits of the economically and politically powerful fossil fuel and utility companies.
- The projected effects are uncertain. Current climate models lead to a wide range in the projected temperature increase and sea-level rise. Thus, there is uncertainty over whether the harmful changes will be moderate or catastrophic. This makes it difficult to plan for avoiding or managing risk. It also highlights the urgent need for more scientific research to reduce the uncertainty in climate models.

There are two main approaches to dealing with the projected harmful effects of global climate change. One, called **mitigation**, is to try to slow down climate change in order to avoid its most harmful effects. The other approach is known as adaptation. To take the adaptation approach is to recognize that some climate change is unavoidable and find ways to adapt to some of its harmful effects. Most analysts call for a combination of both approaches.

Regardless of the approach taken, most climate scientists argue the most urgent priority is to avoid any and all **climate change tipping points**. Tipping points are the estimated thresholds beyond which natural systems could not be reversed. Each of these tipping points would accelerate changes in climate, similar to a ball accelerating down a hill (Figure 16-27). For example, if CO_2 continues to be added to the atmosphere at the current rate, the estimated tipping point marked by 450 ppm of atmospheric CO_2 will be exceeded within a few decades. Such high levels of CO_2 could put Earth's system into a positive feedback loop that may result in climate changes that last for hundreds or perhaps thousands of years. It also would increase the likelihood of exceeding many of the other tipping points.

Many business leaders see climate change as both an economic threat and a global investment opportunity. For example, in the United States, Texas produces more electricity from wind power than all but five of the world's countries. Other countries—notably China, Germany, and Denmark—are building their renewable energy capacities.

Reducing Greenhouse Gas Emission

Climate scientists generally agree that to avoid the most harmful effects of climate change, people need to limit the global average temperature increase to 2°C (3.6°F) over the preindustrial global average.

FIGURE 16-27 ▼
Climate Change Tipping Points This list shows some tipping points that climate scientists are most concerned about.

Atmospheric carbon level of 450 ppm
Melting of all Arctic summer sea ice
Collapse and melting of the Greenland ice sheet
Collapse and melting of most of the West Antarctic ice sheet
Massive release of methane from thawing Arctic permafrost and from the Arctic seafloor
Collapse of part of the Gulf Stream
Severe ocean acidification, collapse of phytoplankton populations, and a sharp drop in the ability of the oceans to absorb CO_2
Massive loss of coral reefs
Severe shrinkage or collapse of Amazon rain forest

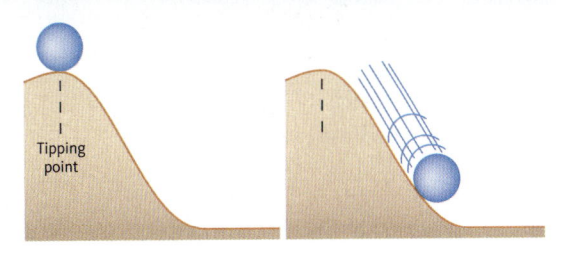

LESSON 16.3

The world's nations and energy companies together hold reserves of fossil fuels that, if burned, would emit nearly five times the amount of CO_2 that climate scientists say is safe. Switching away from fossil fuels to avoid exceeding this climate change tipping point means leaving 82% of the world's coal reserves, 50% of all natural gas, and all Arctic oil reserves in the ground, according to a study from University College of London.

A growing number of scientists recognize that shifting away from fossil fuels is difficult but contend that it can be done. They point out that humans have shifted major energy resources before—first from wood to coal, then to oil, and now to natural gas. Each shift took about 50 years. Humans now have the knowledge and ability to shift to a reliance on sustainable energy over the next 50 years. All that is needed is a political and ethical will to make the shift.

Figure 16-28 lists a number of ways people can slow the rate and degree of atmospheric warming and projected resulting climate change by reducing CO_2 emission and by removing excess CO_2 from the atmosphere.

According to the 2014 IPCC report, there is good news related to reducing human emissions of CO_2 into the atmosphere:

- Hybrid, plug-in hybrid, and electric cars are available. People could switch to these cars and charge their batteries only with electricity produced from renewable energy sources.
- The shift to renewable energy is accelerating as prices for electricity from low-carbon wind turbines and solar cells fall, and investments in these technologies grow.
- Engineers have designed zero-carbon buildings and can reduce the carbon footprints of existing buildings (Figure 16-29).

Cleanup and Geoengineering Some scientists and engineers are designing cleanup strategies for removing some of the CO_2 from the atmosphere or smokestacks and storing it (Figure 16-30 and Engineering Focus 16.3). These strategies are called **carbon capture and storage (CCS)**. Other approaches to carbon cleanup include implementing a massive, global tree planting and forest restoration program and restoring wetlands, both of which remove CO_2 from the atmosphere.

The aim of CCS is to remove excess CO_2 from the atmosphere. Other strategies fall under the umbrella of geoengineering. **Geoengineering** strategies manipulate natural conditions to counter the enhancement of the greenhouse effect. Some of these proposals are shown in Figure 16-31.

Some geoengineering proposals attempt to reflect more sunlight back into space. One proposal calls for injecting sulfate particles into the stratosphere to reflect some of the incoming sunlight. Another calls for placing a series of giant mirrors in orbit above Earth to reflect sunlight.

FIGURE 16-28 ▼

Slowing Climate Change	
Prevention	**Cleanup**
Cut fossil fuel use (especially coal).	Sequester CO_2 by planting trees and preserving forests and wetlands.
Shift from coal to natural gas.	Sequester carbon in soil.
Repair leaky natural gas pipelines and facilities.	Sequester CO_2 deep underground (with no leaks allowed).
Improve energy efficiency.	Remove CO_2 from smokestack and vehicle emissions.
Shift to renewable energy resources.	
Reduce deforestation.	
Use more sustainable agriculture and forestry.	
Put a price on greenhouse gas emissions.	

FIGURE 16-29 ▼
The Beddington Zero Energy Development is a carbon-neutral "eco-village" in London, England.

FIGURE 16-30 ▼
Carbon Capture and Storage (CCS) This illustration shows some proposed CCS schemes for removing some of the carbon dioxide from smokestack emissions and from the atmosphere and storing it in soil, plants, deep underground reserves, and sediments beneath the ocean floor.

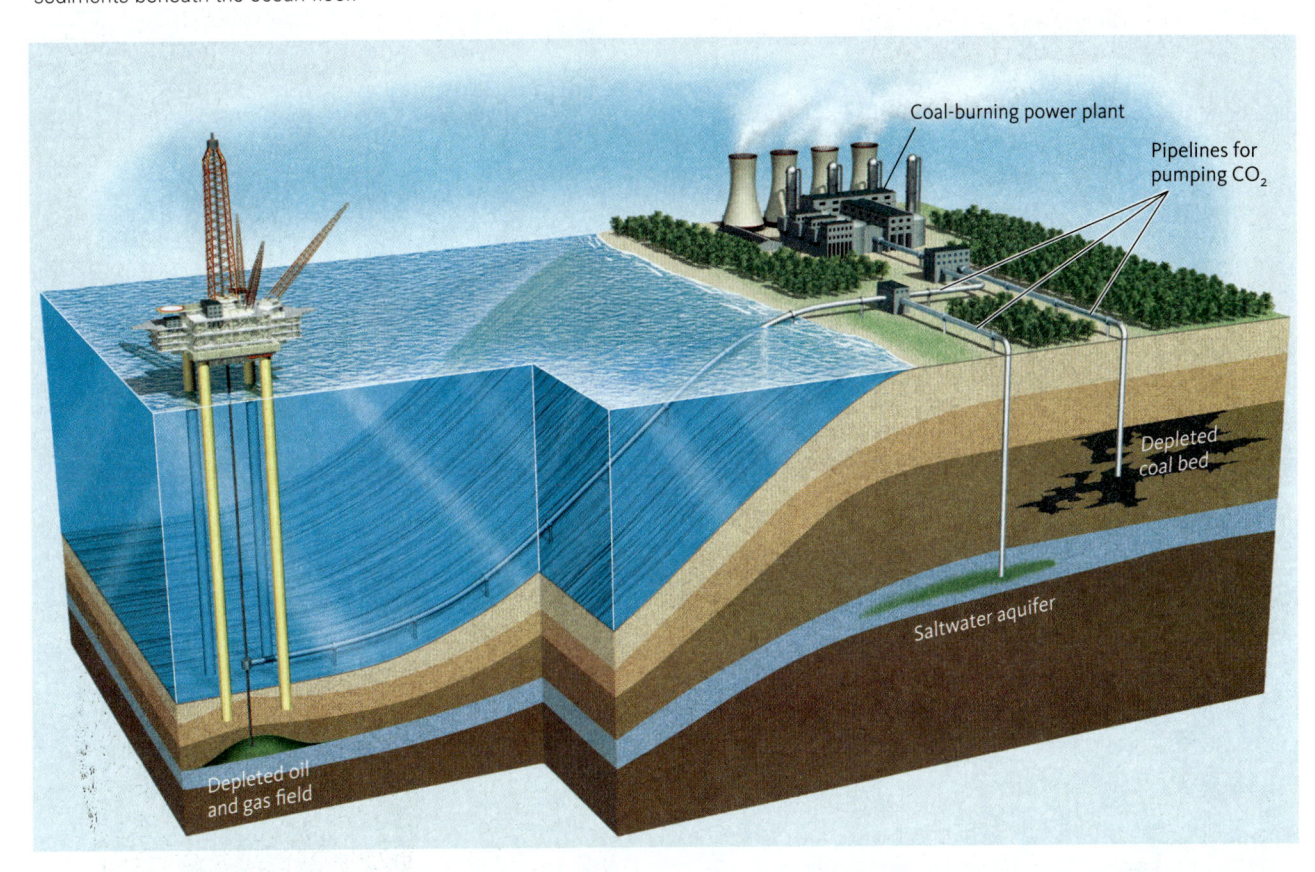

ENGINEERING FOCUS 16.3

CARBON CAPTURE

Carbon capture and storage (CCS) strategies pose a number of challenges. One CCS strategy, shown in Figure 16-30, would remove some of the CO_2 gas from smokestack emissions of coal-burning power and industrial plants or from the atmosphere. It would then convert the CO_2 to a liquid and force it under pressure into underground storage sites.

So far, CCS projects have not been very effective. With current technology, CCS strategies would only remove some of the CO_2 from smokestack emissions and the atmosphere. They would not address the massive amounts of CO_2 from motor vehicle exhaust, food production, and the burning of forests.

In addition, CCS strategies require a lot of energy, which could lead to greater use of fossil fuels and higher emissions of CO_2 and other air pollutants. Also, the CO_2 that is removed would have to remain sealed, or sequestered, from the atmosphere forever. Any large-scale leaks, or a number of smaller continuous leaks, could dramatically increase atmospheric warming in a short time.

Another major challenge is cost. The first large-scale plant designed to convert coal to gas and to capture 65% of the CO_2 produced has experienced large cost overruns and is projected to cost $5.2 billion. It will be one of the most expensive fossil fuel plants ever built in the United States. Such high costs are making U.S. utilities reluctant to build CCS plants.

Thinking Critically
Synthesize Describe three criteria a CCS strategy would need to meet if it were to be successful.

LESSON 16.3   **565**

FIGURE 16-31 ▼

Geoengineering Geoengineering schemes include ways to reduce atmospheric warming by reflecting more sunlight back into space. Which three of the approaches shown here do you think are the most workable? Why?

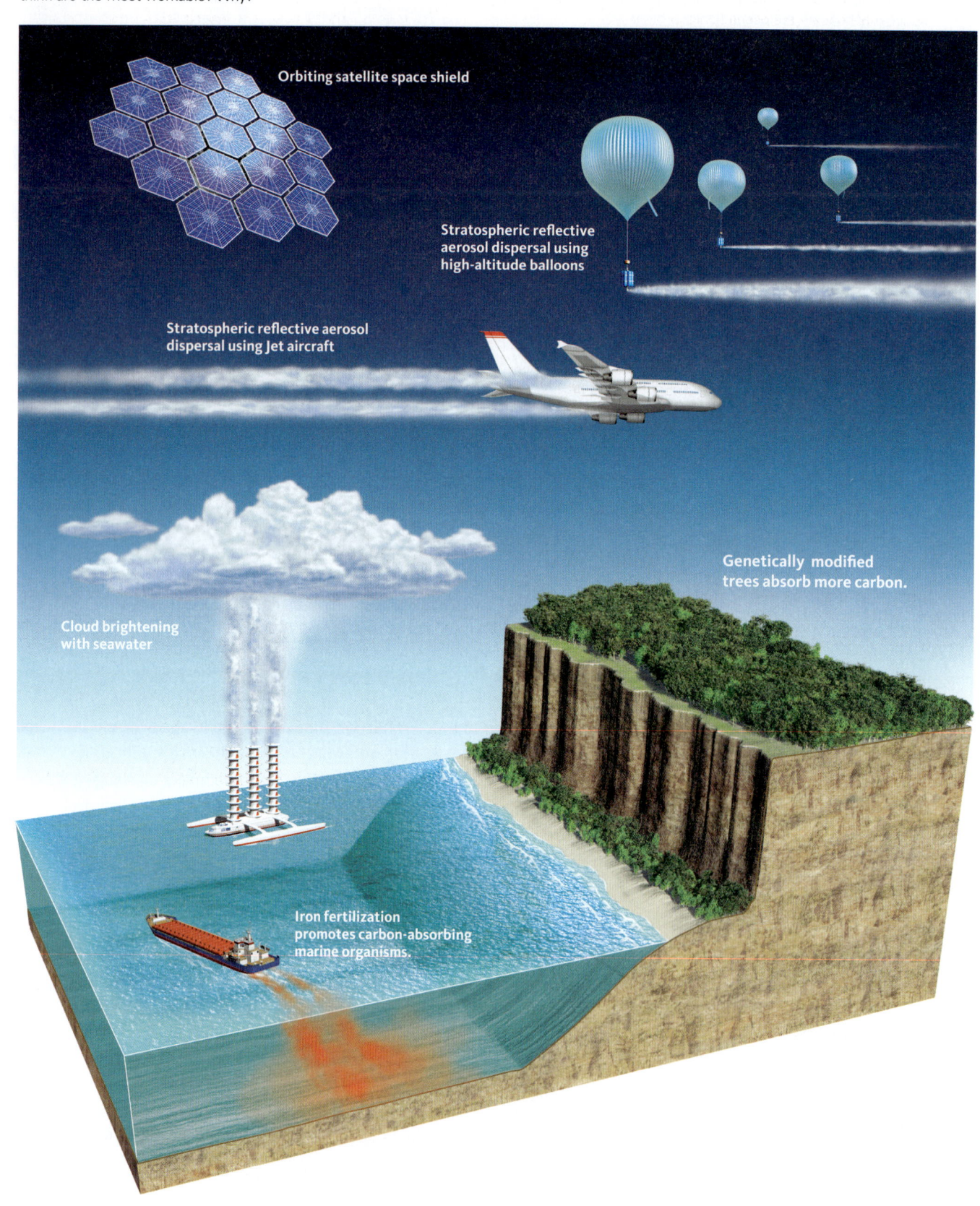

- Orbiting satellite space shield
- Stratospheric reflective aerosol dispersal using high-altitude balloons
- Stratospheric reflective aerosol dispersal using jet aircraft
- Cloud brightening with seawater
- Genetically modified trees absorb more carbon.
- Iron fertilization promotes carbon-absorbing marine organisms.

566 CHAPTER 16 AIR POLLUTION, CLIMATE CHANGE, AND OZONE DEPLETION

Another solution is to deploy a large fleet of computer-controlled ocean-going ships to inject salt water high into the sky in order to make clouds whiter and more reflective (Figure 16-31).

Many scientists reject the idea of launching sulfates into the stratosphere as being too risky because the effects are unknown. If the sulfates reflected too much sunlight, they could reduce evaporation enough to alter global rainfall patterns and worsen the already dangerous droughts in Asia and Africa. In addition, chlorine released by reactions involved in this strategy could speed up the thinning of Earth's vital ozone layer.

A major problem with geoengineered solutions is that if they succeed, they could be used to justify the continued use of fossil fuels. This would allow CO_2 levels in the lower atmosphere to continue building and adding to the serious problem of ocean acidification. In the long run, these solutions might not succeed in slowing atmospheric warming. In that case, atmospheric temperatures could soar at a rapid rate and essentially ensure severe and irreversible climate change, as well as other harmful side effects. Geoengineering strategies may end up delaying a shift from relying on nonrenewable fossil fuels to using improved energy efficiency and a mix of renewable energy resources. Many scientists say we cannot afford such a delay.

Government Strategies Governments can use strategies to tackle climate change. They can:

- Regulate CO_2 and CH_4 as climate-changing pollutants that harm public health.
- Phase out the most polluting coal-burning power plants over the next 50 years and replace them with cleaner natural gas and renewable energy alternatives such as wind power and solar power. The use of next-generation nuclear power plants could also reduce power plant CO_2 emissions.
- Put a price on carbon emissions by phasing in taxes on each unit of CO_2 or CH_4 emitted, or phasing in energy taxes on each unit of any fossil fuel burned (Figure 16-32, top). These tax increases could be offset by reductions in taxes on income, wages, and profits. In 2014, China and 72 other nations, the World Bank, and more than 1,000 corporations called for putting a price on carbon.
- Use a market-based cap-and-trade system to reduce emissions of CO_2 and CH_4 (Figure 16-32, bottom).
- Phase out government subsidies and tax breaks for the fossil-fuel industry and unsustainable industrialized food production. Phase in subsidies and tax breaks for energy-efficient technologies, low-carbon renewable energy development, and more sustainable food production.
- Focus research and development efforts on innovations that lower the cost of clean energy alternatives. That would help these alternatives compete with fossil fuels, which have benefited from government-subsidized research and development for more than 50 years.
- Work out agreements to finance and monitor efforts to reduce deforestation, which accounts for 12–17% of global greenhouse gas emissions, and promote global tree-planting efforts.

Environmental economists and a growing number of business leaders call for putting a price on carbon emissions as the best way to curb them. Fuel prices would include the estimated harmful environmental and health costs of using fossil fuels. This strategy would be in keeping with the full-cost pricing factor of sustainability. Establishing laws and regulations that raise the price of fossil fuels is politically difficult for two reasons:

1. The fossil fuel lobby has immense political and economic power.
2. Reliance on fossil fuels is widespread among consumers and businesses.

In December 1997, delegates from 161 nations met in Kyoto, Japan, to negotiate a treaty to slow atmospheric warming and projected climate change. The first phase of the resulting Kyoto Protocol went into effect in 2005, and 187 of the world's countries (not including the United States) had ratified the agreement by 2009. The 37 participating more-developed countries agreed to cut their emissions of greenhouse gases to certain levels by 2012, when the treaty was to expire. Sixteen of them failed to do so. Less-developed countries, including China, were excused from this requirement because such reductions would have curbed their economic growth.

In 2005, participating countries began negotiating a second phase of the treaty that was to go into effect in 2012. However, these negotiations failed to extend the original agreement. Then in 2015, delegates from 195 countries met in Paris, France, with the goal of achieving a global climate agreement.

In a historic accord, the governments agreed to the following:

- Accept a goal to keep the increase in global average temperatures below 2°C (3.6°F).
- Pledge to reduce greenhouse gas reductions by a set amount.
- Meet every five years to evaluate progress and increase their goals.

Climate scientists applaud this progress but say that it does not do enough to prevent serious environmental and economic problems. Countries are not legally bound to reach their goals. In addition, there was no agreement reached for the wealthier nations to raise a proposed $500 billion by 2020 to assist poorer countries in meeting their goals.

checkpoint In what way is a climate change tipping point similar to a ball balanced on top of a hill?

Facing Climate Change

Some nations are leading others in facing and responding to the challenges of projected climate change. Costa Rica aims to be the first country to become carbon neutral by cutting its net carbon emissions to zero by 2030. The country generates 78% of its electricity with renewable hydroelectric power and another 18% from renewable wind and geothermal energy. Other efforts are being made by state and local governments, colleges and universities, and large corporations. For example, the state of California plans to get 33% of its electricity from low-carbon renewable energy sources by 2030.

Many colleges and universities are taking action. Arizona State University (ASU) boasts the largest collection of solar panels of any U.S. university. ASU was also the first American university to establish a School of Sustainability. The College of the Atlantic in Bar Harbor, Maine, has been carbon neutral since 2007. All its electricity comes from renewable hydropower, and many of its buildings are heated with the use of renewable wood pellets. Students at the University of Washington in Seattle agreed to an increase in their fees to help the school buy electricity from renewable energy sources.

Individuals can also make a difference. You can adjust your lifestyle to reduce your carbon footprint. In addition, you can offset part your carbon footprint by finding ways to reduce CO_2 in the atmosphere. Figure 16-33 lists ways in which you can cut your CO_2 emissions while increasing your beneficial environmental impact.

According to global climate models, the world needs to make a 50–85% cut in greenhouse gas emissions by 2050 to stabilize concentrations of these gases in the atmosphere. This would prevent the atmosphere from warming by more than 2°C (3.6°F) and head off projected rapid changes in the world's climate. Many analysts believe that while people work to slash greenhouse gas emissions, they should also begin to prepare for the harmful effects of projected climate change.

Relief organizations like the International Red Cross are turning their attention to projects to prepare for climate change. These projects include expanding mangrove forests as buffers against storm surges, building shelters on high ground, and planting trees on slopes to help prevent landslides in the face of projected higher levels of precipitation and rising sea levels. Alaska, which is warming quickly, plans to relocate coastal villages at risk from rising sea levels, coastal erosion, and higher storm surges.

Some coastal communities in the United States now require new houses and other construction to be built high enough off the ground or farther back from the current shoreline to survive rising sea levels and storm surges. In anticipation of rising sea levels, Boston elevated one of its sewage treatment plants. Some cities plan to establish cooling centers to

FIGURE 16-32

Carbon and Energy Taxes	
Advantages	**Disadvantages**
Simple to administer	Tax laws can get complex.
Clear price on carbon	Vulnerable to loopholes
Covers all emitters	Doesn't guarantee lower emissions
Predictable revenues	Politically unpopular

Cap-and-Trade Policies	
Advantages	**Disadvantages**
Clear legal limit on emissions	Revenues not predictable
Reward cuts in emissions	Vulnerable to cheating
Record of success	Rich polluters can keep polluting.
Low expense for consumers	Puts variable price on carbon

FIGURE 16-33 ▼

What You Can Do
Calculate your carbon footprint (several helpful websites can assist in this).
Walk, bike, carpool, and use mass transit.
Reduce garbage by reducing consumption, recycling, and reusing more items.
Use energy-efficient appliances and LED light bulbs.
Wash clothes in warm or cold water and hang them up to dry.
Close window curtains to keep heat in or out.
Use a low-flow showerhead.
Eat less meat or no meat.
Heavily insulate your house and seal all air leaks.
Use energy-efficient windows.
Set your hot-water heater to 49°C (120°F).
Plant trees.
Buy from businesses working to reduce their CO_2 emissions.

shelter residents during increasingly intense heat waves. In the low-lying Netherlands, people are dealing with the threat of a rising sea levels by building houses that float.

checkpoint What steps could your school take to reduce its carbon footprint?

16.3 Assessment

1. **Explain** What is a climate change tipping point?
2. **Restate** Describe one thing governments can do to slow atmospheric warming and three things you can do to slow atmospheric warming.
3. **Recall** Describe a CCS strategy and its pros and cons.
4. **Make Judgments** Which geoengineering solutions, if any, do you think are worth pursuing? Weigh the potential benefits and limitations.
5. **Infer** Americans have the largest per capita carbon footprint. What does that mean for the potential of individual Americans to impact climate change?

CROSSCUTTING CONCEPTS

6. **Patterns** Explain how reaching one climate change tipping point could increase the chance of reaching others. Give specific examples.

16.4 How Can People Reverse Ozone Depletion?

CORE IDEAS AND SKILLS

- Explain the causes and effects of stratospheric ozone depletion.
- Describe how people can reverse ozone depletion.

KEY TERM

chlorofluorocarbon (CFC)

The Use of Certain Chemicals Threatens the Ozone Layer

The widespread use of certain chemicals has reduced ozone levels in the stratosphere and allowed more harmful ultraviolet (UV) radiation to reach Earth's surface. The good news is that the problem of ozone depletion can be reversed. To achieve this, people will have to stop producing ozone-depleting chemicals and adhere to the international treaties that ban such chemicals.

As you read in Lesson 16.1, a layer of ozone in the lower stratosphere keeps 95% of the sun's harmful ultraviolet radiation from reaching Earth's surface. This ozone layer is a vital form of natural capital that supports all life on land and in shallow aquatic environments.

Measurements show a considerable seasonal depletion, or thinning, of O_3 concentrations in the stratosphere above Antarctica (Figure 16-34) and the Arctic since the 1970s. Similar measurements reveal slight ozone thinning everywhere except over the tropics. The loss of ozone over Antarctica has been called *the ozone hole*. A more accurate term is *ozone thinning* because the ozone depletion varies with altitude and location. Ozone depletion in the stratosphere poses a serious threat to humans, other animals, and some primary producers (mostly plants).

The origin of this serious environmental threat began in 1930 with the accidental discovery of the first **chlorofluorocarbon (CFC)**, a compound that contains carbon, chlorine, and fluorine. Chemists soon developed similar compounds to create a family of highly useful CFCs, known by their trade name Freons.

These chemically unreactive, odorless, nonflammable, nontoxic, and noncorrosive compounds were thought to be dream chemicals.

LESSON 16.4 **569**

Inexpensive to manufacture, Freons became popular as coolants in air conditioners and refrigerators, propellants in aerosol spray cans, cleansers for electronic parts such as computer chips, fumigants for granaries and ships' cargo holds, and gases used to make insulation and packaging.

It turned out that CFCs actually were too good to be true. Starting in the 1970s, scientists showed that CFCs are persistent chemicals that destroy protective O_3 in the stratosphere. Satellite data, measurements, and models indicate that 75–85% of the observed O_3 losses in the stratosphere since 1976 have been the result of the presence of CFCs and other chemicals in the troposphere beginning in the 1950s.

These long-lived chemicals eventually reached the stratosphere, where they began destroying ozone faster than it was being formed. Formation of O_3 occurs with the interaction of stratospheric oxygen and incoming solar radiation. Ozone depletion is a disruption of one of Earth's most important forms of natural capital. While in the troposphere, CFCs also act as greenhouse gases that raise temperatures in the lower atmosphere.

Why It Matters What makes ozone depletion a cause for concern? Figure 16-35 lists the effects of stratospheric ozone thinning. One effect is that more biologically damaging UV-A and UV-B radiation reaches Earth's surface. This radiation is a likely contributor to rising cases of eye cataracts, damaging sunburns, and skin cancers. Figure 16-36 lists ways to protect yourself from harmful UV radiation.

Another serious threat from ozone depletion and the resulting increase in UV radiation reaching the planet's surface is the possible destruction of phytoplankton, especially in Antarctic waters. These tiny marine plants play a key role in removing CO_2 from the atmosphere, and they form the base of many ocean food webs. By destroying them, we eliminate the vital ecological services they provide. The loss of plankton could accelerate projected atmospheric warming by reducing the capacity of the oceans to remove CO_2 from the atmosphere.

checkpoint How does ozone in the stratosphere protect you from exposure to UV radiation?

People Can Reverse Ozone Depletion

According to researchers in this field, people should immediately stop producing all ozone-depleting chemicals. Models and measurements indicate that, even with immediate and sustained action, it will take at least 60 years for Earth's ozone layer to recover the levels of ozone it had in the 1960s.

FIGURE 16-34

Ozone Depletion These colorized satellite images compare ozone thinning (blue) over Antarctica in September 1979 and September 2021.

Source: National Aeronautics and Space Administration

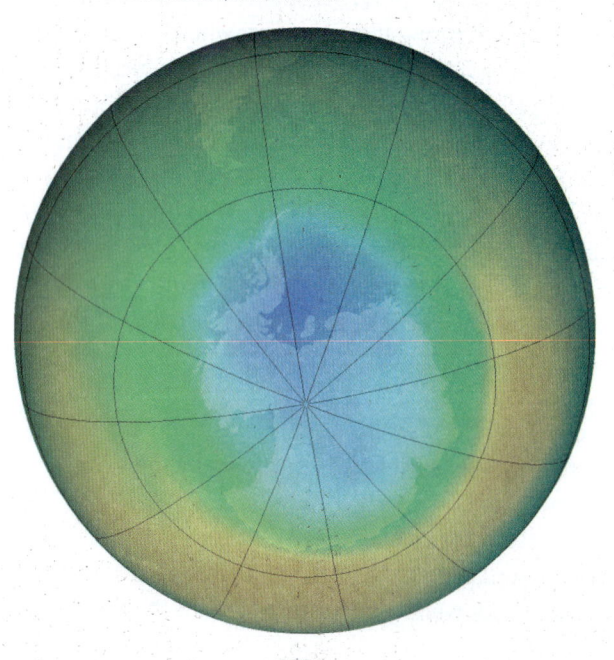

1979

2021

FIGURE 16-35

Effects of CFCs and Ozone Depletion	
Human Health and Structures	**Wildlife**
Worse sunburns	More eye cataracts in some species
More eye cataracts and skin cancers	Shrinking populations of aquatic species sensitive to UV radiation
Immune system suppression	Disruption of aquatic food webs due to shrinking phytoplankton populations
Food and Forests	**Air Pollution and Climate Change**
Reduced yields for some crops	Increased acid deposition
Reduced seafood supplies due to smaller phytoplankton populations	Increased photochemical smog
Decreased forest productivity for UV-sensitive tree species	Degradation of outdoor painted surfaces, plastics, and building materials
	CFCs in troposphere that act as greenhouse gases

In 1987, representatives of 36 nations met in Montreal, Canada, and developed the Montreal Protocol. This treaty's goal was to cut emissions of CFCs by 35% by the year 2000. After learning about seasonal ozone thinning above Antarctica in 1989, representatives of 93 countries continued to meet. In 1992, they adopted the Copenhagen Amendment, which accelerated the phase-out of CFCs and added other key ozone-depleting chemicals to the agreement.

Marking its 25th anniversary, the Montreal Protocol was identified as the world's most successful environmental agreement. This international agreement set an important example because nations and companies worked together and used a prevention approach to solve a serious environmental problem. This approach worked for three reasons:

1. There was convincing and dramatic scientific evidence of a serious problem.
2. The banned chemicals were produced by a small number of international companies, which lead to less corporate resistance in finding a solution.
3. The certainty that CFC sales would decline over a period of years because of government bans motivated the private sector to find more profitable substitute chemicals.

Progress in reducing ozone depletion could be set back by projected climate change. Tropospheric warming makes the stratosphere cooler, slowing down the rate of ozone repair. Thus, O_3 levels may take longer to return to 1960s levels, or they may never recover completely.

However, ozone reduction agreements have lowered the global average temperature by 0.2°C (0.4°F) by reducing emissions of CFC greenhouse gases. The landmark international agreements on stratospheric ozone, now signed by all 195 of the world's countries, are important examples of successful global cooperation in response to a serious global environmental problem.

The impacts of human activities on Earth's atmosphere are greater now than ever. However, the human abilities to model, predict, and manage these impacts are also greater than ever. The six factors of sustainability can help people reduce the harmful effects of air pollution, projected climate change, and stratospheric ozone depletion. People can reduce emissions of pollutants, greenhouse gases, and ozone-depleting chemicals by relying less on fossil fuels. They can recycle and reuse resources much more widely than they do today.

People can also mimic biodiversity by using a

FIGURE 16-36

What You Can Do
Stay out of the sun, especially between 10 a.m. and 3 p.m.
Do not use tanning beds or sunlamps.
When in the sun, wear clothing and sunglasses that protect against UV-A and UV-B radiation.
Be aware that overcast skies do not protect you.
Do not expose yourself to the sun if you are taking antibiotics or birth control pills.
When in the sun, use a sunscreen with an SPF of 15 or higher.

variety of often locally available low-carbon energy resources. People can advance toward these goals by including the harmful environmental and health costs of fossil fuel use in market prices; seeking win-win solutions that will benefit both the economy and the environment; and giving high priority to passing along a healthy atmosphere to future generations.

checkpoint What is the Montreal Protocol and why was it so successful?

16.4 Assessment

1. **Recall** Identify the causes and effects of stratospheric ozone depletion.
2. **Explain** How can people reverse ozone depletion?

SCIENCE AND ENGINEERING PRACTICES

3. **Communicating Information** How does ozone depletion contribute to atmospheric warming? How does atmospheric warming hinder ozone layer repair?

TYING IT ALL TOGETHER STEM
Melting Ice and Sustainability

In this chapter's Case Study, you read about the accelerated melting of ice in Greenland during the summer and the net loss of glacial ice that has resulted. In this exercise, you will analyze and interpret the data presented in Figure 16-1 by estimating the area of ice cover.

Actual ice loss measurements are much more sophisticated. Satellites, such as Europe's CryoSat, use radar to measure ice volume rather than surface area. To measure the volume of ice in Greenland, CryoSat uses 14 million individual data points. Similarly, CryoSat gathers information about a major ice sheet on the other side of the world—in Antarctica. CryoSat computes Antarctica's ice using input from 200 million data points. CryoSat studies suggest that, together, Greenland's and Antarctica's ice sheets are losing about 500 cubic kilometers (120 cubic miles) of ice per year, and this rate is increasing. Other studies that estimate changes in mass also indicate a global net loss of ice.

Use a piece of tracing paper and grid paper to help you complete the steps below.

1. Place the tracing paper over the images and trace the perimeter of both glaciers with a dark line. Label the tracings as *Summer 1982* and *Summer 2012*.
2. Place the grid paper over your tracings. Estimate the relative surface area of each glacier by counting squares. Count complete squares as 1 surface area unit. Count half squares as 0.5 surface area unit. Count squares that fall more than half within the line as 1 surface area unit. Ignore squares that are less than half within the line.
3. Record your numbers in a table. Find the total surface area for each glacier.
4. How much smaller is the summer 2012 glacier than the summer 1982 glacier? Write your answer as a percentage.
5. Supposing the rate of ice loss were steady, by what percentage would ice have decreased per year?
6. Is the data presented here enough to identify a pattern? Explain.
7. How does CryoSat data help scientists identify patterns in ice loss?
8. Work with a partner to evaluate the following statement. Write 1–2 paragraphs that explain the reasoning behind your evaluation. Support your argument with data.
"Melting ice in Greenland doesn't matter because Antarctica is gaining ice."

CHAPTER 16 SUMMARY

16.1 What are the major air pollution problems?

- Air pollution is any gaseous or solid material in the atmosphere that occurs in concentrations high enough to harm organisms, ecosystems, or human-made materials, or to alter climate. Air pollutants come from natural sources such as volcanic eruptions and from human sources such as burning fossil fuels.
- Primary pollutants are emitted directly into the air. Some primary pollutants react with one another and with other natural components of the air to form secondary pollutants.
- The major types of outdoor air pollution include carbon oxides, nitrogen oxides and nitric acid, sulfur dioxide and sulfuric acid, particulates, ozone, and VOCs. Sulfur compounds and suspended particles in the air combine to form industrial smog. Primary and secondary pollutants react with UV radiation to form photochemical smog. Acid deposition occurs when acidic pollutants in the atmosphere fall to Earth's surface.
- Indoor air pollution is more concentrated than outside air pollution and affects more people, especially those in less-developed countries.

16.2 What are the effects of climate change?

- Climate change is a change in weather conditions over a period of at least three decades. Climate change is not new but has accelerated recently. Historical evidence of climate change is found in ice cores, among other sources.
- Oceans play an important role in climate change. Oceans absorb CO_2 and moderate Earth's average surface temperature. The oceans also slow climate change by removing excess heat from the environment, which accounts for a rise in their average temperature since 1970.
- Computers are used to run complex climate models that help scientists make predictions about future average atmospheric temperatures.
- Climate change has serious effects. Melting ice and snow lead to increasing atmospheric temperatures by reducing the amount of sunlight reflected from Earth's surface. Melting permafrost releases large amounts of CH_4 and CO_2 from rotting organic matter. Biodiversity loss is occurring as habitat is lost or rapidly changed.
- Rising sea levels will cause problems such as flooding, disruption of fisheries, contamination of coastal freshwater aquifers, and submersion of low-lying islands. More severe drought and extreme weather are also predicted.

16.3 How can people slow climate change?

- To slow climate change, people must avoid reaching several identified climate change tipping points. Reduction of society's dependence on fossil fuels and shifting to low-carbon sources of energy are steps in the right direction.
- Carbon capture and storage strategies and geoengineering strategies attempt to use technology to slow climate change. Government strategies include carbon or energy taxing and cap-and-trade systems.

16.4 How can people reverse ozone depletion?

- The use of CFCs over several years created thinning in the critical protective ozone layer in the stratosphere.
- Many countries have cooperated to stop and reverse depletion of the ozone layer.

CHAPTER 16 ASSESSMENT

Review Key Terms

Select the key term that best fits each definition. Some terms will not be used.

- air pollution
- acid deposition
- carbon capture and storage (CCS)
- carbon footprint
- chlorofluorocarbon
- climate change
- climate change tipping point
- drought
- geoengineering
- mitigation
- ozone layer
- smog
- temperature inversion

1. A strategy of manipulating natural conditions to counteract atmospheric warming
2. Gaseous or solid materials in the atmosphere in concentrations high enough to harm organisms, ecosystems, and human-made materials, or to alter climate
3. The amount of CO_2 generated by an individual, organization, country, or any other entity over a given period of time
4. Estimated climate change threshold beyond which natural systems could change irreversibly
5. A strategy of trapping carbon dioxide and storing it away from the atmosphere
6. The falling of wet or dry acidic substances to Earth's surface
7. The attempt to slow down or reduce harmful effects
8. Hazy industrial or photochemical air pollution
9. Air pollutant that undergoes a reaction in the atmosphere and damages Earth's ozone layer

Review Key Concepts

10. Why is the ozone layer important?
11. List and briefly explain six natural factors that can worsen outdoor air pollution.
12. What is the most threatening indoor air pollutant in many less-developed countries? What are the three most dangerous indoor air pollutants in more-developed countries?
13. Summarize the use of air-pollution-control laws in the United States and how they could be improved.
14. List three major conclusions of the IPCC and other scientific bodies regarding changes in the temperature of Earth's atmosphere and their effects on climate.
15. List nine pieces of scientific evidence that support the conclusion that human-influenced climate change is happening now.
16. Describe one prevention strategy and one cleanup strategy for slowing projected climate change.

Think Critically

17. Explain how permafrost melt is part of a positive feedback loop. How does positive feedback affect the rate of climate change?
18. Summarize the story of Greenland's melting glaciers and the possible effects of this process. Explain how it fits into projections about climate change during this century.
19. Which of the effects of climate change do you think would be the most difficult to reverse? Explain your reasoning.
20. Suppose someone says that CO_2 should not be classified as an air pollutant because it is a natural chemical that people add to the atmosphere every time we exhale. Do you agree or disagree with the statement? Explain.
21. List three important ways in which your life might be different if citizen-led actions during the 1970s and 1980s had not led to the Clean Air Acts of 1970, 1977, and 1990. Which one or more of the six factors of sustainability were applied by the passage of these acts? Explain.

Chapter Activities

A. Investigate: Climate Models STEM

Climate models are the best—and only—tools scientists have to test scenarios and project the possible results on Earth's atmosphere and climate. As you have seen from this chapter, Earth's atmosphere is such a complex system that it is impossible to make well-founded predictions without the use of models. These models allow scientists to ask critical questions such as: How much of a reduction in carbon emissions would be needed to prevent Earth's atmosphere from reaching climate change tipping points? Such questions and their outputs can inform governments and people about the most efficient ways to take action.

Greatly simplified climate change models are available on the internet, such as the one provided by the National Center for Atmospheric Research: *The Very, Very Simplified Climate Model*. In this activity, you will first run some mathematical calculations of your own. You will then run a computational simulation and compare the results.

1. Look at Figure 16-16. Subtract the atmospheric CO_2 concentration (ppm) in 1960 from that in 2022. Use a ruler to help you estimate the value for 2022. What is the difference in CO_2?

2. How long did it take for that change to occur? Find the rate of change by dividing the difference in CO_2 concentration by the number of years it took to change. Write your answer in ppm/year.

3. If the rate of CO_2 concentration (ppm/year) stays the same, how many years will it take for Earth to reach the climate change tipping point of 450 ppm CO_2?

4. Positive feedback loops, such as those you have read about in this chapter, can accelerate change. Suppose the rate of CO_2 input into the atmosphere doubled. Multiply your rate of change by 2 and run your calculations again. How many years at this rate would it take to reach 450 ppm CO_2?

5. Now use a computer simulation to ask the same questions. Start with a steady rate of 9 gigatons of carbon per year. How many years will it take for Earth to reach the carbon tipping point at this rate? And if the rate is doubled to 18 gigatons of carbon per year?

6. How do the results of the two models compare? How can you account for the differences?

7. Now include temperature data in the computational model. How does changing carbon affect temperature?

8. Draw conclusions from the results. Make projections regarding Earth reaching its carbon tipping point and the implications on temperature, climate, and other Earth systems. Then describe a plausible way people can prevent Earth from reaching this tipping point, using quantitative data in your description.

B. Citizen Science

National Geographic Learning Framework
Attitudes | Curiosity
Skills | Observation
Knowledge | New Frontiers

According to the National Audubon Society, climate change threatens one-fifth of the world's birds, including 314 species in North America. Birds are especially sensitive indicators of climate change because their feeding and migration patterns depend on seasonal temperature-triggered events such as larvae hatching. Suppose the only way you could eat was to use a great amount of energy to travel far across the continent, only to find that your food was not there when you arrived. Some species are adapting by shifting migration patterns northward while other species are in decline.

Join a citizen science group, such as the Audubon's Christmas Bird Count, or another one, to monitor bird species in your area. Meet to discuss your findings with the class.

CHAPTER 17
SOLID AND HAZARDOUS WASTE

FROM EGGSHELLS TO ELECTRONICS, solid waste takes many forms. And all of it must go somewhere. Society faces growing challenges over what to do with its waste. There's a lot of it, and much of it is hazardous. Recycling and reusing solid waste, such as computer parts, can keep it out of landfills. Still, most waste ends up being buried or burned. One goal for managing solid and hazardous wastes is finding ways to produce less of them.

KEY QUESTIONS

17.1 What are problems related to solid and hazardous waste?

17.2 How should society deal with solid waste?

17.3 How should society deal with hazardous waste?

17.4 How can society transition to a low-waste economy?

NATIONAL GEOGRAPHIC | EXPLORERS AT WORK

Waging War Against Food Waste

with National Geographic Explorer Tristram Stuart

Tristram Stuart thinks we should do something revolutionary with food: Eat it. Stuart, a British author and activist, calls the problem of food waste "scandalous and grotesque," and he has the statistics to back him up. One-third of the world's food is wasted from plow to plate, despite the existence of 780 million hungry people across the globe. Just one-quarter of the food wasted in the United States, the United Kingdom, and Europe would go a long way toward lifting these hungry people out of malnourishment. And the water used to irrigate food that ends up being thrown away could be used to meet the domestic water needs of 9 billion people.

Until a few years ago, the huge scale of food waste was largely ignored. Stuart's 2009 book, *Waste: Uncovering the Global Food Scandal*, and the grassroots initiatives he has launched aim to change that. "We want to catalyze a food-waste revolution one person, one town, one country at a time—helping stop needless hunger and environmental destruction across our planet," Stuart declares.

Stuart's passion for his work grew from a personal experience he had at age 15. He raised some pigs to earn extra money, feeding them leftovers from his school kitchen and local shops. Stuart soon realized most of the food the pigs were eating was actually fit for humans to eat—and that supermarket garbage bins were overflowing with fresh food.

"Everywhere I looked, we were hemorrhaging food," Stuart notes. "So I began confronting businesses about the waste and exposing it to the public." His research revealed that most wealthy countries produce between three and four times the food required to meet their citizens' nutritional needs. The effects of food waste are clear to Stuart, who explains, "Producing this huge surplus leads to deforestation, depleted water supplies, massive fossil fuel consumption, and biodiversity loss. Excess food decomposing in landfills accounts for 10 percent of greenhouse gas emissions by wealthy nations."

In 2009, Stuart launched Feedback. Created entirely using food that would have otherwise been wasted, Feedback's flagship event in London was a free feast for 5,000 Londoners that has been replicated around the world. "These events give people a clear, tangible idea of food waste problems and potential solutions right where they live," Stuart says.

Stuart has also campaigned for retailers to relax strict cosmetic standards for produce. As a result, many supermarkets have changed their policies, and "ugly" fruits and vegetables are now the fastest-growing sector in the fresh produce market.

Stuart stresses that ordinary citizens can make a powerful difference: "By rising up and speaking out, we can—and are—making the world's food system less unjust and more sustainable every day."

Thinking Critically
Analyze Cause and Effect According to Stuart, what are some of the results of food waste?

Tristram Stuart sits beside a load of bananas considered too short, too long, or too curved to meet cosmetic standards for produce in the European market. Part of Stuart's war against food waste involves campaigning for the sale of this type of "ugly" produce.

CASE STUDY
E-Waste—An Exploding Problem

What happens to your cell phone, computer, and other electronic devices when they are no longer useful? They become electronic waste, or *e-waste* (as shown in the photo at the beginning of this chapter). E-waste is the fastest-growing solid waste problem in the United States and worldwide.

Only 15% of all U.S. e-waste is recycled. Most ends up in landfills and incinerators. However, much of this e-waste contains gold, rare earths, and other valuable materials that could be recycled or reused. E-waste also contains toxic and hazardous chemicals that can contaminate air, surface water, groundwater, and soil and harm human health.

Much of the e-waste in the United States that is not buried or incinerated is shipped to Asian, African, and Latin American countries for processing. Labor is cheap and environmental regulations are weak in those countries. Workers there—many of them children—dismantle, burn, and treat e-waste with acids to recover valuable metals and reusable parts (Figure 17-1). The work exposes them to toxic metals such as lead and mercury and other harmful chemicals. The remaining scrap is dumped into waterways and fields or burned in open fires, which releases highly toxic chemicals called dioxins.

Transfer of e-waste from more-developed to less-developed countries is banned under the International Basel Convention. The treaty aims to reduce and control shipping of hazardous waste across international borders. Despite the ban, much of the world's e-waste is not officially classified as hazardous waste, or it is illegally smuggled out of some countries. The United States can export its e-waste legally because it is the only industrialized nation yet to approve the treaty.

The European Union (EU) has led the way in dealing with e-waste. The EU requires manufacturers to take back electronic products at the end of their useful lives and repair, remanufacture, or recycle them. In the EU, e-waste is banned from landfills and incinerators. Japan also requires manufacturers to recycle e-wastes. To cover the costs of these programs, consumers pay a recycling tax on electronic products. This is an example of implementing the full-cost pricing factor of sustainability. However, the high cost of complying with these laws has meant much of this waste is still shipped illegally to other countries.

In this chapter, you will read about the challenges of handling e-waste and other types of solid and hazardous waste. You will also consider ways to reduce or reuse this waste in order to limit its harmful impacts on human health and the environment.

As You Read Think about all the waste you generate each day. Do you know what happens to it after you throw it away? How can you create less waste?

◤ FIGURE 17-1 A worker recovers valuable materials from scrapped computers shipped in from the United States. According to the World Health Organization, these "digital dumpsites" expose millions of people to hazardous chemicals and air pollutants.

17.1 What Are Problems Related to Solid and Hazardous Waste?

CORE IDEAS AND SKILLS

- Define and give examples of solid waste.
- Explain what happens to solid waste after its disposal.
- Define and give examples of hazardous waste and understand why hazardous waste requires special handling.

KEY TERMS

solid waste
industrial solid waste
municipal solid waste (MSW)
hazardous waste

Ours Is a High-Waste Society

Think about what you've tossed in the trash today. Perhaps it was leftovers from lunch. Perhaps it was an empty can or bottle. Perhaps it was something you used up or no longer needed. People throw away all sorts of items, and they all add up to huge amounts of waste (Figure 17-2).

In nature, wherever humans are not dominant, there is essentially no waste. The wastes of one organism become nutrients or raw materials for others. This natural nutrient cycling is the basis for one of the three scientific factors of sustainability. Humans violate this factor by producing huge volumes of wastes that either go into landfills or incinerators or litter and pollute the environment.

Scientific studies show that patterning solid waste management after natural cycles could cut the waste of potential resources and the environmental harm from such waste by up to 80%. In their book, *The Upcycle*, William McDonough and Michael Braungart also call for a new view of solid waste and pollution. They challenge society to see these materials and chemicals as potentially useful and economically valuable (Lesson 1.2). Instead of asking, "How do I get rid of this waste?" they say we need to ask, "How much money can I get for these resources?" and "How can I design products that don't end up as waste or pollutants?"

This new way of thinking about waste materials means designing products so they can be recycled or reused, much like nutrients in Earth's biosphere.

FIGURE 17-2
To illustrate the typical waste output from an American household, this family of four in San Diego, California, agreed to pose with all the solid waste they generated in one year. Recyclable items appear on the left and the rest of the solid waste appears on the right.

With this approach, people can think of trashcans and garbage trucks as resource containers. They can view landfills as urban mines packed with materials that should have been recycled, as nature recycles nutrients. People can think of garbage not as something to "throw away" but something to "pass on" for other purposes.

checkpoint Why is there essentially no waste in the natural world except for waste generated by humans?

Solid Waste Is Piling Up

The solid items thrown away in your household fall into the category of **solid waste**—any unwanted or discarded material people produce that is not a liquid or a gas. Solid waste can be classified in one of two ways. The first is **industrial solid waste** produced by mines, farms, and industries that supply people with goods and services. The second is **municipal solid waste (MSW)**, often called garbage or trash. This type of solid waste consists of the combined solid wastes produced by households and workplaces other than factories. Examples of MSW include paper and cardboard, food waste, cans, bottles, yard waste, furniture, plastics, metals, glass, wood, and electronic waste (Case Study). Much of the world's MSW ends up as litter in rivers, lakes, oceans, and natural landscapes.

In more-developed countries, most MSW is collected and buried in landfills or burned in incinerators. For example, in the United States, 50% of MSW is buried in landfills and 12% is incinerated. In many less-developed countries, much of this waste ends up in open dumps, where poor people earn a living finding items they can use or sell (Figure 17-3).

The United States produces more industrial and municipal solid waste than any other country. It also produces the most solid waste per person. With only 4% of the world's people, the United States produces 12% of the world's solid waste. Every year, Americans generate enough MSW to fill a bumper-to-bumper convoy of garbage trucks long enough to circle Earth's Equator almost six times.

According to the EPA, 98.5% of all solid waste produced in the United States is industrial waste from mining (76%), agriculture (13%), and industry (9.5%). The remaining 1.5% is municipal solid waste.

Consider some of the forms of solid waste consumers throw away each year, on average, in the high-waste economy of the United States:

- enough tires to circle Earth's Equator almost three times
- an amount of disposable diapers that, if linked end to end, would reach to the moon and back seven times
- enough carpet to cover the state of Delaware
- enough nonreturnable plastic bottles to form a stack that would reach from Earth to the moon and back about six times
- 100 billion plastic shopping bags—274 million per day, for an average of 3,200 every second
- enough office paper to build a wall 3.5 meters (11 feet) high from New York, New York, to San Francisco, California
- 25 billion plastic-foam cups—enough, if lined up end to end, to circle Earth's Equator 436 times
- $165 billion worth of food

Some resource experts suggest we change the name of the trash we produce from MSW to MWR—*mostly wasted resources*. That's because so much of what is considered "waste" can be useful as a resource if properly managed.

Solid Waste at Sea Trash has made its way into oceans around the world, and the majority of it is made of plastic. Plastic debris can be detected floating on the surface of nearly 90% of open ocean waters. This debris ranges from food wrappers, bags, and bottles to fishing gear and children's toys. Recent analysis suggests that more than 30 times as much plastic might lie at the bottom of the ocean than what floats at the surface.

In 1997, ocean researcher Charles Moore made a disturbing discovery: two huge, slowly rotating masses of plastic and other solid waste floating in the North Pacific Ocean near the Hawaiian Islands.

> **CONSIDER THIS**
>
> **How long does garbage stick around in landfills?**
>
> Most MSW breaks down slowly, if at all, in landfills. Lead, mercury, glass, plastic foam, and most plastic bottles do not break down completely. An aluminum can takes 500 years to disintegrate. Disposable diapers may take 550 years to break down, and a plastic shopping bag may stick around for up to 1,000 years.

FIGURE 17-3
Two men work overnight to pick through a fresh truckload of trash at an open dump in the Philippines. The men are licensed to do this work and hope to scavenge recyclable materials that they can sell to earn a living.

Scientists named these giant masses of trash the Great Pacific Garbage Patch (Figure 17-4). Their trash mainly consists of small particles of plastic floating on or just beneath the ocean's surface. And it keeps on collecting, trapped by a vortex where four rotating ocean currents called gyres meet.

Roughly 80% of this trash comes from land. It is washed or blown off beaches. It pours out of storm drains, and floats down streams and rivers that empty into the sea. Most of it comes from the west coast of North America, the east coast of Asia, and hundreds of Pacific islands. Much of the remaining trash is dumped in the sea by cargo and cruise ships.

According to some estimates, the Great Pacific Garbage Patch covers an area at least the size of Texas. These estimates are difficult to verify because the main ingredient in this garbage soup is very tiny particles of plastic. They are called microplastics and they are suspended just beneath the water's surface. The size of these particles makes the garbage patch difficult to see and measure. In fact, it appears more like a stretch of cloudy water than an island of trash.

While many different types of trash enter the ocean, plastic makes up most of the garbage patch because it is contained in so many disposable products (Figure 17-5). Plus, it does not biodegrade but instead breaks down into smaller and smaller pieces. Research shows the tiny plastic particles can be harmful to marine mammals, seabirds such as albatrosses, and fish and other aquatic species. These animals mistake plastic particles for food and swallow them. Because they cannot digest the plastic, it can cause them to die from starvation or poisoning. Scientists estimate that the amount of plastics entering the ocean will triple by 2040 without immediate action.

Since the Great Pacific Garbage Patch was discovered, five other huge, swirling garbage patches have been found in gyres in the world's other oceans. In total, these garbage patches cover an area of ocean larger than all of Earth's land area. They represent the massive pollution legacy of a throwaway culture.

Engineers have had some success capturing marine debris near the ocean surface as currents pull it into the garbage patch zone. But currently there is no practical or affordable way to clean up this marine litter. Trying to gather the garbage by scooping it out or netting it could actually harm marine life. Any animals similar in size to the bits of trash floating around them could be caught in the scoops and nets.

FIGURE 17-4

The Great Pacific Garbage Patch This garbage patch consists of two vast and slowly swirling subsurface pools of small plastic particles. How long do you think it will take for these garbage patches to decompose?

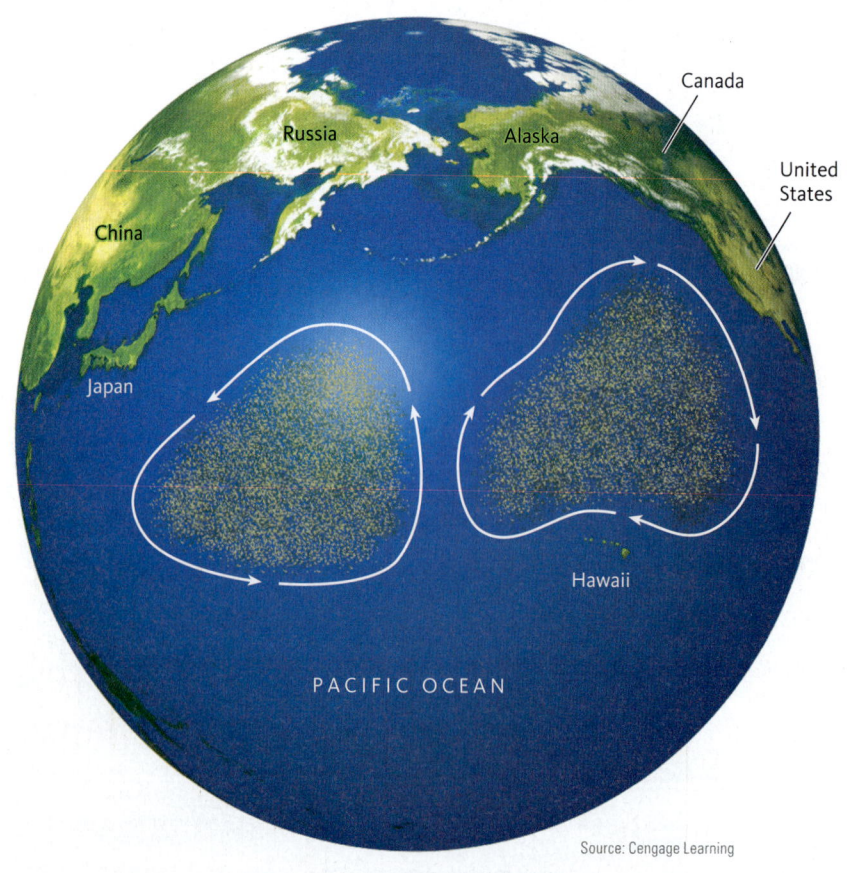

Source: Cengage Learning

FIGURE 17-5
ON ASSIGNMENT National Geographic photographer Frans Lanting came across this Laysan albatross and its chick while visiting Midway Atoll, a Hawaiian island. The bird had brought a variety of plastic items back to its nest from the sea after mistaking the floating trash for food. What items do you recognize?

LESSON 17.1   585

The size of the oceans makes cleanup too time-consuming as well. A more useful approach is to prevent the garbage patch from growing by reducing production of solid waste and keeping it out of oceans in the first place.

checkpoint What is the main source of solid waste generated in the United States?

Hazardous Waste Threatens Human Health and the Environment

Another major category of waste is **hazardous waste**, which is any discarded material or substance that directly threatens human health or the environment. Hazardous waste is identified as having one or more of the following characteristics: It is toxic, flammable, or corrosive, can undergo violent or explosive chemical reactions, or can cause disease.

Hazardous waste falls into two categories. The first includes toxic heavy metals, such as lead, mercury, and arsenic. The second includes synthetic organic compounds, such as pesticides, PCBs, and solvents (liquids or gases used to dissolve or extract other substances). Examples of hazardous waste include paint thinners, medical waste, batteries used in electronic devices, and ash and sludge from incinerators and coal-burning facilities. You might be surprised to learn that chemicals in many household products are deemed hazardous (Figure 17-6). Special rules apply to the disposal of these products.

Another form of extremely hazardous waste is the radioactive waste produced by nuclear power plants and nuclear weapons facilities (Chapter 12). Such waste must be stored safely for 10,000 to 240,000 years, depending on the radioactive isotopes it contains. After 60 years of research, scientists and governments have not found a scientifically and politically acceptable way to safely isolate this dangerous waste for thousands of years.

According to the UN Environment Programme, more-developed countries produce 80–90% of the world's hazardous waste. The United States produces the most hazardous waste. If China continues to industrialize rapidly, largely without adequate pollution controls, it may soon overtake the United States in hazardous waste production.

checkpoint How is hazardous waste identified?

FIGURE 17-6 ▼

What Harmful Chemicals Are in Your Home?
Cleaning Products
Disinfectants
Drain, toilet, and window cleaners
Spot removers
Septic tank cleaners
Paint Products
Paints, stains, varnishes, and lacquers
Paint thinners, solvents, and strippers
Wood preservatives
Artist paints and inks
General
Dry-cell batteries (mercury and cadmium)
Glues and cements
Fire retardants
Gardening
Pesticides
Weed killers
Ant and rodent killers
Flea powders
Automotive
Gasoline
Used motor oil
Antifreeze
Battery acid
Brake and transmission fluid

17.1 Assessment

1. **Recall** What are the two types of solid waste? Give an example of each.
2. **Explain** Why is plastic marine debris so dangerous to ocean life?
3. **Infer** Why do Americans create so much more waste than people in less-developed countries? Explain your answer.

SCIENCE AND ENGINEERING PRACTICES

4. **Communicating Information** Choose one example of U.S. consumer waste from the bulleted list on page 582. Research to find the latest data on the waste described and what is being done to reduce it. Summarize your findings in a paragraph.

17.2 How Should Society Deal with Solid Waste?

CORE IDEAS AND SKILLS

- Understand how waste management, waste reduction, and integrated waste management differ in their approaches to dealing with solid waste.
- Describe the process of landfilling waste, as well as its advantages and disadvantages.
- Describe the process of incinerating waste, as well as its advantages and disadvantages.
- Define the Four Rs approach to dealing with solid waste and identify ways individuals, industries, and communities can use this approach to limit waste and pollution.

KEY TERMS

waste management
waste reduction
integrated waste management
sanitary landfill
primary recycling
secondary recycling
composting

Solid Waste Must Go Somewhere

Society can deal with the solid waste it creates in two ways. One is **waste management**, which focuses on controlling waste in order to limit its environmental harm but does not attempt to seriously reduce the amount of waste produced. This approach begins with the question, "What do we do with solid waste?" It typically involves mixing forms of waste together and then burying them, burning them, or shipping them to another location for processing.

The other option is **waste reduction**—producing much less solid waste and reusing, recycling, or composting what is produced as much as possible. This approach begins with questions such as, "How can we avoid producing so much solid waste?" and "How can we use the forms of waste we produce as resources like nature does?"

Most experts prefer **integrated waste management**—a set of coordinated strategies for waste disposal *and* waste reduction (Figure 17-7).

FIGURE 17-7 ▼

Integrated Waste Management People can reduce waste by refusing or reducing resource use and by reusing, recycling, and composting what we discard. People can manage waste by burying it in landfills or incinerating it. Most countries rely primarily on burial and incineration.

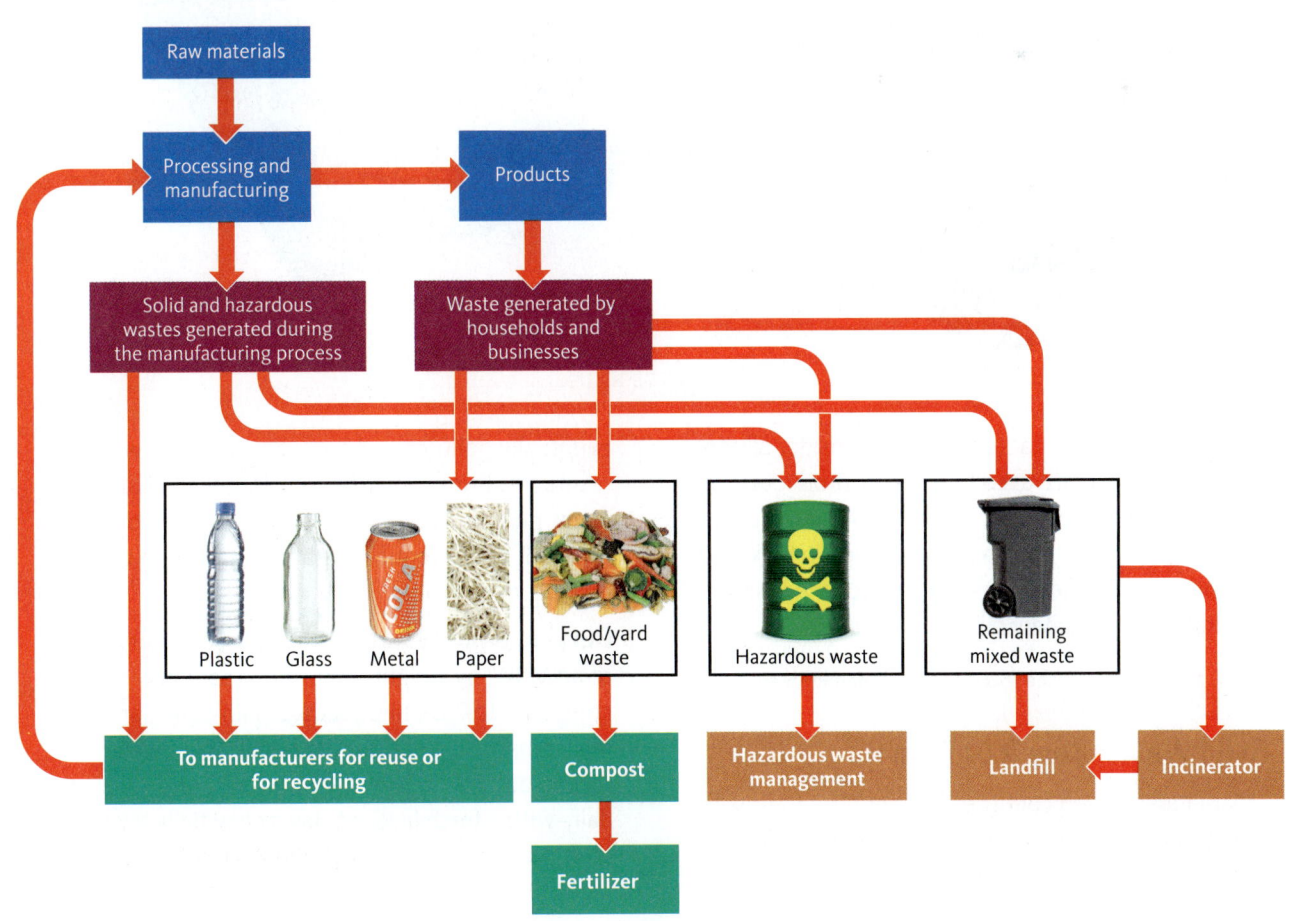

LESSON 17.2   587

FIGURE 17-8 ▼
Burying Solid Waste A sanitary landfill is designed to eliminate or minimize environmental problems that plague older landfills.

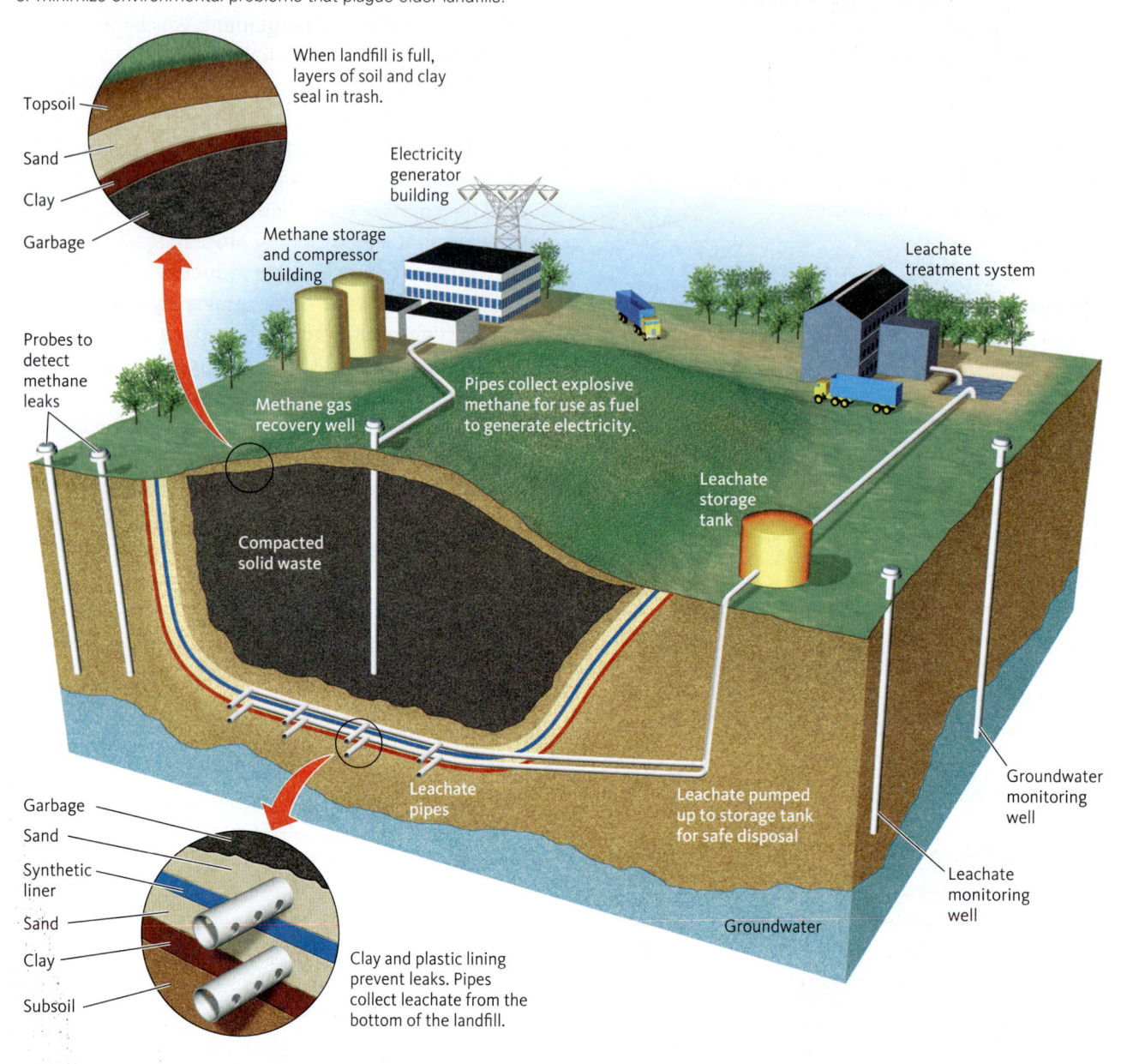

For example, a manufacturer should consider all the components that are used to make its products. Can a product's waste be recycled (plastic, glass, metal, or paper)? If hazardous waste is involved in making a product, what is the best way to reduce the amount generated and to safely dispose of it later?

checkpoint What is the focus of waste management?

Solid Waste: Bury It Or Burn It

Most industrialized countries bury or burn their solid waste. Technologies for burying or burning solid waste are well developed. Burying waste, however, eventually pollutes and degrades land and water resources. Burning waste contributes to air and water pollution and greenhouse gas emissions. Both methods fail to reduce the amount of solid waste produced by society.

Burying Solid Waste In the United States, 50% of all MSW, by weight, is buried in sanitary landfills, compared to 72% in Canada, 2% in Japan, and 5% in Denmark. In a **sanitary landfill**, solid waste is spread out in thin layers, compacted, and covered daily with a fresh layer of clay or plastic foam (Figure 17-8). This process keeps the material dry and cuts

588 CHAPTER 17 SOLID AND HAZARDOUS WASTE

down on odors. It reduces the risk of fire, and keeps out rats and other pest animals. It also helps contain contaminated water called *leachate*, so that it doesn't leak out of the landfill and pollute nearby soil and groundwater supplies.

The bottoms and sides of well-designed sanitary landfills have strong double liners and containment systems that collect any liquids that leach from them. Some landfills also have systems for collecting methane. This potent greenhouse gas is produced when buried waste decomposes in the absence of oxygen. Once collected, the methane can be burned as a fuel to generate electricity.

What gets buried in landfills? Paper products represent the largest percentage of landfill materials. Other common materials include yard waste, plastics, metals, wood, glass, and food waste. Some types of solid waste are not accepted at landfills. For example, tires, waste oil and oil filters, electronics, medical waste, and items that contain mercury such as fluorescent light bulbs are not allowed.

Many people think of landfills as huge compost piles where biodegradable waste decomposes within a few months. Actually, decomposition inside landfills is a very slow process. Buried trash may resist decomposition perhaps for centuries due to conditions in the landfill. Trash is tightly packed and protected from sunlight, water, and air, as well as from bacteria that could digest and decompose most of the biodegradable waste. In fact, researchers have unearthed 50-year-old newspapers that were still readable and hot dogs and pork chops that had not yet decayed. Figure 17-9 lists the advantages and disadvantages of using sanitary landfills to dispose of solid waste.

An open dump is another type of landfill. An open dump is essentially a field or large pit where garbage is deposited and sometimes burned. Open dumps are rare in more-developed countries, but are widely used near major cities in many less-developed countries. China disposes much of its rapidly growing mountains of solid waste in rural open dumps or in poorly designed and poorly regulated landfills. Open dumps pose a variety of health, safety, and environmental threats. The materials they contain may be toxic or explosive. Contaminated liquids leaking from open dumps can pollute soil and groundwater supplies.

Burning Solid Waste Many communities burn their solid waste until nothing remains but fine, white-gray ash, which is then buried in landfills. The heat released by burning trash can be used to generate other forms of energy in facilities called waste-to-energy incinerators (Figure 17-10). Globally, MSW is burned in more than 800 waste-to-energy incinerators, 76 of them located in the United States.

A waste-to-energy incinerator contains a combustion chamber where waste is burned at extremely high temperatures. Heat from the burning material is used to boil water and produce steam. The steam in turn drives a turbine that generates electricity. Combustion also produces waste in the form of ash and gases. The ash must be treated and properly disposed of in landfills. The gases must be filtered to remove pollutants before being released into the atmosphere.

The United States incinerates 12% of its MSW, a fairly low percentage. One reason for this is that in the past, incineration earned a bad reputation as a result of highly polluting and poorly regulated units. Today, incineration also must compete with an abundance of low-cost landfills in many parts of the country. In contrast, Denmark burns just over half its MSW in state-of-the-art waste-to-energy incinerators that far exceed European air pollution standards.

There are benefits and drawbacks to using incinerators to burn solid waste. For example, one EPA study found that landfills actually emit more air pollutants than modern waste-to-energy incinerators. In addition, burning waste in these incinerators reduces its volume by up to 90%. As a result, it takes up less space in landfills.

On the other hand, the toxic ash that is filtered out during incineration must be safely stored. In addition, many members of the public, as well as environmental scientists and local governments, continue to be opposed to solid waste incineration.

FIGURE 17-9

| Burying Solid Waste in Sanitary Landfills ||
Advantages	Disadvantages
Low operating costs	Noise, traffic, and dust
Can handle large amounts of waste	Releases greenhouse gases (methane and carbon dioxide) unless they are collected
Can be used for other purposes once filled	Does not discourage waste production
No shortage of landfill space in many areas	Eventual leaks can contaminate groundwater.

FIGURE 17-10 ▼
Burning Solid Waste A waste-to-energy incinerator burns mixed solid wastes and recovers some of the energy to produce steam for use in heating buildings or producing electricity.

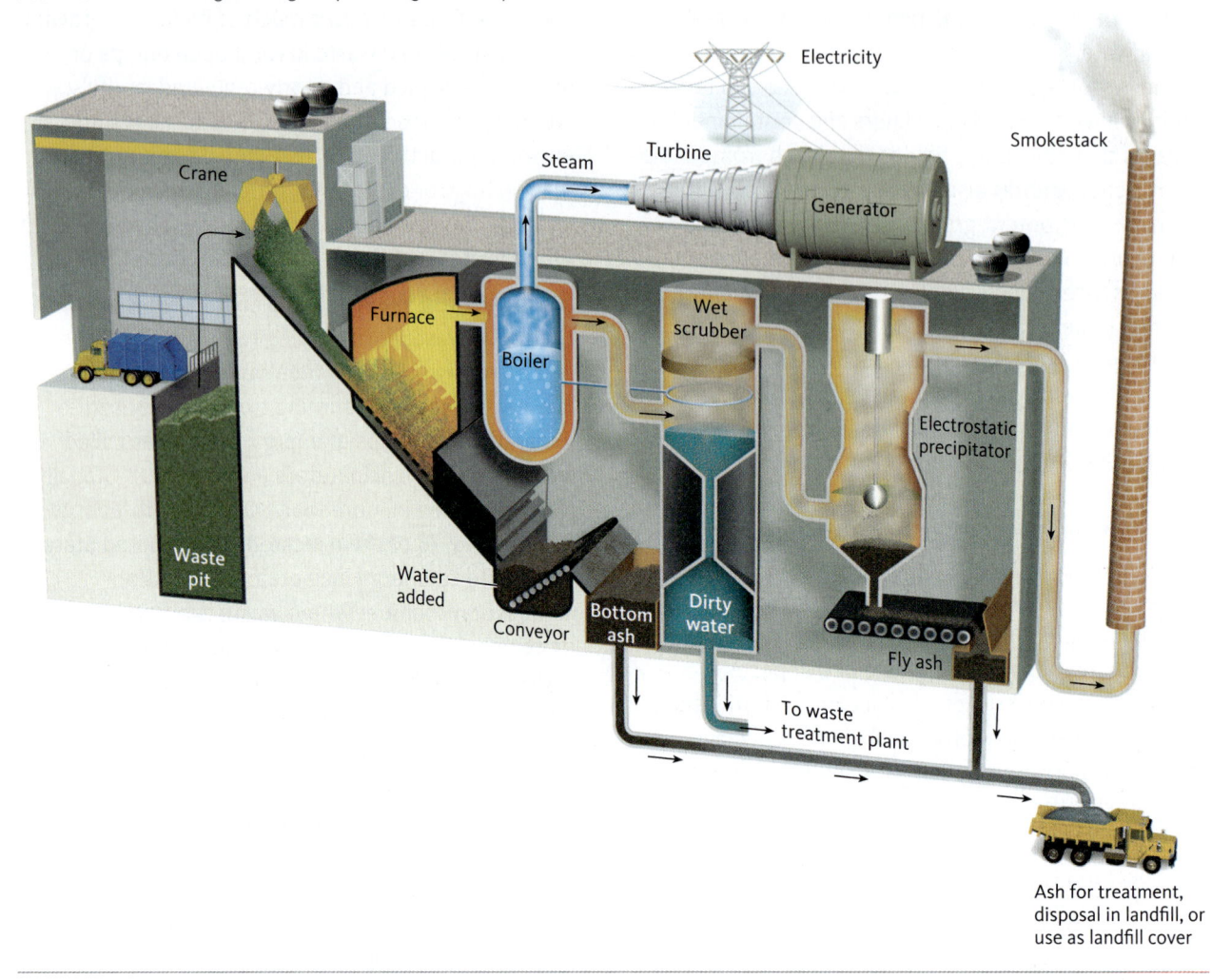

They claim that incineration hurts efforts to increase reuse and recycling because it creates a demand for burnable waste. Incinerators are expensive to build, and as a result they require a large, steady stream of waste to be profitable. Waste incineration also makes it easier for consumers to throw away items that might otherwise be reused or recycled. Figure 17-11 lists the advantages and disadvantages of incinerating solid waste.

checkpoint Explain how conditions in sanitary landfills preserve the materials buried inside them.

Waste Reduction: The "Four R" Strategy

A more sustainable approach to handling solid waste is to first make less of it, then to reuse or recycle it, and finally to safely dispose of what is left. The parts of this strategy, called the "Four Rs" of waste reduction, are listed here in order of priority as suggested by environmental scientists:

- *Refuse*: Don't use it.
- *Reduce*: Use less.
- *Reuse*: Use it over and over.

FIGURE 17-11 ▼

Incinerating Solid Waste	
Advantages	**Disadvantages**
Reduces trash volume	Facilities are expensive to build.
Produces energy	Produces hazardous waste
Concentrates hazardous substances into ash for burial	Emits some CO_2 and other air pollutants
Sale of energy reduces cost	Encourages waste production

FIGURE 17-12

Ways to Apply the Four Rs of Waste Reduction	
Individuals	**Industries and Communities**
Rent, borrow, or barter goods and services instead of buying new ones when possible. Buy secondhand. Donate or sell unused items.	Change industrial processes to eliminate or reduce the use of harmful chemicals.
Buy things that are reusable, recyclable, or compostable. Be sure to reuse, recycle, and compost them.	Redesign manufacturing processes and products to use less material and energy.
Refuse packaging when possible, and choose products with little or no packaging. Recycle packaging as much as possible.	Develop products that are easy to repair, reuse, remanufacture, compost, or recycle.
Avoid disposables such as paper and plastic bags, plates, cups, and utensils, diapers, and razors whenever reusable versions are available.	Eliminate or reduce unnecessary packaging. Use the following hierarchy for product packaging: no packaging, reusable packaging, and recyclable packaging.
Avoid heavily packaged processed foods. Buy products in bulk whenever possible.	Use fee-per-bag waste collection systems that charge consumers for the amount of waste they throw away but provide free pickup of recyclable and reusable items.
Discontinue junk mail as much as possible. Read online newspapers, magazines, and e-books.	Pass laws that require companies to take back various consumer products such as electronic equipment, appliances, and motor vehicles for recycling or remanufacturing.

- *Recycle*: Convert used resources to useful items, and buy products made from recycled materials.

From an environmental standpoint, the first three Rs are preferred. They represent input, or waste-prevention, approaches that tackle the problem of waste production before it occurs. Recycling is important, but it deals with waste after it has been produced. Figure 17-12 describes ways individuals, industries, and communities can apply the Four-R strategy to limit waste and pollution.

By applying the Four Rs, people consume less matter and energy. They reduce pollution and natural capital degradation. They also save money. Some scientists and economists estimate people could eliminate up to 80% of the solid waste produced by following the Four-R strategy. This would mimic Earth's nutrient cycling factor of sustainability.

People in industrialized societies increasingly substitute throwaway items for reusable ones. This habit results in growing masses of solid waste. By applying the Four Rs, society can slow or stop this trend. Individuals can guide and reduce resource consumption by asking questions such as:

- Do I really need this? (refusing)
- How many of these do I actually need? (reducing)
- Is this something I can use more than once? (reusing)
- Can this be converted into something else when I am done with it? (recycling)

As Figure 17-13 illustrates, the United States has yet to prioritize the Four Rs of resource use. However, reuse and recycling do continue to be on the rise.

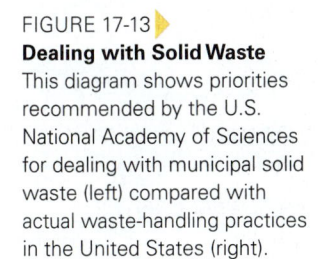

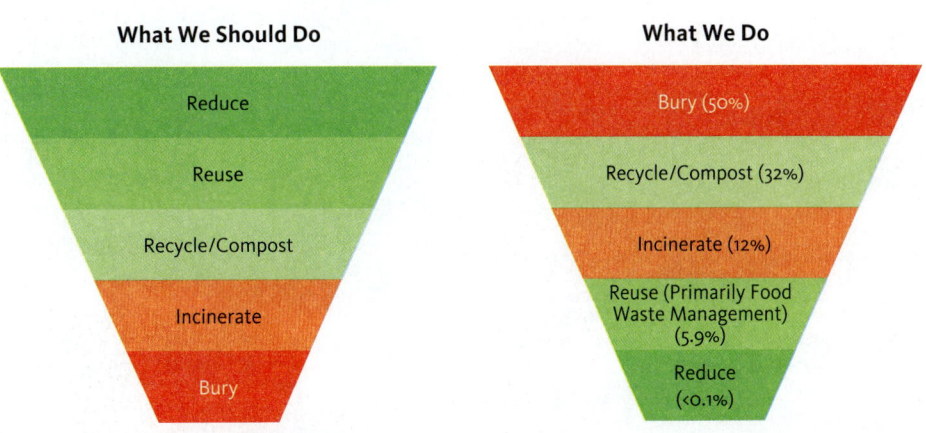

FIGURE 17-13
Dealing with Solid Waste
This diagram shows priorities recommended by the U.S. National Academy of Sciences for dealing with municipal solid waste (left) compared with actual waste-handling practices in the United States (right).

What We Should Do
- Reduce
- Reuse
- Recycle/Compost
- Incinerate
- Bury

What We Do
- Bury (50%)
- Recycle/Compost (32%)
- Incinerate (12%)
- Reuse (Primarily Food Waste Management) (5.9%)
- Reduce (<0.1%)

Sources: U.S. Environmental Protection Agency, U.S. National Academy of Sciences, Columbia University, and *BioCycle*

SCIENCE FOCUS 17.1

BIOPLASTICS

One of the most useful characteristics of plastic—its durability—also happens to be one of its biggest drawbacks. Plastics are made to last, but that means they don't completely break down once they're disposed of. In addition, most of today's plastics are made using petroleum-based chemicals, or petrochemicals. Processing these chemicals creates hazardous waste and causes water and air pollution.

The good news is that some products are now being made from *bioplastic*. This type of plastic is more environmentally friendly because it is made from biologically based chemicals. Bioplastics can be used in the manufacture of a variety of products, including packaging, utensils, and even building materials (Figure 17-14).

Henry Ford, who developed the first Ford car and founded Ford Motor Company, supported research on the development of a bioplastic made from soybeans and another made from hemp. Ford even manufactured a car body using soy bioplastic. However, as oil became cheaper and more widely available, petrochemical plastics took over the market.

Now, evidence shows petroleum use contributes to climate change and other environmental problems. Chemists are stepping up efforts to make biodegradable and more environmentally sustainable plastics in response. These bioplastics can be made from plants such as corn, soy, sugarcane, and switchgrass. They can even be made from chicken feathers and some components of garbage.

Compared with conventional oil-based plastics, properly designed bioplastics are lighter, stronger, and cheaper. Making them requires less energy and produces less pollution per unit of weight. Instead of being sent to landfills, some packaging made from bioplastics can be composted and added to soil to improve its health.

Thinking Critically
Make Judgments Considering the materials used to make bioplastics, would you be willing to buy products made from these alternatives to plastic? Explain your answer.

FIGURE 7-14
Architect Hans Vermeulen stands on a building block manufactured for a house in Amsterdam, Netherlands. The block is made of bioplastic that contains 80% vegetable oil and is recyclable.

FIGURE 17-15
These sandals have soles made from recycled car tires. They are sold in Nairobi and are much more durable and less costly than most other locally sold shoes.

Reuse Programs that encourage reuse are increasingly common. Many coffee shops discount hot drink prices for customers who bring their own refillable containers. Denmark, Finland, and the Canadian province of Prince Edward Island have banned all beverage containers that cannot be reused. In Finland, 95% of all soft drink, beer, wine, and spirits containers are refillable.

The use of rechargeable batteries is cutting toxic waste by reducing the number of conventional batteries that are thrown away. The newest rechargeable batteries come fully charged. They can hold a charge for up to one year when not in use. They can also be recharged up to 500 times.

Globally, more than 2 million single-use plastic bags are used every minute. Each year, 500 billion plastic bags are used in the United States alone. In many countries, the landscape is littered with plastic bags. They can take 400–1,000 years to break down and never disintegrate completely. Huge quantities of plastic bags and other plastic solid waste also end up in ocean waters.

Instead of using throwaway paper or plastic bags to carry groceries and other purchased items, many people now use reusable cloth bags. To encourage people to carry reusable bags, the governments of Denmark, Ireland, Taiwan, and the Netherlands tax plastic shopping bags. In Ireland, a tax of 25 cents per bag cut plastic bag litter by 90%, as people responded to this economic incentive to choose reusable bags. According to the United Nations, as of 2021, 77 countries in the world had passed some sort of full or partial ban on plastic bags. In addition, 32 countries, most located in Europe, had imposed a fee or tax to reduce plastic bag use.

In 2014, California became the first U.S. state to issue a statewide ban on single-use plastic bags. Since then, seven other states have enacted statewide bans. They include Connecticut, Delaware, Hawaii, Maine, New York, Oregon, and Vermont. More than 130 cities or counties have bans as well.

Recycling Five major types of recyclable materials may be produced by households and workplaces.

They are paper products, glass, aluminum, steel, and some plastics. These materials can be reprocessed into new, useful products in two ways. **Primary recycling**, or closed-loop recycling, involves using materials again for the same purpose. Recycling aluminum cans to make more cans is an example of primary recycling. **Secondary recycling** involves converting waste materials to different products, such as turning car tires into sandals (Figure 17-15). During secondary recycling, items may be downcycled into less useful products. They may be upcycled into products that are more useful compared to the original item.

Recycling involves three steps: collection of materials for recycling, conversion of recycled materials to new products, and selling and buying products that contain recycled material. Recycling is successful in environmental and economic terms when all three of these steps are carried out.

Composting is another form of recycling. **Composting** mimics nature by using bacteria to decompose yard trimmings, vegetable food scraps, and other biodegradable organic waste into humus. Humus is an organic component of soil that improves soil fertility. Humus can be added to soil to supply plant nutrients, slow soil erosion, retain water, and improve crop yields. People can compost organic waste several ways. They can use simple backyard containers or low-maintenance composting piles. They can buy composting drums that spin, which eases mixing of wastes and speeds up decomposition.

The United States recycles or composts 32% of its MSW. There is potential to significantly increase this number given that discarded food accounts for 22% of this waste. Here are some of the most commonly recycled materials and the rates at which they are recycled:

- lead-acid batteries (99%)
- newspapers, directories, and newspaper inserts (71%)
- steel cans (67%)
- aluminum cans (67%)

Experts say with education and proper incentives, Americans could recycle and compost at least 80% of their MSW. Doing so would promote the nutrient cycling factor of sustainability.

Currently, only 9%, by weight, of all plastic waste in the United States (and 14% of plastic containers and packaging) is recycled. These percentages are low because many types of plastic resins are difficult to separate from products that contain them. However, progress is being made in the recycling of plastics and in the development of more degradable bioplastics (Science Focus 17.1).

Cities that have higher recycling rates tend to use a single-pickup system for both recyclable and nonrecyclable materials rather than a more expensive dual-pickup system. Successful systems also use a pay-as-you-throw approach. They charge for picking up trash but not for picking up recyclable or reusable materials. They also require citizens and businesses to sort their trash and recyclables by type. San Francisco, California, uses such a system. In 2019, the city recycled, composted, or reused 81% of its MSW.

Whether recycling makes economic sense depends on how we look at its economic and environmental benefits and costs. Critics of recycling programs argue that recycling is costly and adds to the taxpayer burden in communities where recycling is funded through taxation. They say recycling may make economic sense for valuable and easy-to-recycle materials in MSW such as paper and paperboard, steel, and aluminum. But they argue that recycling probably does not make sense for cheap or plentiful resources such as glass.

Those who support recycling say studies show the net economic, health, and environmental benefits of recycling far outweigh the financial costs. For example, the EPA studied the impacts of recycling and composting on U.S. emissions of carbon dioxide.

> **CONSIDER THIS**
>
> **Reusing items is easier than you may think.**
>
> By making a few simple changes, you can significantly reduce the amount of solid waste that you produce.
> - Buy beverages in refillable glass containers.
> - Use reusable lunch containers.
> - Store refrigerated food in reusable containers.
> - Use rechargeable batteries and recycle them when their useful life is over.
> - When eating out, bring your own reusable container for leftovers.
> - Carry groceries and other items in a reusable basket or cloth bags.
> - Buy used furniture, cars, and other items whenever possible.

NATIONAL GEOGRAPHIC | EXPLORERS AT WORK
T.H. Culhane Urban Planner

Thomas "T.H." Culhane gets pretty excited about rotting food. That's because this National Geographic Explorer has found life-changing ways to make use of it. Culhane is an urban planner who specializes in bringing low-cost energy technology to people living in poverty around the world.

Culhane's organization, Solar C.I.T.I.E.S., trains people to build and install rooftop solar water heaters. It has also developed simple devices that turn kitchen scraps and other waste into fuel and fertilizer. These systems are called biogas digesters, and they offer a simple, effective way to recycle solid waste.

Biogas digesters collect methane gas, which is a by-product of the natural anaerobic decomposition of organic materials (Lesson 3.2). This "biogas" is clean burning and can be used just like natural gas. The "digested" liquid material is also collected for use as fertilizer. "You could think of it as a kind of liquid composting tank," says Culhane. "We prefer to call it our domestic dragon, a fire-breathing house pet that eats our food scraps and helps us cook food and make new food again."

Culhane's biogas digesters feature several designs. Most are built for single households. The simplest design includes a sealed plastic container fitted with three pipes. One pipe accepts food waste, one releases liquid fertilizer, and one releases gas. This type of biogas digester can convert one bucket of food waste into roughly two hours of cooking gas.

How does a biogas digester work? First, bacteria-containing material, such as manure or even lake mud, is placed in the digester along with water and left to stew for several weeks. This allows the bacteria that will do the digesting to become established. Then the biogas digester is fed a daily mixture of finely ground organic material and water. Biogas digesters can accept any type of ground-up organic waste, even toilet waste. They break wastes down in as little as 24 hours.

Culhane is honored to help people safely recycle solid waste and produce clean, affordable fuel and fertilizer. As he explains, "You take something that is often considered a terrible health hazard—the organic waste that comes from your kitchen and bathroom and routinely contaminates our air, land and water, causing the untimely death and suffering of women and children the world over—and turn it into life- and health-giving heat and energy and fertilizer."

FIGURE 17-16
T.H. Culhane prepares to install a biogas digester for a community in Kenya.

FIGURE 17-17 ▼

Recycling Solid Waste	
Advantages	**Disadvantages**
Reduces use of energy and mineral resources and air and water pollution	Can cost more than burying in areas with ample landfill space
Reduces greenhouse gas emissions	Reduces profits for landfill and incinerator owners
Reduces solid waste	Inconvenient for some

The EPA estimated that in 2018, the recycling of paper and paperboard alone cut emissions of carbon dioxide by an amount roughly equal to that emitted by 33 million passenger vehicles.

Recycling, reuse, and composting industries create 6 to 10 times as many jobs as landfill and incineration industries. Doubling the U.S. recycling rate would create about 1 million new jobs. However, the recycling industry is not without its challenges. In 2018, China banned the import of certain waste materials for recycling, including post-consumer plastic. This ban has disrupted recycling efforts in the United States and other countries reliant on China for this purpose. The advantages and disadvantages of recycling solid waste are shown in Figure 17-17.

checkpoint What is the difference between primary recycling and secondary recycling?

17.2 Assessment

1. **Compare and Contrast** How is composting solid waste similar to and different from burying it in a landfill?
2. **Recall** Identify the "Four Rs" of solid waste reduction and give an example of how individuals can apply each strategy.
3. **Summarize** What problems do single-use plastic bags pose and what can individuals and governments do to reduce their use?
4. **Evaluate** As city manager of a large city, your job is to recommend to the city council how solid waste should be handled. What technology or technologies would you recommend? Support your recommendations with evidence from the text.

SCIENCE AND ENGINEERING PRACTICES

5. **Engaging in Argument** Do you think a ban on plastic bags is a feasible solution? Use evidence from this lesson and other sources to support your position.

17.3 How Should Society Deal with Hazardous Waste?

CORE IDEAS AND SKILLS

- Describe how hazardous waste can be managed by producing less of it.
- Describe the advantages and disadvantages of recycling, treating, and storing hazardous waste.
- Identify the regulations that apply to hazardous waste.

KEY TERMS

bioremediation
phytoremediation
deep-well disposal
surface impoundment

Hazardous Waste Requires Special Handling

In the United States, most hazardous waste is produced by the chemical and petroleum industries. Special measures must be taken to transport, treat, and dispose of hazardous waste. Handled improperly, hazardous waste may cause injury or death. Hazardous waste may also pollute the environment.

As with solid waste management, the top priority for hazardous waste management should be pollution prevention and waste reduction. Using this approach, industries first try to find substitutes for toxic or hazardous materials. Then they reuse or recycle the hazardous materials that they must use for industrial processes whenever possible. They may also use or sell the hazardous materials as raw materials for making other products.

In keeping with these goals, the U.S. National Academy of Sciences has established three priority levels for dealing with hazardous waste. They are to produce less of this waste, to convert as much of it as possible to less-hazardous substances, and to put the rest in long-term, safe storage. Figure 17-18 illustrates this integrated management approach to dealing with hazardous waste. Unfortunately, most countries do not employ these priorities.

The European Union (EU) has had success with better managing industrial hazardous waste. In the EU, at least 33% of industrial hazardous waste produced is exchanged through clearinghouses, which sell the waste as raw materials for use by other industries. The producers of this waste do not have to pay for their disposal and recipients get low-cost raw materials. In contrast, about 10% of the hazardous waste in the United States is exchanged through

FIGURE 17-18

Hazardous Waste Management The U.S. National Academy of Sciences has suggested these priorities for dealing with hazardous waste.

Produce Less Hazardous Waste	Convert to Less Hazardous or Nonhazardous Substances	Put in Perpetual Storage
■ Change industrial processes to reduce or eliminate hazardous waste production. ■ Recycle and reuse hazardous waste.	■ Natural decomposition ■ Incineration ■ Thermal treatment ■ Chemical, physical, and biological treatment ■ Dilution in air or water	■ Landfill ■ Underground injection wells ■ Surface impoundments ■ Underground salt formations

such clearinghouses. There is significant potential for growth in this area.

Recycling E-Waste As mentioned in the Case Study, e-waste results from the disposal of cell phones, computers, and other electronic devices. E-waste is recyclable. However, most e-waste recycling can create health hazards, especially for workers in some less-developed countries.

Until recently, most of the world's e-waste was shipped to China. Although China officially banned the import of e-waste for processing in 2018, these materials still enter the country illegally. Other countries in Asia as well as Africa and the Latin Americas have stepped in to fill the import gap. Often, e-waste recyclers labor for low wages to extract valuable metals like gold, silver, copper, and various rare earth metals from millions of discarded computers, televisions, and cell phones. They are exposed to toxic chemicals in the process.

Workers usually wear no masks or gloves and often work in rooms with no ventilation. They carry out dangerous activities such as smashing TV picture tubes with large hammers to recover components—a method that releases large amounts of toxic lead dust into the air. They burn computer wires to expose copper. They also melt circuit boards in metal pots over coal fires to extract lead and other metals and douse the boards with strong acid to extract gold. After the metals are removed, leftover parts are burned or dumped into rivers or onto the land.

The World Health Organization estimates that as many as 18 million children and adolescents and 12.9 million women may face health risks related to primitive e-waste recycling practices (Figure 17-19).

The United States is the world's second-largest producer of e-waste and recycles only about 15% of it. As of 2021, 25 states had passed laws mandating statewide e-waste recycling. Most such laws ban the disposal of computers and TVs in landfills and incinerators. Some ban all electronics from landfills. Several states also make manufacturers responsible for recycling most electronic devices.

Some have called for a U.S. federal law requiring manufacturers to take back all electronic devices they produce and recycle them domestically. This regulation could be similar to laws in the European Union, where a recycling fee typically covers the costs of such programs. Without such a law there is little incentive for recycling e-waste and plastics, especially when there is money to be made from illegally sending such materials to other countries.

> **CONSIDER THIS**
>
> **You can take steps to help reduce e-waste.**
>
> If you use electronic devices, consider the following options for extending their lives and for keeping them out of landfills.
>
> - Think carefully before buying new electronic devices. Upgrade the software or hardware you already own if possible.
> - Buy electronics from stores that have trade-in or buy-back programs for these products.
> - Buy a protective case for your device and take good care of it. If it breaks, have it repaired instead of replacing it.
> - Donate used electronic devices if possible.
> - Recycle electronics instead of throwing them away. Contact local recycling centers or visit the U.S. EPA website for information about e-waste recycling.

LESSON 17.3

FIGURE 17-19
A young girl takes apart compact disc players at a family workshope in Guiyu, China. Most of the e-waste that comes to this city for processing and recycling originates in other countries.

The only real long-term solution is a *prevention* approach. In other words, electrical and electronic products must be designed to be manufactured, repaired, remanufactured, or recycled without the use of hazardous materials.

Treating Hazardous Waste Even with successful reduction programs for hazardous waste, there are still large quantities of it that require treatment and disposal. Hazardous waste can be sent to waste management facilities for treatment to remove its toxic materials. Facilities may use chemical, physical, biological, or thermal treatment methods.

Chemical treatment methods involve carrying out chemical reactions. The reaction process converts hazardous chemicals into harmless or less harmful chemicals. Physical treatment methods may involve using charcoal or other materials to filter out harmful solids. Liquid hazardous waste can be boiled to physically remove its toxic chemicals. Boiling produces toxic vapors. The vapors are cooled, causing them to condense so they can be collected. Liquid waste can also be treated in order to separate out harmful chemicals as solids.

Some scientists and engineers consider biological methods for treatment of hazardous waste to be the wave of the future. One such approach is **bioremediation**, in which bacteria and enzymes help to destroy toxic or hazardous substances or convert them to harmless compounds. Bioremediation is often used on contaminated soil. It usually takes a little longer to work than most physical and chemical methods, but it costs much less. (See Engineering Focus 3.2 for more on bioremediation.)

Phytoremediation is another biological method for treating hazardous waste. **Phytoremediation** involves using natural or genetically engineered plants as "pollution sponges." The plants are able to absorb, filter, and remove contaminants from polluted soil and water. Phytoremediation can be used to clean up soil and water contaminated with chemicals such as pesticides, certain solvents, and radioactive or toxic metals. This method is still being evaluated and is slow compared to other alternatives.

Thermal treatment methods use heat to detoxify hazardous waste. Incineration, or burning, is the most common thermal method. It has the same advantages and disadvantages as burning solid waste (Figure 17-11). Incinerating hazardous waste without effective and expensive air pollution controls can release air pollutants such as highly toxic dioxins. It also produces an extremely toxic ash that must be safely and permanently stored in a specially designed landfill or vault.

Plasma gasification is another thermal treatment method. This technology uses arcs of electrical energy to produce very high temperatures in order to vaporize trash in the absence of oxygen. The process reduces the volume of a given amount of waste by 99%. One of its by-products is a gas that can be used as a fuel source. Another by-product is a solid material that safely encases any toxic materials that do not burn. It can be used for construction and paving materials. Plasma gasification is currently very costly, but plasma gasification companies are working to bring prices down.

Storing Hazardous Waste Ideally, hazardous waste managers should use burial on land or storage of hazardous waste in secure vaults only as a last resort. Reduction, recycling, and treatment options should be pursued first. In reality, burial is actually the most widely used method for dealing with hazardous waste in the United States and in most countries, largely because of its lower cost.

The most common form of burial is **deep-well disposal**, in which liquid hazardous waste is pumped under high pressure through a pipe into dry, porous rock formations deep underground. The formations lie far beneath aquifers that are tapped for drinking and irrigation water. Theoretically, the hazardous liquids soak into the porous rock material and are isolated from overlying groundwater by essentially impassible layers of clay and rock.

Deep disposal wells can be built at low cost. In addition, the waste stored at these sites can often be retrieved if problems develop. However, there are a limited number of such sites and limited space within them. Sometimes the waste does leak into groundwater from the well shaft or migrate into groundwater in unexpected ways. Using deep-well disposal also encourages the production of hazardous waste rather than the reduction of such waste.

In the United States, almost two-thirds of all liquid hazardous waste is injected into deep disposal wells. This amount will increase sharply with the country's growing use of hydraulic fracturing, or fracking, to produce natural gas and oil trapped in shale rock. You will recall from Chapter 12 that fracking involves injecting huge volumes of a mixture of water, sand, and chemicals down into a well and out through holes drilled into a horizontal underground pipe.

FIGURE 17-20

Deep-Well Disposal of Hazardous Wastes	
Advantages	**Disadvantages**
Safe if sites are chosen carefully	Leaks possible from corrosion of well casing
Wastes can often be retrieved.	Emits CO_2 and other air pollutants
Low cost	Output approach that encourages waste production

The force of the blast creates cracks in the surrounding shale rock. This allows trapped oil and gas to flow out from the cracks and be pumped to the surface.

Much of the contaminated wastewater that results from fracking is pumped back underground in hazardous waste disposal wells. Not only is the risk of groundwater contamination a concern, but this method of storing hazardous waste is also thought to increase the risk of earthquakes because the fluid places extra pressure on faults in the rock.

Many scientists argue that current regulations for deep-well disposal in the United States are inadequate. Figure 17-20 lists the advantages and disadvantages of this disposal method.

Some liquid hazardous waste is stored in ponds, pits, or lagoons, called **surface impoundments** (Figure 17-21). Some impoundments include liners to contain the waste. As the water evaporates, the waste settles and becomes more concentrated. Where liners are not used and wherever the liners leak, the concentrated waste can seep into groundwater. Because surface impoundments are not covered, harmful chemicals can evaporate and

FIGURE 17-21
Hydraulic fracturing generates toxic wastewater that is usually stored in surface impoundments or injected underground in deep disposal wells. Deep-well disposal of wastewater from hydraulic fracturing may make an area more prone to earthquakes.

FIGURE 17-22

Storing Liquid Hazardous Waste in Surface Impoundments	
Advantages	Disadvantages
Low cost	Water pollution from leaking liners and overflows
Can be retrieved	Air pollution from volatile organic compounds
Can store wastes securely with double liners	Output approach that encourages waste production

pollute the air. In addition, flooding from heavy rains can cause such ponds to overflow.

Studies by the EPA found that 70% of all U.S. hazardous waste storage ponds lack liners and could threaten groundwater supplies. The EPA also warns that all impoundment liners are likely to leak eventually. Figure 17-22 lists the advantages and disadvantages of using surface impoundments.

Liquid or solid hazardous waste may be placed in sealed containers and buried in carefully designed and monitored secure hazardous waste landfills. This is the least-used burial method because of the expense involved.

checkpoint What are the four methods for treating hazardous waste?

Hazardous Waste Regulation in the United States

Several U.S. federal laws help to regulate the management and storage of hazardous waste. The first is called the Resource Conservation and Recovery Act (RCRA, pronounced RECK-ra). In place since 1976 and amended in 1984, RCRA regulates about 5% of all U.S. hazardous waste.

Under RCRA, the EPA sets standards for the management of several types of hazardous waste. It also issues permits to companies to produce and dispose of a certain amount of that waste by approved methods. Permit holders must use a cradle-to-grave system to track hazardous waste. This means tracking its transfer from a point of generation (cradle) to an approved off-site disposal facility (grave) and submitting proof of this disposal to the EPA. RCRA is a good start, but the majority of the hazardous and toxic waste produced in the United States, including e-waste, is not regulated. In most other countries, especially less-developed countries, hazardous waste is even less regulated.

The Toxic Substances Control Act has also been in place since 1976. Its purpose is to regulate and ensure the safety of the thousands of chemicals used to either manufacture many products or as ingredients in many products. Due to limited funding, the EPA so far has used this Act to ban only five of the roughly 85,000 chemicals in use.

Under this law, companies must notify the EPA before introducing a new chemical into the marketplace. However, they are not required to provide data about its safety. A 2016 amendment to the law does give the EPA authority to make an affirmative determination that a new chemical presents no unreasonable risks to human health or the environment before manufacturing can begin. Environmental and health scientists feel manufacturers should bear this burden of proving a new chemical or new product containing a certain chemical is safe before it can be sold.

The Comprehensive Environmental Response, Compensation, and Liability Act (CERCLA) was passed in 1980. It is commonly known as the Superfund Act and is regulated by the EPA. The goals of the act are to identify sites, called Superfund sites, where hazardous waste has contaminated the environment and to clean them up using EPA-approved methods (Figure 17-23). Superfund sites are dealt with on a priority basis. The worst sites—those that represent an immediate and severe threat to human health—are put on a National Priorities List and scheduled for cleanup first.

As of March 2022, the Superfund list included 1,333 existing sites and 43 proposed sites. Nearly 450 sites had been cleaned up and removed from the list. The Waste Management Research Institute estimates at least 10,000 sites should be on the priority list and that cleanup of these sites will cost about $1.7 trillion, not including legal fees. These environmental and economic costs show why it is important to emphasize waste reduction and pollution prevention over the "end-of-pipe" cleanup approach the United States and most countries rely on.

Until 1995, the Superfund Act was funded by a tax on oil and chemical companies. However, the U.S. Congress allowed it to expire due to pressure from polluters. This led to a backlog of unfunded sites. In 2022, the tax was reinstated as part of the Infrastructure Investment and Jobs Act. This law also allocated additional funding to help clear the backlog and accelerate cleanup at other Superfund sites.

FIGURE 17-23
ON ASSIGNMENT National Geographic photographer Fritz Hoffman documented several U.S. Superfund sites, including this one in Picher, Oklahoma. Lead and zinc mines once operated in Picher's backyard. The mountain of waste rock still remains, as does soil and water pollution caused by the mining. People are long gone, however. Residents left after the area was declared a Superfund site in 1983. The government bought most of their homes. Hundreds were torn down, including the ones on these streets.

LESSON 17.3

As part of the Superfund Act, the EPA runs the Toxic Release Inventory website. Anyone can visit the website to find out what toxic chemicals are being stored and released in their communities.

The EPA also runs the Brownfields Program. A brownfield is an industrial or commercial property that is, or may be, contaminated with hazardous pollutants. The program helps states, communities, and other parties to economically redevelop contaminated property such as factories, junkyards, older landfills, and gas stations. The EPA provides help with site assessment, cleanup, or reuse of land designated as a brownfield. Reclaiming brownfields can increase local tax bases and promote job growth.

checkpoint What are the drawbacks of the Toxic Substances Control Act?

17.3 Assessment

1. **Recall** Name the biological methods for treating hazardous materials.
2. **Describe** What dangers are associated with recycling e-waste?
3. **Explain** How does the Superfund Act work?

SCIENCE AND ENGINEERING PRACTICES

4. **Communicating Information** Visit the EPA's Toxic Release Inventory website and use the TRI Toxic Tracker to identify toxic chemicals that may be stored or released in your community. Summarize your findings in a paragraph.

17.4 How Can Society Transition to a Low-Waste Economy?

CORE IDEAS AND SKILLS

- Explain how grassroots action leads to better solid waste management and encourages reuse, recycling, and composting.
- Understand how international treaties have reduced hazardous waste.
- Explain ways to transition to a low-waste economy.

KEY TERM

biomimicry

Governments and Citizens Can Take Action

Shifting to a low-waste economy requires society to make changes at local, national, and global levels. Individuals and businesses will need to find ways to reduce resource use. They will also need to work toward reusing and recycling most solid and hazardous waste.

According to physicist Albert Einstein, "A clever person solves a problem; a wise person avoids it." Many people are taking these words seriously. The governments of Norway, Austria, and the Netherlands have committed to reducing their resource waste by 75%. Many school cafeterias, restaurants, national parks, and corporations are participating in a rapidly growing "zero waste" movement to reduce, reuse, and recycle. Some have lowered their waste outputs by up to 80%, with the ultimate goal of eliminating all waste. They are applying nature's nutrient cycling factor of sustainability.

In the United States, individuals have organized grassroots (bottom-up) campaigns to prevent the construction of hundreds of incinerators, landfills, treatment plants for hazardous and radioactive waste, and chemical plants in or near their communities (Figure 17-24). These campaigns have organized sit-ins, concerts, and protest rallies. They have gathered signatures on petitions and presented them to lawmakers. Health risks from incinerators and landfills, when averaged over the entire country, are quite low. However, the risks for people living near such facilities are higher.

Manufacturers and waste industry officials make the point that something must be done with the toxic and hazardous waste created in the production of certain goods and services. They contend that even if local citizens adopt a NIMBY—"not in my backyard"—approach, the waste will always end up in someone's backyard. Many citizens do not accept this argument. Their view is that the best way to deal with most toxic and hazardous waste is to produce much less of it by focusing on pollution and waste prevention. They argue the goal should be "not in anyone's back yard" (NIABY) or "not on planet Earth" (NOPE).

International Treaties For decades, countries regularly shipped hazardous waste to other countries for disposal or processing. Since 1992, an international treaty known as the Basel Convention has banned participating countries from shipping hazardous waste to or through other countries without their permission. This treaty also applies to e-waste (Case Study).

FIGURE 17-24
People gather outside the Supreme Court building in Denver, Colorado. They are protesting efforts of local oil and gas companies to reverse bans on fracking operations in the area.

By 2021, this agreement had been ratified (formally approved and implemented) by 188 countries. The United States has signed but has not ratified the convention. In 1995, the treaty was amended to outlaw all transfers of hazardous waste from industrial countries to less-developed countries. (Unfortunately, it has not eliminated the highly profitable illegal shipping of hazardous waste.) In 2021, international shipments of most plastic scrap and waste began to be regulated under the treaty. These shipments now require prior consent of the importing country and any transit countries. The goal is to reduce improper disposal of plastic waste.

In 2000, delegates from 122 countries completed a global treaty known as the Stockholm Convention on Persistent Organic Pollutants (POPs). POPs are chemical by-products of manufacturing that persist in the environment. By 2021, 185 countries had ratified the treaty (the United States had not). The treaty currently regulates the use of 30 widely used persistent organic pollutants that build up in the fatty tissues of humans and other animals that occupy high trophic levels in food webs.

These hazardous chemicals can reach levels hundreds of thousands of times higher than levels in the general environment. Because they persist in the environment, POPs also can be transported long distances by wind and water. Based on blood tests and statistical sampling, medical researchers at New York City's Mount Sinai School of Medicine concluded that nearly every person on Earth likely has detectable levels of POPs in their bodies. The long-term health effects of this involuntary global experiment are largely unknown.

checkpoint What are three examples of grassroots action involving hazardous waste?

FIGURE 17-25

Mimicking Nature This ecoindustrial park in Kalundborg, Denmark, reduces waste production by mimicking a natural ecosystem's food web. The wastes of one business become the raw materials for another, thus mimicking the way nature recycles chemicals.

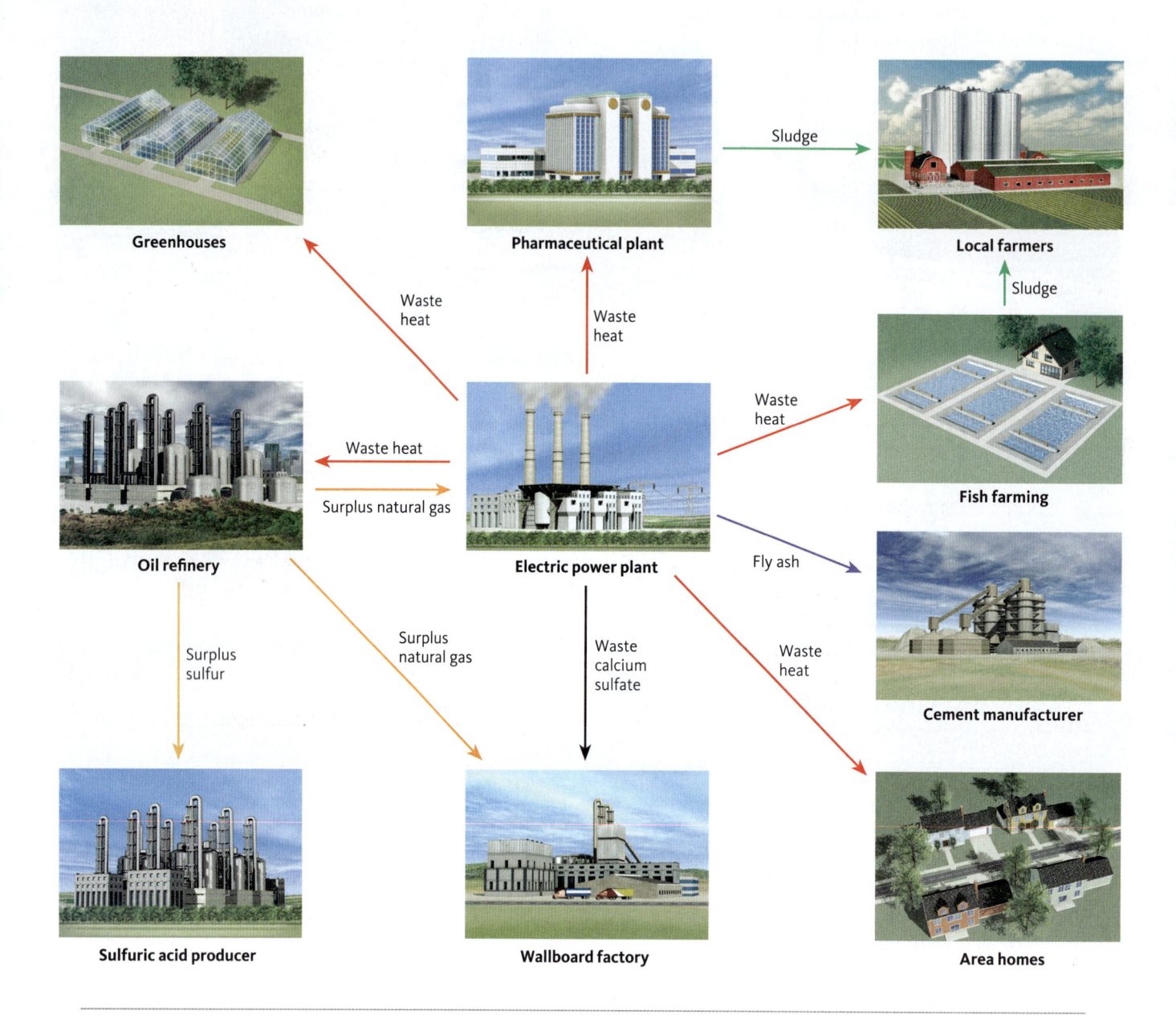

Reuse and Recycling Can Be Encouraged

Why aren't reuse and recycling more common? First, these strategies must compete with the use of cheap, disposable products, which don't include hidden environmental and health costs in their market price. This is a violation of the full-cost pricing factor of sustainability. Second, in most countries, resource extraction industries receive more government tax breaks and subsidies than do reuse and recycling industries. Third, goods made with recycled materials are not given high enough priority to increase demand and lower cost.

How can we encourage reuse and recycling? Proponents say leveling the economic playing field is the best way to start. Governments can *increase* subsidies and tax breaks for reusing and recycling materials and *decrease* subsidies and tax breaks for making items from virgin resources.

Another strategy is to ramp up use of the fee-per-bag waste collection system that charges households

606 CHAPTER 17 SOLID AND HAZARDOUS WASTE

for the trash they throw away but not for their recyclable and reusable waste. When Fort Worth, Texas, launched such a program, the proportion of households recycling trash grew from 21% to 85%. The city went from losing $600,000 in its recycling program to making $1 million a year thanks to increased sales of recycled materials to industries.

Governments can also pass laws requiring companies to take back and recycle or reuse packaging and electronic waste discarded by consumers. Japan and some European Union countries have such laws. Another strategy is to encourage or require government purchases of recycled products to increase demand for and lower prices of these products. Citizens can pressure governments to require product labeling that lists the recycled content of products, as well as the types and amounts of any hazardous materials they contain.

A growing number of people are saving money through reuse by shopping at yard sales, flea markets, and secondhand stores. They are also shopping at online sites that feature used items. One site, run by the Freecycle Network, links people who want to give away their unused household belongings to people who want or need them.

For many, recycling has become a business opportunity. In particular, upcycling, or recycling materials into products of a higher value, is a growing industry. Simple examples include furniture made from shipping pallets and planters made from old tires. Upcycling is one way for individuals to create functional products from "trash."

checkpoint What action can the government take to encourage recycling?

Nature Can Provide a Model

An important goal for a more sustainable society is to make its industrial manufacturing processes cleaner and more sustainable by redesigning them to mimic the way nature deals with waste. This approach is called **biomimicry**. In nature, the waste outputs of one organism become the nutrient inputs of another, so that all of Earth's nutrients are endlessly recycled. This explains why there is essentially no waste in undisturbed ecosystems.

One way for industries to mimic nature is to reuse or recycle most of the minerals and chemicals they use. Industries can set up ecoindustrial parks that operate resource exchange webs. In a resource exchange web, the wastes of one manufacturer become the raw materials for another. In this way, they are similar to food webs in natural ecosystems. An ecoindustrial park can be said to function as an industrial "ecosystem."

One example of an ecoindustrial park is located in Kalundborg, Denmark. There, an electric power plant and nearby industries, farms, and homes are collaborating to save money and to reduce pollution and waste production (Figure 17-25).

Ecoindustrial parks provide many economic benefits for businesses. By encouraging recycling and waste reduction, they cut the costs of managing solid waste, controlling pollution, and complying with pollution regulations. They reduce a company's chances of being sued for damages to people or the environment caused by their actions. Companies improve the health and safety of workers by reducing their exposure to toxic and hazardous materials, thereby bringing down company health insurance costs. Today, more than 300 such parks operate in various places around the world, including the United States.

In recognition of the global importance of ecoindustrial parks, a framework was developed in 2018 that includes an internationally accepted definition for them. The framework also outlines minimum parameters for their environmental, social, and economic performance.

checkpoint Describe the purpose of ecoindustrial parks.

17.4 Assessment

1. **Recall** What is NIMBY and how does it relate to dealing with solid and hazardous waste?
2. **Analyze** What are the obstacles to encouraging reuse and recycling?
3. **Explain** What is biomimicry and how does it relate to transitioning to a low-waste economy?
4. **Evaluate** If the United States ratifies the Basel Convention, what will be the impact on electronics consumers, manufacturers, and processors?

CROSSCUTTING CONCEPTS

5. **Energy and Matter** In nature, the waste outputs of one organism become the nutrient inputs of another, so all of Earth's nutrients are endlessly recycled. Can the same be said of a resource exchange web in an ecoindustrial park? Explain why or why not.

TYING IT ALL TOGETHER STEM
MSW Ecological Footprint

This chapter discussed the impacts a high-waste society can have on the environment and public health. You read about the challenges of safely handling solid and hazardous waste such as e-waste (Case Study) and the need to reduce our waste output. According to conservative estimates, each person in the United States generates 2.2 kg (4.84 lbs) of municipal solid waste (MSW) every day. Figure 17-26 illustrates the breakdown of materials that make up MSW in the United States.

Use Figure 17-26 to answer the questions that follow.

1. Consider the largest component of MSW. How should this material be processed to keep it out of landfills?
2. Think about the items you throw away each day. What category of items do you think makes up the greatest percentage of your daily trash? Why?
3. The latest figures for MSW generated in the United States estimate the annual total at about 265 million metric tons (292 million tons). Calculate the number of metric tons each category of waste represents.
4. Use the data in the pie chart to get an idea of a typical annual MSW ecological footprint for each American. Calculate the total weight in kilograms (and pounds) for each category generated during 1 year (1 kilogram = 2.2 pounds).

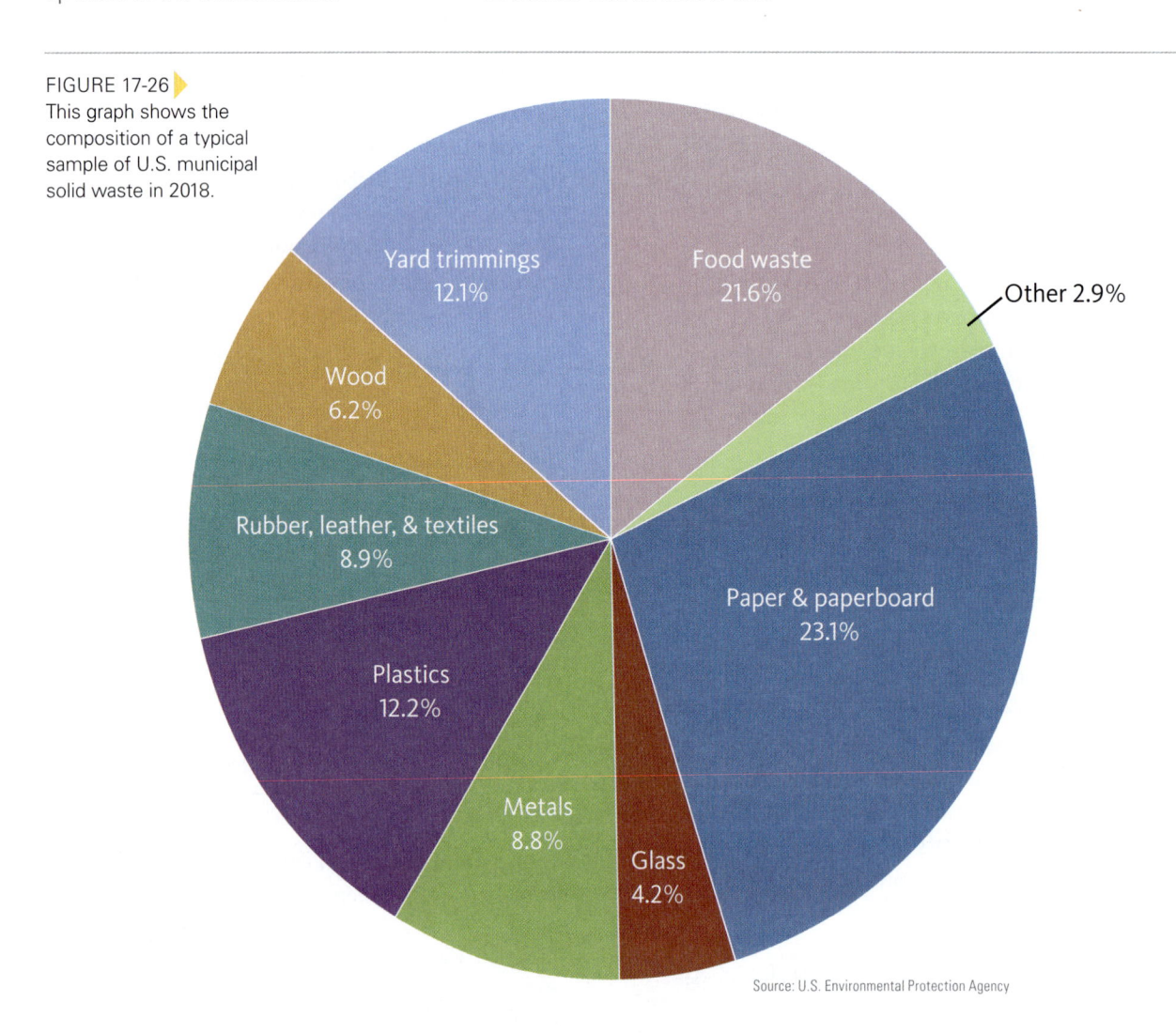

FIGURE 17-26
This graph shows the composition of a typical sample of U.S. municipal solid waste in 2018.

- Yard trimmings 12.1%
- Food waste 21.6%
- Other 2.9%
- Paper & paperboard 23.1%
- Wood 6.2%
- Rubber, leather, & textiles 8.9%
- Plastics 12.2%
- Metals 8.8%
- Glass 4.2%

Source: U.S. Environmental Protection Agency

608 CHAPTER 17 SOLID AND HAZARDOUS WASTE

CHAPTER 17 SUMMARY

17.1 What are problems related to solid and hazardous waste?

- The United States is the world's largest generator of industrial solid waste, municipal solid waste, e-waste, and hazardous waste.
- Plastic marine debris is a serious problem for the world's oceans, especially where it is concentrated in areas called garbage patches.
- Hazardous and toxic waste is defined as any discarded material or substance that threatens human health or the environment because it is poisonous, dangerously chemically reactive, corrosive, or flammable.

17.2 How should society deal with solid waste?

- Solid waste can be handled by burning it (waste-to-energy incineration), burying it in open dumps or landfills, recycling it, or producing less of it. All of these methods have advantages and disadvantages.
- Solid waste can be reduced using the "Four Rs" strategy: refuse, reduce, reuse, and recycle. Integrated waste management is a coordinated strategy for waste disposal and reduction.
- Composting mimics nature by using bacteria to decompose yard waste, vegetable food scraps, and other biodegradable organic waste. The composted material can be used to improve soil fertility.
- Recycling involves three steps: collection of materials, conversion to new products, and marketing recycled products.

17.3 How should society deal with hazardous waste?

- To reduce hazardous waste, society can recycle more of its e-waste.
- Hazardous waste can be detoxified through physical, chemical, biological, and thermal methods.
- Storage of hazardous waste can be accomplished through deep-well disposal or using surface impoundments.
- The U.S. government has passed legislation that deals with hazardous waste: the Resource Conservation and Recovery Act (RCRA); the Toxic Substance Control Act (TSCA); and the Comprehensive Environmental Response, Compensation, and Liability Act (CERCLA), also known as the Superfund Act.

17.4 How can society transition to a low-waste economy?

- Grassroots action involving citizen-led campaigns can influence decisions regarding solid and hazardous waste management.
- Governments can develop and participate in international treaties to reduce and more safely handle solid and hazardous waste.
- Upcycling waste into products of higher value reduces waste and provides economic opportunities.
- Ecoindustrial parks can reduce waste by setting up resource exchange webs, in which the wastes of one manufacturer become the raw materials for another. This mimics natural processes, whereby the waste outputs of one organism become the nutrient inputs of another. Recreating natural processes in this way is called biomimicry.

CHAPTER 17 ASSESSMENT

Review Key Terms

Select the key term that best fits each definition. Not all terms will be used.

biomimicry	phytoremediation
bioremediation	primary recycling
composting	sanitary landfill
deep-well disposal	secondary recycling
hazardous waste	solid waste
industrial solid waste	surface impoundment
integrated waste management	waste management
	waste reduction
municipal solid waste	

1. Practice that mimics nature by using bacteria to decompose yard trimmings, vegetable food scraps, and other forms of biodegradable organic waste into materials than can be used to improve soil fertility

2. The use of bacteria or enzymes to destroy toxic or hazardous substances or convert them to harmless compounds

3. Any discarded material or substance that threatens human health or the environment because it is poisonous, dangerously chemically reactive, corrosive, or flammable

4. Waste disposal site on land in which waste is spread in thin layers, compacted, and covered with a fresh layer of clay or plastic foam each day

5. A variety of coordinated strategies for both waste disposal and waste reduction

6. Process of observing certain changes in nature, studying how natural systems have responded to such changing conditions over many millions of years, and applying what is learned to dealing with some environmental challenge

7. Excess materials produced by mines, farms, and industries that produce goods and services

8. Pumping of liquid hazardous waste under high pressure through a pipe into dry, porous rock formations far beneath aquifers that are tapped for drinking and irrigation water

9. Process by which waste materials are converted into different products

10. The combined solid wastes produced by homes and workplaces other than factories

Review Key Concepts

11. Distinguish among industrial solid waste, municipal solid waste, and hazardous waste, and give an example of each.

12. Explain how and why electronic waste (e-waste) has become a growing solid waste problem.

13. Distinguish among waste management, waste reduction, and integrated waste management.

14. What is the most common way to handle MSW in the United States? Explain the advantages and disadvantages of this method.

15. Distinguish among refusing, reducing, reusing, and recycling in dealing with solid wastes.

16. What is the Great Pacific Garbage Patch and how did it come to be? How does it harm marine life and how can the growth of such patches be prevented?

17. Explain how ecoindustrial parks mimic natural processes.

Think Critically

18. Do you think that manufacturers of computers, televisions, cell phones, and other electronic products should be required to take their products back at the end of their useful lives for repair, remanufacture, or recycling? If so, who do you think should pay to cover the costs of such a take-back program and why?

19. Think of three items that you regularly use once and then throw away such as plastic grocery bags, water bottles, etc. Are there reusable items that you could use in place of these disposable items? Do you think that you could consume less by refusing to buy some of the things you regularly buy? If so, what are three of those things? Do you think that this is something you ought to do? Explain.

20. Imagine there is a proposal to create a storage site in your community for hazardous waste. Citizens have been invited to attend a town hall meeting to ask questions. What legal questions might you ask based on your understanding of how hazardous waste is regulated? What environmental and public health questions might you ask based on your understanding of risks posed by hazardous waste?

Chapter Activities

A. Investigate: Composting STEM

Solid waste is the unwanted material that humans produce and dispose of in a variety of ways. Landfills have long been used to hold much of the solid waste produced. However, other methods of waste management also are used, such as burning and recycling. Composting is one type of recycling that can be used to reduce the amount of solid waste that is put into landfills, while also providing nutrient-rich soil. In this activity, you will design an investigation that compares rates of decomposition of waste when it is composted and when it is landfilled.

Materials

2 clean, empty 2-liter plastic bottles
craft knife
masking tape
marker
rubber gloves
newspaper
organic waste (leaves, grass clippings, shredded paper, fruit and vegetable food waste, eggshells)
other waste materials (plastic, aluminum foil, small metal objects, Styrofoam, etc.)
garden soil
spray bottle of water

1. Working in teams, write a hypothesis predicting the difference in decomposition of materials in a compost bin and in a landfill. Be sure to provide support for your claim.

2. Use a craft knife to carefully cut off the top third of each plastic bottle. Label one of the bottles "Landfill" and the other "Compost Bin." With your team, develop a plan to test your hypothesis. To do this, you will need to simulate the conditions in each environment using the suggested materials. (Use the text and your own research to help you.)

3. Maintain and observe your bottles over the next three weeks. Record any changes you notice.

4. After three weeks, dump out the contents of each bottle onto newspaper and compare the remaining waste materials.

5. What differences did you notice between the organic waste materials in the two bottles? How about the inorganic waste? Why do you think these differences occurred? What conclusions can you draw?

B. Take Action

National Geographic Learning Framework
Attitudes | Responsibility
Skills | Observation
Knowledge | New Frontiers

Investigate ways that your school's cafeteria operations could be modified to generate less waste. What changes can students and staff make to reduce the amount of food and disposable items such as packaging and utensils that ends up in the trash? Start by researching "zero waste" initiatives as they apply to school cafeterias.

Next, work in teams to interview students and staff at your school. Collect information from students about items they typically pack for lunch and how those items are packaged. Collect information from cafeteria staff and other adults about how food is prepared and packaged and how leftover food is handled. Brainstorm ways to reduce waste, such as improving recycling, replacing disposable packaging and utensils, and setting up a collection program to donate unused food to a local food pantry. Research the costs and benefits of these proposed changes, and create a proposal to present to school officials.

CHAPTER 18
ENVIRONMENTAL ECONOMICS, POLITICS, AND WORLDVIEWS

CONSERVATION was a key concern of forward-thinking U.S. President Theodore Roosevelt. In 1903, he designated the first U.S. wildlife refuge. He later created the U.S. Forest Service and signed the Antiquities Act, giving presidents the right to set aside land of historic and scientific value. Later, in 1916, President Woodrow Wilson created the U.S. National Park Service to protect national parks for future generations.

KEY QUESTIONS

18.1 How are economic systems related to the biosphere?

18.2 How can people use economic tools to address environmental problems?

18.3 How can society enact more just environmental policies?

18.4 How can society live more sustainably?

NATIONAL GEOGRAPHIC | EXPLORERS AT WORK

Adventurers and Scientists for Conservation

with National Geographic Explorer Gregg Treinish

Gregg Treinish is the ultimate adventurer. In 2004, he hiked the entire Appalachian Trail, from Georgia to Maine. Next, he spent two years traversing about 12,500 kilometers (7,800 miles) of the rugged Andes Mountains. National Geographic named Treinish Adventurer of the Year in 2008 in recognition of his exploits. Yet he felt dissatisfied. Was this all just adventure for adventure's sake? How could he turn it into something more beneficial to the world?

Treinish decided to pursue a biology degree, which led to fieldwork and the chance to study different types of wildlife. As he recalls, "I still got to hike and explore, but doing it to make a positive difference felt more fulfilling." Fieldwork also highlighted an obvious, yet unmet, need to Treinish. Many outdoor enthusiasts live to explore far-flung regions, but feel guilty and selfish coming home with nothing but personal highs. Meanwhile, scientists desperately need samples, photographs, data, and observations from places too difficult to reach on their shrinking budgets. Why not bring the two worlds together?

In 2011, Treinish did just that by founding Adventure Scientists to connect the scientific and outdoor communities. More than a hundred scientific organizations and a thousand adventurers have already participated. Still, Treinish points out that extreme athletes are only part of his volunteer universe: "We've worked with school kids, teachers, military veterans, and families on vacation. No matter what their skill level in science or the outdoors, they can make a valuable contribution."

So far, Adventure Scientists volunteers have brought back important data from all over the world. Mountain climbers discovered Earth's highest known plant life on Mount Everest, bringing back samples that may help farmers grow crops in extreme conditions. Hundreds of roadkill observations from bikers identified hotspots of risk for animals. Skiers checked glaciers for ice worms to better understand how organisms survive in unforgiving environments. Even six-year-olds have gotten in on the act by sampling for diatoms (single-celled algae) in high mountain lakes. The four new species they found were named after the young adventurers.

It's a win-win for everyone. As Treinish says, "Adventurers tell me these chances to give back have changed their whole perspective. Now, being the strongest or summiting the coolest peak isn't what's important. Trying to contribute and make a difference is what matters."

Thinking Critically
Summarize Why are collaborative expeditions like those sponsored by Adventure Scientists a win-win?

Gregg Treinish examines a research sample gleaned from the forest floor.

CASE STUDY
The United States, China, and Sustainability

The greatest challenge society faces is learning to live more sustainably. Meeting this challenge depends largely on the decisions and actions of the United States and China. These two countries lead the world in resource consumption and production of wastes and pollutants.

From 1940 to 1970, the United States experienced rapid economic growth. The cost of this growth was severe pollution and degradation of air, water, and land. By 1970, public awareness of these problems grew and spurred an environmental movement. Millions of citizens demanded an end to this environmental degradation. The U.S. Congress responded by passing environmental laws that led to improvements in the nation's environmental quality.

Despite this important progress, the United States, with the world's third-largest population, has the world's largest ecological footprint. This is because the average individual in the United States uses more resources than the average individual in any other country. That gives the United States the world's largest per capita (per person) ecological footprint. Figure 18-1 shows how the United States stacks up against China in terms of ecological footprints.

China has the world's largest population and the second largest economy. China's economy grew rapidly for three decades beginning in the late 1970s, after the country partially shifted from a centrally planned economy to a more market-based economy.

As in the U.S. case, China's economic growth has resulted in severe environmental problems that have contributed to its total ecological footprint, the second largest in the world. Also, China's consumer middle class, who lives mostly in its large cities, has grown to roughly 340 million. This number is about equal to the U.S. population. Still, the large majority of Chinese citizens are poor. As a result, the per capita ecological footprint in China is about one-sixth that of the United States.

Since the 1960s, China has cut its birth rate in half. Its population is now growing at a rate slower than that of the United States. However, if China's middle class continues to grow and consume more resources as projected, China could have the world's largest per capita and total ecological footprint within a decade or two.

This chapter discusses the economic and political aspects of environmental problems as well as solutions. It illustrates both aspects and discusses what individuals can do to make a difference. This chapter also considers the future of economic growth and what it will mean to global sustainability.

As You Read Think about your own attitudes and actions regarding sustainability. Has your environmental worldview shifted during this course? If so, in what ways?

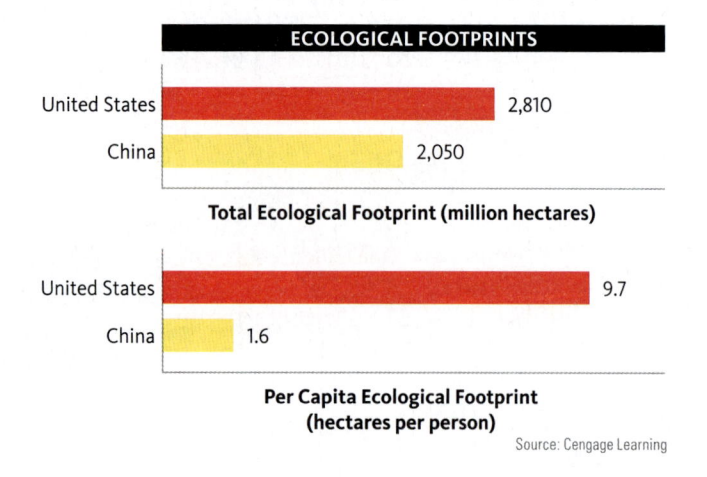

ECOLOGICAL FOOTPRINTS

United States: 2,810
China: 2,050

Total Ecological Footprint (million hectares)

United States: 9.7
China: 1.6

Per Capita Ecological Footprint (hectares per person)

Source: Cengage Learning

FIGURE 18-1
Ecological Footprints The United States' total ecological footprint is larger than China's, and its per capita ecological footprint is much larger.

18.1 How Are Economic Systems Related to the Biosphere?

CORE IDEAS AND SKILLS

- Identify the types of capital used by most economies.
- Describe environmentally sustainable economic development.

KEY TERMS

economics
human capital
manufactured capital
economic growth
economic development

Economic Systems Depend on Natural Capital

Economics is the social science that deals with the production, distribution, and consumption of goods and services to satisfy people's needs and wants. In a market-based economic system, buyers and sellers interact to make economic decisions about how goods and services are produced, distributed, and consumed. In a free-market economic system, all economic decisions are governed by the competitive interactions of supply and demand. Supply is the amount of a good or service producers are willing to offer for sale at a given price. Demand is the amount of a good or service people are both willing and able to buy at a given price. If the demand is greater than the supply, the price rises. If supply is greater than demand, the price falls. Price is the market value of a good or service.

In a true free-market economic system:

- No company or small group of companies controls the prices of any goods or services.
- Market prices include all of the direct and indirect costs (full-cost pricing).
- Consumers have complete information about the beneficial and harmful environmental and health effects of the goods and services.

The economies of the world's countries are not true free-market systems. The combined actions of governments and corporations result in major violations of these three conditions.

Most economic systems use three types of capital, or resources, to produce goods and services. Natural capital includes resources and ecosystem services produced by Earth's natural processes that support all life and all economies. **Human capital** includes the physical and mental talents of the people who provide labor, organizational and management skills, and innovation. **Manufactured capital** includes the machinery, materials, and factories that people create using natural resources.

As the capacity of a nation, state, city, or company to provide goods and services increases, **economic growth** occurs. Sometimes economic growth leads to a high-waste economy (Figure 18-2). Most industrialized countries today depend on this type of economy. A high-waste economy relies on boosting economic growth by increasing the flow of natural matter and energy resources through the economic system to produce more goods and services. Such an economy produces valuable goods and services. However, it also converts large quantities of high-quality matter and energy resources into waste, pollution, and low-quality heat that are transmitted into the air, water, and soil.

Economic development is any set of efforts focused on creating economies that serve to improve human well-being by meeting basic human needs. Human needs include food, shelter, physical and economic security, and good health. The world's countries vary greatly in their levels of economic growth and development.

Economists have debated for more than 200 years whether there are limits to economic growth. Neoclassical economists assume the potential for economic growth is essentially unlimited. They also assume economic growth is necessary to provide profits for businesses and jobs for workers. Neoclassical economists consider natural capital important but not absolutely necessary because they assume people can find substitutes for essentially any resource they might deplete or degrade.

FIGURE 18-2 ▼

High-Waste Economy The high-waste economies of most of the world's more-developed countries rely on continually increasing the flow of resources to promote economic growth. What are three ways in which you regularly add to the throughputs of matter and energy by your daily activities?

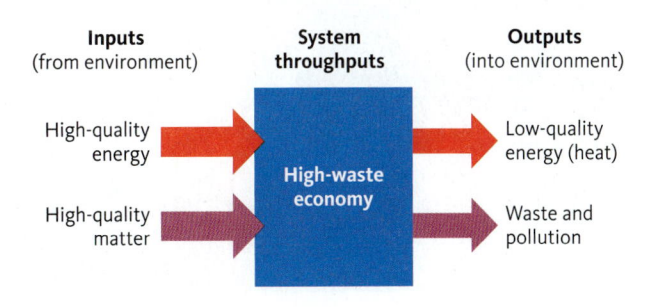

Inputs (from environment)	System throughputs	Outputs (into environment)
High-quality energy	High-waste economy	Low-quality energy (heat)
High-quality matter		Waste and pollution

LESSON 18.1   617

Ecological economists disagree. They point out that there are no substitutes for many vital natural resources such as clean air, water, fertile soil, and biodiversity. They also see no substitutes for crucial ecosystem services such as climate control, air and water purification, pest control, pollination, topsoil renewal, and nutrient cycling. In contrast to neoclassical economists, ecological economists view human economic systems as subsystems of the biosphere that depend heavily on Earth's irreplaceable natural resources and ecosystem services (Figure 18-3).

Environmental economics is closely related to the school of ecological economics. Environmental economists favor adjusting existing economic policies and tools to be more environmentally beneficial over inventing new policies and tools. The debate among these economists is shifting to questions about what kinds of economic growth and development to encourage. Ecological and environmental economists argue that people should promote environmentally sustainable economic development. This approach uses political and economic systems to encourage environmentally beneficial and more sustainable forms of economic growth. It discourages environmentally harmful forms of economic growth that degrade natural capital.

How can economies grow sustainably? The three scientific laws governing matter and energy and the six factors of sustainability suggest the best long-term solution to environmental problems is to shift away from a high-waste economy and develop a low-waste economy. Another goal would be to reuse, recycle, and compost most of our solid wastes.

The drive to improve environmental quality and work toward environmental sustainability has created major growth industries along with profits and large numbers of new green jobs (Figure 18-5). China, the United States, Denmark, the UK, and Germany lead in wind and solar energy development. In 2022, there were more green jobs available in the United States than the available labor force could supply.

FIGURE 18-3
Ecological Economy From an ecological view, all human economies are subsystems of the biosphere and depend on natural resources and services provided by the sun and Earth.

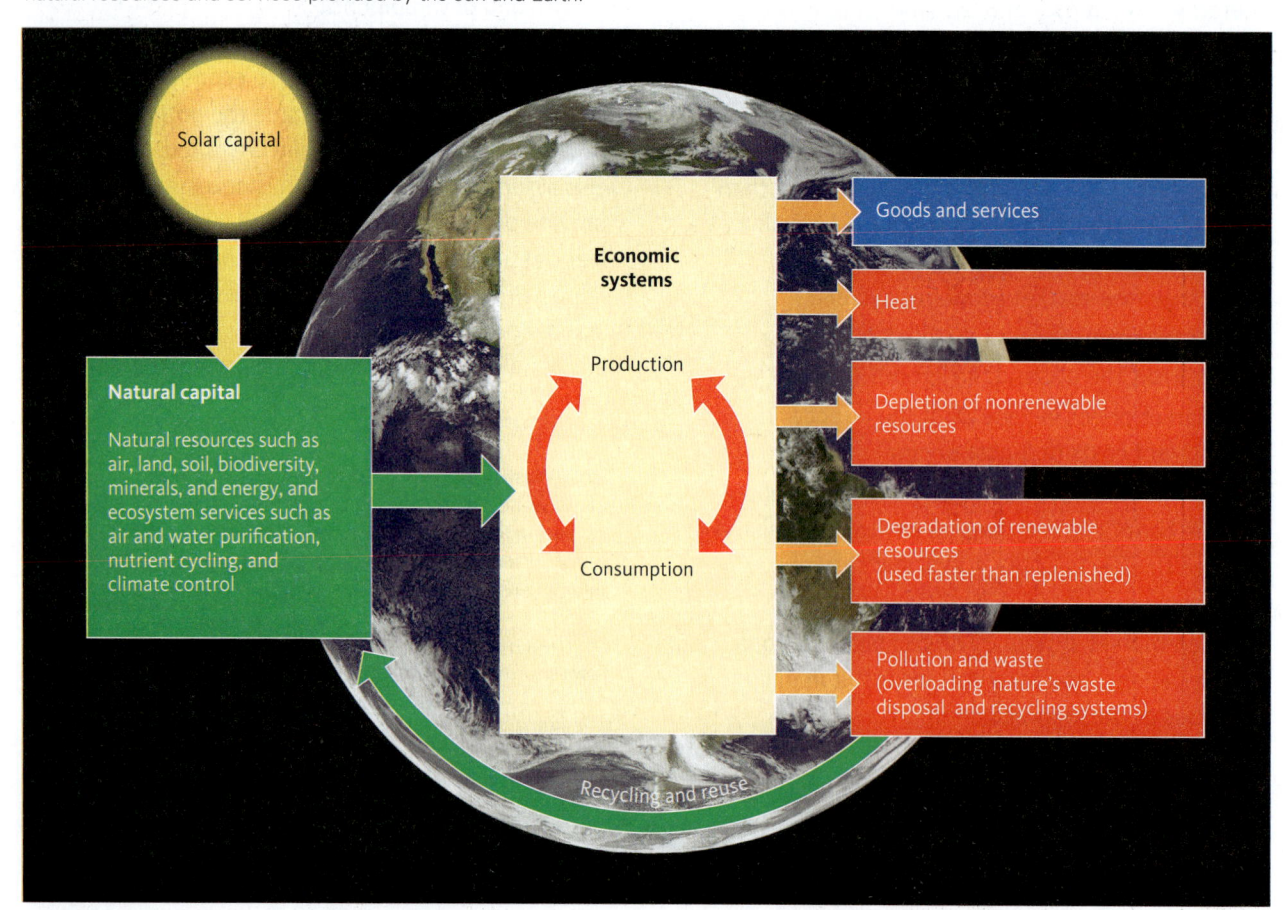

FIGURE 18-4
Biosphere 2, constructed near Tucson, Arizona, was designed to be a self-sustaining life-support system.

SCIENCE FOCUS 18.1

BIOSPHERE 2: A LESSON IN HUMILITY

In 1991, eight scientists were sealed inside Biosphere 2, a $200 million glass and steel enclosure designed to be a self-sustaining life-support system (Figure 18-4). The goal of the project was to increase understanding of Earth's life-support systems.

The sealed system of interconnected domes was built in the desert near Tucson, Arizona. It contained artificial ecosystems including a tropical rain forest, a savanna, a desert, a lake, streams, freshwater and saltwater wetlands, and a mini-ocean with a coral reef.

Biosphere 2 was designed to mimic Earth's natural nutrient cycling systems. Water evaporated from its ocean and other aquatic systems and then condensed to provide rainfall over the tropical rain forest. The precipitation trickled through soil filters into the marshes and back into the mini-ocean before beginning the cycle again.

The facility was stocked with more than 4,000 species of plants and animals, including small primates, chickens, cats, and insects, selected to help maintain life-support functions. Human and animal excrement and other wastes were treated and recycled to help support plant growth. Sunlight and external natural gas-powered generators provided energy. The Biospherians were to be isolated for two years and to raise their own food using intensive organic agriculture. They were to breathe air recirculated by plants and to drink water cleansed by natural nutrient recycling processes.

From the beginning, many unexpected problems cropped up and the life-support system began to unravel. The level of oxygen in the air declined with soil organisms converting it to carbon dioxide. Additional oxygen had to be pumped in from the outside to keep the Biospherians from suffocating. Tropical birds died after the first freeze. An ant species got into the enclosure, proliferated, and killed off most of the system's original insect species. In total, 19 of the Biosphere's 25 small animal species became extinct. Before the two-year period was over, all plant-pollinating insects became extinct, thereby dooming to extinction most of the plant species.

Despite many problems, the facility's waste and wastewater were recycled. With much hard work, the Biospherians were also able to produce 80% of their food supply. However, they suffered from persistent hunger and weight loss.

In the end, an expenditure of $200 million failed to maintain a life-support system for eight people for two years. However, there is a bright side to this experiment: Lessons were learned. Ecologists Joel Cohen and David Tilman, who evaluated the project, concluded, "No one yet knows how to engineer systems that provide humans with life-supporting services that natural ecosystems provide for free."

Thinking Critically
Synthesize What evidence does Biosphere 2 provide in support of an ecological view of natural capital?

FIGURE 18-5

Environmentally Sustainable Businesses and Careers			
Aquaculture	Environmental economics	Fuel-cell technology	Solar-cell technology
Biodiversity protection	Environmental education	Geographic information systems (GIS)	Sustainable agriculture
Biofuels	Environmental engineer	Geothermal geologist	Sustainable forestry
Climate change research	Environmental entrepreneur	Hydrogen energy	Urban gardener
Conservation biology	Environmental health	Hydrologist	Urban planner
Ecotourism management	Environmental writer	Marine science	Waste reduction
Energy-efficient product design	Environmental law	Pollution prevention	Watershed hydrologist
Environmental chemistry	Environmental nanotechnology	Reuse and recycling	Water conservation
Environmental design and architecture	Environmental technologist	Selling services in place of products	Wind energy

Making the shift to more sustainable economies requires governments and industries to increase their spending on research and development—especially in the areas of energy efficiency and renewable energy. The shift toward sustainability requires business leaders and consumers to understand why such a shift is important ecologically and economically.

checkpoint Give two examples of the types of economic growth that an environmental economist would encourage.

18.1 Assessment

1. **Define** Define the three types of capital used by most economies.
2. **Contrast** How do neoclassical economists and ecological economists differ in their views of human economies?
3. **Describe** Describe environmentally sustainable economic development.
4. **Apply** Consider a product you have recently purchased. List the forms of natural, human, and manufactured capital that most likely contributed to the end product.

SCIENCE AND ENGINEERING PRACTICES

5. **Engaging in Argument** Evaluate the following claim, first from a neoclassical economist's perspective and second from an ecological economist's perspective: Freshwater resources can be replaced if they become depleted.

18.2 How Can People Use Economic Tools to Address Environmental Problems?

CORE IDEAS AND SKILLS

- Identify three economic ways in which society can use resources more sustainably.
- Define environmental regulation.
- Describe international goals for poverty reduction and sustainable development.

KEY TERMS

policy   law   regulation

Economic Tools Can Help Society Use Resources More Sustainably

Society can use resources more sustainably by using the tools of full-cost pricing, subsidies for environmentally beneficial goods and services, and taxation of pollution and waste instead of wages and profits.

Full-Cost Pricing Many times the direct price paid for a product or service does not include the indirect or external costs of harm to the environment and human health associated with its production and use. Such costs are often called hidden costs (Figure 18-6).

For example, when someone buys a car, the price includes the direct costs of raw materials, labor, shipping, disposal of solid and hazardous waste, and markup for dealer profit. When using the car, people pay the direct costs for gasoline, maintenance,

FIGURE 18-6 Most of the harmful environmental effects of strip-mining coal and burning it to produce electricity are not included in the cost of electricity.

repairs, and insurance. But the extraction and processing of raw materials uses energy and mineral resources, disturbs land, pollutes the air and water, and releases greenhouse gases into the atmosphere. These are the external costs that have short- or long-term harmful effects on people, economies, and Earth's life-support systems.

Economists and environmental experts call for including these external costs of harm to the environment and human health in the market prices of goods. This practice is called full-cost pricing. According to its proponents, full-cost pricing would reduce resource waste, pollution, and environmental degradation. Putting full-cost pricing into practice would result in some industries and businesses disappearing or remaking themselves. New ones would also appear. This is a normal and revitalizing process in a dynamic and creative capitalist economy. There are three main reasons why full-cost pricing is not used more widely.

First, most producers of harmful products and services would have to charge more for them. This would cause some producers to go out of business. Second, many environmental and health costs are difficult to estimate. Third, many businesses use political and economic power to obtain government subsidies. Subsidies enable them to avoid true free-market competition and retain their economic advantage.

FIGURE 18-7

Environmental Taxes and Fees	
Advantages	**Disadvantages**
They help bring about full-cost pricing.	Low-income groups are penalized unless safety nets are provided.
They encourage businesses to develop environmentally beneficial technologies and goods.	It can be hard to determine optimal level for taxes and fees.
They are easily administered by existing tax agencies.	If they are set too low, wealthy polluters can absorb taxes as costs.

Tradable Environmental Permits	
Advantages	**Disadvantages**
They are flexible and easy to administer.	Wealthy polluters and resource users can buy their way out.
They encourage pollution prevention and waste reduction.	Caps can be too high and not regularly reduced to promote progress.
Permit prices can be determined by market transactions.	Self-monitoring of emissions can allow cheating.

Shift in Subsidies Government subsidies can be beneficial or harmful to the environment. Some subsidies lead to business practices that result in damage to the environment or human health.

FIGURE 18-8
The United States has invested in renewable energy, such as solar energy, as a way to create jobs and stimulate its economy.

Environmental subsidies that end up causing harm to people or the environment are called perverse subsidies. These subsidies cost the world's taxpayers at least $2 trillion a year and cost the average American taxpayer $2,000 per year. Examples of perverse subsidies include tax breaks for extracting minerals and fossil fuels, cutting timber on public lands, irrigating water-thirsty crops, and overfishing commercially valuable aquatic species.

Environmental scientists and economists call for phasing out environmentally perverse subsidies and tax breaks. They encourage phasing in subsidies and tax breaks for environmentally beneficial businesses. These include businesses involved in pollution prevention, waste prevention, sustainable forestry and agriculture, conservation of water supplies, energy efficiency improvements, renewable energy use, and measures to slow projected climate change. A subsidy shift from environmentally harmful to environmentally beneficial subsidies would increase our beneficial environmental impact. Countries including Japan, France, Belgium, as well as the United States have started making such shifts.

Taxing Pollution and Waste One way to discourage pollution and resource waste is to tax it. Governments could levy green taxes on the amount of pollution and hazardous waste produced by farms, businesses, or industries. They could add green taxes to the use of fossil fuels, nitrogen fertilizer, timber, minerals, water, and other resources. This approach would implement the full-cost pricing factor of sustainability and promote a beneficial environmental impact (Figure 18-7).

An environmentally sustainable economic and political system would lower taxes on labor, income, and wealth, and raise taxes on harmful environmental activities. Twenty-five hundred economists, including eight Nobel Prize winners in economics, have endorsed this tax-shifting concept. Proponents list these requirements for the successful implementation of green taxes:

- Phase in taxes over 10–20 years to allow businesses to plan.
- Reduce other taxes by an amount equal to that of the green tax so there is no net increase in taxes.
- Design a safety net for poor and lower-middle class families who would suffer financially from any new taxes on essentials such as fuel, water, electricity, and food.

In Europe and the United States, polls indicate that when tax shifting is explained to voters, 70% of them support the idea. Germany introduced a green tax on fossil fuels in 1999. Germany's green tax reduced pollution and greenhouse gas emissions, created up to 250,000 new jobs, lowered taxes on wages, and greatly increased the use of renewable energy resources. Costa Rica, Sweden, Denmark, Spain, and the Netherlands have raised taxes on several environmentally harmful activities while cutting taxes on income, wages, or both.

checkpoint How can green taxes be used to implement full-cost pricing?

Environmental Laws and Regulations Can Discourage or Encourage Innovation

A major function of the federal government in democratic countries is to develop and implement **policies** for dealing with various issues. The important components of policies are the **laws** passed by the legislative branch, **regulations** instituted by the agencies of the executive branch to put laws and programs into effect, and funding approved by Congress and the president to finance the programs and implement and enforce the laws and regulations (Figure 18-8).

Environmental regulation is a form of government intervention in the marketplace that promotes sustainable resource use. It is used to control or prevent pollution and environmental degradation and encourage efficient use of resources. Regulation involves enacting and enforcing laws that set pollution standards. It addresses the release of toxic chemicals. It also protects slowly replenished resources such as public forests, parks, and wilderness areas from unsustainable use.

checkpoint What is the difference between laws and regulations?

Reducing Poverty Helps the Environment

Poverty is a condition in which people lack money to fulfill their basic needs for food, water, shelter, health care, and education. According to the World Bank, poverty is the way of life for hundreds of millions of people around the world. They struggle to survive on incomes equivalent to less than $1.90 per day or no income at all. Poverty also is responsible for severe health effects such as hunger, malnutrition, infectious disease, and a shorter life span. To reduce poverty, governments, businesses, international lending agencies, and wealthy individuals can undertake the following:

- Provide universal primary school education for all children (Figure 18-9) and the world's nearly 800 million illiterate adults.

FIGURE 18-9
Rwanda's Ministry of Education has been a model in promoting gender equality and high enrollment rates in primary school. The country was able to obtain enrollemnt rates of 95% for females and 92% for males, among the highest rates in Africa.

LESSON 18.2   623

- Mount a massive global effort to combat malnutrition and the infectious diseases that kill millions of people.
- Help less-developed countries to reduce their population growth, mostly by investing in family planning, reducing poverty, and elevating the social and economic status of women.
- Focus on sharply reducing the total and per capita ecological footprints of more-developed countries such as the United States and rapidly growing less-developed countries such as China.
- Make large investments in small-scale infrastructure such as solar-cell power facilities for rural villages and sustainable agriculture projects to help less-developed nations work toward more energy-efficient and environmentally sustainable economies.
- Encourage lending agencies to make small loans to impoverished people who want to increase their income.

For over three decades, an innovation called microlending, or microfinance, has helped people living in poverty deal with the lack of credit and assets. The Grameen Bank in Bangladesh is essentially owned and run by borrowers and the Bangladeshi government. Since it was founded, the bank has provided more than $8 billion in microloans of $50 to $500 at low interest rates to 7.6 million impoverished people in Bangladesh who do not qualify for loans at traditional banks.

Most microloans are used by women to start small businesses. Microloans are used to develop day-care centers, health-care clinics, reforestation projects, drinking water supply projects, literacy programs, and small-scale solar- and wind-power systems in rural villages. The average repayment rate on microloans has been 95% or higher. That is nearly twice the average repayment rate for loans by conventional commercial banks. Banks based on the Grameen microlending model have spread to 58 countries, including the United States, with an estimated 500 million participants.

checkpoint How can microlending benefit the environment?

FIGURE 18-10

Millennium Development Goals Significant progress has been made toward achieving several of the world's Millennium Development Goals.

Goal	Progress
Eliminate extreme poverty and hunger.	In 2000, 50% of the developing world's population lived in extreme poverty (earning less than $1.25 a day). By 2015, 14% remained in extreme poverty.
Achieve universal primary education.	The primary school enrollment rate was raised to 91% in less-developed countries.
Promote gender equality and empower women.	Gender disparity in less-developed regions for primary, secondary, and tertiary education achieved the target goals. In government, women only hold 20% of positions.
Reduce child mortality.	The global mortality rate for children under the age of five was reduced by more than 50%.
Improve the health of pregnant women.	From 1990 to 2015, the maternal mortality ratio declined by 45% worldwide.
Combat HIV/AIDS, malaria, and other diseases.	Between 2000 and 2013, new HIV infections dropped by 40%. Between 1995 and 2013, antiretroviral therapy prevented 7.6 million deaths from AIDS. Globally, the incidence rate of malaria fell by 37% and deaths from malaria dropped by 58% between 2000 and 2015.
Ensure environmental sustainability.	From 1990 through 2015, 98% of ozone-depleting chemicals were eliminated and the ozone layer should be repaired by 2050. The drinking water goal set in 2000 has been met by 147 countries. The sanitation target was met by 95 countries. Between 2000 and 2014, the population living in urban slums in the developing regions fell by 10%.
Establish a global partnership for development.	Between 2000 and 2014, development assistance from developed nations increased by 66% ($135.2 billion). Ninety-five percent of the planet's population is covered by a cellular signal. Internet penetration grew from 6% of the global population in 2000 to 43% in 2015.

Nations Can Work Together to Set Goals

In 2000, the world's nations set goals called the Millennium Development Goals (MDGs). These goals aimed to sharply reduce hunger and poverty, improve health care, achieve universal primary education, empower women, and move toward environmental sustainability by 2015. More-developed countries pledged to donate 0.7%—or $7 of every $1,000—of their annual national income to less-developed countries to help them achieve these goals. The progress toward these goals as of 2015 can be seen in Figure 18-10.

In 2015, the United Nations established a new set of interlinked global goals to build upon the original eight MDGs. These 17 new goals, known as the Sustainable Development Goals (SDGs), are intended to be achieved by 2030. While some progress toward reaching the SDGs has been made, the 2019 UN Global Sustainable Development Report stated that "the world is not on track for achieving most of the 169 targets that comprise the Goals." Critics of the SDGs say that there are too many and sometimes competing goals, and not enough focus on prioritizing the most urgent goals. The global COVID-19 pandemic has also made the achievement of the SDGs within the established timeframe more difficult to accomplish.

checkpoint Which Millennium Development Goals have seen the most progress?

18.2 Assessment

1. **Restate** Describe three ways in which society can use resources more sustainably.
2. **Identify** Give an example of a perverse subsidy.
3. **Explain** How can environmental regulation involve the marketplace?
4. **Draw a Conclusion** Draw a conclusion about the Millennium Development Goals.
5. **Make Judgments** Do you favor phasing out environmentally harmful government subsidies and tax breaks, and phasing in environmentally beneficially ones? Explain. How might such subsidy shifting affect your lifestyle?

SCIENCE AND ENGINEERING PRACTICES
6. **Asking Questions** Write three questions that could help you assess the full-cost price of electricity for your residence.

18.3 How Can Society Enact More Just Environmental Policies?

CORE IDEAS AND SKILLS
- Summarize the politics of environmental law.
- Describe how public land can be protected.
- Identify seven guiding principles of the environmental justice movement.
- Recognize the influence citizens have on environmental policies.

KEY TERMS

politics
lobby

environmental justice

Developing Environmental Policy Is Not Easy

The roles a government plays are determined largely by its policies—the set of laws and regulations it enforces and the programs it funds. **Politics** is the process by which people try to influence or control the policies and actions of governments at local, state, national, and international levels. One important application of this process is the development of environmental policy. This includes the environmental laws, regulations, and programs designed, implemented, and enforced by one or more government agencies.

The United States has a representative democracy. In such a democracy, the government is run by the people through elected officials and representatives. The ideals for a representative democracy are usually embodied in a constitution. This is a document that provides the basis of government authority and limits government power by mandating free elections and the right of free speech. The major function of government in democratic countries is to develop and implement policies for dealing with various issues.

In passing laws, developing budgets, and formulating regulations, government officials must deal with pressure from competing special-interest groups. Each of these groups pushes for the passage of laws, subsidies or tax breaks, or regulations favorable to its cause. Special-interest groups also seek to weaken or repeal laws, subsidies, tax breaks, and regulations unfavorable to their positions.

LESSON 18.3 625

FIGURE 18-11

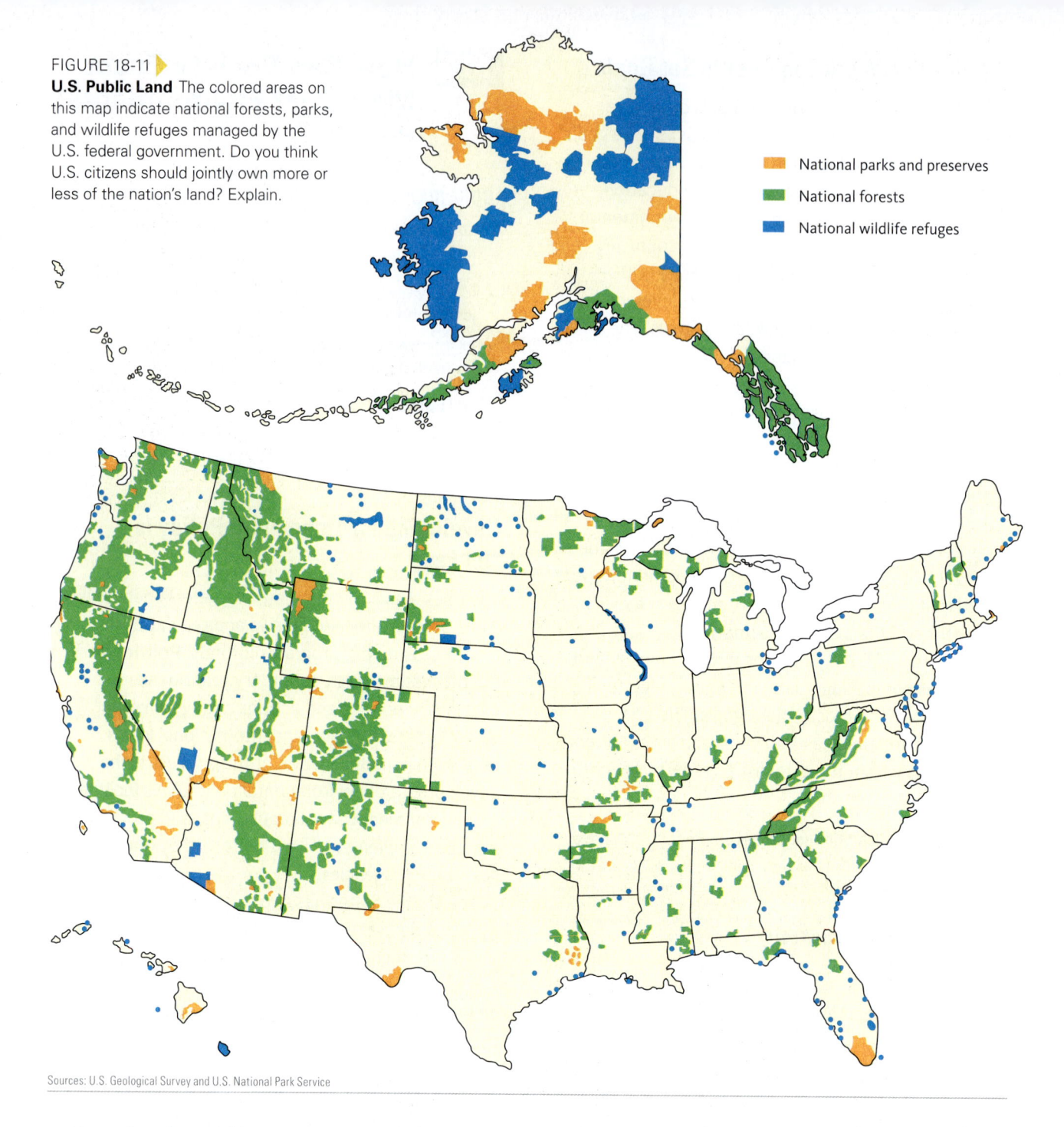

U.S. Public Land The colored areas on this map indicate national forests, parks, and wildlife refuges managed by the U.S. federal government. Do you think U.S. citizens should jointly own more or less of the nation's land? Explain.

- National parks and preserves
- National forests
- National wildlife refuges

Sources: U.S. Geological Survey and U.S. National Park Service

Examples of special-interest groups include:
- Profit-making organizations, such as corporations.
- Nongovernmental organizations (NGOs), most of which are nonprofit organizations such as environmental groups.
- Labor unions representing the interests of workers.
- Trade associations representing various industries.

Passing a law is a complex process. An important factor in this process is lobbying. To **lobby**, an individual or group contacts legislators in person or hires lobbyists to do so, in order to persuade legislators to vote or act in their favor. The opportunity to lobby elected representatives is an important right for everyone in a democracy. However, some critics of the American system believe lobbyists of large corporations have grown too powerful and that their influence overshadows the input of ordinary citizens.

With each group pushing its own agenda, it is difficult for policy makers to see the whole picture. Most politicians tend to focus on short-term, isolated issues rather than on long-term, complex problems. Also, political leaders have hundreds of issues to deal with and are not always aware of how Earth's natural systems work and how those systems support all life, economies, and societies. This can make enacting environmental regulation difficult.

checkpoint What do lobbyists do?

Governments Can Set Aside Land

The U.S. government has set aside much of its land for public use, resource extraction, recreation, or wildlife habitat. About 35% of the country's land is jointly owned by all U.S. citizens and managed by the federal government (Figure 18-11).

The National Forest System consists of 154 national forests and 20 national grasslands. These lands, managed by the U.S. Forest Service (USFS), are used for logging, mining, livestock grazing, farming, oil and gas extraction, recreation, and conservation of watershed, soil, and wildlife resources.

The Bureau of Land Management (BLM) handles 40% of all land managed by the federal government and 13% of the total U.S. land surface. These lands are used mainly for mining, oil and gas extraction, and livestock grazing.

The U.S. Fish and Wildlife Service (USFWS) supervises 560 National Wildlife Refuges. Most refuges protect habitats and breeding areas for waterfowl and big game to provide a harvestable supply for hunters. Approved activities in most refuges include hunting, trapping, fishing, oil and gas development, mining, logging, grazing, farming, and some military activities.

The uses of other public lands are more restricted. The National Park System, under the authority of the National Park Service (NPS), includes 63 major national parks (Figure 18-12) and more than 350 national recreation areas, memorials, battlefields, historic sites, parkways, trails, rivers, seashores, lakeshores, and monuments. Only camping, hiking, sport fishing, and boating can take place in the national parks. Sport hunting, mining, and oil and gas drilling are allowed in national recreation areas.

The most restricted public lands are 803 roadless areas that make up the National Wilderness Preservation System. These areas lie within other public lands and are managed by the agencies in charge of the surrounding lands. Most of these areas are open for recreational activities such as hiking, sport fishing, camping, and non-motorized boating.

Many federal public lands contain valuable oil, natural gas, coal, geothermal, timber, and mineral resources. Since the 1800s, there has been intense controversy over how to use and manage the resources on these public lands.

Most conservation biologists, environmental economists, and many free-market economists believe these principles should govern the use of public lands:

- Protect biodiversity, wildlife habitats, and ecosystems as the top priority.
- Do not provide government subsidies and tax breaks for using or extracting resources on public lands.
- Reimburse the American people with fair compensation for the use of their property.
- Hold all users or extractors of resources on public lands fully responsible for any environmental damage they cause.

There is strong and effective opposition to these ideas. Developers, resource extractors, many economists, and citizens tend to view public lands in terms of their usefulness in providing mineral, timber, and other resources and increasing short-term economic growth. They have succeeded in blocking implementation of the four principles listed above. In recent years, analyses of budgets reveal the government has spent an average of $1 billion per year on subsidies and tax breaks for privately owned interests that use U.S. public lands for mining, fossil fuel extraction, logging, and livestock grazing.

Some developers and resource extractors have sought to go further in opening up more federal lands for economic development. Here are five proposals presented to Congress:

1. Sell public lands or their resources to corporations or individuals. Or, turn over their management to state and local governments.
2. Slash federal funding for enforcement of regulations related to public lands.
3. Cut diverse old-growth forests in the national forestlands for timber and biofuels. Replace them with tree plantations.
4. Open national parks, national wildlife refuges, and wilderness areas to oil drilling, mining, off-road vehicles, and commercial development.

FIGURE 18-12
ON ASSIGNMENT National Geographic Photographer Raul Touzon shot this portrait of El Capitan spotlighted by the last light of sun at Yosemite National Park, California. Carved by ancient glaciers, this formation of granite rises 900 meters (2,900 feet) above the valley below. Yosemite was first protected in 1864 and has been called a "shrine to human foresight."

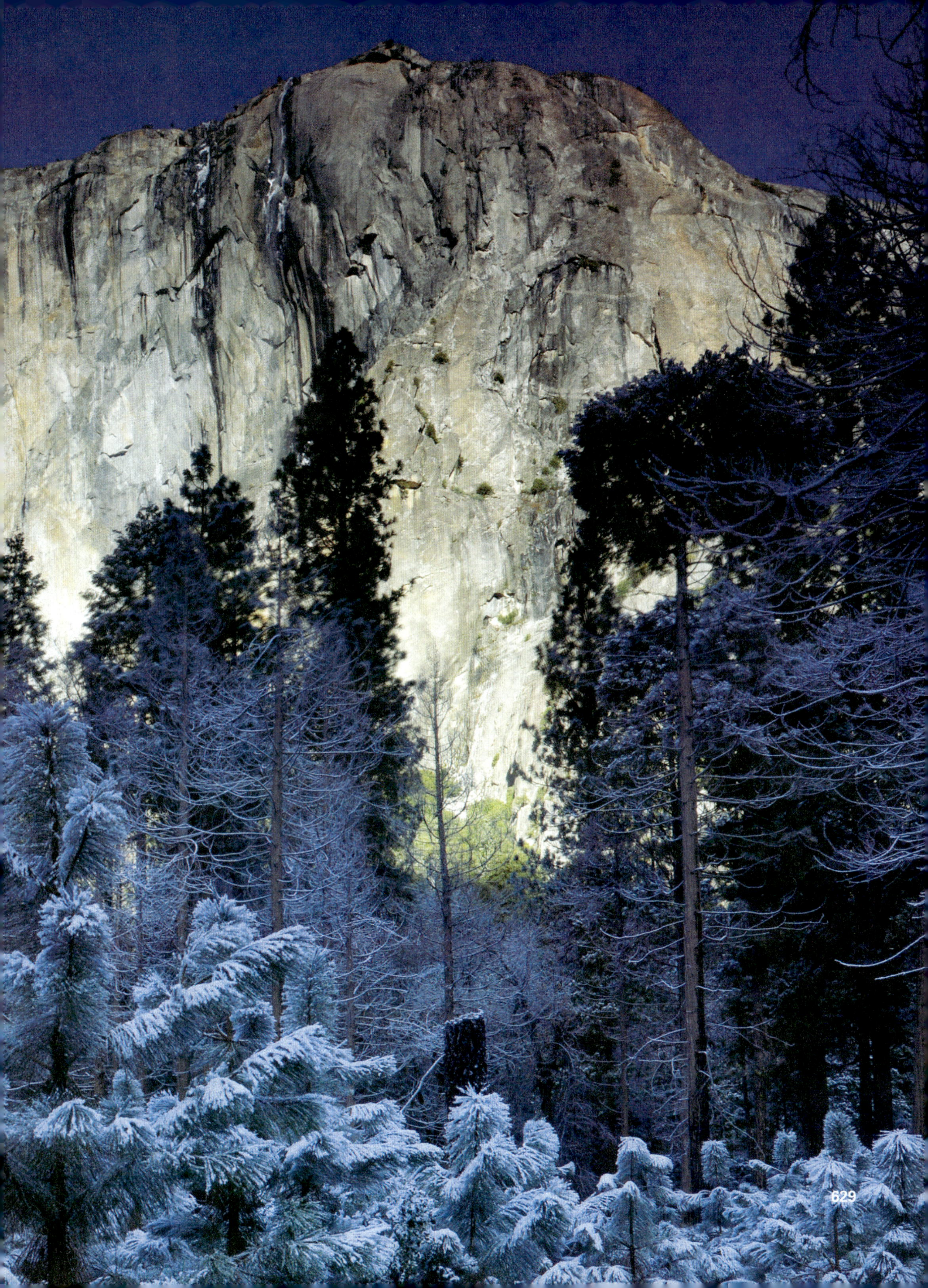

5. Eliminate or take regulatory control away from the National Park Service. Launch a 20-year construction program in the parks to build new concessions and theme parks run by private firms.

checkpoint Which economic view is reflected in the proposals given for opening the use of public lands?

Environmental Policy Considerations

The ideal whereby every person is entitled to protection from environmental hazards regardless of race, gender, age, national origin, income, or social class is called **environmental justice**. Studies show that an uneven share of factories, hazardous waste dumps, incinerators, and landfills in the United States are located in communities populated mostly by minority populations. Other research shows toxic waste sites in predominately white communities are restored faster and more completely than similar sites in minority communities. In China, the problem is worse. Because of intolerable pollution in urban areas, many factories in China are being moved to the countryside. Rates of cancer and other serious human illnesses are rising sharply in these rural areas.

Environmental discrimination in many parts of the world led to a growing grassroots effort known as the environmental justice movement. Supporters of this movement pressure governments, businesses, and environmental organizations to become aware of environmental injustice and act to prevent it. They have made progress toward their goals, but more needs to be done.

Politicians and business representatives suggest economics should be the main factor in deciding where to locate new power plants, freeways, landfills, incinerators, and other potentially disruptive facilities. By locating in undesirable areas, building costs are significantly reduced. Often, these areas are home to low-income residents who have much less political power than developers and corporations. Many analysts argue that environmental justice should carry as much weight as economic factors in such decisions. This political struggle remains unresolved in many areas of the world.

Guiding Principles Several principles can help legislators and individuals evaluate existing or proposed environmental policies:

- *Reversibility principle* Avoid making decisions that cannot be reversed later if they are found to be harmful. Two irreversible actions are the production of toxic coal ash from coal-burning power plants and the production of deadly radioactive wastes from the nuclear power fuel cycle. In both cases, the hazardous wastes must be stored safely for thousands of years.
- *Net energy principle* Prohibit widespread use of energy resources and technologies with low or negative net energy that need subsidies and tax breaks to compete. Examples include nuclear power generation, tar sands, shale oil, ethanol made from corn, and hydrogen fuel.
- *Precautionary principle* When evidence indicates that an activity threatens human health or the environment, take measures to prevent or reduce such harm, even if the evidence is not conclusive.
- *Prevention principle* Make decisions that prevent a problem from occurring or becoming worse.
- *Polluter-pays principle* Use economic tools such as green taxes to ensure polluters bear the costs of dealing with the pollutants and wastes they produce. This stimulates the development of innovative ways to reduce and prevent pollution and wastes. This is in accordance with the full-cost pricing factor of sustainability.
- *Environmental justice principle* Do not allow any group of people to bear an unfair share of the burden created by pollution, environmental degradation, or the execution of environmental laws. This ethical principle also addresses environmental injustices committed by one generation and affecting future generations. Thus, it echoes the ethics factor of sustainability.
- *Holistic principle* Focus on long-term solutions that address root causes of interconnected problems instead of focusing on short-term and often ineffective fixes that treat each problem separately.

Implementing such principles is not easy. It requires policy makers throughout the world to become more environmentally literate. It also requires robust debate among politicians and citizens, mutual respect for diverse beliefs, and a dedication to implementing the win-win factor of sustainability. It requires openness, inclusiveness, innovation, and compromise among political players and other people with divergent views.

checkpoint Summarize the guiding principles of environmental policy evaluation.

You Can Make a Difference

A major theme of this book is that individuals matter. History shows that significant change usually comes from the bottom up when individuals join together to bring about change. Without this grassroots pressure—political action by individual citizens and organized citizen groups—the environment would be more polluted and more of Earth's biodiversity would have been destroyed.

With the growth of the Internet, digital technology, and social media, the number of citizens' groups, national and global action networks, and NGOs focused on environmental and other problems has grown rapidly. People have improved environmental quality in their own neighborhoods, schools, or workplaces. They also inspire actions on regional, national, and global stages. Figure 18-13 lists ways individuals living in democracies can influence government policies.

You can provide environmental leadership in a variety of ways. You can lead by example, using your own lifestyle and values to show others that change is possible. You can use fewer disposable products and eat sustainably raised food. You can walk, ride a bike, or take mass transit to get around. You can reuse and recycle many items and reduce your consumption of goods. You can adjust your lifestyle to reduce your carbon footprint. In addition, you can offset part of your carbon footprint by finding ways to reduce CO_2 in the atmosphere. You can call attention to issues that matter to you by participating in public demonstrations (Figure 18-14).

FIGURE 18-13

What You Can Do
Become informed on environmental issues.
Campaign and vote for informed and environmentally literate candidates.
Write letters to your local school and elected representatives with well-researched recommendations.
Volunteer or intern with a political office that is addressing environmental concerns.
Join nongovernmental organizations (NGOs).
Vote with your wallet. Research the products you buy.
Sign petitions to oppose projects that degrade natural capital for short-term gain.

CONSIDER THIS

A critical mass is all it takes to create change.

Environmentally active citizens and leaders are motivated by two important findings. First, research by social scientists indicates that social change requires active support by only 5–10% of the population. This small percentage is often enough to lead to a political tipping point. Second, experience has shown that reaching such a critical mass can bring about social change much faster than most people think.

U.S. Environmental Laws Concerned citizens have persuaded the U.S. Congress to enact a number of environmental and resource protection laws. Most of them were enacted in the 1970s (Figure 18-16).

U.S. environmental laws have been highly effective, especially in controlling pollution. However, since 1980, a well-organized and well-funded movement has mounted a strong campaign to weaken or repeal existing laws and regulations and change the ways public lands are used. Three major groups strongly opposed to U.S. environmental laws and regulations are:

- Corporate leaders and owners who see regulations as threats to their profits, wealth, and power.
- Citizens who view laws as threatening to their private property rights and jobs.
- State and local government officials who resent having to implement state and federal rules with little or no federal funding, or who disagree with specific regulations.

One major problem working against additional regulations is that the focus of environmental issues has shifted from easy-to-see dirty smokestacks and filthy rivers to complex, long-term, and less visible environmental problems. These include climate change, biodiversity loss, and groundwater pollution. Explaining complex issues to the public and mobilizing support for controversial, long-range solutions to such problems is difficult. Efforts to weaken U.S. environmental laws and regulations has escalated since 2000. However, independent polls show more than 80% of the U.S. public strongly supports environmental laws and regulations.

FIGURE 18-14 More than 400,000 people attended the People's Climate March in New York City to show their support for a 2014 meeting of world leaders on climate change.

One area of current conflict is with the hydraulic fracturing (fracking) industry. Fracking companies currently operate without adequate federal environmental regulations to protect the air, water, and land. Citizens negatively affected by this industry are calling for better protection. Several states have responded to this public pressure by banning fracking until more information is gathered on its environmental impact.

Citizen Environmental Groups The spearheads of the global conservation, environmental, and environmental justice movements are the tens of thousands of nongovernmental organizations (NGOs). These organizations work at the international, national, state, and local levels. The growing influence of NGOs is one of the most important factors in forging environmental decisions and policies.

NGOs range in size from grassroots groups with just a few members to established international organizations. The World Wildlife Fund (WWF) is a 5-million-member global conservation organization that operates in nearly 100 countries, with 1.3 million members in the United States. Other international groups with large memberships include The Nature Conservancy, Conservation International, and the Natural Resources Defense Council (NRDC).

In the United States, more than 8 million citizens belong to more than 30,000 NGOs that deal with environmental issues. Some of the largest U.S. environmental groups include the Sierra Club, the National Wildlife Federation, and the Audubon Society. NGOs are powerful and important forces within political systems. Taken together, a worldwide network of grassroots NGOs is creating bottom-up political, social, economic, and environmental change around the world.

For example, when the Mexican government planned a massive salt plant near Laguna San Ignacio in Baja California Sur, millions of citizens voiced their concern about the impact it would have on the environment. The lagoon provides critical habitat for diverse marine species and is the only remaining breeding site for the eastern Pacific gray whale (Figure 18-17). Pressure from the NRDC, the International Fund for Animal Welfare, and others convinced government officials to amend their plans, preserving the lagoon and saving the gray whale from the brink of extinction. Laguna San Ignacio is now a protected World Heritage Site and part of the largest wildlife refuge in Latin America.

Hundreds of campus environmental groups and many high school environmental groups are also leading the way to make their schools and communities more sustainable.

NATIONAL GEOGRAPHIC | EXPLORERS AT WORK
Jane Goodall Primatologist

Jane Goodall is a pioneering primatologist and conservationist with a Ph.D. from Cambridge University. She is also a National Geographic Explorer-in-Residence Emerita. At age 26, Dr. Goodall began a more-than-50-year career of studying chimpanzee social and family life in the Gombe Stream Game Reserve in what is now the country of Tanzania.

One of Goodall's major scientific discoveries was that chimpanzees make and use tools. She observed some chimpanzees modifying twigs or blades of grass and then poking them into termite mounds. When the termites latched on to these primitive tools, the chimpanzees pulled them out and ate the termites. Goodall and several other scientists have also observed that chimpanzees, including captive chimpanzees, can learn simple sign language, do simple arithmetic, play computer games, develop relationships, and worry about and protect one another.

In 1977, Goodall established the Jane Goodall Institute, an NGO that works to preserve great ape populations and their habitats. And in 1991, Goodall started Roots and Shoots, an environmental education program that is active in more than 130 countries (Figure 18-15).

Goodall has received many awards and prizes for her scientific contributions and conservation efforts. She has written 27 books for adults and children and has been involved with more than 15 films about the lives and importance of chimpanzees.

Goodall spends nearly 300 days a year traveling and educating people throughout the world about chimpanzees and the need to protect the environment. She says, "I can't slow down... if we're not raising new generations to be better stewards of the environment, what's the point?"

FIGURE 18-15
Jane Goodall's Roots and Shoots program empowers young people to take action in their community.

LESSON 18.3

FIGURE 18-16

U.S. Environmental Laws Most of the environmental laws in the United States were enacted between 1969 and 1980.

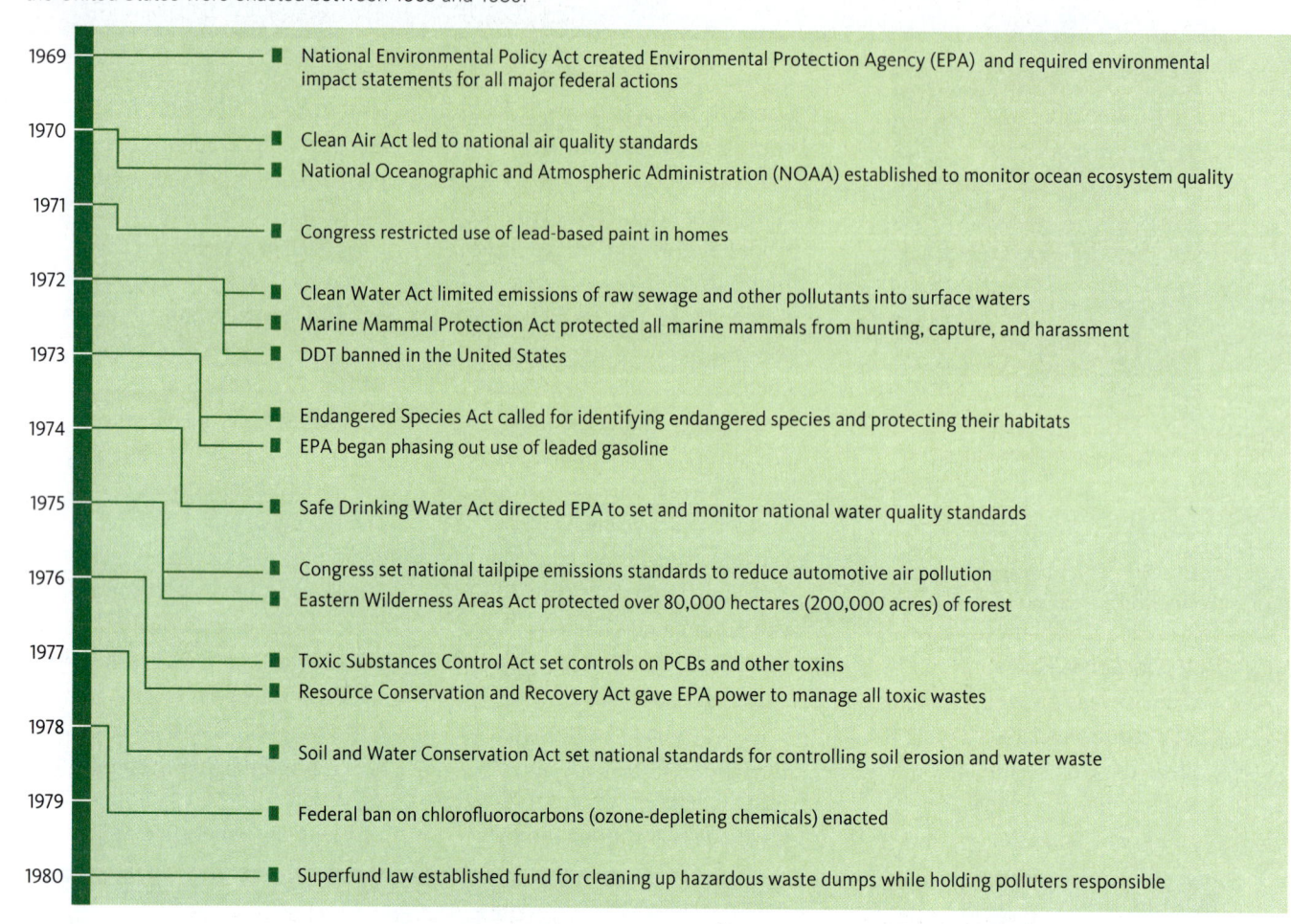

1969	National Environmental Policy Act created Environmental Protection Agency (EPA) and required environmental impact statements for all major federal actions
1970	Clean Air Act led to national air quality standards
	National Oceanographic and Atmospheric Administration (NOAA) established to monitor ocean ecosystem quality
1971	Congress restricted use of lead-based paint in homes
1972	Clean Water Act limited emissions of raw sewage and other pollutants into surface waters
	Marine Mammal Protection Act protected all marine mammals from hunting, capture, and harassment
1973	DDT banned in the United States
1974	Endangered Species Act called for identifying endangered species and protecting their habitats
	EPA began phasing out use of leaded gasoline
1975	Safe Drinking Water Act directed EPA to set and monitor national water quality standards
1976	Congress set national tailpipe emissions standards to reduce automotive air pollution
	Eastern Wilderness Areas Act protected over 80,000 hectares (200,000 acres) of forest
1977	Toxic Substances Control Act set controls on PCBs and other toxins
	Resource Conservation and Recovery Act gave EPA power to manage all toxic wastes
1978	Soil and Water Conservation Act set national standards for controlling soil erosion and water waste
1979	Federal ban on chlorofluorocarbons (ozone-depleting chemicals) enacted
1980	Superfund law established fund for cleaning up hazardous waste dumps while holding polluters responsible

Student groups conduct environmental audits of their campuses or schools. They propose changes to make their campuses or schools more environmentally sustainable while saving money in the process. Such audits focus on improving recycling programs, buying local organic foods, shifting to renewable energy sources, improving the energy efficiency of buildings, and implementing environmental curricula.

checkpoint What influence have grassroots organizations had on developing environmental policy?

18.3 Assessment

1. **Explain** How are policies affected by politics? Give an example.
2. **Relate** Explain how the four principles of public land use in the United States reflect an environmental economist view.
3. **Recall** Describe five ways in which citizens can affect environmental policies.
4. **Make Judgments** Explain why you agree or disagree with each of the seven guiding principles that are recommended by analysts for use in making environmental policy decisions. Which three of these principles do you think are the most important? Why?

SCIENCE AND ENGINEERING PRACTICES

5. **Obtaining and Evaluating Information** Read and summarize two news articles about the same environmental policy that are written from two different viewpoints. Make inferences about the authors' economic views.

FIGURE 18-17
Citizens working through NGOs saved the intelligent and ancient Pacific gray whale by preserving its breeding site.

18.4 How Can Society Live More Sustainably?

CORE IDEAS AND SKILLS
- Describe environmental literacy.
- Identify ways society can live more sustainably.
- Revisit the concept of environmental worldviews.

KEY TERM
environmental literacy

Society Can Increase Its Environmental Literacy

Human populations are degrading the very Earth systems on which they and other species depend. Up to half of the world's species may be wiped out before the century's close. Part of the problem stems from an incomplete understanding of how Earth's life-support system works and how human actions affect it. Improving this understanding begins by grasping three important ideas that form the foundation of **environmental literacy**:

- Natural capital matters because it supports Earth's life and human economies.
- Human ecological footprints are immense and are expanding rapidly.
- Once ecological tipping points are reached, neither wealth nor technology can resolve the consequences that could last for hundreds to thousands of years.

Learning how to live more sustainably requires a foundation of environmental education with the goal of producing environmentally literate citizens.

Learning from Earth Formal environmental education is important, but is it enough? Many analysts say *no*. They call for people to appreciate not only the economic value of nature, but also its ecological, aesthetic, and spiritual values. To these analysts, the problem is not just a lack of environmental literacy but also a lack of intimate contact with nature. People are spending more time connected to electronics and less time outdoors. This lack of connection with nature can reduce people's ability to act responsibly toward Earth.

LESSON 18.4   635

A growing chorus of analysts suggests that people have much to learn from nature. They call for people to acquire a sense of awe, wonder, mystery, excitement, and humility by standing under the stars, exploring a forest, taking in the majesty of the sea, or enjoying a beautiful scene in nature (Figure 18-18). You might pick up a handful of topsoil and try to sense the teeming microscopic life within it that helps to keep people alive by supporting food production. You might look at a tree, a mountain, a rock, or a bee, or listen to the sound of a bird and sense how each of them is connected to you and you to them, through Earth's life-sustaining processes.

Direct experiences with nature can reveal parts of the complex web of life that cannot be bought, recreated with technology, or reproduced with genetic engineering. Understanding and directly experiencing the precious and free gifts we receive from nature can help people to make an ethical commitment to live more sustainably on this planet.

Living Simply and Lightly On a timescale of hundreds of thousands to millions of years, Earth is very resilient and can survive environmental wounds. However, scientists warn that if people continue on their current path, it is very likely that they will be living on a changed planet. Earth is heading toward harsher climates, less dependable supplies of water, more acidic oceans, extensive soil degradation, continued mass extinction, degradation of key ecosystem services, and widespread ecological and economic disruption.

Cultures of excessive consumption feed into these problems. Today's advertising messages encourage people to buy more things as a way to achieve happiness. As American humorist and writer Mark Twain (1835–1910) observed: "Civilization is the limitless multiplication of unnecessary necessities." Not everyone agrees with this. Some people are adopting a lifestyle of voluntary simplicity. This involves living with fewer possessions and using products and services with smaller harmful environmental impacts.

Living simply starts with asking the question: What do I really need? This is not an easy question to answer. People in affluent societies are conditioned to view excessive material possessions as needs. Many people have become addicted to buying more material goods as a way to find meaning in their lives.

Environmental ethicists developed a number of ethical guidelines for living more sustainably by converting environmental concerns, literacy, and wisdom into environmentally responsible actions. Here are five of those guidelines:

- Mimic the ways nature sustains itself, and consider the effects certain activities have on other people and other life forms.

FIGURE 18-18
Kayakers connect with nature on Lake Tahoe in Nevada.

- Protect Earth's natural capital and repair ecological damage human activities have caused.
- Use matter and energy resources as efficiently as possible.
- Celebrate and protect biodiversity.
- Leave Earth in a condition that is as good as what we have enjoyed, or better.

checkpoint What does it mean to improve environmental literacy?

Environmental Worldviews Revisited

People disagree about how serious our environmental problems are, as well as what we should do about them. As discussed in Chapter 1, these conflicts arise mostly from differing environmental worldviews—ways of thinking about how the world works and beliefs that people hold about their roles in the natural world. People with different environmental worldviews can study the same data, be logically consistent in their analysis of those data, and arrive at quite different conclusions. Why? Because each begins with different assumptions and values.

Human-centered worldviews focus primarily on the needs and wants of people. One such worldview, called the planetary management worldview, sees humans as the planet's most important species. It believes the human species can and should dominate and manage Earth. This view holds that other species and parts of nature should be evaluated according to how useful they are to humans. Another human-centered worldview is the stewardship worldview. This view assumes people have an ethical responsibility to be responsible managers, or stewards, of Earth. According to the stewardship view, when people use Earth's natural capital, they are borrowing from Earth and future generations.

Life-centered worldviews hold that all forms of life have value as participating members of the biosphere, regardless of their potential or actual use to humans. Eventually all species become extinct. However, most people who hold a life-centered worldview believe humans have an ethical responsibility to avoid hastening the extinction of any species, for two reasons. One is that each species is a unique part of the genetic storehouse that helps Earth's life to continue by responding to changes in environmental conditions. Another reason is that every species has the potential for providing economic benefits through its participation in providing ecosystem services.

Earth-centered worldviews take this view and expand it to include the entire biosphere, especially ecosystems. In an Earth-centered worldview, humans are part of, not apart from, Earth, and are utterly dependent on Earth's natural capital.

The Sustainability Revolution The Industrial Revolution, which began around the mid 18th century, has been a remarkable global transformation. Now environmental leaders say it is time for another global transformation—a sustainability revolution. Society is beginning to make the shift toward increasing energy efficiency, relying more on renewable energy resources, and preserving forest, ocean, and freshwater resources. People are working toward stabilizing climate, producing food more sustainably, and minimizing waste.

History shows that people can bring about change faster than you might think, once they are willing to leave behind ideas and practices that no longer work. The key to a sustainability revolution is the understanding that individuals matter. Each and every one of your choices and actions makes a difference. Virtually all environmental progress has been made because individuals banded together to insist that humans can do better. This is an exciting time to be alive, as many people's worldviews are expanding, and people are learning to live sustainably within our home, the biosphere.

checkpoint What is a sustainability revolution?

CONSIDER THIS

Earth provides enough.

Indian philosopher and leader Mahatma Gandhi's principle of "enoughness" echoes the living simply ethic:

"The Earth provides enough to satisfy every person's need but not every person's greed… When we take more than we need, we are simply taking from each other, borrowing from the future, or destroying the environment and other species."

18.4 Assessment

1. **Identify** Describe three ways you can live more sustainably.
2. **Synthesize** How are the ethical guidelines for living sustainably in keeping with the six factors of sustainability?
3. **Summarize** Describe how a sustainability revolution could occur in your lifetime.

SCIENCE AND ENGINEERING PRACTICES

4. **Communicating Information** Review the ethical guidelines for living sustainably. Which one do you think is the most important? Develop and carry out a public campaign to promote the guideline you have selected. Who is your target audience? Which form of communication will you use to reach your audience?

TYING IT ALL TOGETHER STEM
A Sustainable World

As you read in the Case Study, the United States and China have the world's largest ecological footprints and the largest economies. As such, these two countries play key roles in determining whether the world can make a transition to a more sustainable future.

Compared to countries with a representative democracy, China has a far more centralized government without the checks and balances of Western democracies. Change there can occur more quickly, because the central government develops policies with little public input. This helped the Chinese economy grow rapidly. However, it also led to major environmental problems.

Because of its strong central government, public protest in China is limited. As China's rapidly expanding middle class gains more economic power, it has begun to put pressure on the government to do something about increasingly intolerable environmental conditions. The internet, and especially social media, has played a critical role in allowing China's citizens to share information about environmental issues. To reduce the growing threat of civil unrest in this country, the government has begun to put greater emphasis on dealing with its serious environmental problems, as the United States did in the 1970s.

In 2014 and 2015, U.S. President Obama met with Chinese President Xi Jinping in an effort to reach a joint vision for addressing climate change in advance of the Paris Climate Change Conference. Despite vastly different political histories, the two presidents were able to form a partnership and help lead the way to a worldwide agreement. China committed to match the United States by pledging $3.1 billion to help developing countries combat and adapt to climate change. Both countries agreed to curb greenhouse gas emissions by 2030 (the United States by 2025), and China launched a national cap-and-trade system in 2021.

Complete the items below.

1. How might politics help or impede each country in reaching its climate goals?
2. Explain how two of the scientific and/or social factors of sustainability can be applied to help each country reach its goals.
3. The United States and China also agreed to improve their transparency. What does "transparency" mean in this context? How is it important to each country's accountability?
4. What is a carbon cap-and-trade system? Do you think China's cap-and-trade system will help it reach its goals? Explain.
5. What role has the internet played in China's environmental policies?
6. In what ways do the internet and social media affect environmental issues in the country where you live?
7. What are the limitations of social media as an agent for environmental action?

CHAPTER 18 SUMMARY

18.1 How are economic systems related to the biosphere?

- Economic systems are related to the biosphere because they depend on natural capital. Natural capital includes the resources provided by Earth's natural systems.
- There is controversy over the sustainability of economic growth because the flow of natural resources is increasingly pushed to produce more goods and services.
- Environmental economists favor adjusting existing economic policies and tools to be more environmentally beneficial over inventing new policies and tools.

18.2 How can people use economic tools to address environmental problems?

- Economic tools to deal with environmental problems include full-cost pricing, subsidies for environmentally beneficial goods and services, and taxation of pollution and waste instead of wages and profits.
- Additional tools include environmental regulation and marketplace innovations, reducing poverty, and reaching development goals.

18.3 How can society enact more just environmental policies?

- Individuals can work together to take part in the political process. Grassroots public pressure—from environmental groups, students, and citizens—drives political leaders to work together. As a result, they develop regulations to protect the environment and improve our sustainability.
- Environmental justice must be addressed when making policies.
- Seven principles can guide people in making environmental policies. These include: reversibility, net energy, precautionary, prevention, polluter-pays, environmental justice, and holistic principles.

18.4 How can society live more sustainably?

- Becoming environmentally literate, learning from nature, and living more simply can lead to living more sustainably.
- Differing environmental worldviews include human-centered, life-centered, and Earth-centered perspectives. The Earth-centered environmental worldview holds that humans and human economies are subsystems of the biosphere.
- Working together, individuals can bring about a sustainability revolution to shift to a more sustainable way of living.

MindTap If you have been provided with access to a MindTap course, additional resources are available at login.cengage.com.

CHAPTER 18 ASSESSMENT

Review Key Terms

Select the key term that best fits each definition. Not all terms will be used.

- economic development
- economic growth
- economics
- environmental justice
- environmental literacy
- human capital
- law
- lobby
- manufactured capital
- policy
- politics
- regulation

1. An increase in the capacity of a nation, state, city, or company to provide goods and services to people
2. Labor, innovation, culture, and organization provided by people
3. Means of putting laws into effect
4. The social science that studies the production, distribution, and consumption of goods and services to satisfy people's needs and wants
5. A rule that is recognized and enforced by a government
6. An ideal whereby every person is entitled to protection from environmental hazards regardless of race, gender, age, national origin, income, social class, or any political factor
7. A set of efforts focused on creating economies that serve to improve human well-being by meeting basic human needs
8. The process by which individuals and groups try to influence or control the policies and actions of governments at local, state, national, and international levels
9. Term for laws and regulations
10. Items made from natural resources, including tools, machinery, equipment, factory buildings, and transportation and distribution facilities

Review Key Concepts

11. How is natural capital the same and different from manufactured capital?
12. How does economic growth differ from economic development? How are they related?
13. Why do products and services actually cost more than most people think?
14. What are perverse subsidies and how do they contribute to environmental problems?
15. List six ways in which governments, businesses, lenders, and individuals can help to reduce poverty.
16. What are four ways in which individuals in democracies can help to influence environmental policy?

Think Critically

17. Suppose that over the next 20 years, the environmental and health costs of goods and services are gradually added to their market prices until those prices more closely reflect their total costs. What harmful effects and what beneficial effects might such a full-cost pricing process have on your lifestyle and on the lives of your children, grandchildren, and great-grandchildren?
18. Describe your own environmental worldview. Which of your views, if any, changed as a result of taking this course?
19. According to the late Steve Jobs, co-founder of Apple, Inc., "The people who are crazy enough to think they can change the world are the ones who do." What did Jobs mean when making this statement? Give an example of someone who fits this statement and explain what they have done to change the world. An example would be Gandhi or Martin Luther King, Jr.
20. Congratulations! You are in charge of the world. Write up a five- to ten-point strategy for shifting the world to more environmentally sustainable economic systems over the next 50 years.

Chapter Activities

A. Investigate: Sustainability STEM

How sustainable is your school's food plan? Some farmers work to produce food in a way that does not harm the environment and preserves the ability of future generations to continue to use the land to grow crops. This is known as sustainable farming. It includes practices that allow crops to grow without the use of chemicals and animals to roam free and grow naturally without the use of hormones or antibiotics. In this activity, you will conduct a sustainability audit of your school's food sourcing, processing, and disposal.

Material
school lunch menu

1. Working in teams, choose a school lunch selection and list the ingredients necessary to make each food item.
2. Find the prices to buy each of the ingredients and add them up to get the total cost of the lunch.
3. Locate a local source that sells sustainable or local organic products and record the price of each item.
4. Compare the prices and determine the difference in cost to prepare the lunch.
5. Combine class data into one spreadsheet.
6. Determine how food is processed and disposed of in your school. Are there ways to reduce waste? Could improved planning or a more efficient system help offset extra costs? Explain.
7. Use this data and what you have learned about the hidden costs of living unsustainably to develop and support an argument for an improved food plan. Discuss the hidden costs to the environment and to students' health. Use quantitative data and estimates if possible.

B. Take Action

National Geographic Learning Framework
Attitudes | Responsibility
Skills | Collaboration
Knowledge | Our Living Planet

Get outside and get involved in a local watershed sustainability project that is going on in your community. The Environmental Protection Agency (EPA) website provides links to help people living anywhere in the United States locate their watershed and connect with local groups taking action to preserve it. Stewardship projects include activities such as building rain gardens, monitoring streams, and trash cleanup. Encourage others to join you and share your experiences with your classmates.

STEM ENGINEERING PROJECT 5
DESIGN A CARBON-CAPTURING DEVICE

Imagine a world in which rooftops are commonly adorned with vast networks of tubes containing a bright green substance. Acting as carbon-capture and storage (CCS) devices, the tubes sequester, or extract, carbon dioxide from the air. Also acting as giant batteries, the tubes provide electricity to the buildings. Carbon dioxide, sunlight, and a steady stream of the building's nutrient-rich wastewater are the only inputs the green tubes require to function.

Although this scenario may seem far-fetched, there is something that can serve all of these functions: algae. The same aquatic plant-like organisms that cause harmful eutrophication of lakes and ponds show promise in engineered solutions. Algae are easy to cultivate on a small or large scale and are efficient at capturing atmospheric carbon and storing it in their bodies. And there is plenty of atmospheric carbon to capture. Since the Industrial Revolution, human activities such as deforestation and the burning of fossil fuels have added a surplus of carbon dioxide to the atmosphere. This surplus of carbon dioxide and other greenhouse gases contributes substantially to atmospheric warming and climate change, affecting every ecosystem on Earth.

Algae is typically composed of several different species when found growing naturally in freshwater and saltwater systems. For example, phytoplankton is often a mixture of diatoms, dinoflagellates, and green algae, as well as bacteria. But different combinations of species need different conditions for survival. Much more data are needed on the cultivation and usefulness of such mixtures for engineered carbon-capture systems. In this challenge, your team will use the practices of engineering to design a solution to a real problem. You will design and test a habitat for cultivating algae and create a carbon-capture and storage (CCS) device.

Engineering DESIGN CYCLE
- Defining problems
- Developing and using models
- Planning and carrying out investigations
- Analyzing data and using math
- Designing solutions
- Forming arguments from evidence
- Obtaining, evaluating, and communicating information

Defining Problems

1. First, identify the larger problem.
   - Explain the problem of carbon dioxide in the atmosphere.
   - How might growing algae help solve the problem?
2. Refer to Engineering Focus 16.3. Consider a large-scale CCS system such as those described in the feature. What criteria would a large-scale, algae-based CCS system need to meet in order to be successful?
3. Large problems can often be broken down into smaller problems. Your team's problem is to determine how to cultivate the algae. What criteria would a successful algae "habitat" need to meet?
4. List your design constraints.

Developing and Using Models

5. Describe your source of algae. Some possibilities include scraping algae from the inside of a fish tank, scooping it from the surface of a pond, or ordering it from a science supply store.
6. Fill three empty bottles with equal amounts of distilled water. Place three marbles in each container to provide additional surface area on which the algae may grow.
7. Add an equal amount of algae to each container and place the containers near a sunny window.

Planning and Carrying Out Investigations

8. Using your three bottles, develop a quantitative test to determine the optimal amount of plant food for your algae mixture. You can use a fine mesh to filter the algae out of the water and measure its volume with a graduated cylinder.

9. Carry out your test and record your results.

Analyzing Data and Using Math

10. Analyze your data. What is the optimal amount of nutrient solution for your algae culture?

11. Prepare a microscope slide with a sample of your algae. Observe your slide under the microscope and sketch your observations. Use the Internet to identify and label as many types of algae as you can.

12. How well does your algae habitat meet each of the criteria you listed in step 3?

Designing Solutions

13. Transfer the algae and marbles to a larger container and add the air pump. You now have a small but fully functioning CCS device. If a carbon dioxide meter is available, use it to test your model's effectiveness.

14. Imagine changing your system to operate on a larger scale. Calculate how many times larger your device would need to be to offset the carbon emissions from using a car. (A typical car emits roughly 3,000 kg CO_2 into the atmosphere per year. Assume your device removes 10 kg CO_2 per year.)

15. What might a larger algae-based CCS system look like? Draw a diagram. Include the system inputs, throughputs, and outputs.

Forming Arguments from Evidence

16. Make a claim about your device and support your claim with evidence.

17. What are the strengths and weaknesses of your design?

18. What further testing is needed?

Obtaining, Evaluating, and Communicating Information

19. Present your team's carbon-capturing device to the class. Describe how your design works and how it was optimized to meet the criteria.

20. Offer thoughtful and specific criticism of other teams' designs.

21. Write a final report. Include clear instructions that explain how to cultivate algae and build your device.

ENGINEERING PROJECT 5

Appendix 1
SCIENCE SAFETY GUIDELINES

Hands-on science can be fun and highly rewarding—as long as it is conducted safely. Whether you are in a laboratory, classroom, or natural setting, safety is the number-one objective. Therefore, it is important that you read and follow all of the safety rules in this appendix. Ignoring safety rules could result in serious injury to yourself or others.

General Lab Rules

1. Never do a laboratory activity without prior approval and supervision.
2. Before you begin, read through all of the instructions at least twice. Ask your teacher if you are unsure about what to do.
3. Be prepared. Create a plan and assemble all of the materials that you will need in advance.
4. Clear any unnecessary items from the work area. Do not place bags or items where you or others might trip on them.
5. Stay focused during the activity. Use the materials only as instructed.
6. Never taste or smell anything unless instructed. Keep your hands away from your face.
7. Know where all safety equipment and exits are located.

Part 1: List of Safety Rules and Symbols

DRESS APPROPRIATELY

Eye Safety	• Goggles must be worn at all times. • Do not wear contact lenses unless you have to.
Skin Safety	• Disposable gloves must be worn at all times.
Lab Clothing	• Wear a lab apron to protect your clothes from chemical spills.
Field Clothing	• Wear appropriate clothing and shoes for the environment you will be in.

CHEMICAL & ELECTRICAL SAFETY

Chemical Safety	• Tie back hair and remove loose jewelry. • Only use chemicals as instructed. • Do not mix chemicals unless instructed. • If a chemical spills on your skin or clothes, wash off immediately and inform your teacher. • If chemicals or fumes get into your eyes, wash in eyewash and inform your teacher immediately. • If chemicals spill on the floor do not attempt to clean up. Inform your teacher immediately. • Always dispose of chemicals as instructed.
Electrical Safety	• Be sure electrical devices are plugged in completely and cords are not exposed. • Use electrical devices only as instructed. • Be sure to turn off devices and unplug them when you are finished. • Avoid touching any heated components.

FIRE & TEMPERATURE SAFETY

Temperature and Fire Safety	• If a fire breaks out, notify your teacher and immediately leave the room. • Tie back hair and remove loose jewelry. • If you or your clothing catch on fire: stop, drop, and roll. • If a classmate's clothing or hair catches on fire, use an emergency extinguisher or fire blanket. Inform your teacher as soon as possible. • Do not place anything in a flame unless instructed to do so. • Upon completion, be sure to extinguish all flames. If you are using gas, turn off gas flow completely.

SAFETY WITH LAB MATERIALS & INSTRUMENTS

Sharps and Scissors Safety	• If using a sharp blade, be sure to use only as instructed. • Be sure to keep fingers away from the object being cut. • If you cut yourself in any way, notify your teacher immediately. • Follow scissor safety and always point down while carrying.
Glassware Safety	• Use glassware carefully and keep glass away from surface edges. • Do not touch heated glass with bare hands. • Use glassware only as instructed. • If glass breaks, notify your teacher immediately.

SAFE & HUMANE HANDLING OF ANIMALS & PLANTS

Live Animal Safety	• Handle and interact with animals only as instructed. • Respect the animal and handle it with care. • Be sure the animal never experiences pain or lack of necessary resources. • Wear appropriate safety gear as instructed. Tie back hair and remove loose jewelry. • Wash your hands after handling any animal.
Live Plant Safety	• Handle all plants carefully. • Only touch the plant if your teacher instructs you to. • Do not taste or eat any plant samples. • Wash your hands after handling any live plants or plant parts.

SAFETY IN THE FIELD

Fieldwork Safety	• Wear appropriate clothing and shoes for the environment you will be in. • Do not wander off alone at any time. • Do not touch any plant or animal that you are not instructed to touch. • Be aware of any poisonous or potentially dangerous species in the area. • Respect the environment and do not litter or make excessive noise.

CLEANUP

Cleanup and Disposal	• Follow all instructions carefully to dispose of materials. • Do not guess on how to handle extra chemicals or other materials. Ask your teacher if you are uncertain. • Wash your hands when finished.

APPENDIX 1

Appendix 1
SCIENCE SAFETY GUIDELINES

Part 2: Chapter Activities

CHAPTER 1
Develop Models: Ecological Footprints

Safety Notes

CHAPTER 2
Experiment: Runoff

Safety Notes

- Wash hands every time after handling soil and plant matter.
- Clean up water spills immediately.

CHAPTER 3
Develop Models: Greenhouse Effect

Safety Notes

- The lamps and glass will be hot to the touch after a short while under the heat lamps.

CHAPTER 4
Investigate: Species Diversity

Safety Notes

CHAPTER 5
Investigate: Populations

CHAPTER 6
Investigate: Leaf Characteristics

Safety Notes

- Wash your hands after handling leaves.
- If you are asked to bring in your own leaves, be careful to avoid poisonous plants. Learn which plants to avoid.

CHAPTER 7
Develop Models: Endangered Species Data

Safety Notes

CHAPTER 8
Investigate: Ocean Acidification

Safety Notes

CHAPTER 9
Develop Models: Fish Consumption

Safety Notes

CHAPTER 10
Develop Models: Oil Spill Cleanup

Safety Notes

CHAPTER 11
Investigate: Product Life Cycles

Safety Notes

CHAPTER 12
Investigate: Fracking

CHAPTER 13
Develop Models: Wind Turbines

Safety Notes

- Be sure to read any instruction on using wiring and motors to avoid electrocution.

CHAPTER 14
Investigate: Transportation Trade-Offs

CHAPTER 15
Investigate: Disease Transmission

Safety Notes

- Do not drink the solution.

CHAPTER 16
Investigate: Climate Models

CHAPTER 17
Investigate: Composting

Safety Notes

CHAPTER 18
Investigate: Sustainability

Part 3: Engineering Projects

PROJECT 1
Design a Method for Treating Contaminated Soil

Safety Notes

- Wear gloves at all times.
- Save unused coal for reuse.
- Carefully read and follow all instructions that come with your lead testing kit.

PROJECT 2
Design a System to Assess a Local Species

Safety Notes

- Conduct all fieldwork on public property and under the supervision of an adult.
- Wear full protective clothing and gloves and avoid touching any plants or animals.
- Collect written permission from caregivers and volunteers.

PROJECT 3
Design a Solar Cooker

Safety Notes

- Use caution when cutting cardboard.
- Be aware that your device may get very hot.
- Always use oven mitts when handling the glass of water and/or cooking pot.

PROJECT 4
Design a Wind-Powered Generator

Safety Notes

- Use caution when cutting plastic.
- Never connect multiple batteries in a row.

PROJECT 5
Design a Carbon-Capturing Device

Safety Notes

- If you are using algae from the field, conduct all fieldwork on your own property or on public property.
- Conduct your fieldwork under the supervision of an adult.

APPENDIX 1

Appendix 2
MEASUREMENT UNITS

Length

Metric
1 kilometer (km) = 1,000 meters (m)
1 meter (m) = 100 centimeters (cm)
1 meter (m) = 1,000 millimeters (mm)
1 centimeter (cm) = 0.01 meter (m)
1 millimeter (mm) = 0.001 meter (m)

English
1 foot (ft) = 12 inches (in)
1 yard (yd) = 3 feet (ft)
1 mile (mi) = 5,280 feet (ft)
1 nautical mile = 1.15 miles

Metric–English
1 kilometer (km) = 0.621 mile (mi)
1 kilometer (km) = 0.54 nautical mile (nmi)
1 meter (m) = 1.094 yards (yd)
1 meter (m) = 3.281 feet (ft)
1 meter (m) = 39.4 inches (in)
1 centimeter (cm) = 0.394 inch (in)

Area

Metric
1 square kilometer (km^2) = 1,000,000 square meters (m^2)
1 square meter (m^2) = 1,000,000 square millimeters (mm^2)
1 square meter (m^2) = 10,000 square centimeters (cm^2)
1 hectare (ha) = 10,000 square meters (m^2)
1 hectare (ha) = 0.01 square kilometer (km^2)

English
1 square foot (ft^2) = 144 square inches (in^2)
1 square yard (yd^2) = 9 square feet (ft^2)
1 square mile (mi^2) = 27,880,000 square feet (ft^2)
1 acre (ac) = 43,560 square feet (ft^2)

Metric–English
1 hectare (ha) = 2.471 acres (ac)
1 square kilometer (km^2) = 0.386 square mile (mi^2)
1 square meter (m^2) = 1.196 square yards (yd^2)
1 square meter (m^2) = 10.76 square feet (ft^2)
1 square centimeter (cm^2) = 0.155 square inch (in^2)

Volume

Metric
1 cubic kilometer (km^3) = 1,000,000,000 cubic meters (m^3)
1 cubic meter (m^3) = 1,000,000 cubic centimeters (cm^3)
1 liter (L) = 1,000 milliliters (mL) = 1,000 cubic centimeters (cm^3)
1 cubic meter (m^3) = 1,000 liters (L)
1 milliliter (mL) = 0.001 liter (L)
1 milliliter (mL) = 1 cubic centimeter (cm^3)

English
1 gallon (gal) = 4 quarts (qt)
1 quart (qt) = 2 pints (pt)

Metric–English
1 liter (L) = 0.265 gallon (gal)
1 liter (L) = 1.06 quarts (qt)
1 liter (L) = 0.0353 cubic foot (ft^3)
1 cubic meter (m^3) = 35.3 cubic feet (ft^3)
1 cubic meter (m^3) = 1.30 cubic yards (yd^3)
1 cubic kilometer (km^3) = 0.24 cubic mile (mi^3)
1 barrel (bbl) = 159 liters (L)
1 barrel (bbl) = 42 U.S. gallons (gal)

Mass

Metric
1 kilogram (kg) = 1,000 grams (g)
1 gram (g) = 1,000 milligrams (mg)
1 gram (g) = 1,000,000 micrograms (µg)
1 milligram (mg) = 0.001 gram (g)
1 microgram (µg) = 0.000001 gram (g)
1 metric ton (mt) = 1,000 kilograms (kg)

English
1 ton (t) = 2,000 pounds (lb)
1 pound (lb) = 16 ounces (oz)

Metric–English
1 metric ton (mt) = 2,200 pounds (lb) = 1.1 tons (t)
1 kilogram (kg) = 2.20 pounds (lb)
1 gram (g) = 0.00220462 pound (lb)
1 gram (g) = 0.035 ounce (oz)

Energy and Power

Metric
1 kilojoule (kJ) = 1,000 joules (J)
1 kilocalorie (kcal) = 1,000 calories (cal)
1 calorie (cal) = 4.184 joules (J)

Metric–English
1 kilojoule (kJ) = 0.949 British thermal unit (Btu)
1 kilojoule (kJ) = 0.000278 kilowatt-hour (kW-h)
1 kilocalorie (kcal) = 3.97 British thermal units (Btu)
1 kilocalorie (kcal) = 0.00116 kilowatt-hour (kW-h)
1 kilowatt-hour (kW-h) = 860 kilocalories (kcal)
1 kilowatt-hour (kW-h) = 3,400 British thermal units (Btu)
1 quad (Q) = 1,050,000,000,000,000 kilojoules (kJ)
1 quad (Q) = 293,000,000,000 kilowatt-hours (kW-h)

Temperature Conversions

Fahrenheit (°F) to Celsius (°C): °C = (°F − 32.0) ÷ 1.80
Celsius (°C) to Fahrenheit (°F): °F = (°C × 1.80) + 32.0

PERIODIC TABLE OF ELEMENTS

Chemists Classify Elements on the Basis of Their Chemical Properties

Matter consists of elements and compounds (Lesson 2.2). The basic unit of each element is a unique atom that is different from the atoms of all other elements. Each atom consists of an extremely small and dense center called its nucleus. The nucleus contains one or more protons and, usually, one or more neutrons, as well as one or more electrons moving rapidly somewhere around the nucleus (Figure 2-7).

Chemists classify the elements in the Periodic Table of Elements according to their chemical behavior. Each row in the table is called a period. Each column, which contains elements with similar chemical properties, is called a group. As you can see from the table, the elements are classified as metals, nonmetals, and metalloids, which have a mixture of metallic and nonmetallic properties.

This Periodic Table also identifies the elements required as nutrients for all or some forms of life (marked by small black squares), and elements that are moderately or highly toxic to all or most forms of life (marked by small red squares). Note that some elements such as copper (Cu) serve as nutrients, but can also be toxic at high doses. Six nonmetallic elements—carbon (C), oxygen (O), hydrogen (H), nitrogen (N), sulfur (S), and phosphorus (P)—make up about 99% of the atoms of all living things.

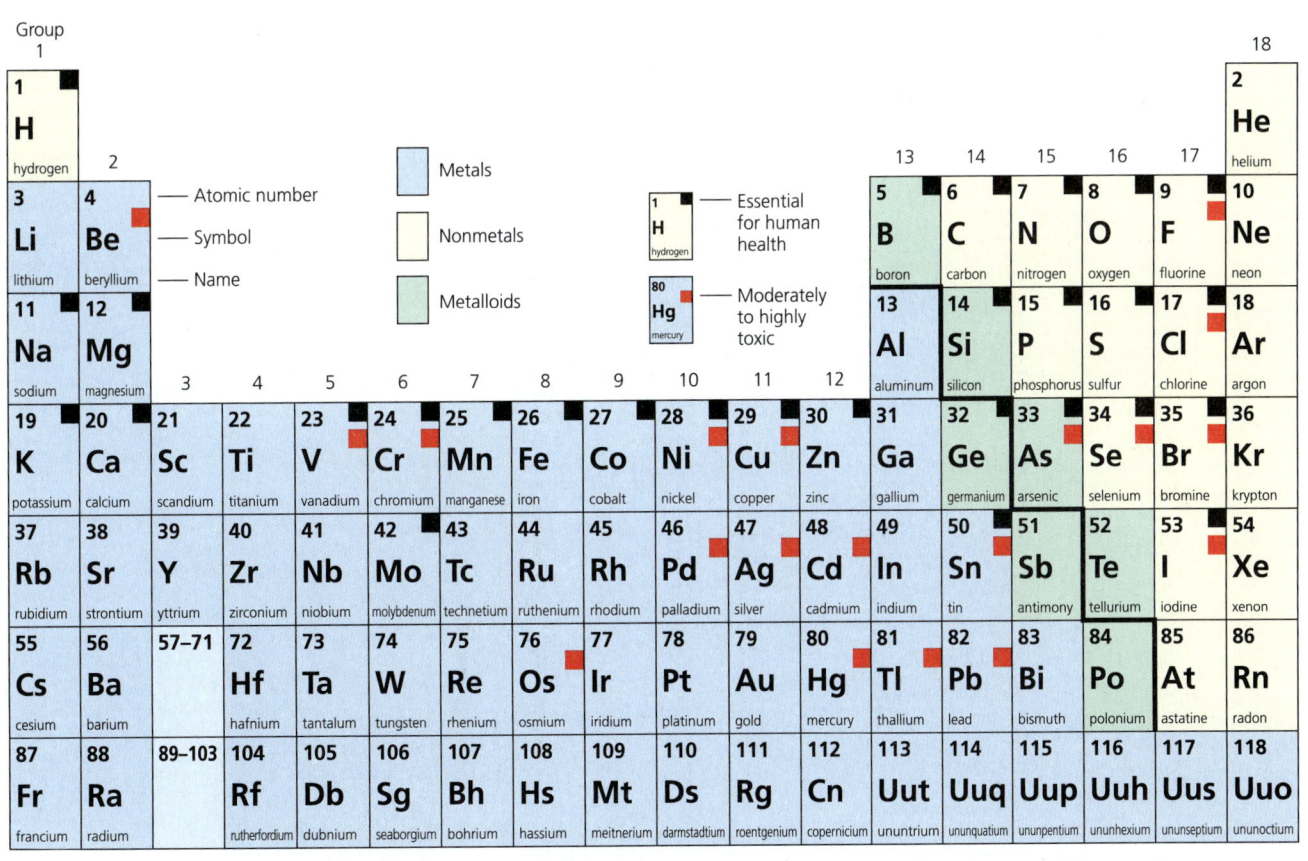

Group 1	2											13	14	15	16	17	18
1 H hydrogen																	2 He helium
3 Li lithium	4 Be beryllium											5 B boron	6 C carbon	7 N nitrogen	8 O oxygen	9 F fluorine	10 Ne neon
11 Na sodium	12 Mg magnesium	3	4	5	6	7	8	9	10	11	12	13 Al aluminum	14 Si silicon	15 P phosphorus	16 S sulfur	17 Cl chlorine	18 Ar argon
19 K potassium	20 Ca calcium	21 Sc scandium	22 Ti titanium	23 V vanadium	24 Cr chromium	25 Mn manganese	26 Fe iron	27 Co cobalt	28 Ni nickel	29 Cu copper	30 Zn zinc	31 Ga gallium	32 Ge germanium	33 As arsenic	34 Se selenium	35 Br bromine	36 Kr krypton
37 Rb rubidium	38 Sr strontium	39 Y yttrium	40 Zr zirconium	41 Nb niobium	42 Mo molybdenum	43 Tc technetium	44 Ru ruthenium	45 Rh rhodium	46 Pd palladium	47 Ag silver	48 Cd cadmium	49 In indium	50 Sn tin	51 Sb antimony	52 Te tellurium	53 I iodine	54 Xe xenon
55 Cs cesium	56 Ba barium	57–103	72 Hf hafnium	73 Ta tantalum	74 W tungsten	75 Re rhenium	76 Os osmium	77 Ir iridium	78 Pt platinum	79 Au gold	80 Hg mercury	81 Tl thallium	82 Pb lead	83 Bi bismuth	84 Po polonium	85 At astatine	86 Rn radon
87 Fr francium	88 Ra radium	89–103	104 Rf rutherfordium	105 Db dubnium	106 Sg seaborgium	107 Bh bohrium	108 Hs hassium	109 Mt meitnerium	110 Ds darmstadtium	111 Rg roentgenium	112 Cn copernicium	113 Uut ununtrium	114 Uuq ununquatium	115 Uup ununpentium	116 Uuh ununhexium	117 Uus ununseptium	118 Uuo ununoctium

Key:
- Metals
- Nonmetals
- Metalloids
- 1 H hydrogen — Essential for human health
- 80 Hg mercury — Moderately to highly toxic
- Atomic number, Symbol, Name

Lanthanides (Rare Earth Elements)

| 57 La lanthanum | 58 Ce cerium | 59 Pr praseodymium | 60 Nd neodmium | 61 Pm promethium | 62 Sm samarium | 63 Eu europium | 64 Gd gadolinium | 65 Tb terbium | 66 Dy dysprosium | 67 Ho holmium | 68 Er erbium | 69 Tm thulium | 70 Yb ytterbium | 71 Lu lutetium |

Actinides

| 89 Ac actinium | 90 Th thorium | 91 Pa protactinium | 92 U uranium | 93 Np neptunium | 94 Pu plutonium | 95 Am americium | 96 Cm curium | 97 Bk berkelium | 98 Cf californium | 99 Es einsteinium | 100 Fm fermium | 101 Md mendelevium | 102 No nobelium | 103 Lr lawrencium |

Source: © 2016 Cengage Learning

APPENDIX 3

Appendix 4
COUNTRIES OF THE WORLD

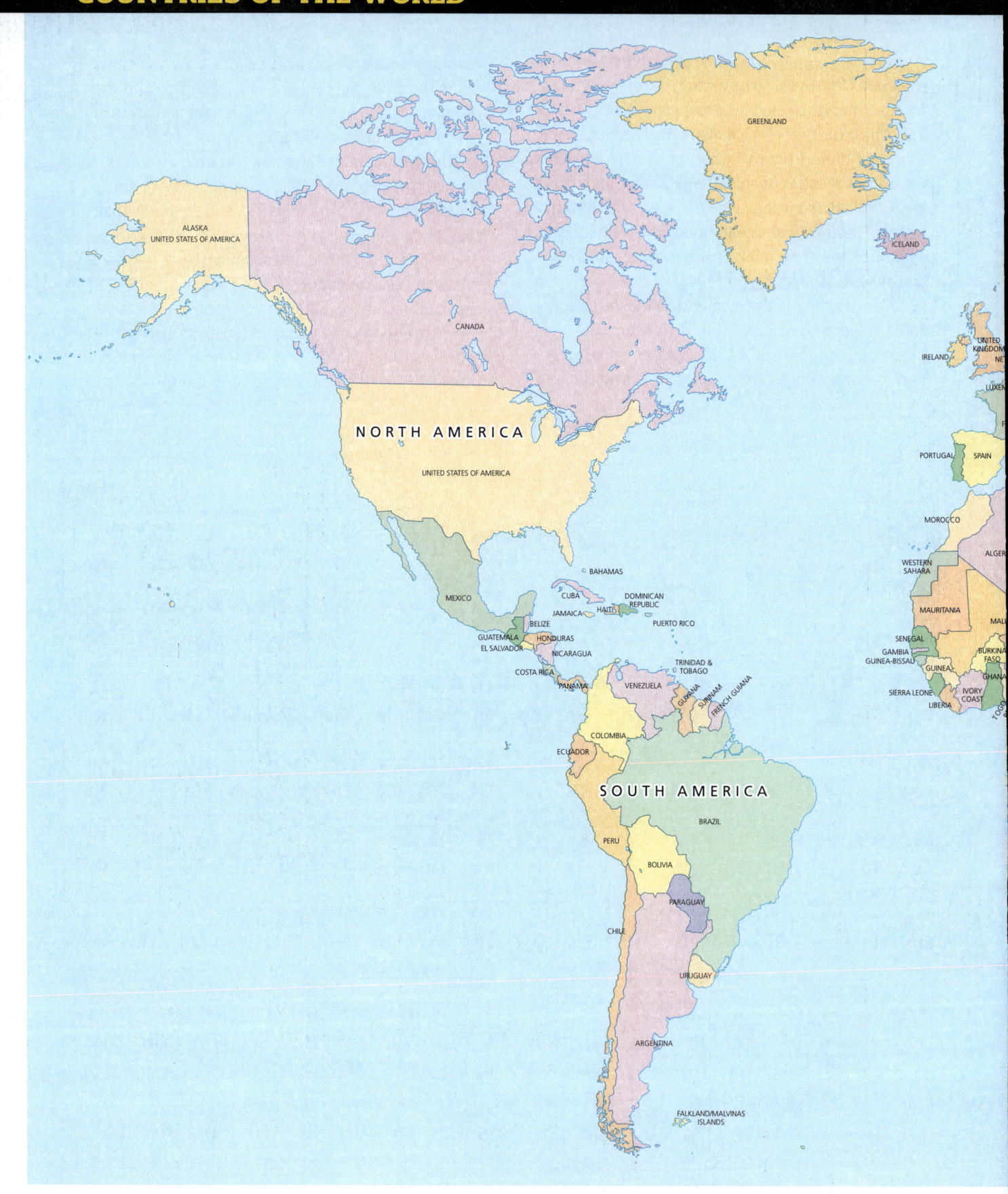

APPENDIX 4

GLOSSARY

A

acid deposition The falling of acids and acid-forming compounds from the atmosphere to Earth's surface. Acid deposition is commonly known as acid rain.

acid rain See *acid deposition*.

active solar heating system System that uses solar collectors to capture energy from the sun and store it as heat for space heating and water heating. Compare *passive solar heating system*.

adaptation Any heritable trait that gives an individual some advantage over other individuals in a given population. See *biological evolution, mutation, natural selection*.

adaptive trait See *adaptation*.

aerobic respiration Complex process that uses oxygen and glucose to produce energy and occurs in the cells of most living organisms. Carbon dioxide and water are the byproducts of this reaction. Compare *photosynthesis*.

age structure Distribution of individuals among various age groups in a population.

agrobiodiversity The genetic variety of animal and plant species used on farms to produce food.

air pollution Any gaseous or solid material in the atmosphere that occurs in concentrations high enough to alter climate or harm organisms, ecosystems, or human-made materials.

anaerobic respiration Form of cellular respiration in which some decomposers get the energy they need through the breakdown of glucose (or other nutrients) in the absence of oxygen.

Anthropocene New epoch in which humans have become major agents of change in the functioning of Earth's life-support system as their ecological footprints have spread over Earth. Compare *Holocene*.

aquaculture Growing and harvesting of fish for human use in freshwater ponds, lakes, reservoirs, and rice paddies, and in underwater cages in coastal and deeper ocean waters.

aquaponics System of growing crops in water fertilized with the waste of farmed fish or other aquatic animals.

aquatic life zone Marine or freshwater portion of the biosphere. Examples include freshwater life zones (such as lakes and streams) and ocean or marine life zones (such as estuaries, coastlines, coral reefs, and the open ocean).

aqueduct Structure used to carry water over land.

aquifer Porous, water-saturated layers of sand, gravel, or bedrock in which groundwater collects.

arboretum Place for the scientific study and public display of various species of trees and shrubs.

artificial selection Process by which humans select one or more desirable genetic traits in the population of a plant or animal species and then use selective breeding to produce populations containing many individuals with the desired traits. Compare *genetic engineering, natural selection*.

asthenosphere Zone within Earth's mantle made up of hot, partly melted, fluid rock.

atmosphere Envelope of gases surrounding Earth. Compare *biosphere, geosphere, hydrosphere, stratosphere, troposphere*.

atom The basic building block of matter. Compare *ion, molecule*.

B

background extinction rate Naturally low rate at which species have disappeared throughout most of Earth's history. Compare *mass extinction*.

bacterium Single-celled organism that can multiply very rapidly. Most are harmless but some may cause diseases such as strep throat and tuberculosis.

bioaccumulation Accumulation of a chemical in an organism's fatty tissues.

biodiversity Variety of life on Earth, including species diversity, genetic diversity, ecological diversity, and functional diversity.

biodiversity hotspot Area rich in highly endangered endemic species.

biofuel Fuel, such as ethyl alcohol or biodiesel, made from plant material (biomass).

biological extinction Complete disappearance of a species from Earth. It happens when a species cannot adapt to survive and reproduce in response to changes in their environment and cannot move to a new environment with more favorable conditions. Compare *speciation*. See also *endangered species, mass extinction, threatened species*.

biological evolution Process by which species change genetically over time.

biomagnification Process in which a chemical becomes more concentrated as it moves up the food chain.

biomass Organic matter produced by plants that can be burned in either solid or gaseous form to produce energy.

biome Geographical area composed of different ecosystems and characterized by a distinct climate and certain species (particularly vegetation) that are able to survive there.

biomimicry Making industrial manufacturing processes cleaner and more sustainable by redesigning them to mimic the way nature deals with wastes.

bioprospector Person trained to search tropical forests and other ecosystems to find plants and animals that can be used to make medicinal drugs.

bioremediation Use of bacteria and enzymes to help destroy toxic or hazardous substances or convert them to harmless compounds.

biosphere Zone of Earth where life is found. It consists of parts of the atmosphere (the troposphere), hydrosphere (mostly surface water and groundwater), and lithosphere (mostly soil and surface rocks and sediments on the bottoms of oceans and other bodies of water). Compare *atmosphere, geosphere, hydrosphere*.

biosphere reserve Area that contains a strictly protected core ecosystem surrounded by a buffer zone where local people can extract resources sustainably.

botanical garden Place that contains living plants on display.

brackish water Mixture of fresh water and salt water.

buffer zone Outer zone surrounding a strictly protected inner core of a reserve where local people can extract resources sustainably.

C

captive breeding Process by which wild individuals of a critically endangered species are collected for breeding in captivity, with the aim of reintroducing the offspring into the wild.

carbon capture and storage (CCS) Process of removing carbon dioxide gas from coal-burning power and industrial plants and storing it somewhere (usually underground or under the seabed). To be effective, carbon dioxide must be stored, essentially forever, so that it cannot be released into the atmosphere.

carbon cycle Cyclic movement of carbon in different chemical forms from the environment to organisms and then back to the environment.

carbon footprint Amount of carbon dioxide generated by an individual, organization, country, or any other entity over a given period of time.

carcinogen Chemical, type of radiation, or virus that can cause or promote cancer. Compare *mutagen, teratogen*.

carnivore Animal that feeds mostly on other animals. Compare *herbivore, omnivore*.

carrying capacity Maximum population of a species that a habitat can sustain indefinitely.

CBD See *Convention on Biological Diversity*.

CCS See *carbon capture and storage*.

CFC See *chlorofluorocarbon*.

chemical change Change in the chemical composition of a substance. Compare *physical change*.

chemical reaction See *chemical change*.

chlorofluorocarbon (CFC) Compound that contains carbon, chlorine and fluorine. Such compounds are harmful to ozone in the stratosphere.

chronic malnutrition Condition where affected individuals do not get enough protein and other key nutrients. It can make individuals more vulnerable to disease and hinder normal physical and mental development of children. Compare *overnutrition*.

CITES See *Convention on International Trade of Endangered Species of Wild Flora and Fauna*.

climate General pattern of atmospheric conditions in a given area over periods ranging from decades to thousands of years. The key factors that influence an area's climate are incoming solar energy, Earth's rotation, global patterns of air and water movement, gases in the atmosphere, and Earth's surface features. Compare *weather*.

climate change Change in the global average atmospheric conditions over a period of at least three decades.

climate change tipping point Point at which an environmental problem reaches a threshold level where scientists fear it could cause irreversible climate disruption.

closed-loop recycling See *primary recycling*.

coal Fossil fuel formed from the remains of land plants that were buried and exposed to intense heat and pressure for 300 to 400 million years.

coastal wetland Land along a coastline that is covered with water for all or part of the year. Examples include marshes and mangrove forests. Compare *inland wetland*.

coastal zone Warm, nutrient-rich, shallow part of the ocean that extends from the high-tide mark on land to the gently sloping, shallow edge of the continental shelf. Compare *open sea*.

coevolution Natural selection process in which changes in the gene pool of one species lead to changes in the gene pool of another species. See *biological evolution, natural selection*.

cogeneration Production of two useful forms of energy, such as high-temperature steam and electricity, from the same fuel source.

commensalism Interaction between organisms of different species in which one type of organism benefits and the other type is neither helped nor harmed to any great degree. Compare *mutualism*.

commercial energy Energy sold in the marketplace.

compost Fertilizer produced when microorganisms break down organic matter such as leaves, crop residues, food wastes, paper, and wood in the presence of oxygen.

composting Form of recycling that mimics nature by using bacteria to decompose yard trimmings, vegetable food scraps, and other biodegradable organic wastes into humus, which is an organic component of soil that improves soil fertility.

GLOSSARY **653**

GLOSSARY

compound Combination of two or more different elements held together in fixed proportions. Compare *element*.

conservation concession Strategy in which governments or private conservation organizations pay governments or landowners in other nations to preserve their land's natural resources for a set amount of time.

consumer Organism that cannot produce its own food and gets its organic nutrients by feeding on the tissues of producers or of other consumers; generally divided into *primary consumers* (herbivores), *secondary consumers* (carnivores), *tertiary* (higher-level) *consumers, omnivores,* and *detritivores* (decomposers and detritus feeders). Compare *producer*.

continental drift Slow movement of the continents across Earth's surface.

Convention on Biological Diversity (CBD) International treaty that commits participating governments to reduce the global rate of biodiversity loss and to share the benefits from use of the world's genetic resources.

Convention on International Trade of Endangered Species of Wild Flora and Fauna (CITES) International treaty that bans hunting, capturing, and selling of threatened or endangered species.

core Inner zone of Earth that consists of a solid inner core and a fluid outer core. Compare *crust, mantle*.

crude oil Gooey liquid consisting of hydrocarbon compounds and other compounds, formed from the remains of ancient organisms over the course of millions of years. Extracted from underground deposits, crude oil is sent to oil refineries, where it is heated to separate it into various fuels and other components.

crust Solid outer zone of Earth. It consists of oceanic crust and continental crust. Compare *core, mantle*.

cultural eutrophication Overnourishment of aquatic ecosystems with plant nutrients (mostly nitrates and phosphates) resulting from human activities such as agriculture, urbanization, and discharges from sewage treatment plants.

D

data Factual information collected by scientists.

debt-for-nature swap Arrangement in which participating countries act as custodians of protected forest reserves in return for foreign aid or debt relief.

decarbonization Global transition away from fossil fuels.

decomposer Consumers that get their nutrients by breaking down nonliving organic matter such as leaf litter, fallen trees, and dead animals. In the process of obtaining their own food, these organisms release nutrients from their waste that return nutrients to the soil and water. Compare *consumer, detritivore, producer*.

deep-well disposal Pumping of liquid hazardous wastes under high pressure through a pipe into dry, porous rock formations far beneath aquifers that are tapped for drinking and irrigation water.

deforestation Temporary or permanent removal of large expanses of forest for agriculture, settlements, or other uses.

delta Area at the mouth of a river built up by deposits of river sediments, often containing estuaries and coastal wetlands.

depletion time Amount of time it takes to use a certain proportion (usually 80%) of the reserves of a mineral at a given rate of use.

desalination Process of removing salt from ocean water or brackish (slightly salty) water in aquifers or lakes.

desertification Condition when the productive potential of topsoil falls by 10% or more due to farming practices such as overgrazing, deforestation, and excessive plowing, combined with drought and climate change. In extreme cases, it leads to desert conditions.

detritivore Consumer organism that feeds on detritus—freshly dead organisms. Examples include earthworms, some insects, hyenas, and vultures. Compare *decomposer*.

district heating Heat distribution system in which heat is sent to multiple buildings from a central plant that sometimes generates power as well.

dose Amount of a harmful chemical that a person has ingested, inhaled, or absorbed through the skin at any one time. Compare *response*.

drip irrigation Irrigation method in which water is delivered directly to plant roots through pipes or tubes.

drought Prolonged dry period in a given region. Atmospheric warming is a major cause of drought.

E

Earth-centered worldview Worldview maintaining that people are part of, and dependent on, nature; that Earth's natural capital exists for all species, not just for humans; that economic success and the long-term survival of cultures and species depend on learning how Earth has sustained itself for billions of years; and lessons from nature should influence how people think and act. Compare *human-centered worldview, life-centered worldview*. See *environmental worldview*.

eco-city Ecologically sustainable city. Residents can walk, bike, or use low-polluting mass transit for most travel. Residents grow their own food and recycle or reuse most of their wastes. Abandoned lots and industrial sites are cleaned up and nearby forests, grassland, wetlands, and farms are preserved.

ecological footprint Amount of land and water needed to supply a population with renewable resources and to absorb and recycle the waste and pollution such resource use produces.

ecological niche Role that a species plays in an ecosystem, encompassing everything that affects its survival and reproduction.

ecological restoration Deliberate alteration of a damaged habitat or ecosystem to a state close to its original one.

ecological succession Typically gradual change in species composition in a given area. See *primary ecological succession, secondary ecological succession.*

ecological tipping point Occurrence in a natural system in which it becomes locked into a positive feedback loop of unchecked change. Beyond this point, the system can change so drastically that it suffers from severe degradation or collapse.

ecology Biological science that studies how living things interact with the living and nonliving parts of their environment.

economic development Any set of efforts focused on creating economies that serve to improve human well-being by meeting basic human needs. Compare *economic growth.*

economic growth Increase in the capacity of a nation, state, city, or company to provide goods and services. Compare *economic development.*

economics Social science that deals with the production, distribution, and consumption of goods and services to satisfy people's needs and wants.

ecosystem One or more communities of different species interacting with one another and with the chemical and physical factors of their nonliving environment.

ecosystem diversity Earth's diversity of biological communities such as deserts, grasslands, forests, lakes, rivers, and wetlands.

ecosystem services Natural services that support life and human economies at no monetary cost. Examples are nutrient cycling, natural pest control, and natural purification of air and water. See *natural capital, natural resource.*

edge effect Tendency for a transition zone between two different ecosystems to have greater species diversity and a higher density of organisms than are found in either of the individual ecosystems.

egg pulling Collecting eggs laid in the wild by critically endangered bird species and then hatching them in zoos or research centers.

electromagnetic radiation Forms of kinetic energy traveling as electromagnetic waves. Examples include radio waves, visible light, and ultraviolet radiation.

element Type of matter with a unique set of properties that cannot be broken down into simpler substances by chemical means. Compare *compound.*

endangered species Species with so few individual survivors that the species could soon become extinct. Compare *threatened species.*

Endangered Species Act (ESA) U.S. law that identifies and protects endangered and threatened species and calls for recovery programs to help species recover to levels where legal protection is no longer needed.

endemic species Species that is found in only one area. Such species are especially vulnerable to extinction.

energy Capacity to do work.

energy efficiency Measure of how much useful work can be extracted from each unit of energy used. See *net energy.*

environment All external conditions, factors, matter, and energy, living and nonliving, that affect any living organism or other specified system.

environmental degradation Depletion, deterioration, or waste of Earth's natural capital.

environmental ethics Study of varying beliefs about what is right or wrong with how people treat the environment.

environmentalism Social movement dedicated to protecting Earth and its resources.

environmental justice An ideal whereby every person is entitled to protection from environmental hazards regardless of race, gender, age, national origin, income, social class, or any political factor.

environmental literacy Knowledge and understanding of the concepts of environmental science.

environmentally sustainable society A society that protects natural capital and lives off its income while also meeting the current and future basic resource needs of its people.

environmental resistance All of the limiting factors that act together to limit the growth of a population. See *limiting factor.*

environmental science Interdisciplinary study of how humans interact with the environment. It includes information and ideas from engineering, natural sciences, and social sciences. The fundamental goals of environmental science are to learn how life on Earth has survived and thrived, understand how humans interact with the environment, and find ways to deal with environmental problems and live more sustainably.

environmental worldview An individual's set of assumptions and values concerning the natural world and what they think their role in managing it should be. See *Earth-centered worldview, human-centered worldview, life-centered worldview.*

ESA See *Endangered Species Act.*

GLOSSARY

estuary Partially enclosed coastal area where a river meets the sea and seawater mixes with fresh water.

eutrophication Process by which a body of water gains nutrients.

evolution See *biological evolution*.

exponential growth Growth in which some quantity, such as population size or economic output, increases at a fixed percentage per unit of time. Exponential growth starts slowly but the quantity soon becomes enormous. When the increase in quantity over time is plotted on a graph, this type of growth yields a curve shaped like the letter J.

extinction See *biological extinction*.

F

feedback loop Process that occurs when an output of matter, energy, or information is fed back into the system as an input and leads to changes in that system.

fertility Number of births in a population.

first law of thermodynamics Scientific law stating that whenever energy is converted from one form to another in a physical or chemical change, no energy is created or destroyed. The total amount of energy does not change. See *second law of thermodynamics*.

fishery Concentration of an aquatic species suitable for commercial harvesting in an ocean area or inland body of water.

fish farming See *aquaculture*.

fishprint Area of ocean needed to sustain the consumption of an average person, a nation, or the world, based on the weight of fish he, she, or they consume annually. Compare *ecological footprint*.

flood irrigation Inefficient irrigation method in which water is pumped into ditches and flows by gravity to crops.

flow See *throughput*.

food chain Sequence of organisms in which each organism is a source of nutrients or energy for the next level of organisms. Compare *food web*.

food insecurity Lack of access to nutritious food. Compare *food security*.

food security Access to enough safe, nutritious foods for a healthy lifestyle. Compare *food insecurity*.

food web Complex network of interconnected food chains. Compare *food chain*.

fossil Preserved remains or traces of prehistoric organisms. Includes mineralized or petrified skeletons, bones, teeth, shells, leaves, seeds, or impressions of such items, as well as impressions of animal activity such as tracks, trails, and burrows.

fracking See *hydraulic fracturing*.

freshwater life zone Aquatic system comprised of water with a dissolved salt concentration of no more than 1,000 ppm. Examples include standing bodies of fresh water such as lakes, ponds, and inland wetlands, and flowing systems such as streams and rivers. Compare *biome*.

functional diversity Variety of processes that occur with ecosystems. Examples include energy flow and cycles of matter. See *biodiversity, ecosystem diversity, genetic diversity, species diversity*.

G

generalist species Species with a broad niche. They can live in many different places, eat a variety of foods, and tolerate a wide range of environmental conditions. Examples include flies, cockroaches, mice, rats, and humans. Compare *specialist species*.

genetic diversity Variety of genes found in a population or in a species. See *biodiversity*. Compare *ecosystem diversity, functional diversity, species diversity*.

genetic engineering Scientific manipulation of genes in order to select desirable traits or eliminate undesirable ones. It allows scientists to alter an organism's genetic material by adding, deleting, or changing segments of its DNA in a process called gene splicing. Compare *artificial selection, natural selection*.

genetic variability Variety in the genetic makeup of individuals in a population.

geoengineering Manipulation of natural conditions to counteract the greenhouse effect.

geographic isolation Separation of populations of a species into different areas. It may occur because of a search for food, a natural event (such as a hurricane, earthquake, or volcanic eruption), or a physical barrier, either natural (such as a mountain or valley) or created by humans (such as a dam or a clearing in a forest).

geosphere Earth's core, mantle, and crust—all the material above and below the surface of Earth that forms the planet's mass. Compare *atmosphere, biosphere, hydrosphere*.

geothermal energy Heat stored in soil, underground rocks, and fluids in the Earth's mantle. It is used to heat and cool buildings and water and to produce electricity.

GPP See *gross primary productivity*.

gray water Wastewater from showers, sinks, dishwashers, and washing machines that can be reused for some purposes such as watering lawns or washing cars.

greenhouse effect Process in which solar energy warms the troposphere as it reflects from Earth's surface (geosphere) and interacts with carbon dioxide, methane, water vapor (from the hydrosphere and biosphere), and other greenhouse gases (atmosphere). This warms Earth and supports life.

greenhouse gas Gas in Earth's lower atmosphere that causes the greenhouse effect. Examples include water vapor, carbon dioxide, methane, and nitrous oxide.

green revolution Process of increasing crop production through industrialized agriculture.

gross primary productivity (GPP) Rate at which an ecosystem's producers convert radiant energy into chemical energy. Compare *net primary productivity*.

groundwater Precipitation that seeps into the soil and collects in an *aquifer*. Compare *runoff, surface water*.

H

habitat Area that provides the abiotic and biotic factors a species needs to survive. Compare *ecological niche*.

habitat fragmentation Breakup of a larger habitat area into smaller areas, usually as a result of human activities.

hazardous waste Any discarded material or substance that threatens human health or the environment.

herbivore Organism that eats mostly green plants or algae. Examples include deer, sheep, grasshoppers, and zooplankton. Compare *carnivore, omnivore*.

Holocene Period of relatively stable climate and other environmental conditions; it has allowed the human population to grow, develop agriculture, and take over a large and growing share of Earth's land and other resources. Compare *Anthropocene*.

horizontal drilling Drilling utilized alongside fracking to mine previously unavailable oil deposits.

human capital People's physical and mental talents that provide labor, organizational and management skills, and innovation. Compare *manufactured capital, natural capital*.

human-centered worldview Worldview that sees the natural world as a support system for human life. It is divided into the planetary management worldview and the stewardship worldview. Compare *Earth-centered worldview, life-centered worldview*.

hydraulic fracturing Method that uses high-pressure pumps to blast huge volumes of a mixture of water, sand, and various chemicals into a well to fracture the porous rock and release natural gas or oil. Also called *fracking*.

hydrogen fuel cell Device that uses hydrogen gas as a fuel to produce electricity when it reacts with oxygen gas in the atmosphere and emits harmless water vapor.

hydrologic cycle Cycle that collects, purifies, and distributes Earth's fixed supply of water.

hydropower Technology that uses the kinetic energy of flowing or falling water to produce electricity.

hydrosphere All of the gaseous, liquid, and solid water on or near Earth's surface. See *hydrologic cycle*. Compare *atmosphere, biosphere, geosphere*.

I

igneous rock Rock formed below or on Earth's surface when magma wells up from the mantle and then cools and hardens. Compare *metamorphic rock, sedimentary rock*. See *rock cycle*.

indicator species Species whose presence or absence indicates the quality or characteristics of certain environmental conditions. Compare *keystone species, native species, nonnative species*.

industrialized agriculture Method of agriculture characterized by the large-scale operation, monoculture, and the goal of steadily increasing each crop's yield, which is the amount of food produced per unit of land. Compare *traditional agriculture*.

industrial solid waste Solid waste produced by mines, farms, and industries that supply people with goods and services. Compare *municipal solid waste*.

inertia Ability of an ecosystem to survive moderate disturbances. Also called *persistence*.

inexhaustible resource Resource available in continuous supply for the conceivable future. Examples include sunlight and the wind and flowing waters that sunlight powers. Compare *nonrenewable resource, renewable resource*.

infant mortality rate Number of babies out of every 1,000 born who die before their first birthday.

infectious disease Disease caused when a pathogen such as a bacterium, virus, or parasite invades the body and multiplies in its cells and tissues. Examples include flu, AIDS, tuberculosis, diarrheal diseases, and malaria. See *transmissible disease*. Compare *nontransmissible disease*.

inland wetland Land away from the coast, such as a swamp, marsh, or small pond, that is covered all or part of the time with fresh water. Compare *coastal wetland*.

input Matter and energy from the environment that is put into a system. See *system*. Compare *output, throughput*.

insurance hypothesis Hypothesis stating that biodiversity ensures ecosystems against a decline in their functioning because many species provide greater guarantees of functioning even if others fail.

integrated pest management (IPM) Program designed to reduce crop damage to an economically tolerable level with minimal use of synthetic pesticides by evaluating each crop and its pests as part of an ecosystem.

integrated waste management Variety of coordinated strategies for both waste reduction and waste management designed to deal with the solid wastes humans produce.

interspecific competition Attempts by members of two or more species to use the same limited resources in an ecosystem.

ion Atom or group of atoms with one or more positive (+) or negative (−) electrical charges. Examples include Na^+ and Cl^-. Compare *atom, molecule*.

GLOSSARY

IPM See *integrated pest management*.

isotope One of two or more forms of a chemical element that have the same atomic number but different mass numbers because they have different numbers of neutrons in their nuclei.

K

keystone species Species that preserves an ecosystem by controlling the populations of prey animals which could otherwise consume enough plant matter to devastate the ecosystem. Examples include wolves, sea otters, alligators, and sharks. Compare *indicator species, native species, nonnative species*.

kinetic energy Energy associated with motion, such as energy in flowing water, a speeding car, or electricity. Compare *potential energy*.

K-selected species Species that tend to do well in competitive conditions when their population size is near the carrying capacity (K) of their environment. They tend to have long lifespans, reproduce later in life, and produce few offspring. Examples include elephants, whales, humans, saguaro cacti, and most tropical rain forest trees. Compare *r-selected species*.

L

law Policy passed by the legislative branch of the U.S. government.

law of conservation of energy See *first law of thermodynamics*.

law of conservation of matter Scientific law stating that whenever matter undergoes a physical or chemical change, no atoms are created or destroyed.

life-centered worldview Worldview holding that all species have value in fulfilling their particular role within the biosphere, regardless of their potential or actual use to society; includes the belief that people have a responsibility to be caring and responsible stewards of the planet. Compare *Earth-centered worldview, human-centered worldview*.

life expectancy Average age a person can be expected to live based on the year and country or region of his or her birth.

limiting factor Factor that is particularly important relative to other factors in regulating the growth of a population.

lithosphere Combination of Earth's crust and the rigid, outermost part of the mantle above the asthenosphere. Compare *crust, geosphere, mantle*.

lobby To influence the vote or action of an elected official.

M

mantle Zone of Earth's interior between its core and its crust. Compare *core, crust*. See *geosphere, lithosphere, asthenosphere*.

manufactured capital Machinery, materials, and factories that people create using natural resources. Compare *human capital, natural resources*.

marine life zone Saltwater environment found in an ocean or its bay, estuary, coastal wetland, shoreline, coral reef, or mangrove forest.

mass extinction Significant rise in extinction rates well above the background extinction rate. Compare *background extinction rate*.

materials revolution Replacement of minerals with other materials for use in industry and technology.

matter Anything that has mass and takes up space. It commonly exists as solid, liquid, or gas.

metamorphic rock Rock formed when an existing rock is subjected to high pressure, high temperatures that cause it to melt partially, or a combination of the two. Compare *igneous rock, sedimentary rock*. See *rock cycle*.

mineral Chemical element or inorganic compound that exists as a solid with a regularly repeating internal structure. See *mineral resource*.

mineral resource Concentration of minerals that is large enough to cover the cost of extracting and processing it into raw materials and useful products.

mitigation Action taken to reduce the severity of a problem.

model Physical or mathematical representation of a structure or system.

molecule Combination of two or more atoms of the same or different elements held together by chemical bonds. Compare *atom, ion*.

monoculture Cultivation of a single crop, usually on a large area of land. Compare *polyculture*.

mortality Number of deaths in a population.

mountaintop removal Type of surface mining that clears large expanses of forests and uses explosives to remove the top of a mountain and expose seams of coal underneath the removed soil and rocks. See *open-pit mining, strip mining, surface mining*. Compare *subsurface mining*.

MSW See *municipal solid waste*.

municipal solid waste (MSW) Combined solid wastes produced by households and workplaces other than factories. See *solid waste*. Compare *industrial solid waste*.

mutagen Toxic agent such as a chemical or form of radiation that causes or increases the frequency of mutations in the

DNA molecules found in cells. See *carcinogen, mutation, teratogen*.

mutation Permanent change in the DNA sequence within a gene in any cell. See *mutagen*.

mutualism Type of species interaction in which two species behave in ways that benefit both by providing each with food, shelter, or some other resource. Compare *commensalism*.

N

nanotechnology Use of science and engineering to manipulate and create materials out of atoms and molecules at the ultra-small scale of less than 100 nanometers.

native species Species that naturally originated in a given ecosystem and have become suited to the environmental conditions there. Compare *indicator species, keystone species, nonnative species*.

natural capital Natural resources and ecosystem services that keep humans and other species alive and support human economies. See *ecosystem services, natural resource*.

natural gas Underground deposits of gases consisting of 50–90% methane and smaller amounts of heavier gaseous hydrocarbon compounds such as propane and butane.

natural income Portion of renewable resource that can be used sustainably.

natural resource Material or energy source in nature that is essential or useful to humans. See *natural capital*.

natural selection Process by which individuals with certain genetic traits are more likely to survive and reproduce under a specific set of environmental conditions, thereby passing these traits on to their offspring. See *adaptation, biological evolution, mutation*.

net energy Amount of high-quality energy available from a given quantity of an energy resource minus the high-quality energy needed to make the energy available.

net primary productivity (NPP) Rate at which producers use photosynthesis to produce and store chemical energy minus the rate at which they use some of this stored chemical energy through cellular respiration. It is used to measure the rate at which producers make chemical energy potentially available to the consumers in an ecosystem. Compare *gross primary productivity*.

nitrogen cycle Cyclic movement of nitrogen in different chemical forms from the environment to organisms and then back to the environment.

nonnative species Species that migrate into an ecosystem or are deliberately or accidentally introduced into an ecosystem by humans. Compare *native species*.

nonpoint source Type of pollution in which pollutants come from many diffuse sources that are hard to pinpoint. Sources include runoff of water and pollutants from cropland, residential areas, clear-cut forests, and construction sites. Compare *point source*.

nonrenewable resource Resource that exists in a fixed amount and takes millions to billions of years to form, so it will be used more quickly than it can be replaced. Examples include copper, aluminum, coal, oil, salt, and sand. Compare *inexhaustible resource, renewable resource*.

nontransmissible disease Disease that is caused by something other than a living organism and does not spread from one person to another. Examples include most cancers, diabetes, cardiovascular disease, and asthma. Compare *transmissible disease*.

NPP See *net primary productivity*.

nuclear fission Method of producing nuclear power by splitting a large nucleus into two or more smaller nuclei. The release of neutrons results in a chain reaction that releases an enormous amount of energy. Compare *nuclear fusion*.

nuclear fusion Method of producing nuclear power in which the nuclei of two isotopes of a light element are forced together at extremely high temperatures until they fuse to form a heavier nucleus, which releases energy in the process. Compare *nuclear fission*.

nutrient cycle Continual movement of the elements and compounds that make up nutrients through air, water, soil, rock, and living organisms within ecosystems. The process is driven by energy from the sun and by Earth's gravity. Some specific nutrient cycles are the carbon, nitrogen, phosphorus, and hydrologic cycles.

O

ocean acidification Rising levels of acidity in ocean waters, occurring because the oceans absorb at least 25% of the carbon dioxide emitted into the atmosphere by human activities, especially the burning of carbon-containing fossil fuels. The carbon dioxide reacts with ocean water to form a weak acid that can endanger entire ecosystems.

ocean current Mass movement of surface water driven by winds and shaped by landforms.

old-growth forest Uncut or regenerated forest that has not been seriously disturbed by human activities or natural disasters for 200 years or more. Also called *primary forest*.

omnivore Animal that can use both plants and other animals as food sources. Examples include pigs, rats, and humans. Compare *carnivore, herbivore*.

open-pit mining Mining technique in which machines dig large holes in Earth's surface and remove metal ores containing copper, gold or other metals, or sand, gravel and stone. See *mountaintop removal, strip mining, surface mining*. Compare *subsurface mining*.

open sea Part of any ocean that lies beyond the continental shelf. Compare *coastal zone*.

GLOSSARY

ore Rock that contains a large enough concentration of a mineral to make it profitable for mining and processing.

organic agriculture Growing crops without the use of synthetic pesticides, synthetic inorganic fertilizers, and genetic engineering; raising livestock on 100% organic feed without the use of antibiotics or growth hormones.

organic compound Compound that contains at least two carbon atoms combined with atoms of one or more other elements. The exception is methane (CH_4), which has only one carbon atom. Known organic compounds number in the millions.

organic farming See *organic agriculture*.

organic fertilizer Organic material such as animal manure, green manure, and compost applied to cropland as a source of plant nutrients. Compare *synthetic inorganic fertilizer*.

output Matter and energy that leaves a system and enters the environment. See *system*. Compare *input, throughput*.

overburden In surface mining, the waste material (vegetation, soil, and rock) that is cleared away in order to access a mineral deposit.

overgrazing Too many animals grazing an area for too long, leading to reduced grass cover, which in turn leads to erosion that compacts soil and reduces its capacity to hold water.

overnutrition Condition in which food energy intake exceeds energy use and causes excess body fat. Overnutrition is caused by consuming too many calories, getting too little exercise, or both. It can lead to obesity, diabetes, and heart disease. Compare *chronic malnutrition*.

ozone layer Stratospheric layer containing much of the atmosphere's ozone. It makes life on land possible by filtering out 95% of the harmful ultraviolet radiation emitted by the sun.

P

parasite Organism that lives on or inside another organism and feeds on it. Parasites can cause serious infectious diseases.

parasitism Relationship in which one organism (a parasite) lives in or on another organism (a host) and benefits at the host's expense. See *parasite*.

passive solar heating system System that absorbs and stores heat from the sun directly within a well-insulated airtight structure. Water tanks as well as walls and floors of concrete, adobe, brick, or stone store much of the collected solar energy as heat that is slowly released. Compare *active solar heating system*.

pasture Managed grassland or enclosed meadow often planted with domesticated grasses or other vegetation to be grazed by livestock.

pathogen Disease-causing agent, such as a bacterium, virus, or parasite. See *bacterium, virus, parasite*.

peak production Point in time, usually after about a decade of pumping, when the pressure in an oil well or oil field drops and its rate of crude oil production starts declining.

peer review Process of scientists reporting the details of the methods they used, the results of their experiments, and the reasoning for their interpretations. Their peers evaluate their work and scientific knowledge advances as scientists question and confirm one another's data and hypotheses.

permafrost Underground soil where captured water can stay frozen for more than two consecutive years.

pest Any species that interferes with human welfare by eating society's food; invading homes, lawns, or gardens; destroying building materials; spreading disease; invading ecosystems; or simply being a nuisance.

petrochemical Usable by-product from crude oil refinement that is utilized to create other chemicals and as an ingredient in products such as plastics, synthetic fibers, paints, medicines, and cosmetics.

petroleum See *crude oil*.

pH Numeric value that indicates the relative acidity or alkalinity of a substance on a scale of 0 to 14, with the neutral point at 7. Acidic solutions have pH values lower than 7; basic or alkaline solutions have pH values greater than 7.

phosphorus cycle Cyclic movement of phosphorus through water, Earth's crust, and living organisms.

photosynthesis Process in which producers change radiant energy (sunlight) into chemical energy. Harnessing the energy of light allows producers to convert inorganic molecules of carbon dioxide and water into organic molecules such as glucose.

photovoltaic cell Device that converts solar energy directly into electrical energy. Also called a *solar cell*.

physical change Process that alters one or more physical properties of matter without changing its chemical composition. Examples include changing the size and shape of a sample of matter (crushing ice or cutting aluminum foil) and changing a sample of matter from one physical state to another (boiling or freezing water). Compare *chemical change*.

phytoremediation Use of natural or genetically engineered plants as "pollution sponges" to absorb, filter, and remove contaminants from polluted soil and water.

plantation agriculture Form of industrialized agriculture used primarily in tropical, less-developed countries to grow cash crops such as bananas, coffee, soybeans, sugarcane, and palm oil. These crops are grown on large monoculture plantations, mostly for export to more-developed countries.

point source Single identifiable source that discharges pollutants into the environment. Examples include the smokestack of a power plant, drainpipe of a meatpacking plant, chimney of a house, or exhaust pipe of an automobile. Compare *nonpoint source*.

policy Set of programs and the laws and regulations through which they are enacted that a government enforces and funds.

politics Process through which individuals and groups try to influence or control government policies and actions that affect local, state, national, and international communities.

pollution Contamination of the environment by any chemical or agent, such as noise or thermal energy, at levels considered harmful to the health, survival, or activities of organisms.

polyculture Growing several different crops on the same plot simultaneously with the use of solar energy and natural fertilizers. Various crops mature at different times, providing year-round food and covering the topsoil, which reduces erosion. Compare *monoculture*.

population Group of interbreeding individuals of the same species, usually living together in a group.

population change Increase or decrease in the size of a population. It is equal to (births + immigration) – (deaths + emigration).

population crash Dieback of a population that has depleted its supply of resources, exceeding the carrying capacity of its environment. See *carrying capacity*.

population density Number of individuals in a population found within a given area or volume.

potential energy Energy stored and potentially available for use. Examples include the water in a reservoir behind a dam and gasoline in a car ready to be burned. Compare *kinetic energy*.

precautionary principle Principle holding that when there is substantial preliminary evidence that an activity, technology, or chemical can harm living things or the environment, decision makers should take action to reduce such harm without waiting for more conclusive evidence.

predation Interaction in which a member of one species (the predator) captures and feeds on all or part of a living organism of another species (the prey) as part of a food web.

predator Organism that captures and feeds on some or all parts of an organism of another species (the prey).

predator-prey relationship Relationship in which the predator species and the prey species evolve to counter one another's defenses and attacks, respectively. See *predator, prey*.

prescribed burn Carefully planned and controlled fire that removes flammable small trees and underbrush in the highest-risk forest areas.

prey Organism that is killed by an organism of another species (the predator) and serves as its source of food.

primary consumer Organism that eats mostly green plants or algae. Compare *detritivore, omnivore, secondary consumer*.

primary ecological succession Ecological succession in an area without soil or bottom sediments. See *ecological succession*. Compare *secondary ecological succession*.

primary recycling Using materials again for the same purpose. Recycling aluminum cans is an example.

producer Organism such as a plant that makes the food it needs from compounds in soil, carbon dioxide, air, and water by using the energy of sunlight. Compare *consumer, decomposer*.

proven oil reserve Known deposit from which oil can be extracted profitably at current prices with current technology.

PV cell See *photovoltaic cell*.

R

rain shadow Semiarid or arid region on the leeward side of a mountain when prevailing winds flow up and over a high mountain, dropping moisture on the windward side.

rangeland Unfenced grassland in temperate and tropical climates that supplies vegetation for grazing (grass-eating) and browsing (shrub-eating) animals. Compare *pasture*.

range of tolerance Range of variations in any physical environment under which a population can survive. See *limiting factor*.

rare earth metal Mineral with superior or unique properties that make it extremely useful in technology products.

reconciliation ecology Method of protecting terrestrial biodiversity by finding ways for humans to share with other species some of the spaces they dominate.

refining Complex process of heating crude oil to separate it into various fuels and other components with different boiling points.

regulation Policies intended to put laws and programs into effect.

reliable surface runoff Surface runoff of water that generally can be counted on as a stable source of fresh water from year to year. It represents one-third of total runoff. See *runoff*.

renewable resource Resource that can be replenished rapidly (in hours to centuries) through natural processes as long as it is not used up faster than it is replaced. Examples include forests, grasslands, wildlife, fertile topsoil, clean air, and fresh water. Compare *nonrenewable resource, inexhaustible resource*. See also *environmental degradation*.

reproductive isolation Halt in the exchange of genes due to the separation of populations. Eventually, members of isolated populations may have very different genetic makeup and no longer be able to interbreed, meaning they have become two distinct species.

GLOSSARY

reserve (mineral) Portion of a mineral resource that is economically and technically feasible for mining.

reservoir Artificial lake created when a river is dammed.

resilience Ability of an ecosystem to be restored through secondary ecological succession after a severe disturbance.

resource partitioning Process in which different species competing for similar scarce resources evolve specialized traits that allow them to "share" the same resources by only using parts of the resources or using them at different times or in different ways. See *ecological niche*.

response Health reaction resulting from exposure to a chemical. See *dose*.

risk Probability of suffering harm from a hazard that can cause injury, disease, death, economic loss, or damage. See *risk analysis, risk assessment, risk management*.

risk analysis Assessing, ranking, and managing risks, as well as communicating information with decision makers and the public about risks. See *risk, risk assessment, risk management*.

risk assessment Process of using statistical methods to estimate how much harm a particular hazard can cause to human health or to the environment. See *risk, risk analysis, risk management*.

risk management Deciding whether and how to reduce a particular risk to a certain level and at what cost. See *risk, risk analysis, risk assessment*.

rock Solid combination of one or more minerals found in Earth's crust. See *mineral*.

rock cycle Interaction of physical and chemical processes that change Earth's rocks from one type to another. In the slowest of Earth's cyclic processes, erosion, melting, and metamorphism produce sedimentary, igneous, and metamorphic rocks, respectively.

***r*-selected species** Species that have a capacity for a high rate of population growth (*r*). They tend to have short life spans and produce many, usually small, offspring, to which they give little or no parental care or protection. Examples include algae, bacteria, and most insects. Compare *K-selected species*.

runoff Surface water that flows into freshwater life zones. See *reliable surface runoff, surface runoff, surface water*. Compare *groundwater*.

S

sanitary landfill Waste disposal site on which waste is spread in thin layers, compacted, and covered with a fresh layer of clay or plastic foam each day.

science Broad field of study focused on discovering how nature works and using that knowledge to describe what is likely to happen in nature. See *data, scientific hypothesis, scientific law, scientific theory*.

scientific hypothesis Possible and testable answer to a scientific question or explanation of what scientists have observed in nature. Compare *scientific law, scientific theory*.

scientific law Description of what scientists find happening in nature repeatedly in the same way, without known exception. See *first law of thermodynamics, law of conservation of matter, second law of thermodynamics*. Compare *scientific hypothesis, scientific theory*.

scientific method Practice used by scientists to ask questions, gather data, and formulate and test hypotheses.

scientific theory Well-tested and widely accepted description of observations that have been repeated many times in a variety of conditions. Compare *scientific hypothesis, scientific law*.

secondary consumer Animal that feeds on primary consumers. Compare *detritivore, omnivore, primary consumer*.

secondary ecological succession Ecological succession in an area in which natural vegetation has been removed or destroyed but some soil or bottom sediment remains. Common sites for secondary ecological succession include abandoned farmland, burned or cut forests, heavily polluted streams, and flooded land. See *ecological succession*. Compare *primary ecological succession*.

secondary recycling Process in which waste materials are converted into different products; for example, used tires can be shredded and turned into rubberized road surfacing. Compare *primary recycling*.

second-growth forest Stand of trees resulting from secondary ecological succession. These forests develop after the trees in an area have been removed by human activities, such as clear-cutting for timber or cropland, or by natural forces such as wildfires and hurricanes.

second law of thermodynamics Scientific law stating that whenever energy is transformed from one type to another in a physical or chemical change, the result is a lower-quality or less usable energy. In any conversion of heat energy to useful work, some of the initial energy input is always degraded to lower-quality, more dispersed, less useful energy—usually low-temperature heat that flows into the environment. See *first law of thermodynamics*.

sedimentary rock Rock formed from sediments such as sand, mud, and organic matter that are laid down in layers, usually at the bottom of lakes and oceans. Layers become rock as they are subjected to increasing weight and pressure from layers above them. Compare *igneous rock, metamorphic rock*. See *rock cycle*.

seed bank Refrigerated, low-humidity storage environment used to preserve genetic information and endangered plant species.

smart growth Set of policies and tools that encourage more environmentally sustainable urban development with less dependence on cars.

smelting Process in which a mineral ore is heated in order to separate a desired metal from the other elements in the ore.

smog Unhealthy combination of air pollutants.

snowpack Mass of accumulated snow that is compressed by its own weight.

soil conservation Methods used to reduce topsoil erosion and restore soil fertility. Examples include terracing, contour planting, strip cropping, and alley cropping.

soil salinization Gradual accumulation of salts in upper soil layers that can stunt crop growth, lower crop yields, and eventually kill plants and ruin the land.

solar thermal system System also known as *concentrating solar power (CSP)* that uses one of several different methods to collect and concentrate solar energy in order to boil water and produce steam for generating electricity.

solid waste Any unwanted or discarded material people produce that is not a liquid or a gas. See *industrial solid waste, municipal solid waste*.

specialist species Species with a narrow ecological niche. They may be able to live in only one type of habitat, tolerate only a narrow range of climatic and other environmental conditions, or eat only one type or a few types of food. Compare *generalist species*.

speciation Formation of a new species from a branch of an existing species through reproductive isolation. Compare *biological extinction*.

species diversity Variety of species present in a specific ecosystem and their abundance within that ecosystem. See *biodiversity*. Compare *genetic diversity*.

spoils Piles of overburden cleared away from a mineral deposit.

stratosphere Layer of the atmosphere between the troposphere and the more distant mesosphere, thermosphere, and exosphere; contains the ozone layer. Compare *troposphere*. See *ozone layer*.

strip mining Any form of mining involving the extraction of mineral deposits that lie in large horizontal beds close to Earth's surface. See *mountaintop removal, open-pit mining, surface mining*. Compare *subsurface mining*.

subsidence Occurs when the land above an aquifer subsides, or sinks.

subsidy Form of government support intended to help a business such as a farm survive.

subsurface mining Extraction of a metal ore or fuel resource such as coal from a deep underground deposit through tunnels and shafts. Compare *mountaintop removal, open-pit mining, strip mining, surface mining*.

surface impoundment Storage for liquid hazardous wastes in ponds, pits, or lagoons.

surface mining Removing vegetation, soil, and rock to extract a mineral deposit in Earth's surface. See *mountaintop removal, open-pit mining, strip mining*. Compare *subsurface mining*.

surface runoff Precipitation that falls on land and flows over land surfaces into streams, rivers, lakes, wetlands, and the ocean, where it can evaporate and repeat the hydrologic cycle.

surface water Precipitation that does not sink into the ground or evaporate; fresh water that flows or is stored in bodies of water on Earth's surface. See *surface runoff*. Compare *groundwater*.

survivorship curve Line graph that shows the percentages of the members of a population surviving at different ages. There are three generalized types of survivorship curves: *late loss, early loss, constant loss*.

sustainability Capacity of Earth's natural systems that support life (including human social systems) to maintain stability or to adapt to changing environmental conditions indefinitely.

synthetic biology Technology that enables scientists to make new sequences of DNA and to use such genetic information to design and create new cells, tissues, organisms, and devices, and to redesign existing natural biological systems.

synthetic inorganic fertilizer Mixture of inorganic plant nutrients applied to the soil to restore fertility and increase crop yields.

synthetic pesticide Chemical used to kill or control populations of organisms considered undesirable.

system Set of components that function and interact in some regular way. Common components of systems are inputs, throughputs, and outputs of matter and energy from the environment. See *input, output, throughput*.

T

tailings Rock wastes that remain after the desired mineral has been extracted. Tailings are put in ponds (where they can be washed out by rain or leak out) or piled up (where the wind can blow them away) and frequently contaminate surface water and groundwater.

GLOSSARY

tectonic plate One of around a dozen pieces of Earth's lithosphere that move slowly across the mantle's flowing asthenosphere and separate, collide, or grind against one another. Tectonic activity is responsible for mountain formation and destruction as well as volcanic eruptions. See *lithosphere*.

temperature inversion Atmospheric condition in which a layer of warm air temporarily lies atop a layer of cooler air nearer the ground. The density of the cooler air keeps it trapped near the ground and pollutants that accumulate in it can rise to harmful, even lethal, concentrations.

teratogen Chemical that harms a fetus or embryo or causes birth defects. Compare *carcinogen*, *mutagen*.

tertiary consumer Consumer that feeds on both primary and secondary consumers. Also called *higher-order consumer*. See *carnivore*. Compare *detritivore, primary consumer, secondary consumer*.

TFR See *total fertility rate*.

thermal energy Energy generated and measured by heat.

threatened species Species that has enough remaining individuals to survive in the short term, but because of declining numbers, is likely to become endangered in the near future. Compare *endangered species*.

throughput Matter and energy flowing through a living system. See *system*. Compare *input, output*.

topsoil Soil underlying all forests, grassland, and croplands. Stores and purifies water and supplies most of the nutrients needed for plant growth. All terrestrial life directly or indirectly depends on topsoil.

total fertility rate (TFR) Average number of children born to women of childbearing age in a population.

toxic chemical Element or compound that can cause temporary or permanent harm or death. See *carcinogen, mutagen, teratogen*.

toxicity Measure of the ability of a substance to cause injury, illness, or death to a living organism.

toxicology Study of the harmful effects of chemicals on humans and other organisms.

toxic waste See *hazardous waste*.

traditional agriculture Method of low-input agriculture characterized by the reliance on human labor and draft animals. Compare *industrial agriculture*.

transmissible disease Infectious disease that can be transmitted from one person to another. Some transmissible diseases are bacterial, such as tuberculosis, meningitis and gonorrhea. Others are viral, such as the common cold, flu, and AIDS. Compare *nontransmissible disease*.

tree farm See *tree plantation*.

tree plantation Managed forest containing only one or two species of trees that are all the same age. Compare *old-growth forest, second-growth forest*.

trophic level Designation for an organism based on its methods of making or finding food and feeding behavior.

troposphere Lowest layer of the atmosphere and the only layer suitable for terrestrial life. Weather occurs in this layer. Compare *stratosphere*.

U

urbanization Trend in which more and more people live in urban areas.

urban sprawl Growth of low-density development on the edges of cities and towns. See *smart growth*.

V

virtual water Water that is not directly consumed but is used to produce food and other products.

virus Infectious agent that is smaller than a bacterium; it works by invading a cell and taking over its genetic machinery to copy itself. It then multiplies and spreads throughout the body, causing a viral disease such as flu or AIDS.

W

waste management Managing wastes to limit their environmental harm without trying to reduce the amount of waste produced. See *integrated waste management*. Compare *waste reduction*.

waste reduction Reducing the amount of waste produced; wastes that are produced are viewed as potential resources that can be reused, recycled, or composted. See *integrated waste management*. Compare *waste management*.

wastewater Water that contains sewage and other wastes from homes and industries.

water cycle See *hydrologic cycle*.

water footprint Rough measure of the volume of water used directly and indirectly to keep a person or group alive and to support their lifestyle(s).

water pollution Any change in water quality that can harm living organisms or make water unfit for human uses such as drinking, irrigation, and recreation.

watershed Land area that delivers runoff, sediment, and dissolved substances to streams, lakes, or wetlands. Watersheds are also called *drainage basins*.

water table Area at the top of the zone of saturation where water rises and falls according to the weather and human intervention. See *zone of saturation*.

weather Set of physical conditions of the lower atmosphere that includes temperature, precipitation, humidity, wind speed, cloud cover, and other factors. Compare *climate*.

wilderness area Area that has not been disturbed by humans and that is protected by law from all harmful human activities.

Z

zone of saturation Zone below the water table where all available pores in soil and rock in Earth's crust are filled by water. See *water table*.

GLOSARIO

A

acidificación oceánica El aumento de los niveles de ácido en las aguas de los océanos; ocurre ya que los océanos absorben al menos el 25% del dióxido de carbono emitido hacia la atmósfera por las actividades humanas, especialmente por la quema de combustibles fósiles que contienen carbono. El dióxido de carbono reacciona con el agua del mar para formar un ácido débil que puede poner en peligro a ecosistemas completos.

acueducto Estructura usada para transportar agua sobre la tierra.

acuicultura Cultivo y cosecha de peces para uso humano que se realiza en lagunas de agua dulce, lagos, reservorios y arrozales y en jaulas bajo el agua en las aguas costeras y profundas del océano.

acuífero Capas porosas y saturadas de agua formadas por arena, grava o roca firme, en donde se acumula agua subterránea.

acuiponía Sistema de producción de cultivos en aguas fertilizadas con los desechos provenientes de la cría de pescado o de otros animales acuáticos.

adaptación Cualquier rasgo hereditario que da a un individuo ciertas ventajas sobre otros individuos en una población determinada. Ver *evolución biológica, mutación, selección natural*.

agricultura de plantación Forma de agricultura industrializada que se usa principalmente en países tropicales y menos desarrollados para sembrar cultivos comerciales, tales como bananas, café, soja, caña de azúcar y aceite de palma. Estos cultivos se siembran en grandes plantaciones de monocultivo, principalmente para ser exportados a países más desarrollados.

agricultura industrializada Método de agricultura caracterizado por una operación a gran escala, monocultivos y el objetivo de aumentar constantemente las cosechas de un cultivo, que es la cantidad de alimento producido por unidad de terreno. Comparar con *agricultura tradicional*.

agricultura tradicional Método de agricultura de baja entrada que se caracteriza por su dependencia en la mano de obra humana y en los animales de tiro. Comparar con *agricultura industrializada*.

agricultura orgánica El cultivo de alimentos sin el uso de pesticidas sintéticos, fertilizantes no orgánicos sintéticos y sin la ingeniería genética; criar ganado con pienso 100% orgánico, sin el uso de antibióticos u hormonas para el crecimiento.

agricultura tradicional de subsistencia Modo de agricultura convencional (no industrializada) que intenta producir suficiente alimento para la supervivencia de una familia que vive en una granja, mientras que a la vez quede una cantidad menor para vender o almacenar. Comparar con *agricultura industrializada*.

agrobiodiversidad La variedad genética de especies de plantas y animales que se usan en las granjas para producir alimento.

agua negruzca Mezcla de agua dulce y agua salada.

agua superficial Precipitación que no es absorbida por el suelo o que no se evapora; agua dulce que fluye o está almacenada en masas de agua en la superficie de la Tierra. Ver *escorrentía de superficie*. Comparar con *aguas subterráneas*.

agua virtual Agua que no es directamente consumida, pero que se usa para producir alimentos y otros productos.

aguas grises Aguas usadas que provienen de las duchas, lavamanos, lavavajillas y lavadoras de ropa que pueden volver a usarse para otros propósitos, como para regar los jardines o lavar los automóviles.

aguas residuales Aguas que contienen desechos de los alcantarillados y otros desechos de hogares e industrias.

aguas subterráneas Precipitación que se filtra hacia el suelo y que se acumula en un *acuífero*. Comparar con *agua superficial, escorrentía*.

aislación geográfica Separación de poblaciones de una especie en distintas áreas. Podría ocurrir a causa de la búsqueda de alimento, un evento natural (como un huracán, terremoto o erupción volcánica) o una barrera física, ya sea natural (como una montaña o valle) o creada por humanos (como una represa o la tala de un bosque).

análisis de riesgos Analizar, organizar y gestionar los riesgos, así como comunicar la información sobre los riesgos a quienes toman las decisiones y con el público. Ver *evaluación de riesgos, gestión de riesgos, riesgo*.

Antropoceno Nueva época en la que los humanos se han convertido en los principales agentes de cambio en el funcionamiento de los sistemas que sustentan la vida en la Tierra a medida que sus huellas ecológicas se han dispersado por todo el planeta. Comparar con *Holoceno*.

arboreto Lugar para el estudio científico y la exhibición pública de distintas especies de árboles y arbustos.

astenosfera Zona dentro del manto terrestre que está compuesta por roca caliente que está parcialmente derretida y que es dúctil.

atenuación Acción tomada para reducir la gravedad de un problema.

atmósfera Capa de gases que rodea la Tierra. Comparar con *biósfera, estratósfera, geósfera, hidrósfera, troposfera*.

átomo Unidad constituyente básica de la materia. Comparar con *ión, molécula*.

B

bacteria Organismo unicelular que puede multiplicarse muy rápidamente. La mayoría no son dañinos, pero algunos pueden causar enfermedades como amigdalitis y tuberculosis.

banco de semillas Ambiente de almacenamiento refrigerado y de baja humedad que se emplea para preservar la información genética de plantas y especies de plantas en peligro de extinción.

bioacumulación Acumulación de un químico en los tejidos adiposos de un organismo.

biocombustible Combustible, tal como el alcohol etílico o biodiesel, hecho a partir de materiales vegetales (biomasa).

biodiversidad Variedad de la vida en la Tierra, incluye la diversidad de especies, la diversidad genética, la diversidad ecológica y la diversidad funcional.

biología sintética Tecnología que permite a los científicos hacer nuevas secuencias de ADN y usar dicha información genética para diseñar y crear nuevas células, tejidos, organismos y aparatos, y para rediseñar los sistemas biológicos naturales existentes.

bioma Área geográfica compuesta por distintos ecosistemas y caracterizada por un clima distintivo y ciertas especies (especialmente vegetación) que son capaces de sobrevivir ahí.

biomagnificación Proceso mediante el cual un químico se hace más concentrado a medida que asciende por la cadena alimenticia.

biomasa Materia orgánica producida por plantas que puede quemarse en su forma sólida o gaseosa para producir energía.

biomímesis Hacer que los procesos de producción sean más limpios y sustentables al rediseñarlos para imitar la manera en que la naturaleza procesa los desechos.

bioprospector Persona entrenada para buscar plantas y animales en bosques tropicales y en otros ecosistemas, de manera que estos puedan ser usados para producir medicamentos.

biorremediación Uso de bacterias y enzimas para ayudar a destruir las sustancias tóxicas o peligrosas o convertirlas en compuestos no dañinos.

biósfera Zona de la Tierra en donde se encuentra la vida. Involucra las partes de la atmósfera (la troposfera), la hidrósfera (principalmente el agua de la superficie y las aguas subterráneas) y la litósfera (principalmente el suelo, las rocas de la superficie y los sedimentos en el lecho de los océanos y otras masas de agua). Comparar con *atmósfera, geósfera, hidrósfera*.

bosque secundario Conjunto de árboles que resultaron de una sucesión ecológica secundaria. Estos bosques se desarrollan después de que los árboles de un área han sido talados para actividades humanas, como la tala de bosques para obtener madera o tierras de cultivo, o por causas naturales, tales como incendios forestales o huracanes.

bosque virgen Bosque no talado o regenerado que no se ha visto impactado seriamente por las actividades humanas o por desastres naturales por 200 años o más. También se conoce como *bosque primario*.

C

CAC Ver *captación y almacenamiento de carbono*.

cadena alimentaria Secuencia de organismos en que cada organismo es una fuente de nutrientes o de energía para el siguiente nivel de organismos. Comparar con *red alimentaria*.

calefacción de distrito Sistema de distribución del calor en que el calor se envía a múltiples edificios desde una planta central que a veces también genera energía.

cambio climático Cambio en las condiciones atmosféricas globales promedio durante un período de al menos tres décadas.

cambio físico Proceso que altera una o más propiedades físicas de la materia sin cambiar su composición química. Ejemplos incluyen cambiar el tamaño y la forma de una muestra de materia (machacar el hielo o cortar el papel de aluminio) y transformar una muestra de materia de un estado físico a otro (hervir o congelar el agua). Comparar con *cambio químico*.

cambio poblacional Aumento o disminución en el tamaño de una población. Es igual a (nacimientos + inmigración) − (fallecimientos + emigración).

cambio químico Cambio en la composición química de una sustancia. Comparar con *cambio físico*.

canje de deuda por naturaleza Acuerdo mediante el cual los países participantes actúan como defensores de reservas de bosques protegidas a cambio de ayuda internacional o de reducción de deudas.

capa de ozono Capa de la estratósfera que contiene gran parte del ozono de la atmósfera. Hace que la vida en la Tierra sea posible, ya que filtra el 95% de la radiación ultravioleta nociva emitida por el Sol.

capa freática Área en la parte superior de la zona de saturación en donde el agua asciende y desciende según el tiempo atmosférico y la intervención humana. Ver *zona de saturación*.

capacidad de carga Población máxima de especies que un hábitat puede sustentar indefinidamente.

capital humano Talentos físicos e intelectuales de las personas que brindan destrezas laborales, organizacionales y de administración e innovación. Comparar con *capital manufacturado, capital natural*.

GLOSARIO

capital manufacturado Maquinarias, materiales y fábricas que las personas han creado usando recursos naturales. Comparar con *capital humano, recurso natural*.

capital natural Recursos naturales y servicios de un ecosistema que mantienen vivos a los humanos y a otras especies y que sustentan las economías humanas. Ver *recurso natural, servicios ecosistémicos*.

captación y almacenamiento de carbono (CAC) Proceso mediante el que se remueve el gas de dióxido de carbono de plantas que queman carbón y plantas industriales y se almacena en algún lugar (usualmente bajo el suelo o bajo el lecho marino). Para que el proceso sea efectivo, el dióxido de carbono debe almacenarse, esencialmente para siempre, para que no pueda liberarse hacia la atmósfera.

carbón Combustible fósil formado a partir de los restos de plantas terrestres que fueron enterradas y expuestas a calor y presión intensas por 300 a 400 millones de años.

carcinógeno Químico, tipo de radiación o virus que podría causar o promover el cáncer. Comparar con *mutágeno, teratógeno*.

carnívoro Animal que se alimenta principalmente de otros animales. Comparar con *herbívoro, omnívoro*.

CDB Ver *Convenio sobre la Diversidad Biológica*.

celda de combustible de hidrógeno Aparato que usa el gas hidrógeno como combustible para producir electricidad cuando reacciona con el gas oxígeno en la atmósfera y emite un vapor de agua no dañino.

celda fotovoltaica Aparato que transforma la energía solar directamente en energía eléctrica. También se conoce como *celda solar*.

celda FV Ver *celda fotovoltaica*.

cénit de producción Punto en el tiempo, usualmente después de una década de bombeo, cuando la presión en un pozo petrolero o campo petrolero baja y su ritmo de producción de petróleo crudo comienza a decaer.

CFC Ver *clorofluorocarbono*.

choque poblacional Declive de una población que ha agotado su suministro de recursos, excediendo la capacidad de carga de su medioambiente. Ver *capacidad de carga*.

ciclo de las rocas Interacción de procesos físicos y químicos que transforman las rocas de la Tierra de un tipo a otro. En los procesos cíclicos más lentos de la Tierra, la erosión, el derretimiento y el metamorfismo producen las rocas sedimentarias, ígneas y metamórficas, respectivamente.

ciclo de los nutrientes Movimiento continuo de los elementos y compuestos que conforman los nutrientes a través del aire, agua, suelo, rocas y organismos vivos de los ecosistemas. El proceso es impulsado por la energía del sol y la gravedad de la Tierra. Algunos ciclos de los nutrientes específicos son el del carbono, nitrógeno, fósforo y el ciclo hidrológico.

ciclo de retroalimentación Proceso que ocurre cuando una salida de materia, energía o información se redirige a la entrada del sistema y causa cambios en ese sistema.

ciclo del agua Ver *ciclo hidrológico*.

ciclo del carbono Movimiento cíclico del carbono en distintas formas químicas desde el medioambiente hacia los organismos y luego de vuelta al medioambiente.

ciclo del fósforo Movimiento cíclico del fósforo a través del agua, la corteza terrestre y los organismos vivos.

ciclo del nitrógeno Movimiento cíclico del nitrógeno en distintas formas químicas desde el medioambiente hacia los organismos y luego de regreso hacia el medioambiente.

ciclo hidrológico Ciclo mediante el cual se reúne, purifica y distribuye el suministro fijo del agua de la Tierra.

ciencia Amplio campo de estudio que se enfoca en descubrir cómo funciona la naturaleza y usa ese conocimiento para describir qué es posible que ocurra en la naturaleza. Ver *datos, hipótesis científica, ley científica, teoría científica*.

ciencias ambientales Estudio interdisciplinario de cómo los humanos interaccionan con el medioambiente. Incluye información e ideas de la ingeniería, las ciencias naturales y las ciencias sociales. Los objetivos fundamentales de las ciencias ambientales son aprender cómo ha sobrevivido y prosperado la vida en la Tierra, comprender cómo humanos interaccionan con el medioambiente y hallar maneras de enfrentar los problemas medioambientales y vivir de una manera más sustentable.

CITES Ver *Convención sobre el Comercio Internacional de Especies Amenazadas de la Flora y Fauna Silvestres*.

clima Patrón general de las condiciones atmosféricas en un área determinada durante períodos que van desde décadas a miles de años. Los factores clave que influencian el clima de un área son la energía solar entrante, la rotación de la Tierra, los patrones del movimiento del aire y el agua, los gases en la atmósfera y las características de la superficie de la Tierra. Comparar con *tiempo*.

clorofluorocarbono (CFC) Compuesto que contiene carbón, cloro y flúor. Estos compuestos son dañinos para el ozono de la estratósfera.

coevolución Proceso de selección natural en el que cambios en el acervo genético de una especie causan cambios en el acervo genético de otra especie. Ver *evolución biológica, selección natural*.

cogeneración Producción de dos formas útiles de energía, tales como vapor de alta temperatura y electricidad, a partir de una misma fuente de combustible.

comensalismo Interacción entre organismos de distintas especies en que un tipo de organismo se beneficia y el otro no sale favorecido ni dañado en ningún grado. Comparar con *mutualismo*.

compactación de nieve Masa de nieve acumulada que se comprime a causa de su propio peso.

competencia interespecífica Intento de los miembros de dos o más especies por usar los mismos recursos limitados en un ecosistema.

competencia medioambiental Conocimiento y comprensión de los conceptos de las ciencias medioambientales.

compost Fertilizante que se produce cuando los microorganismos descomponen la materia orgánica, tales como hojas, residuos de las cosechas, desechos de alimentos, papel y madera en la presencia de oxígeno.

compostación Forma de reciclaje que imita la naturaleza al usar bacterias para descomponer los desechos de los jardines, los restos de alimentos vegetales y otros desechos orgánicos biodegradables en humus, que es un componente orgánico del suelo que mejora la productividad de la tierra.

compuesto Combinación de dos o más elementos diferentes que se mantienen unidos en proporciones fijas. Comparar con *elemento*.

compuesto orgánico Compuesto que contiene al menos dos átomos de carbono combinados con átomos de uno o más elementos externos. La excepción es el metano (CH_4), que solo tiene un átomo de carbono. Hay millones de compuestos orgánicos conocidos.

concesión de conservación Estrategia en la que los gobiernos u organizaciones privadas por la conservación pagan a gobiernos o a quienes tienen tierras en otras naciones para preservar los recursos naturales de sus tierras por una cantidad determinada de tiempo.

conservación del suelo Métodos usados para reducir la erosión del mantillo y restaurar la fertilidad del suelo. Ejemplos incluyen la creación de terrazas, la siembra de contorno, el cultivo en franjas y el cultivo en callejones.

consumidor Organismo que no puede producir su propio alimento y que obtiene sus nutrientes orgánicos al alimentarse de los tejidos de los productores o de otros consumidores; generalmente se dividen en *consumidores primarios* (herbívoros), *consumidores secundarios* (carnívoros), *consumidores terciarios* (de nivel superior), *omnívoros* y *detritívoros* (descomponedores y consumidores de detritus). Comparar con *productor*.

consumidor primario Organismo que se alimenta principalmente de plantas verdes o algas. Comparar con *consumidor secundario, detritívoro, omnívoro*.

consumidor secundario Animal que se alimenta de consumidores primarios. Comparar con *detritívoro, omnívoro, consumidor primario*.

consumidor terciario Consumidor que se alimenta de los consumidores primarios y secundarios. También se conoce como *consumidor de orden superior*. Ver *carnívoro*. Comparar con *consumidor primario, consumidor secundario, detritívoro*.

contaminación del aire Cualquier material gaseoso o sólido en la atmósfera que ocurre en concentraciones lo suficientemente altas como para alterar el clima o causar daño a organismos, ecosistemas o materiales producidos por los humanos.

contaminación hídrica Cualquier cambio en la calidad del agua que puede causar daños a los organismos vivos o hacer que el agua no sea segura para los usos humanos como, por ejemplo, para beber, irrigar y la recreación.

contaminación Polución del medioambiente mediante cualquier químico o agente, tal como el ruido o la energía térmica, a niveles que se consideran dañinos para la salud, la supervivencia o las actividades de los organismos.

Convención sobre el Comercio Internacional de Especies Amenazadas de la Flora y Fauna Silvestres (CITES) Tratado internacional que prohíbe cazar, capturar y vender ejemplares de especies amenazadas.

Convenio sobre la Diversidad Biológica (CDB) Tratado internacional que compromete a los gobiernos participantes a que reduzcan la tasa global de pérdida de biodiversidad y a que compartan los beneficios del uso de los recursos genéticos del mundo.

corriente oceánica Movimiento masivo del agua de la superficie impulsado por los vientos y que toma forma gracias a los accidentes geográficos.

corteza Capa sólida exterior de la Tierra. Está compuesta por la corteza oceánica y la corteza continental. Comparar con *manto, núcleo*.

cosmovisión ambiental Conjunto de supuestos y valores de un individuo con respecto al mundo natural y la gestión que estos piensan que deberían cumplir. Ver *visión global centrada en la Tierra, visión global centrada en los humanos, visión global centrada en la vida*.

crecimiento económico Aumento en la capacidad de una nación, estado, ciudad o compañía para proporcionar bienes y servicios. Comparar con *desarrollo económico*.

crecimiento exponencial Crecimiento en el que alguna cantidad, tal como el tamaño de la población o la productividad económica, aumenta a un porcentaje fijo de unidades de tiempo. El crecimiento exponencial comienza lentamente, pero pronto la cantidad se hace enorme. Cuando ese aumento en cantidad a lo largo del tiempo se representa en una gráfica, este tipo de crecimiento produce una curva con la forma de la letra J.

crecimiento inteligente Conjunto de políticas y herramientas que promueven un desarrollo urbano más sustentable desde el punto de vista ecológico con una menor dependencia en los automóviles.

cuenca Terreno que entrega la escorrentía, sedimentos y sustancias disueltas a arroyos, lagos o humedales. Una cuenca también se conoce como *cuenca hidrográfica*.

GLOSARIO

curva de supervivencia Gráfica lineal que muestra los porcentajes de los miembros de una población que sobreviven en distintas edades. Hay tres tipos generalizados de curvas de supervivencia: *pérdida constante, pérdida tardía, pérdida temprana.*

D

datos Información factual reunida por los científicos.

deforestación Remoción temporal o permanente de amplias extensiones de bosques para la agricultura, asentamientos humanos o para otros usos.

degradación medioambiental Disminución, deterioro o desperdicio del capital natural de la Tierra.

delta Área en la boca de un río que contiene depósitos de sedimentos del río mismo, a menudo contiene estuarios y humedales costeros.

densidad poblacional Número de individuos de una población que se hallan dentro de un área o volumen determinado.

deposición ácida La caída de ácidos y de compuestos que forman ácidos desde atmósfera hacia la superficie de la Tierra. La deposición ácida también se conoce como lluvia ácida.

depredación Interacción en la que un miembro de una especie (el predador) atrapa y se alimenta de un organismo vivo de otra especie o de partes de él (la presa) como parte de una red alimentaria.

deriva continental Movimiento lento de los continentes por la superficie de la Tierra.

desalinización Proceso mediante el cual se remueve la sal del agua del océano o de las aguas negruzcas (levemente saladas) en acuíferos o lagos.

desarrollo económico Cualquier conjunto de esfuerzos que se enfocan en crear economías que mejoran el bienestar de los humanos al satisfacer sus necesidades básicas. Comparar con *crecimiento económico.*

descarbonización Transición global para dejar de usar combustibles fósiles.

descomponedor Consumidores que obtienen nutrientes al descomponer materias orgánicas no vivas, tales como desechos de hojas, árboles caídos y animales muertos. Mientras se alimentan, estos organismos liberan nutrientes de los desechos que devuelven los nutrientes al suelo y al agua. Comparar con *consumidor, detritívoro, productor.*

desechos peligrosos Materiales o sustancias desechadas que amenazan la salud humana o el medioambiente.

desechos sólidos Cualquier material no deseado o desechado que las personas producen que no es líquido o gaseoso. Ver *desechos sólidos industriales, desechos sólidos municipales.*

desechos sólidos industriales Desechos industriales producidos por minas, granjas e industrias que proveen a las personas sus bienes y servicios. Comparar con *desechos sólidos municipales.*

desechos sólidos municipales (DSM) Desechos sólidos combinados producidos por hogares y lugares de trabajo en vez de fábricas. Ver *desechos sólidos.* Comparar con *desechos sólidos industriales.*

desechos tóxicos Ver *desechos peligrosos.*

desertificación Condición en que el potencial productivo del mantillo (capa superior del suelo) decae en un 10% o más a causa de ciertas prácticas agrícolas, tales como el pastoreo excesivo, la deforestación y el arado excesivo, en combinación con la sequía y el cambio climático. En casos extremos, puede causar condiciones desérticas.

desnutrición crónica Condición en que los individuos afectados no obtienen suficientes proteínas y otros nutrientes clave. Puede hacer que los individuos sean más vulnerables a las enfermedades y dificulta el desarrollo físico y mental normal en los niños. Comparar con *hiperalimentación.*

detritívoro Organismo consumidor que se alimenta de detritus (organismos recientemente muertos). Ejemplos incluyen el gusano de tierra, algunos insectos, las hienas y los buitres. Comparar con *descomponedor.*

dispersión urbana Crecimiento de desarrollo de baja densidad en los bordes de las ciudades y pueblos. Ver *crecimiento inteligente.*

diversidad de las especies Variedad de especies presente en un ecosistema específico y su abundancia en dicho ecosistema. Ver *biodiversidad.* Comparar con *diversidad genética.*

diversidad de los ecosistemas Diversidad de los terrenos de comunidades biológicas, tales como los desiertos, pastizales, bosques, lagos, ríos y humedales.

diversidad funcional Variedad de procesos que ocurren en los ecosistemas. Ejemplos incluyen el flujo de energía y los ciclos de la materia. Ver *diversidad de las especies, biodiversidad, diversidad de los ecosistemas, diversidad genética.*

diversidad genética Variedad de genes hallada en una población o especie. Ver *diversidad de las especies, biodiversidad.* Comparar con *diversidad de los ecosistemas, diversidad funcional.*

dosis Cantidad de un químico nocivo que ha sido ingerido, inhalado o absorbido por una persona a través de la piel en un momento determinado. Comparar con *respuesta.*

DSM Ver *desechos sólidos municipales.*

E

ecociudad Ciudad que es sustentable ecológicamente. Los residentes pueden caminar, montar en bicicleta o usar transporte público de baja contaminación para realizar la mayor parte de sus viajes. Los residentes cultivan sus propios alimentos y reciclan o reutilizan la mayoría de sus desechos. Los sitios abandonados e industriales se limpian y se preservan los bosques, pastizales, humedales y granjas cercanas.

ecología Ciencia biológica que estudia cómo interactúan los seres vivos con las partes vivas y no vivas de sus medioambientes.

ecología de reconciliación Método de protección de la biodiversidad terrestre al hallar maneras en que los humanos puedan compartir con otras especies algunos de los espacios que dominan.

ecologismo Movimiento social dedicado a la protección de la Tierra y sus recursos.

economía Ciencia social que trata sobre la producción, distribución y consumo de bienes y servicios para satisfacer las necesidades y deseos de las personas.

ecosistema Una o más comunidades de distintas especies que interaccionan entre sí y con los factores químicos y físicos de su medioambiente abiótico.

efecto de borde Tendencia de que surja una zona de transición entre dos ecosistemas diferentes que tenga una mayor diversidad de especies y una mayor densidad de organismos de lo que se encuentra en cualquiera de los ecosistemas individuales.

efecto invernadero Proceso mediante el cual la energía solar calienta la troposfera a medida que se refleja desde la superficie de la Tierra (geósfera) e interacciona con el dióxido de carbono, metano, vapor de agua (de la hidrósfera y biósfera) y otros gases de efecto invernadero (atmósfera). Esto calienta la Tierra y sustenta la vida.

eficiencia energética Medición de cuánto trabajo útil puede extraerse de cada unidad de energía usada. Ver *energía neta*.

elemento Tipo de materia con un conjunto único de propiedades que no puede dividirse en sustancias más simples mediante procesos químicos. Comparar con *compuesto*.

embalse de superficie Almacenamiento de líquidos peligrosos en estanques, canteras o lagunas.

energía Capacidad de realizar un trabajo.

energía cinética Energía asociada con el movimiento, tal como la energía del agua en movimiento, un automóvil en movimiento o la electricidad. Comparar con *energía potencial*.

energía comercial Energía que se vende en el comercio.

energía geotérmica Calor almacenado en el suelo, en las rocas bajo el suelo y en los fluidos en el manto de la Tierra. Se usa para calefaccionar y enfriar edificios y agua y para producir electricidad.

energía hidráulica Tecnología que usa la energía cinética del agua en movimiento o del agua que cae para producir electricidad.

energía neta Cantidad de energía de alta calidad disponible a partir de una cantidad determinada de un recurso energético menos la energía de alta calidad que se necesita para lograr que esa energía esté disponible.

energía potencial Energía almacenada y potencialmente disponible para el uso. Ejemplos incluyen el agua en un reservorio tras una represa y la gasolina en un coche que está listo para ser usado. Comparar con *energía cinética*.

energía térmica Energía generada y medida por el calor.

enfermedad infecciosa Enfermedad causada cuando un patógeno, por ejemplo, una bacteria, virus o parásito invade el cuerpo y se multiplica en sus células y tejidos. Ejemplos incluyen la gripe, el SIDA, la tuberculosis, las enfermedades con diarrea y la malaria. Ver *enfermedad transmisible*. Comparar con *enfermedad no transmisible*.

enfermedad no transmisible Enfermedad que no es causada por un organismo vivo y que no se contagia de una persona a otra. Ejemplos incluyen la mayoría de los tipos de cáncer, la diabetes, las enfermedades cardiovasculares y el asma. Comparar con *enfermedad transmisible*.

enfermedad transmisible Enfermedad infecciosa que puede transmitirse de una persona a otra. Algunas enfermedades transmisibles son bacterianas, tales como la tuberculosis, la meningitis y la gonorrea. Otras son virales, como el resfrío común, la gripe y el SIDA. Comparar con *enfermedad no transmisible*.

entrada Materia y energía del medioambiente que entran a un sistema. Ver *sistema*. Comparar con *proceso, salida*.

escorrentía Agua de la superficie que fluye hacia áreas de agua dulce. Ver *agua superficial, escorrentía de superficie, escorrentía superficial fiable*. Comparar con *aguas subterráneas*.

escorrentía de superficie Precipitación que cae en el suelo y que fluye sobre las superficies terrestres hacia arroyos, ríos, lagos, humedales y el océano, en donde se evapora. Luego, el ciclo hidrológico se repite.

escorrentía superficial fiable Escorrentía superficial de agua con la que generalmente se puede contar como una fuente estable de agua dulce a través de los años. Representa un tercio de la escorrentía total. Ver *escorrentía*.

especiación Formación de una nueva especie a partir de una rama de una especie existente a través del aislamiento reproductivo. Comparar con *extinción biológica*.

GLOSARIO **671**

GLOSARIO

especie amenazada Especie que tiene suficientes individuos restantes para sobrevivir a corto plazo, pero que a causa de la caída de sus números, es posible que se convierta en una especie en peligro de extinción en el futuro cercano. Comparar con *especie en peligro de extinción*.

especie clave Especie que preserva un ecosistema al controlar las poblaciones de animales presa que de otra manera consumirían suficiente materia orgánica como para devastar el ecosistema. Ejemplos incluyen gusanos, lobos, nutrias marinas, caimanes y tiburones. Comparar con *especie indicadora, especie nativa, especie no nativa*.

especie en peligro de extinción Especie con tan pocos ejemplares sobrevivientes que la especie podría extinguirse pronto. Comparar con *especie amenazada*.

especie endémica Especie que solo se encuentra en un área. Tales especies son especialmente vulnerables a la extinción.

especie especialista Especie con un nicho ecológico reducido. Es posible que solo sea capaz de vivir en un tipo de hábitat, tolerar solo un rango limitado de condiciones climáticas y medioambientales, o solo comer un tipo o unos cuantos tipos de alimento. Comparar con *especie generalista*.

especie generalista Especie con un nicho amplio. Pueden vivir en lugares diferentes, consumir una variedad de alimentos y tolerar una amplia variedad de condiciones medioambientales. Ejemplos incluyen a las moscas, cucarachas, ratones, ratas y humanos. Comparar con *especie especialista*.

especie indicadora Especie cuya presencia o ausencia indica la calidad o características de ciertas condiciones medioambientales. Comparar con *especie clave, especie nativa, especie no nativa*.

especie nativa Especie que se ha originado naturalmente en un ecosistema determinado y que se ha adaptado a las condiciones medioambientales de dicho lugar. Comparar con *especie clave, especie indicadora, especie no nativa*.

especie no nativa Especie que migra hacia un ecosistema o que los humanos introducen intencional o accidentalmente en un ecosistema. Comparar con *especie nativa*.

especies estratega-K Especies que tienden a prosperar en condiciones competitivas cuando el número de su población es cercana a la capacidad de carga (K) de su ambiente. Tienden a tener una larga esperanza de vida, se reproducen más tarde y producen menos crías. Ejemplos incluyen a los elefantes, ballenas, humanos, cactus saguaro y la mayor parte de los árboles del bosque lluvioso. Comparar con *especies estratega-r*.

especies estratega-r Especies que tienen la capacidad de lograr un alto ritmo de crecimiento de sus poblaciones (r). Tienden a tener una expectativa de vida corta y producen mucha descendencia (usualmente pequeñas) a las cuales brindan pocos o nulos cuidados paternos o protección. Ejemplos incluyen algas, bacterias y la mayoría de los insectos. Comparar con *especies estratega-K*.

estéril En la minería de superficie, el material de desecho (vegetación, suelo y roca) que se remueve para obtener acceso a los depósitos minerales.

estratosfera Capa de la atmósfera entre la troposfera y la mesósfera (más distante), la termósfera y la exósfera; contiene la capa de ozono. Comparar con *tropósfera*. Ver *capa de ozono*.

estructura etaria Distribución de los individuos entre distintos grupos de edad en una población.

estuario Área costera parcialmente cerrada en donde un río se une con el mar y el agua de mar se mezcla con el agua dulce.

ética medioambiental Estudio de las distintas creencias sobre qué es correcto o incorrecto con respecto a cómo las personas tratan al medioambiente.

eutrofización Proceso mediante el que una masa de agua obtiene nutrientes.

eutrofización de las aguas Sobrecarga de nutrientes en los ecosistemas acuáticos con nutrientes vegetales (principalmente nitratos y fosfatos) que ocurre como resultado de actividades humanas, tales como la agricultura, la urbanización y los desechos provenientes de las plantas de tratamiento de aguas servidas.

evaluación de riesgos Proceso que consiste en emplear métodos estadísticos para estimar qué tanto daño puede causar un peligro en particular a la salud humana o al medioambiente. Ver *análisis de riesgos, gestión de riesgos, riesgos*.

evolución Ver *evolución biológica*.

evolución biológica Proceso mediante el cual una especie cambia genéticamente a lo largo del tiempo.

expectativa de vida Edad promedio que se puede esperar que viva una persona con base en el año, país y región en que nació.

extinción Ver *extinción biológica*.

extinción biológica Desaparición completa de una especie de la Tierra. Ocurre cuando una especie no puede adaptarse a sobrevivir y reproducirse en respuesta a cambios en su medioambiente y no puede trasladarse a un nuevo medioambiente con condiciones más favorables. Comparar con *especiación*. También ver *especie amenazada, especie en peligro de extinción, extinción masiva*.

extinción masiva Aumento significativo en el ritmo de extinción que es más alto que la tasa de extinción de base. Comparar con *tasa de extinción de base*.

F

factor limitante Factor que es particularmente importante relativo a otros factores en regular el crecimiento de una población.

fertilidad Número de nacimientos en una población.

fertilizante orgánico Material orgánico (como el abono animal y el compost) que se aplica a las tierras de cultivo como una fuente de nutrientes para las plantas. Comparar con *fertilizante sintético inorgánico*.

fertilizante sintético inorgánico Mezcla de nutrientes vegetales inorgánicos que se aplican al suelo para restaurar la fertilidad y aumentar la cantidad de las cosechas.

fisión nuclear Método para producir energía nuclear al dividir un núcleo grande en dos o más núcleos más pequeños. La liberación de neutrones causa una reacción en cadena que libera una enorme cantidad de energía. Comparar con *fusión nuclear*.

fitorremediación Uso de plantas naturales o diseñadas genéticamente como "esponjas para la contaminación" para que absorban, filtren y remuevan los contaminantes del suelo y agua contaminados.

flujo Ver *proceso*.

fósil Restos o huellas preservadas de organismos prehistóricos. Pueden incluir esqueletos, huesos, dientes, conchas, hojas, semillas mineralizados o petrificados o impresiones de estos mismos objetos, así como impresiones de actividad animal, tales como huellas, rastros o madrigueras.

fotosíntesis Proceso mediante el cual los productores transforman la energía radiante (del Sol) en energía química. El emplear la energía de la luz permite que los productores transformen las moléculas inorgánicas de dióxido de carbono y agua en moléculas orgánicas, tales como la glucosa.

fracking Ver *fracturación hidráulica*.

fracturación hidráulica Método que utiliza bombas de alta presión para impulsar enormes cantidades de una mezcla de agua, arena y distintos químicos a un pozo para lograr fracturar la roca porosa y liberar el gas natural o petróleo atrapado en ellas. También se conoce por el anglicismo *fracking*.

fragmentación del hábitat Fragmentación de un hábitat más grande en áreas más pequeñas; ocurre usualmente como resultado de las actividades humanas.

fuente no puntual Tipo de contaminación en que los contaminantes provienen de muchas fuentes difusas que son difíciles de identificar. Entre esas fuentes se incluyen la escorrentía de agua y de contaminantes de las tierras de cultivo, áreas residenciales, bosques talados y sitios de construcción. Comparar con *fuente puntual*.

fuente puntual Fuente única identificable que emite contaminantes hacia el medioambiente. Ejemplos incluyen la chimenea de una planta de energía, una tubería de drenaje de una planta de empaque de carne, la chimenea de una casa o el tubo de escape de un automóvil. Comparar con *fuente no puntual*.

fundición Proceso mediante el que la mena de un mineral se calienta para así separar un metal deseado de los otros elementos de la mena.

fusión nuclear Método para producir energía nuclear en que los núcleos de dos isótopos de un elemento liviano se fusionan a temperaturas extremadamente altas hasta que se fusionan para formar un núcleo más pesado, lo que libera energía en el proceso. Comparar con *fisión nuclear*.

G

gas de efecto invernadero Gas en la parte inferior de la atmósfera de la Tierra que causa el efecto invernadero. Ejemplos incluyen el vapor de agua, dióxido de carbono, metano y óxido nitroso.

gas natural Depósitos subterráneos de gases que consisten en 50–90% metano y cantidades más pequeñas de otros compuestos gaseosos de hidrocarbono, como el propano y el butano.

geoingeniería Manipulación de las condiciones genéticas para contrarrestar el efecto invernadero.

geósfera El núcleo, manto y corteza de la Tierra, todo el material que se encuentra sobre y bajo la superficie de la Tierra y que conforma la masa del planeta. Comparar con *atmósfera, biósfera, hidrósfera*.

gestión de desechos Administración de los desechos para limitar los daños al ambiente sin tratar de reducir la cantidad de desechos producidos. Ver *gestión integral de residuos*. Comparar con *reducción de desechos*.

gestión de riesgos Decidir si es necesario reducir un riesgo en particular a un nivel determinado, cómo hacerlo y qué costos tendrá. Ver *análisis de riesgos, evaluación de riesgos, riesgo*.

gestión integral de residuos Variedad de estrategias coordinadas para la reducción de los residuos y para el manejo de los mismos; diseñada para encargarse de los residuos sólidos que producen los humanos.

granja de árboles Ver *plantación de árboles*.

H

hábitat Área que proporciona todos los factores abióticos y bióticos que una especie necesita para sobrevivir. Comparar con *nicho ecológico*.

herbívoro Organismo que se alimenta principalmente de plantas verdes o algas. Ejemplos incluyen venados, ovejas, saltamontes y zooplancton. Comparar con *carnívoro, omnívoro*.

GLOSARIO

hidrósfera Toda el agua en estado gaseoso, líquido y sólido de la superficie de la Tierra. Ver *ciclo hidrológico*. Comparar con *atmósfera, biósfera, geósfera*.

hiperalimentación Condición en que el consumo de energía proveniente de los alimentos excede el uso de energía y causa un exceso de grasa corporal. La hiperalimentación es causada por consumir demasiadas calorías, no ejercitar lo suficiente o ambas. Puede causar obesidad, diabetes y enfermedades cardíacas. Comparar con *desnutrición crónica*.

hipótesis científica Respuesta posible y que puede ponerse a prueba para contestar una pregunta o explicación científica de lo que los científicos han observado en la naturaleza. Comparar con *ley científica, modelo, teoría científica*.

hipótesis de protección Hipótesis que declara que la biodiversidad asegura los ecosistemas contra una disminución en su funcionamiento, ya que muchas especies proporcionan una mayor garantía de funcionar incluso si otras fallan.

Holoceno Período de clima relativamente estable y otras condiciones medioambientales; permitió que creciera la población humana, que esta desarrollara la agricultura y que ocupara mayores terrenos y otros recursos de la Tierra. Comparar con *Antropoceno*.

huella del carbono Cantidad de dióxido de carbono generada por un individuo, organización, país o cualquier otra entidad durante un período de tiempo determinado.

huella ecológica Cantidad de terreno y agua que se necesita para suministrar los recursos renovables a una población y para absorber y reciclar los desechos y contaminación que el uso de aquellos recursos produce.

huella hídrica Medición del volumen de agua usado directa e indirectamente para mantener viva a una persona o a un grupo y para sustentar sus estilos de vida.

huella piscícola Área del océano necesaria para sustentar el consumo de una persona promedio, una nación o el mundo, con base en la cantidad de pescado (en peso) que consumen anualmente. Comparar con *huella ecológica*.

humedal costero Territorio a lo largo de la costa que está cubierto por agua durante todo el año o parte de él. Ejemplos incluyen marismas y bosques de manglares. Comparar con *humedal interior*.

humedal interior Terrenos alejados de la costa, tales como un pantano, marisma o estanque pequeño que están cubiertos todo el tiempo o parte de él con agua dulce. Comparar con *humedal costero*.

I

inercia Capacidad de un ecosistema de sobrevivir disturbios moderados. También se denomina *persistencia*.

ingeniería genética Manipulación científica de genes para seleccionar características deseables o eliminar las indeseables. Permite que los científicos alteren el material genético de un organismo al agregar, eliminar o cambiar segmentos del ADN en un proceso llamado empalme de genes. Comparar con *selección artificial, selección natural*.

ingreso natural Porción de los recursos renovables que pueden emplearse sustentablemente.

inseguridad alimentaria Falta de acceso a alimentos nutritivos. Comparar con *seguridad alimentaria*.

inversión térmica Condición atmosférica en que una capa de aire cálido se asienta temporalmente sobre una capa de aire más frío cerca del suelo. La densidad del aire más frío la mantiene atrapada cerca del suelo y los contaminantes que se acumulan en ella pueden subir hasta alcanzar concentraciones dañinas e incluso letales.

inyección profunda en pozos Acción de bombear desechos líquidos nocivos a mucha presión a través de una tubería hacia formaciones rocosas secas y porosas que están más allá de los acuíferos que se utilizan para obtener agua potable y para la irrigación.

ion Átomo o grupo de átomos con una o más cargas eléctricas positivas (+) o negativas (−). Ejemplos incluyen Na^+ y Cl^-. Comparar con *átomo, molécula*.

irrigación por goteo Método de irrigación en que el agua se entrega directamente a las raíces de las plantas a través de tuberías o tubos.

isótopo Una de dos o más formas de un elemento químico que tienen el mismo número atómico, pero números de masa distintos, ya que tienen una cantidad distinta de neutrones en sus núcleos.

J

jardín botánico Lugar que contiene una exhibición de plantas vivas.

justicia medioambiental Ideal mediante el cual todo individuo merece protección de riesgos medioambientales sin importar su raza, género, edad, nacionalidad, salario, clase social o cualquier factor político.

L

legislaciones Conjunto de programas y las leyes y regulaciones mediante las cuales se promulgan, que un gobierno hace cumplir y financia.

LEPE Ver *Ley sobre las Especies en Peligro de Extinción*.

ley Política basada en el poder legislativo del gobierno de los Estados Unidos.

ley científica Descripción de lo que los científicos descubren que ocurre en la naturaleza de forma repetida de la misma manera sin excepciones conocidas. Ver *ley de conservación de la materia, primer principio de la termodinámica, segundo principio de la termodinámica*. Comparar con *hipótesis científica, modelo, teoría científica*.

ley de conservación de la energía Ver *primer principio de la termodinámica*.

ley de conservación de la materia Ley científica que declara que cuando la materia pasa por un cambio físico o químico, no se crean ni se destruyen sus átomos.

Ley sobre las Especies en Peligro de Extinción (LEPE) Ley estadounidense que identifica y protege a las especies en peligro de extinción y a las especies amenazadas y que promueve la existencia de programas de restitución para ayudar a que las especies se recuperen hasta alcanzar números en que ya no necesiten protección legal.

litósfera Combinación de la corteza de la Tierra y de la parte rígida y exterior del manto sobre la astenosfera. Comparar con *corteza, geósfera, manto*.

lluvia ácida Ver *deposición ácida*.

M

manejo integrado de plagas (MIP) Programa diseñado para reducir los daños a los cultivos hacia un nivel económicamente tolerable con un uso mínimo de pesticidas sintéticos, ya que evalúa cada cultivo y sus plagas como parte de un mismo ecosistema.

mantillo El suelo que recubre todos los bosques, praderas y tierras de cultivo. Almacena y purifica el agua y proporciona la mayoría de los nutrientes que son necesarios para el crecimiento de las plantas. Toda la vida terrestre depende del mantillo, ya sea directa o indirectamente.

manto Zona del interior de la Tierra, entre el núcleo y la corteza. Comparar con *corteza, núcleo*. Ver *astenosfera, geósfera, litósfera*.

mar abierto Parte del océano que se encuentra más allá de la plataforma continental. Comparar con *zona costera*.

materia Cualquier cosa que tenga masa y que ocupe un espacio. Comúnmente existe en estado sólido, líquido o gaseoso.

medioambiente Todas las condiciones externas, factores, materia y energía (viva y no viva) que afectan a cualquier organismo vivo u otro sistema específico.

mena Roca que contiene una concentración lo suficientemente grande de un mineral como para que sea rentable para la minería y el procesamiento.

metal tierra rara Mineral con propiedades superiores o peculiares que lo hacen extremadamente útil en productos tecnológicos.

método científico Práctica usada por los científicos para formular preguntas, reunir información y formular y poner a prueba sus hipótesis.

mineral Elemento químico o compuesto no orgánico que existe como un sólido con una estructura interna que se repite regularmente. Ver *recurso mineral*.

minería a cielo abierto Cualquier forma de minería que involucre la extracción de depósitos minerales que se encuentran en depósitos horizontales grandes cerca de la superficie de la Tierra. Comparar con *minería de subsuelo, minería de superficie*.

minería a tajo abierto Técnica minera en que las máquinas excavan grandes agujeros en la superficie de la Tierra y remueven menas de metal que contienen cobre, oro u otros metales, o arena, grava y piedra. Comparar con *minería de remoción de cima, minería de subsuelo*.

minería de remoción de cima Tipo de minería de superficie en que se talan grandes áreas boscosas y se usan explosivos para eliminar la cima de una montaña y dejar expuesto el carbón bajo el suelo y roca que se han extraído.

minería de subsuelo Extracción de una mena de metal o recurso de combustible, tal como el carbón, de un depósito que se encuentra en las profundidades del suelo a través de túneles y pasajes. Comparar con *minería de superficie*.

minería de superficie Remover la vegetación, el suelo y las rocas para extraer un depósito mineral en la superficie de la Tierra. Ver *minería de remoción de cima, minería a tajo abierto, minería a cielo abierto*. Comparar con *minería de subsuelo*.

MIP Ver *manejo integrado de plagas*.

modelo Representación física o matemática de una estructura o sistema.

molécula Combinación de dos o más átomos del mismo o de distintos elementos que se mantienen unidos por enlaces químicos. Comparar con *átomo, ion*.

GLOSARIO

momento crítico del cambio climático Punto en el que un problema medioambiental alcanza un umbral en donde los científicos temen que podría causar trastornos climáticos irreversibles.

monocultivo Cultivo de una sola cosecha, usualmente en un terreno amplio. Comparar con *policultivo*.

mortalidad Número de muertes en una población.

mutación Cambio permanente en la secuencia de ADN en un gen en cualquier célula. Ver *mutágeno*.

mutágeno Agente tóxico tal como un químico o una forma de radiación que causa o aumenta la frecuencia de mutaciones en las moléculas de ADN que se encuentran en las células. Ver *carcinógeno, mutación, teratógeno*.

mutualismo Tipo de interacción entre las especies en que dos especies se comportan de maneras en que ambas se benefician al proporcionarles a ambas alimento, refugio u otro recurso. Comparar con *comensalismo*.

N

nanotecnología Uso de las ciencias y de la ingeniería para manipular y crear materiales en base a átomos y moléculas en una escala ultra pequeña de menos de 100 nanómetros.

nicho ecológico Papel que tiene una especie en un ecosistema y que incluye todo lo que afecta su supervivencia y reproducción.

nivel trófico Designación para un organismo en base a sus métodos para producir o hallar alimento y alimentarse.

núcleo Parte central de la Tierra que consiste en un núcleo interno sólido y un núcleo externo líquido. Comparar con *corteza, manto*.

O

omnívoro Animal que puede alimentarse tanto de plantas como de otros animales. Ejemplos incluyen a los cerdos, ratas y humanos. Comparar con *carnívoro, herbívoro*.

P

parasitismo Relación en la que un organismo (el parásito) vive sobre o dentro de otro organismo (el huésped) y se beneficia a costa de él. Ver *parásito*.

parásito Organismo que vive dentro de o sobre otro organismo y que se alimenta de él. Los parásitos pueden causar graves enfermedades infecciosas.

partición de recursos Proceso en el que distintas especies que compiten por escasos recursos parecidos evolucionan y desarrollan rasgos especializados que les permiten "compartir" los mismos recursos al solo usar partes de esos recursos o al usarlos en distintos momentos o de maneras diferentes. Ver *nicho ecológico*.

pastadero Pradera sin cercar en climas templados y tropicales que proporciona vegetación para animales de pastoreo (que se alimentan de pasto) y para los que pacen (se alimentan de arbustos). Comparar con *pastizal*.

pastizal Praderas controladas o prados cercados que suelen contener pastos domesticados u otros tipos de vegetación para que paste el ganado.

patógeno Agente que causa enfermedades, como las bacterias, virus o parásitos. Ver *bacteria, parásito, virus*.

perforación horizontal Perforación que se realiza junto con la fracturación hidráulica para minar depósitos petroleros anteriormente no disponibles.

permafrost Capa de subsuelo en donde el agua puede permanecer congelada por más de dos años consecutivos.

pesticida sintético Químicos usados para matar o controlar las poblaciones de organismos que se consideran indeseables.

petróleo Ver *petróleo crudo*.

petróleo crudo Líquido viscoso que consiste en compuestos de hidrocarburo y otros compuestos, y que se formó a partir de los restos de organismos antiguos a lo largo de millones de años. El petróleo crudo, que se extrae de depósitos subterráneos, luego se envía a refinerías de petróleo, en donde se calienta para separarlo en distintos tipos de combustibles y otros componentes.

petroquímico Producto utilizable derivado del refinamiento del petróleo crudo que se emplea para crear otros químicos y como ingrediente en productos, tales como plásticos, fibras sintéticas, pinturas, medicinas y cosméticos.

pH Valor numérico que indica la acidez o alcalinidad relativa de una sustancia en una escala del 0 al 14, su punto neutro es 7. Las soluciones ácidas tienen un valor de pH menores que 7; las soluciones básicas o alcalinas tienen un pH mayor que 7.

piscicultura Ver *acuicultura*.

placa tectónica Una entre aproximadamente una docena de fragmentos de litósfera de la Tierra que se mueven lentamente por el manto sobre la astenosfera. Los fragmentos se separan, chocan o se rozan entre sí hasta pulverizar sus bordes. La actividad tectónica es responsable por la formación y destrucción de las montañas así como por la actividad volcánica. Ver *litósfera*.

plaga Cualquier especie que interfiere con el bienestar de los humanos al consumir el alimento de la sociedad; invadir hogares, céspedes o jardines; destruir materiales de construcción; diseminar enfermedades; invadir ecosistemas o simplemente al ser una molestia.

plantación de árboles Bosque administrado que solo contiene una o dos especies de árboles de la misma edad. Comparar con *bosque secundario, bosque virgen*.

población Grupo de individuos de la misma especie, que se reproducen entre sí y que suelen vivir juntos.

policultivo Cultivar distintos sembradíos en el mismo terreno de forma simultánea con el uso de la energía solar y de fertilizantes naturales. Los distintos cultivos maduran en épocas diferentes, lo que proporciona alimento a lo largo de todo el año y cubre el mantillo, lo que reduce la erosión. Comparar con *monocultivo*.

política Proceso mediante el que los individuos y grupos tratan de influenciar o controlar las políticas del gobierno y las acciones que afectan a las comunidades locales, estatales, nacionales e internacionales.

PPB Ver *producción primaria bruta*.

PPN Ver *producción primaria neta*.

predador Organismo que atrapa y se alimenta de un organismo vivo de otra especie o de partes de él (la presa).

presa Organismo que es atrapado por un organismo de otra especie (el predador) y que le sirve como fuente de alimento.

presionar Influenciar el voto o acciones de un funcionario elegido.

primer principio de la termodinámica Ley científica que declara que cuando la energía se transforma de una forma a otra mediante un cambio físico o químico, la energía no se crea ni se destruye. La cantidad total de energía no cambia. Ver *segundo principio de la termodinámica*.

principio precautorio Principio que sostiene que cuando hay evidencia preliminar significativa de que una actividad, tecnología o químico puede dañar a los seres vivos o al medioambiente, quienes tienen el poder para tomar decisiones, deben entrar en acción para reducir dicho daño sin esperar a obtener más pruebas concluyentes.

proceso La materia y energía que fluyen por un sistema vivo. Ver *sistema*. Comparar con *entrada, salida*.

producción primaria bruta (PPB) Ritmo al cual los productores de un ecosistema convierten la energía radiante en energía química. Comparar con *producción primaria neta*.

producción primaria neta (PPN) Ritmo al que los productores usan la fotosíntesis para producir y almacenar la energía menos el ritmo al que usan parte de esta energía química almacenada a través de la respiración celular. Se usa para medir el ritmo al que los productores hacen que la energía química esté potencialmente disponible a los consumidores de un ecosistema. Comparar con *producción primaria bruta*.

productor Organismo, tal como una planta, que produce el alimento que necesita a partir de los compuestos en el suelo, dióxido de carbono, aire y agua al usar la energía del Sol. Comparar con *consumidor, descomponedor*.

punto crítico de biodiversidad Área rica en especies endémicas en peligro de extinción.

punto ecológico de no retorno Ocurrencia en un sistema natural que queda estancado en un ciclo de retroalimentación de cambios sin comprobar. Más allá de este punto, el sistema puede cambiar tan drásticamente que sufre de una degradación o colapso severo.

Q

quema controlada Incendio cuidadosamente planificado y controlado que elimina los árboles pequeños e inflamables y la maleza en áreas boscosas de alto riesgo.

químico tóxico Elemento o compuesto que puede causar daños temporarios o permanentes o la muerte. Ver *carcinógeno, mutágeno, teratógeno*.

R

radiación electromagnética Formas de energía cinética que viajan como ondas electromagnéticas. Ejemplos incluyen ondas de radio, la luz visible y la radiación ultravioleta.

rango de tolerancia Rango de variaciones de cualquier medioambiente físico en el que puede sobrevivir una población. Ver *factor limitante*.

rasgo adaptativo Ver *adaptación*.

reacción química Ver *cambio químico*.

reciclaje de circuito cerrado Ver *reciclaje primario*.

reciclaje primario Usar los materiales nuevamente con el mismo propósito. El reciclaje de las latas de aluminio es un ejemplo de esto.

reciclaje secundario Proceso mediante el que materiales de desecho se convierten en distintos productos; por ejemplo, las llantas usadas pueden triturarse y convertirse en asfalto goma. Comparar con *reciclaje primario*.

recurso inexhaustible Recurso disponible en un suministro constante durante el futuro concebible. Ejemplos incluyen la luz del sol, el viento y las aguas en movimiento impulsadas por la luz solar. Comparar con *recurso no renovable, recurso renovable*.

recurso mineral Concentración de minerales que es lo suficientemente amplia como para cubrir el costo de extraerla y procesarla para crear materia prima y productos útiles.

recurso natural Fuente material o energética de la naturaleza que es esencial o útil para los humanos. Ver *capital natural*.

GLOSARIO

recurso no renovable Recurso que existe en una cantidad fija y que toma de millones a billones de años en formarse, de manera que será usado más rápidamente de lo que puede reemplazarse. Ejemplos incluyen el cobre, aluminio, carbón, petróleo, sal y arena. Comparar con *recurso inexhaustible, recurso renovable*.

recurso renovable Recurso que puede reemplazarse rápidamente (en horas o siglos) a través de procesos naturales, siempre y cuando no se use más rápido de lo que es reemplazado. Ejemplos incluyen bosques, pastizales, la vida silvestre, el mantillo fértil, el aire puro y el agua dulce. Comparar con *recurso inexhaustible, recurso no renovable*. Ver también *degradación medioambiental*.

red alimentaria Red compleja de cadenas alimentarias interconectadas. Comparar con *cadena alimentaria*.

reducción de desechos Reducción de la cantidad de desechos producidos; los desechos producidos se ven como recursos potenciales que podrían reutilizarse, reciclarse o usarse para compost. Ver *gestión integral de residuos*. Comparar con *gestión de desechos*.

refinar Proceso complejo en que el crudo se calienta para luego separarse en distintos tipos de combustibles y otros componentes mediante el uso de distintos puntos de ebullición.

reglamentación Políticas que buscan poner en efecto leyes y programas.

relación predador-presa Relación en que tanto la especie de predador y la especie de presa evolucionan para contrarrestar las defensas y ataques del otro, respectivamente. Ver *predador, presa*.

relave Desechos de roca que quedan después de que se han extraído los minerales deseados. El relave se coloca en estanques (en donde se lo puede llevar el agua de lluvia o puede filtrarse) o puede apilarse (donde el viento puede llevárselo) y con frecuencia contamina las aguas superficiales y las subterráneas.

relleno sanitario Sitio de eliminación de residuos en donde la basura se dispone en capas delgadas, se compacta y se cubre con una capa fresca de arcilla o de espuma plástica cada día.

remoción de huevos Recolectar los huevos que especies en peligro crítico de extinción han puesto en su hábitat natural para que luego eclosionen en zoológicos o en centros de investigación.

reproducción aislada Detención en el intercambio de genes a causa de la separación de las poblaciones. Eventualmente, los miembros de las poblaciones aisladas podrían tener una composición genética diferente y ya no serían capaces de reproducirse entre sí, lo que quiere decir que se han convertido en dos especies diferentes.

reproducción en cautiverio Proceso mediante el cual se reúne a individuos silvestres de una especie en peligro crítico de extinción para que se reproduzcan en cautiverio, con el objetivo de luego introducir a sus crías en su hábitat silvestre.

reserva (mineral) Porción de un recurso mineral que es económica y técnicamente viable para la minería.

reserva de la biósfera Área que contiene un ecosistema nuclear estrictamente protegido rodeado por una zona neutral en donde los habitantes locales pueden extraer recursos de manera sustentable.

reserva probada de petróleo Depósito conocido desde el que puede extraerse petróleo de forma rentable a los precios actuales con la tecnología actual.

reservorio Lago artificial que se crea cuando se hace una represa en un río.

resiliencia Capacidad de un ecosistema de restaurarse mediante la sucesión ecológica secundaria después de trastornos graves.

resistencia medioambiental Todos los factores limitantes que actúan en conjunto para frenar el crecimiento de una población. Ver *factor limitante*.

respiración aeróbica Proceso complejo que utiliza el oxígeno y glucosa para producir energía; ocurre en las células de la mayoría de los organismos vivos. El dióxido de carbono y el agua son los derivados de esta reacción. Comparar con *fotosíntesis*.

respiración anaeróbica Forma de respiración celular a través de la cual algunos descomponedores obtienen la energía que necesitan mediante la descomposición de glucosa y otros nutrientes en la ausencia de oxígeno.

respuesta Reacción de la salud que resulta de la exposición a un químico. Ver *dosis*.

restauración ecológica Alteración intencional de un hábitat o ecosistema que ha sido dañado a un estado que se asemeja a su estado original.

revisión por pares Proceso en que los científicos reportan los detalles de los métodos que utilizan, los resultados de sus experimentos y el razonamiento para sus interpretaciones. Sus pares evalúan su trabajo y avances en el conocimiento científico a medida que cuestionan y confirman sus datos e hipótesis.

revolución de los materiales Reemplazo de los minerales con otros materiales para su uso en la industria y tecnología.

Revolución Verde Proceso mediante el cual se aumenta la producción de los cultivos a través de la agricultura industrializada.

riego por inundación Método de irrigación no eficiente en que el agua se bombea hacia zanjas y desde allí fluye hacia los cultivos, gracias a la gravedad.

riesgo Probabilidad de sufrir daños a causa de un peligro que podría causar lesiones, enfermedades, muerte, pérdida económica o daños. Ver *análisis de riesgos, evaluación de riesgos, gestión de riesgos*.

rimero Pilas de estéril que se remueven de un depósito mineral.

roca Combinación sólida de uno o más minerales que se encuentra en la corteza de la Tierra. Ver *mineral*.

roca ígnea Roca que se forma bajo o sobre la superficie de la Tierra cuando el magma surge desde el manto y luego se enfría y endurece. Comparar con *roca metamórfica, roca sedimentaria*. Ver *ciclo de las rocas*.

roca metamórfica Roca que se forma cuando la roca existente se coloca bajo mucha presión y altas temperaturas, lo que causa que se derrita parcialmente o una combinación de ambas. Comparar con *roca ígnea, roca sedimentaria*. Ver *ciclo de las rocas*.

roca sedimentaria Roca formada por sedimentos tales como arena, lodo y materia orgánica que se depositan en capas, usualmente en el fondo de lagos y océanos. Estas capas se convierten en roca a medida que se ven expuestas a un mayor peso y presión de las capas que están sobre ellas. Comparar con *roca ígnea, roca metamórfica*. Ver *ciclo de las rocas*.

S

salida Materia y energía que salen de un sistema y que ingresan en el medioambiente. Ver *sistema*. Comparar con *entrada, proceso*.

salinización del suelo Acumulación gradual de sales en las capas superiores del suelo que puede dificultar el crecimiento de los cultivos, disminuir la cantidad de las cosechas y eventualmente matar las plantas y arruinar el terreno.

sector pesquero Concentración de una especie acuática apta para el cultivo comercial en un océano o masa de agua tierra adentro.

segundo principio de la termodinámica Ley científica que declara que cuando la energía se transforma de un tipo a otro durante un cambio físico o químico, el resultado es un tipo de energía de menor calidad o menos utilizable. Durante cualquier conversión de energía calórica a energía utilizable, parte de la energía inicial entrante siempre se degrada a un tipo de energía de menor calidad, más dispersa y menos útil—usualmente calor de baja temperatura que fluye por el medioambiente. Ver *primer principio de la termodinámica*.

seguridad alimentaria Acceso a suficientes alimentos seguros y nutritivos para tener un estilo de vida saludable. Comparar con *inseguridad alimentaria*.

selección artificial Proceso mediante el cual los humanos seleccionan uno o más rasgos genéticos deseables en la población de una especie vegetal o animal y luego emplean la reproducción selectiva para producir poblaciones que contengan muchos individuos con los rasgos deseados. Comparar con *ingeniería genética, selección natural*.

selección natural Proceso mediante el cual es más probable que sobrevivan y se reproduzcan los individuos con ciertas características genéticas bajo un conjunto específico de condiciones medioambientales, transmitiendo esas características a su descendencia. Ver *adaptación, evolución biológica, mutación*.

sequía Período seco prolongado en una región determinada. El calentamiento atmosférico es una causa principal de la sequía.

servicios ecosistémicos Servicios naturales que sustentan la vida y las economías humanas sin costo monetario. Ejemplos de esto son el ciclo de los nutrientes, el control natural de plagas y la purificación natural del aire y del agua. Ver *capital natural, recurso natural*.

siembra orgánica Ver *agricultura orgánica*.

sistema Conjunto de componentes que funcionan e interaccionan de manera regular. Los componentes comunes de los sistemas son las entradas, procesos y salidas de materia y energía del medioambiente. Ver *entrada, salida, proceso*.

sistema de calefacción solar activo Sistema que usa colectores solares para capturar la energía del Sol y almacenarla como calor para calefaccionar el ambiente y el agua. Comparar con *sistema de calefacción solar pasivo*.

sistema de calefacción solar pasivo Sistema que absorbe y almacena el calor del sol directamente de una estructura bien aislada y hermética. Los tanques de agua, así como los muros de concreto, adobe, ladrillo o piedra, almacenan gran parte de la energía solar recolectada y liberan lentamente el calor. Comparar con *sistema de calefacción solar activo*.

sistema térmico solar Sistema que también se conoce como *energía solar de concentración (ESC)* que usa uno de distintos métodos para recolectar y concentrar la energía solar para así hervir el agua y producir vapor para generar electricidad.

smog Combinación no saludable de contaminantes en el aire.

sobrepastoreo Ocurre cuando demasiados animales han pastado en un área por demasiado tiempo, lo que causa que se reduzca la cobertura de pasto en un área. A su vez, esto causa erosión, lo que compacta el suelo y reduce su capacidad para retener el agua.

sociedad medioambientalmente sostenible Sociedad que protege su capital natural y que vive de sus beneficios al mismo tiempo que satisface las necesidades básicas de recursos del presente y del futuro de sus habitantes.

sombra orográfica Región semiárida o árida en el lado de sotavento de una montaña; ocurre cuando los vientos prevalentes soplan hacia la parte superior y sobre una montaña alta, dejando caer humedad en el costado a barlovento.

GLOSARIO

subsidencia Ocurre cuando el terreno sobre un acuífero se hunde.

subsidiar Forma de apoyo gubernamental que busca ayudar a la subsistencia de un negocio, por ejemplo, una granja.

sucesión ecológica Cambio típicamente gradual en la composición de las especies de un área determinada. Ver *sucesión ecológica primaria, sucesión ecológica secundaria*.

sucesión ecológica primaria Sucesión ecológica en un área sin suelo o sedimentos de base. Ver *sucesión ecológica*. Comparar con *sucesión ecológica secundaria*.

sucesión ecológica secundaria Sucesión ecológica en un área en donde la vegetación natural ha sido removida o destruida, pero en donde aún queda parte del suelo o de los sedimentos de fondo. Sitios comunes de sucesión ecológica secundaria incluyen tierras agrícolas abandonadas, bosques quemados o talados, arroyos seriamente contaminados y territorios inundados. Ver *sucesión ecológica*. Comparar con *sucesión ecológica primaria*.

sustentabilidad Capacidad de los sistemas naturales de la Tierra que sustentan la vida (incluyendo los sistemas sociales humanos), para mantener la estabilidad o adaptarse a las condiciones medioambientales cambiantes indefinidamente.

T

tasa de extinción de base Tasa naturalmente baja según la cual las especies han desaparecido a través de la mayor parte de la historia de la Tierra. Comparar con *extinción masiva*.

tasa de mortalidad infantil Número de recién nacidos de cada 1,000 que mueren antes de su primer cumpleaños.

tasa total de fertilidad (TTF) Número promedio de hijos que tiene una mujer en edad reproductiva dentro de una población.

teoría científica Descripción de observaciones bien probadas y ampliamente aceptadas que han sido repetidas muchas veces en una variedad de condiciones. Comparar con *hipótesis científica, ley científica*.

teratógeno Químico que daña a un feto o embrión o que causa defectos de nacimiento. Comparar con *carcinógeno, mutágeno*.

tiempo Conjunto de condiciones físicas en la parte inferior de la atmósfera. El tiempo incluye la temperatura, precipitación, humedad, velocidad del viento, cobertura de nubes y otros factores. Comparar con *clima*.

tiempo de agotamiento Cantidad de tiempo que toma utilizar cierta proporción (usualmente el 80%) de las reservas de un mineral a una tasa de empleo determinada.

toxicidad Medición de la capacidad de una sustancia de causar daños, enfermedades o la muerte a los organismos vivos.

toxicología Estudio de los efectos dañinos de los químicos en los humanos y en otros organismos.

troposfera Capa más baja de la atmósfera y la única capa capaz de sustentar la vida terrestre. El tiempo atmosférico tiene lugar en esta capa. Comparar con *estratósfera*.

TTF Ver *tasa total de fertilidad*.

U

urbanización Tendencia en que más y más personas deciden vivir en áreas urbanas.

V

variación genética Variedad en la composición genética de individuos en una población.

virus Agente infeccioso más pequeño que una bacteria; actúa al invadir a una célula y al tomar control de su maquinaria genética para producir copias de sí mismo. Luego, se multiplica y propaga por todo el cuerpo, causando una enfermedad viral como el resfrío común o el SIDA.

visión global centrada en la Tierra Visión global que sostiene que las personas somos parte de la naturaleza y que dependemos de ella; que el capital natural de la Tierra existe para todas las especies, no solo para los humanos; que el éxito económico y la supervivencia a largo plazo y de las culturas y especies dependen de aprender cómo la Tierra se ha sustentado a sí misma durante miles de millones de años; y que las lecciones de la naturaleza deben influenciar cómo las personas piensan y actúan. Comparar con *visión global centrada en los humanos, visión global centrada en la vida*. Ver *cosmovisión ambiental*.

visión global centrada en la vida Visión global que sostiene que todas las especies tienen valor al cumplir con su papel particular en la biósfera, independientemente de su uso potencial o real en la sociedad; incluye la creencia de que las personas tienen la responsabilidad de cuidar y de ser guardianes responsables del planeta. Comparar con *visión global centrada en la Tierra, visión global centrada en los humanos*.

visión global centrada en los humanos Visión global que ve al mundo natural como el sistema que sustenta la vida humana. Se divide en la visión global de administración del planeta y en la visión global de protección. Comparar con *visión global centrada en la Tierra, visión global centrada en la vida*.

Z

zona costera   Parte cálida, rica en nutrientes y poco profunda del océano que se extiende desde la marca de marea alta en tierra hasta el borde poco profundo y levemente inclinado de la plataforma continental. Comparar con *mar abierto*.

zona de saturación   Zona bajo la capa freática en donde todos los poros del suelo y de las rocas de la corteza terrestre están llenos de agua. Ver *capa freática*.

zona de vida acuática   Sistema acuático compuesto por agua con una concentración de sal disuelta de no más de 1,000 ppm (partes por millón). Ejemplos incluyen masas de agua dulce como lagos, estanques y humedales tierra adentro y sistemas que fluyen, como arroyos y ríos. Comparar con *bioma*.

zona neutral   Zona periférica que rodea el núcleo interior estrictamente protegido de una reserva en donde las personas pueden extraer recursos de manera sustentable.

zona silvestre   Área que no ha sido intervenida por los humanos y que es protegida por la ley de toda actividad humana perjudicial.

zonas de vida acuática   Porción marina o de agua dulce de la biósfera. Ejemplos incluyen las zonas de vida de agua dulce (como en lagos y arroyos) y las zonas de vida oceánica marina (como estuarios, zonas costeras, arrecifes de coral y el mar abierto).

zonas de vida marina   Medioambiente de agua salada que se encuentra en el océano o en su bahía, estuario, humedal costero, costa, arrecife de coral o bosque de manglares.

INDEX

Note: Page numbers in **boldface** type indicate key terms. Page numbers followed by italicized *f* or *b* indicate figures and boxes.

A

AAAS. *See* American Association for the Advancement of Science (AAAS)
Abyssal zone, 181, 182*f*–185*f*
Acid deposition, **540**–543, 541*f*–542*f*
Acidification of oceans, 158, **184**, 267–270, 270*f*
Acidity, pH to measure, 56
Acid rain, 405. *See also* Acid deposition
Acoustic research, 532–533*f*
Active solar heating systems, **434**, 434*f*, 436*f*
Adaptation, **122**, 169, 563
Adventure Scientists, 614
Aerobic respiration, **80**
Aeroponics, 491, 491*f*
Aerosols
  as air pollutants, 551
  atmospheric, 95*f*
Affluence, unsustainable resource use and, 30–31
African savanna, 138, 172–173*f*
"Age of Man," 214*b*. *See also* Anthropocene epoch
Age structure of human population, **145**, 479–480, 479*f*, 509*f*
Agribusiness, 280*f*–281*f*. *See also* Agriculture
Agricultural chemicals, 227–228, 228*f*
Agricultural wastes, as biomass source, 443
Agriculture, 290–294, 291*f*–293*f*
  air pollution, climate change and, 307
  biodiversity reduction and, 307–308, 308*f*
  climate change impacts, 560
  energy requirements, 302
  fish/shellfish production, 299–301, 300*f*–301*f*, 309–311, 310*f*
  for feed and fuel, 296*f*, 307*b*
  food production increases, 290
  green revolutions, 294–295, 294*f*–295*f*
  importance of pollinators for, 212, 223
  improved varieties, 295*f*, 296–299, 296*f*–298*f*
  meat production, 299, 299*b*, 309, 309*f*
  urban farming, 491, 491*f*
  pest control and, **311**, 312*f*
  soil and, 302–307, 303*f*–307*f*,
  314–318, 315*f*–317*f*
  sustainable practices, 282, 282*b*, 283*f*. *See also* Sustainable practices
  synthetic pesticides, 312–314, 313*f*
Agrobiodiversity, **307**–308
Agroecology, 306
Agroforestry, 315
Air circulation, climate and, 162–164, 162*f*–164*f*
Air pollution, 530–575
  acid deposition, 540–543, 541*f*–542*f*
  agriculture and, 307
  atmospheric layers, 535, 535*f*
  in coastal cities, 530*f*–531*f*
  factors decreasing, 538
  factors increasing, 538
  fracking and, 404
  glacier research, 532–533*f*
  Greenland ice, melting of, 534, 534*f*, 572
  indicator species and, 118, 537
  indoor, 544, 544*f*
  from industrialization, 37
  laws and regulations on, 543–545, 545*f*
  from mining, 379
  in National Parks, 257
  natural and human sources of, 397, 405–406, **536**–538, 536*f*–537*f*, 538*b*
  purification of, as ecosystem service, 21
  smog, 539–540, 539*f*–540*f*
Al Bagr Towers, Abu Dhabi, United Arab Emirates, 50*f*
Algae, 444–445*f*, 642
Algae blooms, 152, 347–348
Alkalinity, pH to measure, 56
Allen, Will, 284, 284*f*, 287
Alley cropping, 315
Amazonia, 66*f*–67*f*, 176*f*–177*f*
American alligator, as keystone species, 117
American Association for the Advancement of Science (AAAS), 546–547, 551
American Water Works Association, 332*f*, 340
Amino acids, 57
Amphibian species
  declines in, 106–108*f*, 128
  as pollution indicators, 118, 118*b*
Anaerobic respiration, **80**
Anglerfish, 184*f*–185*f*
Animal waste, 309
Antarctic ice, melting of, 555, 572
Anthony, Katey Walter, 392–393*f*
Anthropocene epoch, **95**, 214*b*
Antibacterial cleansers, 505, 513
Antibiotics
  bacterial genetic resistance to, 122–123, 122*f*, 505
  in livestock production, 309, 505
Antiquities Act, 612
Apatite mineral, 56*f*
Appalachian Trail, 614–615*f*
Appliances, energy-efficient, 433
Aquaculture, **301**, 301*f*, 321
  advantages and disadvantages of, 310*f*
Aquaponics, **321**
Aquatic biodiversity
  difficulties in protecting, 270–271, 271*f*
  ecosystem approach to, 271
  human impact on, 266–270, 267*f*–269*f*
  reversal of damage to, 271*b*
  sustaining, 271*f*
Aquatic life zones, **180**
Aqueducts, **336**
Aquifers. *See also* Groundwater; Hydrologic cycle
  depletion of, 334–335, 334*f*, 335*f*
  description of, **84**, 331
Arboretum, **236**
Arctic National Wildlife Refuge, 236*f*
Area strip mining, 378
Arizona-Sonora Desert Museum (Tucson, AZ), 201, 201*f*
Artificial selection, **127**
Ashoka Trust for Research in Ecology and the Environment (India), 264
Asian carp, 224, 224*f*
Asthenosphere, **366**
Atlantic Forest (Brazil), 68–69*f*
Atmosphere, **71**–73, 73*f*
Atmospheric layers, 535, 535*f*
Atoms, **55**, 55*f*, 74*f*
Automobiles, advantages and disadvantages of, 489–490, 490*f*

B

Baby boom, 479, 479*b*
Background extinction rate, **125**, 213–214
Bacteria, 79*f*, 122–123, 122*f*, **504**–505
Ballard, Robert, 377, 377*f*, 387
Bangladesh, climate change and 555, 558*f*–559*f*
Basel Convention, 604
Bathyal zone, 181, 182*f*–183*f*
Batrachochytrium dendrobatidis fungus, 128
Batteries
  for electric cars, 385
  mineral resources for, 362–363*f*
  rechargeable, 593
Bawa, Kamaljit St., 264, 264*f*
BDFFP (Biological Dynamics of Forest Fragments Project), 94
Becker, Vitor, 68–70, 69*f*, 98, 109
Beddington Zero Energy Development (UK), 564*f*
Bicycle lanes, 490, 490*f*
Big Bend National Park (Texas), 260*f*
Bioaccumulation, **227**, 517
BioBlitz events, 51, 51*f*, 465*f*–467*f*
Biodiesel fuel, 444
Biodiversity
  agricultural reduction in, 307–308, 308*f*
  bird species and, 244–245*f*
  climate change impacts on, 560, 560*f*
  of Coral Triangle, 14*f*
  Costa Rica, conservation in, 246, 246*f*
  defined, **109**
  loss of, 95*f*
  Lovejoy on, 94
  in natural capital, 136
  polyculture for, 291
  rain forest disappearance and, 70
  sustainability and, 20*b*
  threats to, 208*f*–209*f*, 220, 221*f*
Biodiversity, aquatic
  difficulties in protecting, 270–271, 271*f*
  ecosystem approach to, 271
  human impact on, 266–270, 267*f*–269*f*
  reversal of damage to, 271*b*
Biodiversity, ecosystem approach to sustaining
  biodiversity hotspots, **262**, 263*f*
  damaged ecosystem restoration, 263–264
  ecosystem services, protecting, 263
  five-point plan, 262–263, 262*f*
  sharing ecosystems, 264–265, 265*b*
Biodiversity, evolution and. *See also* Ecosystem dynamics
  amphibian species declines, 106–108, 128
  extinction, 125–127, 213–214
  genetic variability and natural selection, 120–123
  organism changes over time, 120
  speciation, 124
  species in ecosystems, 115–120
  variety as, 109–115, 112*f*–113*f*
Biodiversity, forests and
  age and structure of, 247–249, 248*f*
  demand for harvesting, 253–254, 254*b*, 254*f*
  fire protection and, 252–253

682 INDEX

life supported by, 247
management of, 249–252, 250f–253f
tropical deforestation, 254
tropical loss rates, 272
Biodiversity, grasslands and
buffer zone approach, 260
overgrazing, 255, 256f
rangeland protection, 255–257
U.S. National Parks, 257, 258f–259f
wilderness areas, 261
Biodiversity hotspots, **262**, 263f, 274
Bioengineering, 79
Biofuels, 127, 307–309, **443**, 444–445
Biogas digesters, 595, 595f
Biological Dynamics of Forest Fragments Project (BDFFP), 94
Biological extinction, **125**–127, 125f, 213–214
Biological hazards, 503. *See also* Human health, environmental hazards and
Biological pest control, 313–314, 313f
Bioluminescence, 184f–185f
Biomagnification, **227**, 228f
Biomass and biofuels, **443**–445, 444f–445f
Biomes, 110, 111f, 168f, 193
Biomimicry approach, **607**
Bioplastics, 592, 592f
Bioprospectors, **216**, 220
Bioremediation, **599**
Biosphere, **73**–74, 74f
Biosphere 2 experiment, 619, 619f
Biosphere reserve, **260**
Biotechnology, 500
Bird species, sustaining, 244–245f, 575
Birth and fertility rates, 477–478, 477f–478f
Birth defects, chemical hazards causing, 512f
Bisphenol A (BPA), 342b, 514, 514f
"Black smokers" (hydrothermal vents), in oceans, 375f
Blanding's Turtle Restoration Project (Chicago, IL), 198
Bleaching, in sewage treatment, 353
Blue Planet Prize, 94
"Blue Zones," 524, 526
Body systems, chemical effects on, 512–513
Boreal forests, 175
Bormann, F. Herbert, 48–49, 56
Botanical gardens, **236**
BP *Deepwater Horizon* oil rig explosion, 183f, 351

BPA. *See* Bisphenol A (BPA)
Brackish water, **180**
Brady, George "Duke," 46
Braungart, Michael, 581
British Royal Society, 547
Broken Record (Jay), 233f
Brookhaven National Laboratory (New York), 384f
Buettner, Dan, 524, 524f, 526
Buffer zones, **260**, 261f
Buildings, energy efficient, 429–433, 430f–432f
Bureau of Land Management (BLM), 627
Burmese pythons, 225–226f
Burney, Jennifer, 282–283f
Bus rapid-transit (BRT) system, Curitiba, Brazil, 492, 492f
Bycatch, from overfishing, 309f

C

Caddo Lake (Texas-Louisiana border), 188f
CAFOs (concentrated animal feeding operations), 298f, 299, 309
advantages and disadvantages of, 309f
Calcium losses in forests, 62f
California State Water Project, 336
Cambridge University (UK), 500, 633
Camels, humps for water storage, 170b, 171f
Camouflage, in predator-prey relationships, 138–139
Cancer, 220, 512
Cap-and-trade system, 567–568, 638
advantages and disadvantages of, 568f
Captive breeding programs, **236**–237, 236f
Carbohydrates, 57
Carbon-based molecules, 57f
Carbon capture and storage (CCS) strategies, 564–565, 565f, 642–643, 643f
Carbon cycle, **87**–89, 88f, 95f, 134
Carbon dioxide (CO_2), 307, 397, 537, 550–551, 550f, 567
Carbon monoxide (CO), 537
Carbon footprint, **550**, 564, 568–569
Carlsbad Desalination Project (California), 337f
Carcinogen, **512**
Carnivores, **75**
Carrère, Alizé, 175, 175f
Carrying capacity, **147**, 147f, 474
Cars, advantages and disadvantages of, 489–490, 490f, 564

Carson, Rachel, 37–38f, 312
Casita Volcano (Nicaragua), 422
CBD (Convention on Biological Diversity), **232**
CCD. *See* Colony collapse disorder, of honeybees
CCS. *See* Carbon capture and storage
CDC. *See* Centers for Disease Control and Prevention
Cells, as level of organization, 74f
Cellular respiration, **80**
Cellulose, as organic compound, 57f
Cellulosic ethanol, for biofuels, 444–445
Center for Science and Mathematics Education, University of Utah, 46
Centers for Disease Control and Prevention (CDC), 505, 514, 519
CERCLA. *See* Comprehensive Environmental Response, Compensation, and Liability Act (CERCLA)
CFC (chlorofluorocarbon compounds), **569**–570
Change the Course campaign, 341, 359
Charcoal, as biomass source, 443
Chemical change, **57**
Chemical defenses, in predator- prey relationships, 139
Chemical hazards. *See also* Solid and hazardous waste
body systems affected by, 512–513
cancers, mutations, and birth defects from, 511–512, 512f
diagnostic tools for, 500f–501f
endocrine system affected by, 513–515, 513f–514f
overview, 503
precautionary principle for, **520**–521, 520b, 521f
testing, 517–518, 518f
toxicity of, 515–517, 516f–517f
trace levels of, 519–520, 519f
Chemical pollution, 95f
Chemical reactions of matter, **57**, 57f
Chernobyl (Ukraine) Nuclear Power Plant accident (1986), 413, 413f
Child labor, 477, 477f, 597–598f
Children & Nature Network, 16
China
air pollution and, 539, 540f
ivory and, 229
minerals and, 373, 385
population growth and, 480b

sustainability in, 616, 616f, 618, 630, 638
Chinese Academy of Sciences, 362
Chlorinated solvents, 79f
Chlorination, in sewage treatment, 353
Chlorofluorocarbon (CFC) compounds, **569**–570
Chromium, 374f, 383
Chronic effects, of chemical hazards, 517
Chronic malnutrition, **285**
Chytrid disease of amphibians, 106–107f, 128
Cigarette smoking, 522–524, 522f–523f
CITES (Convention on International Trade in Endangered Species of Wild Fauna and Flora), **231**–232
Cities. *See also* Human population
air pollution in, 1940–1970, 37
as artificial environments, 24
business subsidies and, 34
freshwater ecosystems and, 189
freshwater supplies and, 334, 338–339
growing populations and, 29f
microclimates in, 166
recycling programs, 518
ocean pollution, 350f
sustainability, 482–493
transportation alternatives for, 429, 490, 492
water usage by, 332
wetlands sewage treatment in, 354b, 354f
Cities, in less-developed countries
food insecurity, 285
water pollution, 344
Citizen action, 604–605, 605f, 631–634, 631b, 631f–634f
Citizen science, 65, 99, 131, 155, 193, 275, 458f–467f, 529, 575
City/FARM project (MIT), 491, 491f
Clean Air Act, 405, 541
Clean Water Act, 351, 353, 405
Clear-cutting, 31, 249b, 250f
Climate
air circulation and, 162, 163f
arid, native plants and, 340
in biomes, 110
cities and, 485, 488
cloud cover and, 551
desert, 170, 170f, 171f, 304
Earth's rotation and, 163
Earth's spheres and, 71
Earth's surface features and, 166, 166f
ecosystem services and, 21, 22f
factors influencing, 161–166
forest, 174–175, 174f, 247, 250
freshwater scarcity and dry, 333
geological processes and, 367

INDEX **683**

grassland, 172, 172f, 255
greenhouse gases and, 165b, 213, 228, 392, 547, 550
Hadley cells and, 163–164, 163f
mass extinctions and, 214f
mathematical models and, 94, 552–553
microclimates, 166, 246
in mountains, 178
ocean currents and, 162, 162f
solar energy and, 58, 163f
tectonic plate movement and, 126b
terrestrial ecosystems and, 167–168, 167f–168f
topsoil organisms and, 302
water circulation and, 87b, 162, 162f
weather vs., **161**
Climate change
agriculture and, 295, 307
amphibians and, 119
atmospheric warming and, **546**–553, 546f–549f
carbon cycle and, 95f
droughts expected from, 333, 336, 338, 556
flooding and, 557, 558f–559f
greenhouse gases and, 165b, 213, 228, 392, 547, 550
kelp forests and, 134
models to predict, 552–553, 552f–553f, 575
species extinction from, 228, 557
tipping points, 534, **563**, 563f
Climate change, impacts of, 551–562
biodiversity loss, 557, 560, 560f
droughts expected from, 333, 336, 338, 556
economies, 562
flooding, 557, 558f–559f
food production, 560
Greenland ice melt, 534, 534f, 572
human health, environmental hazards, 562
permafrost melt, 554, 556f
sea level and, 530f–531f, 554–555, 555f, 558f–559f
snow and ice melt, 534, 534f, 572, 554
weather extremes, 556–557, 556f–557f
Climate change, slowing
cleanup and geoengineering, 564–567, 564f–566f
government strategies, 567–568
greenhouse gas reduction, 563–564
overview, 562–563, 563f, 564f
response to challenge, 568–569, 568f–569f
Coal
advantages and disadvantages of, 407f
smog from burning of, 537
trends in use of, 416, 416f
Coal ash, 405–406
Coal mining

description of, **405**–407, 405f–407f
in India, 380f–381f
mountaintop removal, 360f–361f
Coal sludge, 100
Coastal areas, water pollution in, 347
Coastal cities, threat of sea level rise, 530f–531f, 556f
Coastal wetlands, **180**, 181f–182f
Coastal zones, **180**
Cobalt, 374f
Coevolution, **141**
Cogeneration of energy, **426**
Cohen, Joel, 619
Cold deserts, 170, 170f
Cold forests, 174f, 175
Cold grasslands, 172, 172f
Colony collapse disorder (CCD), of honeybees, 212, 223, 238
Colorado River, 86f, 326f–327f, 328–329, 329f–330f, 341, 356
Commensalism, **141**
Commercial energy, 390, **395**. See also Energy, nonrenewable resources of
Commercial forest, **247**, 248f. See also Tree plantation
Communities, as level of organization, 74f
Community-supported agriculture (CSA) programs, 288, 320f
Commuter trains, 429
Competition, **137**–138
Compost, 318
Composting, **594**
Composting toilet systems, 353
Compounds, **57**, 57f
Comprehensive Environmental Response, Compensation, and Liability Act (CERCLA), 601, 602f–603f
Computers, energy-efficient, 433
Concentrated animal feeding operations (CAFOs), 298f, 299, 309, 309f
Concentrating solar power (CSP), 436
Condensation, 85f
Conservation, 246, 256–263, 612f–613f, 614f–615f, 627
Conservation concessions, **254**
Conservation easements, 256
Conservation International, 228, 632
Conservation-tillage farming, 302, 315
Constant-loss population, 151
Consumer Electronics Association, 597
Consumers in ecosystems, 75–77, 75f, 82f
Contaminated soil treatment, 100–101, 599
Continental drift, **366**–367
Contour planting, 280f–281f, 315

Contour strip mining, 378
Convection, 164, 164f
Convention on Biological Diversity (CBD), **232**
Convention on International Trade in Endangered Species of Wild Fauna and Flora (CITES), **231**–232
Cook, Diane, 430
Copenhagen Amendment to Montreal Protocol on CFC emissions, 571
Copper, 376b, 379f
Coral reefs, 156f–157f, 159f, 160, 160f, 190, 561, 561f
Coral Triangle, 14f
Corcovado National Park (Costa Rica), 246f
Core, of geosphere of Earth, **366**
Coriolis effect, 163
Corn for ethanol, 307b
Cortes Bank (California), 132f–133f
Costanza, Robert, 248
Costa Rica
bird diversity in, 244–245f
buffer zones in, 260
carbon neutrality of, 568
conservation in, 246, 246f
Countries of the world (Appendix 4), 650–651
COVID-19, 508, 625
Coyne, Michael, 558
Crop rotation, 318
Crops, sustainable, 318–319, 319f, 322. See also Agriculture; Sustainable practices
Crossbreeding, 127, 297
Crown fires, in forests, 249
Crude oil (petroleum), **397**–402, 398f–402f, 416, 416f
advantages and disadvantages of, 400f
Cruise ships, ocean pollution by, 353
Crust, of Earth's geosphere, 366
CryoSat satellite (European Union), 572
CSA (community-supported agriculture) programs, 288, 320f
CSP (concentrating solar power), **436**
Culhane, T. H., 595, 595f
Cultural eutrophication, **345**, 345f
Cultural hazards. See Human health, environmental hazards and
Curitiba (Brazil), as eco-city, 492, 492f
Currents. See Ocean currents

D

Damaged ecosystem restoration, 263–264
Dams, benefits and problems of, 334–336, 336f
Darwin, Charles, 122
Data collection, **49**, 52f–53f

DDT (pesticide), 227–228, 228f, 312
Dead zones, ocean, 270, 299b, 350
Death rates, 478
Debt-for-nature swap, **254**
Decarbonization, **451**
Decomposers
in ecosystems, 75f, **78**–80, 78f–80f
insects as, 114, 114f
Deep-sea trawling, 266, 268f–269f
Deepwater Horizon oil rig (BP) explosion, 183f, 351
Deep-well disposal, of hazardous waste, **599**–600
advantages and disadvantages of, 600f
Deforestation
description of, **250**
forest types and, 175
malaria and, 510b
overview, 76f–77f
rates of, 272, 272f
reducing, 253–254
tropical, 179, 251f–252f, 254
Delta, at river mouths, **187**, 187f
Demographic transition, 480
Denali National Park and Preserve (AK), 242f–243f
Dengue fever, 562
Depletion allowance, for mineral mining, 374
Depletion time, for minerals, **371**, 373f
Desalination, 337f, 338
Desert ecosystems, 167f–171f, 169–171
Desertification, **304**–305f, 318
Detritivores, **78**, 78f, 114
Dewan, Leslie, 412, 412f
Dimick, Dennis, 36, 36f
Dirty dozen foods, 294
Disrupters, endocrine, 513
Distillation, for desalination, 338
District heating, **449**
DNA (deoxyribonucleic acid), 122, 127–128
DOE. See U.S. Department of Energy (DOE)
"Doomsday Vault" (Svalbard Global Seed Vault), 236–237f, 308, 308f
Doses, of chemical hazards, **515**, 517f
Doubilet, David, 149
Doubling time, in population growth, 30b
Dow Chemical Company, 426
Dredging operations, 266–267, 268f–269f
Drip irrigation, **339**, 339f
Drought, **556**

E

Earle, Sylvia, 561, 561f
Earth
composition of, 365, 365f
crust elements of, 56b

spheres of, 71f. See also Ecosystem dynamics
surface features of, climate and, 166
Earth-centered worldview, **35**, 636
Earth Day (April 24), 37
Earthquakes
  fracking and, 404, 419, 600, 600f
  Fukushima nuclear accident and, 413
  geographic isolation from, 124
  geologic processes and, 126b, 365, 365f, 367
  geothermal energy drilling and, 448
  as natural health hazard cause, 503
Ebola virus, 508
E-cigarettes, 522–524
Eco-cities, 492–493, 492f, 564f
Ecoindustrial park, 620
Ecological economy, 618, 618f
Ecological footprint, **26**–28, 27f, 43, 476, 484, 493, 608, 616, 616f
Ecological inertia, **143**
"Ecological insurance policy," 261
Ecological niche, **115**–116, 115f, 116f
Ecological resilience, **143**
Ecological restoration, **263**
Ecological succession, **142**–144, 143f, 144f
Ecological tipping point, **62**, 70, 95, 143, 178, 635
Ecology, **19**
Economic development, **617**
Economic growth, **617**
Economics
  climate change effects and, 562, 567
  of coal, 407f
  conservation, 612f–613f, 614–615f
  of crude oil, 401f
  of forests, 247f
  of fuel, 428
  full-cost pricing, 20f, 34, 248b, 396, 620–621, 621f
  innovation encouragement, 623
  of mineral depletion, 371–373
  Millennium Development Goals and, 624–625, 624f
  natural capital impact on, **617**–620, 617f–620f
  of natural gas, 403f
  poverty reduction, 623–624, 623f–624f
  slowing population growth and, 480
  subsidy shifts, 621–622, 622f
  sustainability and, 20b, 34, 616, 616f, 638
  taxing pollution and waste, 622–623
Ecosystem diversity, **110**
Ecosystem dynamics, 66–99. See also Biodiversity and evolution
  atmosphere, **71**–73

carbon cycle, **87**–89
cellular respiration in, **80**
consumers in, **75**–77
decomposers in, **78**–80
energy in, 81–83
geosphere, **71**
greenhouse effect, **73**
hydrologic cycle, **84**–87
hydrosphere, **73**
matter in ecosystems, **84**–91
nitrogen cycle, **89**–90
phosphorus cycle, **90**–91
producers in, **74**–75
rain forest and, 70, 70f, 96
at Serra Bonita, 68–69f
species in, 109f, 115–120
study of, 92–95
Ecosystems. See also entries for specific ecosystem types
  defined, **19**
  reversal of damage to, 271b
  coral reefs, 159f, 160, 160f, 190
  freshwater, 186–189, 186f–188f
  greenhouse gases and, 165, 165f
  marine/ocean, 158–159, 159f, 180–185, 181f–185f
  terrestrial. See Terrestrial ecosystems
Ecosystem services
  forests, 247f
  human impacts on, 26
  insects in, 114, 114f
  oceans, 180
  overview, **21**, 22f
  predator-prey relationship in, 141
  protecting, 263
  topsoil, **302**
  value of, 248
Ecosystem services, species-saving, 208–241
  captive breeding programs, 236–237, 236f
  extinction. See Extinction of species
  habitat destruction, 221–222f
  honeybees, 210–212, 211f, 212f, 223, 223f
  human population, 226–227f, 227–228, 228f
  invasive species, 222, 222f, 224–226, 224f, 225b, 226f
  overexploitation, 228–231, 229f–230f
  protected lands, 234, 235f
  questions about, 238
  seed storage and cultivation, 236, 237f
  treaties and laws on endangered species, 231–234, 233f
Ecotourism, 32f–33f, 220, 260, 264
Edge effect, **168**
Education. See Nature museums and preserves
Efficiency, energy. See Energy efficiency
Egg pulling, **236**
Ehrlich, Paul, 27
Einstein, Albert, 604

Electrical grid, 426–427, 427f
Electric cars, 362, 364f, 385
Electricity production
  average usage, 40
  energy efficient, 427–429, 427b, 427f
  nonrenewable energy resources for, 394–396
  renewable energy resources for, 60, 422–423, 433–449, 456–457
Electromagnetic radiation, **58**–59f
Electronic waste
  recycling, 34f, 576f–577f, 580, 580f, 598f
  special handling for, 597–599, 597b
Electrons, in atoms, 55f
Elements, in matter, **55**–56b
Elevation and latitude, 167–168
El Niño-Southern Oscillation (ENSO), 163
Emissions trading, 541–543
Emphysema, 523f
Empowerment, 481f, 482
Endangered species
  American alligator, 117
  data on, 241, 241f
  defined, **215**
  giant panda, 115
  orangutan, 216
  overview, 37
  primates, 218f–219f
  snow leopards, 208f–209f
  treaties and laws on, 231–234
Endangered Species Act (ESA), **232**
Endemic species, **125**, 262
Endocrine system, effects of chemicals on, 513–515, 513f–514f
Energy
  defined, **58**
  in agriculture, 302, 303f–305f
  ecosystem dynamics and, 81–83
  energy pyramid, 82f
Energy, nonrenewable sources of, 390–419
  coal mining, 405–407, 405f–407f
  crude oil (petroleum), **397**–402, 398f–402f
  fracking for oil and gas, 394, 394f, 403–405, 404f
  methane leaks in Arctic, 392f–393f
  natural gas, **402**–403, 402f–403f
  net energy, 396
  nuclear power. See Nuclear power
  power plants, 390f–391f
  source breakdown, 395, 395f
  subsidies for, 396
  trends in, 416, 416f
Energy, renewable resources of, 420–455
  biomass and biofuels, **443**–445, 444f–445f
  efficiency. See Energy efficiency

for electricity production, 433–434
geothermal, 422–423f, **445**–448, 446f–448f
hydrogen fuel, 448–449
hydropower, 60f, **440**–442, 441f–442f
solar, **434**–440, 434f–439f, 622f
transitioning to, 449–451, 449b, 450f–451f
wind, 420f–421f, 424, 424f, 442–443, 443f, 452, 452f
Energy efficiency
  attaining, 449b, 450–451, 451f
  in buildings, 429–433, 430f–432f
  in electrical utilities, 426–427, 427b, 427f
  in industry, 426
  overview, **425**
  in people's use of, 425–426, 425f
  in transportation, 427–429, 429f
Energy Star Program, U.S. EPA, 433
Engineering
  carbon capture, 565, 565f
  description of, 50
  design cycle of, 100f, 194f, 276f, 456f, 642f
  pollution cleanup, 79
  practices in, 50b
  Solar Sewage Treatment Plant (Rhode Island), 354, 354f
ENSO (El Niño–Southern Oscillation), 163
Environment, **19**. See also Sustainability
Environmental degradation, **24**, 31
Environmental engineering, 79
Environmental ethics, **34**–35
Environmental impacts, 26–28. See also Sustainability
Environmentalism, **19**
Environmental justice movement, **630**
Environmental literacy, **635**
Environmentally sustainable society, **38**
Environmental permits, advantages and disadvantages of, 621f
Environmental policies, 623–626, 634f. See also Economics
Environmental Protection Agency. See U.S. Environmental Protection Agency (EPA)
Environmental resistance, **147**
Environmental science, **19**
Environmental taxes and fees, advantages and disadvantages of, 621f
Environmental Working Group (EWG), 294b, 309
Environmental worldview, **34**–35, 637, 637b
EPA. See U.S. Environmental Protection Agency (EPA)

INDEX **685**

Epidemics, 504
Epidemiological studies, 472, 518
Erosion. *See* Soil conservation; Soil erosion
ESA (Endangered Species Act), 232
Essick, Peter, 317
Estes, Jim, 134–135*f*
Estuaries, **180**
Ethanol, 306*b*, 444
Ethics, sustainability and, 19–20, 20*b*, 34–35, 630
Euphotic zone, 181, 182*f*–183*f*
European Union (EU), 524, 580, 596
Eutrophication, **187**, 345*f*, 642
Evaporation, 85*f*
Everglades National Park (Florida), 199, 199*f*
Evolution, 139–140. *See also* Biodiversity and evolution
E-waste. *See* Electronic waste
EWG (Environmental Working Group), 294*b*, 309
Exotic pet trade, 228–229
Explorers at Work
　Anthony, Katey Walter, 392–393*f*
　Ballard, Robert, 377, 377*f*
　Bawa, Kamaljit St., 264, 264*f*
　Becker, Vitor, 68–69*f*
　Buettner, Dan, 524, 524*f*
　Burney, Jennifer, 282–283*f*
　Carrère, Alizé, 175, 175*f*
　Culhane, T. H., 595, 595*f*
　Dewan, Leslie, 412, 412*f*
　Dimick, Dennis, 36, 36*f*
　Earle, Sylvia, 561, 561*f*
　Estes, Jim, 134–135*f*
　Francis, John, 51, 51*f*
　Glover, Jerry, 306, 306*f*
　Golden, Christopher, 472–473*f*
　Goodall, Jane, 633, 633*f*
　Guo, Yu-Guo, 362–363*f*
　Harper, Caleb, 491, 491*f*
　Hogan, Zeb, 146, 146*f*
　Hinojosa Huerta, Osvel, 328–329*f*
　Lovejoy, Thomas E., 94, 94*f*
　Martinez, Juan, 16–17*f*
　Nadkarni, Nalini, 46*f*–47*f*
　Pettit, Erin, 532–533*f*
　Postel, Sandra, 341, 341*f*
　Ruzo, Andrés, 422–423*f*
　Sala, Enric, 158–159, 159*f*
　Sartore, Joel, 217, 217*f*
　Savage, Anna, 106–107*f*
　Şekercioğlu, Çağan, 244–245*f*
　Sindi, Hayat, 500–501*f*
　Stuart, Tristram, 578–579*f*
　Treinish, Gregg, 614–615*f*
　Varma, Anand, 210–211*f*
　Wilson, Edward O., 121, 121*f*
　Zheng, Xiaolin, 436, 436*f*
Exponential population growth, **28**, 30*f*, 147–150, 147*f*–150*f*

Extinction of species
　biodiversity and ecosystem services affected by, 216, 220
　documenting, 217*f*–219*f*
　in history, 213
　overview, **125–127**, 125*f*
　rates of, 213–215, 214*f*, 215*f*
　vulnerability to, 215, 216*f*

F

Family planning, 482
Farming. *See* Agriculture
Farm-to-City Market Basket program (Milwaukee, WI), 284
Federal Insecticide, Fungicide, and Rodenticide Act (FIFRA), 312
Feedback loops, **61**–62, 61*f*, 165*f*
Feeding the 5000, 578
Feedlots, 298*f*, 299, 309
　advantages and disadvantages of, 309*f*
Fertility rates, **476**–478
Fertilizer
　air pollution and, 307
　algae blooms and, 152, 185*f*
　animal excrement as, 309, 317*f*, 321
　annual crops and, 306
　compost as, 355
　fossil fuels to produce, 444
　green tax on, 622
　human excrement as, 353
　integrated pest management and, 314
　kitchen waste, 595*b*
　manure as, 309
　natural capital degradation and, 22
　nitrogen cycle and, 89*f*, 90–91, 91*f*
　nitrous oxide emissions from, 537
　as nonpoint water pollution source, 342, 346*f*
　as ocean pollution source, 350–351, 350*f*
　organic, **318**
　phosphate, 373*f*, 542*f*
　planetary boundaries and, 95*f*
　as population limiting factor, 147
　potash in, 378*f*
　soil conservation and, 315
　synthetic inorganic, **318**
　use in agriculture, 227, 270, 290–291, 294–295, 298, 302, 304
Fiber-optic cables, 383
FIFRA (Federal Insecticide, Fungicide, and Rodenticide Act), 312
Finke, Brian, 298
Fire ants, 224–225
Fires, forest, 249, 252–253
First law of thermodynamics, **59**
Fish consumption, 325, 325*f*
Fish/shellfish production
　environmental issues in, 309–311, 310*f*
　harvest rates, 325*f*
　overview, 299–301, 300*f*–301*f*

　sustainable, 300*f*, 321, 321*b*
Fisheries, **266**
Fishing trawlers, 266–270, 268*f*–269*f*
Fishprint, **267**
Fission, in nuclear power, **408**–410, 408*f*–410*f*
Flooding, 343, 555, 557, 558*f*–559*f*
Flood irrigation, **339**
"Flu" (influenza), 506–507
Food. *See also* Agriculture
　food security/insecurity, 285–289, 286*f*–289*f*
　industrialized production, 296, 296*f*
　urban food oasis, 284, 284*f*
　wasted, 578–579*f*
Food and Agriculture Organization (FAO), United Nations, 249, 288, 307, 309, 311
Food chains and food webs, **81**, 81*f*–83*f*
Food deserts, 286*f*, 287*f*
Food Quality Protection Act, 294*b*
Ford, Henry, 592
Ford Motor Company, 426, 592
Forests
　age and structure of, 247–249, 248*f*
　clearing of, 22–23*f*
　demand for harvesting and, 253–254, 254*b*, 254*f*
　economic and ecosystem services of, 247*f*, 248
　ecosystems of, 174–177, 174*f*–177*f*
　experimentation in, 48, 62
　fire protection, 252–253
　life supported by, 247
　loss rates of, 272, 272*f*
　management of, 249–252, 250*f*–253*f*
　tropical deforestation, 254
Forest Stewardship Council (FSC), 252
Fossil fuels. *See* Air pollution; Energy, nonrenewable resources of
Fossils, **120**, 120*f*, 125–126, 213
"Four R" strategy for reducing waste
　overview, 590–592, 591*f*
　recycling, 591, 592*f*–593*f*, 595*f*–596*f*
　reducing, 590
　reusing, 590, 593–594*b*
Fowler, Cary, 237*f*
Fracking for oil and gas, 394, 394*f*, 400, 403–405, 404*f*, 600, 600*f*, 632
Fragmentation of habitat, 221
Francis, John, 51, 51*f*
Free-range chicken farming, 319
Freshwater ecosystems, 186–189, 186*f*–188*f*
Freshwater life zones, **186**, 187*f*
Freshwater purification, 347
Freshwater use, 95*f*

Frogs, 118, 118*b*, 119*f*. *See also* Amphibian species, declines in
FSC (Forest Stewardship Council), 252
Fuel economy in cars, 428
Fuelwood crisis, 254
Fukushima Daiichi Nuclear Power Plant (Japan) accident (2011), 413–414
Full-cost pricing, 20, 34, 248, 252, 309, 318, 339, 353, 396, 407, 428, 442, 449, 451, 489, 493, 567, 606, 617, 620–621, 621*f*
Functional diversity, **110**
Fusion, as nuclear power source, 415

G

Gas Guzzler Tax, 428
Gasoline taxes, 428
Generalist species, **115**–116
Gene splicing, 127
Genetically modified organisms (GMOs), 127, 297–299, 297*f*, 299*b*
Genetic diversity, **110**, 110*f*
Genetic engineering, **127**
Genetic resistance, 122, 122*f*, 505
Genetic variability, 120–123, **122**, 128*f*
Geoengineering, **564**–567, 564*f*–566*f*
Geographic information systems (GIS), 92
Geographic isolation, **124**
Geology, 126, 126*f*. *See also* Mineral resources
Geosphere, **71**
Geothermal energy, 422–423*f*, **445**–448, 446*f*–448*f*
Ghandi, Mahatma, 640
Giant pandas, as endangered species, 115
Giant sequoia, 44*f*–45*f*
GIS (geographic information systems), 92
Glacier National Park (Montana), 554, 612*f*–613*f*
Glaciers, research on, 532–533*f*
Global Coral Reef Monitoring Network, 160
Global ecological deficit, 28*f*
Global positioning systems (GPS), 92
Global warming, 70. *See also* Climate change
Global Water Policy Project, 341
Glover, Jerry, 306, 306*f*
Glucose, 57*f*, 74
GMOs (genetically modified organisms), 127, 297–299, 297*f*, 299*b*
Golden Age of Conservation (1901–1909), 37
Golden, Christopher, 472–473*f*
Goldsmith, Greg, 93*f*

686 INDEX

Gombe Stream Game Reserve (Tanzania), 633
Goodall, Jane, 633, 633*f*
Gorongosa National Park (Mozambique), 121
GPP (gross primary productivity), **81**
GPS (global positioning systems), 92
Gran, Sally, 320*f*
Graphene, 383, 383*f*, 386
Grasshopper effect, in air pollution, 536
Grasslands
  buffer zone approach to protecting, 260
  ecosystems in, 172, 172*f*–173*f*
  overgrazing, 255, 256*f*
  rangeland protection, 255–257
Gray water, recycling, **340**
Great Pacific Garbage Patch, 350, 584, 584*f*
Great Smoky Mountains National Park, 257
Green architecture, 429–431, 430*f*
Green Belt Movement, 254*b*
Green building certification, 431
Greenhouse effect, **73**, 99, 165, 165*f*, 550–551, 550*f*
Greenhouse Gas Equivalency Calculator (EPA), 40
Greenhouse gases
  from agriculture, 307
  from Arctic methane seeps, 392
  climate affected by, **165**, 165*f*, 546–547
  habitat loss from, 228
  reduced in Curitiba, Brazil, 492, 492*f*
  reduction of, 563–564
Greening of America, 18, 40
Greenland ice, melting of, 534, 534*f*, 555, 572
Green landscaping, 18*f*
Green manure, 318
Green revolutions, **294**–295, 294*f*–295*f*
Griffiths, Annie, 439
Grob, Marco, 507
Gross primary productivity (GPP), **81**
Groundwater. *See also* Aquifers; Hydrologic cycle
  defined, **84**
  description of, 331–332
  flood irrigation use of, 339
  pollution of, 346–347, 346*f*, 348*f*–349*f*
Growing Power program (Milwaukee, WI), 284, 287, 293
Gulf of Mexico, 182*f*–183*f*
Guo, Yu-Guo, 362–363*f*, 387

H
Habitats
  climate change impact on, 228, 554–555, 557
  destruction of, 221–222*f*
  in forests, 247
  fragmentation of, **221**
  overview, **115**
Hadley cells, 163–164, 163*f*
Hammond, Robin, 292, 486
Hardin, Garrett, 24
Harper, Caleb, 491, 491*f*
Harvesting fisheries, 267*f*, 325*f*
Harvesting forests
  description of, 253–254, 254*b*, 254*f*
  harmful effects of, 250*b*, 252*b*
  impacts of, 247–250, 248*f*, 250*f*
  methods of, 250*f*
  reduced demand for, 254*b*
Hazardous waste. *See also* Solid and hazardous waste
  electronics, 597–599, 597*b*, 598*f*
  household, 510*f*
  overview, 596–597, 597*f*
  regulations on, 601, 604
  storing, 599–601, 600*f*–601*f*, 602*f*–603*f*
  treating, 599
  types of, **586**
HBV (hepatitis B virus), 508
Health, 31–34. *See also* Human health, environmental hazards and
Health and Ecosystems: Analysis of Linkages (HEAL), 472
Health risks, 522–524, 522*f*–523*f*
Heat (thermal energy), **58**
Heating of Earth's surface, 162–163*f*
Heat island effect, of urban areas, 166, 485
Heat loss in buildings, preventing, 431–432, 431*f*–432*f*
Heavy oil, 401
  advantages and disadvantages of, 402*f*
Heavy metals, 100–101. *See also* Human health, environmental hazards and; Mineral resources
Hedges, Barrett, 258*f*–259*f*
Hemis National Park (India), 209*f*–210*f*
Hepatitis B virus (HBV), 508
Herbivores in ecosystems, **75**
Hero for the Planet aware (*Time* magazine), 561
HHAs (hormonally active agents), 513–515
Hidden water use, 332*f*
High-input agriculture, **290**
High Line park (New York), 484*f*
High-quality energy, 59
High-waste society, 581–582, 581*f*, 608, 617, 617*f*
Hill, Julia "Butterfly," 39*f*
Himalayas: Mountains of Life (Bawa), 264
Hinojosa Huerta, Osvel, 328, 329*f*
HIV/AIDS, 482, 508, 509*f*
Hoffman, Fritz, 602
Hogan, Zeb, 146, 146*f*

Hoh Rain Forest (Olympic National Park), 196
Holdren, John, 27
Holistic principle, 630
Holocene epoch, **95**, 214*b*
Homeostasis, 169
Honeybees, 210–212, 211*f*, 212*f*, 223, 223*f*
Hong Kong, population growth in, 29*f*
Horizons of soil, 303, 303*f*
Horizontal drilling, in fracking, 394, 394*f*, **400**
Hormonally active agents (HAAs), 513–515
Hubbard Brook Experimental Forest (New Hampshire), 48, 62
Hubbard Medal (National Geographic Society), 121
Human capital, **617**
Human-centered worldview, **35**, 636
Human evolution, 123, 123*b*
Human health, environmental hazards and, 498–529
  chemical hazards. *See* Chemical hazards
  climate change impacts, 562
  diagnostic tools for, 500*f*–501*f*
  infectious diseases, **504**–506, 505*f*–506*f*, 511, 511*b*
  life expectancy estimation, 526
  risks to, 522–524, 522*f*–523*f*, 525*f*
  toxic waste exposure, 498*f*–499*f*, 502, 502*f*
  viruses and parasites, 506–510, 507*f*, 509*f*–510*f*
Human impacts. *See also* Climate change
  on aquatic biodiversity, 266–270, 267*f*–270*f*
  BP *Deepwater Horizon* oil rig explosion, 183*f*, 351
  on carbon cycle, 88
  extinction rates increased by, 215
  flooding as pollution source, 343, 343*f*
  on freshwater ecosystems, 189
  on hydrologic cycle, 85–87
  on marine ecosystems, 184–185*f*
  on nitrogen cycle, 90
  on phosphorus cycle, 91
  on terrestrial ecosystems, 178–179, 178*f*, 179*f*
  UV radiation increases, 118–119
Human population, 470–497
  age structure, 479–480, 479*f*, 509*f*
  analyzing growth of, 494, 494*f*
  birth and fertility rates, 477–478, 477*f*–478*f*
  changes in, 475–476
  death rates, 478
  growth rate of, 30*f*, 226–227*f*, 227–228, 228*f*, 474–475, 474*f*, 475*f*
  in Madagascar, 472–473*f*
  migration rates, 479
  slowing growth of, 477, 480–482, 480*b*, 480*f*, 481*f*
  slums, 485–488, 485*b*, 486*f*–488*f*
  sustainable cities, 490, 493, 493*f*
  urbanization of. *See* Urbanization
Hunger, 285. *See also* Poverty
Hybrid cars, 428, 564
Hydraulic fracturing, 394, 394*f*, **400**, 403–405, 404*f*, 600, 600*f*, 632
Hydrocarbons, 57
Hydrogen fuel cells, **429**, 448
Hydrogen fuel energy, 448–449
Hydrologic cycle, **84**–87, 85*f*, 178
Hydropower, 60*f*, **440**–442, 441*f*–442*f*
  advantages and disadvantages of, 442*f*
Hydrosphere, **73**, 73f
Hydrothermal vents, 375*f*, 377
Hypercities, 483

I
Ice ages, 546*f*
Ice core analysis, 532, 547, 547*f*
Iceland, renewable energy use in, 447*f*, 450, 450*f*
Igneous rock, **370**, 372*f*
Iguaçu National Park (Brazil), 76*f*–77*f*
Illegal trade in wild species, 229–231, 229*f*–230*f*
Inbreeding, diversity loss by, 237
Incineration, to treat hazardous waste, 512–514, 514*f*, 599
  advantages and disadvantages of, 590*f*
Indicator species, **118**, 537
Indoor air pollution, 544, 544*f*
Industrialized food production, 280*f*–281*f*, **290**, 291*f*, 296, 296*f*
Industrial Revolution, 95, 550, 637
Industrial solid waste, **582**
Industry, energy efficiency in, 426
Inertia, ecological, **143**
Inexhaustible natural resources, **21***b*, 58, 395, 442
Infant mortality rate, **478**, 482
Infectious diseases, 128*f*, **504**–506, 505*f*–506*f*, 511, 511*b*
Infiltration, 85*f*
Influenza, 506–507
Inland wetlands, **187**–189, 188*f*
Innovation, encouragement of, 623
Inputs to systems, **61***f*
Insecticides
  frog exposure to, 108
  honeybees and, 223*b*
  malarial protozoa and, 509–510
  Neem tree as, 220

INDEX **687**

spiders *vs.*, 311
   synthetic, 312–313
Insect predators, 311–312*f*
Insects, in ecosystem services, 114
Insecurity, food, **285**–289, 286*f*–289*f*
Institute of Chemistry, Chinese Academy of Sciences, 362
Insurance hypothesis, **111**
Integrated pest management (IPM), **314**, 320*f*
Integrated waste management, **587**–588, 587*f*
Intergovernmental Panel on Climate Change (IPCC), 547, 554, 557, 560, 564
International Center for Technology Assessment, 428
International Crane Foundation (Wisconsin), 204, 204*f*
International Fund for Animal Welfare, 632
International Red Cross, 568
International Union for Conservation of Nature (IUCN), 215, 256, 262, 270
Interspecific competition, **137**
Invasive species
   Asian carp, 224, 224*f*
   Burmese python, 225, 226*f*
   control of, 225–226
   fire ants, 224–225
   Kudzu vine, 224
   in National Parks, 257
   native species *vs.*, 116
   in oceans, 158
   overview, 222, 222*f*
   stopping spread of, 225*b*
   zebra mussel, 225
Ions, **55**–56
IPAT model of environmental impact, 27
IPCC (Intergovernmental Panel on Climate Change), 547, 554–555, 557, 560, 564
IPM (integrated pest management), **314**, 320*f*
Irrigation, 306–307, 307*f*, **339**, 339*f*–340*f*
Isolation, geographic, **124**
Isolation from nature, 34, 261*b*, 635
Isotopes, of elements, **55**
IUCN (International Union for Conservation of Nature), 215, 256, 262, 270
Ivanpah Solar Electric Generating System (California), 435*f*

J
Jane Goodall Institute, 633
Japan,
   urban population, 475, 483
   minerals, 385
JASON organization, 377
Jay, Asher, 233*f*
Jensen, Vicki, 507*f*
Jenshel, Len, 430
Jinping, Xi, 638

Jobs, Steve, 640
Joubert, Beverly, 32*f*–33*f*
Joubert, Dereck, 32*f*–33*f*

K
Kadur, Sandesh, 264
Kalundborg ecoindustrial park (Denmark), 606*f*, 607
Kariega Game Reserve (South Africa), 230*f*
Kelp forests, 134
Kendrick, Robb, 381*f*
Kerogen, shale oil from, 401
Keystone species, **116**–117, 216
Kilauea volcano (Hawai'i), 142
Kinetic energy, **58**, 60*f*, 456
King, Martin Luther, Jr., 640
K-selected species, **150**–151*f*
Kudzu vine, 224
Kyoto Protocol, on climate change, 567

L
Lactose intolerance, 123*b*
Laguna Atascosa National Wildlife Refuge (Texas), 205, 205*f*
Lakes, 186–187, 186*f*, 344–346
Laman, Tim, 52, 140
Landfills, 582*b*, **588**–589, 588*f*–589
   advantages and disadvantages of, 589*f*
Land for public use, 627–630, 628*f*–629*f*
Lands, protected, 234
Land trusts, 256
Land use, 95*f*
Lanthanum (mineral), 364, 364*f*, 370
Lanting, Frans, 76*f*–77*f*, 549, 585
Latitude and elevation, 167–168
Law of conservation of matter, 57–**58**
Laws and regulations, **623**, 631–632, 634*f*. See also entries for specific laws
Leachate, 589
Leach fields, for sewage treatment, 355
Leaf characteristics, 193
Leaks, water loss from, 340*b*
LEED (U.S. Green Building Council's Leadership in Energy and Environmental Design), 431
Leen, Sarah, 399
Leopold, Aldo, 37–38*f*
Less-developed countries, 26
Levels of organization, 74*f*
Lianas, 176–177*f*, 252*b*
Lichens, as air pollution indicators, 537*b*, 537*f*
Life-centered worldview, **35**, 636
Life expectancy, **478**, 526. See also Human population
Life raft ecosystems, 263
Lifestyle choices, 522*f*, 524. See also Human health, environmental hazards and

Lighting, energy-efficient, 433
Liittschwager, David, 113
Likens, Gene, 48–49, 56
Limits to population growth
   growth, shrinkage, and stability, 145–146
   limiting factors, **147**–150
   reproductive patterns, 150–151
Lipids, 57
Liquefied petroleum gas (LPG), 402–403
Liquid water, uniqueness of, 86
Lithium, 385
Lithosphere, of geosphere of Earth, **366**
Lobby, **324**
Logistic population growth, 147, 147*f*
Lovejoy, Thomas E., 92, 94, 94*f*
Low-quality energy, 59
Low-waste society, transitioning to biomimicry approach, 607
   government and citizen action, 604–605, 605*f*
   reuse and recycling, 605–606, 606*f*
LPG (liquefied petroleum gas), 402–403
Lyme disease, 562

M
Maathai, Wangari, 254*b*
MacKenzie River Valley (Canada), 161*f*
Macronutrients, 72, 287*f*
Madagascar, 175, 472–473*f*
Madagascar Health and Environmental Research (MAHERY), 472
Malaria, 508–510, 509*f*–510*f*, 510*b*
Malnutrition, **285**, 287
Manganese, 374*f*
Mantle, of geosphere of Earth, **366**
Manufactured capital, **617**
Manú National Park (Peru), 72*f*
Manure, 316*f*–317*f*, 318
Marine biology, 51
Marine ecosystems, 180–185, 181*f*–185*f*
Marine life zones, **180**
Marine pollution, 158
Marine protected areas (MPAs), 270–271, 561
Marine reserves, 271
Marine snow, 181
Marine Stewardship Council (MSC), 321, 321*b*
Market prices for minerals, 373–376, 373*f*–375*f*, 376*b*
Marriage, average age at, 477
MARS (Magnetic Acoustic Resonance Sensor), 500
Marsh, George Perkins, 36
Martinez, Juan, 16–17*f*, 34, 42
Mass extinctions, **125**, 213–215, 214*f*

Mass transit, 429, 490
Materials revolution, **383**
Matter in ecosystems
   carbon cycle, **87**–89
   hydrologic cycle, **84**–87
   ladder of, 74*f*
   movement of, 75*f*
   nitrogen cycle, **89**–90
   phosphorus cycle, **90**–91
McBride, Peter, 86*f*
McDonough, William, 581
McKibben, Bills, 249*b*
Mead, Margaret, 39
Measurement units (Appendix 2), 648
Meat production
   environmental issues in, 309, 309*f*
   impacts of, 299, 299*b*
   sustainable, 319–321, 320*f*
   water consumption in, 334*b*
Medicinal properties of plants, 220, 220*f*
Megacities, 483, 488
Megafish Project, 146
Megareserves, in Costa Rica, 260
Mercury, toxic effects of, 502, 502*f*, 512–513, 512*f*, 521*f*
Metamorphic rock, **370**, 372*f*
Methane, 392*f*–393*f*, 404, 404*f*, 536
Mexico City, Sustainable Transportation Award, to (2013), 488
Microbes, pollution cleanup by, 79, 79*f*, 599
Microhydropower generators, 441
Micronutrients, 72, 285, 287*f*
Migration rates, 479
Millennium Development Goals, 624–625, 624*f*
Millennium Ecosystem Assessment (United Nations), 26, 90
Millennium Seed Bank (UK), 236
Milton-Union School District (West Milton, OH), 18, 18*f*, 40
Minamata Convention, of United Nations, 520
Mineral deficiencies, in diets, 285
Mineral resources, 72, 360–389
   for batteries, 362–363*f*
   defined, **370**
   geological processes, 365–369, 365*f*–369*f*
   market prices of, 373–376, 373*f*–375*f*, 376*b*
   mining techniques for, 376–382, 377*f*–381*f*, 382*f*
   in mountains, 360*f*–361*f*
   nonrenewability of, 371–373, 372*f*–373*f*
   ore processing, 382
   rare earth, 364, 364*f*, 386
   rock cycle, 370–371, 371*f*
   substitutions for, 383, 383*f*
   sustainable use of, 383–385, 384*f*
   uses of, 373*f*

Mining
  heavy metals released by, 100–101
  techniques of, 376–382, 377f–381f, 382b
Ministry of Education, Rwanda, 623f
Minnesota Pollution Control Agency, 118b
Mission Blue, 561
Mitigation approach, **563**
Models, **49**
Molecules
  carbon-based, 57f
  description of, **55**
  level of organization, 74f
  of life, 57
Molten salt reactors, for nuclear power, 412
Monoculture, **290**
Montane cloud forests (Costa Rica), 93f
Monterey Bay Aquarium (California), 202–203f, 321b
Monteverde Cloud Forest Reserve (Costa Rica), 47f, 125
Montreal Protocol on CFC emissions (1987), 571
More-developed countries, 26
Mortality, **476**, 478
Motu Motiro Hiva Marine Park (Chile), 158
Mountains
  ecosystems in, 178
  glacier losses in, 554
  mineral resources in, 360f–361f
  mountaintop removal (surface mining), **378**
MPAs (marine protected areas), 270–271, 561
MSC (Marine Stewardship Council), 321
Muir, John, 36–37f
Municipal solid waste (MSW), **582**, 608, 608f
Mutagens, **512**
Mutations, **122**, 512f
Mutualism, 140f, **141**, 160

N

Nadkarni, Nalini, 46f–47f
Nanotechnologies, 362, **383**–384, 384f, 386
Narrow-spectrum synthetic pesticides, 312
NAS (U.S. National Academy of Sciences), 312–314, 519, 547, 555, 562
NASA (National Aeronautics and Space Administration), 491, 547
National Advertising Council, 252
National Audubon Society, 575, 632
National Center for Atmospheric Research, 575
National Environmental Policy Act, 405
National Forensics Lab, U.S. Fish and Wildlife Service (USFWS), 235, 235f
National Forest System, 626f, 627
National Geographic Explorers
  Anthony, Katey Walter, 392–393f
  Ballard, Robert, 377, 377f
  Bawa, Kamaljit St., 264, 264f
  Becker, Vitor, 68–70, 69f
  Buettner, Dan, 524, 524f
  Burney, Jennifer, 282–283f, 287
  Carrère, Alizé, 175, 175f
  Culhane, T. H., 595, 595f
  Dewan, Leslie, 412, 412f
  Dimick, Dennis, 36, 36f
  Dollar, Luke, 472
  Earle, Sylvia, 561, 561f
  Estes, Jim, 134–135f
  Francis, John, 51, 51f
  Glover, Jerry, 306, 306f
  Golden, Christopher, 472–473f
  Goldsmith, Greg, 93f
  Goodall, Jane, 633, 633f
  Guo, Yu-Guo, 362–363f
  Harper, Caleb, 491, 491f
  Hinojosa Huerta, Osvel, 328–330, 329f
  Hogan, Zeb, 146, 146f
  Jay, Asher, 233f
  Joubert, Beverly, 32f–33f
  Joubert, Dereck, 32f–33f
  Kadur, Sandesh, 264
  Lovejoy, Thomas E., 92, 94, 94f
  Martinez, Juan, 16–17f, 34, 42
  Nadkarni, Nalini, 46f–47f
  Pettit, Erin, 532–533f
  Postel, Sandra, 341, 341f
  Ruzo, Andrés, 422–423f
  Sala, Enric, 158–159, 159f
  Sartore, Joel, 217, 217f, 219f
  Savage, Anna, 106–107f
  Şekercioğlu, Çağan, 244–245f
  Sindi, Hayat, 500–501f
  Stuart, Tristram, 578–579f
  Treinish, Gregg, 614–615f
  Varma, Anand, 210–211f
  Wilson, Edward O., 121, 121f
  Zheng, Xiaolin, 436, 436f
National Geographic magazine (NGM), 36, 217
National Geographic's Committee for Research and Exploration, 264
National Interagency Biodefense Campus (Fort Detrick, Maryland), 507
National Marine Fisheries Service (NMFS), 232–234
National Park Service (NPS), 51, 257, 612, 627. See also Nature museums and preserves
National Park System, 626f, 627
National Renewable Energy Laboratory (NREL), 424
National Wilderness Preservation System, 261, 627
National Wildlife Federation, 632
National Wildlife Refuge System, 234, 626f
National Woman's Hall of Fame, 561
Native species, **116**
Natural capital
  biodiversity in, 111, 136
  degradation of, 23–24f
  economic impact of, 617–620, 617f–620f
  sustainability and, **20**–23, 22f
Natural gas, 40, **402**–403, 402f–403f, 416, 416f
  advantages and disadvantages of, 404f
Natural hazards. See Human health, environmental hazards and
Natural income, **39**
Natural Leaders Network, Children & Nature Network, 16
Natural resources
  categories of, **20**, 21b
  in natural capital, 22f
  unsustainable use of, 30–31
Natural Resources Defense Council (NRDC), 632
Natural selection, **120**–123
Nature, 16–17f, 34
Nature (television series), 94
Nature Conservancy, The, 234, 632
Nature museums and preserves, 196–205. See also National Park Service
  Arizona-Sonora Desert Museum (Tucson, AZ), 201, 201f
  Everglades National Park (Florida), 199, 199f
  International Crane Foundation (Wisconsin), 204, 204f
  Laguna Atascosa National Wildlife Refuge (Texas), 205, 205f
  Monterey Bay Aquarium (California), 202, 203f
  Olympic National Park (Washington), 198, 198f
  overview, 196f–197f, 197
  Peggy Notebaert Nature Museum (Chicago, IL), 200, 200f
  Smithsonian's National Zoo (Washington, D.C.), 202, 202f
Negative feedback loops, 61f–62
Neonicotinoid pesticides, 223
Nesjevallir power plant (Iceland), 447f
Net energy, **396**
Net energy principle, 630
Net primary productivity (NPP), 81–82, 96, 96t, 123
Neurotoxins, 512
Neutrons, in atoms, 55f
New Guinea, 15f, 104f–105f
NGM (National Geographic magazine), 36, 217
NGOs (nongovernmental organizations), 631–633, 635
Niche, ecological, **115**–116, 116f
Nitrate concentrations in water, 48f
Nitric acid, 537
Nitrogen cycle, **89**–90, 89f, 95f
Nitrogen oxides, 537
NMFS (National Marine Fisheries Service), 232–234
NOAA (U.S. National Oceanic and Atmospheric Administration), 547, 561
Nobel Peace Prize, 254b
Noise pollution, 485
Nongovernmental organizations (NGOs), 631–633, 635
Nonnative species, **116**
Nonpoint sources of pollution, **26**
Nonpoint water pollution sources, 342–344, 342b, 343f, 344f, 351, 353
Nonrenewability of mineral resources, 371–373, 372f–373f
Nonrenewable energy, 318, 394–395, 395f, 397–415
Nonrenewable natural resources, **21b**, 22f, 71, 371–373
Nontransmissible diseases, **504**
NPP (net primary productivity), **81**–82, 96, 96f, 123
NPS (National Park Service), 51, 257, 612, 627
NRDC (Natural Resources Defense Council), 632
NREL (National Renewable Energy Laboratory), 424
Nuclear power
  accidents and uncertainty, 411–414, 412f–413f
  advantages and disadvantages of, 414f
  consumption trends for, 416, 416f
  environmental challenges of, 410, 410f
  fission, **408**–410, 408f–410f
  fusion, **415**
  government subsidies, 414
  safety concerns, 414–415, 414f
Nuclear weapons, 414–415
Nucleic acids, 57
"Nurse logs," downed trees as, 253f
Nutrient cycles
  carbon cycle, **87**–89
  description of, 72, **84**
  ecosystem components linked by, 80, 80f
  as ecosystem service, 21, 78, 180, 186, 247, 255, 303, 618
  history shown in, 90b
  hydrologic cycle, **84**–87
  importance of, 20, 72, 618–619
  nitrogen cycle, **89**–90
  phosphorus cycle, **90**–91
  sustainability and, 20b, 321, 385, 493, 591–592, 604

O

Obama, Barack, 638
Obesity, 285–286, 522
Ocean currents
 changes in, 546
 climate and, 162–164, 162f, 164f, 172, 552b
 CO_2 transfer by, 551
 continental drift and, 367
 Great Pacific Garbage Patch and, 584
 sea-level rise and, 555
 solar energy and, 58
Oceans. *See also* Marine ecosystems
 acidification of, 95f, **184**, 267–270, 270f, 275
 climate affected by, 162, 162f
 climate change and, 551
 dead zones in, 270, 350
 ecosystems and climate in, 158–159, 159f
 energy in, 441–442
 mineral mining in, 374–377, 375f
 Mission Blue for protection of, 561
 open-ocean aquaculture, 321
 plate tectonics in, 377, 377f
 pollution of, 350–351, 350f, 353, 353b
 trash in, 582–586, 584f–585f
Ogallala Aquifer, 335, 335f
Oil. *See* Crude oil (petroleum)
Oil spills, 37, 351–352f, 359
Okavango Delta (Botswana), 2–11
Old-growth forest, **247**
Old Weather project, 193
Olympic National Park (Washington), 198, 198f, 257
Omnivores in ecosystems, **75**
OPEC (Organization of Petroleum Exporting Countries), 401
Open-ocean aquaculture, 321
Open-pit mining, **376**, 379f
Open sea, **180**–181
Orangutans, 216, 219f, 220
Orchards, agroforestry for, 315
Ores, **371**, 382
Organic agriculture, **291**–294, 294f, 299b, 318
Organic compounds, **57**, 57f
Organic fertilizers, 316f–317f, **318**
Organisms, level of organization, 74f
Organization of Petroleum Exporting Countries (OPEC)., 401
Outputs of systems, 61f
Overbrowsing, 257
Overburden, in surface mining, **376**, 378
Overexploitation, 228–231
Overfishing, 158, 189, **266**–267, 267f, 309f
Overgrazing, **255**, 256f, 305f
Overharvesting, 228
Overnutrition, **285**–286

Overpumping, 334
Oxygen-depleted zones, of oceans, 248, 350
Ozone
 depletion of, 570–572, 570f–571f
 ozone layer, 71, 95f, **535**, 535f
 in smog, 536, 538

P

Pacific Remote Islands Marine National Monument, 158
Pangaea supercontinent, 126, 126f
Paper, tree-free, 253
Parasites, 118, **504**, 506–510, 507f, 509f–510f
Parasitism, **141**, 152
Paris Climate Change Conference, 567–568, 638
Particulates, in air pollution, 520–521f, 538
Passive solar heating systems, **434**, 434f, 436f
Pastures, **255**
Pathogens, **503**
Peak production, of oil wells, **397**
Peel-and-stick solar cells, 436
Peer review, in science, **54**
Peggy Notebaert Nature Museum (Chicago, IL), 200, 200f
Pelican Island National Wildlife Refuge (Florida), 234
Penguin Lifelines project, 155
Percolation, 85f
Periodic Table of Elements (Appendix 3), 649
Permafrost, **172**
Permafrost melt, 554, 556f
Persistence of chemical hazards, 517
Pesticides. *See* Pest management
Pest management, 37, 311–314
 neonicotinoid pesticides, 223
 pest control in nature, **311**, 312f
 synthetic pesticides, 294, 294b, 312–314, 313f
Petcoke, 397
Petrochemicals, **397**
Pettit, Erin, 532–533f
pH, of solutions, **56**, 56f
Phosphorus cycle, **90**–91, 91f, 95f
Photo Ark project, 217, 219f
Photochemical smog, 537–538
Photosynthesis, **74**, 134
Photovoltaic cells (solar cells), **436**–437, 437f, 438f–439f, 440
Phthalates, 513–515, 513f
Physical changes in matter, **57**
Phytoremediation, 100–101, **599**
Pinchot, Gifford, 36
Planetary boundaries for systems, 95, 95f, 476
Planet Earth (BBC television), 341
Plankton, 99
Plantation agriculture, 290, 292f–293f

Plants, medicinal properties of, 220, 220f
Plasma gasification, to treat hazardous waste, 599
Plastic bags, single-use, 517, 598
Plastic waste, 25f, 342b, 343, 598, 599f
Plate tectonics, 377, 377f
Platinum, 374f, 383
Plug-in electric hybrid cars, 362, 428–429f, 564
Point and nonpoint water pollution sources, 342–344, 342b, 343f, 344f, 351, 353
Point sources of pollution, **26**, 342–343f
Policy, **623**
Political science, sustainability and, 20b
Politics, 333, **625**–634
Pollination, 21, 114, 212, 212f, 223f
Polluter-pays principle, 630
Pollution. *See also* Human health, environmental hazards and; Water pollution
 chemical, 95f
 chlorinated solvents, 79f
 eutrophication accelerated by, **187**
 indicator species and, 118
 marine, 158
 pesticide use, 37, 227
 plastic waste, 25f
 prevention, 26
 sources of, **26**
 taxing, 622–623
 in urban areas, 485
Pollution Prevention Pays (3P) program, of 3M Company, 520b
Polyculture, **291**, 318
Polymers, 57
Population
 biological evolution of, 122
 crash, **150**
 defined, **145**
 density of, **147**
 level of organization, 74f
 growth of, 28–30, 29f
Population change, 145–146, 155, **475**–**476**
Population crash, 150f
Population control
 growth, shrinkage, and stability, 145–146, 475–476
 limiting factors, **147**–150, 148f–149f
 limits to population growth, 145–151, 476
 reproductive patterns, 150–151
Positive feedback loops, 62, 165f, 554
Postel, Sandra, 329f, 341, 341f
Potential energy, 58, 60f
Poverty
 environment and health affected by, 31
 food insecurity and, 285
 health risks from, 522f

 life expectancy affected by, 478
 population growth and, 476
 reduction in, 623–624, 623f–624f
 slums, 485–488, 485b, 486f–488f
 urban, 483
Power outages, 427b
Power plants, 60f, 390f–391f, 406f, 409f, 427f, 446–447f, 621f
Precautionary principle, **520**–521, 520b, 521f, 630
Precipitation, 84, 85f. *See also* Hydrologic cycle
Predation, **138**–141, 138f
Predator-prey relationship, **138**
Predators, 81f, **138**, 311–312f
Prescribed burns, 253
Prevailing winds, 163
Prevention principle, 630
Prey, **138**
Prices, environment and health costs *vs.*, 31–34
Pricing, full-cost, 620–621, 621f
Primary consumers in ecosystems, **75**, 75f, 82f
Primary ecological succession, **142**–143f
Primary recycling, **594**
Primary sewage treatment, 353
Pristine Seas project, 158–159
Producers in ecosystems, **74**–75, 75f, 82f
Product life cycles, 389
Protected lands, 234, 256–257, 260–261, 260f–261f
Proteins, 57
Protons, in atoms, 55f
Proven oil reserves, 398f–399f, **400**, 401f
"Pulse flow," of water, 328
Purification, freshwater, 347

Q

Quadrant studies, 131
Quality of energy, 59
Quality of life, 23–24

R

Rainforest Alliance, 254
Rain forests, 70, 70f, 96, 174, 174f, 176f–177f, 179
Rain shadow, **166**, 166f
Rainwater harvesting system, 18
Random sampling, 131
Rangeland protection, **255**–257
Range of tolerance, of population, **145**, 145f
Rapid laser pulse test for malaria, 510
Rare earth mineral resources, 364, 364f, **370**, 386
Reconciliation ecology, **264**
Recycling
 advantages and disadvantages of, 596f
 electronics, 34f, 576f–577f, 580, 580f

in "four R" strategy, 591,
592f–593f, 605–606, 606f
gray water, 340
minerals and metals, 385–386
to save energy, 426
in urban areas, 484
Red imported fire ants, 224–225
Red List, of International Union
for Conservation of Nature
(IUCN), 215
Reducing waste, 590
Refining oil, 397, 400f
Regulations, **623**, 631–632, 634f.
*See also entries for specific
laws*
Reliability, of scientific data, 54
Reliable surface runoff, **332**
Renewable energy, 318, 395f,
420–455, 493, 563–564,
622–623
Renewable natural resources, **21***b*,
22, 22f, 24f, 26, 39, 255, 303,
318, 395f, 622–623
"Renting biodiversity," 254
Reproductive isolation, **124**
Reproductive patterns, 150–151
Reserves of minerals, **371**
Reserves of oil, 398f–399f
Reservoirs
dams to create, **334**, 336f
pollution of, 344–346
Resilience, ecological, **143**
Resistance, environmental, **147**
Resource Conservation and
Recovery Act (RCRA), 601
Resource partitioning, **137**–138,
137f
Resource scarcity, 376b
Response, to chemical hazards,
517, 517f
Restoration of damaged ecosys-
tems, 254, 263–264
Reusing, 590,
593–594b, 605–606, 606f
Reverse osmosis, for desalination,
338
Reversibility principle, 630
Ribeiroia ondatrae (flatworm),
118–119f
Richardson, Bill, 426
Richardson, Jim, 320
Riparian zones, **255**
Risk analysis, **522**
Risk assessment, **503**, 503f
Risk management, **503**, 503f
Risks
health, 522–524, 522f–523f
irrational evaluation
of, 524–525, 525f
overview, **503**
Rivers, 187, 344, 344f–345f
Rock, **370**
Rock cycle, **370**–371, 371f, 372f
Rockefeller University, 529
Rockström, Johan, 95, 178
Rooftop active solar collectors, 434
Roosevelt, Theodore, 36–37, 37f,
234, 612
Roots and Shoots program, 633f

Rotation, of Earth, 163
r-selected species, **150**
"Rule of 72," to calculate population growth, 30b
Runoff
defined, **186**
source of fresh water, 84,
331–332, 334
nutrients in, 49, 65,
prevention, 315
source of pollution, 26, 90–91,
160, 180, 249, 270, 309, 379
Ruzo, Andrés, 422–423f

S

Safe Water Drinking Act, 404
Sahara desert, 171f
Sala, Enric, 158–159, 159f
Salinization, soil, **307**–308, 308f,
318
Saltwater ecosystems, 266
Salt water in aquifers, 334
Sand County Almanac, A (Leopold),
37
Sanitary landfills, **588**–589, 588f,
588f–589f
Sanitation, 31
Sartore, Joel, 217, 217f, 218f,
219f, 348
Savage, Anna, 118, 128
Scavenger species, 257
Science, **49**
Science safety guidelines
(Appendix 1), 644–647
"Scientific American 52", 341
Scientific factors of sustainability,
19–20, 20b
Scientific hypothesis, **49**
Scientific law, **54**
Scientific Method, **49**
Scientific practices, 49b
Scientific theories, **49**, 54
Scripps Institution of
Oceanography, 158
Scrubbers, on coal burning power
plants, 405–406
Seagrass beds, on coastlines,
180–181f
Sea level rise, 530f–531f, 534,
551, 554–555, 556f
Seamounts Marine Managed Area
(Costa Rica), 158
Sea otters, 134–136, 136f, 152,
152f
Secondary consumers in ecosys-
tems, 75, 75f, 82f
Secondary ecological succession,
142–143, 143f, 144f
Secondary recycling, **594**
Secondary sewage treatment, 353
Second-growth forests, **247**, 249b
Secondhand smoke, 523
Second law of thermodynamics,
59
Security, food, **285**, 287–288, 560
Sedimentary rock, **370**, 372f
Seed banks, **236**, 308
Seed storage and cultivation, 236,
237f, 308

S¸ekerciog˘lu, Çag˘an, 244–245f
Selective breeding, 297
Selective cutting, of forests, 250f
Selective synthetic pesticides, 312
SELF (Solar Electric Light Fund),
282
Self Employed Women's
Association, 439
Selinda Spillway, Botswana,
32f–33f
Sequoia National Park, 44f–45f
Serra Bonita Reserve (Brazil),
68–69f, 109
Sewage treatment, 353–355, 354f
Shanghai, population growth in,
470f–471f
Shale oil, 401
Sharing cars, 490
Sharks, as keystone species,
117–118f
Shellfish. *See* Fish/shellfish pro-
duction
Shelterbelts, on farms, 315
Sierra Club, 36, 632
Silent Spring (Carson), 37, 312
Sindi, Hayat, 500–501f
Skerry, Brian, 300f
Slash-and-burn agriculture, 251f
Slums, 485–488, 485b, 486f–488f
"Smart" electrical grid, 426–427,
427f
Smart growth, in eco-cities, **493**,
493f
Smelting, 382
Smithsonian's National Zoo
(Washington, D.C.), 202, 202f
Smog, 536, **539**–540, 539f–540f
Smoking tobacco, 522–524,
522f–523f
Snowpacks, **338**, 554
Social factors of sustainability,
20, 20b
Society for Conservation Biology,
94
Sodium, as nutrient, 72f
Soil conservation, **315**
Soil erosion, 31, 304
Soil, 100–101, 302–303f
Solar cells, 436–437f, 440
advantages and disadvantages
of, 440f
Solar C.I.T.I.E.S., 595
Solar cookers, 276–277f, 436
Solar Electric Light Fund (SELF),
282
Solar energy
description of, 58, 424,
sustainability and, 20b, 21, 21b
72, 276, 354, 395, 431, 434–
440, 493, 550, 554, 622f
Solar heating systems, 434, 434f
advantages and disadvantages
of, 436f
Solar irrigation systems, 282–283f
Solar Sewage Treatment Plant
(Rhode Island), 354, 354f
Solar thermal systems, 18, 435f,
436,

advantages and disadvantages
of, 436f
Solid and hazardous waste,
576–611. *See also* Hazardous
waste
animal, 309
burying or burning, 588–590,
588f–590f
electronic, 576f–577f, 580, 580f
food, 578–579f
"four R" strategy for reducing,
590–596, 591f–593f, 594b,
595b, 595f–596f
growth of, 582–586, 582b,
583f–586f
high-waste society, 581–582,
581f, 608
management of, **587**–588, 587f
nuclear, 410–411, 410f
taxing, 622–623
transitioning to low-waste society, 604–607, 605f–606f
Solubility of chemical hazards, 517
Sonoran Desert (Arizona), 199
Southern sea otters, 134–136,
136f, 152, 152f
Specialist species, **115**–116
Speciation, **124**, 124f
Species, 194–195, 195f,
216. *See also* Ecosystem
services, saving species and
Species diversity, **109**, 131
Species interaction
commensalism, **141**
competition, **137**–138, 137f
mutualism, 140f, 141
parasitism, **141**
predation, **138**–141, 138f–141f
sea otters, 134–136
Spoils, in surface mining, **376**,
378f
Steffen, Will, 95
STEM projects
carbon-capturing
device, 642–643, 643f
species assessment, 194–195,
195f
treating contaminated soil,
100–101
wind-powered genera-
tor, 456–457, 456f–457f
Stockholm Convention on
Persistent Organic
Pollutants (POPs) of 2002,
605
Stockholm Resilience Centre,
95, 178
Stratosphere, **71**, 535, 535f
Straw bale construction, 431
Streams, 186f, 187
Streams and rivers, 344–345,
344f–345f
Strip cropping, 280f–281f, 315
Strip-cutting, of forests, 250f
Strip mining, **376**, 378
Stuart, Tristram, 578–579f
Subsidence, **334**, 379
Subsidies
agricultural, 567
energy, 396

INDEX **691**

environmentally harmful, 34
farming, **287**–288, 288*f*, 335
fossil fuel, 567
mineral supplies controlled by, 374
nuclear power, 414
shifts in, 621–622, 622*f*
Subsurface mining, **379**, 380*f*–381*f*
Succession. *See* Ecological succession
Succulent plants, 168
Sulfur dioxide (SO$_2$), 405, 537
Sulfuric acid, 537
Sunflowers, in phytoremediation, 100–101*f*
Superfund Act, 601, 602*f*–603*f*
Superinsulation, 431
Surface features of Earth, 166
Surface impoundments, **600**
advantages and disadvantages of, 601*f*
Surface mining, 376–378, 376*f*–378*f*
Surface Mining Control and Reclamation Act, 378
Surface runoff, **84**, 85*f*, 332. *See also* Hydrologic cycle
Surface water, **186**, 332
Survivorship curves, 150*f*–**151**
Sustainability, 14–43. *See also* Economics; Nature museums and preserves
achieving, 38–39
affluence and unsustainable resource use, 30–31
in cities, 493, 493*f*
environmental impacts, 26–28
environmental worldviews and ethics, 34–35
greening of American mindset, 40
greening of American schools, 18
isolation from nature, 34
natural capital and, 20–23
natural connections in, 19
overview, **20**
pollution sources, 26
population growth, 28–30
poverty impact on environment and health, 31
prices *vs.* environment and health costs, 31–34
quality of life, impacts of, 23–24
reconnecting people with nature, 16–17*f*
scientific factors of, 19–20, 20*b*
social factors of, 20, 20*b*
unsustainable living, 24–26
in U.S. and China, 616, 616*f*, 638
U.S. environmental protection and, 36–38
Sustainability in Prisons Project, Washington State Department of Corrections, 46
Sustainable cities, 490, 493, 493*f*
Sustainable practices
in agriculture, 282–283*f*
crops that promote, 318–319, 319*f*, 322

in fish/shellfish production, 300*f*, 321, 321*b*
in forestry, 252, 252*f*, 254*b*
in meat production, 319–321, 320*f*
mineral resource use, 383–385, 384*f*
Sustainable society, 635–637, 635*f*–636*f*
Sustainable Transportation Award, to Mexico City (2013), 488
Svalbard Global Seed Vault, 236–237*f*, 308, 308*f*
Swamps, 188*f*
Sweating, as heat adaptation, 169
Synthetic biology, **127**
Synthetic inorganic fertilizers, **318**
Synthetic pesticides, 312–314, 313*f*
advantages and disadvantages of, 312*f*
System, **61**
Systematic sampling, 131

T

TableTop Farm (Iowa), 320*f*
Taft, William Howard, 613
Tailings, from mining, 348*f*–349*f*, **382**
Tar sands (oil sands), 401–402, 402*f*
Taxing pollution and waste, 622–623
Taxing carbon and energy, advantages and disadvantages of, 568*f*
Tectonic plates, 126, 126*f*, **366**–369, 366*f*, 368*f*–369*f*
Teen Conservation Leader program (Monterey Bay Aquarium, California), 202
Teen Conservation Leadership Corps (Arizona), 199
Temperate deserts, 170, 170*f*
Temperate forests, 174, 174*f*
Temperate grassland, 172, 172*f*
Temperate rain forests, 196
Temperature inversion, **538**, 539*f*
Teratogens, **512**
Terracing, for soil conservation, 315, 315*f*
Terrestrial ecosystems
climate and, 167–168, 168*f*
desert, 167*f*–171*f*, 169–171
forest, 19, 174–177, 174*f*–177*f*
human impacts on, 178, 178*f*
Terrestrial organisms
climate and, 167
troposphere and, 71
Tertiary consumers in ecosystems, **75**, 82*f*
TFR (total fertility rate), **477**–478, 478*f*, 479*b*
Theft, resource scarcity enticing, 376*b*
Thermal energy, **58**
Thermodynamics, first and second laws of, **59**
Third-growth forests, 249*b*

Thirdhand smoke, 523
Thometz, Nicole, 152
Threatened species, **215**
3-D conducting nanonetworks, 362
3M Company, 520*b*
Three Mile Island (TMI, Pennsylvania) nuclear power plant accident (1979), 411–413
Throughputs in systems, 61*f*
Thumbs, opposable, 123
Tidal energy dams, 442
Tiered pricing, for water, 356
Tilman, David, 619
Time, organism changes over, 120
Time magazine, 561
Tipping points. *See* Climate change; Ecological tipping point
Tobacco ringspot virus, 223
Todd, John, 354, 354*f*
Topsoil,
overview, **302**, 304, 304*f*
renewal of, 21, **302**
Total fertility rate (TFR), **477**–478, 478*f*, 479*b*
Touzon, Raul, 628
Toxic chemicals, **511**. *See also* Chemical hazards
Toxicity of chemical hazards, **515**–517, 516*f*–517*f*
Toxicology, **515**. *See also* Chemical hazards
Toxic Release Inventory website, 604
Toxic Substances Control Act, 601
Toxic waste exposure, 498*f*–499*f*, 502, 502*f*
Trace levels of chemical hazards, 519–520, 519*f*
Trade winds, 163*b*
Traditional agriculture, **290**–291
Trampling, for rangeland management, 255
Transatomic Power, Inc., 412
Transfers of water, 336–338
Transition zones, 168
Transmissible diseases, **504**
Transpiration, 84, 85*f*. *See also* Hydrologic cycle
Transportation
energy efficient, 427–429, 429*f*
in urban areas, 489–490, 490*f*, 492, 492*f*
Tree plantation, **247**, 249, 249*b*
Treinish, Gregg, 614–615*f*
Trophic levels, **74**, 82*f*
Tropical deforestation, 179, 251*f*–252, 254
Tropical deserts, 170, 170*f*
Tropical forests, 174, 176*f*–177*f*, 254*b*, 272, 272*f*
Tropical grassland, 172, 172*f*
Tropical rain forests, 174, 174*f*, 176*f*–177*f*, 179, 272
Troposphere, **71**, 535, 535*f*
Tuberculosis, 504–506, 505*f*–506*f*

U

UNESCO Goodwill Ambassador, 500
Union for Conservation of Nature, 108
United Nations
on clean water, 333
on electronic waste, 597
Environment Programme (UNEP), 94, 249, 254*b*, 302, 586
Food and Agriculture Organization (FAO), 249, 288, 307, 309, 311
on forest cover, 249
human population estimate, 226
on irrigation, 339
Millennium Ecosystem Assessment, 26, 90, 255, 263
Minamata Convention of, 520
Population Division, 482
Unruh, Jon, 175
Unsustainable living, 24–26
Unsustainable resource use, 30–31
Upcycle, The (McDonough and Braungart), 581
Urban farming, 491, 491*f*
Urban food oasis, 284, 284*f*
Urbanization, **483**, 483*f*
advantages and disadvantages of, 484–485
overview, 470*f*–471*f*
transportation in, 489–490, 490*f*, 492, 492*f*
trends in, 482–484, 483*f*
Urban sprawl, **484**, 489
U.S. Agency for International Development (USAID), 228
U.S. Census Bureau, 479*b*
U.S. Congress, National Wilderness Preservation System and, 261
U.S. Department of Agriculture (USDA), 127, 297, 309
U.S. Department of Defense, 562
U.S. Department of Energy (DOE)
on electrical grid, 427
on mining wastes, 385
on plug-in hybrid cars, 428
on wind farms, 424, 442
U.S. Department of the Interior, 16
U.S. Environmental Protection Agency (EPA)
air quality standards of, 541
creation of, 36–38
delayed implementation of air pollution standards and, 520–521
Energy Star Program, 433
Federal Insecticide, Fungicide, and Rodenticide Act (FIFRA), 312
on geothermal heat pump systems, 446
Greenhouse Gas Equivalency Calculator, 40
on methane emissions, 404
MyEnvironment website of, 419
Resource Conservation and Recovery Act (RCRA), 601
on surface mining, 378

on toxic chemicals, 511
Toxic Release Inventory website, 604
User-pays approach, 489–490
U.S. Fish and Wildlife Service (USFWS), 136, 222, 234–235, 235f, 627
U.S. Food and Drug Administration (FDA), 309, 513–514
U.S. Forest Service (USFS), 36, 252, 612, 627
U.S. Geological Survey (USGS), 152, 372, 385
U.S. Green Building Council's Leadership in Energy and Environmental Design (LEED), 431
U.S. Library of Congress, 561
U.S. National Academy of Sciences (NAS), 312–314, 519, 547, 555, 562
U.S. National Oceanic and Atmospheric Administration (NOAA), 547, 561
U.S. National Ocean Service, 271
U.S. National Park System, 257, 258f–259f, 261b. *See also* Nature museums and preserves
U.S. Surgeon General, 524
U.S. Wilderness Society, 37
UV filtering effect of ozone layer, 535

V

Varma, Anand, 210–211f
Varroa mite, 210–211f, 223
Vegetation, climatic variations in, 167–168, 167f–168f
Vermeulen, Hans, 592
Vertebrate extinctions, 215f
Virtual water use, in United States, **332**, 332f
Viruses and parasites, **504**, 506–510, 507f, 509f–510f

Vitamins, 72, 285
Volatile organic compounds (VOCs), 536
Volatile organic hydrocarbons (VOCs), 351
Volcanoes, 126, 142b, 367, 367f
Volcanoes National Park (Hawai'i), 51, 51f

W

Wallace, Alfred Russel, 122
Warblers, resource partitioning among, 137f
Washington State Department of Corrections, 46
Waste. *See* Solid and hazardous waste
Waste reduction, **587**
Waste-to-energy incinerator, 589–590, 590f
Wastewater, 343f, **344**, 346
Water, 326–359
  aquifer depletion, 334–335, 334f, 335f
  brackish, **180**
  camels' humps for, 170b
  clean drinking, 31
  climate affected by circulation of, 162–164, 162f–164f
  Colorado River delta, 326f–327f, 328–329, 329f–330f, 356
  dams, benefits and problems of, 334–336, 336f
  deforestation impact on, 251
  desalination of, 337f, **338**
  electrical power from, 60f, 326f–327f, 440–443, 441f
  freshwater purification, 347
  freshwater use, 95f
  in hydrologic cycle, **84**–87
  indicator species and, 118
  management of fresh, 331–332, 332f
  pure, 56
  purification of, as ecosystem service, 21
  reducing losses of, 338–341, 339f, 340f, 341f

  shortages of, 333, 333f
  soil erosion by, 48
  transfers of, 336–338
  urban problems with, 485
Water int, **332**
Waterlogging, 307
Water pollution, 37, **342**. *See also* Pollution
  bacteria to clean, 79f
  eliminating, 355, 355f
  fracking pollutants, 404, 404f
  groundwater, 346–347, 346f, 348f–349f
  lakes and reservoirs, 344–346
  mining as source of, 379, 382b
  nitrate concentrations and, 48f
  oceans, 347, 350–351, 350f
  point and nonpoint sources, 342–344, 342b, 343f, 344f
  reduction of, 351–355, 352f, 353b
  rivers and streams, 344–345f
Water-saving appliances, 340
Water scarcity hotspots, 333f
Watershed, 48, **186**–187, 330, 332, 334, 641
Water table, **331**
Water use, U.S. daily, 332f
Wave energy, 442
Wavelengths of electromagnetic radiation, 58
Weather, climate *vs.*, **161**
Weather extremes, 556–557, 556f–557f
West Nile virus, 508, 562
Wetlands, 85, 116f, 117, 180, 182f, 186–189, 188f, 199, 204, 221, 270, 328
Wild Bird Conservation Act, 229
Wilderness areas, **261**
Wilson, Edward O., 121, 121f, 271, 271b
Wilson, Woodrow, 612
Windbreaks, on farms, 315
Wind energy, 420f–421f, 424, 424f, 442–443, 443, 452, 452f, 456–457, 456f–457f

  advantages and disadvantages of, 443f
Wind turbines, 18f, 455
Wood, as biomass source, 443–444
World Bank, 31, 334, 567
World Commission on Water for the 23st Century, 344
World countries (Appendix 4), 650–651
World Economic Forum, 562
World Green Building Council, 431
World Health Organization (WHO)
  on air pollution, 31, 536
  on clean water, 331, 333
  on Fukushima Daiichi Nuclear Power Plant (Japan) accident (2011), 414
  on infectious diseases, 511
  on iodine, 288
  on obesity, 286, 288
  on tobacco use, 522
World Heritage Sites, 632
World Meteorological Organization, 557
World Resources Institute (WRI), 250
Worldwatch Institute, 442
World Wildlife Fund (WWF), 30–31, 94, 216, 632
Wrangell-St. Elias National Park (Alaska), 548f–549f

Y

Yellowstone National Park (Wyoming), 257, 258f–259f
Yosemite National Park (California), 36, 628f–629f

Z

Zebra mussel, 225
Zebras, 173f
Zheng, Xiaolin, 436, 436f
Zika virus, 508
Zone of saturation, **331**

ACKNOWLEDGMENTS

Photo Credits:
Front Matter: iii ©Carsten Peter/National Geographic Creative. **vi** ©David Doubilet/National Geographic Creative. **vii** Tim Laman/National Geographic Creative. **viii-ix** ©Edmund Lowe Photography/Moment Open/Getty Images. **xi** ©Gregory Rec/Portland Press Herald/Getty Images. **xii** ©Ed Kashi/National Geographic Creative. **xiii** NPS/Tim Rains. **xvi** (tl) ©Jen Shook. (tcl) ©Mark Thiessen/National Geographic Creative. (tc) ©Sandesh Kadur. (tcr) ©Robin Moore. (tr) ©Sora Devore/National Geographic Creative. (bl) ©Mark Thiessen/National Geographic Creative. (bcl) ©Rebecca Hale/National Geographic Creative. (bc) Courtesy of Alizé Carrère. (bcr) ©Mark Thiessen/National Geographic Creative. (br) ©Sora Devore/National Geographic Creative. **xvii** (top left to bottom right) (tl) ©PRNewsFoto/Land O›Lakes, Inc./AP Images. (tlc) ©Rebecca Hale/National Geographic Creative. (tc) ©Brian Hatfield. (tcr) ©Mark Thiessen/National Geographic Creative. (tr) ©Rebecca Drobis/National Geographic Creative. (cl) ©Mark Thiessen/National Geographic Creative. (ccl) ©Michael Nichols/National Geographic Creative. (c) ©HU JIN-SONG/National Geographic Creative. (ccr) ©Sora Devore/National Geographic Creative. (cr) ©Rebecca Hale/National Geographic Creative. (bl) ©Rebecca Hale/National Geographic Creative. (bcl) ©Sue Cunningham Photographic/Alamy Stock Photo. (bc) ©Rebecca Hale/National Geographic Creative. (bcr) ©Minden Pictures/SuperStock. (br) ©Rebecca Drobis/National Geographic Creative. **xviii** (tl) ©Mark Thiessen/National Geographic Creative. (tlc) Courtesy of Andres Ruzo. (tc) ©Mark Thiessen/National Geographic Creative. (tcr) ©Mark Thiessen/National Geographic Creative. (tr) ©Anna Savage. (cl) ©Rebecca Hale/National Geographic Stock. (ccl) ©Benjamin Grimes/National Geographic Creative. (c) ©Rebecca Hale/National Geographic Creative. (ccr) ©Rebecca Drobis/National Geographic Creative. (cr) ©Rebecca Drobis/National Geographic Creative. (bl) ©Mark Thiessen/National Geographic Creative. (bcl) ©Rebecca Hale/National Geographic Creative. (bcr) ©Michael Dwyer/Alamy Stock Photo. (br) ©Rebecca Drobis/National Geographic Creative.

02-03 ©Gaston Piccinetti/AGE Fotostock. **04** ©Shah Selbe. **05** ©Chris Schmid/Aurora Photos. **06-07** ©Shah Selbe. **09** Frans Lanting/National Geographic Creative. **10** ©David Doubilet/National Geographic Creative. **11** (cl) (bl) (r) ©Shah Selbe. **12-13** ©JoannaPerchaluk/Shutterstock. **14-15** ©David Doubilet/National Geographic Creative. **17** ©Juan Martinez. **18** ©Victoria Hathaway. **21** ©Alaska Stock. **23** JAMES L. STANFIELD/National Geographic Creative. **25** ©Kevin Schafer/The Image Bank/Getty Images. **28** ©NASA Goddard Space Flight Center. **29** ©Design Pics/National Geographic Creative. **30** ©Cengage Learning. **32-33** ©Beverly Joubert/National Geographic Creative. **34** PETER ESSICK/National Geographic. **36** AP Images. **37** (tl) ©Library of Congress/SuperStock.com.(br) Library of Congress.Prints and Photographs Division.LC-DIG-ppmsca-36426. **38** (bl) Library of Congress. Prints and Photographs Division.LC-USZ62-102274. (tr) U.S. Fish and Wildlife Service. **39** John Storey/Getty Images. **44-45** ©Michael Nichols/National Geographic Creative. **47** ©Minden Pictures/SuperStock. **50** ©FocusTechnology/Alamy Stock Photo. **51** (t) ©Mark Thiessen/National Geographic Creative. (b) ©Chris Johns/National Geographic Creative. **52-53** ©Tim Laman/National Geographic Creative. **56** ©Melissa Carroll/iStock/Getty Images. **60** ©Imaginechina/Corbis. **66-67** ©Paul Harris/AWL Images/Getty Images. **69** (t, bl, br) ©Robin Moore. **70** (bl, br) U.S. Department of the Interior U.S. Geological Survey. **72** ©Frans Lanting/National Geographic Creative. **76-77** ©FRANS LANTING/National Geographic Creative. **79** ©Eye of Science/Science Source. **81** ©Roy Toft/National Geographic Creative. **86** ©Peter McBride/National Geographic Creative. **93** ©2011 Drew Fulton. **94** Sue Cunningham Photographic/Alamy Stock Photo. **95** MODIS/Terra/NASA Images. **99** Jim Richardson/National Geographic Creative. **102-103** ©Gilles Barbier/Getty Images. **104-105** ©Tim Laman/National Geographic Creative. **107** (t) ©Anna Savage. (bl, br) ©Dennis Caldwell. **108** ©Joel Sartore/National Geographic Creative. **109** ©Ralph H. Bendjebar/Danita Delimont/Alamy Stock Photo. **110** ©James Carmichael Jr/NHPA/Photoshot/Newscom. **111** (b) (l to r) ©cdrin/

Shutterstock.com. ©Karl Weatherly/PhotoDisc/Getty Images. ©Richard Broadwell/Alamy Stock Photo. ©Olivier Le Queinec/Shutterstock.com. ©Niv Koren/Shutterstock.com. ©Creatas/Jupiter Images. **112-113** ©David Liittschwager/National Geographic Creative. **114** ©Joel Sartore/National Geographic Creative. **118** ©Paul Hilton/epa european pressphoto agency b.v./Alamy Stock Photo. **119** ©Anand Varma/National Geographic Creative. **120** ©Ira Block/National Geographic Creative. **121** ©Joel Sartore/National Geographic Creative. **125** ©JOEL SARTORE/National Geographic. **132-133** Brian J.Skerry/National Geographic Creative. **135** ©Norman Smith. **136** David Courtenay/Oxford Scientific/Getty Images. **138** BEVERLY JOUBERT/National Geographic Creative. **140** ©Tim Laman/National Geographic Creative. **146** (tl) ©Rebecca Hale/National Geographic Creative. (b) ©Brant Allen. **148-149** ©David Doubilet/National Geographic Creative. **151** ©Frans Lanting/National Geographic Creative. **156-157** ©MAURICIO HANDLER /National Geographic Creative. **159** ©Manu San Felix/National Geographic Creative. **160** ©2011 BRIAN J. SKERRY /National Geographic Image Collection. **161** ©John Foster/Masterfile/Corbis. **165** Nightman1965/Shutterstock.com. **169** ©Alex Saberi/National Geographic Creative. **171** ©Daniel Gilbey/Hemera/Getty Images **173** ©Frans Lanting/National Geographic Creative. **175** ©Alize Carrre. **176-177** ©FERNANDO G. BAPTISTA/National Geographic Creative. **179** ©Michael Nichols/National Geographic Creative. **181** ©Bertie Gregory/National Geographic Creative. **182-183** ©National Geographic Creative. **184-185** ©Norbert Wu/Minden Pictures. **188** Keith Kapple/Alamy Stock Photo. **195** ©Leonardo Patrizi/iStock/Getty Images. **193** ©Edmund Lowe Photography/Moment Open/Getty Images. **196-197** ©Edmund Lowe Photography/Moment Open/Getty Images. **198** ©Frank Pali/All Canada Photos/Getty Images. **199** ©Masa Ushioda/Age Fotostock. **200** ©Chicago Academy of Sciences/Peggy Notebaert Nature Museum. **201** ©Richard J. Green/Science Source. **202** ©Jason Knauer/500px. **203** ©Lacey Ann Johnson/Aurora Photos. **204** ©Thomas Wiewandt/Danita Delimont/Alamy Stock Photo. **205** Rolf Nussbaumer Photography/Alamy Stock Photo. **206-207** ©leekris/Getty Images. **208-209** ©Steve Winter/National Geographic Creative. **211** (t, bl, br) ©Anand Varma/National Geographic Creative. **212** ©Darlyne A. Murawski/National Geographic Creative. **214** ©In Green/Shutterstock.com. **217** (t) ©Joel Sartore/National Geographic Creative. (b) ©Joel Sartore/National Geographic Creative. **218-219** ©Joel Sartore/National Geographic Creative. **223** ©Virigina W. Mason/National Geographic Creative. **224** AP Images/Nerissa Michaels/Illinois River Biological Station via the Detroit Free Press. **226** Dan Callister/Alamy Stock photo. **229** ©Bill Curtsinger/National Geographic Creative. **230** ©Neil Aldridge/npl/Minden Pictures. **223** ©Cengage/National Geographic Creative. **235** ©Lynn Johnson/National Geographic Creative. **236** ©Michio Hoshino/Minden Pictures. **237** ©Jim Richardson/National Geographic Creative. **242-243** ©Thomas Mangelsen/Minden Pictures. **245** ©Marco Grob/National Geographic Creative. **246** ©H Lansdown/Alamy Stock Photo. **248** ©145/Wesley Hitt/Ocean/Corbis. **251** ©Inaki Relanzon/npl/Minden Pictures. **253** ©Bryan Mullennix/500px. **256** ©Photoshot License Ltd/Alamy Stock Photo. **258-259** ©Barrett Hedges/Natonal Geographic Creative. **260** ©Inge Johnsson/Alamy Stock Photo. **264** ©Sandesh Kadur. **268-269** ©Don Foley. **270** ©National Geographic Maps/National Geographic Creative. **277** ©Rebecca Hale/National Geographic Creative. **278-279** ©zhihao/Getty Images. **280-281** ©R Hamilton Smith/Agstockusa/Age Fotostock. **283** (t) ©Marshall Burke. (bl) ©Zacharie Sero Tamou. (br) ©Marshall Burke. **284** ©Darren Hauck/The New York Times/Redux. **286** ©Virginia W Mason/National Geographic Creative. **287** ©Virigina W Mason/National Geographic Creative. **289** ©Amy Toensing/National Geographic Creative. **291** ©Jim Richardson/National Geographic Creative. **292-293** ©Robin Hammond/National Geographic Creative. **294** ©Wally Eberhart/Visuals Unlimited/Corbis. **295** ©kataleewan intarachote/ Shutterstock.com. **296** ©Virginia W Mason/National Geographic Creative. **297** ©Cheryl Ravelo/REUTERS. **298** ©Brian Finke/National Geographic Creative. **300** ©Brian J. Skerry/National Geographic Creative. **301** ©NGM Art/National Geographic Creative. **304** Lynn Betts, USDA Natural Resources Conservation Service.

305 ©Dirk Ercken/Shutterstock.com. **306** ©Mark Thiessen/National Geographic Creative. **307** ©2010/JIM RICHARDSON/National Geographic Stock. **308** ©Jim Richardson/National Geographic Creative. **310** ©2007 BRIAN J. SKERRY/National Geographic Image Collection. **312** ©Cathy Keifer/Shutterstock.com. **313** Scott Bauer/U.S. Department of Agriculture (USDA). **315** ©Lim Yong Hian/Shutterstock.com. **316-317** ©Peter Essick/National Geographic Creative. **319** ©Virigina W Mason/National Geographic Creative. **320** ©Jim Richardson/National Geographic Creative. **326-327** ©Pete McBride/National Geographic Creative. **329** ©Cheryl Zook/National Geographic Creative. **337** ©Patrick T. Fallon/Bloomberg/Getty Images. **341** (t) ©Mark Thiessen/National Geographic Creative. (b) ©CHERYL ZOOK/National Geographic Creative. **343** ©Dragana Gerasimoski/Shutterstock.com. **345** ©Li Yuanbo/ZUMApress/Newscom. **348-349** ©Joel Sartore/National Geographic Creative. **352** ©Photodisc/Getty Images. **354** ©Raf Makda/VIEW Pictures Ltd/Alamy Stock Photo. **360-361** ©Robb Kendrick/National Geographic Creative. **363** ©HU.JIN-SONG/National Geographic Creative. **364** ©SPL/Science Source. **367** ©Carsten Peter/National Geographic Image Collection. **368-369** ©Alex Mustard/npl/Minden Pictures. **371** (l) ©Larry Stepanowicz/Visuals Unlimited.Inc. (r) ©John Cancalosi/Nature Picture Library/Corbis. **372** (tl) Dwight Smith/Shutterstock.com.(c) ©LesPalenik/Shutterstock.com. (tr) Bragin Alexey/Shutterstock.com. **374** (tl) ©Scientifica/Corbis. (tr) ©Russ Lappa/Science Source. (cl) ©Theodore Gray/Visuals Unlimited.Inc. (cr) ©De Agostini/C. Bevilacqua/Science Source. **375** ©Woods Hole Oceanographic Institution/Visuals Unlimited.Inc. **377** (t) ©Mark Thiessen/National Geograpic Creative. (b)©Emory Kristof/National Geograhic Creative. **378** ©Martin Siepmann/image BROKER/Alamy Stock Photo. **379** David R. Frazier Photolibrary.Inc./Alamy Stock Photo. **380-381** ©Robb Kendrick/National Geographic Creative. **383** Vincenzo Lombardo/Getty Images. **384** ©Science Source. **390-391** ©Robb Kendrick/National Geographic Creative. **393** ©Mark Thiessen/National Geographic Creative. **398-399** ©Sarah Leen/National Geographic Creative. **402** (t) U.S. Department of Energy. (b) Daniel Barnes/Getty Images. **404** ©Mark Thiessen/National Geographic Creative. **405** (tl) ©www.sandatlas.org/Shutterstock.com. (tl)©JIANG HONGYAN/Shutterstock.com. (c)©farbled/Shutterstock.com. (tr)©macrowildlife/Shutterstock. **410** (bl) U.S. Department of Energy. (br) Nuclear Regulatory Commission/U.S. Department of Energy. **412** ©Lynn Johnson/National Geographic Creative. **413** ©Gerd Ludwig/National Geographic Creative. **420-421** ©Alexander W Helin/Moment Open/Getty Images. **423** (cl, bl, r) ©Sofía Ruzo and Andrés Ruzo. **427** chombosan/Shutterstock.com. **429** Ellen Isaacs/AGE Fotostock. **430** ©Diane Cook & Len Jenshel/National Geographic Creative. **431** ©Science Source. **435** Steve Proehl/Getty Images. **437** ©Kat Keene Hogue/National Geographic Creative. **438-439** ©ANNIE GRIFFITHS/National Geographic Creative. **441** nesneJkraM/Getty Images. **443** ©Greg Girard/National Geographic Creative. **445** ©Robert Clark/National Geographic Creative. **447** ©Martin Thomas/500px. **450** ©iStockphoto.com/Johan Sjolander. ©**458-459** ©Gregory Rec/Portland Press Herald/Getty Images. **460** ©Stephen Dalton/Minden Pictures. **461** CycloneCenter.org. **462** ©NHPA/Photoshot / Photoshot. **463** (l) HAYKIRDI/Getty Images. (r) ©Alex Norton/Eyewire. **464** ©Karine Aigner/Meet Your Neighbours/iLCP. **465** ©Karine Aigner/Meet Your Neighbours/iLCP. **466** The Photolibrary Wales/Alamy Stock Photo. **467** ©Ken Gillespie/All Canada Photos/Getty Images. **468-469** ©SAM YEH/Getty Images. **470-471** ©Prisma Bildagentur AG/Alamy Stock Photo. **473** ©Jon Betz/National Geographic Creative. **474** ©BartlomiejMagierowski/Shutterstock.com. **477** ©Anadolu Agency/Getty Images. **478** ©Jenny Matthews/Alamy Stock Photo. **481** ©Jake Lyell/Alamy Stock Photo. **483** ©AFP/Getty Images. **484** ©Diane Cook & Len Jenshel/National Geographic Creative. **486-487** ©Robin Hammond/National Geographic Creative. **488** ©Jess Kraft/Shutterstock.com. **490** ©Diane Cook & Len Jenshel/National Geographic Creative.

491 ©Lynn Johnson/National Geographic Creative. **492** ©Marcio Jose Bastos Silva/Shutterstock.com. **498-499** ©Ed Kashi/National Geographic Creative. **501** ©Rebecca Hale/National Geographic Creative. **502** ©Robert and Jean Pollock/Science Source. **506** ©Simon Fraser/Science Source. **507** ©Marco Grob/National Geographic Creative. **510** Olivier Asselin/Alamy Stock Photo. **512** ©Robin Treadwell/Science Source. **513** YanLev/Shutterstock.com. **514** David McNew/Getty Images. **516** ©Graeme Robertson/eyevine/Redux. **523** ©Mark Thiessen/National Geographic Creative. **524** ©Arthur Glauberman/Science Source. **525** ©Alexandra Lande/Shutterstock.com. **530-531** NASA. **533** ©Adam Taylor. **534** (bl) NASA/Goddard Space Flight Center Scientific Visualization Studio.(br) NASA/Goddard Space Flight Center Scientific Visualization Studio. **537** ©Alex Hyde/NPL/Minden Pictures. **539** ©Ash Lindsey Photography/Getty Images. **540** ©VCG/Getty Images. **547** ©ARCTIC IMAGES/Alamy Stock Photo. **548-549** ©FRANS LANTING/National Geographic Creative. **555** ©Patrick J. Endres/Getty Images. **556** ©Brandt Meixell.USGS. **558-559** ©Michael Coyne/National Geographic Creative. **560** ©Peter Essick/National Geographic Creative. **561** (t) ©Tyrone Turner/National Geographic Creative. (b) ©BATES LITTLEHALES /National Geographic Creative. **564** ©Global Warming Images/Alamy Stock Photo. **570** (bl) (br) NASA Ozone Watch. **576-577** ©Peter Essick/National Geographic Creative. **579** ©Brian Finke/National Geographic Creative. **580** ©Peter Essick/National Geographic Creative. **581** ©Jose Azel /Aurora Photos. **583** ©Craig Pulsifer/Aurora Photos. **585** ©Frans Lanting/National Geographic Creative. **587** (left to right) ©Mariyana M/Shutterstock.com. ©donatas1205/Shutterstock.com. ©Oleksiy Mark /Shutterstock.com. ©PhotoDisc/Getty Images. ©vilax/Shutterstock.com. ©MrGarry/Shutterstock.com. ©Le Do/Shutterstock.com. **592** ©Robert Clark/National Geographic Creative. **593** ©Simon Maina/AFP/Getty Images. **595** Courtesy of TH Culhane. **598** ©Jim Xu/Getty Images. **600** ©Lynn Johnson/National Geographic Creative. **602-603** ©Fritz Hoffmann/National Geographic Creative. **605** ©Cyrus McCrimmon/The Denver Post via Getty Images. **612-613** NPS/Tim Rains. **615** ©Explorer Gregg Treinish/Alexandria Bombach. **618** NASA Goddard Space Flight Center. **619** Library of Congress.Prints & Photographs Division.photograph by Carol M. Highsmith LC-DIG-highsm-04715. **621** ©Andreas Reinhold/Shutterstock.com. **622** ©Elena Elisseeva/Shutterstock.com. **623** ©William Campbell/Corbis Historical/Getty Images. **628-629** ©Raul Touzon/National Geographic Creative. **632** ©Peter Foley/epa european pressphoto agency b.v./Alamy Stock Photo. **633** (t) ©Bertrand Guay/AFP/Getty Images. (b) ©Chris Dickinson/Jane Goodall Institute. **635** ©Flip Nicklin/Minden Pictures. **636** ©Justin Bailie/Aurora Photos. **643** ©Robert Clark/National Geographic Creative.

Illustrations:
Unless otherwise indicated, all illustrations are owned by Cengage Learning.

Maps:
Unless otherwise indicated, all maps are created by National Geographic.